Oxford Resources for IB
Diploma Programme

2023 EDITION

CHEMISTRY

COURSE COMPANION

Sergey Bylikin
Gary Horner
Elisa Jimenez Grant
David Tarcy

OXFORD
UNIVERSITY PRESS

OXFORD
UNIVERSITY PRESS

Great Clarendon Street, Oxford, OX2 6DP, United Kingdom

Oxford University Press is a department of the University of Oxford. It furthers the University's objective of excellence in research, scholarship, and education by publishing worldwide. Oxford is a registered trade mark of Oxford University Press in the UK and in certain other countries.

British Library Cataloguing in Publication Data
Data available

9781382016469

9781382016506 (ebook)

10 9 8 7 6 5 4 3

Paper used in the production of this book is a natural, recyclable product made from wood grown in sustainable forests.

The manufacturing process conforms to the environmental regulations of the country of origin.

Printed in China by Shanghai Offset Printing Products Ltd

Acknowledgements

The "In cooperation with IB" logo signifies the content in this textbook has been reviewed by the IB to ensure it fully aligns with current IB curriculum and offers high-quality guidance and support for IB teaching and learning.

The authors have the following acknowledgements and thanks:
Sergey Bylikin: I would like to thank Dr Natalia Kalashnikova for her support and suggestions.
Gary Horner: To my friends and colleagues for their support throughout my teaching career, my sister Susan for her unwavering friendship, care and professional advice, my inspirational parents Myrtle and Dennis: I dedicate this book in their loving memory.
Elisa Jimenez Grant: To Miljan.
In memory of David Tarcy

The Publisher would like to thank the following members of the DP Science 2023 Research Panel for sharing their insights, expertise, and feedback:
Sandy Hawkinson, Kathryn Russell, Ilias Liakatas, Nerissa Puntawe, Vicki Boyd, Cendrella Kettaneh, Astha Acharya, Natalie Parker, Chandan Bhosale, Deepa Sathya and Síle-Caitríona O'Callaghan.

The Publisher wishes to thank the International Baccalaureate Organization for permission to reproduce their intellectual property.

The publisher and authors would like to thank the following for permission to use photographs and other copyright material:

Cover: www.flickr/Getty Images. **Photos; p2:** Copyright by Boonchet Ch./Getty Images; **p3:** With permission of Cornell University; **p5(t):** Jonak Photography/Getty Images; **p5(b):** Masterpics / Alamy Stock Photo; **p6(l):** mirecca/Getty Images; **p6(m):** Jrgen Wambach / EyeEm/Getty Images; **p6(r):** RHJPhtotos/Shutterstock; **p7(l):** Remigiusz Gora/Getty Images; **p7(m):** MARTYN F. CHILLMAID / SCIENCE PHOTO LIBRARY; **p7(r):** stockcreations/Shutterstock; **p12:** Andrea Izzotti/Shutterstock; **p13(t):** Jan Halaska / Alamy Stock Photo; **p13(b):** KPG_Payless/Shutterstock; **p14:** Alexey Kljatov/Shutterstock; **p15:** Wayne Eastep/Getty Images; **p16:** © BIPM; **p22:** one-image photography / Alamy Stock Photo; **p23:** IBM RESEARCH / SCIENCE PHOTO LIBRARY; **p26:** Artiom Photo/Shutterstock; **p28(t):** PATRICK LANDMANN / SCIENCE PHOTO LIBRARY; **p28(b):** Science History Images / Alamy Stock Photo; **p31:** FRANCOIS GUILLOT/AFP via Getty Images; **p35:** Denis Belitsky/Shutterstock; **p36:** Fotomaton / Alamy Stock Photo; **p37(t):** H.S. Photos / Alamy Stock Photo; **p37(b):** Zern Liew/Shutterstock; **p63:** CHARLES D. WINTERS / SCIENCE PHOTO LIBRARY; **p66:** Nneirda/Shutterstock; **p67(l):** Ashley Cooper/Getty Images; **p67(r):** Ludovic Caritey / 500px/Getty Images; **p68:** Mysid/Wikimedia Commons; **p69:** Wirestock Creators/Shutterstock; **p76:** Sergey Bylikin; **p90:** DAVID PARKER / SCIENCE PHOTO LIBRARY; **p94:** Pi-Lens/Shutterstock; **p96(l):** KKStock/Getty Images; **p96(m):** Marina Kryuchina/Shutterstock; **p96(r):** Nneirda/Shutterstock; **p100:** MARTYN F. CHILLMAID / SCIENCE PHOTO LIBRARY; **p104:** KARL GAFF / SCIENCE PHOTO LIBRARY; **p108:** Nadezda Boltaca/Shutterstock; **p111:** au_uhoo/Shutterstock; **p112:** Peky/Shutterstock; **p113:** ggw/Shutterstock; **p118(l):** Pixel B/Shutterstock; **p118(m):** maurobeltran/Shutterstock; **p118(r):** Anthony Bradshaw/Getty Images; **p126:** bszef/Shutterstock; **p127:** Michelangelus/Shutterstock; **p129:** pirita/Shutterstock; **p130:** Ben Rout **p135(bl):** Ambelrip/Shutterstock; **p135(bm):** Mopic/Shutterstock; **p136(l):** DENNIS KUNKEL MICROSCOPY / SCIENCE PHOTO LIBRARY; **p136(r):** Jennifer Sophie/Shutterstock; **p137(l):** ANDREW LAMBERT PHOTOGRAPHY / SCIENCE PHOTO LIBRARY; **p137(r):** Pi-Lens/Shutterstock; **p138:** Africa Studio/Shutterstock; **p139:** TURTLE ROCK SCIENTIFIC / SCIENCE SOURCE / SCIENCE PHOTO LIBRARY; **p146(t):** Wirestock Creators/Shutterstock; **p146(bl):** Alexey Kljatov/Shutterstock; **p146(br):** JanMiko/Getty Images; **p146(r):** Andrey Armyagov/Shutterstock; **p161(tl):** Admin/Shutterstock; **p161(tr):** NASA EARTH OBSERVATORY / OZONE HOLE WATCH / SCIENCE PHOTO LIBRARY; **p161(b):** Brooks Kraft LLC/Corbis via Getty Images; **p163(tr):** SMDSS/Shutterstock; **p163(m):** Fotokostic/Shutterstock; **p163(r):** DUSAN ZIDAR/Shutterstock; **p164:** Boris15/Shutterstock; **p169:** PHIL DEGGINGER / SCIENCE PHOTO LIBRARY; **p170:** PHIL DEGGINGER / SCIENCE PHOTO LIBRARY; **p186:** © 2014 Garfinkel et al; **p188:** S-F/Shutterstock; **p190(l):** Yermolov/Shutterstock; **p190(r):** Westend61 GmbH / Alamy Stock Photo; **p192:** TURTLE ROCK SCIENTIFIC / SCIENCE PHOTO LIBRARY; **p193:** Adam J/Shutterstock; **p194:** Adrienne Bresnahan/Getty Images; **p196:** ANDREW LAMBERT PHOTOGRAPHY / SCIENCE PHOTO LIBRARY; **p197(l):** simonkr/Getty Images; **p197(r):** aditya_frzhm/Shutterstock; **p199:** sezer66/Shutterstock; **p203:** GrayMark/Shutterstock; **p206(l):** Gurgen Bakhshetyan/Shutterstock; **p206(r):** Tyler Olson/Shutterstock; **p207(l):** David Sanger/Photographer›s Choice/Getty Images; **p207(r):** Dario Sabljak/Shutterstock; **p209:** JUAN GAERTNER / SCIENCE PHOTO LIBRARY; **p211:** HAGLEY MUSEUM AND ARCHIVE / SCIENCE PHOTO LIBRARY; **p212:** Khomulo Anna/Shutterstock; **p213:** JAMES KING-HOLMES / SCIENCE PHOTO LIBRARY; **p215(t):** Rich Carey/Shutterstock; **p215(b):** STEVE GSCHMEISSNER / SCIENCE PHOTO LIBRARY; **p216(t):** Sne Tak; **p216(b):** STEVE GSCHMEISSNER / SCIENCE PHOTO LIBRARY; **p220:** Matej Kastelic/Shutterstock; **p222:** Leonid Andronov/Shutterstock; **p228:** Tom Grill/Getty Images; **p231:** SCIENCE SOURCE / SCIENCE PHOTO LIBRARY; **p234:** magnetix/Shutterstock; **p239:** Albert Russ/Shutterstock; **p240:** SCIENCE PHOTO LIBRARY; **p260:** AtWaG/Getty Images; **p278:** Nigel Cattlin / Alamy; **p287:** ANDREW LAMBERT PHOTOGRAPHY / SCIENCE PHOTO LIBRARY; **p296:** smereka/Shutterstock; **p298:** SCIENCE PHOTO LIBRARY; **p308:** SHEILA TERRY / SCIENCE PHOTO LIBRARY; **p311:** WH CHOW/Shutterstock; **p313(t):** Ron Kloberdanz/Shutterstock; **p313(b):** Tobias / Alamy; **p314:** MarcelClemens/Shutterstock; **p317:** Volodymyr Nahaiets/Shutterstock; **p318:** Bjoern Wylezich/Shutterstock; **p319:** Starring Lab / Alamy Stock Photo; **p321:** Cultura Creative RF / Alamy Stock Photo; **p325:** Sorapop Udomsri/Shutterstock; **p327:** Oxford University Press ANZ; **p332:** Rattiya Thongdumhyu/Shutterstock; **p335:** MARTYN F. CHILLMAID / SCIENCE PHOTO LIBRARY; **p338:** SCIENCE PHOTO LIBRARY; **p339:** TED KINSMAN / SCIENCE PHOTO LIBRARY; **p341(t):** Andrea Obzerova / Alamy Stock Photo; **p341(b):** Mark Lorch/Shutterstock; **p342:** Reprinted (adapted) with permission from J. Chem. Educ. 2016, 93, 7, 1249–1252 Copyright 2016 American Chemical Society; **p343:** Stu Shaw/Shutterstock; **p351(t):** Julien_N/Shutterstock; **p351(b):** Ron Kloberdanz/Shutterstock; **p352(tl):** Oxford University Press; **p352(tr):** chromatos/Shutterstock; **p352(bl):** MARTYN F. CHILLMAID / SCIENCE PHOTO LIBRARY; **p352(br):** photong/Shutterstock; **p386:** timandtim/Getty Images; **p388(tl):** John and Tina Reid/Getty Images; **p388(tr):** Dr Ajay Kumar Singh/Shutterstock; **p388(b):** Valery Lisin/Shutterstock; **p389:** keerati/Shutterstock; **p390:** CHARLES D. WINTERS / SCIENCE PHOTO LIBRARY; **p408:** ilyankou/Shutterstock; **p426:** tab1962/Getty Images; **p427:** R Kawka / Alamy Stock Photo; **p431:** Xinhua / Alamy Stock Photo; **p434:** Richard Liu; **p439:** REUTERS / Alamy Stock Photo; **p443(t):** Boltzmann›s Tomb photos by Thomas D. Schneider (https://alum.mit.edu/www/toms/images/boltzmann); **p443(b):** Mariyana M/Shutterstock; **p452:** SCIENCE PHOTO LIBRARY; **p453:** Reprinted (adapted) with permission from J. Chem. Educ. 2012, 89, 5, 675–677. Copyright 2012 American Chemical Society; **p460:** Richard Liu; **p468:** sciencephotos / Alamy Stock Photo; **p470(l):** Heritage Image Partnership Ltd / Alamy Stock Photo; **p470(r):** The Granger Collection / Alamy Stock Photo; **p477(l):** MAXIMILIAN STOCK LTD / SCIENCE PHOTO LIBRARY; **p477(r):** ASHLEY COOPER / SCIENCE PHOTO LIBRARY; **p481:** nagelestock.com / Alamy Stock Photo; **p484:** ANDREW LAMBERT PHOTOGRAPHY / SCIENCE PHOTO LIBRARY; **p497:** Kenneth Brown / EyeEm; **p499:** dpa picture alliance / Alamy Stock Photo; **p509:** bluesnote/Shutterstock; **p513:** CHARLES D. WINTERS / SCIENCE PHOTO LIBRARY; **p520:** ANDREW LAMBERT PHOTOGRAPHY / SCIENCE PHOTO LIBRARY; **p526:** Sankei Archive via Getty Images; **p536:** STEVE HORRELL / SCIENCE PHOTO LIBRARY; **p538:** CHARLES D. WINTERS / SCIENCE PHOTO LIBRARY; **p543:** SCIENCE PHOTO LIBRARY; **p553:** ANDREW LAMBERT PHOTOGRAPHY / SCIENCE PHOTO LIBRARY; **p554:** CHARLES D. WINTERS/SCIENCE PHOTO LIBRARY; **p555:** David R. Frazier Photolibrary, Inc. / Alamy Stock Photo; **p557:** SCIENCE PHOTO LIBRARY; **p571:** ANDREW LAMBERT PHOTOGRAPHY / SCIENCE PHOTO LIBRARY; **p581:** Judy Kennamer/Shutterstock; **p583:** MARTYN F. CHILLMAID / SCIENCE PHOTO LIBRARY; **p591:** MARTYN F. CHILLMAID / SCIENCE PHOTO LIBRARY; **p592:** Colin Hawkins/Getty Images; **p594(l):** SCIENCE PHOTO LIBRARY; **p594(r):** SHEILA TERRY / SCIENCE PHOTO LIBRARY; **p599:** Dorling Kindersley/Getty Images; **p608(l):** © Corbis; **p608(r):** ggw/Shutterstock; **p619:** REUTERS / Alamy Stock Photo; **p632:** Pictorial Press Ltd / Alamy Stock Photo; **p637:** ANDREW LAMBERT PHOTOGRAPHY / SCIENCE PHOTO LIBRARY.

Artwork by Q2A Media, Aptara Inc., GreenGate Publishing Services, Six Red Marbles, Barking Dog Art, IFA Design, Phoenix Photosetting, Thomson Digital, Tech-Set Ltd, Wearset Ltd, HL Studios, Peter Bull Art Studio, Tech Graphics, James Stayte, Trystan Mitchell, Clive Goodyer, Jeff Bowles, Roger Courthold, Mike Ogden, Jeff Edwards, Russell Walker, Clive Goodyer, Jamie Sneddon, David Russell, Mark Walker, Erwin Haya, Paul Gamble, Sergey Bylikin, Elisa Jimenez Grant and Oxford University Press. Index by James Helling.

Although we have made every effort to trace and contact all copyright holders before publication this has not been possible in all cases. If notified, the publisher will rectify any errors or omissions at the earliest opportunity.
Links to third party websites are provided by Oxford in good faith and for information only. Oxford disclaims any responsibility for the materials contained in any third party website referenced in this work.

Contents

Structure 1. Models of the particulate nature of matter	2
Structure 1.1 Introduction to the particulate nature of matter	3
Structure 1.2 The nuclear atom	20
Structure 1.3 Electron configurations	34
Structure 1.4 Counting particles by mass: The mole	63
Structure 1.5 Ideal gases	80

Structure 2. Models of bonding and structure	94
Structure 2.1 The ionic model	95
Structure 2.2 The covalent model	117
Structure 2.3 The metallic model	187
Structure 2.4 From models to materials	197

Structure 3. Classification of matter	228
Structure 3.1 The periodic table: Classification of elements	229
Structure 3.2 Functional groups	256

Tools for chemistry	308
Tool 1: Experimental techniques	309
Tool 2: Technology	342
Tool 3: Mathematics	350

Reactivity 1. What drives chemical reactions?	386
Reactivity 1.1 Measuring enthalpy changes	387
Reactivity 1.2 Energy cycles in reactions	404
Reactivity 1.3 Energy from fuels	424
Reactivity 1.4 Entropy and spontaneity (AHL)	442

Reactivity 2. How much, how fast and how far?	460
Reactivity 2.1 How much? The amount of chemical change	461
Reactivity 2.2 How fast? The rate of chemical change	480
Reactivity 2.3 How far? The extent of chemical change	512

Reactivity 3. What are the mechanisms of chemical change?	536
Reactivity 3.1 Proton transfer reactions	537
Reactivity 3.2 Electron transfer reactions	580
Reactivity 3.3 Electron sharing reactions	622
Reactivity 3.4 Electron-pair sharing reactions	628

Cross-topic exam-style questions	652
The inquiry process (authored by Maria Muñiz Valcárcel)	655
The internal assessment (IA) (authored by Maria Muñiz Valcárcel)	668
Index	686
Periodic Table	708

Answers: www.oxfordsecondary.com/ib-science-support

Introduction

The diploma programme (DP) chemistry course is aimed at students in the 16 to 19 age group. The curriculum seeks to develop a conceptual understanding of the nature of science, working knowledge of fundamental principles of chemistry and practical skills that can be applied in familiar and unfamiliar contexts. As with all the components of the DP, this course fosters the IB learner profile attributes (see page viii) in the members of the school community.

Nature of science

Nature of science (NOS) is concerned with methods, purposes and outcomes that are specific to science. NOS is a central theme that is present across the entire course. You will find suggested NOS features throughout the book and are encouraged to come up with further examples of your own as you work through the programme.

NOS can be organized into the following eleven aspects:

- **Observations and experiments**
  Sometimes the observations in experiments are unexpected and lead to serendipitous results.

- **Measurements**
  Measurements can be qualitative or quantitative, but all data are prone to error. It is important to know the limitations of your data.

- **Evidence**
  Scientists learn to be sceptical about their observations and they require their knowledge to be fully supported by evidence.

- **Patterns and trends**
  Recognition of a pattern or trend forms an important part of the scientist's work whatever the science.

- **Hypotheses**
  Patterns lead to a possible explanation. The hypothesis is this provisional view and it requires further verification.

- **Falsification**
  Hypotheses can be proved false using other evidence, but they cannot be proved to be definitely true. This has led to paradigm shifts in science throughout history.

- **Models**
  Scientists construct models as simplified explanations of their observations. Models often contain assumptions or unrealistic simplifications, but the aim of science is to increase the complexity of the model, and to reduce its limitations.

- **Theories**
  A theory is a broad explanation that takes observed patterns and hypotheses and uses them to generate predictions. These predictions may confirm a theory (within observable limitations) or may falsify it.

- **Science as a shared activity**
  Scientific activities are often carried out in collaboration, such as peer review of work before publication or agreement on a convention for clear communication.

- **Global impact of science**
  Scientists are responsible to society for the consequences of their work, whether ethical, environmental, economic or social. Scientific knowledge must be shared with the public clearly and fairly.

Syllabus structure

Topics are organized into two main concepts: structure and reactivity. This is shown in the syllabus roadmap below. The skills in the study of chemistry are overarching experimental, technological, mathematical and inquiry skills that are integrated into the course. Chemistry is a practical subject, so these skills will be developed through experimental work, inquiries and investigations.

Skills in the study of chemistry			
Structure Structure refers to the nature of matter from simple to more complex forms		Reactivity Reactivity refers to how and why chemical reactions occur	
Structure determines reactivity, which in turn transforms structure			
Structure 1. Models of the particulate nature of matter	Structure 1.1 — Introduction to the particulate nature of matter	Reactivity 1. What drives chemical reactions?	Reactivity 1.1 — Measuring enthalpy changes
	Structure 1.2 — The nuclear atom		Reactivity 1.2 — Energy cycles in reactions
	Structure 1.3 — Electron configurations		Reactivity 1.3 — Energy from fuels
	Structure 1.4 — Counting particles by mass: The mole		Reactivity 1.4 — Entropy and spontaneity (Additional higher level)
	Structure 1.5 — Ideal gases		
Structure 2. Models of bonding and structure	Structure 2.1 — The ionic model	Reactivity 2. How much, how fast and how far?	Reactivity 2.1 — How much? The amount of chemical change
	Structure 2.2 — The covalent model		Reactivity 2.2 — How fast? The rate of chemical change
	Structure 2.3 — The metallic model		Reactivity 2.3 — How far? The extent of chemical change
	Structure 2.4 — From models to materials		
Structure 3. Classification of matter	Structure 3.1 — The periodic table: Classification of elements	Reactivity 3. What are the mechanisms of chemical change?	Reactivity 3.1 — Proton transfer reactions
			Reactivity 3.2 — Electron transfer reactions
	Structure 3.2 — Functional groups: Classification of organic compounds		Reactivity 3.3 — Electron sharing reactions
			Reactivity 3.4 — Electron-pair sharing reactions

Chemistry concepts are thoroughly interlinked. For example, as shown in the roadmap above, "Structure determines reactivity, which in turn transforms structure". You are therefore encouraged to continuously reflect on the connections between new and prior knowledge as you progress through the course. Linking questions will help you explore those connections. In assessment tasks, you will be expected to identify and apply the links between different topics. On page 652, there are three examples of DP-style exam questions that link several different topics in the course.

How to use this book

The aim of this book is to develop conceptual understanding, aid in skills development and provide opportunities to cement knowledge and understanding through practice.

Feature boxes and sections throughout the book are designed to support these aims, by signposting content relating to particular ideas and concepts, as well as opportunities for practice. This is an overview of these features:

Developing conceptual understanding

Guiding questions

Each topic begins with a guiding question to get you thinking. When you start studying a topic, you might not be able to answer these questions confidently or fully, but by studying that topic, you will be able to answer them with increasing depth. Hence, you should consider these as you work through the topic and come back to them when you revise your understanding.

These boxes in the margin will direct you to other parts of the book where a concept is explored further or in a different context. They may also direct you to prior knowledge or a skill you will need, or give a different way to think about something.

Linking questions

Linking questions within each topic highlight the connections between content discussed there and other parts of the course.

Nature of science

These illustrate NOS using issues from both modern science and science history, and show how the ways of doing science have evolved over the centuries. There is a detailed description of what is meant by NOS and the different aspects of NOS on the previous page. The headings of NOS feature boxes show which of the eleven aspects they highlight.

Theory of knowledge

This is an important part of the IB Diploma course. It focuses on critical thinking and understanding how we arrive at our knowledge of the world. The TOK features in this book pose questions for you that highlight these issues.

AHL

Parts of the book have a coloured bar on the edge of the page or next to a question. This indicates that the material is for students studying at DP Chemistry Higher Level. AHL means "additional higher level".

Developing skills

ATL Approaches to learning

These ATL features give examples of how famous scientists have demonstrated the ATL skills of communication, self-management, research, thinking and social skills, and prompt you to think about how to develop your own strategies.

Chemistry skills

These contain ways to develop your mathematical, experimental or inquiry skills, especially through experiments and practical work. Some of these can be used as springboards for your Internal Assessment.

Tools for chemistry, the inquiry process and internal assessment

These three section of the book are full of reference material for all the essential mathematical and experimental tools required for DP Chemistry, details on data analysis and modelling chemistry, as well as guidance on how to use the inquiry process in the study of the subject and to work through your Internal Assessment. Flick to this section as your working through the rest of the book for more information. Links in the margin throughout the book will direct you towards it too.

Practicing

Worked examples

These are step-by-step examples of how to answer questions or how to complete calculations. You should review these examples carefully, preferably after attempting the question yourself.

Practice questions

These are designed to give you further practice at using your chemistry knowledge and to allow you to check your own understanding and progress.

Data-based questions

Part of your final assessment requires you to answer questions that are based on the interpretation of data. Use these questions to prepare for this. They are also designed to make you aware of the possibilities for data acquisition and analysis for day-to-day experiments and for your IA.

Activity

These give you an opportunity to apply your chemistry knowledge and skills, often in a practical way.

End-of-topic questions

Use these questions at the end of each topic to draw together concepts from that topic and to practise answering exam-style questions.

Course book definition

The IB Diploma Programme course books are resource materials designed to support students throughout their two-year Diploma Programme course of study in a particular subject. They will help students gain an understanding of what is expected from the study of an IB Diploma Programme subject while presenting content in a way that illustrates the purpose and aims of the IB. They reflect the philosophy and approach of the IB and encourage a deep understanding of each subject by making connections to wider issues and providing opportunities for critical thinking.

The books mirror the IB philosophy of viewing the curriculum in terms of a whole-course approach; the use of a wide range of resources, international mindedness, the IB learner profile and the IB Diploma Programme core requirements, theory of knowledge, the extended essay, and creativity, activity, service (CAS).

Each book can be used in conjunction with other materials and, indeed, students of the IB are required and encouraged to draw conclusions from a variety of resources. Suggestions for additional and further reading are given in each book and suggestions for how to extend research are provided.

In addition, the course companions provide advice and guidance on the specific course assessment requirements and on academic honesty protocol. They are distinctive and authoritative without being prescriptive.

IB mission statement

The International Baccalaureate aims to develop inquiring, knowledgeable and caring young people who help to create a better and more peaceful world through intercultural understanding and respect.

To this end, the organization works with schools, governments and international organizations to develop challenging programmes of international education and rigorous assessment.

These programmes encourage students across the world to become active, compassionate and lifelong learners who understand that other people, with their differences, can also be right.

The IB Learner Profile

The aim of all IB programmes to develop internationally minded people who work to create a better and more peaceful world. The aim of the programme is to develop this person through ten learner attributes, as described below.

Inquirers: They develop their natural curiosity. They acquire the skills necessary to conduct inquiry and research and snow independence in learning. They actively enjoy learning and this love of learning will be sustained throughout their lives.

Knowledgeable: They explore concepts, ideas and issues that have local and global significance. In so doing, they acquire in-depth knowledge and develop understanding across a broad and balanced range of disciplines.

Thinkers: They exercise initiative in applying thinking skills critically and creatively to recognize and approach complex problems, and to make reasoned, ethical decisions.

Communicators: They understand and express ideas and information confidently and creatively in more than one language and in a variety of modes of communication. They work effectively and willingly in collaboration with others.

Principled: They act with integrity and honesty, with a strong sense of fairness, justice and respect for the dignity of the individual, groups and communities. They take responsibility for their own action and the consequences that accompany them.

Open-minded: They understand and appreciate their own cultures and personal histories, and are open to the perspectives, values and traditions of other individuals and communities. They are accustomed to seeking and evaluating a range of points of view, and are willing to grow from the experience.

Caring: They show empathy, compassion and respect towards the needs and feelings of others. They have a personal commitment to service, and to act to make a positive difference to the lives of others and to the environment.

Risk-takers: They approach unfamiliar situations and uncertainty with courage and forethought, and have the independence of spirit to explore new roles, ideas and strategies. They are brave and articulate in defending their beliefs.

Balanced: They understand the importance of intellectual, physical and emotional ballance to achieve personal wellbeing for themselves and others.

Reflective: They give thoughtful consideration to their own learning and experience. They are able to assess and understand their strengths and limitations in order to support their learning and personal development.

A note on academic integrity

It is of vital importance to acknowledge and appropriately credit the owners of information when that information is used in your work. After all, owners of ideas (intellectual property) have property rights. To have an authentic piece of work, it must be based on your individual and original ideas with the work of others fully acknowledged. Therefore, all assignments, written or oral, completed for assessment must use your own language and expression. Where sources are used or referred to, whether in the form of direct quotation or paraphrase, such sources must be appropriately acknowledged.

How do I acknowledge the work of others?

The way that you acknowledge that you have used the ideas of other people is through the use of footnotes and bibliographies.

Footnotes (placed at the bottom of a page) or endnotes (placed at the end of a document) are to be provided when you quote or paraphrase from another document or closely summarize the information provided in another document. You do not need to provide a footnote for information that is part of a 'body of knowledge'. That is, definitions do not need to be footnoted as they are part of the assumed knowledge.

Bibliographies should include a formal list of the resources that you used in your work.

'Formal' means that you should use one of the several accepted forms of presentation. This usually involves separating the resources that you use into different categories (e.g. books, magazines, newspaper articles, internet-based resources, and works of art) and providing full information as to how a reader or viewer of your work can find the same information. A bibliography is compulsory in the Extended Essay.

What constitutes malpractice?

Malpractice is behaviour that results in, or may result in, you or any student gaining an unfair advantage in one or more assessment component. Malpractice includes plagiarism and collusion.

Plagiarism is defined as the representation of the ideas or work of another person as your own. The following are some of the ways to avoid plagiarism:

- words and ideas of another person to support one's arguments must be acknowledged

- passages that are quoted verbatim must be enclosed within quotation marks and acknowledged

- email messages, and any other electronic media must be treated in the same way as books and journals

- the sources of all photographs, maps, illustrations, computer programs, data, graphs, audio-visual and similar material must be acknowledged if they are not your own work

- when referring to works of art, whether music, film dance, theatre arts or visual arts and where the creative use of a part of a work takes place, the original artist must be acknowledged.

Collusion is defined as supporting malpractice by another student. This includes:

- allowing your work to be copied or submitted for assessment by another student

- duplicating work for different assessment components and/or diploma requirements.

Other forms of malpractice include any action that gives you an unfair advantage or affects the results of another student. Examples include, taking unauthorized material into an examination room, misconduct during an examination and falsifying a CAS record.

Experience the future of education technology with Oxford's digital offer for DP Science

You're already using our print resources, but have you tried our digital course on Kerboodle?

Developed in cooperation with the IB and designed for the next generation of students and teachers, Oxford's DP Science offer brings together the IB curriculum and future-facing functionality, enabling success in DP and beyond. Use both print and digital components for the best blended teaching and learning experience.

Learn anywhere with mobile-optimized onscreen access to student resources and offline access to the digital Course Book

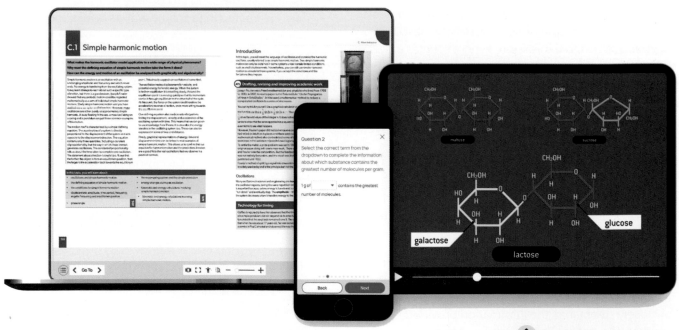

Encourage motivation with a variety of engaging content including interactive activities, vocabulary exercises, animations, and videos

Embrace independent learning and progression with adaptive technology that provides a personalized journey so students can self-assign auto-marked assessments, get real-time results and are offered next steps ↓

Deepen understanding with intervention and extension support, and spaced repetition, where students are asked follow-up questions on completed topics at regular intervals to encourage knowledge retention ↓

Enhance reporting with rich data collected to support responsive teaching at an individual and class level

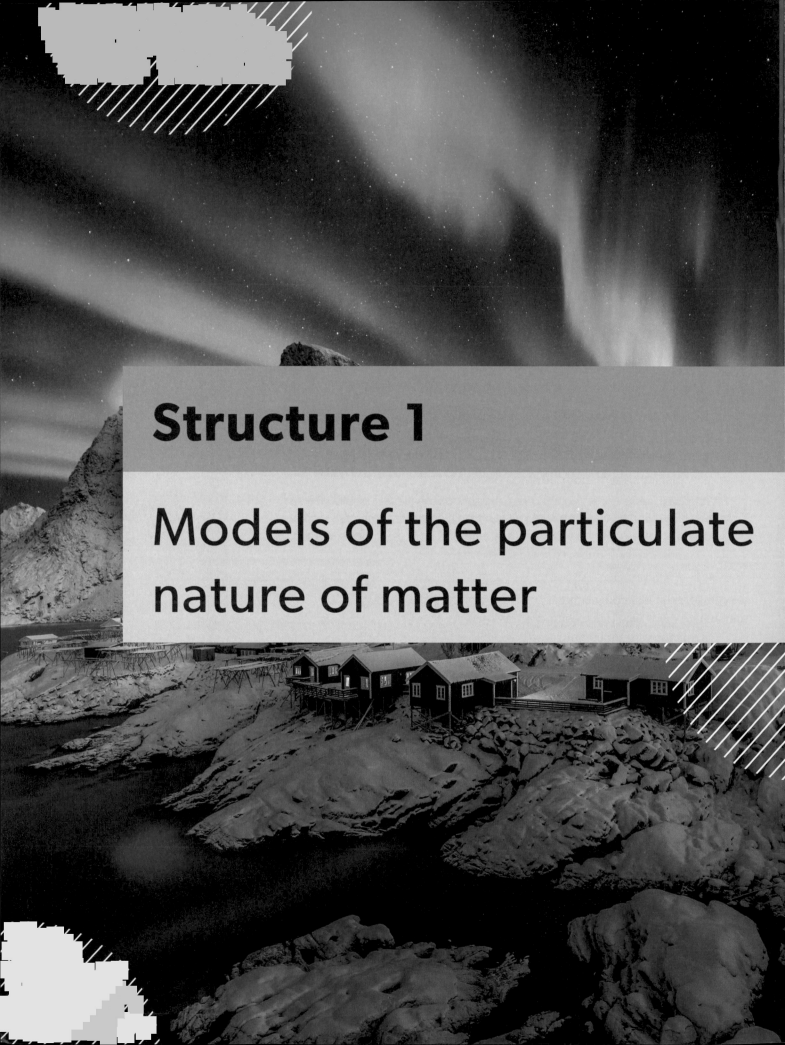

Structure 1

Models of the particulate nature of matter

Structure 1.1 Introduction to the particulate nature of matter

How can we model the particulate nature of matter?

The universally accepted idea that all matter is composed of atoms came from experimental evidence that could only be explained if matter were made of particles.

Early classical theory suggested that all matter was composed of earth, air, fire, and water. However, this theory lacked predictive power and could not account for the great variety of chemical compounds, so it was eventually abandoned. The systematic study of chemical changes led to the discovery of many chemical elements that could not be broken down into simpler substances. The fact that these elements could only combine with one another in fixed proportions suggested the existence of atoms. It was this way of processing knowledge through observation and experimentation which led to the modern **atomic theory**.

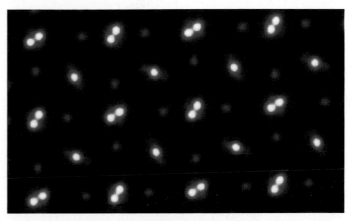

▲ Figure 1 In 2021, scientists at Cornell University captured the most detailed picture of atoms to date. What do models show us that microscope images cannot?

Understandings

Structure 1.1.1 — Elements are the primary constituents of matter, which cannot be chemically broken down into simpler substances.

Compounds consist of atoms of different elements chemically bonded together in a fixed ratio.

Mixtures contain more than one element or compound in no fixed ratio, which are not chemically bonded and so can be separated by physical methods.

Structure 1.1.2 — The kinetic molecular theory is a model to explain physical properties of matter (solids, liquids, and gases) and changes of state.

Structure 1.1.3 — Temperature (in K) is a measure of average kinetic energy (E_k) of particles.

The composition of matter (*Structure 1.1.1*)
Matter and energy

Chemistry is the study of **matter** and its composition. Matter is everywhere. We are made up of matter, we consume it, it surrounds us, and we can see and touch many forms of matter. Air is a form of matter that we know is there, though we cannot see it. The universe is made of matter and chemistry seeks to expand our understanding of matter and its properties. The characteristics of matter are shown in figure 2.

In contrast, energy is anything that exists but does not have these properties. Matter and energy are closely associated with each other, and energy is often considered as a property of matter, such as the ability to perform work or produce heat.

Chemical reactions are introduced in *Reactivity 1.1*.

Although mass and energy can be converted into one another (for example, in nuclear reactors or inside stars), chemistry studies only those transformations of matter where both mass and energy are conserved. In **chemical reactions**, the products have the same mass as starting materials, and the energy is transformed from one form to another rather than created or destroyed.

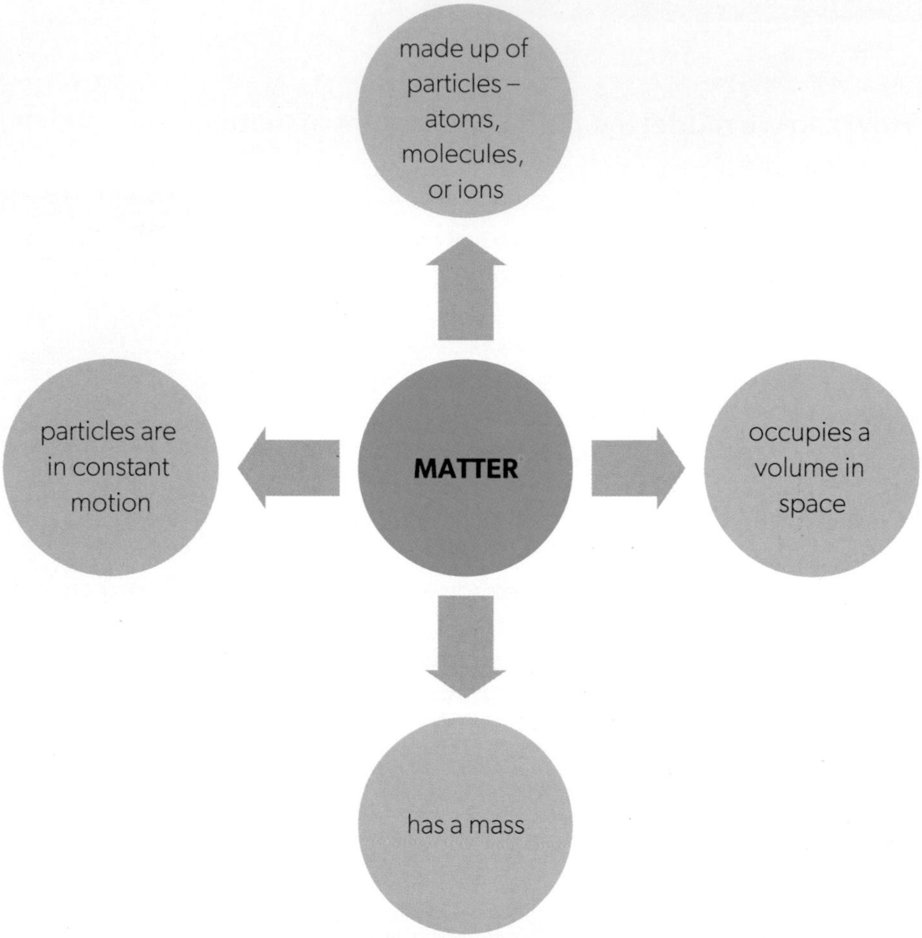

▶ **Figure 2** The characteristics of matter

made up of particles – atoms, molecules, or ions

particles are in constant motion

MATTER

occupies a volume in space

has a mass

ATL Thinking skills

The famous Einstein equation, $E = mc^2$, shows that mass (m) and energy (E) are interconvertible. However, the energy released or absorbed in chemical reactions is relatively small while the speed of light (c) is very large ($3.00 \times 10^8 \, m \, s^{-1}$). As a result, the loss or gain in mass caused by chemical changes is negligible.

This example demonstrates the importance of approximation in science: if the effect of a certain factor is minor, it can often be ignored in calculations without compromising the final result.

What other examples of negligible effects have you encountered in chemistry?

The atomic theory

The law of conservation of mass and the observation that certain substances always combine in definite proportions led to the idea that matter was composed of **elements**. It was theorized that elements combined to form other substances but could not be broken down chemically. Hydrogen and oxygen can react to form water, and experiments showed that the mass of hydrogen and oxygen consumed equalled the mass of water formed. Other experiments showed that 1.0 g of carbon would react with 1.33 g of oxygen through combustion to form carbon monoxide, and with 2.66 g of oxygen to form carbon dioxide.

It was proposed that elements, such as hydrogen, oxygen or carbon, are the primary constituents of matter, and they cannot be chemically broken down into simpler substances. The idea of definite proportions suggested that particles of one element, called **atoms**, would combine with atoms of another element in a fixed, simple ratio, and that atoms of one element have a different mass than atoms of a different element. This, and other experimental evidence, led to the atomic theory.

The atomic theory states that all matter is composed of atoms. These atoms cannot be created or destroyed, but they are rearranged during chemical reactions. Physical and chemical properties of matter depend on the bonding and arrangement of these atoms.

The internal structure and characteristics of atoms will be discussed in *Structure 1.2*.

Evidence

Ancient atomists, among them the Indian sage Uddālaka Aruni and the Greek philosophers Democritus and Leucippus, reasoned that matter was made up of tiny, indivisible particles. They postulated that changes in the natural world are due to interactions between these particles.

In 8th century BCE, Āruni proposed that "particles too small to be seen mass together into the substances and objects of experience". He called the particles "kana". Similarly, in 5th century BCE, Democritus is said to have observed that one could successively snap a seashell into increasingly smaller parts until producing powder composed of indivisible units, known as "atomos", "not splittable", that could not be broken any further.

The next stage in the development of atomic theory, over 2000 years later, is credited to John Dalton. Dalton drew from mass conservation experiments to propose that atoms could be classified into different types known as "elements", based on their masses.

Scientific knowledge must be supported by verifiable evidence. What evidence was used to develop these atomic theories? What is evidence? Is evidence shaped by our perspective?

▲ **Figure 3** Top: Āruni lived in what is now modern day Northern India, by the Ganges river. Bottom: Democritus is depicted in a Renaissance-era painting

Chemical symbols

In modern chemistry, atoms and elements are represented by the same **symbols**, which consist of one or two letters and are derived from the element names. For example, the chemical symbol for hydrogen is H (the first letter of hydrogen), and the chemical symbol for iron is Fe (the first two letters of the Latin *ferrum* "iron"). Common chemical elements and their symbols are listed in table 1; the full list is given in the data booklet and in the periodic table at the end of this book.

Symbol	Name
H	hydrogen
C	carbon
O	oxygen
Na	sodium
Mg	magnesium
S	sulfur
Cl	chlorine
Fe	iron

▲ Table 1 Common chemical elements

Atoms are the smallest units of matter that still possess certain chemical properties. While atoms can exist individually, they tend to combine together and form chemical substances. **Elementary substances** contain atoms of a single element, while **chemical compounds** contain atoms of two or more elements bound together by chemical forces. For example, magnesium metal is an elementary substance, as it contains only one type of atom, Mg. Similarly, sulfur (S) is another elementary substance composed of sulfur atoms only. In contrast, magnesium sulfide (MgS) is a chemical compound, as it consists of two different, chemically bound atomic species, Mg and S (figure 4). MgS is the **chemical formula** of magnesium sulfide.

▲ Figure 4 Magnesium (left), sulfur (middle) and magnesium sulfide (right)

Pure substances and mixtures

Matter can be classified as a pure substance or a mixture, depending on the type of particle arrangement (figure 5).

matter – any substance that occupies space and has mass

pure substance – has a definite and uniform chemical composition

mixture – a combination of two or more pure substances that retain their individual properties

element – composed of one kind of atoms, e.g., magnesium (Mg), sulfur (S)

compound – composed of two or more kinds of atoms in a fixed ratio, e.g., magnesium sulfide (MgS), water (H_2O)

homogeneous – has uniform composition and properties throughout, e.g., sea water, metal alloy

heterogeneous – has non-uniform composition and varying properties, e.g., paint, salad dressing

▲ Figure 5 How matter is classified according to the arrangement of particles

Pure substances cannot be separated into individual constituents without a chemical reaction, which alters their physical properties. In contrast, mixtures can be separated into individual components that retain their respective physical properties.

Data-based questions

A student had two pure substances, A and B. They were heated in separate crucibles and some qualitative and quantitative observations were made and recorded in table 2.

▲ Substance A

▲ Substance B

▲ Appearance after heating each of the two substances

Substance	Observations before heating	Mass of crucible and substance / g	Mass of crucible and contents after heating / g	Change in mass / g	Observations after heating
A	Red colour	26.12 ± 0.02	26.62 ± 0.02		Black colour
B	Green colour	27.05 ± 0.02	25.76 ± 0.02		Black colour

▲ Table 2 Results from heating substances A and B

1. Calculate the change in mass for substances A and B.

2. State a qualitative observation from the experiment performed on A and B.

3. Melting ice is a physical change while rusting iron is a chemical change. Explain, using the observations, whether the changes to substances A and B represented a physical change or a chemical change.

4. A and B were both pure substances, not mixtures. Discuss whether the experiment shows that A and B are elements.

5. Both A and B turned black on heating. Can it be concluded that the heating of these two substances produced the same substance?

Melting point determination

Melting point data can be used to assess the purity of a substance. Pure substances have sharp melting points, which means they melt at a specific temperature that closely matches the theoretical value. The presence of impurities in a substance lowers its melting point and causes melting to occur over a temperature range.

Relevant skills
- Tool 1: Melting point determination
- Inquiry 2: Identify and record relevant qualitative observations and sufficient relevant quantitative data

Materials
- Melting point apparatus
- Capillary tubes
- Samples of two known organic solids, for example, aspirin and salol (phenyl 2-hydroxybenzoate)

Safety
- Wear eye protection.
- Note that the melting point apparatus gets very hot.
- You teacher will give you further safety precautions, depending on the identity of the solids being analysed (for example, salol and aspirin are irritants and environmentally hazardous).

Method
(Your teacher will provide specific instructions, depending on the identity of the solids being analysed.)
1. Obtain samples of two organic solids (A and B) for analysis.
2. Prepare samples of each solid in two separate capillary tubes.
3. Following your teacher's instructions, mix small amounts of the two solids together.
4. Prepare, in a third capillary tube, a small sample of the mixture of the two solids.
5. Determine the melting point of your three samples (A, B and the mixture).

Questions
1. Record relevant qualitative and quantitative data in an appropriate format.
2. Comment on the results, comparing the melting points of pure substances with impure substances.
3. Research the structural formulas of A and B and use this information to explain the difference in their melting points.
4. To what extent could melting point data be used to analyse the success of an organic synthesis?

Methods for determining the melting point of a substance are discussed in the *Tools for chemistry* chapter.

The most common homogeneous mixtures, aqueous solutions, will be discussed in *Reactivity 3.1*, and the properties of metal alloys in *Structure 2.4*.

Mixtures contain more than one element or compound in no fixed ratio, which are not chemically bonded and so can be separated by physical methods. Mixtures can be **homogeneous**, in which the particles are evenly distributed. Air is a mixture of nitrogen, oxygen, and small amounts of other gases. Air is a homogeneous mixture, and its composition of roughly 80% nitrogen and 20% oxygen is consistent regardless of where air is sampled.

If the particles are not evenly distributed, such as in a mixture of two solids, then the mixture is referred to as **heterogeneous**. Natural milk will have the cream rise to the top, which reveals that milk is a heterogeneous mixture.

Each component of a mixture maintains its physical and chemical properties. For example, hydrogen, H_2, is explosive, and oxygen, O_2, supports combustion. When these substances are present in a mixture, their properties stay the same. In contrast, water, H_2O, is not a mixture of hydrogen and oxygen but a chemical compound formed by bonding two hydrogen atoms with one oxygen atom. The new substance has none of the properties of hydrogen or oxygen. It is not a gas, is not explosive, and it does not support combustion. It is a pure substance with its own properties and the hydrogen and oxygen cannot be separated from water without a chemical reaction, which creates new substances.

Separating mixtures

Mixtures can be separated by physical means because each component of the mixture has unique properties. A mixture of iron and sulfur powders can be separated using a magnet. Iron is magnetic while sulfur is not. This difference in property is used to separate them. The compound iron(II) sulfide, FeS, is not magnetic and does not have a sulfurous smell. It maintains none of the properties of the components as it is a new, individual pure substance.

Two solids can usually be separated if we understand their intermolecular forces. Sand can be separated from sugar because sugar will dissolve in water, due to the intermolecular attractions between sugar and water.

The solid mixture of sand and sugar is placed in water and the sugar dissolves. The solution can then be poured through filter paper placed inside a funnel, a process called **filtration** (figure 6). The large sand particles will not pass through and remain on the filter paper, whereas the sugar dissolved in the water will pass through the filter paper. The wet sand is dried, and the water evaporates leaving behind the pure sand. The sugar can be obtained by evaporating the water from the **filtrate** — the solution which passed through the filter paper. Sugar crystals will form in this **crystallization** process (figure 7).

Intermolecular forces are discussed in *Structure 2.2.*

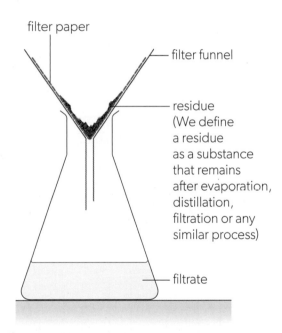

filter paper

filter funnel

residue
(We define a residue as a substance that remains after evaporation, distillation, filtration or any similar process)

filtrate

▲ Figure 6 Filtration apparatus

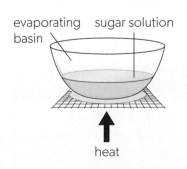

evaporating basin   sugar solution

heat

solution from evaporating basin

cold tile

leave for a few days for sugar to crystallize

◄ Figure 7 The crystallization process

Distillation can be used to separate **miscible** liquids with different boiling points, such as ethanol and water. Ethanol has a lower boiling point and will evaporate first. Once the vapours rise up a cooling column, they can be condensed to a liquid. As shown in figure 8, cold water surrounds the condenser and allows the vapours to condense to liquid ethanol. The water remains mostly in the distillation flask.

▶ Figure 8 Distillation apparatus

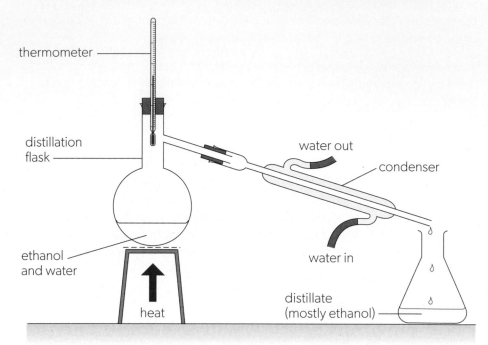

Paper chromatography will be discussed in more detail in *Structure 2.2.*

Paper chromatography can be used to separate substances such as components in inks. A piece of chromatography paper is spotted with the mixture. The bottom of the paper, below the spot, is placed in a suitable solvent as in figure 9(a).

The substances in the mixture have different affinities for the solvent (the **mobile phase**) and the paper (the **stationary phase**). The affinity depends on the **intermolecular forces of attraction** between the pure substances in the mixture and the solvent or the paper. Figure 9(c) shows a mixture that was composed of five pure substances.

▶ Figure 9 The stages in 2D paper chromatography

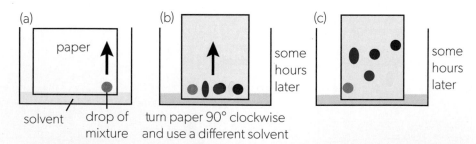

Data-based questions

Look at figure 9.
1. Which colour dot had the strongest affinity for both solvent 1 and solvent 2?
2. Which colour dots had a stronger affinity for solvent 1 than solvent 2?
3. Which had a stronger affinity for solvent 2 than solvent 1?

Table 3 shows a summary of the separation techniques discussed.

Technique	Description	Components	
		removed	left
filtration	mixture is poured through a paper filter or other porous material	liquid(s)	solid(s)
dissolution (solvation)	mixture is added to water or an organic solvent	soluble substance(s)	insoluble substance(s)
crystallization	mixture is dissolved in hot water or an organic solvent, the solution cools down, and the crystals formed are isolated by filtration	more soluble substance(s)	less soluble substance(s)
evaporation or distillation	mixture is heated up until one or more of its components vaporize(s)	volatile liquid(s)	solid(s) and/or non-volatile liquid(s)
paper chromatography	mixture is placed on a piece of paper; one side of the paper is submerged in water or a solvent; components move along the paper	more soluble component(s) move(s) faster	less soluble component(s) move(s) slower or stay(s) in place

▲ Table 3 Summary of separation techniques

 Activity

Suggest a suitable method for separating each of the following mixtures:

a. salt and pepper
b. several water-soluble dyes
c. sugar and water
d. iron and copper filings

For each mixture, describe the separation technique and outline how each component is isolated.

▲ Figure 10 An advanced filtration technique called reverse osmosis extracts salt from seawater, providing fresh water for millions of people. However, this process requires vast amounts of energy, most of which is currently provided by fossil fuels. Why might it be important to consider alternative energy sources?

Planning experiments and risk assessments

Relevant skills
- Tool 1: Separation of mixtures
- Tool 1: Addressing safety of self, others and the environment

Instructions
1. Using the ideas in this chapter, devise a method that would allow you to separate a mixture containing sand, salt, iron filings and powdered calcium carbonate. In doing so, you must consider the physical and chemical properties of each of these four substances.
2. Once you have decided on a method, identify the hazards and complete a risk assessment protocol in which you:
   - Identify the **hazards**
   - Assess the level of **risk**
   - Determine relevant **control measures**
   - Identify suitable disposal methods aligned with your school's health and safety policies.
3. If you have time, try it out! Remember that your teacher should validate your methodology and risk assessment beforehand.

Extension
You could evaluate the effectiveness of your method by comparing the mass of each component (sand, salt, iron filings, and calcium carbonate) before and after the separation. Measure the mass of each component prior to mixing them together. Then mix them together, carry out your separation, make sure the components are all dry, and measure the mass of each again. Compare the masses before and after to calculate the percentage recovery of each component.

 Linking questions

What factors are considered in choosing a method to separate the components of a mixture? (Tool 1)

How can the products of a reaction be purified? (Tool 1)

How do intermolecular forces influence the type of mixture that forms between two substances? (Structure 2.2)

Why are alloys generally considered to be mixtures, even though they often contain metallic bonding? (Structure 2.3 and Structure 2.4)

States of matter (*Structure 1.1.2*)
Solids, liquids and gases

Matter is composed of particles. The types of interactions between these particles determine the state of matter of a substance: solid, liquid or gas. All substances can exist in these three states, depending on the temperature and pressure.

The states of matter of substances are shown by letters in brackets after the formula: (s) for solid, (l) for liquid and (g) for gas. For example:

* Water is a solid below 0 °C: $H_2O(s)$

* Water is a liquid between 0 and 100 °C: $H_2O(l)$

* Water is a gas above 100 °C: $H_2O(g)$.

A special symbol, (aq), is used for molecules or other species in aqueous solutions. For example, the expression "NaCl(aq)" tells us that sodium chloride is dissolved in water while "NaCl(s)" refers to the pure compound (solid sodium chloride). The properties of the three states of matter are summarized in figure 11.

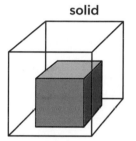
solid

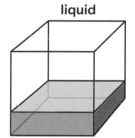
liquid

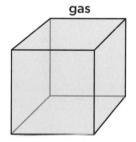
gas

solid	liquid	gas
• fixed volume	• fixed volume	• no fixed volume
• fixed shape	• no fixed shape	• no fixed shape
• cannot be compressed	• cannot be compressed	• can be compressed
• attractive forces between particles are strong	• attractive forces between particles are weaker than those in solids	• attractive forces between particles are negligible
• particles vibrate in fixed positions but do not move around	• particles vibrate, rotate, and move around	• particles vibrate, rotate, and move around faster than in a liquid

▲ **Figure 11** Steam, liquid water and ice are the three states of water

Changes of state

Substances change their states of matter as they absorb or release energy. Solid ice will absorb energy as it is heated. The particles continue to vibrate in fixed positions, but more violently, until a temperature known as the melting point is reached. At this point, the ice melts (changes its state from solid to liquid). A further increase in temperature accelerates the movement of particles, and eventually the water vaporizes and becomes a gas. The decrease in temperature reverses these changes of state.

Under certain conditions, solid substances can turn into gases directly, without melting. This change of state, known as **sublimation**, is typical for dry ice (solid carbon dioxide, $CO_2(s)$, figure 12), which is commonly used for refrigerating ice cream and biological samples.

▲ **Figure 12** Sublimation of dry ice

▲ Figure 13 A snowflake, the product of deposition of water

The process opposite to sublimation is called *deposition*. At low temperatures, water vapour in the air solidifies and forms snowflakes of various shapes and sizes (figure 13).

When a substance changes from a more condensed state to a less condensed state, energy is absorbed by the particles from the surroundings. This happens when a solid becomes a liquid or a gas, and when a liquid becomes a gas. These are **endothermic** processes.

When a substance changes from a less condensed state to a more condensed state, the particles lose energy to the surroundings and, for a molecular substance, the intermolecular forces become stronger. This happens when a gas becomes a liquid or a solid, and when a liquid becomes a solid. The process of releasing energy to the surroundings is an **exothermic** process.

The changes of state occurring in these transformations are shown in figure 14.

Non-Newtonian fluids

Some substances, known as **non-Newtonian fluids**, do not behave like typical liquids. The **viscosity** of non-Newtonian fluids varies depending on the force applied to them. You will make a non-Newtonian fluid commonly known as maize starch slime or "oobleck", and explore its properties.

Relevant skills
- Inquiry 1: Identify dependent and independent variables
- Inquiry 2: Identify and record relevant qualitative observations

Safety
Wear eye protection.

Materials
- Spoon or large spatula
- 250 cm³ beaker
- Powdered maize starch
- Water

Method
1. Add three or four heaped spoons of maize starch to the beaker. Note its appearance and consistency.

2. Slowly add water to the maize starch and mix. Continue adding water until the mixture achieves a thick consistency. Adjust by adding more maize starch or more water, as needed.
3. Spend some time exploring the properties of your mixture. It should harden if tapped, and flow smoothly if stirred slowly.

Questions
1. Describe the properties and identify the state of matter of each of the following:
   - powdered maize starch
   - water
   - the maize starch–water mixture.
2. Suppose you were asked to develop a research question relating to a maize starch–water mixture. Consider possible independent and dependent variables.
3. Research non-Newtonian fluids and identify other examples of these substances.
4. How has this experience changed the way you think about states of matter and their properties? Reflect on this, completing the following sentence starters:
   - I used to think...
   - Now, I think...

 Linking questions

Why are some substances solid while others are fluid under standard conditions? (Structure 2.4)

Why are some changes of state endothermic and some exothermic? (Structure 2, Reactivity 1.2)

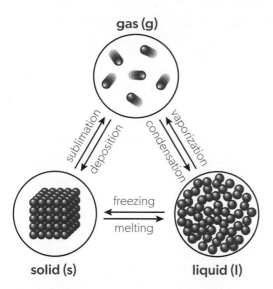

▲ Figure 14 Endothermic and exothermic changes of state

▲ Figure 15 Orange growers spray their fruit with water on cold nights. Freezing of water is an exothermic process that releases energy (in the form of heat) to the fruit, protecting it against cold

Kelvin temperature scale (*Structure 1.1.3*)

As temperature rises, the energies of particles increase. **Temperature** is a measure of the average kinetic energy of particles. As substances absorb energy, particles of a solid vibrate in the lattice more, particles in a liquid vibrate more and move faster, while in a gas they move faster.

When water is heated, there is no temperature change during the periods when a solid changes to a liquid and when a liquid changes to a gas (figure 16). The added energy is used to disrupt the solid lattice and overcome the intermolecular forces between molecules in the liquid.

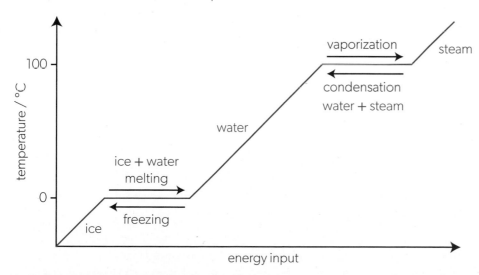

▲ Figure 16 Graph of the heating curve for water

There were many attempts to measure relative temperature, but the first widely accepted temperature scale was introduced by the Polish-born Dutch physicist Daniel Gabriel Fahrenheit.

The kelvin is the base unit of temperature measurement in the International System of Units (SI). There are seven base units, and all other units of measurements can be derived from these (figure 17).

▲ Figure 17 The seven base SI units are kilogram (kg) for mass, meter (m) for length, second (s) for time, ampere (A) for electric current, kelvin (K) for temperature, mole (mol) for amount of substance, and candela (cd) for luminous intensity. All units of measurement can be derived from these seven base units

You will learn more about the mole in *Structure 1.4*.

Measurement

Making, recording, and communicating measurements greatly benefits from agreed upon scales. The International Bureau of Weights and Measures (BIPM, from the French *Bureau international des poids et mesures*), established in the late 19th century, is an international organisation which seeks to set up and continuously refine measurement standards.

The International System of Units (SI, from the French *Système international d'unités*) is the most commonly used system of measurement. Its building blocks are the seven base units: length (metre, m), mass (kilogram, kg), time (second, s), electric current (ampere, A), temperature (kelvin, K), amount of substance (mole, mol) and luminous intensity (candela, cd). All other units, such as those of volume (m^3), density ($kg\,m^{-3}$), energy (joule, J, where $1\,J = 1\,kg\,m^2\,s^{-2}$) and so on, are derived from the seven base units.

The base units are defined according to seven constants, including several that you will recognize, such as the Boltzmann constant, k; speed of light, c; the Avogadro constant, N_A; and the Plank constant, h.

The use of universal and precisely defined units is very important, as it allows scientists from different countries to understand one another and share the results of their studies. What other advantages are there to internationally shared and continuously updated measurement systems in the natural sciences? You might want to look up the Mars Climate Orbiter.

▲ Figure 18 A platinum–iridium cylinder in the US was used to define a kilogram of mass. This standard became obsolete in 2019, when the kilogram and all other SI units were redefined as exact quantities based on physical constants

(ATL) **Thinking skills**

Throughout history, several universal temperature scales have been developed, each with different reference points. Some of these are summarized in table 4.

Scale	Date	Reference points
Newton	1700s	H_2O freezing point = 0° Human body temperature = 12°
Fahrenheit	1700s	H_2O freezing point = 32° H_2O boiling point = 212°
Delisle	1700s	H_2O freezing point = 150° H_2O boiling point = 0°
Celsius	1700s	H_2O freezing point = 0° H_2O boiling point = 100°
Kelvin	1800s	Absolute zero = 0
CGPM	1950s	Triple point of water = 273.16 K
BIPM	2018	Kelvin defined in terms of the Boltzmann constant, k.

▲ Table 4 Examples of various temperature scales

Temperature is related to thermal energy and as such it could be expressed in the unit for energy, joules (J), which are in turn defined in terms of the base units kg, m and s. It has been decided to keep kelvin as an SI base unit "for historical and practical reasons". What do you think some of these historical and practical reasons could be?

Look carefully at table 4 above. Identify one thing you see, one thing it makes you think about, and one thing it makes you wonder. Share your ideas with your class.

Kelvin temperature is proportional to the average kinetic energy of particles and is considered an absolute scale.

Absolute zero (0 K) implies that at this temperature the particles cannot transfer any kinetic energy on collisions. Matter at absolute zero cannot lose heat and hence cannot get any colder. An increase in temperature of 1 kelvin is equivalent to an increase in temperature of 1 degree Celsius. 0 °C is equal to 273.15 K. Under normal pressure, water boils at 100 °C, so that makes the boiling point of water 373.15 K. Absolute zero on the Celsius scale is −273.15 °C.

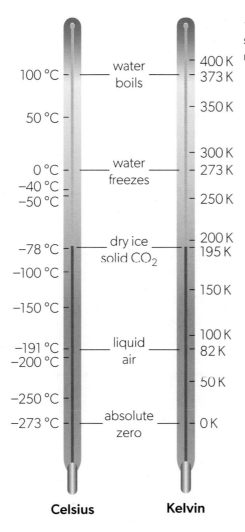

◄ Figure 19 The Celsius and Kelvin scales for temperature (all values are rounded to whole numbers)

You will learn more about the kinetic energy of particles in *Reactivity 2.2*.

Celsius   **Kelvin**

 Linking questions

What is the graphical distribution of kinetic energy values of particles in a sample at a fixed temperature? (Reactivity 2.2)

What must happen to particles for a chemical reaction to occur? (Reactivity 2.2)

End-of-topic questions

Topic review

1. Using your knowledge from the *Structure 1.1* topic, answer the guiding question as fully as possible:

   How can we model the particulate nature of matter?

Exam-style questions

Multiple-choice questions

2. Which of the following are examples of homogeneous mixtures?

   I.   Air

   II. Steel

   III. Aqueous potassium manganate(VII), $KMnO_4(aq)$.

   A. II only

   B. III only

   C. I and II only

   D. I, II and III

3. What correctly describes the sublimation of dry ice (carbon dioxide)?

	Exothermic or endothermic?	Equation describing the process
A	exothermic	$CO_2(s) \rightarrow CO_2(g)$
B	exothermic	$CO_2(s) \rightarrow C(g) + O_2(g)$
C	endothermic	$CO_2(s) \rightarrow CO_2(g)$
D	endothermic	$CO_2(s) \rightarrow C(g) + O_2(g)$

4. Which of the following methods could be used to obtain solid sodium chloride from a solution of sodium chloride in water?

   I.   evaporation

   II. filtration

   III. distillation

   A. I only

   B. I and II only

   C. I and III only

   D. I, II and III

5. Which changes of state are opposite to each other?

   A. melting and condensation

   B. vaporization and deposition

   C. deposition and sublimation

   D. sublimation and freezing

6. Which of the following statements is incorrect?

   A. solids and liquids are almost incompressible

   B. particles in both solids and liquids are mobile

   C. liquids and gases have no fixed shape

   D. particles in solids, liquids and gases can vibrate

7. Which elements can be separated from each other by physical methods?

   A. oxygen and nitrogen in air

   B. hydrogen and oxygen in water

   C. carbon and oxygen in dry ice

   D. magnesium and sulfur in magnesium sulfide

8. Which change in temperature on the Celsius scale is equivalent to the increase in temperature by 20 K?

   A. decrease by 20 °C

   B. increase by 20 °C

   C. decrease by 293.15 °C

   D. increase by 293.15 °C

Extended-response questions

9. Explain why the Kelvin temperature is directly proportional to average kinetic energy but the Celsius temperature is not, even though a 1-degree temperature increment is the same in each scale? [2]

10. Ionic salts can be broken down in electrolysis. The unbalanced ionic equation for the electrolysis of molten lead(II) bromide is:

    $Pb^{2+} + Br^- \rightarrow Pb + X$

    a. One of the products is lead, Pb. State the formula of product X. [1]

    b. Balance the equation. [1]

    c. The electrolysis of molten lead(II) bromide is carried out at 380 °C. With reference to melting point and boiling point data, deduce the state of matter of each of the species in the equation at this temperature. Write state symbols in the balanced equation you gave in (b). [2]

11. The kinetic energy of particles is equal to half of their mass × the square of the velocity of the particles: $E_k = \frac{1}{2}mv^2$. Determine how much the speed of molecules in a pure gaseous substance will increase when the Kelvin temperature is doubled. [2]

12. Pure caffeine is a white powder with melting point 235 °C.

   a. State the melting point of caffeine in kelvin. [1]

   b. A chemist is investigating the efficacy of three caffeine extraction methods. The theoretical yield in all three cases is 0.960 g. She uses each method once and collects the following data for the yield and melting point of the product:

	Method 1	Method 2	Method 3
Mass of caffeine obtained / g	0.229	0.094	0.380
Melting point of caffeine product / °C	190–220	229–233	188–201

     i. Calculate the mean and range of the mass of caffeine obtained. [2]

     ii. Calculate the percentage yield of Method 1. Give your answer to an appropriate number of significant figures. [2]

     iii. Determine, giving a reason, which method gave the purest caffeine product. [1]

   c. Suggest one way to minimize the random error in this experiment. [1]

13. A student prepares a copper(II) sulfate solution by reacting dilute sulfuric acid with excess copper(II) oxide. Copper(II) oxide is insoluble in water.

The word equation for this reaction is as follows:

sulfuric acid + copper(II) oxide → copper(II) sulfate + water

   a. Write a balanced chemical equation, including state symbols, for this reaction. [2]

   b. The acid was heated, then copper(II) oxide powder was added until it was in excess and could be observed suspended in the solution, quickly sinking to the bottom of the beaker. Suggest, giving a reason, a method the student could use to remove the excess copper(II) oxide. [2]

   c. Once the excess copper(II) oxide had been removed, the student needed to figure out how to obtain pure crystals of copper(II) sulfate from the solution. Describe a method the student could follow to obtain pure, dry copper(II) sulfate crystals. [3]

14. Study the figure below.

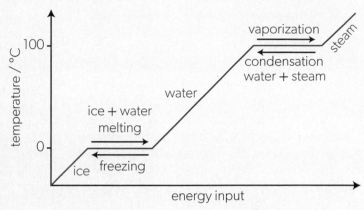

   a. Explain why, in spite of the increasing energy input, the temperature of the sample remains constant at 0 °C for a period of time. [2]

A solution of 5.00 g of sodium chloride in 100.0 g of pure water (at standard atmospheric temperature and pressure) has the following properties:

• melting point: –3 °C

• boiling point: 101 °C

   b. Sketch a graph similar to the one in figure 16 to show the heating curve for a sample of this sodium chloride solution. [2]

15. Elemental iodine exists as diatomic molecules, I_2. At room temperature and pressure, it is a lustrous purple-black solid that readily forms violet fumes when heated gently. When cooled, gaseous iodine deposits on cold surfaces without condensing. Under increased pressure, solid iodine melts at 114 °C to form a deep-violet liquid.

   a. Formulate equations that represent all changes of state mentioned above. [3]

   b. State the melting point of iodine in kelvin. [1]

   c. Suggest how liquid iodine can be obtained from gaseous iodine. [1]

Structure 1.2 The nuclear atom

How do nuclei of atoms differ?

The answer to this question was obtained by over 100 years of brilliant research. Sometimes, the question of *how we know* is more fascinating than the question of *what is known*.

In the late 1800s, the idea that matter was composed of atoms that were indivisible and rearranged in chemical reactions (known as the atomic theory) was gaining popularity. The discovery of electricity and radioactivity allowed scientists to study the structure of the atom itself.

Understandings

Structure 1.2.1 — Atoms contain a positively charged, dense nucleus composed of protons and neutrons (nucleons). Negatively charged electrons occupy the space outside the nucleus

Structure 1.2.2 — Isotopes are atoms of the same element with different numbers of neutrons.

Structure 1.2.3 — Mass spectra are used to determine the relative atomic masses of elements from their isotopic composition.

The structure of the atom (*Structure 1.2.1*)

An atom contains a positively charged nucleus, which itself contains protons and neutrons (collectively known as nucleons). Atoms also contain electrons, which occupy the vast region outside of the nucleus. The protons, neutrons and electrons are known as subatomic particles.

The key factors of the nucleus are:

1. It is very small in comparison to the atom itself.

2. It is a highly dense structure containing virtually all the mass of the atom.

3. It has a positive charge.

In an experiment designed by Ernest Rutherford in 1911, positively charged radioactive alpha particles were fired toward a sheet of gold foil. The main observations made are given in figure 1.

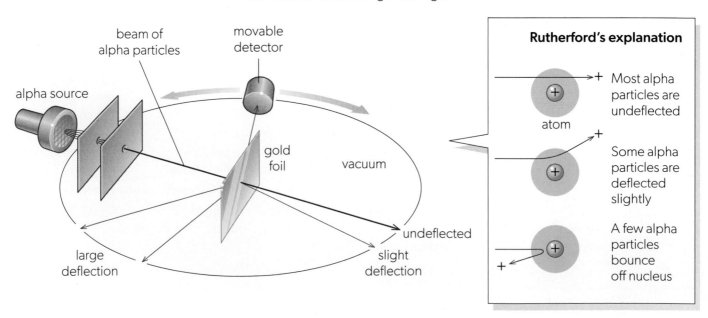

▲ Figure 1 Rutherford's gold foil experiment

Falsification

The gold foil experiment **falsified** the atomic model that preceded it, namely the "plum-pudding model". The plum-pudding model suggested that the atom was an amorphous positively charged blob with electrons present throughout. If this were the case, all alpha particles fired at the gold foil would have gone through its atoms undeflected. Rutherford's results contradicted the existing model, paving the way for the development of a new model of the atom.

Scientific claims are falsifiable. This means that they are vulnerable to evidence that contradicts them. A scientific claim that stands up to severe testing is strong but can never be proven true with absolute certainty. Scientific knowledge is therefore always accompanied by a degree of uncertainty. The provisional nature of scientific knowledge means that further evidence can steer it in new directions.

Can a single counterexample falsify a claim?

Activity

The lists below show the observations in the gold foil experiment and the properties of the nucleus. Determine which observation is explained by which property.

Observation	Property
Nearly all the alpha particles went straight through the gold foil.	The nucleus has a positive charge.
Occasionally, some of the alpha particles bounced straight back.	The nucleus is very small in comparison to the size of the atom.
The alpha particles are repelled when closely approaching the nucleus.	The nucleus is very dense, containing virtually all the mass of the atom.

In 1911, Rutherford summarized the results of his experiments by proposing the **planetary model of the atom**, also known as the **Rutherford model** (figure 2). In this model, negatively charged electrons orbit the positively charged atomic nucleus in the same way as planets orbit the Sun. Just as the Sun contains 99.8% of the solar system's mass, the atomic nucleus contains over 99.9% of the mass of the entire atom. However, instead of by gravity, the electrons are held around the nucleus by electrostatic attraction.

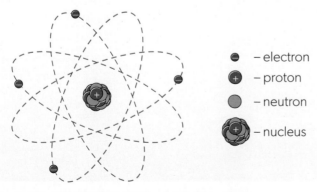

▲ Figure 2 The Rutherford model of the atom

21

Models

Scientists use models to represent natural phenomena. All models have limitations, which should be identified and understood. Consider the depiction of the atom in figure 2. The size of the nucleus is exaggerated but it serves as a useful model of the nuclear atom.

The vast space in the atom compared to the tiny size of the nucleus is hard to fully appreciate. Rutherford's native New Zealand is a great rugby-playing nation. Imagine being at Eden Park stadium (figure 3) and looking down at the centre of the pitch from the top row of seats. If a golf ball were placed at the centre of the field, the distance between you and the golf ball would represent the distance between the electron and the nucleus.

The relative volume of open space in the atom is vast, and our simple representation of Rutherford's atomic model in figure 2 is obviously unrealistic. The nucleus occupies a tiny volume of the atom and the diameter of an atom is approximately 100 000 times the diameter of the nucleus.

Atoms themselves are extremely small. The diameter of most atoms is in the range 1×10^{-10} to 5×10^{-10} m. The unit used to describe the dimensions of atoms is the **picometre, pm**:

$1 \, pm = 10^{-12} \, m$

In X-ray crystallography a commonly used unit for atomic dimensions is the angstrom, symbol Å:

$1 \, \text{Å} = 10^{-10} \, m$

For example, the atomic radius of the fluorine atom is 60×10^{-12} m (60 pm). To convert this to Å we can use **dimensional analysis**, using the conversion factors given above:

$$60 \, pm \times \frac{10^{-12} \, m}{1 \, pm} \times \frac{1 \, \text{Å}}{10^{-10} m} = 0.60 \, \text{Å} = 6.0 \times 10^{-1} \, \text{Å}$$

In spite of its limitations, Rutherford's work has formed the basis of much of our thinking on the structure of the atom. Rutherford is rumoured to have said to his students:

All science is either physics or stamp collecting!

▲ Figure 3 Eden Park, Auckland, New Zealand. If the atom were the size of the stadium, the nucleus would look like a golf ball in the centre of the field

TOK

All the models we have discussed assume that atoms are real. However, it could be argued that objects are only "real" when they can be seen. In 1981 two physicists, Gerd Binnig and Heinrich Rohrer, working at IBM in Zurich, Switzerland invented the scanning tunnelling microscope (STM), an electron microscope that generates three-dimensional images of surfaces at the atomic level. This gave scientists the ability to observe individual atoms directly. The Nobel Prize in Physics in 1986 was awarded to Binnig and Rohrer for their groundbreaking work.

You can find an atomic scale film created by IBM called *A Boy and his Atom* on the internet.

▲ Figure 4 A still from *A Boy and his Atom*

Has technology extended human's capacity to make observations of the natural world?

How important are material tools in the production or acquisition of knowledge?

Other experiments have shown that the nucleus also contains a neutral subatomic particle, the neutron, with nearly the same mass as the proton. The relative masses and charges of the subatomic particles are shown in table 1.

Particle	Relative mass	Relative charge	Location
proton	1	+1	nucleus
neutron	1	0	nucleus
electron	negligible	−1	outside nucleus

◀ Table 1 Relative masses and charges for the proton, neutron and electron

The electric charge carried by a single electron is known as the **elementary charge** (e) and it has a value of approximately 1.602×10^{-19} C. The charges of subatomic particles are commonly expressed in elementary charge units. For example, the charge of an electron can be represented as $-e$, and the charge of a proton as $+e$. The symbol e is often omitted, so it is customary to say that electrons and protons have charges of −1 and +1, respectively.

The actual masses and charges of these particles can be found in the data booklet.

How small is small?

Relevant skills
- Tool 3: Apply and use SI prefixes and units
- Tool 3: Use and interpret scientific notation

Instructions

1. A variety of small lengths are shown in table 2. Without looking at their lengths, but rather based on what you know about each item, list these objects in order of size, from smallest to largest.

Item	Length
proton, charge radius	0.84 fm
sheet of paper, thickness	0.10 mm
onion cell, diameter	250 µm
iodine-iodine bond, length	267 pm
printed full stop, diameter	0.30 mm
carbon atom, diameter	150 pm
C_{60} fullerene, diameter	0.71 nm

▲ Table 2 Lengths of various small items

2. Convert the length values into metres and state them in standard form to two significant figures. Refer to the following conversion factors:
   - milli, m: 10^{-3}
   - micro, µ: 10^{-6}
   - nano, n: 10^{-9}
   - pico, p: 10^{-12}
   - femto, f: 10^{-15}
3. List the length values in table 2 in order of increasing size. Was the list you gave for question 1 correct?
4. Conduct a web search to find three more values to add to the list: one smaller than the values given in table 2, one larger, and one intermediate.

5. Provide the full reference for your information sources in question 4, following your school's citing and referencing system.

Atomic number and the nuclear symbol

As of 2023, there are 118 known elements, given atomic numbers 1 to 118. The atomic number of an element is also the number of protons in the nucleus of that atom. Gold, atomic number 79, has 79 protons, while carbon, atomic number 6, has 6 protons. As all the relative mass is in the nucleus, the difference between the atomic number and mass number is the number of neutrons in the element. Gold has atomic number 79 and mass number 197. Therefore, it has 197 – 79 = 118 neutrons. Each element is neutral, with no charge, so the number of electrons in a neutral atom must equal the number of protons.

Activity

Determine the missing values from the table.

Atomic symbol	Atomic number	Mass number	Protons	Neutrons	Electrons
O				8	
	13	27			
		85	37		
		80			35
	27			32	
				120	80
Pb		207			
			69	100	

Chemists frequently use nuclear symbol notation, $^{A}_{Z}X$, to denote the number of neutrons, protons and electrons in an atom. A represents the mass number of the isotope, Z is the atomic number, and X is the chemical symbol (figure 5). Gold, for example, with mass number 197 and atomic number 79, would have a nuclear symbol notation of $^{197}_{79}Au$.

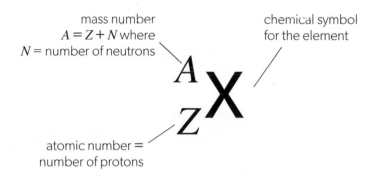

mass number
$A = Z + N$ where
N = number of neutrons

chemical symbol
for the element

$$^{A}_{Z}X$$

atomic number =
number of protons

▲ Figure 5 The nuclear symbol notation

Atoms form compounds by sharing or transferring electrons. As a result, these atoms sometimes are no longer neutral, having more or fewer electrons than protons. For example, magnesium atoms react with oxygen atoms to produce the ionic compound magnesium oxide. Magnesium loses two electrons to form a magnesium **ion** with a 2+ charge, as the number of positively charged protons in the nucleus (12) is two greater than the number of negatively charged electrons remaining (10).

The resulting charge is also displayed in the nuclear symbol notation below:

mass number: 24
(12 protons + 12 neutrons)

charge: 2+
(12 protons – 10 electrons)

$$^{24}_{12}Mg^{2+}$$

atomic number: 12
(12 protons)

chemical element: Mg
(magnesium)

The oxygen atom gains the two electrons lost by magnesium to produce an oxide ion with a 2– negative charge. The nuclear symbol for the oxide ion is $^{16}_{8}O^{2-}$.

The overall **chemical equation** for the reaction between magnesium and oxygen is

$$Mg + \frac{1}{2}O_2 \rightarrow Mg^{2+} + O^{2-}$$

$Mg^{2+} + O^{2-}$ is more commonly written as MgO, as the opposite charges on the two ions result in a force of attraction between them known as an **ionic bond**. Ionic bonds hold the ions together to form solid magnesium oxide.

Ionic bonding is discussed further in *Structure 2.1*.

 Activity

Deduce the nuclear symbol notation for an ion with 24 protons, 21 electrons, and 28 neutrons.

 Linking questions

What determines the different chemical properties of atoms? (Structure 1.3)

How does the atomic number relate to the position of an element in the periodic table? (Structure 3.1)

Isotopes (*Structure 1.2.2*)

Isotopes are different atoms of the same element with a different number of neutrons. As a result, they have different mass numbers, A, but the same atomic number, Z. Chlorine, for example, has two isotopes: one with mass number 35, $^{35}_{17}Cl$, and one with mass number 37, $^{37}_{17}Cl$. They have similar chemical properties, as they are both chlorine atoms with the same number of electrons, but different physical properties, such as density, because atoms of one isotope are heavier than atoms of the other.

Naturally occurring hydrogen consists of two stable isotopes, hydrogen-1 (protium) and hydrogen-2 (deuterium). The third isotope of hydrogen, tritium (figure 6), is radioactive, so it does not occur in nature in significant quantities.

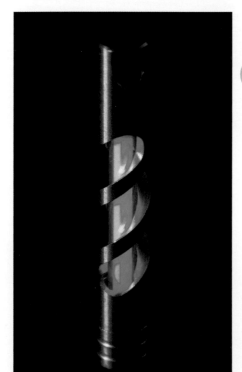

▲ **Figure 6** A portable tritium light source. The radioactive decay of tritium produces high-energy electrons (beta particles). These electrons hit a fluorescent material and make it glow in the dark

 Activity

Copy the table below and complete it by deducing the nuclear symbols and/or composition of these isotopes.

Isotope	Nuclear symbol	Z	N	A
hydrogen-1 (protium)	$^{1}_{1}H$			
hydrogen-2 (deuterium)			1	
hydrogen-3 (tritium)				3

Atomic numbers of isotopes are often omitted in nuclear symbol notation. For example, the isotope of chlorine with mass number 37 can be written as ^{37}Cl. 'Cl' tells you the isotope is chlorine and therefore must have an atomic number of 17, so including the atomic number is not necessary. These isotopes can also be written with a hyphen, such as chlorine-37, or Cl-37. The **relative atomic mass**, A_r, listed for each element on the periodic table is not a whole number because it is the weighted average of all isotopes of that element.

Natural abundance (NA) of an isotope is the percentage of its atoms among all atoms of the given element found on our planet. If we know the natural abundances for all isotopes of an element, we can calculate the average A_r of that element. The opposite task (calculation of natural abundances from A_r) is possible only if the element is composed of two known isotopes.

Worked example 1

Calculate the A_r for iron using the values in the following table.

Isotope	Natural abundance (NA) / %
^{54}Fe	5.845
^{56}Fe	91.754
^{57}Fe	2.119
^{58}Fe	0.282

Solution

We know A_r = average of the natural abundance of each isotope multiplied by their mass numbers. The natural abundance values add up to 100% so we divide by 100 to obtain the average.

Therefore:

$$A_r = \frac{54 \times 5.845 + 56 \times 91.754 + 57 \times 2.119 + 58 \times 0.282}{100} = 55.91$$

Worked example 2

There are two stable isotopes of chlorine: Cl-35 and Cl-37. Calculate the natural abundance (NA) of each isotope given that A_r for chlorine is 35.45.

Solution

$$A_r = \frac{(A \text{ of isotope 1} \times NA \text{ of isotope 1}) + (A \text{ of isotope 2} \times NA \text{ of isotope 2})}{100}$$

Therefore:

$$\frac{(35 \times NA \text{ of Cl-35}) + (37 \times NA \text{ of Cl-37})}{100} = 35.45$$

Let $x = NA$ of Cl-35, then $100 - x = NA$ of Cl-37.

Substituting in the above equation gives:

$$\frac{35x + 37(100 - x)}{100} = 35.45$$

Expanding the brackets and resolving the x terms gives:

$$\frac{3700 - 2x}{100} = 35.45$$

Then rearrange in terms of x:

$$x = \frac{3700 - 3545}{2}$$

$x = 77.5$ and $100 - x = 22.5$. Therefore, the natural abundance of Cl-35 is 77.5% and Cl-37 22.5%.

The actual natural abundances of ^{35}Cl and ^{37}Cl are 75.8 and 24.2%, respectively. The results of our calculations are slightly different because we used mass numbers, which are rounded values for the actual masses of the ^{35}Cl and ^{37}Cl atoms.

Average A_r values for all elements are given in the data booklet and in the periodic table at the end of this book.

▲ Figure 7 A pellet of enriched uranium used as fuel in nuclear reactors

Isotopes have the same nuclear charge and differ only in the number of neutrons. Since neutrons have no effect on the electron configuration (*Structure 1.3*), the chemical properties of isotopes are nearly identical. However, their physical properties differ noticeably, especially in the case of compounds containing isotopes of hydrogen. For example, the density, melting point and boiling point of heavy water (deuterium oxide, ^2H_2O or D_2O) are all higher than those of normal water (hydrogen oxide, ^1H_2O), as shown in table 3.

Compound	Density at 4 °C / g cm^{-3}	Melting point / °C	Boiling point / °C
^1H_2O	1.000	0.00	100.0
^2H_2O	1.106	3.82	101.4

▲ Table 3 Physical properties of normal and heavy water

Naturally occurring uranium consists of two main isotopes, ^{235}U and ^{238}U. The differences in physical properties of these isotopes are used for the **enrichment** (increase in the proportion of ^{235}U over ^{238}U) of nuclear fuel (figure 7), as most nuclear reactors require uranium with at least 3% of ^{235}U, while natural uranium contains only 0.72% of this isotope.

Enriching one type of isotope in a particular substance can also make it possible to track the mechanisms and progress of reactions. This is often referred to as isotope labelling.

▲ Figure 8 Austrian-Swedish physicist Lise Meitner in 1906

 ## Global impact of science

Developments in science and their applications may have ethical, environmental, political, social, cultural and economic consequences. Nuclear fission, which involves splitting up the nuclei of large atoms releasing colossal amounts of energy, is one such development. It has led to the development of nuclear energy, as well as the atomic bomb.

Element 109, meitnerium (Mt), is named after Lise Meitner, the second woman in history to receive a physics doctorate from the University of Vienna (figure 8). Her work with Otto Frisch led to the discovery of nuclear fission, published in *Nature* in 1939. In later years, Meitner was invited to work on atomic bomb technology being developed in the US. She declined, famously stating "I will have nothing to do with a bomb!"

Can you think of other scientific developments that have had important ethical implications?

Practice questions

1. State the nuclear symbols for potassium-39 and copper-65. Deduce the numbers of protons and neutrons in the nucleus of each isotope.
2. Naturally occurring sulfur has four isotopes with the following natural abundances: ^{32}S (95.02%), ^{33}S (0.75%), ^{34}S (4.21%) and ^{36}S (0.02%). Calculate the average A_r value for sulfur.
3. The actual A_r value of sulfur is 32.07. Suggest why your answer to the previous question differs from this value.

Linking question

How can isotope tracers provide evidence for a reaction mechanism? (Reactivity 3.4)

Mass spectrometry (*Structure 1.2.3*)

The **mass spectrometer** (figure 9) is an instrument used to detect the relative abundance of isotopes in a sample.

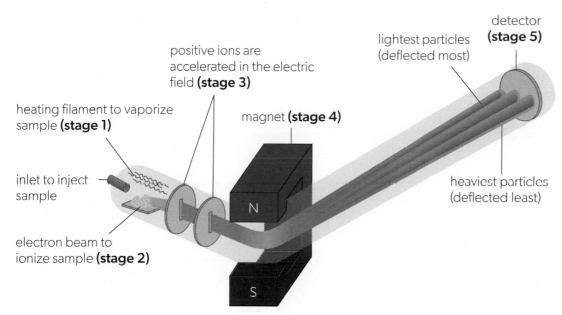

▲ Figure 9 Schematic diagram of a mass spectrometer

The sample is injected into the instrument and vaporized (stage 1). The atoms within the sample are then bombarded with high-energy electrons (stage 2). As a result, the atoms lose some of their electrons to form positively charged ions, known as **cations**. For example, copper atoms can be ionized as follows:

$$Cu(g) + e^- \rightarrow Cu^+(g) + 2e^-$$

The resulting ions are then accelerated by an electric field (stage 3) and deflected by a magnetic field (stage 4). The degree of deflection depends on the mass to charge ratio (m/z **ratio**). Particles with no charge are not affected by the magnetic field and therefore never reach the detector. The species with the lowest m and highest z will be deflected the most. When ions hit the detector (stage 5), their m/z values are determined and passed to a computer. The computer generates the **mass spectrum** of the sample, in which relative abundances of all detected ions are plotted against their m/z ratios (figure 10).

▶ Figure 10 Mass spectrum of a sample of copper

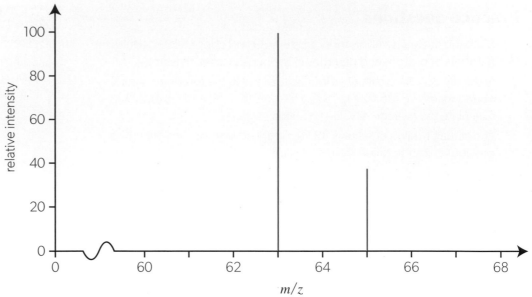

The operational details of the mass spectrometer will not be assessed in examination papers.

Worked example 3

Figure 11 shows a mass spectrum from a sample of boron. Calculate the relative atomic mass, A_r, of boron from this mass spectrum.

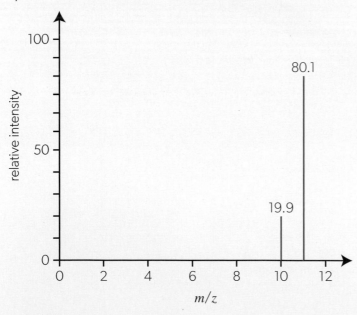

▲ Figure 11 Mass spectrum of boron

Solution

First, we need to derive the information from graph. The peak at $m/z = 10$ represents an isotope with a mass number of 10, which has a relative abundance of 19.9%. The peak at $m/z = 11$ represents an isotope with a mass number of 11, which has a relative abundance of 80.1%.

We can then calculate A_r by finding the sum of the relative abundance of each isotope multiplied by its mass number. The relative abundance values add up to 100%, so we divide the result by 100 to obtain the average.

$$\frac{11 \times 80.1 + 10 \times 19.9}{100} = 10.8$$

Data-based questions

1. Estimate the relative abundance of each isotope from figure 12. Use your estimates to calculate the relative atomic mass, A_r, for this element and identify the element.

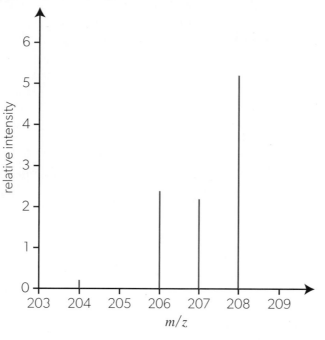

◀ **Figure 12** Mass spectrum of unknown element

2. Mass spectrometry is used for discovering the presence of specific elements in geological samples, including those of cosmic origin. For example, cobalt and nickel are common components of iron meteorites (figure 14).

   Cobalt and nickel have similar properties and nearly identical relative atomic masses. However, the isotopic compositions of these two metals are very different, so they can easily be distinguished by mass spectrometry (figure 13).

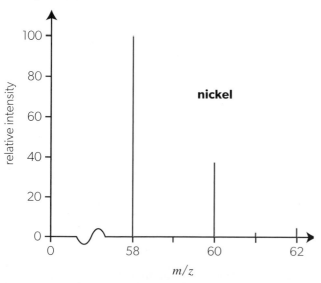
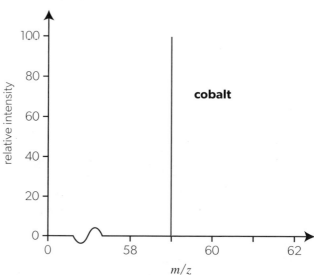

▲ **Figure 13** Mass spectra of cobalt (left) and nickel (right)

   Estimate the relative abundance of each isotope for nickel. Use your estimates to calculate its relative atomic mass, A_r and hence deduce whether cobalt or nickel has the larger A_r.

3. The actual A_r value for nickel is 58.69. Suggest why your result in question 2 is different.

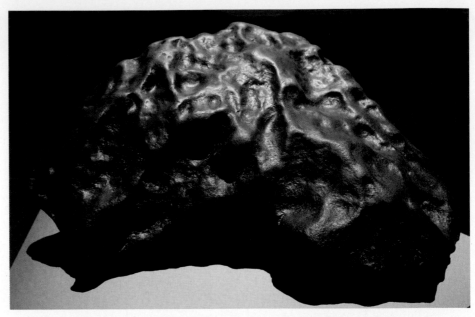

▲ Figure 14 Tamentit iron meteorite, found in 1864 in the Sahara Desert

 Mass spectra

Mass spectra can be found in various databases on the internet, giving you a chance to practice calculating average atomic mass values from authentic data.

Relevant skills
- Tool 2: Identify and extract data from databases
- Tool 3: Percentages

Instructions
1. Using a database of your choice, search for the mass spectra of three different elements.

2. From the mass spectra, calculate the relative atomic mass of each element.

3. Compare your calculated relative atomic mass to that stated in the data booklet. Comment on any differences you observe.

 Linking question

How does the fragmentation pattern of a compound in the mass spectrometer help in the determination of its structure? (Structure 3.2)

End-of-topic questions

Topic review

1. Using your knowledge from the *Structure 1.2* topic, answer the guiding question as fully as possible:

   How do nuclei of atoms differ?

Exam-style questions

Multiple-choice questions

2. What is correct for $^{63}_{29}Cu^{2+}$?

	Protons	Neutrons	Electrons
A	29	34	27
B	29	34	31
C	34	63	31
D	34	29	27

3. Which values are the same for both $^1H_2$ and $^2H_2$?

   I.   boiling point

   II. ΔH of combustion

   III. number of protons

   IV. density

   A. I and III only

   B. I and IV only

   C. II and III only

   D. I, II and III

4. The naturally occurring isotopes of lithium are ^6Li and ^7Li. Which shows the correct approximate percentage abundances for lithium?

	Percentage abundance of ^6Li	Percentage abundance of ^7Li
A	75	25
B	50	50
C	35	65
D	10	90

5. Which of the following statements are correct?

   I.   Nearly all mass of the atom is contained within its nucleus.

   II. The mass number shows the number of protons in an atomic nucleus

   III. Isotopes of the same element have equal numbers of protons.

   A. I and II only

   B. I and III only

   C. II and III only

   D. I, II and III

6. Which of the following species contain equal numbers of neutrons in their nuclei?

   A. cobalt-58 and nickel-58

   B. cobalt-58 and nickel-59

   C. cobalt-59 and nickel-58

   D. cobalt-58 and cobalt-59

Extended-response questions

7. The gold foil experiment involved firing alpha particles at gold foil. This experiment is depicted in figure 1 on page 20.

   a. An alpha particle is a helium nucleus. State the nuclear symbol for an alpha particle. [1]

   b. Suggest the results of the gold foil experiment that would have been observed in each of the following alternative scenarios:

   i. Atoms are instead hard, dense, solid balls of positive charge. [1]

   ii. Atomic nuclei are instead negatively charged. [1]

8. There are two stable isotopes of potassium: ^{39}K and ^{41}K. The A_r of potassium is 39.10. Use this information to determine the relative abundances of the two isotopes and sketch the mass spectrum of potassium metal. [3]

9. "Dutch metal" is an alloy composed of 86% copper and 14% zinc. This alloy closely resembles gold, so it is often used for making costume jewellery. Explain how Dutch metal can be distinguished from gold using mass spectrometry. [2]

AHL

How can we model the energy states of electrons in atoms?

This question is complex with many layers. What are electrons? How do we know they exist in energy states? What various models about these energy states are there?

According to modern views, electrons are quantum objects that behave as both particles and waves. Although such behaviour has no analogues in our everyday life, we can visualize electrons in atoms as fuzzy clouds. The shapes and sizes of these clouds depend on the energies of electrons, which can have only certain, predefined values.

Understandings

Structure 1.3.1 — Emission spectra are produced by atoms emitting photons when electrons in excited states return to lower energy levels.

Structure 1.3.2 — The line emission spectrum of hydrogen provides evidence for the existence of electrons in discrete energy levels, which converge at higher energies.

Structure 1.3.3 — The main energy level is given an integer number, n, and can hold a maximum of $2n^2$ electrons.

Structure 1.3.4 — A more detailed model of the atom describes the division of the main energy level into s, p, d and f sublevels of successively higher energies.

Structure 1.3.5 — Each orbital has a defined energy state for a given electron configuration and chemical environment, and can hold two electrons of opposite spin. Sublevels contain a fixed number of orbitals, regions of space where there is a high probability of finding an electron.

Structure 1.3.6 — In an emission spectrum, the limit of convergence at higher frequency corresponds to ionization.

Structure 1.3.7 — Successive ionization energy data for an element give information about its electron configuration.

AHL

Emission spectra (*Structure 1.3.1*)

Much of our understanding of electron configurations in atoms has come from studies involving interaction with light. In the 1600s, Sir Isaac Newton showed that sunlight can be broken down into different coloured components using a prism. This generates a **continuous spectrum** (figure 1a). This type of spectrum contains light of all **wavelengths**, and appears as a continuous series of colours, in which each colour merges into the next, and no gaps are visible. The classic example of a continuous spectrum is the rainbow. The wavelength of visible light ranges from 400 nm to 700 nm.

A pure gaseous element subjected to a high voltage under reduced pressure will glow — in other words, it will emit light. When this light passes through a prism, it produces a series of lines against a dark background. This is known as an **emission spectrum** (figure 1b). In contrast, when a cold gas is placed between the prism and a source of visible light of all wavelengths, a series of dark lines within a continuous spectrum will appear. This is known as an **absorption spectrum** (figure 1c).

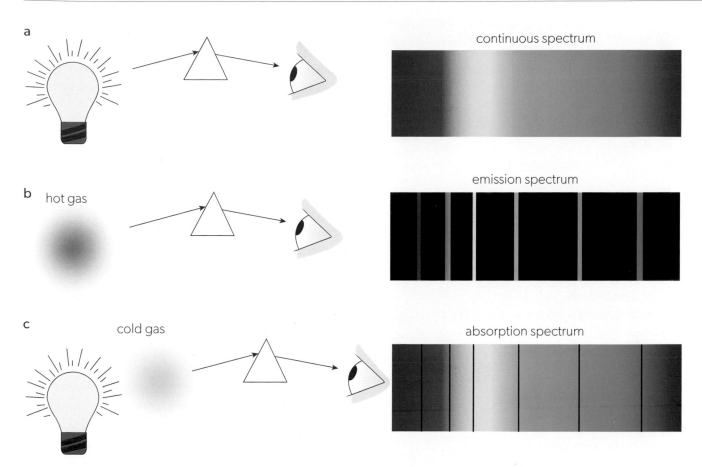

▲ Figure 1. The spectra generated from (a) visible light of all wavelengths (b) a heated gas (c) visible light of all wavelengths passing through a cold gas

▲ Figure 2 The aurora borealis (Northern Lights) in Lapland, Sweden. Charged high-energy particles from the Sun are drawn by the Earth's magnetic field to the polar regions, where they excite atoms and molecules of atmospheric gases, causing them to emit light

Emission spectra

Emission spectra can be observed through a simple handheld spectroscope by holding it up to a light source. Discharge lamps contain low-pressure gases which are ionized when a voltage is applied.

Relevant skills

- Tool 3: Construct graphs and draw lines of best fit
- Inquiry 2: Identify and record relevant qualitative observations and sufficient relevant quantitative data.
- Inquiry 2: Identify and describe patterns, trends and relationships
- Inquiry 2: Assess accuracy

Safety

- Wear eye protection.
- The discharge lamps will get very hot. Handle them with care.
- Further safety precautions will be given by your teacher, depending on the exact nature of the discharge lamps.

Materials

- Discharge lamps
- Handheld spectroscope

Method

1. Observe natural light through the spectroscope. Note down the details of the spectrum you observe.
2. Observe artificial light from a computer screen or LED. Note down the details of the spectrum you observe.
3. Observe light from various discharge lamps. Note down the details of the emission lines you observe, including colours, wavelengths and number of lines.

Questions

1. Sketch the spectra you observed.
2. Describe each as a continuous, emission or absorption spectrum.
3. Look up the emission spectra of the elements in the discharge lamps you observed. Compare the theoretical and observed emission lines, commenting on the number, colours and positions of the emission lines.
4. Next, you will compare the theoretical and observed wavelengths of the emission lines. Construct a graph of theoretical wavelength vs observed wavelength. Draw a line of best fit through your data.
5. Comment on the relationship shown in your graph.
6. Comment on the accuracy of the observed wavelength data.

Each element has its own characteristic line spectrum, which can be used to identify the element. For example, excited sodium atoms emit yellow-orange light with wavelengths of 589.0 and 589.6 nm (figure 3, right). The same yellow-orange colour appears in a flame test of any sodium-containing substance. Like barcodes in a shop that can be used to identify products, line emission spectra can be used to identify chemical elements.

▲ Figure 3 Sodium streetlights (left) and the line emission spectrum of sodium (right)

Observations

Chemists often generate data from observing the properties of matter. Observations can be made directly through the human senses (often sight), or with instruments. Advancements in technology expand the boundaries of our observations, revealing otherwise imperceptible features or detail. Sodium vapour lamps emit orange-yellow light. As seen in figure 3, observing the light through a spectroscope reveals a strong emission in the yellow region of the spectrum. The light from helium lamps is also orange to the naked eye but the emission spectrum of helium is more complex (figure 4).

What is the difference between observing a natural phenomenon directly and with the aid of an instrument?

▲ Figure 4 Helium emission spectrum

Flame tests

Flame testing is an analytical technique that can be used to identify the presence of some metals. The principle behind flame tests is atomic emission. Electrons are promoted to a higher energy level by the heat of the flame. When they fall back to a lower energy level, photons of certain wavelengths are emitted. Some of these photons are in the visible region of the spectrum.

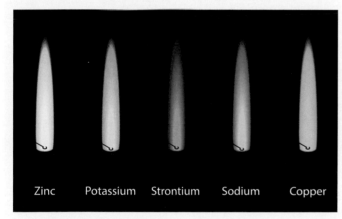

Zinc   Potassium   Strontium   Sodium   Copper

▲ Figure 5 Flame test colours for different elements

Relevant skills
- Inquiry 2: Identify and record relevant qualitative observations

Safety
- Wear eye protection.
- Take suitable precautions around open flames.
- Dilute hydrochloric acid is an irritant.
- A variety of different chloride salts will be used, some of which are irritants — avoid contact with the skin.
- Dispose of all substances appropriately.
- Further safety precautions will be given by your teacher, depending on the identity of the salts being analysed.

Materials
- Flame test wire (platinum or nichrome)
- Small portion of dilute hydrochloric acid
- Bunsen burner and heatproof mat
- Small samples of various metal salts (e.g. LiCl, NaCl, KCl, $CaCl_2$, $SrCl_2$, $CuCl_2$)

Method
1. Clean the end of the flame test wire by dipping it into the HCl solution and placing it in a non-luminous Bunsen burner flame. Repeat until no flame colour is observed.
2. Dip the end of the flame test wire into one of the salt samples, and place it in the edge of the non-luminous Bunsen burner flame, noting down the identity of the metal in the salt and the colour(s) you observe.

3. Clean the wire again and repeat with other salt samples.
4. Clear up as instructed by your teacher.

Questions:
1. Look up the emission spectra of the metals you tested. Compare these to the colours you observed. Comment on any similarities and differences.
2. Explain why the different metals show different flame colours.

TOK

One of the ways knowledge is developed is through reasoning. Reasoning can be deductive or inductive.

Inductive reasoning involves drawing conclusions from experimental observations. Inductive arguments are "bottom up": they take specific observations and build general principles from them.

inductive reasoning ("bottom-up" approach):

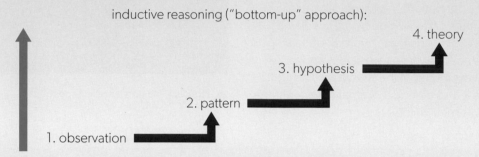

For example, you might make the following observations about lithium salts:

Lithium chloride gives a red flame test.

Lithium sulfate gives a red flame test.

Lithium iodide gives a red flame test.

From these observations, you can make the conclusion that all lithium salts give red flame tests.

Deductive arguments are "top down": they infer specific conclusions from general premises. You do this all the time when asked to apply your scientific knowledge in a new context.

deductive reasoning ("top-down" approach):

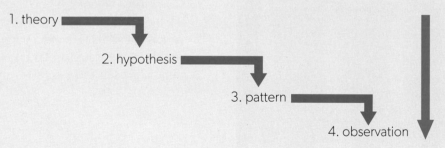

For example, suppose your scientific knowledge includes the following existing premises:

Lithium bromide is a lithium salt.

Lithium salts give red flame tests.

From this, you could propose that lithium bromide gives a red flame test.

What are the advantages and disadvantages of each type of reasoning?

Can reasoning always be neatly classified into these two types?

On what grounds might we doubt a claim reached through inductive reasoning?

On what grounds might we doubt a claim reached through deductive reasoning?

Visible light is one type of **electromagnetic (EM) radiation**. In addition to visible light, microwaves, infrared radiation (IR), ultraviolet (UV), X-rays and gamma rays are all part of the electromagnetic spectrum.

The energy of the radiation is inversely proportional to the wavelength, λ:

$$E \propto \frac{1}{\lambda}$$

Electromagnetic waves all travel at the **speed of light,** c**,** in a vacuum. The speed of light is approximately equal to $3.00 \times 10^8 \, m \, s^{-1}$. Wavelength is related to the **frequency** of the radiation, f, by the following equation:

$$c = f \times \lambda$$

High energy EM waves, such as gamma rays, have short wavelengths and high frequencies while low energy waves, such as microwaves, have long wavelengths and low frequencies.

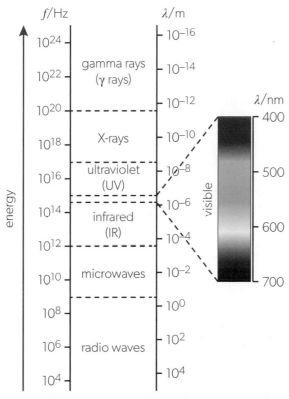

▲ **Figure 6** The wavelength (λ) of electromagnetic radiation is inversely proportional to both frequency and energy of that radiation

Activity

Compare the colours red and green in figure 6. Determine which colour has:

a. the highest wavelength

b. the highest frequency

c. the highest energy

Data-based questions

Look at the spectra below. Explain how we know that stars are partly composed of hydrogen.

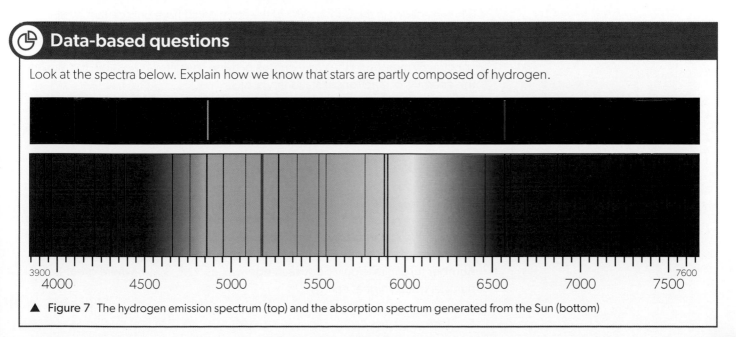

▲ **Figure 7** The hydrogen emission spectrum (top) and the absorption spectrum generated from the Sun (bottom)

The line emission spectrum of hydrogen (*Structure 1.3.2 and 1.3.3*)

Each line in the emission spectrum of an element has a specific wavelength, which corresponds to a specific amount of energy. This is called **quantization**: the idea that electromagnetic radiation comes in discrete packets, or quanta. A **photon** is a quantum of energy, which is proportional to the frequency of the radiation as follows:

$$E = h \times f$$

Where E = the specific energy possessed by the photon, expressed in joules, J

h = Planck's constant, 6.63×10^{-34} J s

f = frequency of the radiation, expressed in hertz, Hz, or inverse seconds, s^{-1}

In 1913, Niels Bohr proposed a model of the hydrogen atom based on its emission spectra. The main postulates of his theory were:

1. The electron can exist only in certain stationary orbits around the nucleus. These orbits are associated with **discrete energy levels**.

2. When an electron in the orbit with the lowest energy level absorbs a photon of the right amount of energy, it moves to a higher energy level and remains at that level for a short time.

3. When the electron returns to a lower energy level, it emits a photon of light. This photon represents the energy difference between the two levels.

Bohr's theory was the first attempt to overcome the main problem of the Rutherford model of the atom (*Structure 1.2*). Classical electrodynamics predicted that orbiting electrons would radiate energy and quickly fall into the nucleus, making any prolonged existence of atoms impossible. Bohr postulated that electrons did not radiate energy when staying in stationary orbits.

Since electrons in the Bohr model of the atom could have only certain, well-defined energies, their transitions between stationary orbits could absorb or emit photons of specific wavelengths, producing characteristic lines in the atomic spectra. By measuring the wavelengths of these lines, it was possible to calculate the energies of electrons in stationary orbits.

For a hydrogen atom, the electron energy (E_n) in joules could be related to the energy level number (n) by a simple equation:

$$E_n = -R_H \frac{1}{n^2}$$

where $R_H \approx 2.18 \times 10^{-18}$ J is the Rydberg constant. This equation clearly represents the quantum nature of the atom, where the energy of an electron can have only discrete, **quantized** values. These values are characterized by integer or half-integer parameters, known as **quantum numbers**. The **principal quantum number** (n) can take only positive integer values (1, 2, 3, …), where greater numbers mean higher energy.

The most stable state of the hydrogen atom is the state at $n = 1$, where the electron has the lowest possible energy. This energy level is known as the **ground state** of the atom. In contrast, the energy levels with $n = 2, 3, \ldots$ are called **excited states**. Atoms in excited states are unstable and spontaneously return to the ground state by emitting photons of specific wavelengths (figure 8).

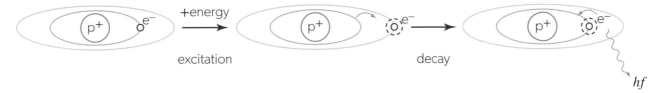

▲ Figure 8 Electrons returning to lower energy levels emit a photon of light, hf

Energy levels in atoms resemble ladders with varying distances between the rungs. Electrons cannot exist between energy levels, much like how you cannot stand between the rungs of a ladder. Jumping up each rung or level requires a specific, discrete amount of energy, and jumping down a rung or level releases the same amount of energy.

An electron can be excited to any energy level, n, and return to any lower energy level. Electrons returning to $n = 2$ will produce distinct lines in the visible spectrum of hydrogen (figure 9).

Note that the red line has a longer wavelength and lower frequency than the violet line. The energy of the photon released is lower when an electron falls from $n = 3$ to $n = 2$, than from $n = 6$ to $n = 2$. In both cases, it represents the difference between two of the allowable energy states of the electron in the hydrogen atom.

colour	violet	blue	cyan		red
wavelength / nm	410	434	486		656
transition from	$n = 6$	$n = 5$	$n = 4$		$n = 3$

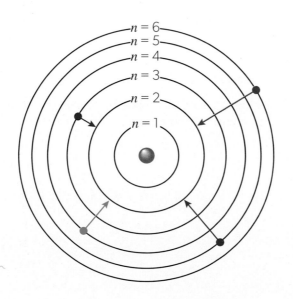

◄ Figure 9 The visible lines in the emission spectrum of hydrogen show electrons returning from higher energy levels to energy level $n = 2$

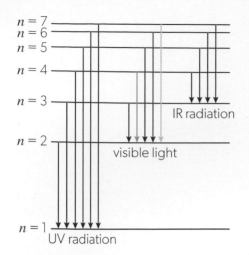

▲ Figure 10 Electron transitions for the hydrogen atom. Notice how the allowable energy levels get closer together when the electron moves further away from the nucleus. The energy difference between $n = 3$ and $n = 2$ is much smaller than that between $n = 2$ and $n = 1$

Electron transitions to the ground state, $n = 1$, release higher energy, shorter wavelength ultraviolet light, while electrons returning to $n = 3$ produce lines in the infrared region of the electromagnetic spectrum (figure 10).

It is important to note that electrons will absorb or release only the exact energy required to move between allowable energy states. Any excess will not be absorbed, and if an insufficient amount of energy is supplied the electrons will not move.

Energy levels closer to the nucleus hold fewer electrons. The maximum number of electrons in any energy level, n, is $2n^2$. For example, the energy level with $n = 1$ holds up to two electrons, at $n = 2$ there could be a maximum of eight electrons, $n = 3$ has a maximum of 18 electrons, and $n = 4$ has a maximum of 32 electrons.

(ATL) Communication skills

When explaining concepts, we sometimes use diagrams, graphs or images to help us convey our ideas more clearly.

Prepare a written explanation of atomic emission that does not include any diagrams. Exchange it with a partner. Give each other feedback, concentrating on:

- Use of scientific vocabulary
- Order in which ideas are given
- Whether any important concepts are missing from the explanation.

When you have shared each other's feedback, spend some time using the feedback to make improvements to your work. Finally, choose a graph, image or diagram to accompany your explanation. Discuss why you chose it and whether or not it adds to the explanation.

Linking questions

What qualitative and quantitative data can be collected from instruments such as gas discharge tubes and prisms in the study of emission spectra from gaseous elements and from light? (Inquiry 2)

How do emission spectra provide evidence for the existence of different elements? (Structure 1.2)

How does an element's highest occupied main energy level relate to its period number in the periodic table? (Structure 3.1)

The quantum mechanical model of the atom (*Structure 1.3.4*)

The **Bohr model** was an attempt to explain the energy states of electrons in atoms. It was based on quantization: the idea that electrons existed in discrete energy levels. According to Bohr, the emission spectra of hydrogen consisted of narrow lines because the wavelengths of these lines corresponded to the differences in allowable energy levels. However, this model was limited by several problems and incorrect assumptions:

1. The model could not predict the emission spectra of elements containing more than one electron. It was only successful with the hydrogen atom.
2. It assumed the electron was a subatomic particle in a fixed orbit about the nucleus.
3. It could not account for the effect of electric and magnetic fields on the spectral lines of atoms and ions.
4. It could not explain molecular bonding and geometry.
5. Heisenberg's uncertainty principle states that it is impossible to precisely know the location and momentum of an electron simultaneously. Bohr's model stated that electrons exhibited fixed momentum in specific circular orbits.

> The principles behind molecular bonding and geometry are explained in *Structure 2.2*.

Because of these limitations, the Bohr theory has been eventually superseded by the modern quantum mechanical model of the atom.

TOK

The modern quantum mechanics combines the idea of quantization with the following key principles.

Heisenberg's uncertainty principle states that it is impossible to determine accurately both the momentum and the position of a particle simultaneously. This means that the more we know about the position of an electron, the less we know about its momentum, and vice versa. Although it is not possible to pinpoint the location or predict the trajectory of an electron in an atom, we can calculate the **probability** of finding an electron in each region of space.

> *One aim of the physical sciences has been to give an exact picture of the material world. One achievement ... has been to prove that this aim is unattainable.*
> Jacob Bronowski (1908–1974)

What are the implications of this uncertainty principle on the boundaries of knowledge?

What are the limits of human knowledge?

Wave–particle duality is the ability of electrons and other subatomic species to behave as both particles and waves. Certain characteristics of these species, such as mass, momentum and the tendency to be absorbed or released as discrete entities suggest their particulate nature. However, photons, electrons, and even whole atoms and small molecules, are capable of interference (combination of waveforms), diffraction (bending around obstacles) and tunnelling (passing through obstacles), all of which are characteristic to waves.

> *We have two contradictory pictures of reality; separately neither of them fully explains the phenomena of light, but together they do.*
> Albert Einstein (1879–1955)

The wave–particle duality of the electron is quantitatively described by the **Schrödinger equation**, which was formulated in 1926 by the Austrian physicist Erwin Schrödinger (1887–1961). Solutions to the Schrödinger equation give a series of three-dimensional mathematical functions, known as **wave functions**, which describe the possible states and energies of electrons in atoms.

The concept of wave–particle duality illustrates the fact that objects of study do not always fall neatly into the discrete categories we have developed. What is the role of categorisation in the construction of knowledge?

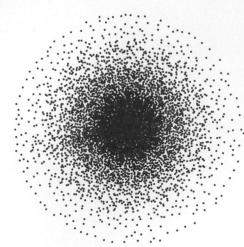

▲ Figure 11 An s orbital is spherical. The sphere represents the boundary space where there is a 99% probability of finding an electron. The s orbital can hold two electrons

Schrödinger's wave functions describe the electrons in atoms in terms of their probability density, using Heisenberg's idea that the momentum and position of electrons are uncertain. Instead of saying that electrons follow a defined travel path, this theory gives the probability that an electron will be found in a specific region of space at a certain distance from the nucleus. An **atomic orbital** is a region in space where there is a high probability of finding an electron.

There are several types of atomic orbitals, and each orbital can hold a maximum of two electrons. Each orbital has a characteristic shape and energy. The first four atomic orbitals, in order of increasing energy are labelled **s, p, d,** and **f.** Subsequent orbitals are **theoretical**, and these are labelled alphabetically (g, h, i, k and so on).

The principal quantum number, n, introduced by the Bohr model represents the main energy levels. These energy levels are split into sublevels comprised of atomic orbitals. For example, for $n = 1$, 2 and 3, the s atomic orbitals are 1s, 2s and 3s. As n increases, the s orbitals are further distanced from the nucleus.

Figure 12 shows that, for 1s, there is a high probability of finding electrons close to the nucleus and this probability never reaches zero when we move further away from the nucleus. For 2s, the highest probability is somewhat further away, although there is a small probability that an electron could be found closer to the nucleus. There is zero probability of finding the electron between the two peaks. The same is true for 3s, with the highest probability at an even greater distance from the nucleus and two regions of zero probability.

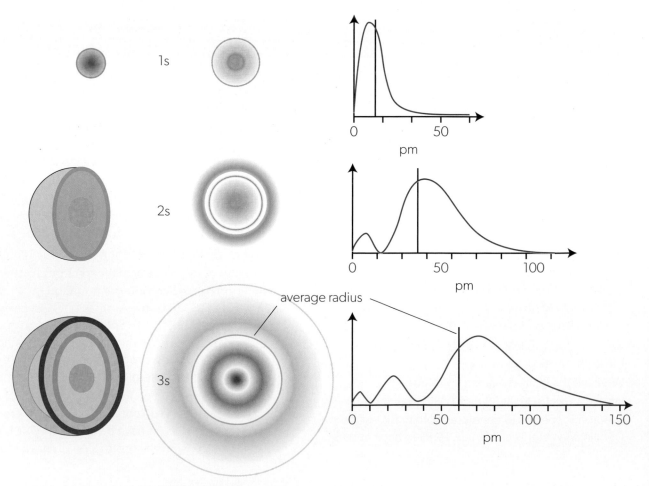

▲ Figure 12 The plots of the wavefunctions for the first three s orbitals

Imagine that you are a student waiting for your DP chemistry lesson to begin at 8.00am. At 8.15am, there is still no sign of your teacher, so you wonder where they could be. Some students from your class suggest that the teacher:

- is possibly in the staff room, the chemistry laboratory, or the library
- could be in the school principal's office or in the school car park
- may be at their house in the town centre
- might perhaps be at the airport
- might even have gone to the North Pole!

Although the exact location of the teacher is unknown, it is possible to draw a three-dimensional cluster of dots showing areas where there is a high probability of finding the teacher. A boundary surface could be drawn around this cluster to define a region of space where there is a 99% chance of finding them. This might be the school perimeter, or the town where your school is located, or a certain region around the town that includes the airport. Similarly, an atomic orbital represents the region of space with a high probability of finding an electron (figure 13).

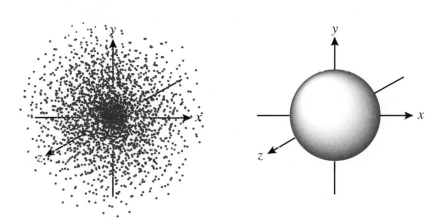

◀ Figure 13 Representation of a 1s atomic orbital as a cluster of dots (left) and a sphere that encloses 99% of the dots (right)

A **p orbital** is dumbbell shaped. There are three p orbitals, each described with orientations parallel to the x, y and z axes (figure 14). These are labelled p_x, p_y and p_z. These shapes all describe boundaries with the highest probability of finding electrons in these orbitals.

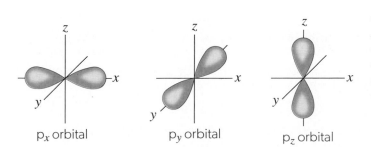

p_x orbital p_y orbital p_z orbital

◀ Figure 14 The three p atomic orbitals are dumbbell shaped, aligned along the x, y and z axes. There is zero probability of finding the electron at the intersection of the axes between the two lobes of the dumbbell. Each of the p orbitals can hold two electrons

Theories and models

Current atomic theory evolved from previous models, each superseding the one that came before. Theories are comprehensive systems of ideas that model and explain an aspect of the natural world. Contrary to the use of the word "theory" in everyday language, scientific theories are substantiated by vast amounts of observations and tested hypotheses, which are amassed, documented and communicated by a large number of scientists.

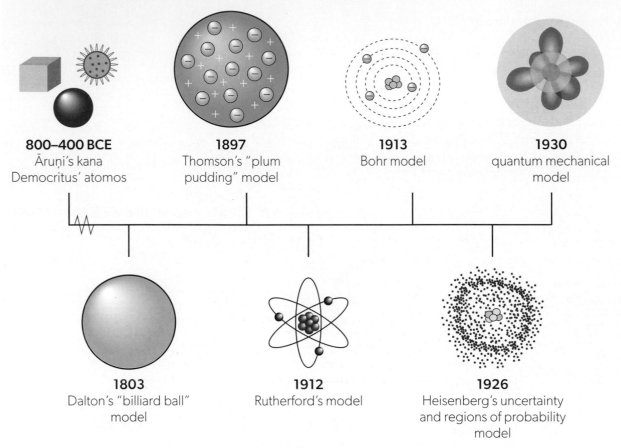

800–400 BCE
Āruṇi's kana
Democritus' atomos

1897
Thomson's "plum pudding" model

1913
Bohr model

1930
quantum mechanical model

1803
Dalton's "billiard ball" model

1912
Rutherford's model

1926
Heisenberg's uncertainty and regions of probability model

▲ **Figure 15** The atomic theory has seen the idea of atoms evolve from indestructible spheres to the quantum mechanical model where electrons have specific energies and are found in regions of high probability

What other examples of theories can you think of?

 Linking question

What is the relationship between energy sublevels and the block nature of the periodic table? (Structure 3.1)

Electron configurations (*Structure 1.3.5*)

Each atomic orbital type has a characteristic shape and energy. The s orbital is spherical and it has the lowest possible energy. There are three p orbitals, each oriented differently. There are five d orbitals and seven f orbitals, and these are higher in energy than s or p.

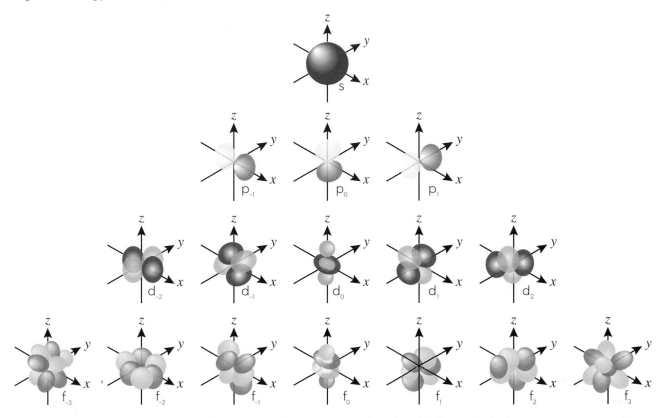

▲ Figure 16 The shapes of the s, p, d and f orbitals. Only the shapes of s and p orbitals need to be known

Each energy level defined by the principal quantum number, n, can hold n types of orbitals (table 1). For $n = 1$, only the s orbital exists. For $n = 2$, there are two types of orbital: s and p. For $n = 3$, there are three types: s, p, and d. For $n = 4$, there are four types: s, p, d, and f.

Principal quantum number (n)	Type of sublevel	Number of orbitals per type	Total number of orbitals per energy level (n^2)	Maximum number of electrons within energy level ($2n^2$)
1	s	1	1	2
2	s	1	4	8
	p	3		
3	s	1	9	18
	p	3		
	d	5		
4	s	1	16	32
	p	3		
	d	5		
	f	7		

◀ Table 1 Each energy level, defined by n, can hold $2n^2$ electrons. The number of sublevels, or atomic orbital types, is equal to n. For $n = 4$ there are four types of orbitals (s, p, d, and f) with 16 atomic orbitals in total occupied by a maximum of $2(4)^2 = 32$ total electrons

Activity

State the following for the energy level with $n = 5$:

a. the sublevel types

b. the number of atomic orbitals in each sublevel

c. the total number of atomic orbitals

d. the maximum number of electrons at that energy level.

Orbital diagrams

For convention, an "arrow in box" notation called an **orbital diagram** is used to represent how electrons are arranged in atomic orbitals (figure 17). The arrangement of electrons in orbitals is called **electron configuration**.

▶ Figure 17 In orbital diagrams, each box represents an orbital. This diagram shows the number of orbitals for each sublevel. Arrows are drawn in the boxes to represent electrons. A maximum of two electrons can occupy each orbital, so each box has a maximum of two "arrows"

s sublevel (one box representing an s orbital)

p sublevel (three boxes representing the three p orbitals p_x, p_y, and p_z)

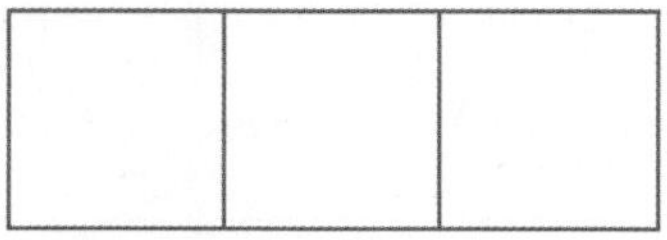

d sublevel (five boxes representing the five d orbitals)

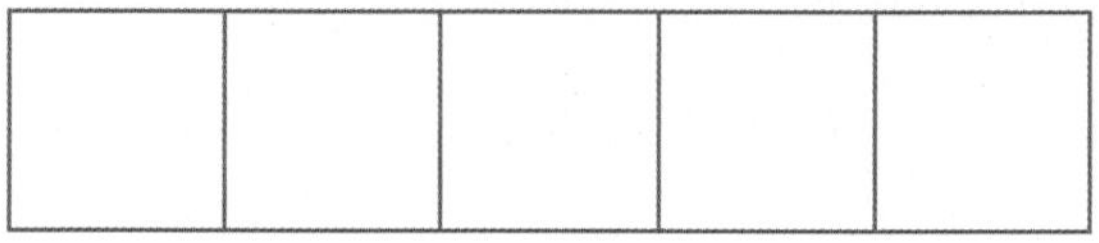

f sublevel (seven boxes representing the seven f orbitals)

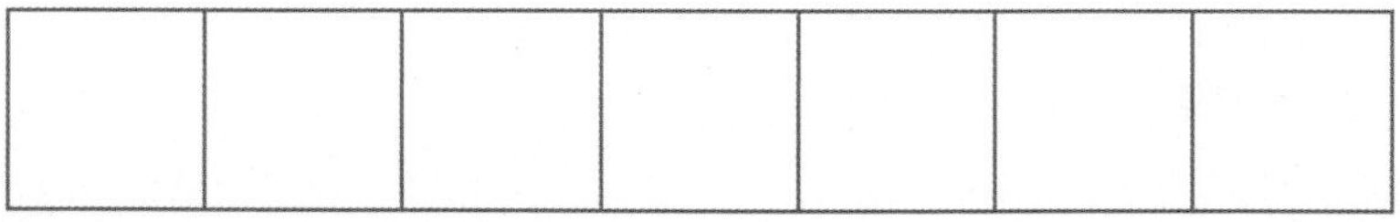

Atomic orbitals are regions of space where there is a high probability of finding electrons. Electrons are charged negatively, and like charges repel each other, so two electrons should not be able to occupy the same region of space. Quantum mechanics solves this problem by using a ± spin notation for each electron. A pair of electrons with opposite spins behave like magnets facing in opposite directions. Hence each orbital box is shown with an upwards half-arrow, ↿, and one downwards half-arrow, ⇂ (figure 18). This is known as the **Pauli exclusion principle**:

Only two electrons can occupy the same atomic orbital and those electrons must have opposite spins.

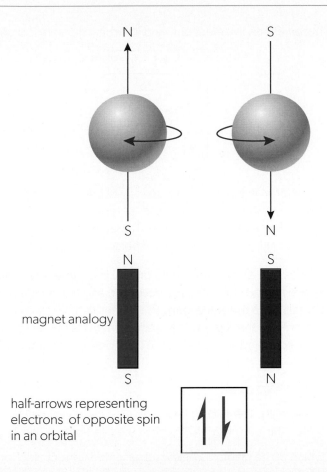

magnet analogy

half-arrows representing electrons of opposite spin in an orbital

TOK

Electron spin is often interpreted as the rotation of the electron around its own axis. However, this interpretation has no physical basis: electrons in atoms behave like waves, and a wave cannot rotate. Unfortunately, neither the spin nor the wave-like behaviour of electrons can be visualized in any way, as they have no analogues in our everyday life and can be expressed only in mathematical form. This lack of visualization does not undermine the quantum theory but rather shows the limits of human perception and, at the same time, the power of mathematics as the language of science.

To what extent does mathematics support knowledge development in the natural sciences?

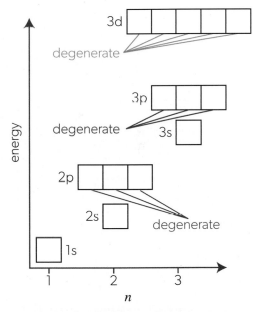

▲ Figure 19 The three 2p orbitals are degenerate as they have the same energy. These three degenerate atomic orbitals have lower energy than the three 3p orbitals

Each of the atomic orbitals of the same type in one sublevel are of equal energy. Orbitals with the same energy are referred to as **degenerate** orbitals (figure 19).

An atom of boron (B) has five electrons, and its orbital diagram is drawn as follows:

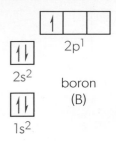

The single 2p electron in boron can occupy any of the three orbitals, as they have equal energies. The degenerate 2p orbitals are represented by boxes joined together to show their energy equivalence. Traditionally, the half-arrow is drawn in the leftmost box, although it is a matter of personal preference.

Hund's rule states that every **degenerate orbital in a sublevel is singly occupied before any orbital is doubly occupied** and that **all electrons in singly occupied orbitals have the same spin**. This means that the three p orbitals must have one electron with the same spin in each of them before any orbital can become doubly occupied with an electron of opposite spin (figure 20).

Practice questions

1. Look at figure 20. The 1s and 2s orbitals are fully occupied by electrons. Why do you think the 1s and 2s orbitals are filled before the 2p orbitals?

2. State which of the diagrams below represents a correct electron configuration based on Hund's rule and the Pauli exclusion principle. State the reason for the four incorrect diagrams being wrong.

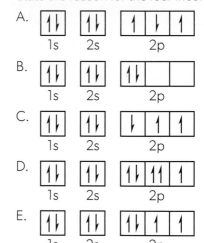

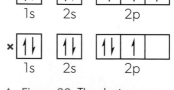

▲ Figure 20 The electrons are evenly distributed across the three degenerate 2p orbitals in nitrogen before an orbital is doubly occupied

The **Aufbau principle** states that as electrons are added to atoms, **the lowest available energy orbitals fill before higher energy orbitals do**. The third and fourth energy levels contain d and f orbitals (figure 21). These orbitals are typically filled after the s orbitals of the following levels because they are higher in energy. As shown in figure 21, the 3d sublevel has a higher energy than 4s but lower than 4p, so 4s is filled with electrons first, followed by 3d and finally 4p. For the same reason, 4d orbitals are filled after 5s, and 4f orbitals are filled only after 6s.

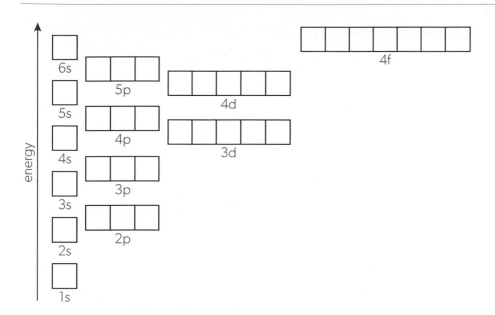

◀ **Figure 21** The 4s sublevel has a lower energy and will fill before the 3d sublevel

This is consistent with experimental data that show that potassium, K, and calcium, Ca have electrons in the 4s sublevel, not in 3d.

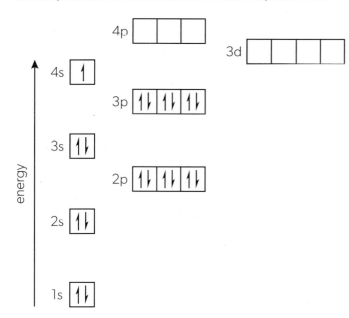

◀ **Figure 22** Potassium orbital filling diagram showing the outermost electron in the 4s orbital because 3d orbitals are higher in energy

Generally, the following order is observed:

1s < 2s < 2p < 3s < 3p < 4s < 3d < 4p < 5s < 4d < 5p < 6s < 4f < 5d < 6p ...

In the IB Diploma Programme, only the electron configurations of atoms and ions up to $Z = 36$ will be assessed. Electrons in these species can fill sublevels up to 4p.

Electron sharing and transfer are fundamental to understanding chemical reactions, so it is important to know the electron configuration of an atom or an ion. There are three ways to show the electron configuration:

1. Full electron configuration

2. Condensed electron configuration

3. Orbital filling diagram ("arrows in boxes" notation)

The orbital filling diagram for potassium is given in figure 22.

Activity

Copy the orbital diagram from figure 21 and complete it for the following elements in their ground states:

a. aluminium, Al

b. chlorine, Cl

c. iron, Fe

Refer to the periodic table at the back of this book to deduce the number of electrons in each atom.

Full electron configurations

To write a full electron configuration, we use the periodic table, and "build up" the electrons in successive orbitals according to the Aufbau principle, Hund's Rule and the Pauli exclusion principle.

> ### Worked example 1
>
> Determine the full electron configuration for the calcium atom.
>
> ### Solution
>
> The Aufbau principle states that as electrons are added to atoms, the lowest available energy orbitals fill before higher energy orbitals do. From the Pauli exclusion principle, we know that each orbital will have a maximum of two electrons.
>
> The atomic number of calcium is 20. Let's split the 20 electrons evenly across each orbital, starting with the lowest energy first. When writing electron configurations, the number of electrons within each sublevel is given in superscript, next to the sublevel:
>
> - The 1s orbital has two electrons: $1s^2$
> - The 2s orbital also has two electrons: $2s^2$
> - The three 2p orbitals have two electrons each, six in total: $2p^6$
> - The 3s orbital has two electrons: $3s^2$
> - The three 3p orbitals have three electrons each, six in total: $3p^6$
>
> This brings us up to 18 electrons, with two left over to go into the orbital with the next lowest energy, 4s: $4s^2$
> So, for calcium, the full electron configuration is $1s^2\,2s^2\,2p^6\,3s^2\,3p^6\,4s^2$.

Practice question

3. Determine the full electron configuration for the phosphorus atom.

Condensed electron configurations

As the atomic number of an element increases, the full electron configuration gets longer and it can be time-consuming to write. The chemistry of atoms and ions is mostly determined by their **valence electrons**, that is, the outermost electrons, rather than the **inner core electrons.** A more convenient way of writing electron configurations is to highlight the valence electrons and represent the inner core electrons as having the same electron configuration as the previous group 18 (known as the **noble gases**) element in the periodic table:

Condensed electron configuration = [previous noble gas] + valence electrons

Table 2 shows some more examples of full and condensed electron configurations.

▶ Table 2 Examples of full and condensed electron configurations for selected elements

Element	Atomic number	Full electron configuration	Condensed electron configuration
O	8	$1s^2\,2s^2\,2p^4$	$[He]\,2s^2\,2p^4$
Ne	10	$1s^2\,2s^2\,2p^6$	$[Ne]$
Mn	25	$1s^2\,2s^2\,2p^6\,3s^2\,3p^6\,4s^2\,3d^5$	$[Ar]\,4s^2\,3d^5$
Br	35	$1s^2\,2s^2\,2p^6\,3s^2\,3p^6\,4s^2\,3d^{10}\,4p^5$	$[Ar]\,4s^2\,3d^{10}\,4p^5$

Worked example 2

Determine the condensed electron configuration for the calcium atom.

Solution

In worked example 1, we determined the full electron configuration of calcium to be $1s^2 2s^2 2p^6 3s^2 3p^6 4s^2$.

The previous noble gas in the periodic table is argon, which has an atomic number of 18. Argon has an electron configuration of $1s^2 2s^2 2p^6 3s^2 3p^6$, and we can therefore write the condensed electron configuration of calcium as $[Ar] 4s^2$.

Orbital diagrams can also sometimes be shortened by using a condensed electron configuration. The condensed orbital diagrams for oxygen and manganese are shown below.

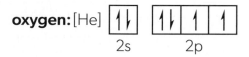

oxygen: [He] 2s   2p

manganese: [Ar] 4s   3d

Self-management skills

The ideas in this topic span a range of concepts and skills: from the quantum mechanical model of the atom, to how to write electron configurations.

Write a chapter summary, no longer than a sheet of A4 paper.

Write three key takeaways from the chapter.

List the key vocabulary you should know from this chapter.

Write five brief questions that test your understanding of the ideas in this chapter. Make an answer key, then try them out on one of your peers.

Exceptions to the Aufbau principle

The Aufbau principle correctly predicts the order of filling of atomic orbitals for most elements. However, when atoms lose electrons to form ions, the electrons in the sublevel with the highest principal quantum number (n) are lost first.

So, for Mn, with electron configuration $[Ar] 4s^2 3d^5$, the 4s electrons will be lost first. This gives the manganese ion, Mn^{2+}, with electron configuration $[Ar] 3d^5$, **not** $[Ar] 4s^2 3d^3$.

All elements with 3d valence electrons tend to lose two 4s electrons to form 2+ ions. These are known as the 3d **transition elements**, or **transition metals**. If you look at the periodic table at the back of this book, these elements are from scandium (Sc) to copper (Cu). These transition metals can also have **variable oxidation states** in compounds.

There are some exceptions. With only one electron in its 3d orbital, scandium readily forms only Sc^{3+} ions, by losing this 3d electron and the two 4s electrons.

Practice question

4. Determine the condensed electron configuration for the phosphorus atom.

The periodic table is structured according to the type of sublevel that valence electrons of elements appear in. This is discussed further in *Structure 3.1*.

Ionization and oxidation are covered in *Structure 2.1* and *Structure 3.1*.

Activity

Deduce the electron configuration of the Cu^{2+} cation.

The ground state configurations of copper and chromium are also different from those predicted by Aufbau principle.

The predicted electron configuration of copper is $[Ar]\,4s^2\,3d^9$, as the Aufbau principle suggests that the lower-energy 4s orbital should be filled first. However, the observed ground-state electron configuration for copper is $[Ar]\,4s^1\,3d^{10}$ (figure 23). For chromium, the predicted configuration is $[Ar]\,4s^2\,3d^4$ and the observed is $[Ar]\,4s^1\,3d^5$ (figure 23). In each case, promoting a 4s electron to a 3d level leads to a more stable electron configuration. In the case of copper, this gives a full d sublevel, and in the case of chromium, there are no paired electrons but rather six half-occupied orbitals, each containing an electron with the same spin.

▶ Figure 23 The expected and observed electron configurations of copper and chromium

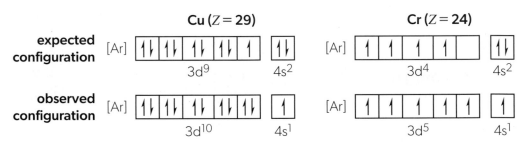

The electron configurations of chromium (Cr) and copper (Cu) are the only two exceptions that you need to know. In all other elements up to $Z = 36$, the sublevels are filled with electrons according to the general order.

Ionization energy (*Structure 1.3.6 and 1.3.7*)

The quantum mechanical model of the atom helps to explain the trends and discontinuities in the first **ionization energies (*IE*)** of elements. Ionization energy is the minimum energy required to eject an electron out of a neutral atom or molecule in its ground state.

$$X(g) + energy \rightarrow X^+(g) + e^-$$

The columns in the periodic table are known as groups, and the rows are known as periods. Going across the periodic table, the groups are numbered from 1 to 18. The periodic table can be shown as four blocks corresponding to the four sublevels s, p, d, and f (figure 24). The sublevels holding the outermost valence electrons for each element are also shown.

First ionization energy (IE_1) generally decreases down the groups of the periodic table and increases across the periods.

Ionization energy and periodic table trends are discussed further in *Structure 3.1*.

Going down a group, the number of sublevels increases. The outermost electrons are shielded from the pull of the nucleus by the electrons in the lower energy sublevels (so-called "inner electrons"). The more sublevels, the greater the shielding and therefore less energy is required to remove electrons from the outermost sublevel.

Going across a period, the number of protons in the nucleus increases, so the outermost electrons are held closer to the nucleus by the **increased nuclear charge**. At the same time, the shielding effect remains nearly constant because the number of inner electrons does not change.

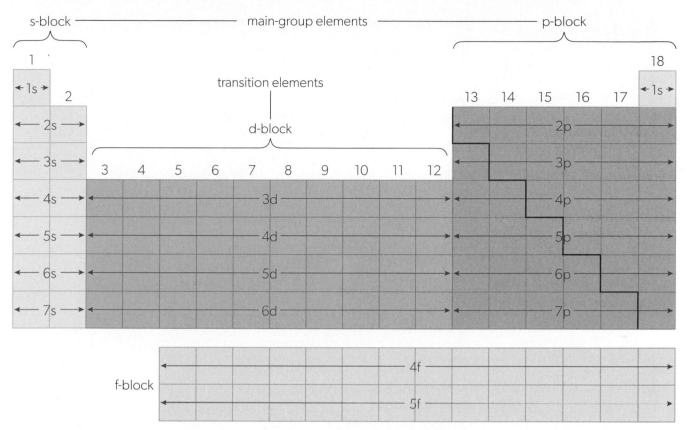

▲ Figure 24 The blocks of the periodic table correspond to the sublevels s, p, d and f

Therefore, more energy is required to remove outermost electrons, so ionization energy increases across the period.

The general trend of decreasing ionization energy down a group and increasing across a period is shown in figure 25:

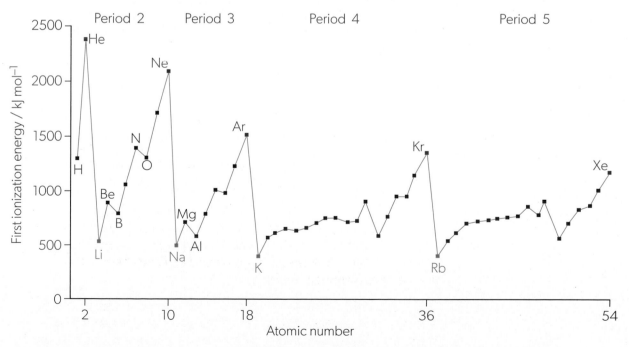

▲ Figure 25 Plot of first ionization energy against atomic number for the elements from hydrogen to xenon

There are two clear discontinuities across the period:

1. **Between the group 2 and group 3 elements**

The valence electron configuration of beryllium is $2s^2$ while for boron it is $2s^2 2p^1$. The paired $2s^2$ electrons shield the single $2p$ electron in boron from the nucleus, making the electron slightly easier to remove.

Patterns and trends

Scientists look out for patterns and trends in the data they collect. The presence of discrepancies — results that do not fit the overall pattern — allows further conclusions to be drawn.

What can be inferred from the patterns in successive ionization energies?

The same trend can be observed in comparing group 2 to group 3 elements in any period. For example, the $3s^2$ electrons shield the lone $3p^1$ electron in aluminium, so the first ionization energy of aluminium is lower than that of magnesium.

Suppose you have a two-story building and you need to remove one floor to meet new height regulations. Which floor would you remove? Obviously, it will be the top floor, as the building would collapse otherwise! The same reasoning can be applied to the ionization of atoms — electrons are removed first from the highest occupied energy level, and from the highest energy sublevel within that level.

2. **Between the group 15 and 16 elements**

From group 15 to 16 there is also a drop in ionization energy. The electron configuration of nitrogen is $1s^2 2s^2 2p^3$ while for oxygen it is $1s^2 2s^2 2p^4$.

Nitrogen has a more stable electron configuration than oxygen as it has a half-filled p sublevel, and therefore more energy is required to remove an electron from nitrogen (figure 26). This is because the paired electrons in oxygen occupy the same region of space and have increased repulsion. However, in nitrogen, the three electrons in the 2p orbitals do not come into close proximity.

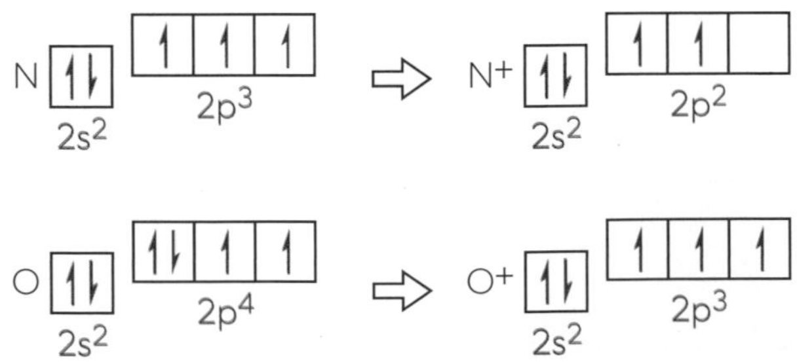

▶ **Figure 26** A half-filled p sublevel is more stable than p sublevels with 2 or 4 electrons

The most stable p orbital configuration is p^6, a completely filled p sublevel, followed by p^3, a half-filled sublevel. This is generally true for other sublevels. For example, d^{10} and d^5 electron configurations are also stable, which partly explains why chromium and copper do not obey the Aufbau principle (figure 23).

Calculating ionization energy from spectral data

As the principal quantum number of energy levels increases, the distance between the levels converges to a continuum. This can be observed by spectral lines converging in the hydrogen emission spectrum, shown in figure 27.

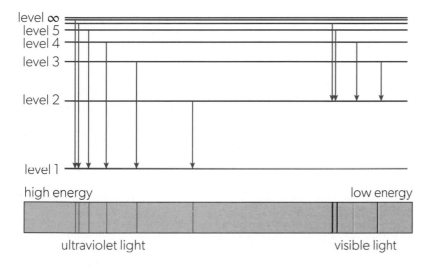

◀ **Figure 27** Ultraviolet and visible light transitions in hydrogen and the resulting emission spectrum

The spectral lines in the hydrogen emission spectrum converge at 9.12×10^{-8} m, or 912 Å (figure 28). This represents the wavelength of light at which the hydrogen atom is ionized.

This wavelength can be used to calculate the first ionization energy of hydrogen.

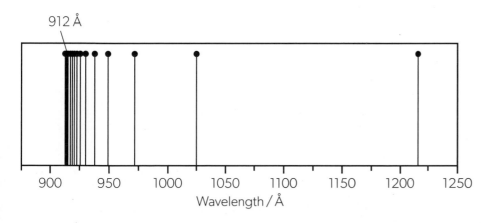

◀ **Figure 28** Hydrogen is ionized at the wavelength where the spectral lines converge in the emission spectrum

Worked example 3

Spectral lines converge at 9.12×10^{-8} m in the emission spectrum of the hydrogen atom. Calculate the first ionization energy of hydrogen in kJ mol^{-1}.

Solution

First, calculate the frequency of radiation using $c = f \times \lambda$, where c is the speed of light, approximately equal to 3.00×10^8 m s^{-1}.

$$3.00 \times 10^8 \text{ m s}^{-1} = f \times 9.12 \times 10^{-8} \text{ m}$$
$$f = 3.29 \times 10^{15} \text{ Hz (s}^{-1})$$

Then, calculate the energy using Planck's constant and the equation $E = h \times f$

$$E = 6.63 \times 10^{-34} \text{ J s} \times 3.29 \times 10^{15} \text{ s}^{-1}$$
$$= 2.18 \times 10^{-18} \text{ J}$$

Alternatively, these two steps can be merged into one by using the equation $E = \dfrac{h \times c}{\lambda}$

This represents the energy of a single photon of light which would be absorbed in exciting the electron in a hydrogen atom to the convergence level, or removing one electron from the atom.

Ionization energies are usually given in kJ mol^{-1}. You can convert the ionization energy value to kJ mol^{-1} using Avogadro's constant (N_A, the number of atoms in 1 mol) and the following equation:

The ionization energy in kJ mol^{-1}

$$= \frac{\text{(energy needed to remove one electron from an atom)} \times N_A}{1000}$$
$$= \frac{2.18 \times 10^{-18} \text{ J} \times 6.02 \times 10^{23} \text{ mol}^{-1}}{1000}$$
$$= 1.31 \times 10^3 \text{ kJ mol}^{-1}$$

The values of the speed of light, Planck's constant and Avogadro's constant are given in the data booklet. The mole and Avogadro's constant are discussed further in *Structure 1.4*.

Worked example 4

The first ionization energy of Na is 496 kJ mol^{-1} as given by the IB data booklet. Calculate the wavelength of convergence for the sodium atom spectrum in Å.

Solution

First, find the energy of ionization for one atom by converting the given value from kJ to J and dividing it by Avogadro's constant.

$$496\,000 \text{ J mol}^{-1} / 6.02 \times 10^{23} \text{ mol}^{-1} = 8.24 \times 10^{-19} \text{ J}$$

Then calculate the wavelength of light using $E = \dfrac{h \times c}{\lambda}$

$$8.24 \times 10^{-19} \text{ J} = \frac{6.63 \times 10^{-34} \text{ J s} \times 3.00 \times 10^8 \text{ m s}^{-1}}{\lambda}$$
$$\lambda = 2.41 \times 10^{-7} \text{ m}$$
$$= 2410 \text{ Å}.$$

This corresponds to the UV region in the electromagnetic spectrum.

Practice questions

5. In the emission spectrum of the helium atom, the spectral lines converge at 5.05×10^{-8} m. Calculate the first ionization energy, in kJ mol^{-1}, of helium.

6. The first ionization energy of calcium is 590 kJ mol^{-1}. Calculate the wavelength of convergence, in Å, for the calcium atomic spectrum.

Successive ionization energies

It requires more energy to remove the second and **successive** electrons from an atom because the number of protons exceeds the number of remaining electrons while the electron–electron repulsion decreases.

As a result, electron clouds are pulled closer to the nucleus and held tighter by the increased electrostatic attraction. Once all the valence electrons are removed so that only the stable noble gas configuration remains, the energy required to remove the next electron increases sharply, as shown in figure 29.

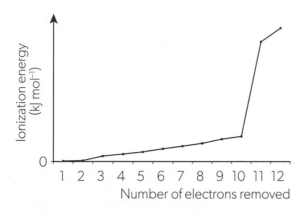

◀ **Figure 29** Removing 10 electrons from magnesium gives the noble-gas configuration $1s^2$ or [He]. There is a considerable increase in energy required to remove the 11th electron

Worked example 5

The first five successive ionization energies for an unknown element X have the following values: 403, 2633, 3860, 5080 and 6850 kJ mol^{-1}. Deduce the group of the periodic table in which element X is likely to be found.

Solution

The largest increase in energy occurs from the first ionization (403 kJ mol^{-1}) to the second (2633 kJ mol^{-1}). This means that the second electron is likely to be removed from a stable noble gas configuration of the atom. Therefore, the outermost energy level of the element contains one electron, so the element belongs to group 1 of the periodic table.

Practice question

7. The first five successive ionization energies of an unknown element have the following values: 801, 2427, 3660, 25 026 and 32 827 kJ mol^{-1}. Deduce the group of the periodic table in which this element is likely to be found.

Data-based question

Using figure 30 and the periodic table, explain the two large jumps in the successive ionization energies for sodium.

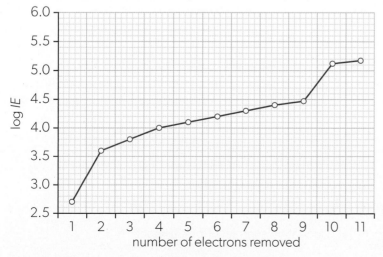

▲ **Figure 30** Successive ionization energies for sodium

AHL

 Ionization energy data

Relevant skills

- Tool 2: Extract data from databases
- Tool 2: Use spreadsheets to manipulate data and represent data in graphical form
- Tool 3: Construct and interpret graphs

Instructions

Part 1: Data collection

1. Identify a database that contains successive ionization energy data for the elements (for example, WebElements).

2. Choose one of the following elements: sulfur, chlorine, argon, potassium or calcium.

3. Collect successive ionization energy data in a spreadsheet, labelling the columns as follows:

	A	B
1	**element name:**	
2		
3	**ionization**	**ionization energy/ kJ mol^{-1}**
4	1	
5	2	
6	etc	

Part 2: Graphing successive ionization energies

4. Plot a line graph of ionization energy vs ionization. Make sure that you present your graph suitably, with axis labels, suitable scales and a descriptive title.

5. Answer the following questions:
   a. Identify the *IE* values that correspond to the innermost and outermost electrons.
   b. Explain why ionization energy increases with each successive electron.
   c. Explain how the graph provides evidence for the existence of main energy levels in the atom.

Part 3: Graphing the logarithm of the ionization energies

6. Title the third column in your spreadsheet "log (ionization energy)" as shown below:

	A	B	C
1	**element name:**		
2			
3	**ionization**	**ionization energy/ kJ mol^{-1}**	**log (ionization energy)**
4	1		
5	2		
6	etc		

7. Compute the logarithm of each ionization energy using the spreadsheet LOG (or LOG10) function.

8. Construct a graph showing the logs of successive ionization energies by plotting log (ionization energy) vs ionization number.

9. Answer the following questions:
   a. Identify the large increases in ionization energy that indicate a change in main energy level.
   b. Why is it useful to plot the logs of the ionization energies?

Part 4: Evidence for the existence of sublevels

10. Construct a graph that will allow you to closely examine the electrons in energy level $n = 2$. Enlarge the graph and "zoom in" to inspect the increases closely.

11. Can you see any unusually large increases in ionization energy? Explain how they relate to the existence of sublevels.

12. Experimental data is often organized into tables and later transformed into graphical forms. What is the role of graphical representations in the advancement of scientific knowledge? Are graphical representations employed in other subject areas?

TOK

 Linking questions

How does the trend in *IE* values across a period and down a group explain the trends in properties of metals and non-metals? (Structure 3.1)

 Why are log scales useful when discussing [H$^+$] and ionization energies? (Tool 3, Reactivity 3.1)

AHL How do patterns of successive ionization energies of transition elements help to explain the variable oxidation states of these elements? (Structure 3.1)

End-of-topic questions

Topic review

1. Using your knowledge from the *Structure 1.3* topic, answer the guiding question as fully as possible:

   How can we model the energy states of electrons in atoms?

Exam-style questions
Multiple-choice questions

2. Which row is correct for the following regions of the electromagnetic spectrum?

	Ultraviolet (UV)		Infrared (IR)	
A.	high energy	short wavelength	low energy	low frequency
B.	high energy	low frequency	low energy	long wavelength
C.	high frequency	short wavelength	high energy	long wavelength
D.	high frequency	long wavelength	low frequency	low energy

3. Which of the following sources of light will produce a line spectrum when placed behind a prism?

   I.   a gas discharge tube

   II. an incandescent lamp

   III. an alkali metal salt placed in a Bunsen burner flame

   A. I and II only

   B. I and III only

   C. II and III only

   D. I, II and III

4. An electron transition between energy levels $n = 4$ and $n = 2$ in an isolated atom produces a line in the visible spectrum. Which electron transition in the same atom is likely to produce a line in the UV spectrum?

   A. from $n = 4$ to $n = 1$

   B. from $n = 4$ to $n = 3$

   C. from $n = 3$ to $n = 2$

   D. from $n = 5$ to $n = 3$

5. What is the maximum possible number of electrons in the third energy level?

   A. 3

   B. 6

   C. 9

   D. 18

6. What is the electron configuration of chromium ($Z = 24$) in the ground state?

   A. $[Ar]\, 3d^7$

   B. $[Ar]\, 4s^2\, 3d^4$

   C. $[Ar]\, 4s^1\, 3d^5$

   D. $[Ar]\, 4s^1\, 4p^5$

7. Which of the following is correct?

   A. $IE_3 > IE_4$

   B. Molar ionization energies are measured in kJ.

   C. The third ionization energy of the atom X represents the process:
   $$X^{2+}(g) \rightarrow X^{3+}(g) + e^-$$

   D. Ionization energies decrease across a period going from left to right.

8. Which statement about the first ionization energies of nitrogen and oxygen atoms is correct?

   A. $IE_1(N) < IE_1(O)$ because oxygen has two paired electrons in its partly filled sublevel

   B. $IE_1(N) > IE_1(O)$ because oxygen has two paired electrons in its partly filled sublevel

   C. $IE_1(N) < IE_1(O)$ because oxygen loses an electron from a higher sublevel than nitrogen

   D. $IE_1(N) > IE_1(O)$ because oxygen loses an electron from a higher sublevel than nitrogen

9. The first five successive ionization energies for an unknown element are 578, 1817, 2745, 11577 and 14842 kJ mol^{-1}. In which group of the periodic table is this element is likely to be found?

   A. 1

   B. 2

   C. 13

   D. 14

AHL

Extended-response questions

10. Explain, in your own words, why gaseous atoms produce line spectra instead of continuous spectra. [3]

11. State the full and condensed electron configurations for the following species in their ground states:
    a. titanium atom [1]
    b. selenium atom [1]
    c. silicon atom [1]
    d. Ti^{3+} cation [1]
    e. S^{2-} anion [1]

12. Determine which of the configurations below is impossible. Explain why it cannot exist. [2]
    $1s^2 2s^2 2p^7 3s^2 3p^5$
    $1s^2 2s^2 2p^6 3s^2 3p^6$

13. Deduce and explain, which of the following electron configurations represents a ground state. [2]
    $1s^2 2s^2 2p^6 3s^2 3p^6 4s^2 3d^{10} 4p^5 5s^1$
    $1s^2 2s^2 2p^6 3s^2 3p^6 4s^2 3d^{10} 4p^6 5s^1$

14. Sketch the shape of an s orbital. [1]

15. The diagram below (not to scale) represents some of the electron energy levels in the hydrogen atom.

    _____ $n = 7$
    _____ $n = 6$

    _____ $n = 5$

    _____ $n = 4$

    _____ $n = 3$

    _____ $n = 2$

    _____ $n = 1$

    Draw an arrow on the diagram to represent the lowest energy transition in the visible region of the emission spectrum of hydrogen. [1]

16. Sketch an orbital filling diagram for Al and deduce the number of unpaired electrons. [2]

17. A transition element ion, X^{3+}, has the electron configuration [Ar] $3d^5$. Determine the atomic number of element X. [1]

18. Sketch the condensed orbital filling diagram for germanium and deduce the total number of p orbitals containing one or more electrons. [2]

19. Describe, in your own words, how the first ionization energy of an atom can be determined from its emission spectrum. [2]

20. Using the data booklet, explore the first ionization energies of the period 3 elements, from sodium to argon. Explain the general trend and discontinuities in these energies. [3]

21. The first four successive ionization energies for an unknown element X are given in table 3. Deduce the group of the periodic table in which element X is likely to be found. [1]

n	IE_n / kJ mol^{-1}
1	738
2	1451
3	7733
4	10543

▲ Table 3 Successive ionization energies for element X

AHL

Structure 1.4 | Counting particles by mass: the mole

How do we quantify matter on the atomic scale?

Atoms are extremely small, so any physical object contains a huge number of these particles. There are more atoms in a glass of water than glasses of water in all of the oceans combined. The unit of the amount of substance, the mole, enables chemists to deal comfortably with large numbers of very small particles. At the same time, the concepts of molar, relative atomic and relative molecular masses allow the use of small numbers for expressing masses of atomic species.

Understandings

Structure 1.4.1 — The mole (mol) is the SI unit of amount of substance. One mole contains exactly the number of elementary entities given by the Avogadro constant.

Structure 1.4.2 — Masses of atoms are compared on a scale relative to ^{12}C and are expressed as relative atomic mass (A_r) and relative formula mass (M_r).

Structure 1.4.3 — Molar mass (M) has the units $g\,mol^{-1}$.

Structure 1.4.4 — The empirical formula of a compound gives the simplest ratio of atoms of each element present in that compound. The molecular formula gives the actual number of atoms of each element present in a molecule.

Structure 1.4.5 — The molar concentration is determined by the amount of solute and the volume of solution.

Structure 1.4.6 — Avogadro's law states that equal volumes of all gases measured under the same conditions of temperature and pressure contain equal numbers of molecules.

The mole (*Structure 1.4.1*)

Atoms and molecules are so small that their masses cannot be measured directly. Even a million atoms of lead, Pb, the heaviest stable element, would have a mass of only 3.4×10^{-16} g. This is too small to be weighed even on the most sensitive analytical balance. At the same time, the number of Pb atoms in 1 g of lead is huge, about 2.9×10^{21}, which is hard to imagine, let alone count. Therefore, chemists need a unit that allows them to work comfortably with both very small masses and very large numbers of atoms. This unit, the **mole**, was devised in the 19th century and quickly became one of the most useful concepts in chemistry.

The mole (with the unit "mol") is the SI unit of amount of substance that contains $6.02214076 \times 10^{23}$ elementary entities of that substance. An elementary entity can be an atom, a molecule, an electron or any other species. In this book, we will use the rounded value of the mole: $1\,mol = 6.02 \times 10^{23}$.

▲ Figure 1 One mole quantities of different substances (left to right): aluminium, water, copper, sucrose and sodium chloride

Prefix	Symbol	Factor
pico	p	10^{-12}
nano	n	10^{-9}
micro	μ	10^{-6}
milli	m	10^{-3}
centi	c	10^{-2}
deci	d	10^{-1}
kilo	k	10^{3}
mega	M	10^{6}
giga	G	10^{9}

▲ Table 1 Decimal prefixes

Avogadro's constant (N_A) is the conversion factor linking the number of particles and **amount of substance** in moles. It has the unit of mol^{-1}:

$$N_A = 6.02 \times 10^{23}\,mol^{-1}$$

In chemical calculations, Avogadro's constant is used in the same way as any other conversion factor (table 1). For example, to convert kilograms into grams, we need to multiply the mass in kg by 1,000. Similarly, to convert the amount of substance (n) into the number of atoms or any other structural units (N), we need to multiply that amount by N_A:

$$N = n \times N_A$$

In chemistry texts, the term "amount of substance" is often abbreviated to just "amount".

Worked example 1

Calculate the amount of lead (Pb), in mol and mmol, in a sample containing 2.9×10^{21} atoms of this element.

Solution

To find n, we can rearrange the equation $N = n \times N_A$ as follows:

$$n = \frac{N}{N_A}$$

Therefore, $n(\text{Pb}) = \dfrac{2.9 \times 10^{21}}{6.02 \times 10^{23}} \approx 4.8 \times 10^{-3}\,mol$

According to table 1, $1\,mmol = 10^{-3}\,mol$, so $4.8 \times 10^{-3}\,mol = 4.8\,mmol$. In this example, both answers ($4.8 \times 10^{-3}\,mol$ and $4.8\,mmol$) have been rounded to two significant figures, the same as in the least precise value used in the division (2.9×10^{21}).

The use of correct **significant figures** is discussed in the *Tools for chemistry* chapter.

Activity

Calculate:

a. the number of atoms in 2.5 mol of copper metal

b. the number of molecules in 0.25 mol of water

c. the number of atoms in 0.25 mol of water

ATL Research skills

The mole is a huge number, and it is useful for counting particles because they are so small. Measuring amounts of everyday objects in moles can help use to convey just how large this number is.

Choose one of the following and conduct the necessary research to reach an approximate answer.

- How many moles of grains of sand are in a desert of your choice?
- How many moles of water molecules are in a large sea or ocean of your choice?
- One mole of human cells represents roughly how many people?
- What is the age of the universe, in moles of seconds?
- How tall is a stack of one mole of sheets of paper?
- How many moles of air are in your school building?

Relative molecular mass and molar mass (*Structure 1.4.2 and 1.4.3*)

In *Structure 1.2*, we introduced the concept of **relative atomic mass**, A_r, which is the ratio of the mass of a certain atom to one-twelfth of the mass of a carbon-12 atom. Similarly, **relative molecular mass**, M_r, is the ratio of the mass of a molecule or other multiatomic species to one-twelfth of the mass of a carbon-12 atom. Both A_r and M_r are ratios, so they have no units.

To find the M_r of a molecule, we need to add together the A_r values for all atoms in that molecule.

Worked example 2

Calculate the M_r for a molecule of water.

Solution

Water, H_2O, is composed of two hydrogen atoms ($A_r = 1.01$) and one oxygen atom ($A_r = 16.00$). Therefore $M_r(H_2O) = 2 \times 1.01 + 16.00 = 18.02$.
You should always use the actual (not rounded) values of A_r, which are given in the data booklet and the periodic table at the end of this book. Similarly, keep all significant figures in calculated M_r values and never round them to the nearest integer number.

If a substance is composed of ions instead of molecules, the M_r for that substance is calculated using the smallest **formula unit**. For example, calcium chloride ($CaCl_2$) is an ionic compound that consists of many calcium cations (Ca^{2+}) and twice as many chloride anions (Cl^-). Its smallest formula unit contains one Ca^{2+} and two Cl^- ions. The ions have approximately the same masses as neutral atoms because the masses of electrons are negligible.

Therefore, $M_r(CaCl_2) = A_r(Ca) + 2 \times A_r(Cl) = 40.08 + (2 \times 35.45) = 110.98$.

Many ionic compounds form **hydrates**: compounds in which water molecules form coordination bonds (Structure 2.2) with the ions. One of the most common hydrates is copper(II) sulfate pentahydrate, $CuSO_4 \bullet 5H_2O$. Copper(II) sulfate pentahydrate forms large, clear, deep-blue crystals (figure 2). The **stoichiometric coefficient** "5" before "H_2O" means that one formula unit of copper(II) sulfate is bound with five molecules of water. Therefore, the M_r value for this hydrate can be calculated as follows:

$$M_r(CuSO_4 \bullet 5H_2O) = A_r(Cu) + A_r(S) + 4 \times A_r(O) + 5 \times M_r(H_2O)$$

$$= 63.55 + 32.07 + (4 \times 16.00) + (5 \times 18.02)$$

$$= 249.72$$

The composition and structure of ionic compounds will be discussed in *Structure 2.1*.

 Activity

Calculate the M_r values for the following species:

a. ammonia, NH_3

b. sulfuric acid, H_2SO_4

c. sodium sulfate decahydrate, $Na_2SO_4 \bullet 10H_2O$

▲ **Figure 2** Crystals of copper(II) sulfate pentahydrate, $CuSO_4 \cdot 5H_2O$

Molar mass, M, of a chemical substance is the mass of 1 mol of that substance. Molar mass is numerically equal to relative molecular mass (for substances with molecular and ionic structures) or relative atomic mass (for substances with atomic structure). For example, $M(Na) = 22.99\,g\,mol^{-1}$ and $M(H_2O) = 18.02\,g\,mol^{-1}$.

 ## Science as a shared endeavour

A shared understanding of common terminology helps scientists to communicate effectively. This terminology is constantly being updated.

Historically, the mole was defined as the amount of substance that contained as many elementary entities (atoms, molecules, ions, electrons or other particles) as there were atoms in 0.012 kg (or 12 g) of carbon-12. However, the numerical value of the mole (approximately 6.02×10^{23}) had to be revised frequently, as the improvements in instrumentation allowed scientists to measure mass with greater precision.

On 16 November 2018, scientists from more than 60 countries met at the *General Conference on Weights and Measures* in Versailles, France. It was agreed here that all SI base units, including the mole, were defined in terms of physical constants instead of physical objects. Following these changes, one mole of a substance is now defined exactly as $6.02214076 \times 10^{23}$ elementary entities of that substance.

The 2018 redefinition of the mole means that the mass of 1 mol of carbon-12 no longer equals 12 g exactly. As a result, the numerical values of M (defined through two exact SI quantities, the kilogram and the mole) no longer match the numerical values of their respective A_r or M_r (defined through the experimentally determined mass of a carbon-12 atom). However, the differences between these numerical values are so small (approximately $4 \times 10^{-8}\,\%$) that they can be ignored for all practical purposes.

Why are constants and values continuously being revised and updated?

How do scientists achieve a shared understanding of changes made to existing definitions?

The amount (n), mass (m) and molar mass (M) of any substance are related as follows:

$$n = \frac{m}{M}$$

This is probably the most common expression in chemistry, as it is used in almost all stoichiometric calculations. Although the base SI unit of mass is the kilogram, the masses of chemical substances are traditionally expressed in grams, and molar masses in $g\,mol^{-1}$.

Worked example 3

Table sugar is often sold in the form of cubes that are made almost entirely of sucrose. Sucrose is an organic compound with the molecular formula $C_{12}H_{22}O_{11}$. Calculate:

a. the molar mass of sucrose

b. the amount of sucrose in one cube (2.80 g) of sugar

c. the number of oxygen atoms in one cube of sugar

Solution

a. $M_r(C_{12}H_{22}O_{11}) = 12 \times 12.01 + 22 \times 1.01 + 11 \times 16.00 = 342.34$

$M(C_{12}H_{22}O_{11}) = 342.34\,g\,mol^{-1}$

b. $n = \frac{m}{M}$

$n(C_{12}H_{22}O_{11}) = \frac{2.80\,g}{342.34\,g\,mol^{-1}} \approx 0.00818\,mol$

c. One mole of sucrose contains 11 mol of oxygen atoms, so

$n(O) = 11 \times n(C_{12}H_{22}O_{11})$

$\quad = 11 \times 0.00818\,mol \approx 0.0900\,mol$

$N(O) = n(O) \times N_A = 0.0900\,mol \times 6.02 \times 10^{23}\,mol^{-1} \approx 5.42 \times 10^{22}$

 Activity

Calculate:

a. the molar mass of sulfuric acid, H_2SO_4

b. the amount of substance in 1.00 g of sulfuric acid

c. the number of hydrogen atoms in 1.00 g of sulfuric acid

▲ Figure 3 There are more oxygen atoms in one sugar cube than the estimated total insect population on Earth (10^{19}) and total grains of sand on Earth's beaches (10^{21})

Empirical formula, molecular formula and chemical analysis (*Structure 1.4.4*)

The composition of a chemical substance with a molecular structure can be represented by a **molecular formula**, which shows the actual number of atoms of each element in the molecule of that substance. In contrast, the **empirical formula** shows the simplest ratio of atoms of the different elements that are present in the substance. The molecular and empirical formulas of the same substance can be identical or different (table 2). For ionic compounds, the empirical formula is the same as the formula unit, which represents the simplest ratio of ions in the compound (figure 4).

Substance	Molecular formula	Empirical formula
oxygen	O_2	O
ozone	O_3	O
water	H_2O	H_2O
hydrogen peroxide	H_2O_2	HO
butane	C_4H_{10}	C_2H_5
glucose	$C_6H_{12}O_6$	CH_2O
sucrose	$C_{12}H_{22}O_{11}$	$C_{12}H_{22}O_{11}$

▲ **Table 2** Molecular and empirical formulas of selected substances

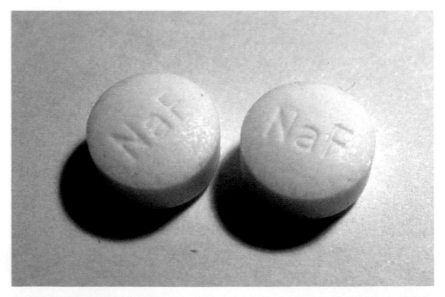

▲ **Figure 4** Sodium fluoride is an ionic compound with the empirical formula NaF. It is used in some countries as a food supplement to prevent tooth decay

The number of atoms of a certain element is proportional to the amount of that element in mol ($N = n \times N_A$). Therefore, the empirical formula also shows the **mole ratio** of elements in a chemical compound. For example, one molecule of water, H_2O, contains two atoms of hydrogen and one atom of oxygen, so the atomic ratio of hydrogen to oxygen in water is 2:1. Similarly, one mole of water contains two moles of hydrogen atoms and one mole of oxygen atoms, so the mole ratio of hydrogen to oxygen in water is also 2:1.

The elemental composition of a compound is often expressed in percent by mass, which is commonly referred to as the **percentage composition, ω**. The mole ratio can be used to calculate the percentage composition of a compound.

Worked example 4

Calculate the percentage composition of water.

Solution

Let $n(H_2O) = 1\,mol$, then $n(H) = 2\,mol$ and $n(O) = 1\,mol$. Using $m = n \times M$:

$m(H) = 2\,mol \times 1.01\,g\,mol^{-1} = 2.02\,g$

$m(O) = 1\,mol \times 16.00\,g\,mol^{-1} = 16.00\,g$

$m(H_2O) = 1\,mol \times 18.02\,g\,mol^{-1} = 18.02\,g$

$\omega(H) = \dfrac{2.02\,g}{18.02\,g} \times 100\% \approx 11.2\%$

$\omega(O) = 100\% - 11.2\% = 88.8\%$

In practice, chemists more often face the opposite problem of deducing the empirical formula for a compound from its percentage composition or other experimental data. The percentage composition can be determined by destruction analysis, in which the compound is combusted or decomposed, and the masses of the combustion or decomposition products are measured.

The mass percentages of elements in a sample can be determined by various analytical techniques, such as fully automated combustion elemental analysis. In a typical experiment, the sample is burned in excess oxygen, and the volatile combustion products are trapped and weighed. These weights are then converted into mass percentages of chemical elements in the original sample.

Practice question

Calculate the percentage composition of sulfuric acid, H_2SO_4.

Worked example 5

Iron and oxygen form several compounds (iron oxides). Deduce the empirical formula of an oxide that contains 72.36% of iron.

Solution

If $\omega(Fe) = 72.36\%$, then $\omega(O) = 100\% - 72.36\% = 27.64\%$.

Let $m(Fe_xO_y) = 100\,g$, then $m(Fe) = 72.36\,g$ and $m(O) = 27.64\,g$

Use $n = \dfrac{m}{M}$ to determine the amount of each element:

$n(Fe) = \dfrac{72.36\,g}{55.85\,g\,mol^{-1}} \approx 1.296\,mol$

$n(O) = \dfrac{27.64\,g}{16.00\,g\,mol^{-1}} \approx 1.728\,mol$

The mole ratio $x:y = 1.296:1.728 \approx 1:1.333 \approx 3:4$

Therefore, the empirical formula of the oxide is Fe_3O_4.

▲ Figure 5 Fe_3O_4 is the main component of the mineral magnetite, a common iron ore

Worked example 6

Hydrocarbons are organic compounds of carbon and hydrogen. An unknown hydrocarbon has undergone combustion in excess oxygen to produce 26.41 g of carbon dioxide, CO_2, and 13.52 g of water, H_2O. Deduce the empirical formula of the hydrocarbon.

Solution

$M(CO_2) = 12.01 + 2 \times 16.00 = 44.01\,g\,mol^{-1}$

$n(CO_2) = \dfrac{26.41\,g}{44.01\,g\,mol^{-1}} \approx 0.6001\,mol$

$n(C) = n(CO_2) = 0.6001\,mol$

$M(H_2O) = 2 \times 1.01 + 16.00 = 18.02\,g\,mol^{-1}$

$n(H_2O) = \dfrac{13.52\,g}{18.02\,g\,mol^{-1}} \approx 0.7503\,mol$

$n(H) = 2 \times n(H_2O) = 2 \times 0.7503\,mol \approx 1.501\,mol$

All carbon and hydrogen atoms in the combustion products originate from the hydrocarbon, C_xH_y, so:

The mole ratio $x : y = 0.6001 : 1.501 \approx 1 : 2.5 = 2 : 5$

Therefore, the empirical formula of the hydrocarbon is C_2H_5.

We express empirical formulas as whole number ratios. Whole numbers are also known as **integers**. In worked example 5, the ratio we initially calculated was comprised of two non-integer values: 1.296 and 1.728. To convert it to a whole number ratio, you divide each term in the ratio by the smallest number in the ratio. This gives a ratio of 1 : 1.333. Then, you can use trial and error to determine a factor by which you should multiply the ratio to obtain the whole number ratio. Multiplying this ratio by 3, and then subsequently rounding the result, gives a whole number ratio of 3 : 4.

The molecular formula of a compound can be deduced from the empirical formula if we know the molar mass of the compound. For example, you might determine experimentally that the molar mass of the hydrocarbon in worked example 6 is 58.12 g mol⁻¹. The molar mass of the empirical formula can be calculated:

$(12.01 \times 2) + (1.01 \times 5) = 29.07\,g\,mol^{-1}$

The value of 29.07 is roughly half of 58.12, therefore the molecular formula must have twice the number of atoms as the empirical formula: C_4H_{10}.

Table 2 suggests that this hydrocarbon could be butane, C_4H_{10}. However, we cannot be sure about it without further analysis, as there is another hydrocarbon, methylpropane, with the same molecular formula. Butane and methylpropane can be distinguished by measuring their boiling points (*Structure 1.1*) or comparing their infrared spectra (*Structure 3.2*).

Determining the molar masses of gaseous substances is discussed in *Structure 1.5*.

Practice questions

1. Deduce the empirical formulas of the following compounds:
   a. an oxide of manganese that contains 36.81% of oxygen
   b. a hydrocarbon that produces 5.501 g of carbon dioxide and 2.253 g of water upon complete combustion
2. Deduce the molecular formula of the hydrocarbon from 1b if its molar mass is 42.09 g mol⁻¹.

Experimental determination of empirical formula

Relevant skills

- Tool 1: Measure mass
- Tool 3: Carry out calculations involving decimals and ratios
- Tool 3: Use approximation and estimation
- Tool 3: Construct and interpret graphs
- Inquiry 3: Explain realistic and relevant improvements to an investigation

Safety

- Wear eye protection.
- Take suitable precautions around open flames.
- The equipment will get very hot. Take suitable precautions around it and do not touch it while it is hot.
- Magnesium burns with a very bright light. Do not look directly at it.

Materials

- crucible and lid
- balance (±0.01 g)
- pipeclay triangle
- tripod
- heat-proof mat
- tongs
- magnesium ribbon
- Bunsen burner

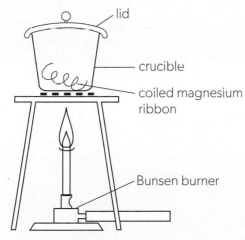

▲ Figure 6 The experimental set-up

Instructions

1. Weigh a clean, dry crucible.
2. Obtain a piece of magnesium ribbon (between 0.3 g and 1.0 g) from your teacher. Measure its exact mass.
3. Twist the magnesium into a loose coil and place it inside the crucible.
4. Heat the crucible, with its lid on, over a roaring Bunsen flame. Periodically lift the crucible lid to allow air to enter the crucible.
5. Continue heating until the magnesium no longer lights up. Then, remove the heat source and allow the crucible to cool for a few minutes.
6. When the crucible is cool, weigh it.
7. Heat the crucible and its contents strongly for an additional minute. Allow to cool and re-weigh. Repeat this heating-cooling-weighing cycle until the mass is constant.

Questions

1. Process the data to determine the empirical formula of magnesium oxide.
2. Compare your experimental empirical formula to the actual one.
3. Obtain mass data from other members of your class. Plot a graph of mass of magnesium oxide vs mass of magnesium.
4. Identify any anomalies (if applicable) and draw a best fit line on the graph.
5. Explain what the graph shows about the composition of magnesium oxide.
6. Explain why you repeatedly heated and weighed the crucible until a constant mass was achieved.
7. Identify and explain two major sources of error in this procedure.
8. Suggest realistic improvements to the methodology that could minimize the sources of error you have identified.
9. Reflect on the role of approximation and rounding in empirical formula calculations. When is it suitable to round to the nearest whole number? When is it not? Can you come up with a rule of thumb of when to round and when not to round?

Measurement

Atoms, molecules and ions are so small that counting them directly is virtually impossible. The concept of the mole is powerful because it relates number of particles to mass, which can be easily measured.

As with all measurements, mass has an uncertainty associated with it. Consider the mass of a sample of calcium carbonate, $CaCO_3$, is found to be 3.500 g ± 0.001 g. This means the mass measurement can be inaccurate by up to 0.001 g in either direction. This is clearly a minuscule mass. How many moles does it represent? How many particles does it represent? Do a quick calculation and find out. You will see that in moles the uncertainty is tiny, but in terms of particles, it is quite large.

Is a measurement uncertainty ever negligible? If so, when? Think about these questions as you proceed through the DP chemistry course, particularly when doing experiments that involve making measurements.

Solutions and concentration (*Structure 1.4.5*)

Many chemical reactions are carried out in solutions. Solutions are easier to handle and mix than solids or gases. Sometimes a solvent is used because it can affect the properties of dissolved substances or participate in chemical reactions.

Solutions are homogeneous mixtures of two or more components. Each solution consists of a **solvent** and one or more **solutes**. The solvent is usually the major component of the solution, so the properties of the whole solution are similar to the properties of the solvent. The other components of the solution are called solutes. For example, a solution of sugar in water is more like water (clear colourless liquid) than sugar (white crystalline powder), so water is the solvent while sugar is the solute. In this topic, we will consider only **aqueous solutions** (from the Latin *aqua* meaning "water"), in which the solvent is water.

Homogeneous and heterogeneous mixtures are discussed in *Structure 1.1.*

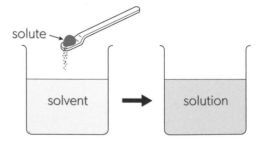

▲ **Figure 7** How a solution is formed

In some cases, the identity of the solvent is unclear: for example, if we mix ethanol and water, each of these liquids can be called a solvent. However, if water is present in the mixture, it is traditionally regarded as the solvent, even if it is not the major component. For example, we say "96% solution of ethanol in water" rather than "4% solution of water in ethanol".

Solutions are often classified according to the mass or mole ratio between the solute and solvent. A **concentrated solution** contains a large proportion of solute, and so has a high ratio of solute to solvent, while a **dilute solution** has a small proportion of solute, and so has a low ratio of solute to solvent. Generally, the term "concentrated" refers to solutions with much more than 10 g of the solute per 100 g of the solvent, and the term "dilute" refers to solutions with much less than 10 g of the solute per 100 g of the solvent.

TOK

Some words do not have precise definitions and their choice and interpretation is context dependent. The terms "concentrated" and "dilute" are not precisely defined and should be used with care. For example, most chemists would call a solution of 5 g of sulfuric acid (H_2SO_4) in 100 g of water "dilute", as much higher proportions of sulfuric acid to water are commonly used in laboratories. At the same time, a solution of 5 g of potassium permanganate ($KMnO_4$) in 100 g of water would be considered very concentrated by any medical worker, as typical concentrations of potassium permanganate in antiseptic solutions are less than 0.1 g per 100 g of water.

The concentrations in the examples above could be expressed numerically. To what extent does expressing a quantity numerically help or hinder the communication of knowledge?

Quantitatively, the composition of solutions is expressed in terms of concentration. **Molar concentration**, c, also known as **molarity**, is the ratio of the amount of a solute to the volume of the solution:

$$c_{solute} = \frac{n_{solute}}{V_{solution}}$$

The most common units for molar concentration are $mol\,dm^{-3}$ (which is the same as $mol\,L^{-1}$). For very dilute solutions, smaller units ($mmol\,dm^{-3}$ or $\mu mol\,dm^{-3}$) can also be used:

$1\,mmol\,dm^{-3} = 1 \times 10^{-3}\,mol\,dm^{-3}$

$1\,\mu mol\,dm^{-3} = 1 \times 10^{-6}\,mol\,dm^{-3}$

The units of molar concentrations are sometimes abbreviated as M (for $mol\,dm^{-3}$) or mM (for $mmol\,dm^{-3}$). For example, the expression "2.5 M NaOH" means that each dm^3 of the solution contains 2.5 mol of sodium hydroxide.

Note that the term "molar concentration" refers to a specific substance, not the whole solution. For example, it is incorrect to say that "the concentration of a sodium chloride solution is $1.0\,mol\,dm^{-3}$", as it is not clear whether we are talking about the concentration of sodium chloride or water. The correct statement would be "the concentration of sodium chloride in a solution is $1.0\,mol\,dm^{-3}$".

Molar concentration is often represented by square brackets around the solute formula. For example, the expression $[NH_3] = 0.5\,M$ refers to a $0.5\,mol\,dm^{-3}$ solution of ammonia. Similarly, the expression $[Cl^-]$ refers to the molar concentration of chloride ions in a solution.

Worked example 7

Calculate the molar concentration of sodium chloride, in $mol\,dm^{-3}$, in a solution prepared by dissolving 3.60 g of NaCl(s) in water to make $25.0\,cm^3$ of the final solution.

Solution

First, calculate the molar mass of sodium chloride:

$$M(NaCl) = 22.99 + 35.45 = 58.44\,g\,mol^{-1}$$

Then use $n = \dfrac{m}{M}$ to calculate the amount of solute:

$$n(NaCl) = \frac{3.60\,g}{58.44\,g\,mol^{-1}} \approx 0.0616\,mol$$

Convert the volume to dm^3 by dividing by 1,000:

$$V(\text{solution}) = 25.0\,cm^3 = 0.0250\,dm^3$$

Use $c = \dfrac{n}{V}$ to calculate the concentration:

$$c(NaCl) = \frac{0.0616\,mol}{0.0250\,dm^3} \approx 2.46\,mol\,dm^{-3}.$$

The composition of a solution is sometimes expressed as the **mass concentration**, ρ_{solute}, of the solute, which is the ratio of the mass of the solute to the volume of the solution:

$$\rho_{\text{solute}} = \frac{m_{\text{solute}}}{V_{\text{solution}}}$$

Worked example 8

Calculate the mass concentration of sodium chloride in the solution from worked example 7.

Solution

If we know the mass of the solute and the volume of the solution, we can calculate the mass concentration as follows:

$$\rho(NaCl) = \frac{3.60\,g}{0.0250\,dm^3} = 144\,g\,dm^{-3}$$

Alternatively, the mass concentration of NaCl can be found from its molar concentration and molar mass, using the relationship $\rho_{\text{solute}} = c_{\text{solute}} \times M_{\text{solute}}$:

$$\rho(NaCl) = c(NaCl) \times M(NaCl) = 2.46\,mol\,dm^{-3} \times 58.44\,g\,mol^{-1} \approx 144\,g\,dm^{-3}$$

The most common units for mass concentration are $g\,dm^{-3}$ and $g\,cm^{-3}$. Mass concentration and molar concentration of the same solute are related by molar mass, as follows:

$$c_{\text{solute}} = \frac{\rho_{\text{solute}}}{M_{\text{solute}}}$$

Activity

Calculate the mass of sulfuric acid, H_2SO_4, in $50.0\,cm^3$ of a solution where $[H_2SO_4] = 1.50\,mol\,dm^{-3}$.

Activity

Calculate the molar concentration of sulfuric acid, in $mol\,dm^{-3}$ in a solution with $\rho(H_2SO_4) = 0.150\,g\,cm^{-3}$.

Worked example 9

A standard solution was prepared by dissolving 6.624 g of sodium carbonate, Na_2CO_3, in deionized water using a 250 cm³ volumetric flask. An analytical pipette was used to transfer 10.0 cm³ sample of this solution to a 100 cm³ volumetric flask, and the flask was topped up to the graduation mark with deionized water. Calculate the concentration, in mol dm⁻³, of sodium carbonate in the new solution.

Solution

First, we need to find the concentration of sodium carbonate in the standard solution:

$$M(Na_2CO_3) = 2 \times 22.99 + 12.01 + 3 \times 16.00 = 105.99 \, g \, mol^{-1}$$

$$n(Na_2CO_3) = \frac{6.624 \, g}{105.99 \, g \, mol^{-1}} \approx 0.06250 \, mol$$

$$V_{standard} = 250 \, cm^3 = 0.250 \, dm^3$$

Note that the accuracy of a typical volumetric flask is three significant figures.

$$c_{standard}(Na_2CO_3) = \frac{0.06250 \, mol}{0.250 \, dm^3} = 0.250 \, mol \, dm^{-3}$$

Then we need to calculate the concentration of sodium carbonate in the new solution.

First, calculate the amount of Na_2CO_3 in the sample. Remember to convert all volumes to dm³.

$$V_{sample} = 10.0 \, cm^3 = 0.0100 \, dm^3$$

$$c_{standard}(Na_2CO_3) = c_{sample}(Na_2CO_3) = 0.250 \, mol \, dm^{-3}$$

$$n_{sample}(Na_2CO_3) = 0.250 \, mol \, dm^{-3} \times 0.0100 \, dm^3 = 0.00250 \, mol$$

When the sample is diluted with deionized water to produce the new solution, the amount of solute does not change. Therefore

$$n_{sample}(Na_2CO_3) = n_{new}(Na_2CO_3) = 0.00250 \, mol$$

Now you can work out the concentration of Na_2CO_3 in the new solution by dividing the amount of Na_2CO_3 by the volume of the new solution:

$$V_{new} = 100 \, cm^3 = 0.100 \, dm^3$$

$$c_{new}(Na_2CO_3) = \frac{0.00250 \, mol}{0.100 \, dm^3} = 0.0250 \, mol \, dm^{-3}$$

It is a common practice to store chemicals in the form of concentrated solutions (so-called **stock solutions**) and dilute them to the required concentration when needed. Stock solutions with a known concentration of the solute are called **standard solutions**.

To determine the concentration of the standard solution in worked example 9, we did the following two calculations:

1. $n_{sample} = c_{sample} \times V_{sample}$

2. $c_{new} = \dfrac{n_{new}}{V_{new}}$

You know that $n_{sample} = n_{new}$, so you can substitute equation 1 into equation 2. This gives the following expression:

$$c_{new} = \frac{c_{sample} \times V_{sample}}{V_{new}}$$

Therefore, to calculate the concentration of a solute in a new solution, you just need to know the original concentration of the solute, the volume of the original solution, and the volume of the new solution. In summary:

$$c_1 \times V_1 = c_2 \times V_2$$

Practice question

3. A standard solution was prepared by dissolving 2.497 g of copper(II) sulfate pentahydrate, $CuSO_4 \cdot 5H_2O$, in deionized water using a 100 cm³ volumetric flask. A 5.00 cm³ sample of this solution was diluted to 250.0 cm³. Calculate the concentration, in mol dm⁻³, of copper(II) sulfate in the final solution.

The process for preparing standard solutions is discussed in the *Tools for chemistry* chapter.

Case study: spectrophotometry and calibration curves

Spectrophotometry is an analytical technique based on the measurement of the intensity of visible, ultraviolet and near-infrared radiation. This technique is commonly used for determining concentrations of coloured substances in solutions.

A spectrophotometer produces light of a certain wavelength, which passes through a small sample of the studied solution. The photodetector measures the intensity of the transmitted light and converts it into the **absorbance**. Absorbance is a value describing the amount of light absorbed by the sample. Initially, several standard solutions of the studied substance are prepared by serial dilution (figure 9), and their absorbances are measured. These absorbances are plotted against concentrations, producing a **calibration curve** (figure 8). The calibration curve is then used for determining the unknown concentration of the coloured substance in the studied solution.

In the general case, a calibration curve relates a measurable property (such as absorbance, pH or electrical conductivity) of the solution to the concentration of the solute. The unknown concentration can be found by measuring that property and plotting the result of the measurement on the calibration curve.

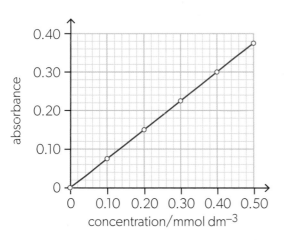

▲ Figure 8 A typical calibration curve

▲ Figure 9 A series of standard solutions of potassium permanganate

🕒 Data-based question

The calibration curve in figure 8 was obtained using a series of standard solutions of potassium permanganate, $KMnO_4$. A solution with unknown concentration of $KMnO_4$ has an absorbance of 0.285. Determine the concentration of $KMnO_4$ in that solution.

Ideally, the calibration curve should be linear, pass through the origin and have a tilt of approximately 45°. If the curve does not meet any of these requirements, it should be constructed again using a slightly different wavelength of light and/or different set of standard solutions. Sometimes linearity can only be achieved within a narrow range of concentrations. In this case, the studied solution can be diluted, so the concentration of the studied substance falls within the range of calibration curve. In the last case, some additional calculations will be required to relate the concentrations of the studied substance in the diluted and original solutions.

Another technique, **colorimetry**, is based on the same principles as spectrophotometry but limited to visible light. The terms "colorimetry" and "spectrophotometry" are often used interchangeably, which is not entirely correct but very common.

Concentration uncertainty of a standard solution

A standard solution is a solution of known concentration. In this activity, you will prepare two standard solutions of copper(II) sulfate, each by using different equipment. By propagating the measurement uncertainties, you will assess the precision of the concentration values. You will determine the actual concentration of your solutions using a colorimeter. This will then allow you to assess the accuracy of the concentrations.

Relevant skills
- Tool 1: Measuring volume and mass
- Tool 1: Standard solution preparation
- Tool 3: Calculate and interpret percentage error and percentage uncertainty
- Tool 3: Express quantities and uncertainties to an appropriate number of significant figures
- Tool 3: Record measurement uncertainties
- Tool 3: Propagate uncertainties
- Inquiry 2: Assess accuracy and precision

Safety
- Wear eye protection.
- Solid copper(II) sulfate is an irritant and toxic to the environment
- Dispose of all solutions appropriately.

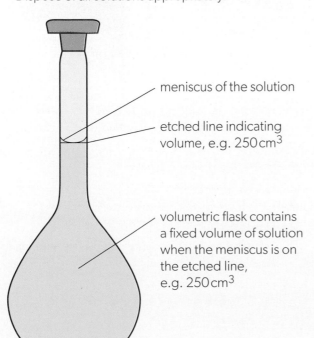

meniscus of the solution

etched line indicating volume, e.g. 250 cm³

volumetric flask contains a fixed volume of solution when the meniscus is on the etched line, e.g. 250 cm³

Materials
- Wash bottle containing distilled water
- Weighing boats (2)
- 100 cm³ beakers (2)
- Stirring rods (2)
- Funnels (2)
- Pipettes
- Spatula
- Reagent bottles (2)
- Blank labels
- Colorimeter
- Cuvettes
- Calibration curve relating concentration of copper(II) sulfate and absorbance
- Copper(II) sulfate pentahydrate, $CuSO_4 \cdot 5H_2O$

Additional equipment for solution 1:
- 100 cm³ volumetric flask
- Milligram balance (three decimal places)

Additional equipment for solution 2:
- 100 cm³ measuring cylinder
- Centigram balance (two decimal places)

Instructions
1. Use the equipment provided to prepare two copper(II) sulfate standard solutions, both with concentration 0.020 mol dm⁻³. When preparing solution 1, you should use the volumetric flask and milligram balance. For solution 2, use the measuring cylinder and centigram balance.
2. Record the measurements you make along the way, including their uncertainties.
3. Following your teacher's instructions on how to use the colorimeter, measure the absorbance of your solutions.
4. Refer to the calibration curve to determine the actual concentration of your solutions.

Questions
1. Determine the uncertainty of the concentrations of solutions 1 and 2.
2. Calculate the percentage error of the concentrations of solutions 1 and 2.
3. Assess the precision and accuracy of the concentrations of solutions 1 and 2.

4. Consider the way you have presented your calculations for the questions above. Do you think they convey your thinking? Do you think a reader would be able to easily follow your thought process? How could you improve the presentation of your calculations? You may want to look through some of the worked examples in this textbook for ideas.

5. The construction of calibration curves involves preparing samples of solutions that cover a range of concentrations. Instead of measuring and dissolving a certain mass of solute the way you have done here, chemists often start with a stock solution and perform a **serial dilution**. Discuss the advantages and disadvantages of using a serial dilution in the preparation of samples for a calibration curve.

Avogadro's law (*Structure 1.4.6*)

In 1811, Amedeo Avogadro suggested that equal volumes of any two gases at the same temperature and pressure contain equal numbers of molecules. This hypothesis has been confirmed in many experiments and is now known as **Avogadro's law**.

Since the amount of a substance and the number of particles are proportional to each other, the amount of a gas is proportional to its volume. Therefore, the volumes of two reacting gaseous species measured under the same conditions are proportional to the amounts of these species:

$$\frac{n_1}{n_2} = \frac{V_1}{V_2}$$

In turn, the amounts of reactants and products are proportional to their stoichiometric coefficients in a balanced chemical equation. As a result, if we know the volume of any gas consumed or produced in the reaction, the volumes of other gaseous substances can be found without calculating their amounts.

Worked example 10

The combustion of hydrogen sulfide, H_2S, proceeds as follows:

$$2H_2S(g) + 3O_2(g) \rightarrow 2H_2O(l) + 2SO_2(g)$$

Calculate the volumes of oxygen, $O_2(g)$, consumed and sulfur dioxide, $SO_2(g)$, produced if the volume of hydrogen sulfide combusted was $0.908\,dm^3$. All volumes are measured under the same conditions.

Solution

The ratio of the stoichiometric coefficients of H_2S and O_2 is $2:3$. Therefore, you can multiply the volume of combusted H_2S by $\frac{3}{2}$ to find the volume of combusted O_2:

$$V(O_2) = \frac{3}{2}\,V(H_2S) = \frac{3}{2} \times 0.908\,dm^3 \approx 1.36\,dm^3$$

The ratio of the stoichiometric coefficients of H_2S and SO_2 is $1:1$. Therefore, the volume of combusted H_2S is the same as the volume of produced SO_2:

$$V(SO_2) = V(H_2S) = 0.908\,dm^3$$

Note that the volume of liquid water cannot be found in the same manner, as Avogadro's law applies to gases only.

Practice question

4. Incomplete combustion of hydrogen sulfide produces elemental sulfur instead of sulfur dioxide:

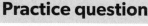

$$2H_2S(g) + O_2(g) \rightarrow 2H_2O(l) + 2S(s)$$

Calculate the volume of combusted hydrogen sulfide if the volume of oxygen consumed in this reaction was $1.25\,dm^3$.

 Linking question

Avogadro's law applies to ideal gases. Under what conditions might the behaviour of a real gas deviate most from an ideal gas? (Structure 1.5)

End-of-topic questions

Topic review

1. Using your knowledge from the *Structure 1.4* topic, answer the guiding question as fully as possible:

   How do we quantify matter on the atomic scale?

Exam-style questions

Multiple-choice questions

2. What is the number of oxygen atoms in 0.400 mol of copper(II) sulfate pentahydrate, $CuSO_4 \cdot 5H_2O$?

   A. 3.60

   B. 9

   C. 2.16×10^{24}

   D. 5.40×10^{24}

3. A sample containing 0.70 g of calcium nitrate, $Ca(NO_3)_2$, is dissolved in water to a volume of 200 cm³. What is the concentration of NO_3^- ions in this solution?

   A. $3.5 \, g \, dm^{-3}$

   B. $7.0 \, g \, dm^{-3}$

   C. $0.021 \, mol \, dm^{-3}$

   D. $0.043 \, mol \, dm^{-3}$

4. For which molecule is the empirical formula the same as the molecular formula?

   A. $CH_3CH_2CH_2OH$

   B. $CH_3COOCH_2CH_3$

   C. $CH_3CH_2CH_2CH_3$

   D. CH_3COOH

5. Which volume of a $5.0 \, mol \, dm^{-3}$ sulfuric acid (H_2SO_4) stock solution is required to prepare 0.50 dm³ of a solution whose concentration of hydrogen ions is $0.10 \, mol \, dm^{-3}$?

   A. 0.010 cm³

   B. 0.0050 cm³

   C. 5.0 cm³

   D. 10 cm³

6. A student obtained the following data during an experimental determination of the empirical formula of an oxide of tin:

   - Mass of tin before heating = 1.78 g

   - Mass of oxide of tin after heating to a constant mass = 2.26 g

   According to these data, what is the correct formula of the oxide of tin?

   A. SnO

   B. SnO_2

   C. SnO_3

   D. SnO_5

Extended-response questions

7. Alums are salt hydrates of the general formula $XAl(SO_4)_2 \cdot 12H_2O$, where X is an alkali metal or other singly-charged cation. When heated, most alums decompose as follows:

   $$XAl(SO_4)_2 \cdot 12H_2O(s) \rightarrow XAl(SO_4)_2(s) + 12H_2O(l)$$

Calculate:

   a. The molar mass of potassium alum, $KAl(SO_4)_2 \cdot 12H_2O$. [1]

   b. The amount of substance, in mol, in 1.00 g of potassium alum. [1]

   c. The total number of atoms in 1.00 g of potassium alum. [1]

   d. The amount of water, in mol, that can be produced by complete decomposition of 1.00 g of potassium alum. [1]

   e. The percentage composition, by mass, of potassium alum. [2]

8. To visualize the mole, a chemistry student decided to pile up 6.02×10^{23} grains of sand. Estimate the time needed to complete this project if an average grain of sand weighs 5 mg, and the student can shovel 50 kg of sand per minute. [3]

9. Deduce the empirical formulas for the following compounds:

   a. an oxide of sulfur that contains 59.95% of oxygen [1]

   b. an oxygen-containing organic compound, 5.00 g of which produces 9.55 g of carbon dioxide and 5.87 g of water upon complete combustion. [2]

10. A standard solution of potassium sulfate, K_2SO_4, was prepared from 8.714 g of the solid salt using a 250 cm³ volumetric flask. Calculate the mass concentration, in $g \, dm^{-3}$, and the molar concentration, in $mol \, dm^{-3}$, of potassium sulfate in the final solution. [2]

11. The calibration curve in figure 7 was constructed using five standard solutions, in which the concentration of potassium permanganate, $KMnO_4$, varied from 0.100 to $0.500 \, mmol \, dm^{-3}$. Describe how you would prepare these solutions using serial dilution. [3]

12. Carbon monoxide, CO, is a toxic gas. Its combustion produces carbon dioxide, $CO_2(g)$.

   a. Deduce the balanced equation for the combustion of carbon monoxide. [1]

   b. Calculate the volumes, in dm³, of consumed carbon monoxide and oxygen if the combustion produced 2.00 dm³ of carbon dioxide. All volumes are measured under the same conditions. [2]

Structure 1.5 Ideal gases

How does the model of ideal gas behaviour help us to predict the behaviour of real gases?

As with any theoretical model, the concept of an ideal gas is a simplification that has its advantages and limitations. In many cases, it predicts the properties of real gases with a precision sufficient for most practical purposes. However, at low temperatures and high pressures the behaviour of real gases deviates significantly from the prediction, so the ideal gas model cannot be used under these conditions.

Understandings

Structure 1.5.1 — An ideal gas consists of moving particles with negligible volume and no intermolecular forces. All collisions between particles are considered elastic.

Structure 1.5.2 — Real gases deviate from the ideal gas model, particularly at low temperature and high pressure.

Structure 1.5.3 — The molar volume of an ideal gas is a constant at a specific temperature and pressure.

Structure 1.5.4 — The relationship between the pressure, volume, temperature and amount of an ideal gas is shown in the ideal gas equation $pV = nRT$ and the combined gas law $\dfrac{p_1 V_1}{T_1} = \dfrac{p_2 V_2}{T_2}$.

Assumptions of the ideal gas model (*Structure 1.5.1*)

The ideal gas model states that an ideal gas conforms to the following five assumptions:

1. **Molecules of a gas are in constant random motion**

   This means that gas molecules are not stationary. They move in straight lines until they collide with another gas molecule or the side of a container.

2. **Collisions between molecules are perfectly elastic**

   In inelastic collisions of larger objects, energy can be transferred as heat or sound. However, the collisions between molecules in an ideal gas are perfectly elastic and no energy is lost from the system.

3. **The volume occupied by gas molecules is negligible compared to the volume of the container they occupy**

   Vaporized water occupies 1600 times the volume of liquid water at 273.15 K (0 °C) and 100 kPa pressure (**standard temperature and pressure, STP**). Nitrogen gas occupies about 650 times the volume of liquid nitrogen under the same conditions. In both cases, the number of molecules in liquid and gaseous phase is the same and the size of individual molecules has not changed, but the volume of the gas is >99.9% empty space. This is the space in which the gas molecules are free to move.

4. **There are no intermolecular forces between gas particles**

   For an ideal gas, the intermolecular forces are negligible compared to the kinetic energy of the molecules. As such, an ideal gas will not condense into a liquid.

Intermolecular forces are studied in *Structure 2.2*.

5. **The kinetic energy of the molecules is directly proportional to Kelvin temperature**

This relationship is studied in *Reactivity 2.2*.

Pressure–volume relationships

Robert Boyle (1627–1691) established that, at constant temperature, the pressure of a given amount of a gas is inversely proportional to its volume. This relationship, now known as **Boyle's law**, can be expressed as follows:

$$p \propto \frac{1}{V} \quad \text{or} \quad pV = k \text{ (a constant)} \quad \text{or} \quad p_1 V_1 = p_2 V_2$$

In figure 1, the molecules of a gas are constantly striking and bouncing off the walls of the container. The force of these impacts produces a measurable pressure. If the volume is halved, there are twice as many molecules in each unit of space, so every second there are twice as many impacts with the container walls, so the pressure is doubled (figure 2).

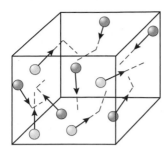

▲ Figure 1 An ideal gas consists of particles that collide elastically, have no intermolecular forces and occupy negligible volume when compared to the volume of the gas (the container)

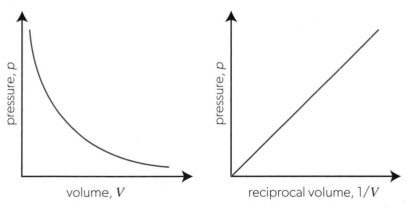

volume halved

pressure doubled

▲ Figure 2 Halving the volume of a container doubles the pressure

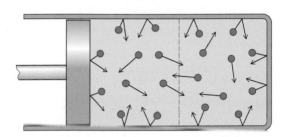 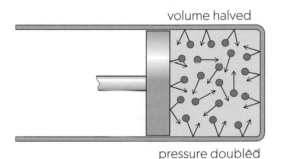

▲ Figure 3 Graphs showing the inverse relationship between pressure and volume of an ideal gas

The SI unit of pressure is the **pascal (Pa)**, where $1\,Pa = 1\,N\,m^{-2} = 1\,J\,m^{-3}$. Many other units of pressure are commonly used in different countries, including the atmosphere (atm), millimetres of mercury (mm Hg), bar, and pounds per square inch (psi). Standard temperature and pressure conditions (STP) are frequently found in databases for comparative purposes. STP for gases is 0 °C or 273.15 K temperature and 100.0 kPa pressure.

TOK

Models are simplified representations of natural phenomena. The ideal gas model is built on certain assumptions related to the behaviour of ideal gases.

What is the role of assumptions in the development of scientific models?

What are the implications of not acknowledging a model's limitations?

Worked example 1

A weather balloon filled with 32.0 dm³ of helium at a pressure of 100.0 kPa is released at sea level. The balloon reaches an altitude of 4500 m, where the atmospheric pressure is 57.7 kPa. Calculate the volume, in dm³, of the balloon at that altitude. Assume that the temperature and the amount of helium in the balloon remain constant.

Solution

From Boyle's law, it follows that $V_2 = \dfrac{V_1 \times P_1}{P_2}$, so:

$$V_2 = \frac{32.0 \, \text{dm}^3 \times 100 \, \text{kPa}}{57.7 \, \text{kPa}} \approx 55.5 \, \text{dm}^3$$

Practice question

1. At a certain altitude, a weather balloon has a temperature of −35.0 °C and a volume of 0.250 m³. Calculate the pressure, in kPa, inside the balloon if contains 16.0 g of helium, He(g).

Real gases vs ideal gases (*Structure 1.5.2*)

When the volume of a real gas decreases significantly, the molecules may begin to occupy a large proportion of the container. With so little space to move, intermolecular forces become significant. This decreases the number of collisions, reducing the pressure. This means that, for a real gas, the inverse relationship between pressure and volume no longer applies. Figure 4 shows a graph of pressure against volume for a real gas and an ideal gas. For the real gas, doubling the pressure no longer halves the volume.

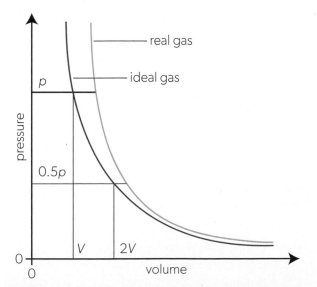

▲ **Figure 4** Doubling the pressure halves the volume for an ideal gas but not for a real gas

For a gas to deviate from ideal gas behaviour, there must be detectable intermolecular forces and/or a significant volume of the gas must be occupied by the molecules themselves. This commonly occurs at a low temperature and high pressure.

Low temperature: At low temperature, the kinetic energy of the gas molecules is reduced. As they collide with one another, intermolecular forces of attraction form and molecules may not necessarily rebound elastically.

High pressure: At high pressure, there are more molecules in a reduced space. The volume of the molecules becomes a significant part of the volume of the gas. As molecules themselves cannot be compressed, only the space between them, the relationship between pressure and volume is no longer inverse, so the gas is not considered to be an ideal gas.

Ideal gas conditions keep molecules far apart and prevent interaction between them. The conditions for an ideal gas behaviour are low pressure and high temperature. At low pressure, there are very few molecules per unit of volume in the container, so the space occupied by the molecules themselves is negligible. At high temperature, the molecules are moving too fast to allow for intermolecular forces of attraction to form.

 Activity

1. Outline the main assumptions behind the ideal gas model.

2. Discuss what conditions of pressure and temperature are likely to lead to deviations from ideal behaviour.

3. Consider how each of the following might affect the validity of the ideal gas model:

   a. Strong intermolecular forces

   b. Large molecular volume

4. For each of the following pairs, predict which is more likely to exhibit ideal behaviour and give a reason:

   a. gas at low pressure
      or
      gas at high pressure

   b. gas at low temperature
      or
      gas at high temperature

   c. hydrogen fluoride, $HF(g)$
      or
      hydrogen bromide, $HBr(g)$

   d. methane, $CH_4(g)$
      or
      decane, $C_{10}H_{22}(g)$

   e. propanone, $CH_3COCH_3(g)$
      or
      butane, $C_4H_{10}(g)$

Real gases

Gases that deviate from the ideal gas model are known as real gases.

Relevant skills

- Tool 2: Use spreadsheets to manipulate data.
- Inquiry 1: Select sufficient and relevant sources of information.
- Inquiry 1: Demonstrate creativity in the designing, implementation or presentation of the investigation.

Instructions

The relationship between pressure, volume, amount and temperature of real gases is modelled by the van der Waals equation:

$$\left(p + a\left(\tfrac{n}{V}\right)^2\right)\left(V - nb\right) = nRT$$

measured pressure — correction for forces between molecules — correction for volume of molecules — measured volume

Parameter a corrects for intermolecular force strength and parameter b corrects for molecular volume. Values of a and b for various gases are shown in table 1.

Substance	$a \;/\; \times 10^{-1}\ \mathrm{Pa\,m^6\,mol^{-2}}$	$b \;/\; \times 10^{-3}\ \mathrm{m^3\,mol^{-1}}$
ammonia, NH_3	4.225	0.0371
argon, Ar	1.355	0.0320
butane, C_4H_{10}	13.89	0.1164
butan-1-ol, C_4H_9OH	20.94	0.1326
chloromethane, CH_3Cl	7.566	0.0648
ethane, C_2H_6	5.580	0.0651
ethanol, C_2H_5OH	12.56	0.0871
helium, He	0.0346	0.0238
hydrogen bromide, HBr	4.500	0.0442
hydrogen chloride, HCl	3.700	0.0406
hydrogen fluoride, HF	9.565	0.0739
krypton, Kr	5.193	0.0106
methane, CH_4	2.303	0.0431
methanol, CH_3OH	9.476	0.0659
neon, Ne	0.208	0.0167
pentane, C_5H_{12}	19.09	0.1449
pentan-1-ol, $C_5H_{11}OH$	25.88	0.1568
propane, C_3H_8	9.39	0.0905
propan-1-ol, C_3H_7OH	16.26	0.1079
water, H_2O	5.537	0.0305
xenon, Xe	4.192	0.0516

▲ Table 1 Van der Waals parameters, a and b, for a selection of gases

1. Use a selection of the data in table 1 to explore some of the factors affecting the values of *a* and *b*. For instance, you could look at:

   • intermolecular force strength and the value of *a*

   • molar mass and the value of *b*

   • the effect of volume on the deviation from ideal gas behaviour.

You will need to decide how much data to select, and how to analyse it. Depending on which option you choose to explore, you may need to perform calculations and/or look up additional data.

 2. Consider how you could present your data graphically. Prepare a one-page summary of your exploration to share with your class.

 Linking question

Under comparable conditions, why do some gases deviate more from ideal behaviour than others? (Structure 2.2)

The molar volume of an ideal gas (*Structure 1.5.3*)

> Avogadro's law is covered in *Structure 1.4*.

Avogadro's law states that equal volumes of any two gases at the same temperature and pressure contain equal numbers of particles. The molar volume of an ideal gas is a constant at specified temperature and pressure. For example, at STP, the **molar volume of an ideal gas**, V_m, is equal to 22.7 dm³ mol⁻¹.

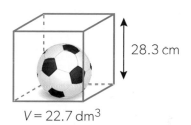

28.3 cm

$V = 22.7$ dm³

▲ Figure 5 Molar volume of an ideal gas compared with a soccer ball

H_2 He CH_4 O_2 Cl_2

2.02 g mol⁻¹   4.00 g mol⁻¹   16.05 g mol⁻¹   32.00 g mol⁻¹   70.90 g mol⁻¹

▲ Figure 6 Molar volume of any gas is identical at a given temperature and pressure

Worked example 2

A 2.00 dm³ sample of an unknown gas at STP has a mass of 2.47 g. Determine the molar mass, in g mol⁻¹, of the gas.

Solution

$$n = \frac{V}{V_m} = \frac{2.00\,\text{dm}^3}{22.7\,\text{dm}^3\,\text{mol}^{-1}} = 0.0881\,\text{mol}$$

$$M = \frac{m}{n} = \frac{2.47\,\text{g}}{0.0881\,\text{mol}} = 28.0\,\text{g mol}^{-1}$$

Practice question

2. Determine the molar mass, in g mol⁻¹, of an elemental gaseous substance that has a density of 3.12 g dm⁻³ at STP. Identify the substance if its molecules are diatomic.

Hypotheses

Amedeo Avogadro postulated that equal volumes of different gases would contain equal numbers of particles under the same conditions of temperature and pressure. This became known as Avogadro's hypothesis.

A hypothesis is a tentative and falsifiable explanation or description of a natural phenomenon, from which predictions can be deduced. Predictions can then be used to test the hypothesis.

What predictions might be derived from Avogadro's hypothesis?

Experimental determination of the molar mass of a gas

The ideal gas equation can be used to determine the molar mass of a gas by collecting a known volume of it under known conditions of temperature and pressure. In this practical you will experimentally determine the molar mass of butane found in disposable plastic lighters.

Relevant skills

- Tool 3: Record uncertainties in measurements as a range to an appropriate precision and propagate uncertainties in processed data.
- Tool 3: Calculate and interpret percentage error.
- Inquiry 2: Assess accuracy and precision.
- Inquiry 3: Identify and discuss sources and impacts of systematic and random error.
- Inquiry 3: Evaluate the implications of methodological weaknesses, limitations and assumptions on conclusions.
- Inquiry 3: Explain realistic and relevant improvements to an investigation.

Safety

- Wear eye protection.
- Butane gas is flammable. Keep away from open flames and sparks.

Materials

- Disposable plastic lighter
- Large container, for example a large plastic trough
- 100 cm³ measuring cylinder
- Balance (±0.01 g)
- Clamp and stand
- Thermometer
- Barometer (if available)

Instructions

1. Measure ambient pressure with a barometer. Alternatively, you can search local weather data for atmospheric pressure in your geographic location on the day you do the experiment.

2. Half-fill the plastic trough with water. Measure the temperature of the water.

3. Fill the measuring cylinder to the brim with water and invert it in the trough so that its mouth is under water. If done correctly, the measuring cylinder should be full of water. Hold it in this position with a clamp (figure 7).

4. Submerge the lighter in water, then take it out again and dry it thoroughly with a paper towel. Weigh the lighter.

5. Hold the lighter under water and press the button on the lighter to release the gas so that it bubbles up inside the measuring cylinder (figure 7). Continue until you have collected around 100 cm³ of gas. Record the exact volume.

6. Release the gas in a well-ventilated area.

7. Dry the lighter as thoroughly as possible and reweigh it.

8. If you have time, repeat to get three sets of results.

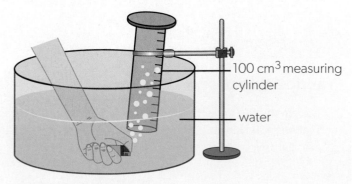

▲ Figure 7 Experiment apparatus

100 cm³ measuring cylinder

water

Questions

1. Design a suitable results table for your data.

2. Process your data to obtain an experimental value for the molar mass of butane.

3. Propagate the uncertainties.

4. Compare your experimental value to the theoretical value by calculating the percentage error.

5. Assess the accuracy and precision of your data.

6. Discuss the relative impacts of systematic and random errors on your results.

7. Comment on at least two major sources of experimental error.

8. Suggest realistic improvements to this method for determining the molar mass of a gas.

9. The pressure inside the measuring cylinder is in fact the sum of the vapour pressure of water and the partial pressure of the butane. How could you adjust your data processing to account for this? What additional data and information do you need to research?

10. Consider alternative methods for determining the molar mass of a gas that could be done in a school laboratory. If you have time, show your ideas to your teacher and try them out.

Linking question

Graphs can be presented as sketches or as accurately plotted data points. What are the advantages and limitations of each representation? (Tools 2 and 3, Reactivity 2.2)

Pressure, volume, temperature and amount of an ideal gas (*Structure 1.5.4*)

There are four variables of an ideal gas that affect each other:

1. The pressure exerted by the gas, p

2. The volume the gas occupies, V

3. The absolute temperature of the gas, T

4. The amount of the gas, n

The effect of any two of these variables on each other can be investigated by keeping the other two constant. This is what Robert Boyle did when he came up with Boyle's law: he performed an experiment where the amount and the temperature of a gas were kept constant, but the volume of the container was changed. He observed that the pressure and volume of the gas were inversely proportional.

Graphing the gas laws

Online simulations allow you to easily explore the relationships between pressure and volume, volume and temperature, and pressure and temperature for a fixed amount of gas. In this task, you will collect data from an ideal gas simulation, which will allow you to practice spreadsheet data analysis skills, as well as reinforce ideas about direct and inverse proportionality.

Relevant skills

- Tool 2: Generate data from simulations.
- Tool 2: Use spreadsheets to manipulate data.
- Tool 3: Understand direct and inverse proportionality.
- Inquiry 1: Identify dependent, independent and controlled variables.

Materials

- Simulation that allows you to change pressure, volume and temperature for an ideal gas. It must have an option to hold one variable constant and vary the other two.
- Spreadsheet software

Instructions

1. Using the simulation, vary the temperature at a constant volume and record the resulting pressure for a certain amount of gas. Collect data for at least five different temperatures in a suitable table in your spreadsheet.

2. Construct a graph of p vs T.

3. Using the simulation, vary the temperature at a constant pressure and record the resulting volume for a certain amount of gas. Collect data for at least five different temperatures in a suitable table in your spreadsheet.

4. Compute the temperature values in both °C and K.

5. Construct two graphs of V vs T; one with T in °C and the other with T in K.

6. Using the simulation, vary the volume at a constant temperature and record the resulting pressure for a certain amount of gas. Collect data for at least five different volumes in a suitable table in your spreadsheet.

7. Construct a graph of p vs V.

8. Use your spreadsheet to compute values for $\frac{1}{V}$.

9. Construct a graph of p vs $\frac{1}{V}$.

Questions

1. What were your dependent and independent variables in each case? Which variables were controlled?

2. Describe the relationship shown in each graph as direct proportionality, inverse proportionality, or other.

3. When studying gases, it is important to convert all temperature values into SI units (kelvin). Discuss why this is the case for temperature, whereas pressure and volume units can vary depending on the source.

The combined gas law

We have seen that pressure is inversely proportional to volume and directly proportional to absolute temperature.

$$p \propto \frac{1}{V}; p \propto T$$

Combining the two relationships gives:

$$pV \propto T \text{ or } \frac{pV}{T} = k \text{ (a constant) or}$$

$$\frac{p_1V_1}{T_1} = \frac{p_2V_2}{T_2}$$

This equation is known as the **combined gas law**.

Experiments

The gas laws arose from experiments in which certain variables were controlled, while others were carefully manipulated. Inspect the apparatus shown in figure 8. What might be explored with this set-up? What is the independent variable? What variables must be controlled? What is the purpose of each of the items depicted?

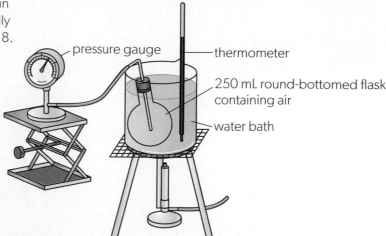

▶ Figure 8 Apparatus for conducting an experiment into the behaviour of a gas

Worked example 3

A weather balloon filled with 32.0 dm³ of helium at 25 °C and a pressure of 100.0 kPa is released at sea level. The balloon eventually reaches an altitude of 35,000 m, where the pressure is 475 Pa and the temperature is −50 °C. Calculate the volume, in m³, of the gas in the balloon under these conditions.

Solution

List the conditions of the gas in the weather balloon.

$$p_1 = 100.0 \text{ kPa}$$
$$V_1 = 32.0 \text{ dm}^3$$

Remember to convert temperature to kelvin:

$$T_1 = 25 + 273.15 = 298.15 \text{ K}$$

Then, list the conditions of the gas in the weather balloon at 35,000 m. Remember to make sure the units are consistent with the initial conditions of the balloon.

$$p_2 = 0.475 \text{ kPa}$$
$$V_2 = \text{unknown}$$
$$T_2 = -50 + 273.15 = 223.15 \text{ K}$$

Substitute the numbers into the combined gas law:

$$\frac{p_1 V_1}{T_1} = \frac{p_2 V_2}{T_2}$$

$$\frac{100.0 \text{ kPa} \times 32.0 \text{ dm}^3}{298.15 \text{ K}} = \frac{0.475 \text{ kPa} \times V_2}{223.15 \text{ K}}$$

Rearranging the expression in terms of V_2 gives:

$$V_2 \approx 5.04 \times 10^3 \text{ dm}^3 = 5.04 \text{ m}^3$$

Practice question

3. A sample of an ideal gas has a volume of 1.00 dm³ at STP. Calculate the volume, in dm³, of that sample at 50.0 °C and 50.0 kPa.

TOK

Throughout this chapter you have explored models related to the particulate nature of matter. Many of these concepts were developed through observations of the natural world or obtained through experimentation: from the interaction of alpha particles with gold atoms, to the manipulation of gases in the gas laws, to explorations of subatomic particles at CERN.

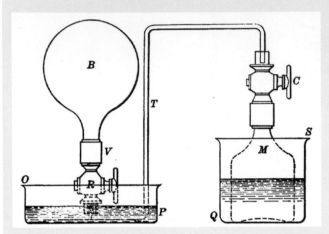 

▲ **Figure 9** The set-up used by Joseph Louis Gay-Lussac to investigate the thermal expansion of gases (left)
The ATLAS detector at CERN used to investigate elementary particles (right)

How do scientists investigate the behaviour of particles that are too small to be observed directly? How have advances in technology influenced scientific research into what matter is made up of?

Ideal gas equation

The combined gas law suggests that for any given gas, the change in one of the three parameters, p, V or T, affects the other two in such a way that the expression $\dfrac{pV}{T}$ remains constant. The exact value of that constant must be proportional to the amount of the gas, n:

$$\frac{pV}{T} \propto n \quad \text{or} \quad \frac{pV}{T} = nR$$

where R is the **universal gas constant**, or simply **gas constant**. The last expression is known as the **ideal gas equation**, which is traditionally written as follows:

$$pV = nRT$$

The value and units of R depend on the units of p, V, T and n. If all four parameters are expressed in standard SI units (p in Pa, V in m^3, T in K and n in mol), then $R \approx 8.31\,\mathrm{J\,K^{-1}\,mol^{-1}}$.

The same value and units of R can be used if pressure is expressed in kPa and volume in dm^3, as the two conversion factors (10^3 for kPa to Pa and 10^{-3} for dm^3 to m^3) cancel each other out.

 Linking question

How can the ideal gas law be used to calculate the molar mass of a gas from experimental data? (Tool 1, Inquiry 2)

Worked example 4

A 3.30 g sample of an unknown organic compound was vaporized at $T = 150\,°C$ and $p = 101.3\,kPa$ to produce 1.91 dm^3 of a gas. The gas was combusted in excess oxygen to produce 3.96 g of water, 2.49 dm^3 of carbon dioxide and 1.25 dm^3 of nitrogen at STP.

Determine the following for the compound:

a. molar mass

b. empirical formula

c. molecular formula

Solution

a. To determine the molar mass, we need to find out the amount of the compound using the ideal gas equation:

$$n = \frac{pV}{RT}$$

$T = 150 + 273.15 = 423.15\,K$

$$n = \frac{101.3\,kPa \times 1.91\,dm^3}{8.31\,J\,K^{-1}\,mol^{-1} \times 423.15\,K} \approx 0.0550\,mol$$

Therefore, $M = \dfrac{3.30\,g}{0.0550\,mol} = 60.0\,g\,mol^{-1}$

b. All carbon, hydrogen and nitrogen atoms in the combustion products originate from the organic compound, so the amounts of these elements in carbon dioxide, water and nitrogen are the same as those in the original sample:

$$n(H_2O) = \frac{m}{M} = \frac{3.96\,g}{18.02\,g\,mol^{-1}} \approx 0.220\,mol$$

$n(H) = 2 \times n(H_2O) = 2 \times 0.220\,mol = 0.440\,mol$

$$n(CO_2) = \frac{V}{V_M} = \frac{2.49\,dm^3}{22.7\,dm^3\,mol^{-1}} \approx 0.110\,mol$$

$n(C) = n(CO_2) = 0.110\,mol$

$$n(N_2) = \frac{V}{V_M} = \frac{1.25\,dm^3}{22.7\,dm^3\,mol^{-1}} \approx 0.0551\,mol$$

$n(N) = 2 \times n(N_2) = 2 \times 0.0551 = 0.110\,mol$

The original compound could also contain oxygen. To check this, we need to compare the total mass of the three elements (hydrogen, carbon and nitrogen) with the mass of the original sample:

$m(H) = 0.440\,mol \times 1.01\,g\,mol^{-1} \approx 0.444\,g$

$m(C) = 0.110\,mol \times 12.01\,g\,mol^{-1} \approx 1.32\,g$

$m(N) = 0.110\,mol \times 14.01\,g\,mol^{-1} \approx 1.54\,g$

$m(total) = 0.444\,g + 1.32\,g + 1.54\,g \approx 3.30\,g$

Therefore, the organic compound did not contain oxygen, so its formula can be represented as $C_xH_yN_z$.

$x:y:z = 0.110:0.440:0.110 = 1:4:1$

The empirical formula of the compound is CH_4N.

c. $M(CH_4N) = 12.01 + 4 + 1.01 + 14.01 = 30.06\,g\,mol^{-1}$. This value is half the experimental value (60.0 g mol^{-1}), so the molecular formula of the compound will have twice the number of atoms of each element: $C_2H_8N_2$.

End-of-topic questions

Topic review

1. Using your knowledge from the *Structure 1.5* topic, answer the guiding question as fully as possible:

   How does the model of ideal gas behaviour help us to predict the behaviour of real gases?

Exam-style questions

Multiple-choice questions

2. Which of the following are assumptions of the ideal gas model?

   I. The volume occupied by the gas particles is negligible

   II. There are no intermolecular forces between gas particles

   III. Zero particle movement

   A. I and II only

   B. I and III only

   C. II and III only

   D. I, II and III

3. The temperature of an ideal gas is 27 °C. What is the temperature of the gas after the pressure is doubled and the volume is tripled?

   A. 162 °C

   B. 450 °C

   C. 1527 °C

   D. 1800 °C

4. A gas syringe contains 40 cm³ of an ideal gas at 27 °C. What will the volume of the gas be after it is warmed to 57 °C at constant pressure?

   A. 1.9 cm³

   B. 3.6 cm³

   C. 44 cm³

   D. 84 cm³

5. A 0.58 g sample of an ideal gas at 100 kPa and 100 °C has a volume of 250 cm³. Which of the following expressions is equal to the molar mass of the gas?

   A. $\dfrac{0.58 \times 8.31 \times 100}{100 \times 10^3 \times 250 \times 10^{-6}}$

   B. $\dfrac{0.58 \times 8.31 \times 373}{100 \times 10^3 \times 250 \times 10^{-6}}$

   C. $\dfrac{0.58 \times 100 \times 10^{-3} \times 250 \times 10^6}{8.31 \times 100}$

   D. $\dfrac{0.58 \times 100 \times 10^{-3} \times 250 \times 10^6}{8.31 \times 373}$

6. Which graph correctly shows the relationship between the pressure and volume of an ideal gas, at constant temperature?

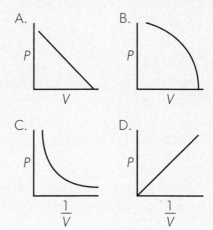

7. Which of the following balloons contains the largest number of hydrogen atoms, at constant temperature and pressure?

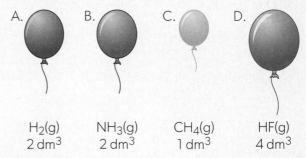

A.	B.	C.	D.
$H_2(g)$	$NH_3(g)$	$CH_4(g)$	$HF(g)$
2 dm³	2 dm³	1 dm³	4 dm³

8. What are the conditions for the ideal gas behaviour of real gases?

   A. Low temperature and low pressure.

   B. Low temperature and high pressure.

   C. High temperature and low pressure.

   D. High temperature and high pressure.

9. Which of the following statements about an ideal gas are correct?

   I.   At constant temperature, $\frac{p}{V}$ = constant

   II. At constant volume, $\frac{p}{T}$ = constant

   III. At constant pressure, $\frac{V}{T}$ = constant

   A. I and II only

   B. I and III only

   C. II and III only

   D. I, II and III

Extended-response questions

10. Explain, in your own words, why real gases deviate from ideal behaviour at low temperatures and high pressures. [2]

11. A car tyre inflated to 2.50 bar (250 kPa) at 10 °C contains 12.0 dm³ of compressed air. After a long journey, the tyre temperature increases to 25 °C and the pressure to 261 kPa. Determine the tyre volume under these conditions. Assume that there was no air loss during the journey. [2]

12. Ammonium carbonate, $(NH_4)_2CO_3(s)$, decomposes readily when heated:

    $$(NH_4)_2CO_3(s) \rightarrow 2NH_3(g) + CO_2(g) + H_2O(l)$$

    Determine the volumes, in dm³ at STP, of the individual gases produced on decomposition of 2.25 g of ammonium carbonate. [2]

13. The gases produced in question 12 were transferred to a sealed vessel with a volume of 1.50 dm³, and the vessel was heated up to 200 °C. Calculate the pressure in the vessel at that temperature. Assume that the gases do not react with each other. [2]

14. Carbon forms several gaseous compounds with fluorine. Deduce the empirical and molecular formulas for these compounds using the data from the table below. [3]

Compound	Carbon / mass %	Mass of 1.00 dm³ at STP / g
X	13.65	3.88
Y	24.02	4.41
Z	17.40	6.08

15. An organic compound A contains 54.5% of carbon, 9.1% of hydrogen and 36.4% of oxygen by mass. A vaporized sample of A with a mass of 0.230 g occupies a volume of 0.0785 dm³ at $T = 95$ °C and $p = 102$ kPa.

    a. Determine the empirical formula of A. [2]

    b. Determine the relative molecular mass of A. [1]

    c. Using your answers to parts a and b, determine the molecular formula of A. [1]

16. A closed steel cylinder contains 0.32 mol of hydrogen gas and 0.16 mol of oxygen gas. The volume of the cylinder is 25 dm³ and the initial temperature of the gas mixture is 25 °C.

    a. Calculate the initial pressure, in kPa, of the gas mixture in the cylinder. [1]

    b. When the gas mixture is ignited, both reactants are consumed completely and the temperature inside the cylinder rises to 800 °C. Calculate the pressure inside the cylinder at that moment. [2]

17. An unknown gas X has a density of 2.82 g dm⁻³ at STP. A 4.00 g sample of X was combusted in excess oxygen to produce 2.50 g of hydrogen fluoride and 2.84 dm³ of carbon dioxide at STP.

    Determine the following for X:

    a. molar mass [1]

    b. empirical formula [2]

    c. molecular formula [1]

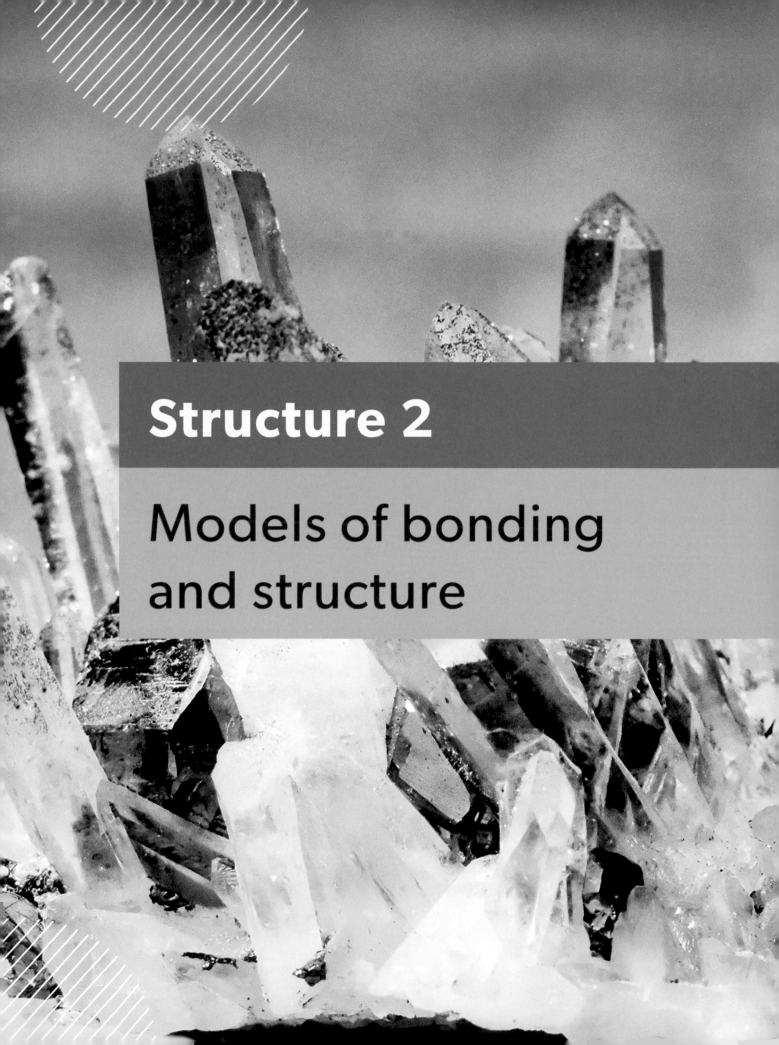

Structure 2

Models of bonding and structure

Structure 2.1 The ionic model

What determines the ionic nature and properties of a compound?

Ionic compounds are characterized by the presence of positive and negative ions, which attract each other electrostatically. In solid ionic compounds, these ions are arranged in rigid crystalline lattices. Melting these solids requires a large amount of thermal energy due to the strong electrostatic attractions between oppositely charged ions. Once liquid, however, ionic compounds are electrical conductors due to the presence of mobile ions. Due to their charge, ions interact strongly with polar water molecules, so ionic compounds are often water-soluble.

Understandings

Structure 2.1.1 — When metal atoms lose electrons, they form positive ions called cations. When non-metal atoms gain electrons, they form negative ions called anions.

Structure 2.1.2 — The ionic bond is formed by electrostatic attractions between oppositely charged ions.

Binary ionic compounds are named with the cation first, followed by the anion. The anion adopts the suffix "ide".

Structure 2.1.3 — Ionic compounds exist as three-dimensional lattice structures, represented by empirical formulas.

Introduction to bonds and structure

Atoms rarely exist in isolation. They are connected together in several different ways. Atoms can be bonded to atoms of the same type, or to atoms of different elements. The varying arrangements of atoms and features of the bonds between them give rise to certain different properties. For example, 78% of the air around us is nitrogen, N_2. However, in agriculture, nitrogen fertilizers are added to soils to help crops grow. This is because the structure and bonding of nitrogen in air are different to that of the nitrogenous compounds found in fertilizers.

Atoms are held together by **chemical bonds**. This chapter discusses three different bonding models: **ionic**, **covalent** and **metallic**. These lead to four types of structure: ionic, **molecular covalent**, **covalent network** and metallic. You may be wondering why there are four types of structure, given that there are only three types of bonds. This is because covalent substances can be found in two arrangements: a continuous 3D network, or discrete groups of atoms known as molecules.

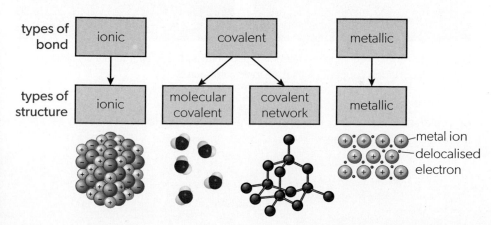

◄ Figure 1 There are three types of bonds and four types of structure

Models

Structure 2.1, 2.2 and *2.3* discuss models of bonding and structure. Scientific models simplify and represent complex phenomena. Sometimes models help us to visualize things that we cannot observe directly. This is one of the reasons bonding models are so useful.

All models have limitations. This does not necessarily make the models inadequate, but it is important to understand the weaknesses of a model. As you work through these sections, identify some of the strengths and limitations of the various bonding models.

Chemical bonds

Chemical bonds are strong forces of attraction that hold atoms or **ions** together in a substance. All chemical bonds occur due to electrostatic attractions between positively charged species and negatively charged species. The type of bonding depends on which species are involved (table 1).

Type of bonding	The electrostatic attraction between...	
	Positively charged species	**Negatively charged species**
ionic	cations	anions
covalent	atomic nuclei	shared pair of electrons
metallic	cations	delocalized electrons

▲ Table 1 All bonding types involve a positively charged species and a negatively charged species that are electrostatically attracted to each other

Ions (*Structure 2.1.1*)

Sodium chloride and copper(II) sulfate are examples of **ionic compounds**. They are crystalline and brittle, which are properties characteristic of ionic compounds. Ionic compounds are also poor electrical conductors when solid, but good electrical conductors when molten or dissolved. The reactions and properties of these ionic compounds are very different to those of their constituent elements. For instance, sodium chloride, the main ingredient in table salt, is water-soluble. However, elemental sodium is a soft metal that reacts violently with water, and chlorine is a poisonous gas.

▲ Figure 3 Sodium chloride crystals on a tree branch and copper(II) sulfate crystals. Sodium chloride and copper(II) sulfate are ionic compounds

▲ Figure 2 Offshore oil platform in California, USA. What examples of structure and bonding are present in the photo?

Before discussing ionic bonds and the characteristics of ionic structures, we will first look into what ions are.

Cations and anions

Sodium chloride contains sodium ions, not sodium atoms. Sodium atoms and sodium ions have different numbers of electrons, and therefore behave differently.

You will notice three differences between Na and Na^+:

1. number of electrons

2. electron arrangement

3. charge.

Sodium atoms are neutral. Sodium ions have a 1+ charge, indicated by a superscript + sign next to the symbol: Na^+.

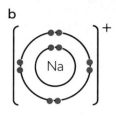

▲ **Figure 4** (a) sodium atom (b) sodium ion

Worked example 1

Determine the number of subatomic particles to show that

a. sodium atoms are neutral

b. sodium ions have a 1+ charge.

Solution

a. In a sodium atom there are:

   11 protons (charge = 11+)

   11 electrons (charge = 11−)

   Overall charge is 11 − 11 = 0

b. In a sodium ion there are:

   11 protons (charge = 11+)

   10 electrons (charge = 10−)

   Overall charge is 11 − 10 = 1+

> In *Structure 1.2*, you learned that protons have a 1+ charge and electrons have a 1− charge. You can ignore neutrons in ionic charge calculations as these are uncharged.

Worked example 2

Deduce the electron configuration of a sodium atom and a sodium ion.

Solution

Na: $1s^2\,2s^2\,2p^6\,3s^1$

Na^+: $1s^2\,2s^2\,2p^6$

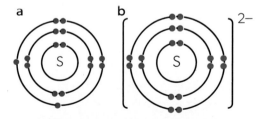

▲ **Figure 5** (a) sulfur atom (b) sulfide ion

Cations are ions with more protons than electrons. This means that cations are positively charged, as the combined positive charge of protons is greater than the combined negative charge of electrons. As sodium ions have 11 protons and 10 electrons, the overall charge is 1+.

Anions are negatively charged ions. They contain a greater number of electrons than protons. Figure 5 shows a sulfur atom and a sulfide ion. The sulfide ion has a 2− charge, denoted by the superscript in the symbol S^{2-}. Note that anions adopt a slightly different name: the first part corresponds to the name of their parent atom. This is followed by the suffix –**ide.**

For now, we will only consider monatomic ions. You will look at charged groups of atoms (called **polyatomic ions**) in a later section.

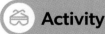 **Activity**

Show that the sulfur atom is neutral and the sulfide ion has a charge of 2− by counting their subatomic particles. Determine the electron configuration of the sulfur atom and sulfide ion.

Predicting the charge of an ion

The main group elements are in periodic table groups 1, 2, 13, 14, 15, 16, 17 and 18. The electron configurations for some main group element atoms and their corresponding cations are shown below:

$$Na \xrightarrow{-1e^-} Na^+$$
$$1s^2\,2s^2\,2p^6\,3s^1 \qquad\qquad 1s^2\,2s^2\,2p^6$$

Na^+ has the same electron configuration as neon, Ne. Two different species with same electron configuration are called **isoelectronic**. Therefore, Na^+ and Ne are isoelectronic.

$$Li \xrightarrow{-1e^-} Li^+ \qquad Ca \xrightarrow{-2e^-} Ca^{2+}$$
$$1s^2\,2s^1 \qquad\qquad 1s^2 \qquad 1s^2\,2s^2\,2p^6\,3s^2\,3p^6\,4s^2 \qquad\qquad 1s^2\,2s^2\,2p^6\,3s^2\,3p^6$$

Li^+ is isoelectronic with helium, He. Ca^{2+} is isoelectronic with argon, Ar.

The resulting cations all have noble gas configurations. Noble gases have full (or "closed") sublevels. When main group elements form ions, they often achieve this noble gas electron configuration. The atoms above have all done so by losing their outermost valence electrons. As they have lost electrons, and electrons are negatively charged, the resulting cations are positively charged. Cation formation is an example of **oxidation** because it involves the loss of electrons.

Anions are formed when atoms gain electrons. Look at the examples below where the parent atoms gain electrons in order to achieve a noble gas electron configuration:

$$Cl \xrightarrow{+1e^-} Cl^- \qquad O \xrightarrow{+2e^-} O^{2-}$$
$$1s^2\,2s^2\,2p^6\,3s^2\,3p^5 \qquad\qquad 1s^2\,2s^2\,2p^6\,3s^2\,3p^6 \qquad 1s^2\,2s^2\,2p^4 \qquad\qquad 1s^2\,2s^2\,2p^6$$

Cl^- is isoelectronic with argon, Ar. O^{2-} is isoelectronic with neon, Ne.

Atoms that gain electrons become anions. As **reduction** is the gain of electrons, the formation of anions is a reduction process.

The formation of an ionic compound from its elements is a **redox** reaction. Consider the formation of sodium chloride from its elements:

$$2Na(s) + Cl_2(g) \longrightarrow 2NaCl(s)$$

Sodium chloride, NaCl, is made up of sodium cations, Na^+, and chloride anions, Cl^-. The half equations are shown below. The first is an oxidation and the other is a reduction and therefore the formation of NaCl from its elements is a redox reaction.

$$2Na \longrightarrow 2Na^+ + 2e^- \qquad \text{Electron loss} = \text{oxidation}$$
$$Cl_2 + 2e^- \longrightarrow 2Cl^- \qquad \text{Electron gain} = \text{reduction}$$

Once you have learned about **oxidation states** (*Structure 3.1*), you should also be able to see that the sodium is undergoing oxidation because its oxidation state increases (from 0 to +1) and the chlorine is reduced because its oxidation state decreases (from 0 to −1).

To obtain a noble gas configuration, a chlorine atom gains an electron. Chlorine would also have a noble gas configuration if it lost the seven outermost electrons. However, the removal of so many electrons from the attractive pull of the positively charged nucleus requires a large amount of energy while the addition of a single electron releases energy. This is why chlorine instead will gain an electron to become a chloride ion. The energetics of these processes, called **ionization energy** and **electron affinity**, are discussed in *Structure 3.1* and relevant in the construction of Born–Haber cycles (*Reactivity 1.2*).

Atoms tend to achieve a noble gas electron configuration through gaining, losing, or, as we will see in *Structure 2.2*, sharing electrons. This is often referred to as the **octet rule**. It is called the octet rule because most noble gases have eight electrons in their outer shell.

There exists a relationship between the charge of the ion formed by a main group element and its periodic table group. In general:

- Elements in groups 1, 2 and 13 form 1+, 2+ and 3+ ions, respectively

- Elements in groups 15, 16 and 17 form 3–, 2– and 1– ions, respectively

- Elements in group 18 (noble gases) do not form ions

The relationship between periodic table group and ionic charge is illustrated in figure 6.

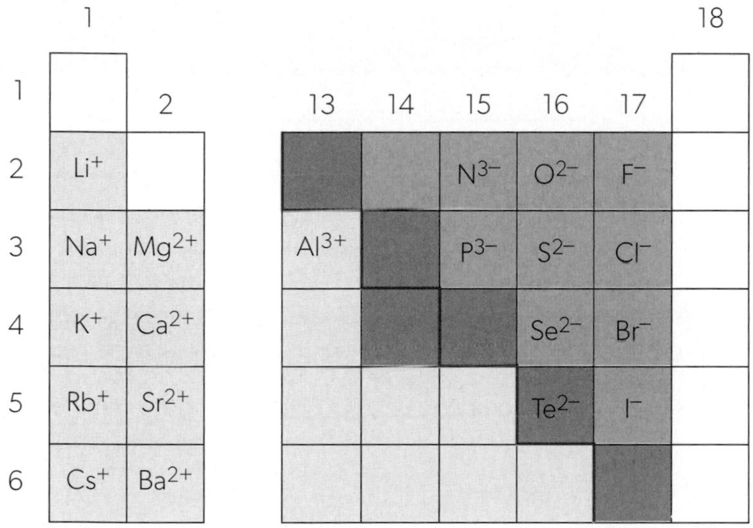

1								18
1	2	13	14	15	16	17		
2	Li^+				N^{3-}	O^{2-}	F^-	
3	Na^+	Mg^{2+}	Al^{3+}		P^{3-}	S^{2-}	Cl^-	
4	K^+	Ca^{2+}				Se^{2-}	Br^-	
5	Rb^+	Sr^{2+}				Te^{2-}	I^-	
6	Cs^+	Ba^{2+}						

▲ Figure 6 The charges of some common ions

Hydrogen atoms have only one electron in the 1s sublevel. They form ions by either losing that electron or gaining one. Electron loss leads to the formation of H^+, which is simply a hydrogen nucleus: a proton with no electrons surrounding it. The charge density of a H^+ ion is therefore very high, so these cations readily combine with other species. One such example is the formation of acidic hydronium ions, H_3O^+, formed when hydrogen cations bond with water.

Hydrogen atoms can also gain an electron to achieve a noble gas configuration, thus forming **hydride** anions, H^-.

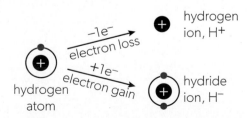

▲ Figure 7 The formation of H^+ and H^-

The electron configuration of carbon, $1s^2 2s^2 2p^2$, suggests that carbon atoms could lose or gain four electrons in order to achieve a noble gas configuration. This would result in the formation of C^{4+} or C^{4-} ions, respectively. Although this is possible, carbon more commonly forms compounds through a process called covalent bonding, which does not involve ion formation. Covalent bonding is discussed in *Structure 2.2*.

Hydride anions are very strong bases. You will learn more about bases in *Reactivity 3.1*.

Practice questions

1. Determine the charge of the ion formed by each of the following elements.
   a. lithium, Li
   b. magnesium, Mg
   c. aluminium, Al
   d. fluorine, F
   e. nitrogen, N
   f. selenium, Se
   g. barium, Ba

2. State the name of ions d, e and f above.

3. Complete the table:

Name	Symbol	Number of protons	Number of electrons	Electron configuration	Charge
beryllium					0
	K⁺				
		8			2–
		15	18		
	H⁺				

4. For each electron configuration given, identify three isoelectronic species:
   a. $1s^2\,2s^2\,2p^6\,3s^2\,3p^6$
   b. $1s^2$

5. Explain why noble gases do not form ions.

A **transition element** is an element with a partially filled d sublevel. In contrast to main group elements, a transition element can form multiple ions with different charges. For example, iron commonly forms Fe^{2+} and Fe^{3+} ions (figure 8).

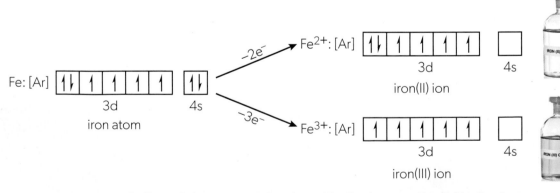

▲ Figure 8 Iron atoms can form ions with a 2+ charge and ions with a 3+ charge

Consider the electron configurations of the first-row transition elements.

Most of them contain two 4s electrons, which are lost when the M^{2+} ions are formed. This helps to explain why most of these elements commonly form 2+ cations.

As seen in chapter *Structure 1.3*, the 4s sublevel fills up before the 3d sublevel.

◀ Table 2 Electron configurations of the first-row transition elements

Symbol	Element	Electron configuration	
		Atom	2+ ion
Sc	scandium	$[Ar]4s^2 3d^1$	$[Ar]3d^1$
Ti	titanium	$[Ar]4s^2 3d^2$	$[Ar]3d^2$
V	vanadium	$[Ar]4s^2 3d^3$	$[Ar]3d^3$
Cr	chromium	$[Ar]4s^1 3d^5$	$[Ar]3d^4$
Mn	manganese	$[Ar]4s^2 3d^5$	$[Ar]3d^5$
Fe	iron	$[Ar]4s^2 3d^6$	$[Ar]3d^6$
Co	cobalt	$[Ar]4s^2 3d^7$	$[Ar]3d^7$
Ni	nickel	$[Ar]4s^2 3d^8$	$[Ar]3d^8$
Cu	copper	$[Ar]4s^1 3d^{10}$	$[Ar]3d^9$

When the first row transition elements are ionized, the 4s electrons are lost before the 3d electrons. Further successive ionizations occur in many of these elements because the 3d sublevel is similar in energy to the 4s sublevel.

Transition elements have variable oxidation states (*Structure 3.1*). This characteristic can be explored by examining successive ionization energy data. Let's focus on the transition elements in period 4. They have variable oxidation states because the 4s and 3d sublevels are close together in energy, as shown by successive ionization energy data (figure 9). It is important to realize that ionization rarely happens in isolation. Ionization absorbs energy, but this energy investment is usually offset by other processes that release energy, such as lattice formation. If a certain ionization only requires a small amount of additional energy compared to the previous ionization, then it could be energetically favourable if it leads to a subsequent **exothermic** process.

Practice questions

6. Deduce the abbreviated electron configuration of each of the following:
   a. Mn^{2+}
   b. V^{3+}
   c. Cu^+
   d. Cu^{2+}
7. Zinc only forms 2+ ions.
   a. Deduce the full electron configuration of Zn^{2+}.
   b. Explain why zinc is not a transition element.
8. The ion of a transition metal has mass number 55, electron configuration $[Ar]3d^5$ and a charge of 2+.
   a. Write its symbol using nuclear notation.
   b. Identify a 1+ ion that has the same electron configuration as the above.

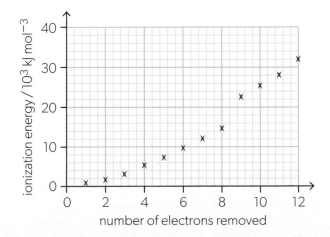

▲ Figure 9 Data for the first 12 ionization energies of iron. As you can see, the 4s and 3d electrons are very close together in energy. The large jump between the 8th and 9th electrons occurs because the 9th electron is removed from the 3p energy level, which is closer to the nucleus

Linking questions

How does the position of an element in the periodic table relate to the charge of its ion(s)? (Structure 3.1)

How does the trend in successive ionization energies of transition elements explain their variable oxidation states? (Structure 1.3)

Ionic bonds (*Structure 2.1.2*)

Cations and anions are **electrostatically attracted** to each other because of their opposite charges. This attraction results in the formation of ionic bonds. Therefore, if a given element forms cations, and another forms anions, they can bond ionically to form an ionic compound.

Electronegativity (χ)

One way to estimate whether a bond between two given elements is ionic is to look at the difference in electronegativity between the two. **Electronegativity** (χ) is a measure of the ability of an atom to attract a pair of covalently bonded electrons. Within the periodic table, electronegativity increases across the periods and up the groups. This means that fluorine is the most electronegative element, so it has a high tendency to attract pairs of covalently bonded electrons.

One of the electronegativity scales used by chemists is called the **Pauling scale**. Values in the Pauling scale are dimensionless and range from 0.8 to 4.0. Fluorine has an electronegativity value of 4.0. The electronegativity of caesium, one of the least electronegative elements, is 0.8. Noble gases are generally not assigned electronegativity values.

The larger the difference in electronegativity between two elements in a compound, the greater the ionic character of the bond between them. Ionic bonding is assumed to occur when the difference in electronegativity is greater than 1.8 (figure 10). In reality, bonding occurs across a continuum, so above 1.8 the main type of bonding in the compound is ionic, but there may be other types of bonding present.

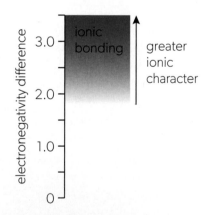

▲ **Figure 10** If two elements have an electronegativity difference greater than 1.8, the bonding between them will have a high ionic character

Electronegativity and other periodic trends are discussed in greater detail in *Structure 3.1*.

Data-based question

Predict which of the compounds in table 3 will have ionic structure.

Compound	Difference in electronegativity ($\Delta \chi$)
sodium fluoride, NaF	$\chi(Na) = 0.9$ and $\Delta\chi(F) = 4.0$ $\Delta\chi = 3.1$
sodium chloride, NaCl	$\chi(Na) = 0.9$ and $\chi(Cl) = 3.2$ $\Delta\chi = 2.3$
aluminium chloride, AlCl$_3$	$\chi(Al) = 1.6$ and $\chi(Cl) = 3.2$ $\Delta\chi = 1.6$

▲ **Table 3** Electronegativity differences for selected metal chlorides

Periodic table position

You can qualitatively approximate how ionic a compound will be by looking at the positions of its constituent elements in the periodic table. Elements with large differences in electronegativity are generally found at a greater horizontal distance from each other.

Worked example 3

Compare the ionic character of bonding in the following pairs of compounds:

a. caesium fluoride, CsF, and caesium iodide, CsI

b. magnesium oxide, MgO, and carbon monoxide, CO

Solution

a. Qualitative comparison:

Cs and F are a greater distance from each other than Cs and I are in the periodic table. Therefore, the difference in electronegativity is larger between Cs and F, meaning the bond between them is more ionic.

Quantitative comparison:

$\chi(Cs) = 0.8$ and $\chi(F) = 4.0$

$\Delta\chi(CsF) = 3.2$

$\chi(Cs) = 0.8$ and $\chi(I) = 2.7$

$\Delta\chi(CsI) = 1.9$

Both $\Delta\chi$ values are greater than 1.8, so the bonds in both compounds are ionic. However, CsF has a higher percentage ionic character than CsI.

b. Qualitative comparison:

In the periodic table, Mg and O are further away from each other than C and O are. Therefore, the electronegativity difference between Mg and O must be larger than that between C and O, and the bond between Mg and O must be more ionic.

Quantitative comparison:

$\chi(Mg) = 1.3$ and $\chi(O) = 3.4$

$\Delta\chi(MgO) = 2.1$

$\chi(C) = 2.6$ and $\chi(O) = 3.4$

$\Delta\chi(CO) = 0.8$

Mg and O bond ionically because $\Delta\chi$ is greater than 1.8 for this compound. C and O do not bond ionically because $\Delta\chi$ is lower than 1.8.

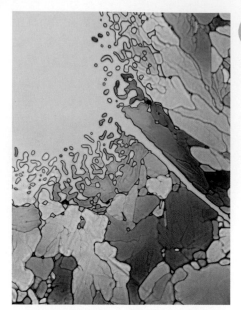

▲ **Figure 11** Polarized light micrograph of ammonium nitrate crystals. Ammonium nitrate contains two polyatomic ions: NH_4^+ and NO_3^-. Its uses include fertilizers and rocket propellants

Name	Formula
ammonium	NH_4^+
hydroxide	OH^-
nitrate	NO_3^-
hydrogencarbonate	HCO_3^-
carbonate	CO_3^{2-}
sulfate	SO_4^{2-}
phosphate	PO_4^{3-}

▲ **Table 4** Common polyatomic ions

Name	Formula
potassium fluoride	KF
magnesium fluoride	MgF_2
calcium carbonate	$CaCO_3$
barium hydroxide	$Ba(OH)_2$
iron(III) oxide	Fe_2O_3
silver(I) sulfide	Ag_2S

▲ **Table 5** Names and formulas of some ionic compounds

Activity

Determine whether the following pairs of elements are likely to bond ionically using the following two methods:

    i.   look at their positions in the periodic table

    ii. refer to their electronegativity values in the data booklet.

a.   Li and F   d.   As and S

b.   Rb and Ga   e.   P and Cl

c.   Ca and I   f.   Ag and Br

It is often incorrectly said that only ionic bonds form when a metallic element and a non-metallic element bond together. There are substances, such as aluminium chloride, $AlCl_3$, that do not fit this description. Aluminium is a metal and chlorine is a non-metal, so you would expect them to bond ionically. But the compound has properties that are characteristic of covalent compounds, such as a low melting point and high volatility. The electronegativity difference between these two elements (1.6) suggests they do not bond ionically.

Polyatomic ions

Some ionic compounds contain more than two elements. For instance, NH_4Cl, which is made up of NH_4^+ cations and Cl^- anions. Ammonium ions, NH_4^+, are **polyatomic ions**. As their name suggests, polyatomic ions are ions that contain several atoms.

You are expected to know the names and formulas of the polyatomic ions shown in table 4.

ATL Self-management skills

You will need to spend some time memorizing the names and formulas of the polyatomic ions in table 4. Some students like to use flashcards, others make up mnemonics. What strategies will you use? How will you make sure you actively engage with them?

Naming ionic compounds

Consider the list of ionic compounds shown in table 5. Can you notice any patterns in their names?

You should notice that, in the names of ionic compounds:

- the cation name is given first and is followed by the anion

- cations adopt the name of the parent atom and the name remains unchanged

- monatomic anions adopt the first part of the name of the parent atom, followed by the suffix -*ide*. If the anion is polyatomic, refer to table 4

- the name of the compound does not reflect the number of ions in the formula.

Practice questions

9. State the name of each of the following compounds:
   a. RbF
   b. Al_2S_3
   c. AlN
   d. $Sr(OH)_2$
   e. $BaCO_3$
   f. NH_4HCO_3

> Anions are **conjugate bases** of common acids. The strength of acids and stability of their anions can be compared quantitatively using their dissociation constants, K_a, which will be introduced in *Reactivity 3.1*.

The formulas of ionic compounds

The name of an ionic compound tells you what elements it contains, but not the ratio of the ions in it. The basis for working out the formula of an ionic compound is remembering that the net charge of the compound is zero, so the positive charges and negative charges must cancel out. First, determine the charge of the anion and the cation, then work out how many of each ion you need to reach a total charge of zero.

Worked example 4

Deduce the formulas of the following ionic compounds:

a. calcium oxide
b. calcium nitride
c. sodium carbonate
d. aluminium nitrate

Solution

a. To deduce the formula of calcium oxide, work through the following steps.

Step 1: Determine the charges of the cation and the anion

Calcium has an electron configuration of $1s^2 2s^2 2p^6 3s^2 3p^6 4s^2$. Calcium atoms have two outer shell electrons, so they form ions with a 2+ charge.

Oxygen has an electron configuration of $1s^2 2s^2 2p^4$. Oxygen atoms have six outer shell electrons, so they form ions with a 2– charge.

Therefore, calcium ions = Ca^{2+} and oxide ions = O^{2-}.

Step 2: Determine how many of each ion are needed in order to achieve a net charge of zero

There are two methods you can use for this step. The first is the **bar diagram** method. Write out the ions as blocks equal to the number of charges on each individual ion:

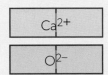

The bar diagram contains one calcium ion and one oxide ion, so the ratio of calcium to oxide in the compound is 1:1.

$$Ca_1O_1$$

The second method is the **criss-cross rule**. Swap the charges and turn them into the other ion's subscript, ignoring the sign:

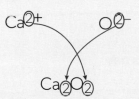

Then, simplify the ratio:

$$Ca_1O_1$$

Step 3: Check that the net charge is zero

You check your working by adding up the charges of each individual ion. If you did Step 2 correctly, the charges will add to zero:

$$\text{Total positive charge} = \overset{Ca^{2+}}{2+}$$

$$\text{Total negative charge} = \overset{O^{2-}}{2-}$$

Net charge = $2 - 2 = 0$

Step 4: Write the formula

This is a straightforward example where the magnitude of charge is equal for the cation and anion, and hence the formula is CaO.

b. Calcium nitride is a more complex example as the ions have different charges. Work through the steps as before.

Step 1: Determine the charges of the cation and the anion

$$Ca: 1s^2\,2s^2\,2p^6\,3s^2\,3p^6\,4s^2$$

Calcium atoms have two outer shell electrons, so they form ions with a 2+ charge.

$$N: 1s^2\,2s^2\,2p^3$$

Nitrogen atoms have five outer shell electrons, so they form ions with a 3– charge.

$$\text{Calcium ions} = Ca^{2+} \text{ and nitride ions} = N^{3-}$$

Step 2: Determine how many of each ion are needed in order to achieve a net charge of zero

Bar diagram method

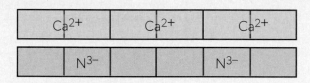

The bar diagram contains three calcium ions and two nitride ions, so the ratio of calcium to nitride ions in the compound is 3:2

$$Ca_3N_2$$

Criss-cross rule

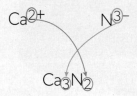

There is no need to simplify the ratio here because it is already in its simplest form.

Step 3: Check that the net charge is zero

Ca^{2+}	
Ca^{2+}	N^{3-}
Ca^{2+}	N^{3-}

Total positive charge = 6+   Total negative charge = 6–

Net charge = 6 – 6 = 0

Step 4: Write the formula

$$Ca_3N_2$$

c. Sodium carbonate contains a polyatomic ion. You must not split up or change the ratio of atoms in the polyatomic cluster. Treat it like an indivisible entity and draw brackets around it if the formula contains more than one such ion.

Step 1: Determine the charges of the cation and the anion

$$Na: 1s^2\,2s^2\,2p^6\,3s^1$$

Sodium atoms have one outer shell electron, so they form ions with a 1+ charge.

Carbonate ions, CO_3^{2-}, have a 2– charge.

Sodium ions = Na^+ and carbonate ions = CO_3^{2-}

Step 2: Determine how many of each ion are needed in order to achieve a net charge of zero

Bar diagram method

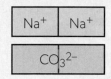

The bar diagram contains two sodium ions and one carbonate ion, so the ratio of sodium to carbonate in the compound is 2:1

$$Na_2CO_3$$

Criss-cross rule

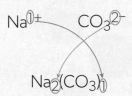

Remember, you can draw brackets around the polyatomic ion to remind yourself that its formula does not change. Again, there is no need to simplify the ratio here.

Step 3: Check that the net charge is zero

| Na^+ | |
| Na^+ | CO_3^{2-} |

Total positive charge = 2+   Total negative charge = 2–

Net charge = 2 – 2 = 0

Step 4: Write the formula

The formula is Na_2CO_3. Note that in the final answer there are no brackets around the polyatomic ion because the formula contains only one carbonate ion.

d. The final example, aluminium nitrate, also contains a polyatomic ion. Follow the same steps as before.

Step 1: Determine the charges of the cation and the anion

Al: $1s^2\,2s^2\,2p^6\,3s^2\,3p^1$

Aluminium atoms have three outer shell electrons, so they form ions with a 3+ charge.

Nitrate ions, NO_3^-, have a 1– charge.

Al ions = Al^{3+} and nitrate ions = NO_3^-

> If you do not remember the charge on a polyatomic ion, revise table 4. Make sure that you learn these formulas and charges off by heart.

Step 2: Determine how many of each ion are needed in order to achieve a net charge of zero

Bar diagram method

	Al^{3+}	

NO_3^-	NO_3^-	NO_3^-

The bar diagram contains one aluminium ion and three nitrate ions, so the ratio of aluminium to nitrate in the compound is 1:3.

$$Al_1(NO_3)_3$$

Criss-cross rule

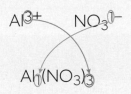

Again, there is no need to simplify the ratio here.

Step 3: Check that the net charge is zero

$\underline{Al^{3+}}$ $\quad$ $\begin{array}{l}NO_3^-\\NO_3^-\\\underline{NO_3^-}\end{array}$

Total positive charge = 3+   Total negative charge = 3–

Net charge = 3 – 3 = 0

Step 4: Write the formula

The formula is $Al(NO_3)_3$. Note that brackets are used to indicate the presence of more than one nitrate ion. The formula should not be written as AlN_3O_9.

Practice questions

10. Deduce the formula of each of the following compounds:

   a. magnesium oxide
   b. strontium chloride
   c. sodium sulfide
   d. lithium nitride
   e. lithium nitrate
   f. barium hydrogencarbonate
   g. ammonium phosphate.

> Names of ionic compounds that contain transition elements have the oxidation number of the transition metal ion in brackets. This is covered in *Structure 3.1*.

Linking questions

Why is the formation of an ionic compound from its elements a redox reaction? (Reactivity 3.2)

 How is formal charge used to predict the preferred structure of sulfate? (Structure 2.2)

 Polyatomic anions are conjugate bases of common acids. What is the relationship between their stability and the conjugate acid's dissociation constant, K_a? (Reactivity 3.1)

Ionic lattices and properties of ionic compounds (*Structure 2.1.3*)

Lattices

Within **ionic crystals**, the ions are arranged in a **lattice structure**. Lattices are continuous, three-dimensional networks of repeating units of positive and negative ions. The exact arrangement of ions in a lattice depends on the size and charge ratio of the ions.

▶ Figure 12 Ionic compounds are made of ions arranged in a lattice structure

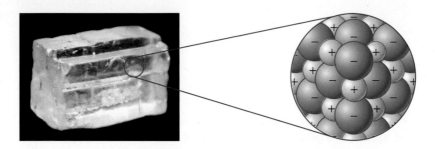

The formula of ionic compounds is an empirical formula: it indicates the ratio of each type of ion in the structure. A single grain of sodium chloride, NaCl, can easily contain a quadrillion ions arranged in a continuous lattice. The formula indicates that the Na^+ and Cl^- ions are present in the lattice in a 1:1 ratio.

ATL Thinking skills

Research often involves finding information but also evaluating its usefulness and reliability.

Consider the statement "each grain of NaCl can easily contain a quadrillion ions".

- Does this sound reasonable to you?
- What information would you need to fact-check the statement?
- How could you reliably find this information?
- Come up with your own estimate of the number of ions in a grain of salt and compare it to this one. How do they compare? Why might they be different? Is the difference between the two values significant?

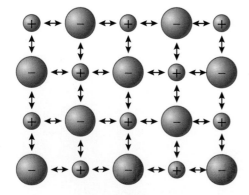

▲ Figure 13 Ionic bonding is non-directional. Each ion attracts all oppositely charged ions around it

Ionic bonds are **non-directional**. This means that an ion will attract all oppositely charged species surrounding it, with the attraction being equal in all directions. Because of this non-directional quality, each cation in the ionic lattice attracts the all the surrounding anions, and vice versa. This means the forces of attraction in an ionic lattice are very strong. Figure 13 shows a 2D representation of the forces in an ionic lattice, but remember, actual lattices are 3D.

Lattice enthalpy

Lattice enthalpy values tell us how strong the ionic bonds are in particular ionic lattice. Lattice enthalpy, $\Delta H^{\ominus}_{\text{lattice}}$, is the standard enthalpy change that occurs on the formation of gaseous ions from one mole of the solid lattice. It is a measure of the strength of an ionic bond because, in order for the ions to become gaseous, all the electrostatic forces of attraction between cations and anions in the lattice need to be overcome. A general equation for the lattice dissociation process is shown below:

$$MX(s) \longrightarrow M^+(g) + X^-(g) \qquad \Delta H^{\ominus}_{\text{lattice}} > 0$$

The process is **endothermic**. Experimental values of lattice enthalpy at 298 K for some compounds can be found in the data booklet. Lattice enthalpies are often quoted as negative values that represent the exothermic formation of the lattice from gaseous ions—the opposite process to that shown in figure 14. However, in this book we shall consider only the endothermic formation of gaseous ions from a lattice, which is consistent with the definition given in the data booklet.

Lattice enthalpy increases as the energy required to overcome the electrostatic forces of attraction between ions increases. Two factors affecting lattice enthalpy are **ionic radius** and **ionic charge**. The strength of the electrostatic attraction between oppositely charged ions:

- increases with increasing ionic charge

- decreases with increasing ionic radius.

Look at the variations in lattice enthalpy of the compounds in table 6.

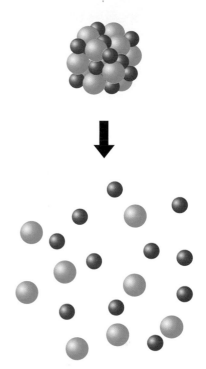

▲ Figure 14 Lattice enthalpy is the energy required to overcome the electrostatic forces of attraction holding ions together in the lattice

Ionic compound	$\Delta H^{\ominus}_{\text{lattice}}$/kJ mol^{-1}	Ionic radius/10^{-12} m		Ionic charge	
		Cation	Anion	Cation	Anion
KF	829	138	133	+1	−1
NaF	930	102	133	+1	−1
CaF$_2$	2651	100	133	+2	−1

▲ Table 6 Lattice enthalpies of selected compounds

NaF has a greater lattice enthalpy than KF because the cations in NaF are smaller, and therefore the electrostatic attraction between Na$^+$ ions and F$^-$ ions is greater. The lattice enthalpy of CaF$_2$ is considerably larger than that of KF. This is in part due to the smaller ionic radius of Ca^{2+} compared to K$^+$. However, it is mainly due to the greater charge on the Ca^{2+} cation, which results in greater electrostatic attraction between the cations and anions in CaF$_2$.

Factors affecting the lattice enthalpy of the group 1 chlorides

Charge density is a term used to describe charge per unit volume. In this task, you will explore the charge density of the group 1 cations and relate this to the trend in lattice enthalpies of the group 1 chlorides.

Relevant skills

- Tool 2: Use spreadsheets to manipulate data

- Tool 3: General mathematics

- Inquiry 1: State and explain predictions

- Inquiry 2: Identify, describe and explain patterns, trends and relationships

Instructions

Part 1: Prediction

1. For the group 1 cations, predict the relationship between:
   a. ionic radius and lattice enthalpy of their chlorides
   b. charge density and lattice enthalpy of their chlorides.

2. Sketch the graphs you expect to obtain for the relationships above.

Part 2: Data collection

3. Collect the following data for the group 1 chlorides: ionic charge, ionic radius, lattice enthalpy. Possible sources of information include the data booklet and online databases. Cite each source appropriately.

4. Input your data into a spreadsheet and organize it into a suitable table.

Part 3: Analysis

5. Calculate the volume of each cation.

6. Calculate the charge density of each cation.

7. Plot two graphs: one graph showing the relationship between ionic radius and lattice enthalpy, and the other, between charge density and lattice enthalpy.

8. Describe and explain the trends shown in the graphs.

9. Discuss the differences between the two graphs.

10. Evaluate your prediction, including a comparison of the graphs you obtained in 7 and the sketched graphs you obtained in 2.

11. Consider possible extensions to this investigation: what other aspects of ionic radius, charge density and lattice enthalpy could you explore?

Practice questions

1. Write equations, including state symbols, that represent the lattice enthalpies of KBr, CaO and $MgCl_2$.

2. State and explain whether you expect KF or K_2O to have a lower lattice enthalpy value.

3. State and explain which of the following ionic compounds you expect to have the greatest lattice enthalpy value: NaCl, $MgCl_2$, Na_2O or MgO.

4. Describe and explain the trend in lattice enthalpy of the group 1 chlorides down the group from LiCl to CsCl.

Properties of ionic compounds

The properties of ionic compounds are due to their structural features: they contain cations and anions held together by strong electrostatic attractive forces in a lattice.

Volatility

Volatility (from the Latin *volare*, to fly) refers to the tendency of a substance to vaporize (turn into a gas). For an ionic compound to turn into a gas, the strong electrostatic forces of attraction holding the ions together must be overcome. The volatility of ionic compounds is therefore very low: they are said to be "non-volatile". This also means they have high boiling points.

Ionic compounds typically have high melting points too. The melting point of sodium chloride is approximately 1075 K. Magnesium oxide, frequently used in furnaces due to its ability to withstand high temperatures, melts at around 3098 K.

Electrical conductivity

In order to conduct electricity, substances must contain charged particles that are able to move. Ionic compounds contain charged particles, cations and anions. In a solid ionic lattice, cations and anions can vibrate around a fixed point, but they cannot change position. Solid ionic compounds do not conduct electricity because ions in a solid lattice are not mobile. When molten or aqueous, both cations and anions are free to move past one another, allowing them to conduct electricity when a potential difference is applied.

Global impact of science

Some ionic compounds have uncharacteristically low melting points and can be used as solvents. Such **ionic liquids** can be described as "green solvents" because they are non-volatile. This means they can be more easily contained and, often, recycled. This, however, does not necessarily mean they are harmless. Their manufacturing, disposal and transportation can have significant environmental impacts.

▲ Figure 15 These are waste separation bins in Jakarta, Indonesia. The leftmost bin (red) is for batteries. Electrolytes in batteries conduct electricity because they contain mobile ions. Used batteries are frequently separated from other types of waste to prevent them from ending up in landfill. This is because batteries contain valuable metals and other substances, which can be recycled. Do you separate your used batteries from other household waste?

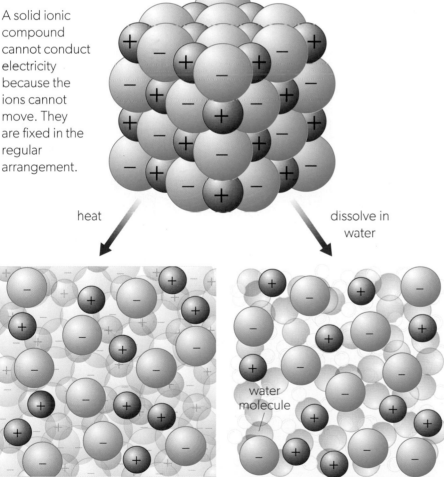

A solid ionic compound cannot conduct electricity because the ions cannot move. They are fixed in the regular arrangement.

heat

dissolve in water

water molecule

When an ionic compound is heated strongly and melts, the ions can move around and the molten compound conducts electricity.

When an ionic compound is dissolved in water, it can conduct electricity because its ions can move among the water molecules.

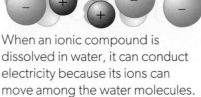

◀ Figure 16 Molten and aqueous ionic compounds are electrical conductors

Solubility

Ionic compounds are typically soluble in **polar solvents** such as water, and insoluble in **non-polar solvents** such as hexane.

Water is a polar solvent. The difference in electronegativity between the oxygen and hydrogen atoms, combined with the bent geometry of the water molecule, result in the water molecule having a partial negative charge on the oxygen atom and partial positive charges on the hydrogen atoms.

Imagine an ionic compound being added to water. The water molecules position themselves so that their partial negative charges point towards the cations, and their partial positive charges point towards the anions. As a result, individual ions are pulled out of the lattice and become surrounded by water molecules. In the case of a non-polar solvent, there is no attraction between the ions of the ionic compound and the solvent molecules, so the cations and anions remain within the lattice.

> Polarity is discussed in greater detail in *Structure 2.2*.

Activity

Draw a labelled diagram explaining why ionic compounds conduct electricity when molten or dissolved, but not when solid.

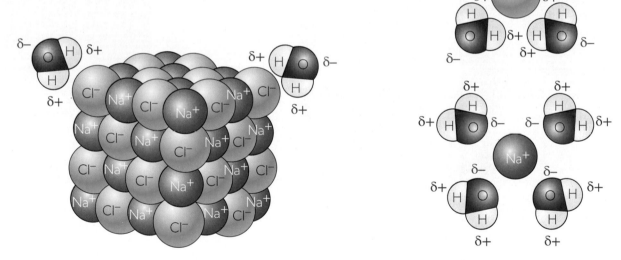

▲ **Figure 17** The dissolution of ionic compounds in water involves interactions between ions and water molecules

Not all ionic compounds dissolve in water. This is because there are two competing forces of attraction present:

- ionic bonds between cations and anions in the lattice

- the association between the ions and the partial charges of water molecules

Ionic compounds are insoluble when the electrostatic attractions between the cations and anions in the lattice are stronger than the association between the ions and water molecules. Examples of ionic compounds that are insoluble in water include calcium carbonate and silver chloride.

Research skills

Heavy metal ions, such as lead and nickel, often form insoluble salts. Some wastewater treatment processes take advantage of this property, removing heavy metals out of industrial effluents through precipitation.

- Use the internet to research other examples of precipitation reactions and their uses.
- Describe and explain the changes that are observed when a precipitate is formed.

Linking questions

What experimental data demonstrate the physical properties of ionic compounds? (Tool 1, Inquiry 2)

How can lattice enthalpies and the bonding continuum explain the trend in melting points of metal chlorides across period 3? (Structure 3.1)

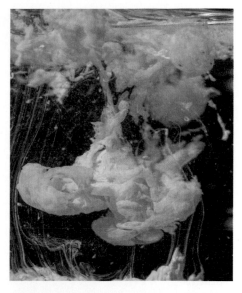

▲ Figure 18 Close-up photograph of the formation of a lead(II) chromate precipitate in the reaction between aqueous solutions of lead(II) nitrate and potassium chromate

▲ Figure 19 Salt flats at Salar de Uyuni in Bolivia, which are mainly made of halite, the mineral form of sodium chloride. The brine below the rock salt crust is rich in dissolved metal ions, particularly lithium. Global demand for lithium is increasing due to its use in batteries. Lithium-ion batteries can be used to power mobile phones, laptops and electric vehicles

Solubility of ionic salts

The patterns in aqueous solubility of several common ionic compounds are often referred to as **solubility rules**. We can use these differences in solubility to help us deduce the identity of an unknown ionic compound. In this task you will mix different solutions of ionic compounds and observe whether an insoluble product (known as a **precipitate**) is produced.

Relevant skills

- **Tool 1:** Recognize and address relevant safety and environmental issues in an investigation
- **Inquiry 2:** Interpret qualitative data

Safety

- Wear eye protection.
- Dilute calcium nitrate and silver nitrate solutions are irritants.
- You should take care when handling all solutions because you do not know exactly which is which. They are all potential irritants.
- Collect and retain any precipitates formed.
- Dispose of waste solutions and precipitates according to your school's guidelines.

Materials

- Clear plastic sheet on a black background
- Pipettes
- Small piece of copper wire (~0.5 cm long)
- Dilute acid solution
- Samples of solutions labelled A, B, C and D

Instructions

Part 1: Solubility rules

General solubility rules can be inferred from the data in table 7. For example:

- All nitrates are soluble.
- Sulfates are generally insoluble, except group 1 sulfates and ammonium sulfate.

Study table 7 and infer at least three more general solubility rules.

Part 2: Identification of ionic compounds

You will be provided with samples of solutions labelled A, B, C, and D. These solutions are potassium carbonate, sodium chloride, silver nitrate and calcium nitrate, but you do not know which is which. Your job is to identify each solution, using the materials listed and the solubility rules. You may also draw from your knowledge of other areas of chemistry.

Note you do not have to mix the solutions inside test tubes. You can prepare small-scale mixtures of solutions by mixing a drop of each on a plastic sheet. If a precipitate is formed, the solution will become opaque. This will be easily observable, particularly if you lay the sheet on a black background.

Devise a method, present it clearly, and show it to your teacher. If they approve it and if you have time, try it out!

		Cations					
		group 1 cations (Li⁺, Na⁺, K⁺)	ammonium, NH_4^+	barium, Ba^{2+}	calcium, Ca^{2+}	silver, Ag^+	lead, Pb^{2+}
Anions	**nitrate, NO_3^-**	soluble	soluble	soluble	soluble	soluble	soluble
	ethanoate, CH_3COO^-	soluble	soluble	soluble	soluble	soluble	soluble
	chloride, Cl^-	soluble	soluble	soluble	soluble	insoluble	insoluble
	hydroxide, OH^-	soluble	soluble	soluble	slightly soluble	insoluble	insoluble
	sulfate, SO_4^{2-}	soluble	soluble	insoluble	slightly soluble	slightly soluble	insoluble
	carbonate, CO_3^{2-}	soluble	soluble	insoluble	insoluble	insoluble	insoluble

▲ Table 7 Aqueous solubility of common ionic compounds

End-of-topic questions

Topic review

1. Using your knowledge from the *Structure 2.1* topic, answer the guiding question as fully as possible:

   What determines the ionic nature and properties of a compound?

2. Explain why ionic substances are always compounds.

Exam-style questions

Multiple-choice questions

3. The elements in group 17 generally form ions with which charge?

   A. 7+

   B. 1+

   C. 1−

   D. 7−

4. The ion of element X has a 2+ charge and contains 20 protons. Give the full electron configuration of this ion.

   A. $1s^22s^22p^63s^23p^6$

   B. $1s^22s^22p^63s^23p^64s^2$

   C. $1s^22s^22p^63s^23p^63d^2$

   D. $1s^22s^22p^63s^23p^64s^23d^2$

5. Which statement about ionic compounds is correct?

   A. Ionic bonding results from the electrostatic attraction between cations and anions.

   B. Calcium fluoride is made up of CaF_2 molecules.

   C. Ionic structures contain delocalized electrons when molten or dissolved, but not when solid.

   D. Ions are held together because anions transfer electrons to cations.

6. What is the name of $CaSO_4$?

   A. carbon sulfite

   B. calcium sulfite

   C. carbon sulfate

   D. calcium sulfate

7. What is the formula of sodium nitrate?

   A. $NaNO_2$

   B. $NaNO_3$

   C. Na_3N

   D. S_3N_2

8. Which compound has the largest value of lattice enthalpy?

   A. CaS

   B. CaO

   C. K_2S

   D. K_2O

9. Which statement is correct?

   A. Lattice enthalpy increases when the radii of the component ions increase.

   B. Lattice enthalpy represents the energy needed to transfer an electron from a cation to an anion.

   C. Lattice enthalpy decreases when the charge of the component ions increases.

   D. Lattice enthalpy increases when the charge density of the component ions increases.

10. Which equation correctly represents the lattice enthalpy of potassium oxide?

    A. $K_2O(s) \rightarrow 2K^+(g) + O^{2-}(g)$

    B. $K_2O(s) \rightarrow 2K(s) + \frac{1}{2}O_2(g)$

    C. $K_2O(s) \rightarrow K_2^{2+}(g) + O^{2-}(g)$

    D. $K_2O(s) \rightarrow 2K(g) + \frac{1}{2}O_2(g)$

11. List the lithium halides in order of **increasing** lattice enthalpy.

    A. LiF, LiCl, LiBr, LiI

    B. LiF, LiBr, LiCl, LiI

    C. LiI, LiBr, LiCl, LiF

    D. LiI, LiCl, LiBr, LiF

12. Which substance has an ionic structure?

	Melting point / °C	Solubility in water	Electrical conductivity when molten	Electrical conductivity when solid
A	36	high	none	none
B	186	low	none	none
C	1083	high	good	none
D	1710	low	good	good

Extended-response questions

13. Calcium fluoride has applications in the field of optics.

    a. Deduce the formula for calcium fluoride. [1]

    b. Describe the structure and bonding in solid calcium fluoride. [3]

    c. Explain why solid calcium fluoride is a poor electrical conductor, but molten calcium fluoride can conduct electricity. [2]

    d. The lattice enthalpy of calcium fluoride is $2\,651\,kJ\,mol^{-1}$.

       i. Write an equation, including state symbols, that shows the process associated with the lattice enthalpy of calcium fluoride. [2]

       ii. Explain why the process in part (i) is endothermic. [1]

       iii. The lattice enthalpy of calcium oxide, CaO, is $3\,401\,kJ\,mol^{-1}$. Explain why the lattice enthalpy of calcium oxide is greater than that of calcium fluoride. [2]

14. Certain types of breathalyser contain the bright orange ionic compound potassium dichromate(VI), $K_2Cr_2O_7$.

    a. Write the full electron configuration of a potassium ion. [1]

    b. Explain why the difference in mass between a potassium atom and a potassium ion is negligible. [1]

    c. Deduce the charge of the dichromate(VI) ion. [1]

    d. Potassium dichromate(VI) contains chromium. Chromium is a transition element that commonly forms form Cr^{2+} ions and Cr^{3+} ions among others.

       i. Write the abbreviated electron configuration of a chromium atom. [1]

       ii. Copy the diagram below and draw arrows in the boxes to represent the electron configuration in the 3d and 4s orbitals of a Cr^{2+} ion. [1]

       4s   3d

       iii. Write the full electron configuration of a Cr^{3+} ion. [1]

15. The equation below shows the formation of lithium fluoride from its elements under standard conditions.

    $Li(s) + F_2(g) \rightarrow LiF(s)$

    a. Balance the equation. [1]

    b. Identify the charge of the lithium ion. [1]

    c. Identify the oxidized species and the reduced species in this reaction. [1]

    d. Sketch a diagram showing the structure of lithium fluoride. [2]

Structure 2.2 | The covalent model

What determines the covalent nature and properties of a substance?

Covalent bonds lead to a vast range of different substances. From water to diamond to nitrogen gas, from oils to plastics to polyatomic ions, these species contain atoms held together by strong covalent bonds. Covalent bonds lead to the formation of two different types of structure: **covalent network structures** (also known as **giant covalent structures**) and **molecular covalent structures**.

In general, covalent substances are poor electrical conductors.

Substances with covalent network structures are also characterized by high melting points and boiling points, low volatility and poor solubility in water. Substances with molecular structures, on the other hand, generally have low melting points and boiling points. Their solubility and volatility vary greatly depending on their intermolecular forces.

In *Structure 2.2*, you will learn what a covalent bond is, how we represent molecules as well as how we describe and explain their shapes and intermolecular forces. You will also learn about covalent network structures.

Understandings

Structure 2.2.1 — A covalent bond is formed by the electrostatic attraction between a shared pair of electrons and the positively charged nuclei.

The octet rule refers to the tendency of atoms to gain a valence shell with a total of 8 electrons.

Structure 2.2.2 — Single, double and triple bonds involve one, two and three shared pairs of electrons respectively.

Structure 2.2.3 — A coordination bond is a covalent bond in which both the electrons of the shared pair originate from the same atom.

Structure 2.2.4 — The Valence Shell Electron Pair Repulsion (VSEPR) model enables the shapes of molecules to be predicted from the repulsion of electron domains around a central atom.

Structure 2.2.5 — Bond polarity results from the difference in electronegativities of the bonded atoms.

Structure 2.2.6 — Molecular polarity depends on both bond polarity and molecular geometry.

Structure 2.2.7 — Carbon and silicon form covalent network structures.

Structure 2.2.8 — The nature of the force that exists between molecules is determined by the size and polarity of the molecules. Intermolecular forces include London (dispersion), dipole–induced dipole, dipole–dipole and hydrogen bonding.

Structure 2.2.9 — Given comparable molar mass, the relative strengths of intermolecular forces are generally:

London (dispersion) forces < dipole–dipole forces < hydrogen bonding.

Structure 2.2.10 — Chromatography is a technique used to separate the components of a mixture based on their relative attractions involving intermolecular forces to mobile and stationary phases.

Structure 2.2.11 — Resonance structures occur when there is more than one possible position for a double bond in a molecule.

Structure 2.2.12 — Benzene, C_6H_6, is an important example of a molecule that has resonance.

Structure 2.2.13 — Some atoms can form molecules in which they have an expanded octet of electrons.

Structure 2.2.14 — Formal charge values can be calculated for each atom in a species and used to determine which of several possible Lewis formulas is preferred.

Structure 2.2.15 — Sigma bonds (σ) form by the head-on combination of atomic orbitals where the electron density is concentrated along the bond axis.

Pi bonds (π) form by the lateral combination of p orbitals where the electron density is concentrated on opposite sides of the bond axis.

Structure 2.2.16 — Hybridization is the concept of mixing atomic orbitals to form new hybrid orbitals for bonding.

AHL

Covalent bonds and molecules (*Structure 2.2.1*)

Covalent bonds are formed when atoms share pairs of valence electrons. A covalent bond results from the electrostatic attraction between a shared pair of electrons and the positively charged nuclei of the atoms involved in the bond.

Diatomic hydrogen, H_2, is the simplest covalent molecule. It consists of two hydrogen atoms held together by a covalent bond made from one electron from each hydrogen atom (figure 1). Atoms can share one, two or three pairs of electrons, forming single, double or triple bonds, respectively. Figure 1 also shows the formation of the double covalent bond in diatomic oxygen, O_2.

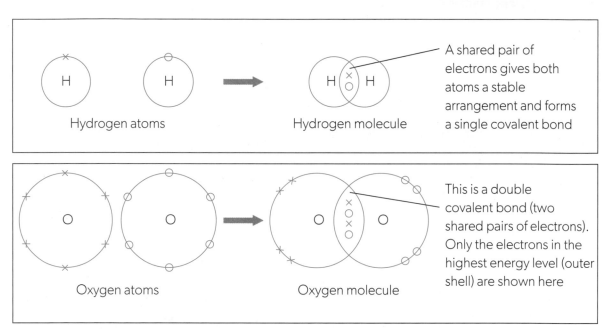

A shared pair of electrons gives both atoms a stable arrangement and forms a single covalent bond

Hydrogen atoms   Hydrogen molecule

This is a double covalent bond (two shared pairs of electrons). Only the electrons in the highest energy level (outer shell) are shown here

Oxygen atoms   Oxygen molecule

▲ Figure 1 Covalent bond formation in diatomic hydrogen, H_2, and diatomic oxygen, O_2

▲ Figure 2 Examples of substances that contain covalent bonds: plastics, graphite (in pencil leads), fizzy water, and the carbon dioxide bubbles in it

Activity

Draw a diagram to show the triple covalent bond in diatomic nitrogen, N_2.

Electronegativity

Covalent bonds generally form between atoms of relatively high electronegativity, typically non-metals. In *Structure 2.1*, we saw that ionic bonds form when the electronegativity difference between two atoms is greater than 1.8. When the difference in electronegativity between the atoms is less than 1.8 (figure 3), the bond between them is predominantly covalent.

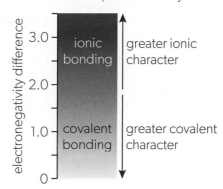

▲ **Figure 3** Electronegativity differences that are lower than 1.8 suggest that covalent bonding is present

Unlike ionic substances, which are always compounds, covalent substances can be either elements or compounds. When two non-metal atoms of the same element bond together, the electronegativity difference is zero, so they form a covalent bond between them.

Electronegativity increases across periods and up groups. If two atoms of relatively high electronegativity are found close to each other in the periodic table, the bonds they form are likely to be covalent. For example, oxygen and fluorine form covalent bonds with each other.

As discussed in *Structure 2.1*, there is no sharp borderline between ionic and covalent bonding. Bonding occurs along a continuum, with greater ionic character at one end and greater covalent character at the other. *Structure 2.4* addresses this continuum in more detail with the **bonding triangle**.

Practice questions

1. Determine whether the following pairs of elements bond covalently or ionically, by referring to their electronegativity values:
   a. carbon and oxygen
   b. sodium and oxygen
   c. carbon and hydrogen
   d. chlorine and oxygen
   e. iodine and iodine
   f. aluminium and fluorine

Lewis formulas

Groups of atoms that are covalently bonded together are called **molecules**. The way valence electrons are arranged in a molecule can be shown using **Lewis formulas**. In Lewis formulas, pairs of valence electrons are represented as dashes, pairs of dots or pairs of crosses, or a combination of all three. Consider the various Lewis formulas of fluoroamine, NH_2F, in figure 4.

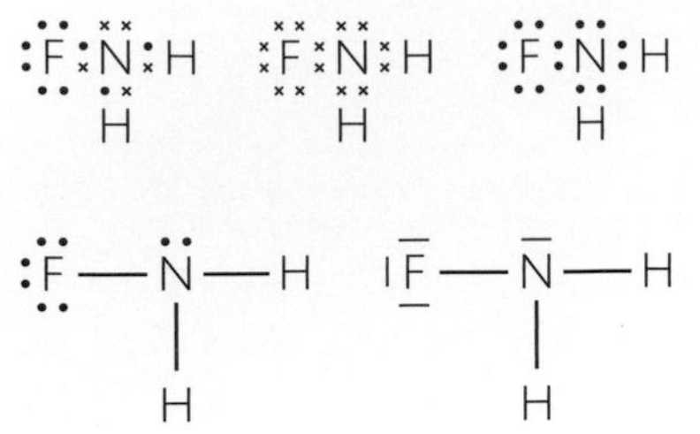

▲ **Figure 4** These are all acceptable Lewis formulas of fluoroamine, NH_2F

Here are the rules for drawing Lewis formulas regardless of whether dots, crosses or dashes are used:

1. Only valence electrons are shown.

2. Electrons are arranged in pairs.

3. Each pair of electrons shared between two atoms represents a covalent bond.

4. Electrons in a bond (termed **bonding electrons**) are positioned in the region between the two atoms involved in the bond.

5. **Non-bonding electrons** (also referred to as **lone pairs**) are positioned away from the region between the two atoms involved in the bond.

Atoms in Lewis formulas generally have noble gas electron configurations. This is known as the **octet rule** because noble gases often have eight valence electrons. Noble gases already have full octets and do not readily gain, lose or share electrons. Hence, they are usually found in an unbonded, monatomic form.

The octet rule is a useful rule of thumb, but it has limitations. It does not explain why some noble gases form bonds in certain situations. Sometimes atoms can form stable molecules even with fewer than an octet of electrons. Conversely, larger atoms can form **expanded octets** (this is seen at AHL). Species with odd numbers of valence electrons are also exceptions to the octet rule.

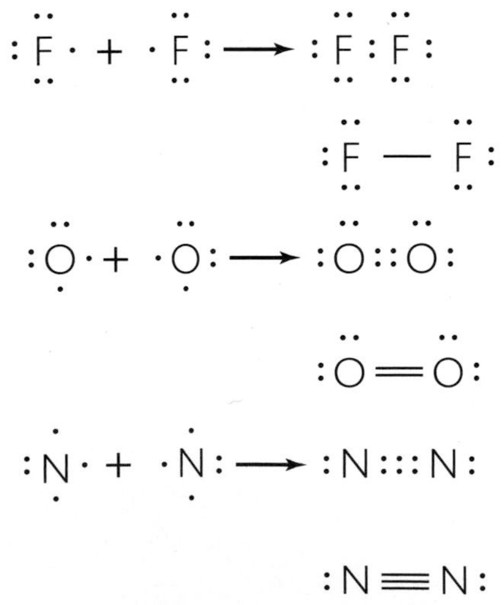

Following these steps will help you draw Lewis formulas in most cases:

1. Work out the total number of the valence electrons for each atom in the molecule.

2. Divide the total number of valence electrons by two to work out how many pairs of electrons there are.

3. Arrange the atoms by drawing their symbols on the page. The element with the least number of atoms is usually found in the centre. Hydrogen atoms always surround the central atom(s).

4. Bond the central and peripheral atoms together by drawing single bonds between them. Each single bond represents an electron pair.

5. Assign non-bonding pairs of electrons to the peripheral atoms. Keep going until they achieve noble-gas configurations.

6. Assign any remaining electron pairs to the central atom(s).

7. Check that the central atom has a full octet. If it does not, try the following two methods:

    •   Reassign non-bonding pairs on the peripheral atoms to become additional bonds to the central atom

    •   Check that the molecule you are looking at is not an exception to the octet rule (see page 123 for examples)

▲ **Figure 5** By sharing a pair of electrons, two fluorine atoms can obtain a full octet of valence electrons. Oxygen atoms can do this by sharing two pairs of electrons, and nitrogen atoms share three pairs

Worked example 1

Draw the Lewis formulas of each of the following molecules:

a. water, H_2O b. nitrogen trichloride, NCl_3 c. carbon dioxide, CO_2

Solution

Follow the steps above for each molecule:

	a. water, H_2O	b. nitrogen trichloride, NCl_3	c. carbon dioxide, CO_2
Step 1 Count valence electrons	hydrogen: $1 \times 2 = 2$ oxygen: 6 Total: $2 + 6 = 8$	nitrogen: 5 chlorine: $7 \times 3 = 21$ Total: $5 + 21 = 26$	carbon: 4 oxygen: $6 \times 2 = 12$ Total: $4 + 12 = 16$
Step 2 Calculate the number of electron pairs	$\frac{8}{2} = 4$ pairs	$\frac{26}{2} = 13$ pairs	$\frac{16}{2} = 8$ pairs
Step 3 Arrange the atoms	H O H Hydrogens on the periphery. There is only one oxygen, so it is likely to be in the centre.	Cl Cl N Cl There is only one nitrogen, so it is likely to be in the centre.	O C O There is only one carbon, so it is likely to be in the centre.
Step 4 Draw the single bonds	H —— O —— H 2 pairs used so far…	Cl | Cl —— N —— Cl 3 pairs used so far…	O —— C —— O 2 pairs used so far…
Step 5 Put non-bonding pairs on peripheral atoms	The peripheral hydrogen atoms already have noble gas configurations because the first energy level only holds up to two electrons. They do not need any more electrons.	:Cl: | :Cl —— N —— Cl: The chlorine atoms now have full octets. 12 pairs used so far…	:Ö —— C —— Ö: The oxygen atoms now have full octets. 8 pairs used so far…
Step 6 Put any remaining electron pairs on the central atom	H —— Ö —— H The 2 remaining electron pairs are assigned to the oxygen.	:Cl: | :Cl —— N —— Cl: The remaining electron pair is assigned to the nitrogen.	We have used all 8 available electron pairs. None are available for the central carbon.
Step 7 Check that the central atom has a full octet	Oxygen has 4 electron pairs in the Lewis formula, therefore a full octet.	Nitrogen has 4 electron pairs in the Lewis formula, therefore a full octet.	⚠ :Ö —— C —— Ö: The carbon atom has only 2 electron pairs in this Lewis formula. We need to reassign two of the non-bonding pairs on the oxygen atoms to form double bonds: :O ═══ C ═══ O: Now the carbon atom has 4 electron pairs and therefore a full octet.

Polyatomic ions are charged groups of covalently bonded atoms. The magnitude of the charge indicates how many electrons have been lost or gained. Lewis formulas of polyatomic ions are enclosed in square brackets, with the charge indicated with a superscript outside the brackets. When counting the valence electrons, you also need to factor in the charge: for every additional negative charge, you add an electron, and for every additional positive charge, you subtract an electron. This is illustrated in the next set of worked examples.

Worked example 2

Draw the Lewis formulas of the following ions:

a.   hydroxide ion, OH^- b.   carbonate ion, CO_3^{2-} c.   hydronium ion, H_3O^+

Solution

Follow the steps above for each ion:

	a hydroxide ion, OH^-	b carbonate ion, CO_3^{2-}	c hydronium ion, H_3O^+
Step 1 Count valence electrons	oxygen: 6 hydrogen: 1 This polyatomic ion has a 1– charge, meaning that it has one additional electron. Total: $6 + 1 + 1 = 8$	carbon: 4 oxygen: $6 \times 3 = 18$ This polyatomic ion has a 2– charge, meaning that it has two additional electrons. Total: $4 + 18 + 2 = 24$	hydrogen: $1 \times 3 = 3$ oxygen: 6 This polyatomic ion has a 1+ charge, meaning that it has one fewer electron. Total: $3 + 6 - 1 = 8$
Step 2 Calculate the number of electron pairs	$\dfrac{8}{2} = 4$ pairs	$\dfrac{24}{2} = 12$ pairs	$\dfrac{8}{2} = 4$ pairs
Step 3 Arrange the atoms	O H	O C O O	H O H H
Step 4 Draw the single bonds	O —— H 1 pair used so far...	O | C O⁄ ⁀O 3 pairs used so far...	H —— O —— H | H 3 pairs used so far...
Step 5 Put non-bonding pairs on peripheral atoms	The hydroxide ion has only two atoms in it, so you do not need to distinguish between central and peripheral atoms here. Skip this step.	:Ö: | C :Ö̤⁄ ⁀Ö̤: The oxygen atoms now have full octets. 12 pairs used so far...	The peripheral hydrogen atoms have noble gas configurations and therefore do not need any more electrons.
Step 6 Put any remaining electron pairs on the central atom	:Ö —— H ¨ The remaining 3 electron pairs are assigned to the oxygen atom. The hydrogen already has a noble gas configuration.	We have used all 12 available electron pairs. None are available for the central carbon.	H —— Ö —— H | H The fourth electron pair is assigned to the oxygen atom.

	a hydroxide ion, OH⁻	b carbonate ion, CO₃²⁻	c hydronium ion, H₃O⁺
Step 7 Check that the central atom has a full octet	$\left[:\ddot{O}-H\right]^{-}$ All atoms have noble-gas configurations. Note the square brackets and charge, which are used in the Lewis formulas of polyatomic ions.	The carbon atom has only 3 pairs in this Lewis formula. We need to reassign one of the non-bonding pairs on the oxygen atoms to form a double bond: Now the carbon atom has 4 electron pairs and therefore a full octet. Brackets and charge are added to complete the Lewis formula.	$\left[H-\ddot{O}-H\right]^{+}$ All atoms have noble gas configurations. Brackets and charge are added to complete the Lewis formula.

Sometimes molecules contain atoms that do not follow the octet rule. At SL, you should be aware of Lewis formulas that contain atoms with fewer than eight valence electrons. Consider the arrangement of the 12 electron pairs of boron trifluoride, BF_3 in the Lewis formula shown in figure 6. Here, the boron atom has only three pairs of electrons around it. The boron atom is **electron deficient**.

Similar to boron, many other elements of groups 2 and 13 form stable electron-deficient molecules. These include beryllium, magnesium and aluminium. In Lewis formulas of such molecules, group 2 elements (Be and Mg) have only two bonding electron pairs while group 13 elements (B and Al) have three bonding electron pairs.

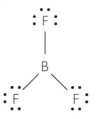

▲ Figure 6 Lewis formula of boron trifluoride

Practice questions

2. Draw the Lewis formula of each of the following:
    a. BH_3
    b. BCl_3
    c. $BeCl_2$
    d. $AlCl_3$
    e. CH_3^+

3. Draw the Lewis formula of each of the following:
    a. HBr
    b. OF_2
    c. O_2
    d. O_3
    e. HCN
    f. CH_3Cl
    g. CH_2O
    h. N_2H_4

4. Draw the Lewis formula of each of the following:
    a. NH_4^+
    b. NO_3^-
    c. NO_2^+
    d. NO_2^-
    e. OCl^-

Linking questions

 What are some of the limitations of the octet rule?

Why do noble gases form covalent bonds less readily than other elements? (Structure 1.3)

Why do ionic bonds only form between different elements while covalent bonds can form between atoms of the same element? (Structure 2.1)

Bond order (*Structure 2.2.2*)

The hydrogen molecule, H_2, shown in figure 7, contains a **single bond**. The oxygen molecule, O_2, contains a **double bond**. Diatomic nitrogen, N_2, contains a **triple bond.**

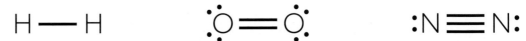

▲ **Figure 7** Hydrogen, oxygen and nitrogen molecules with differing bond order

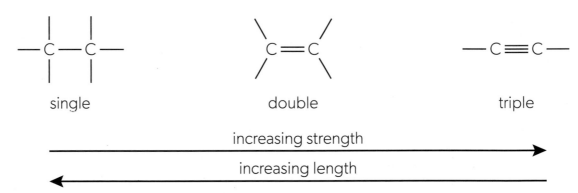

single   double   triple

increasing strength
→

increasing length
←

▲ **Figure 8** The relationship between bond order, bond length and bond strength

The number of bonded electron pairs between two atoms is referred to as **bond order**. Single bonds (bond order 1), double bonds (bond order 2) and triple bonds (bond order 3) differ in their strength and length. Double bonds are stronger than single bonds, and triple bonds are stronger still. Triple bonds hold atoms closer together than double bonds, and hence triple bonds are shorter than double bonds. Double bonds are, in turn, shorter than single bonds. Figure 8 shows this pattern for carbon–carbon bonds.

Bond enthalpies are discussed in *Reactivity 1.2*.

Bond strength and length can be quantified. The stronger a bond is, the larger its bond enthalpy.

Bond length is defined as the average distance between two bonded nuclei. Table 1 shows bond enthalpy and bond length data for carbon–carbon single, double and triple bonds. The carbon–carbon triple bond is the shortest but also the strongest.

Bond	Bond enthalpy (at 298 K) / kJ mol^{-1}	Bond length / 10^{-12} m
C–C	346	154
C=C	614	134
C≡C	839	120

◀ Table 1 Bond enthalpies and bond lengths of carbon–carbon bonds

 Data-based question

The bond enthalpy of a carbon–nitrogen triple bond is 890 kJ mol^{-1} and its bond length is 116×10^{-12} m. Use these data and the information in table 1 to predict the bond enthalpies and bond lengths of single and double carbon–nitrogen bonds. Then compare your predictions with the values in the data booklet.

ATL **Communication skills**

Graphs often complement written explanations. The paragraph below explains why a covalent bond forms between two hydrogen atoms by describing the relationship between potential energy and the distance between two hydrogen atoms. Read it and try to represent what it is describing by sketching a graph of potential energy (*y*-axis) vs distance between hydrogen nuclei (*x*-axis).

Two hydrogen atoms have no effect on each other if they are separated by a sufficiently large distance. As the two atoms approach each other, the electrostatic attraction between the hydrogen nuclei and each other's electrons increases. This process leads to a decrease in potential energy. As the atoms get closer, they reach a point where the two nuclei attract the two electrons in the pair, effectively sharing them. This arrangement occurs at a potential energy minimum, meaning that the molecule is energetically stable. If the atoms were to get any closer together, the resulting repulsion between the two positively charged nuclei would outweigh the attraction for the shared pair of electrons, leading to a rise in potential energy.

When you have finished, compare your sketch graph to a potential energy curve. You can find one by typing "hydrogen molecule potential energy curve" into a search engine.

What can graphical representations add to explanations such as the one above? To what extent is a graphical representation better than a textual description?

 Linking question

How does the presence of double and triple bonds in molecules influence their reactivity? (Reactivity 2.2)

The formation of coordination bonds is a feature of electron pair sharing reactions, as discussed in *Reactivity 3.4*.

Coordination bonds (*Structure 2.2.3*)

We have seen that a covalent bond is formed when each of the two atoms in the bond contributes an electron to the bond. Sometimes both the electrons in the covalent bond come from the same atom. The resulting bonds are called **coordination bonds.** For example, when a hydrogen cation encounters a water molecule, a coordination bond is formed, leading to the formation of a hydronium ion (figure 9). Coordination bonds are often indicated with an arrow along the bond. The direction of the arrow shows which atom donated the electron pair to the bond, and which atom accepted the electron pair.

▶ Figure 9 Different representations of the hydronium ion, showing the coordination bond

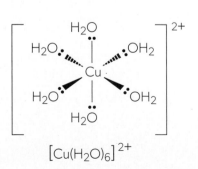

coordination bond

Coordination bonds help to explain how bonding occurs in certain molecules. Once formed, coordination bonds are indistinguishable from any other covalent bond.

Practice questions

5. Draw the Lewis formulas of the following molecules, indicating the coordination bonds clearly.
    a. ammonium ion, NH_4^+
    b. ozone, O_3
    c. carbon monoxide, CO
    d. ammonia boron trifluoride, NH_3BF_3

The properties of complex ions and their reactions are discussed in *Reactivity 3.4*.

Linking question

Why do Lewis acid–base reactions lead to the formation of coordination bonds? (Reactivity 3.4)

Coordination bonds in transition metal complexes

Transition metals can form complex ions such as the one shown in Figure 10. These complex ions contain coordination bonds, which hold together the central metal cation and the surrounding atoms or groups of atoms, called **ligands**.

A common feature of all ligands is a lone pair of electrons, which can be used to form the coordination bond to the metal ion.

$[Cu(H_2O)_6]^{2+}$

▲ Figure 10 Water molecules form coordination bonds with copper(II) cations, leading to the formation of a blue hexaaquacopper(II) complex ion

The valence shell electron pair repulsion model (VSEPR) (*Structure 2.2.4*)

You might have come across a Lewis formula for water that shows the two hydrogen atoms at an angle. Such formulas better illustrate the geometry of the molecule. This can also be shown by ball-and-stick and space-filling models (figure 11).

Two-dimensional representations of molecules such as Lewis formulas do not reflect the three-dimensional arrangement of atoms in a molecule. The three-dimensional shape of a molecule is termed **molecular geometry**, and it is an essential feature of molecules which contributes to their properties. For example, the molecular geometry within a structure known as a beta-lactam ring is key to its antibiotic properties. This is shown in figure 12.

Molecular geometry can be explored using the **VSEPR** model:

Valence Shell Electron Pair Repulsion

The VSEPR model is based on the following premises:

1. Electron pairs repel each other and therefore arrange themselves as far apart from each other as possible.

2. Non-bonding (or lone) electron pairs occupy more space than bonding pairs (single bonds).

3. Double and triple bonds occupy more space than single bonds.

The term **electron domain** will be useful in these discussions. An electron domain is a region of high electron density due to the presence of electron pairs. An electron domain can be:

- a non-bonding pair of electrons (known as a **lone pair**)

- a bonding pair of electrons (single bond)

- a double or triple bond (which involve multiple pairs of electrons).

The Lewis formula in figure 13 illustrates this point: the central sulfur atom has three electron domains, of which one is a non-bonding domain, and two are bonding domains. One of the bonding domains is a double bond, the other is a single bond.

▲ Figure 11 Space-filling model of a water molecule

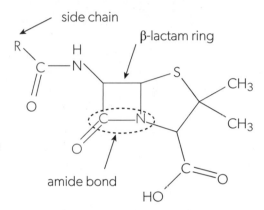

▲ Figure 12 The beta-lactam ring (in red above) is a common feature of antibiotic molecules such as penicillins. The geometry of this structure is vital to its reactivity. The amide bond hydrolyses readily due to the strain caused by the 90° bond angles. This means that it can attach to enzymes responsible for building bacterial cell walls.

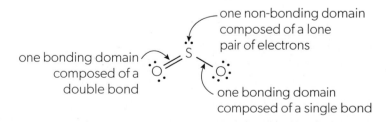

◀ Figure 13 The sulfur atom in sulfur dioxide has three electron domains

We will now explore geometries involving two, three and four electron domains using VSEPR. Predicting the VSEPR shape of a molecule involves two steps:

1. Count the number of electron domains around a central atom to deduce the electron domain geometry.

2. Determine how many of these are **bonding domains** and how many are **non-bonding domains**.

▲ **Figure 14** Linear geometry has two electron domains at a 180° angle to each other

Two domains: Linear geometry

If there are two electron domains, the electron pairs in those domains repel each other. They therefore adopt positions at 180° (figure 14). The angle between bonding pairs in a molecule is known as the **bond angle**. This **electron domain geometry** is called **linear** to illustrate that the central atom and both domains are on a straight line.

Molecules with this electron domain geometry also have a linear **molecular geometry**. Examples include beryllium chloride, $BeCl_2$, carbon dioxide, CO_2, and ethyne, C_2H_2. These are shown in figure 15.

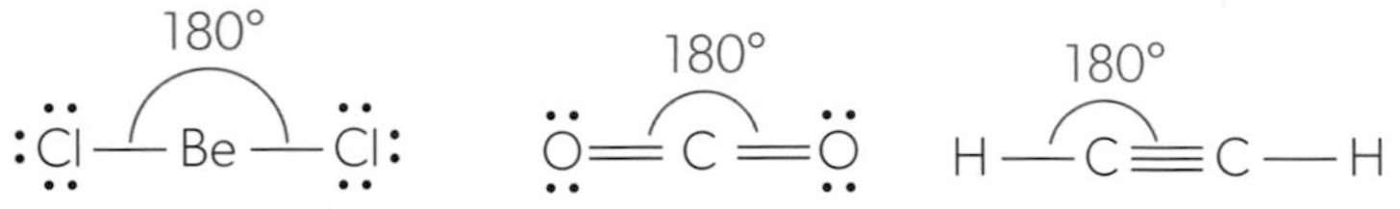

▲ **Figure 15** Examples of linear molecules, in which the central atom has two electron domains. Note that double and triple bonds count as one domain

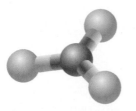

▲ **Figure 16** Trigonal planar geometry has three electron domains at 120° angles to each other

Three domains: Trigonal planar geometry

If there are three bonding domains, the electron pairs adopt positions at 120° from each other. This electron domain geometry is called **trigonal planar**. Trigonal, because the domains form a triangle, and planar, because the atoms lie flat on a plane.

A trigonal planar electron domain geometry can have two possible molecular geometries, depending on the presence of non-bonding domains:

- When all three domains are bonding domains, the molecule has **trigonal planar** geometry.

- When only two of the three domains are bonding domains, and one is a non-bonding domain, the molecule has **bent** (or **V-shaped**) geometry.

Examples of each type are shown in figure 17.

(a) :Cl:
        | 120°
    :Cl—B—Cl:

(b) N
      O⁻   <120°   O

▲ **Figure 17** A trigonal planar electron domain geometry can lead to (a) trigonal planar molecular geometry if all three domains are bonding domains, or (b) bent (V-shaped) molecular geometry if one of the domains is a lone pair of electrons

The bond angle in NO_2^- is smaller than the predicted 120°. This is because the non-bonding pair (or lone pair) exerts a stronger repulsion than the bonding domains, and therefore takes up more space.

Four domains: Tetrahedral geometry

If there are four bonding domains, the electron pairs adopt positions at 109.5° from each other. This electron domain geometry is called **tetrahedral** because the ends of the domains form the corners of a tetrahedron.

You might expect four electron domains to arrange themselves in a square configuration. However, this would result in angles of 90° between the domains. The domains arrange themselves tetrahedrally to maximize the bond angles and distances between them.

A tetrahedral electron domain geometry gives rise to three possible molecular geometries, depending on the presence of non-bonding domains:

- When all four domains are bonding domains, the molecule has **tetrahedral** geometry.

- When three of the four domains are bonding domains and one is a non-bonding domain, the molecule has **trigonal pyramidal** geometry.

- When only two of the four domains are bonding domains and two are non-bonding domains, the molecule has **bent** (or **V-shaped**) geometry.

Examples of each type are shown in figure 20.

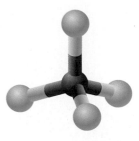

▲ Figure 18 Tetrahedral geometry has four electron domains at 109.5° angles to each other

▲ Figure 19 Tetrahedral food packages are convenient because they can be easily filled, stacked and packed. Their invention revolutionized the food packaging industry in the mid-20th century

tetrahedral   trigonal pyramidal   bent or V-shaped

◀ Figure 20 A tetrahedral electron domain geometry leads to tetrahedral, trigonal pyramidal or bent (V-shaped) molecular geometries

Non-bonding pairs occupy more space than bonding domains, which leads to decreased bond angles. Methane, CH_4, has no non-bonding pairs, and therefore the bond angle corresponds to the predicted 109.5°. Ammonia, NH_3, has one non-bonding pair, and therefore has a smaller bond angle (107°). Water, H_2O, has two non-bonding pairs and therefore has an even smaller bond angle of 104.5°.

The angles and geometries are shown in figure 21. Note that some bonds are represented using **wedges** (▬►) and **dashes** (llllll···) to better convey their 3D shape. Wedges represent bonds that are coming out of the plane of the page at an angle and dashes represent bonds that are going into the plane of the page.

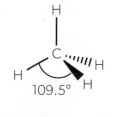
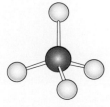

◀ Figure 21 In methane, CH_4, ammonia, NH_3, and water, H_2O, the bond angles are all slightly different due to the stronger repulsion exerted by the lone pairs

Multiple bonds

Multiple bonds (double and triple bonds) count as one domain. A double bond represents one domain, but it contains two electron pairs. Similarly, a triple bond is one domain composed of three electron pairs. Since multiple bonds contain more than one pair of electrons, they exert a greater repulsion than single bonds. The increased repulsion causes the bond angles in the molecule to deviate from predicted values.

For example, each carbon atom in ethene has three bonding electron domains and hence its molecular geometry is trigonal planar. This would suggest 120° bond angles. However, the H–C–H bond angle is 117° and the H–C=C bond angle is 121.5°, as the greater repulsion exerted by the double C=C bond pushes away both C–H bonding domains. This is shown in figure 22.

Table 2 shows a summary of the geometry of molecules with two, three and four electron domains.

▲ Figure 22 Multiple bonds generally exert a greater repulsion than single bonds

Electron domain geometry	Number of domains	Number of bonding domains	Number of non-bonding domains	Molecular geometry	Bond angle	Examples
linear	2	2	0	linear	180°	$BeCl_2$ CO_2 HCN
trigonal planar	3	3	0	trigonal planar	120°	BF_3, NO_3^-, SO_3
		2	1	bent (or V-shaped)	<120°	O_3, SO_2, NO_2^-
tetrahedral	4	4	0	tetrahedral	109.5°	CH_4, NH_4^+, ClO_4^-
		3	1	trigonal pyramidal	<109.5°	NH_3, H_3O^+, PBr_3
		2	2	bent (or V-shaped)	<109.5°	H_2O, NH_2^-, SF_2

▲ Table 2 Summary of the geometries for two, three and four electron domains

(ATL) Thinking skills

You can model electron domains using balloons. If you hold several balloons at a central point by their knots, they will arrange themselves as far away from each other as possible. The central point where the balloons meet represents the central atom in the molecule. How else might you model electron domains?

VSEPR

Unlike Lewis formulas, molecular models represent the 3D geometry of molecules. You will explore the models in a VSEPR database to develop your understanding of molecular geometry.

Relevant skills

- Tool 1: Digital molecular models
- Tool 2: Identify and extract data from a database

Instructions

1. Write a list of the molecular geometries that you need to know.

2. Access a VSEPR database (such as those suggested in the *Tools for chemistry* chapter) and extract two examples of species belonging to each molecular geometry. If possible, rotate the images to get a sense of their three-dimensional shape.

3. For each, write down the following:

   - electron domain geometry
   - molecular geometry
   - molecular formula
   - bond angle data (if available)
   - 3D sketch of the molecule, including bonding and non-bonding domains.

4. Draw the Lewis formula of each. Then deduce the number of bonding domains and non-bonding domains, and whether the molecule is polar or non-polar.

5. Organize your data into a suitable table.

6. If bond angle data is available, search for examples that illustrate:

   - the effect of non-bonding domains on the bond angle
   - the effect of multiple bonds (e.g. double or triple bonds) on bond angle.

Practice questions

6. Identify the electron domain geometry and molecular geometry given the following numbers of bonding and non-bonding domains.

   a. four bonding domains
   b. two bonding domains and one non-bonding domain
   c. two bonding domains only
   d. two bonding domains and two non-bonding domains
   e. three bonding domains and one non-bonding domain.

7. Identify the number of bonding and non-bonding domains around atoms with the following molecular geometries.

   a. linear
   b. trigonal pyramidal
   c. tetrahedral
   d. trigonal planar
   e. bent (there are two possible answers here).

8. Deduce the electron domain geometry and molecular geometry of each of the following:

   a. NF_3
   b. CH_2Cl_2
   c. BF_3
   d. BeH_2
   e. HCN
   f. SO_3
   g. SF_2
   h. NO_2^-

9. Predict the bond angles in each of the species in question 8.

10. Methanal is an organic compound with the molecular formula CH_2O.

    a. Draw the Lewis formula for methanal.
    b. Deduce the molecular geometry of methanal.
    c. Suggest values for the bond angles in methanal. Explain your reasoning.

Linking question

How useful is the VSEPR model at predicting molecular geometry?

Bond polarity (*Structure 2.2.5*)

Polarity is related to the way electrons are distributed within bonds and molecules. The shared pair of electrons in a covalent bond is not necessarily shared equally between the two atoms in the bond. **Bond polarity** results from the difference in the electronegativities of the bonded atoms.

In the case of identical atoms, such as the two fluorine atoms in F_2, there is an equal sharing of the electrons in the bonding pair. This is not the case, however, in HF. The fluorine atom has a much greater electronegativity than the hydrogen atom, so it pulls the shared electron pair more strongly than hydrogen does. This leads to what we describe as a **polar covalent bond**, with one atom (fluorine) adopting a partial negative charge, $\delta-$, and the other atom (hydrogen) adopting a partial positive charge, $\delta+$.

The separation of charge between two non-identical bonded atoms is called a **dipole moment**. The dipole moment can be represented by a vector (figure 23). This vector is shown as an arrow along the bond. The arrow starts as a plus sign at the less electronegative atom in the bond, and points towards the more electronegative atom in the bond.

The shared pair of electrons is shifted towards the more electronegative atom in the covalent bond. The greater the difference in electronegativity between the two atoms, the greater this shift.

If the two atoms involved in the formation of the covalent bond are identical, the bond is said to be a **pure covalent bond.** The charge is distributed evenly about the covalent bond, meaning that it is non-polar and has no dipole moment.

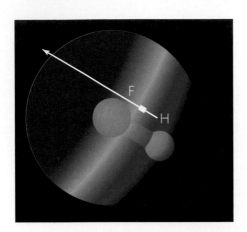

▲ **Figure 23** Representation of the molecular electrostatic potential of the polar covalent bond in HF. Note that the bond dipole in HF can be represented as a vector (arrow)

As seen in *Structure 2.1*, if the difference of electronegativity between the two atoms is very large (greater than 1.8), then the bonding pair of electrons is shifted almost entirely towards the more electronegative atom. Such bonds are considered ionic because electrons are transferred, not shared.

Practice questions

11. Rank the following bonds in order of increasing polarity:

C–H, C–C, C–F, C–O

12. Explain why Br_2 is non-polar, whereas HBr is polar.

 Linking question

What properties of ionic compounds might be expected in compounds with polar covalent bonding? (Structure 2.1)

Molecular polarity (*Structure 2.2.6*)

Molecular polarity is similar to bond polarity, but it describes the electron distribution throughout the whole molecule. A molecule is polar when the electron distribution leads to a partial negative charge on one end of the molecule, and a partial positive charge on the other. This results in a **dipole moment, μ.** The value of the dipole moment is often reported in the unit debye, D. Polarity is an important trait of molecules because it gives rise to many characteristic properties such as volatility, solubility and boiling point.

Molecular polarity depends on both bond polarity and molecular geometry. Some examples are given in table 3.

Molecule	Lewis formula	Polarity	Dipole moment / D
water, H_2O	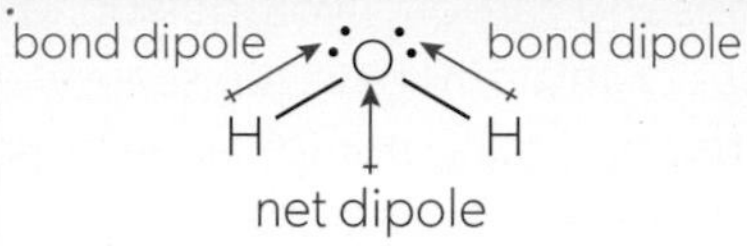	polar	1.85
trichloromethane, $CHCl_3$	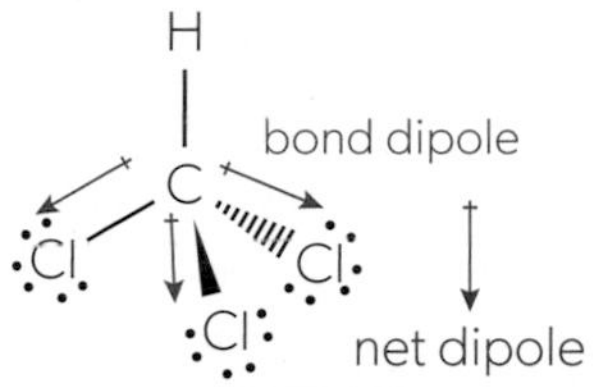	polar	1.01
boron trifluoride, BF_3	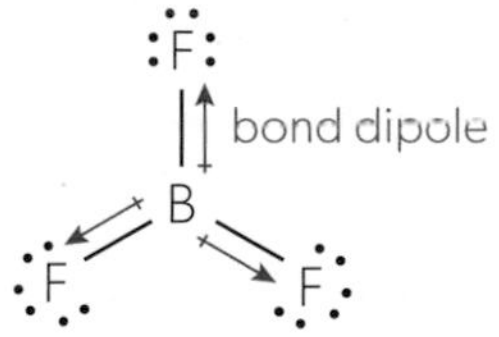	non-polar	0
carbon dioxide, CO_2	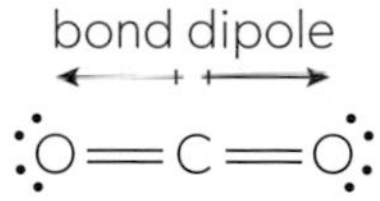	non-polar	0
ethane, C_2H_6	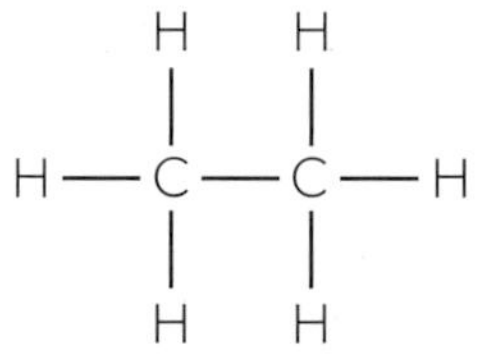	non-polar	0

▲ Table 3 Examples of polar and non-polar molecules

Molecules are polar when their bond dipoles do not cancel each other out. This may occur due to the geometry of the molecule, or because the bonds have different polarities. Both water and trichloromethane (table 3) are polar because they contain polar bonds that do not cancel out. This results in a net dipole across the molecule.

Molecules are non-polar when their bond dipoles cancel each other out. This happens when the bond dipoles are equal and their arrangement results in no net dipole. Examples include boron trifluoride and carbon dioxide. Both molecules contain polar bonds because of the difference in electronegativity of the atoms involved. These polar bonds are positioned in such a way that they cancel each other out. This means that boron trifluoride and carbon dioxide are both non-polar, despite the presence of polar bonds.

Molecules are non-polar when all their bonds are non-polar. Hydrocarbons (compounds of hydrogen and carbon) such as ethane contain two types of bonds: carbon–carbon bonds and carbon–hydrogen bonds. The carbon–carbon bonds are non-polar because both atoms have the same electronegativity. The carbon–hydrogen bonds are virtually non-polar because of the small electronegativity difference between the two atoms. In addition, the tetrahedral geometry around each carbon atom results in a net dipole of zero, or very close to zero. Therefore, hydrocarbons are generally non-polar.

So far, we have only discussed small molecules. Long molecules can have a polar region and a non-polar region, leading to many interesting applications including emulsifiers, soaps and detergents, as well as the phospholipid bilayer in cell membranes. Figure 24 shows a typical detergent molecule. Can you identify the polar and non-polar regions in the molecule?

Practice questions

13. Determine whether each of the following molecules is polar or non-polar:
    a methane, CH_4
    b ammonia, NH_3
    c hydrogen cyanide, HCN
    d hexane, C_6H_{14}
14. Explain why CO_2 is non-polar.
15. State and explain whether PH_3 is polar or non-polar.

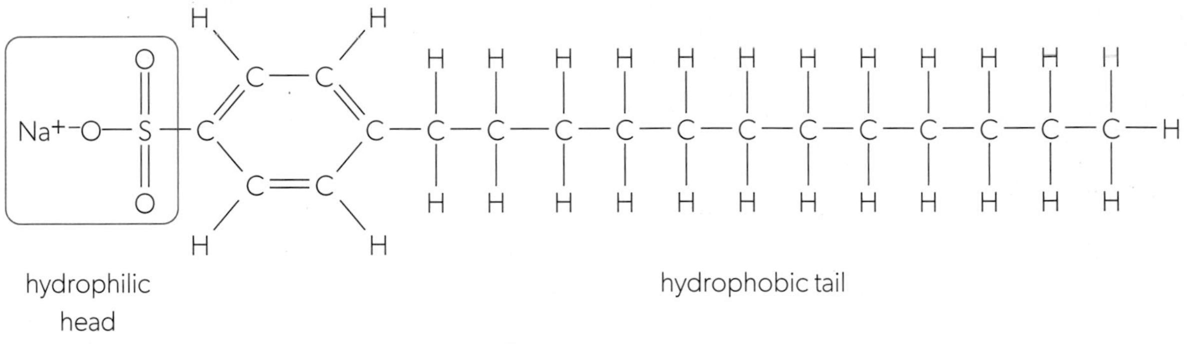

hydrophilic head

hydrophobic tail

▲ Figure 24 A diagram of a synthetic detergent molecule, showing the hydrophobic and hydrophilic regions

Linking question

AHL

What features of a molecule make it "infrared (IR) active"? (Structure 3.2)

Covalent network structures (*Structure 2.2.7*)

Covalent bonds give rise to two different types of structure: **molecular** and **covalent network** (also known as **giant covalent**). The atoms in both cases are joined by covalent bonds, but the two structures are very different in terms of their properties. Molecular substances consist of discrete groups of covalently bonded atoms called molecules. By contrast, covalent network structures contain atoms that are held together by covalent bonds in a continuous three-dimensional lattice. Examples of covalent network structures include silicon, silicon dioxide, and most of the allotropes of carbon, such as diamond and graphite.

Allotropes of carbon

Some elements have different structural forms known as **allotropes**. Carbon is one such element, and its allotropes include diamond, graphite, graphene, and a group of substances known as fullerenes. They are all composed of carbon atoms but have different chemical and physical properties due to their different structural arrangements.

Chemists talk about two types of stability: kinetic (*Reactivity 2.2*) and thermodynamic (*Reactivity 1*), and diamond illustrates the difference well. At high pressures, such as those occurring deep inside the Earth's crust, diamond is the most stable allotrope. At standard temperature and pressure, graphite is the most thermodynamically stable allotrope of carbon. Under these conditions, the transformation of diamond into graphite is thermodynamically spontaneous, but is so slow that diamonds effectively last forever. Diamond is kinetically stable but thermodynamically unstable (we describe it as being metastable). You can read more about spontancity in *Reactivity 1.4*.

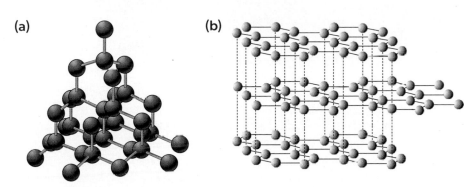

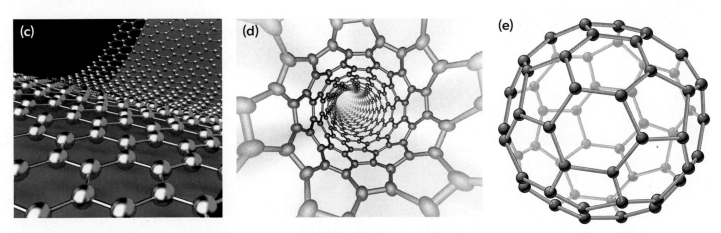

▲ Figure 25 The allotropes of carbon: (a) diamond, (b) graphite, (c) graphene, (d) carbon nanotubes and (e) buckminsterfullerene, C_{60}

In **diamond**, each carbon atom is bonded to four other atoms in a tetrahedral arrangement. It is one of the hardest substances known. For this reason, it is often used in heavy-duty cutting tools such as saws, polishing tools and dental drills. Diamond's high refractive index and durability mean that it is also used to make jewelry. Diamond is a poor electrical conductor because it has no mobile charged particles—the electrons are all localized in the bonds. It is however an excellent thermal conductor: vibrational energy carriers called phonons travel well through the highly regular lattice and strong covalent bonds.

▶ Figure 26 The Big Hole: a former diamond mine in Kimberley, South Africa

Graphite is composed of layers of sheets made of carbon atoms. Each carbon atom is bonded to three other carbon atoms in a hexagonal arrangement where the geometry around each carbon atom is trigonal planar. The carbon atoms in graphite are bonded such that one electron per carbon atom is delocalized. These delocalized electrons are free to move in the planes above and below each sheet and therefore graphite is a good electrical conductor. While the covalent bonds between the carbon atoms within the sheets are strong, the forces of attraction (called **London (dispersion) forces**) between the sheets are weak. This means that the sheets can be separated easily, making graphite a good lubricant, as well as an ideal material for pencil leads. When you use a pencil to write, the force applied causes parts of the graphite sheets to come off, leaving a mark on the paper.

▲ Figure 27 A microscope image of a dentist's drill showing the diamond chips

Graphene is essentially a single sheet of graphite. Graphene is thus one-atom thick and is therefore said to be two-dimensional. Like graphite, it is an excellent electrical conductor. It is also flexible, lightweight, transparent and, at the same time, extraordinarily strong. First isolated in 2004, graphene is a promising new material, with a vast range of potential applications, from desalination technology to bendable electronic displays.

Fullerenes are a group of carbon allotropes with atoms arranged in interlinking hexagonal and pentagonal rings. Some fullerenes form long hollow cylinders: these are known as **carbon nanotubes**. Due to their size, carbon nanotubes are utilized in **nanotechnology**. Nanotechnology involves the use of atoms, molecules and objects with dimensions of less than 100 nm (about 1000 atoms or less across).

Carbon nanotubes are used to reinforce composite materials. Their exceptional strength is due to the strong covalent bonds holding the carbon atoms together. Carbon nanotubes have potential applications in electronics because, like graphite and graphene, the presence of delocalized electrons means that they are good electrical conductors.

Buckminsterfullerenes, or **buckyballs**, have a covalent molecular structure. With the formula C_{60}, the atoms in buckminsterfullerene are arranged in hexagons and pentagons to suggest a very familiar shape: a football. Along with other spherical fullerenes, buckyballs could have exciting new roles in medicine as drug carriers. Because of its molecular structure, C_{60} has a low boiling point: overcoming the weak intermolecular forces of attraction does not require much thermal energy.

Silicon and silicon dioxide

Silicon forms a three-dimensional lattice where each silicon atom is bonded to four other silicon atoms in a tetrahedral arrangement. This arrangement is similar to that of the carbon atoms in diamond.

The extensive covalent bonds in the lattice result in high strength, as well as high melting point and boiling point. Table 4 contrasts some of the features of carbon and silicon. The Si–Si bond is weaker than its C–C counterpart as silicon has a larger atomic radius. Therefore, the Si–Si bond is more reactive than the C–C bond. This difference in strength also helps to explain the different melting and boiling points.

	C (graphite)	Si
Atomic radius / 10^{-12} m	75	114
X–X bond enthalpy / kJ mol^{-1}	346	226
Melting point / °C	3500	1414
Boiling point / °C	4827	3265
First ionization energy / kJ mol^{-1}	1086	787
X–O bond enthalpy	358	466

▲ Table 4 A selection of features of carbon and silicon, where X = Si or C

Interestingly, despite belonging to the same periodic table group, the properties of carbon and silicon are not as similar as one might anticipate. For example, carbon is a non metal, while silicon is a metalloid. Diamond is a poor electrical conductor, while silicon is a semiconductor. Double and triple bonds are common between carbon atoms but unusual between silicon atoms.

Comparing the Si–O and C–O bond strengths (table 4) shows that single bonds between Si and O are very strong. Silicon is one of the most abundant elements in the Earth's crust and much of it is found in species containing Si–O bonds. **Silica**, or **silicon dioxide**, is a mineral that forms a covalent network containing Si and O in a 1:2 ratio, hence the formula SiO_2. Each silicon atom is bonded covalently to four oxygen atoms and each oxygen atom is bonded covalently to two silicon atoms. The crystalline form of silicon dioxide is quartz (figure 28). Other forms of silica include sand and glass.

The properties of silicon dioxide are those we associate with sand: it is hard, insoluble in water and has a high melting point. In its solid crystalline form, quartz, it is a poor conductor of heat and electricity.

Silicon and carbon are found in group 14 hence they have some similar properties. Other patterns in the periodic table are explored in *Structure 3.1*.

▲ Figure 28 Structure (left) and crystals of silicon dioxide, SiO_2, known as quartz. The photograph also contains crystals of iron disulfide, FeS_2, commonly known as "fool's gold"

 Linking question

Why are silicon–silicon bonds generally weaker than carbon–carbon bonds? (Structure 3.1)

Practice questions

16. Explain why graphite and graphene conduct electricity, but diamond does not.
17. Explain why diamond and silicon dioxide have high melting and boiling points.

Activity

Summarize the properties of some of the covalent substances using a table like the one below.

	Diamond	Graphite	Graphene	C_{60} fullerene	Silicon dioxide
Element(s)					
Arrangement of atoms					
Electrical conductivity					

▲ Figure 29 Intermolecular forces hold the molecules together in water and inside the ice cubes

Intermolecular forces (*Structure 2.2.8*)

So far, we have been discussing forces of attraction that hold atoms and ions together, which are collectively referred to as **bonding**. We will now turn our attention to the forces of attraction between molecules, known as **intermolecular forces.** As their name suggests, intermolecular forces apply to molecular substances. The type and strength of intermolecular forces depend on the size and polarity of the molecules. Types of intermolecular forces include **London (dispersion) forces, dipole–induced dipole forces, dipole–dipole forces** and **hydrogen bonding**. Collectively the first three intermolecular forces are termed **van der Waals forces**.

Consider a glass of water. You know that the water molecules contain two hydrogen atoms each, bonded to a central oxygen atom through strong covalent bonds. But what forces cause the water molecules to stay close to one another? How about the water molecules inside ice cubes? What exactly is being overcome when some of the water molecules in ice break away from the cube and join the liquid water surrounding it? The answer in each case is intermolecular forces. Intermolecular forces are weaker than chemical bonds, but they affect physical properties such as volatility, solubility and boiling point of molecular substances.

Intermolecular forces are weak electrostatic forces of attraction that occur between molecules. Figure 30 highlights the difference between intermolecular and **intramolecular** forces.

▶ Figure 30 Intramolecular forces, such as covalent bonds, occur within molecules. Intermolecular forces occur between molecules

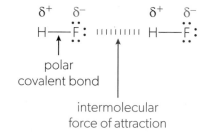

When molecular substances melt, boil or sublime, the intermolecular forces between the molecules are overcome. Therefore, melting and boiling points are often used as indicators of intermolecular force strength. The covalent bonds in the molecule do not break during phase changes (figure 31).

In **Structure 1.5,** we discussed the ideal gas law, which carries the assumption that no intermolecular forces are present in the gas. This is generally true at low pressures and high temperatures, for which the ideal gas equation can frequently be used. When a gas deviates from ideal behaviour, we must correct for the presence of intermolecular forces using a **real gas** model. Real gas models also take into account the actual volume of gas particles, which is considered negligible in the ideal gas model.

weak intermolecular interactions *between I₂ molecules* break on changing state

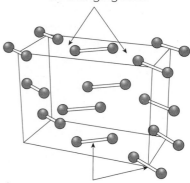

strong covalent bonds *between atoms* in I₂ molecule *do not* break on changing state

▲ Figure 31 Intermolecular forces of attraction are overcome when a molecular substance changes state, but the covalent bonds remain intact

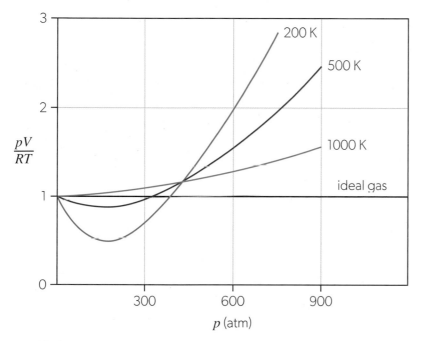

▲ Figure 32 Gases deviate more from ideal gas behaviour at lower temperatures. The horizontal black line represents an ideal gas. The curves show the deviations from ideal gas behaviour at different temperatures for a real gas

London (dispersion) forces (LDFs)

All molecules experience London (dispersion) forces (or LDFs), which are intermolecular forces resulting from temporary **instantaneous dipoles**. A molecule has a dipole when one side carries a partial positive charge, and the other a partial negative charge. London (dispersion) forces involve induced dipoles, which means that one molecule causes (or **induces**) a temporary dipole in another molecule. These induced dipoles occur due to the random movement of electrons around the molecule.

▲ Figure 33 Intermolecular forces of attraction are overcome when iodine sublimes, and are formed when iodine is deposited on a surface

In a simple non-polar molecule such as hydrogen, H_2, the electron distribution is on average symmetrical (figure 34). However, electrons are constantly moving around within the molecule. If we could somehow freeze time and take a photo of electrons in a molecule, the electron distribution would very unlikely be perfectly symmetrical. Instead, we would see a somewhat unequal electron density, with one region of the molecule having more electrons, rendering it slightly negative ($\delta-$). There would also be a region of lower electron density elsewhere in the molecule with a positive partial charge ($\delta+$).

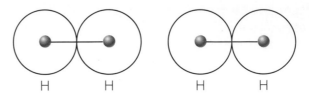

▲ Figure 34 On average, the electron distribution in H_2 is symmetrical

This temporary dipole then induces further temporary dipoles in the molecules around it. Due to electrostatic repulsion, a region of partial negative charge ($\delta-$) on one molecule will repel the electrons in neighbouring molecules. This temporarily creates regions of partial positive charge ($\delta+$). The resulting electrostatic attraction between the $\delta-$ region on one molecule and the $\delta+$ region on the next is termed a London dispersion force. However, this arrangement is temporary. In the next moment of time a different pattern of induced dipoles will emerge.

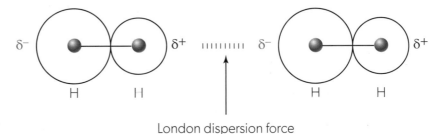

London dispersion force

▲ Figure 35 At any given moment in time, a temporary dipole is present in a molecule, inducing a dipole in another. A weak London dispersion force forms between the opposite partial charges

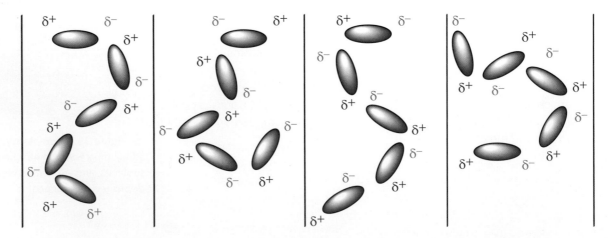

▲ Figure 36 Different arrangements of LDFs between molecules, which result from interactions between an instantaneous dipole on one molecule and an induced dipole on an adjacent molecule

There are two main factors that increase the strength of LDFs: the number of electrons and the molecular shape. They affect the **polarizability** of the molecule: how easily the electron distribution is distorted by an electric field. The greater the polarizability, the stronger the dispersion forces.

Consider the group 17 elements, all of which form non-polar diatomic molecules. The number of electrons increases going down the group. As a result, the electron clouds become larger and less attracted to the nuclei. This means that the electrons become more easily polarized and the strength of LDFs increases, leading to higher boiling points as you descend the group (table 5).

Substance	Boiling point / °C
fluorine, F_2	−188.1
chlorine, Cl_2	−34.04
bromine, Br_2	58.78
iodine, I_2	184.4

Increasing LDF strength

▲ Table 5 Boiling point increases down group 17 due to an increasing number of electrons, resulting in stronger LDFs

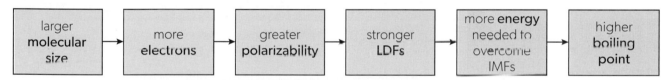

▲ Figure 37 The effect of molecular size on LDF strength and boiling point

The boiling point also increases for successive members of a **homologous series** of organic compounds: each molecule has one more CH_2 than the one before it, hence greater molecular size and a larger number of electrons. Molecular size can be quantified in terms of mass, so we often refer to molecular mass when comparing molecules in terms of their LDFs.

Homologous series and isomers of organic compounds are discussed in *Structure 3.2*.

Patterns and trends

Scientists look for patterns and try to understand relationships between variables. For example, we often say LDF strength increases with increasing molecular mass, due to an increasing number of electrons. This may seem contradictory because electrons have negligible mass. But here the relationship between molecular mass and the number of electrons is not **causal**. Rather, the greater mass of larger molecules is *accompanied by* a greater number of electrons. The greater number of protons and neutrons is responsible for the increase in molecular mass. However, each proton requires an electron to maintain the overall neutrality of the molecule. Therefore, any increase in the number of protons always leads to a proportional increase in the number of electrons.

Compare the boiling points of the two isomers of C_5H_{12}: pentane, $CH_3CH_2CH_2CH_2CH_3$ and 2,2-dimethylpropane $(CH_3)_4C$ (table 6 and figure 38).

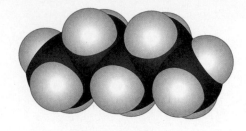 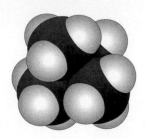

▶ Figure 38 Space-filling models show that pentane (top) has a greater area of contact for LDFs than 2,2-dimethylpropane (bottom)

Isomer	Boiling point / °C
pentane	36.1
2,2-dimethylpropane	9.5

▲ Table 6 Boiling points of the two isomers of pentane

Both isomers contain the same number of electrons and are non-polar. The difference is in their shapes. Pentane molecules are long and are therefore able to interact with each other across the full length of the molecule; that is, there is a large area of interaction because of the better contact between molecules of pentane. As a result, pentane has a relatively high boiling point and is a liquid at room temperature. Molecules of 2,2-dimethylpropane are rounder and more compact, so they pack together less efficiently. As the interaction between the molecules is limited, the LDFs in 2,2-dimethylpropane are weaker, so it is a gas at room temperature.

Data-based questions

1. Boiling point data for the first four alkanes are shown below. Describe and explain the boiling point trend you observe.

Alkane	Boiling point / °C
methane, CH_4	−161
ethane, C_2H_6	−88
propane, C_3H_8	−42.1
butane, C_4H_{10}	−0.5

2. The boiling points of 2,2-dimethylpropane and pentane are 9.5 °C and 36.1 °C, respectively. Predict a value for the boiling point of 2-methylbutane, $CH_3CH_2CH(CH_3)_2$. Explain your answer.

LDFs occur between all types of molecules because all molecules contain electrons. If we have a substance made up of only non-polar molecules, only LDFs will occur.

Dipole–induced dipole forces

LDFs are forces of attraction between temporary, or instantaneous, dipoles. **Dipole–induced dipole forces** are a type of related intermolecular force occurring between a polar molecule and a nearby non-polar molecule. The presence of a permanent dipole in the polar molecule induces the formation of a temporary dipole in the neighbouring non-polar molecule. For example, this type of intermolecular force attracts non-polar oxygen molecules, O_2, to polar water molecules. Dipole–induced dipole forces are weak, which explains why the aqueous solubility of oxygen is relatively low.

Activity

Compare and contrast London (dispersion) forces and dipole-induced dipole forces.

Dipole–dipole forces

Dipole–dipole forces involve permanent dipoles, whereas LDFs result from temporary dipoles. When a molecule is polar, it has a permanent dipole and therefore experiences dipole–dipole forces of attraction with neighbouring polar molecules (figure 39).

Hydrogen chloride, HCl, and diatomic fluorine, F_2, have similar sizes and comparable molecular masses (table 7). They both therefore experience LDFs of a similar strength. However, HCl molecules are polar and therefore experience dipole–dipole forces in addition to LDFs. The intermolecular forces between HCl molecules are stronger than those between F_2 molecules, so HCl has a higher boiling point.

$$\overset{\delta+}{H}\!-\!\overset{\delta-}{Cl} \longleftrightarrow \overset{\delta+}{H}\!-\!\overset{\delta-}{Cl}$$

▲ Figure 39 Hydrogen chloride, HCl, molecules are polar and attract each other due to dipole–dipole forces

Molecule	Molecular mass	Boiling point / °C	Intermolecular forces
F_2	38.00	−188.1	London (dispersion)
HCl	36.46	−85.05	London (dispersion) and dipole–dipole

▲ Table 7 Diatomic fluorine and hydrogen chloride have similar molecular masses but different boiling points

Practice questions

18. Which of the following species experience dipole–dipole forces?
    A. oxygen, O_2
    B. carbon dioxide, CO_2
    C. carbon monoxide, CO
    D. carbonate ion, CO_3^{2-}
19. The molecular masses of ICl and Br_2 are very similar, while their boiling points are 97.4 °C and 58.8 °C, respectively. Explain this difference.
20. Draw a diagram to show how propanone molecules, CH_3COCH_3, might arrange themselves when near one another. Label any dipole–dipole forces that occur.

Hydrogen bonding

Hydrogen bonds are strong intermolecular forces that form when a molecule contains a strong dipole involving hydrogen. When a hydrogen atom is covalently bonded to a highly electronegative atom such as oxygen, nitrogen or fluorine, the bond between them is very polar. The electrons in the covalent bond are drawn towards the more electronegative atom, resulting in a considerable partial positive charge ($\delta+$) on the hydrogen. This hydrogen atom can then form a strong electrostatic interaction—a hydrogen bond—with the electrons of another electronegative atom. This is usually found on a different molecule, but intramolecular hydrogen bonds can also exist in certain situations.

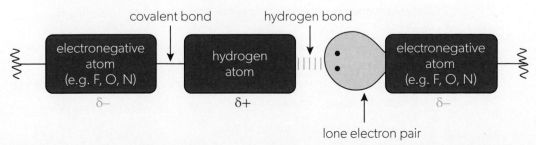

◀ Figure 40 Hydrogen bonds form between the hydrogen in a very polar bond and electrons on a highly electronegative atom

Hydrogen bonds occur, for example, between:

1. water molecules

2. ammonia (NH_3) molecules

3. hydrogen fluoride (HF) molecules

4. water molecules and dimethyl ether molecules, $(CH_3)_2O$

These hydrogen bonds are depicted with blue dashes in figure 41.

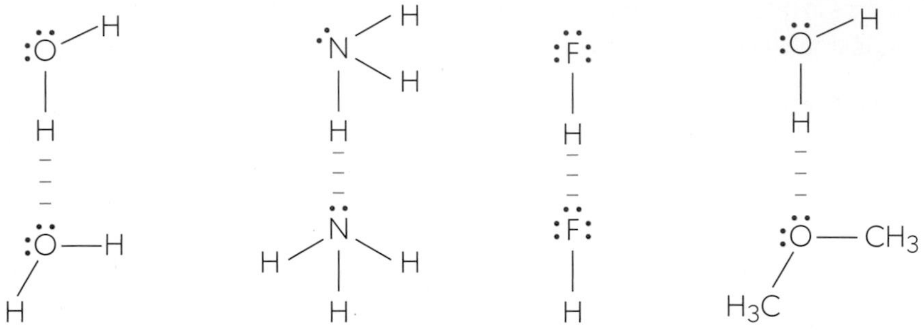

▶ Figure 41 Hydrogen bonds can form between a variety of different molecules

Despite the name suggesting otherwise, hydrogen bonds are not chemical bonds: they are intermolecular forces. They are generally stronger than other types of intermolecular force, but much weaker than covalent bonds.

Bond or intermolecular force type	Typical bond or intermolecular force enthalpy / kJ mol⁻¹
Hydrogen bond	20–40
Single covalent bonds	150–600

▶ Table 8 Enthalpies of hydrogen bonds and single covalent bonds compared

A good way to remember that hydrogen bonds are weaker than chemical bonds is to look at what happens when water boils. At 100 °C, the energy available is sufficient to overcome the hydrogen bonds between water molecules (~20 kJ mol⁻¹), but not the covalent bonds between oxygen and hydrogen atoms in the water molecules themselves (463 kJ mol⁻¹).

Practice questions

21. Which **two** of the following substances can form hydrogen bonds between their molecules?

    A. hydrogen fluoride, HF

    B. fluorine, F_2

    C. methanol, CH_3OH

    D. fluoromethane, CH_3F

22. Draw a diagram to show a hydrogen bond between two methanol molecules, CH_3OH.

23. Draw a diagram to show a hydrogen bond between a water molecule and a methanamine molecule, CH_3NH_2.

24. Draw a diagram to show a hydrogen bond between a water molecule and a methanal molecule, CH_2O.

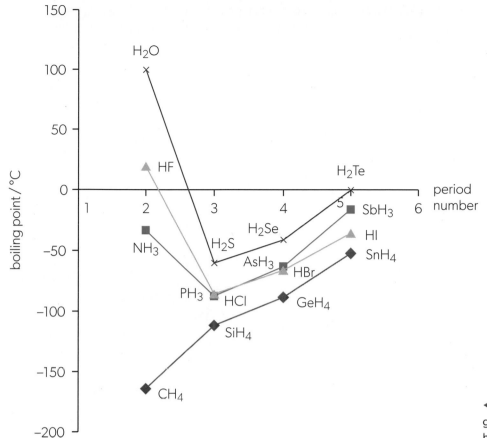

How do the boiling points of the group 14 hydrides (CH_4, SiH_4, GeH_4 and SnH_4) change as you go down the group? The molecular mass increases down the group, increasing the number of electrons and polarizability of the electron cloud. Descending the group, there are stronger LDFs, and therefore an increasing boiling point is expected. This is precisely what the data show in the bottom curve of figure 42.

Now look at the trend of the group 15 hydrides, NH_3, PH_3, AsH_3 and SbH_3, in the next curve up. We see the expected rise in boiling point from phosphine, PH_3, to stibane, SbH_3, due to increasing molecular mass. Extrapolating the trend to ammonia predicts a boiling point of around –130 °C, when in fact it is 100 °C higher, approximately –30 °C. The higher-than-expected boiling point is due to the presence of hydrogen bonds between ammonia molecules.

The trends in the group 16 and group 17 hydrides are similar to that of group 15: the period 2 hydrides, water, H_2O, and hydrogen fluoride, HF, have very high boiling points. Again, this is due to the presence of hydrogen bonds between their molecules.

A single water molecule can form up to four hydrogen bonds (figure 43) and therefore the boiling point of water is very high: 100 °C.

▲ Figure 43 Water molecules can form up to four hydrogen bonds each

Hydrogen bonds give water its notable properties. Most substances have higher densities when solid than when liquid. Solid water (ice) floats on liquid water because it is less dense than liquid water. This unusual property leads to the formation of layers of ice on lake surfaces during the winter. Insulation from this ice layer prevents lakes freezing into solid blocks of ice, allowing aquatic ecosystems to survive.

The hydrogen bonding in ice results in the formation of a regular, very ordered open-cavity network of molecules (figure 44). In the liquid phase, the hydrogen bonding is more random, which results in a higher density compared to ice. Note that O–H covalent bonds in water are shorter than H - - - O hydrogen bonds between water molecules.

(a)

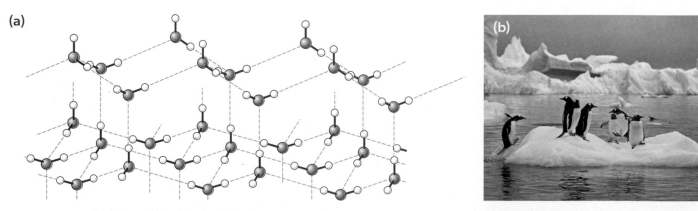

▲ Figure 44 (a) The open-cavity structure of ice causes it to be less dense than water (b) Ice floats on water because of this

▲ Figure 45 The hexagonal arrangement of water molecules in the ice lattice leads to the characteristic shape of snowflakes

▲ Figure 46 Extensive hydrogen bonding is responsible for the high surface tension of water, which allows small insects to walk on the surface without sinking or becoming even partly submerged

TOK

DNA (deoxyribonucleic acid) molecules store genetic information. A highly accurate DNA-copying mechanism is essential for this information to be passed along between generations. DNA has a double-helix structure composed of two strands of organic molecules held together by hydrogen bonds. DNA replication involves breaking these hydrogen bonds to "unzip" the two strands, allowing copies to be made.

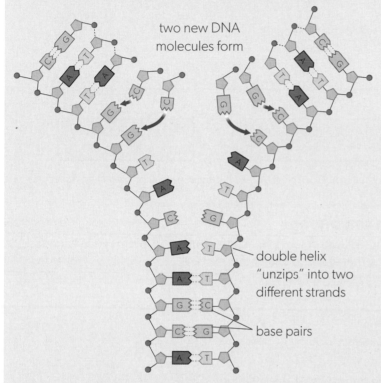

two new DNA molecules form

double helix "unzips" into two different strands

base pairs

▲ Figure 47 The DNA double helix is held together by hydrogen bonds, which are broken during replication. Hydrogen bonds are shown by black dashes

What other types of information storage systems exist in the natural world, in science and in other areas of knowledge?

In summary, intermolecular forces are electrostatic attractions that keep molecules together. We have discussed the four types of intermolecular forces, summarised in table 9.

	Type of intermolecular force	Where does it occur?	
van der Waals forces	London (dispersion) forces (LDFs)	between all molecules	increasing strength
	dipole–induced dipole forces	between a polar molecule and a non-polar molecule	
	dipole–dipole forces	between polar molecules	
	hydrogen bonding	between a highly electronegative atom (F, O or N) and a hydrogen covalently bonded to another highly electronegative atom	

▲ Table 9 Summary of types of intermolecular forces

Figure 48 shows the overall method for determining the strongest intermolecular force present in a substance.

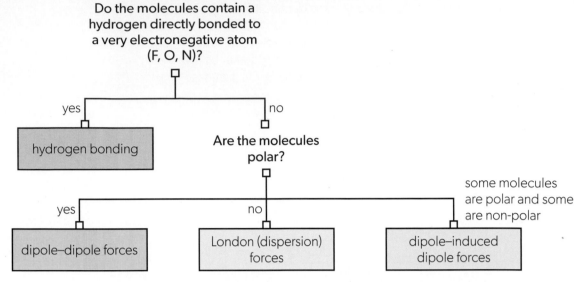

Do the molecules contain a hydrogen directly bonded to a very electronegative atom (F, O, N)?

yes → hydrogen bonding

no → **Are the molecules polar?**

yes → dipole–dipole forces

no → London (dispersion) forces

some molecules are polar and some are non-polar → dipole–induced dipole forces

▲ **Figure 48** Figuring out the strongest intermolecular force present between a pair of interacting molecules

Worked example 3

For each of the following substances, determine the intermolecular forces that attract the molecules to each other.

a. butane, $CH_3CH_2CH_2CH_3$
b. ethoxyethane, $CH_3CH_2OCH_2CH_3$
c. carbon tetrafluoride, CF_4
d. ethanamine, $CH_3CH_2NH_2$
e. fluoroethane, CH_3CH_2F
f. chloromethane, CH_3Cl, and methane, CH_4
g. water, H_2O, and hydrogen fluoride, HF

Solution

All the substances in this worked example form London (dispersion) forces.

a. Butane is non-polar, so it only forms London (dispersion) forces.
b. Ethoxyethane is polar, so it forms dipole–dipole forces.
c. Carbon tetrafluoride is non-polar, so it only forms LDFs.
d. Ethanamine can form hydrogen bonds because it contains a hydrogen atom directly bonded to nitrogen, which is a very electronegative atom. It is also polar, so it forms dipole–dipole forces.
e. Fluoroethane is polar, so it forms dipole–dipole forces. Note that it does not form hydrogen bonds. The structural formula might suggest that fluorine is directly bonded to hydrogen, but this is not the case. If you draw the Lewis formula, you will see that fluorine is bonded to a carbon atom.
f. Chloromethane is polar while methane is non-polar. They would therefore form dipole–induced dipole forces.
g. Hydrogen bonds are formed between water and hydrogen fluoride molecules because they both contain a highly electronegative atom that is directly bonded to a hydrogen. They are both polar, so they also experience dipole–dipole forces.

Practice questions

25. For each of the following substances, determine the intermolecular forces that attract molecules to each other:

    a.   propane, $CH_3CH_2CH_3$

    b.   hydrogen bromide, HBr

    c.   hydrogen fluoride, HF

    d.   buckminsterfullerene, C_{60}

    e.   ammonia, NH_3, and diatomic oxygen, O_2

    f.   ethanal, CH_3CHO

26. State and explain which of the following species can form hydrogen bonds with water molecules:

    a.   ammonia, NH_3

    b.   propane, $CH_3CH_2CH_3$

    c.   ethanoic acid, CH_3COOH

 Linking questions

To what extent can intermolecular forces explain the deviation of real gases from ideal behaviour? (Structure 1.5)

 How do the terms "bonds" and "forces" compare? (Structure 1.1, Structure 2.1, Structure 2.3)

 How can advances in technology lead to changes in scientific definitions, e.g. the updated International Union of Pure and Applied Chemistry (IUPAC) definition of the hydrogen bond?

The properties of covalent substances (*Structure 2.2.9*)

Relative strength of intermolecular forces

Given comparable molar mass, the relative strengths of intermolecular forces are generally:

> London (dispersion) forces < dipole–dipole forces < hydrogen bonding.

Table 10 shows typical enthalpy ranges associated with overcoming each of these forces. The trend in enthalpy reflects the trend in strength. The strength of dipole–induced dipole forces lies somewhere in-between London (dispersion) and dipole–dipole forces.

Intermolecular force type	Typical enthalpy / kJ mol⁻¹
London (dispersion) forces	1 to 5
sipole–dipole forces	3 to 25
hydrogen bonds	20 to 40

▲ Table 10 Comparison of the strength of various intermolecular forces

Data-based questions

The relative strengths of intermolecular forces are best illustrated by comparing molecules that have similar molecular masses. Table 11 shows molecules with molecular masses in the range of 44.0 to 46.1 and wildly different boiling points.

Name and molecular formula	Structural formula	Molecular mass	Molecular dipole moment / D	Boiling point / °C	Intermolecular forces		
					London (dispersion)	Dipole–dipole	Hydrogen bond
propane C_3H_8		44.11	0.084	−42.1	yes	no	no
methoxymethane C_2H_6O		46.08	1.30	−24.8	yes	yes	no
ethanal C_2H_4O		44.06	2.69	20.2	yes	yes	no
ethanol C_2H_6O		46.08	1.69	78.2	yes	yes	yes
methanoic acid CH_2O_2		46.03	1.41	100.7	yes	yes	yes

▲ **Table 11** Polarity, boiling point and intermolecular force data for organic compounds of similar molecular masses

Interpreting and explaining data are important skills in science. What do the data in table 11 suggest? For example, they show that all molecules experience London (dispersion) forces because they all contain electrons that move around, causing temporary dipoles that then induce further temporary dipoles in the surrounding molecules.

Write down four more statements describing the relationship between molecular structure and intermolecular forces that are shown by the data in table 11. Attempt to explain each statement using your knowledge from this topic.

Physical properties of covalent substances

There are two types of covalent structures: covalent network and molecular covalent. These two structures have very different properties. Most of the properties of molecular substances are governed by intermolecular forces. Substances with a covalent network structure are not comprised of molecules, so their properties are dictated by their lattice features.

We will look at physical properties of covalent substances including volatility, electrical conductivity and solubility.

Volatility

Substances with a covalent network structure are solids at room temperature and pressure. Vaporizing them requires a lot of energy because of the strong covalent bonds holding the structure together. They are therefore non-volatile and have very high melting and boiling points.

In order to vaporize molecular substances, the intermolecular forces between the molecules need to be overcome. Since intermolecular forces are relatively weak, the energy required to overcome them is low and therefore molecular substances are generally volatile.

Due to the variety of sizes and intermolecular forces, there is, however, a large variation in the volatility of molecular substances. Many of the molecular substances we have considered in this chapter consist of small molecules, for example, N_2, CO_2, H_2O and some hydrocarbons. Many of these are gases or liquids at room temperature.

However, substances that consist of much larger molecules have lower volatility and higher melting and boiling points. This is because larger molecules have stronger LDFs between each other. Examples include buckminsterfullerene (C_{60}), which sublimes at 500–600 °C, and a component of candle wax, tetracosane ($C_{24}H_{50}$), which boils at 391 °C.

Electrical conductivity

Covalent substances, both network and molecular, are usually not electrical conductors. To conduct electricity, a substance needs to contain charged particles that are free to move. Covalent substances generally do not contain such particles, as their electrons are "locked up" in localized covalent bonds, and they do not contain ions.

There are some notable exceptions, such as graphite, which is a good electrical conductor due to the presence of delocalized electrons. Silicon is a semiconductor, meaning that its intermediate conductivity places it between conductors and insulators.

Solar panels consist of photovoltaic cells, which convert solar energy into electricity. The cells contain semiconductors, such as silicon.

▲ Figure 49 Molecular covalent substances vaporize when the intermolecular forces are overcome

Practice questions

27. Identify which species in each pair has the highest boiling point.
    a. He or Kr
    b. CH_4 or NH_3
28. The boiling point of carbon tetrachloride, CCl_4, is 76.7 °C. The boiling point of fluoromethane, CH_3F, is −78.2 °C. Comment on this difference using your knowledge of intermolecular forces.

▲ Figure 50 Solar panels containing silicon provide this satellite with electrical power

Solubility

When a substance dissolves, forces of attraction are formed between the substance (known as the solute) and the solvent. Substances with covalent network structures are insoluble in most solvents because of the strong covalent bonds holding their atoms together.

A molecular substance is likely to dissolve in a solvent if the intermolecular forces between the solute and solvent are stronger than the attraction between the solute molecules.

It is often said in chemistry that "like dissolves like". In other words, non-polar solutes are likely to dissolve in non-polar solvents. Similarly, polar solutes are likely to dissolve in polar solvents. Dissolving is unlikely to occur if the solute is polar and the solvent is non-polar, or vice versa.

TOK

"Like dissolves like" is a useful rule of thumb, which is helpful in many cases but not all, and it has no explanatory power. In a small group, discuss the advantages and disadvantages of having rules like this. You may want to consider: what are the alternatives? Do such rules have an "expiry date"? What rules of thumb do you use in other areas of knowledge?

Iodine, I_2, is non-polar. It dissolves readily in non-polar solvents such as hexane. Iodine is almost insoluble in water because it cannot associate strongly with water molecules. Substances that can form hydrogen bonds, such as ethanol, are more likely to dissolve in water (figure 51). Ethanol and water are said to be **miscible**, as they can form a solution when mixed in any proportion.

Methanol, ethanol and propan-1-ol readily dissolve in water. The solubility of the primary alcohols, however, decreases with increasing carbon chain length (table 12).

Primary alcohol	Structural formula	Solubility g / 100 g of H_2O
ethanol	CH_3CH_2OH	miscible
butan-1-ol	$CH_3(CH_2)_3OH$	6.8
nonan-1-ol	$CH_3(CH_2)_8OH$	0.014

▲ Table 12 The aqueous solubility of primary alcohols decreases with increasing hydrocarbon chain length

Hydrocarbon chains are non-polar and do not form hydrogen bonds with water. Fats and oils are largely non-polar and therefore do not dissolve in water. This is easily observable when vinegar (an aqueous solution of ethanoic acid) is mixed with oil to make salad dressing. The two liquids form two layers and are said to be **immiscible**, which means they do not mix, due to their differing polarities.

Long hydrocarbon chains, such as those in fats and oils, are said to be **hydrophobic** (water-hating). Soap and detergent molecules often have a hydrophobic tail and a **hydrophilic** (water-loving) head. They work because they associate with both water and the non-polar fats and oils frequently found in greasy stains (figure 52). Molecules with a hydrophobic tail and hydrophilic head are called **surface-active agents**, or **surfactants**.

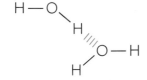

(a) between water molecules

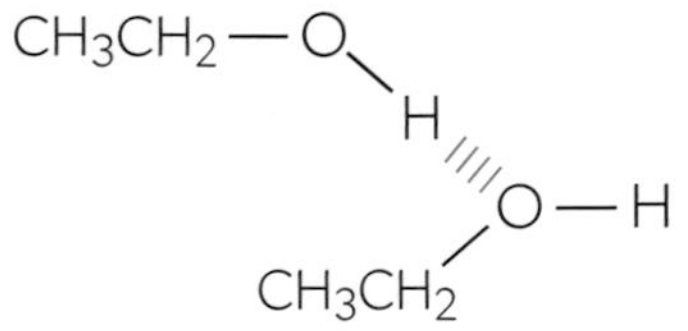

(b) between ethanol molecules

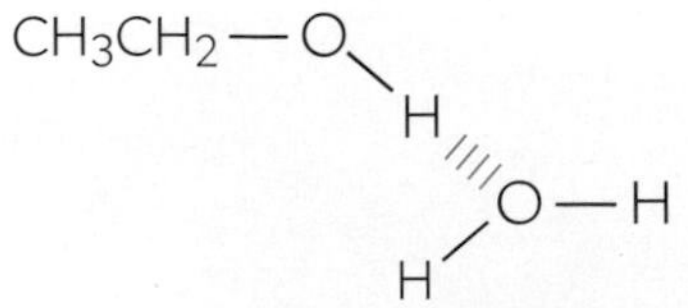

(c) between ethanol and water

▲ Figure 51 Ethanol and water can form hydrogen bonds. Therefore, ethanol is water soluble

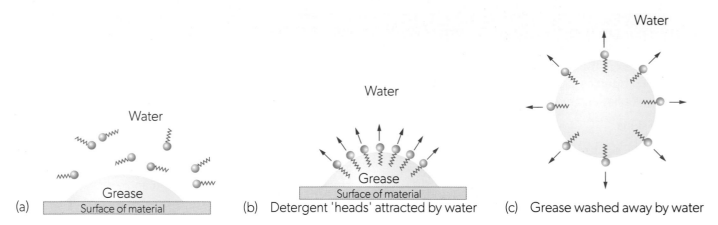

▲ **Figure 52** In a surfactant molecule, the hydrophilic (water-loving) head and a hydrophobic (water-hating) tail lift greasy stains off materials

Global impact of science

Morphine and its derivatives, known as opiates, are strong painkillers with a high potential for misuse and addiction. The effect they have on the body depends on two factors:

1. their ability to cross the blood-brain barrier
2. their ability to bind to opioid receptors in the brain.

The blood-brain barrier consists of a series of cell membranes that coat the blood vessels in the brain and prevent polar molecules from entering the central nervous system. These membranes are lipophilic, which means they are permeable to lipid molecules. The presence of one amino and two hydroxyl groups in the morphine molecule (figure 53) makes it sufficiently polar to be soluble in water but at the same time reduces its solubility in lipids and therefore limits its ability to reach the opioid receptors in the brain.

In diamorphine, also known as heroin, both hydroxyl groups are substituted with ester groups, which greatly reduces the polarity of the molecule. Diamorphine is soluble in lipids and can easily cross the blood-brain barrier. In the brain, diamorphine is quickly metabolized into morphine, which binds to the opioid receptor.

This mechanism of action makes diamorphine about five times more potent an analgesic than morphine. At the same time, diamorphine has more severe side effects, including tolerance, addiction and central nervous system depression. In most countries the use of diamorphine is either banned or restricted to terminally ill patients with certain forms of cancer or central nervous system disorders.

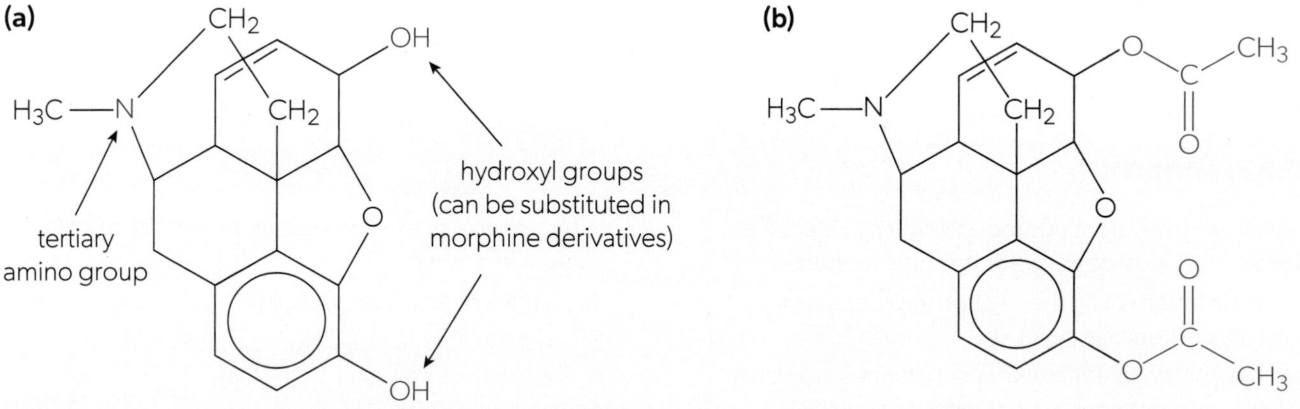

(a)

H₃C —N

CH₂

CH₂

OH

tertiary amino group

hydroxyl groups (can be substituted in morphine derivatives)

OH

(b)

H₃C —N

CH₂

CH₂

O—C—CH₃

O

O—C—CH₃

O

▲ **Figure 53** The structures of (a) morphine and (b) diamorphine (heroin)

Physical properties of ionic and covalent substances

You can figure out the structure of a substance by testing its physical properties such as electrical conductivity, aqueous solubility and melting point.

Relevant skills

- Tool 1: Recognize and address the relevant safety, ethical or environmental issues in an investigation.

- Inquiry 1: Demonstrate independent thinking, initiative or insight.

- Inquiry 2: Identify and record relevant qualitative observations.

Instructions

You are presented with three bottles labelled A, B and C. Each bottle contains a crystalline solid. One is sodium chloride, another is sugar, and the third is white sand, but you do not know which is which. Design a safe method that could be used to identify A, B and C.

Possible equipment and substances you could use might include:

spatulas, beakers, distilled water, pipettes, power pack, leads, ammeter, lamp, conductivity probe, electrodes, crucible, Bunsen burner, pipeclay triangle, heat-proof mat, mortar and pestle, melting point apparatus, fume hood, combustion spoon.

Complete a risk assessment and show this and your proposed method to your teacher for approval. If you have time, try it out.

Extension

What if the ionic substance were not water soluble? How would you modify your method?

To summarize, substances with covalent network structures are typically:

- non-volatile because they are held together by strong covalent bonds between atoms
- poor electrical conductors (exceptions include graphite and graphene) because they do not contain mobile charged particles
- insoluble in all solvents because their dissolution would require breaking the strong covalent bonds.

Substances with molecular covalent structures are typically:

- volatile because their molecules are held together by intermolecular forces, which are weak
- poor electrical conductors because they do not contain mobile charged particles
- of varying solubility, depending on the strength of the intermolecular forces they form with solvent molecules.

Linking questions

What experimental data demonstrate the physical properties of covalent substances? (Tool 1, Inquiry 2)

To what extent does a functional group determine the nature of the intermolecular forces? (Structure 3.2)

Practice questions

29. Explain why covalent compounds are not electrical conductors, except for graphite and graphene.

30. Describe and explain the relationship between volatility and boiling point.

31. Explain why covalent network substances are non-volatile, and molecular substances are volatile, although they both contain covalent bonds.

32. Explain why ethanol, CH_3CH_2OH, is miscible with water.

33. Which **two** of the following substances are readily soluble in water?
    A. ethanoic acid, CH_3COOH
    B. octane, $CH_3(CH_2)_6CH_3$
    C. octan-1-ol, $CH_3(CH_2)_6CH_2OH$
    D. propanone, CH_3COCH_3

34. Explain why ionic substances, such as sodium chloride, do not dissolve in oil.

35. Explain why diatomic bromine, Br_2, is more soluble in hexane than in water.

Chromatography (*Structure 2.2.10*)

The components of a mixture can often be separated and identified using **chromatography**. All forms of chromatography (figure 54) have the same underlying principle: the existence of a **mobile phase** and a **stationary phase**. In most forms of chromatography, the components of the mixture are separated based on their different affinities for each of the two phases. These affinities involve intermolecular forces.

The mobile phase moves through the stationary phase, carrying the components of the mixture. The components all start at the same point but are transported through the stationary phase at different rates due to their differing affinities for each of the two phases. The components are eventually separated.

> Methods for separating mixtures are discussed in *Structure 1.1.*

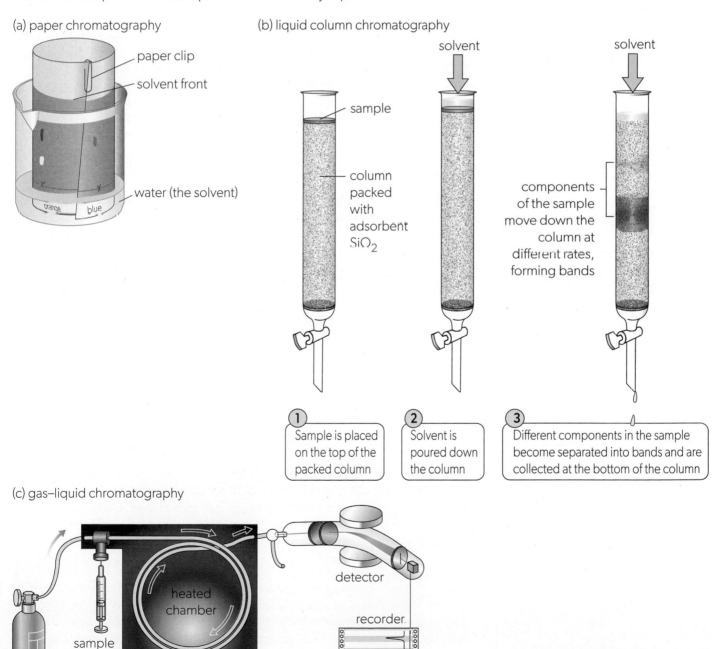

(a) paper chromatography

paper clip

solvent front

water (the solvent)

orange   blue

(b) liquid column chromatography

solvent   solvent

sample

column packed with adsorbent SiO$_2$

components of the sample move down the column at different rates, forming bands

1 Sample is placed on the top of the packed column

2 Solvent is poured down the column

3 Different components in the sample become separated into bands and are collected at the bottom of the column

(c) gas–liquid chromatography

detector

recorder

heated chamber

sample injected here

carrier gas

column – capillary tube with liquid-coated walls

◀ **Figure 54** All types of chromatography involve a mobile phase and a stationary phase

155

Chromatographic methods can be classified into categories. They can be grouped according on their format (for example, **planar** vs **column chromatography**), or the mechanism of separation (for example, partition, adsorption, size exclusion or ion exchange chromatography). In this chapter, we will focus on two types of planar chromatography: paper and thin-layer. The former separates mixtures based on the principle of partition, whereas the latter is based on adsorption. Research at least one other method of chromatography (such as gas chromatography) and compare and contrast the different methods. Organize your notes in a Venn diagram.

Paper chromatography

Separating mixtures of pigments is a simple way to demonstrate **paper chromatography** (figure 55). The stationary phase is a rectangular piece of chromatography paper (made of hydrated cellulose) and the mobile phase is a suitable solvent. The mixture to be separated is dotted onto a start line near the bottom of the paper. The paper is then placed in a chamber such as a beaker, containing a small amount of solvent at the bottom. Placing a lid on top of the beaker allows the atmosphere in the chamber to become saturated with the solvent vapour and prevents solvent loss through evaporation.

▼ Figure 55 Simple paper chromatography experiment using different inks. You can see that the purple ink at the right-hand side is a mixture containing red, yellow and blue dyes

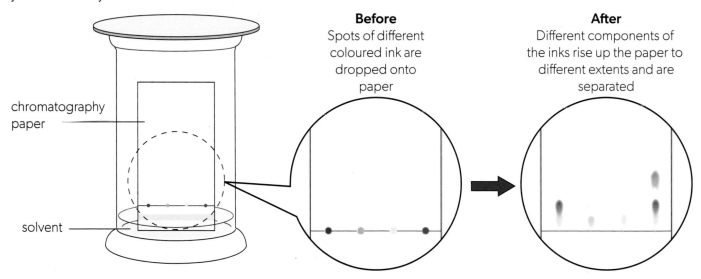

The solvent is allowed to move up the paper until it reaches a point near the top. The components of the mixture move different distances up the paper according to their relative affinities for the stationary and mobile phases. Components with a greater affinities for the solvent will dissolve more readily in it and are therefore transported further up the paper. The results of a chromatography experiment are known as **chromatogram.**

Chromatography paper is composed of hydrated cellulose, which contains many –OH groups. These groups are very polar and attract water molecules tightly, saturating the surface of cellulose with water. When coupled with a less polar organic solvent, this technique can be used to investigate mixtures such as leaf pigments or amino acid mixtures. The mixture components are **partitioned** between the water layer and the solvent: less polar components dissolve in the solvent and get carried further up the paper, whereas more polar components stay predominantly in the water layer. The separation can be optimized by using different solvents, or mixtures of solvents.

Thin layer chromatography (TLC)

Thin layer chromatography (TLC) is more expensive than its paper counterpart, but it offers greater sensitivity. In TLC, the stationary phase is a rectangular plate made of glass or metal coated with silica (silicon dioxide, SiO_2) or alumina (aluminium oxide, Al_2O_3). The silica or alumina surfaces are very polar and contain many hydroxyl groups. The mixture is spotted onto the plate and placed into an eluting chamber containing a non-polar organic solvent. The polar substances in the mixture travel up the plate slowly because their tendency to adsorb onto the silica or alumina surface is greater than their tendency to dissolve in the solvent. Conversely, non-polar components dissolve in the solvent and travel further up the plate.

Experiments

Experimental techniques are adapted to suit different purposes. Coloured substances are easily located on a chromatogram. When a substance is colourless, an additional step is needed. For example, when a mixture contains amino acids, the chromatogram can be sprayed with a locating agent. A locating agent is a substance that reacts with the spots to form coloured products. Other colourless compounds can be located in different ways. For example, some might show up under UV light. Can you think of any other examples of how existing scientific methods were adapted to solve a problem?

Retardation factor (R_F)

The results of a chromatography experiment can be quantified by calculating the **retardation factor** (R_F). The R_F value for a spot on a chromatogram is the ratio of the distance travelled by the spot (b) to the distance travelled by the solvent (a):

$$R_F = \frac{b}{a}$$

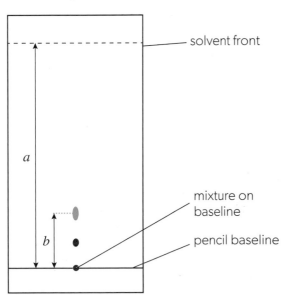

◀ **Figure 56** Retardation factors, R_F, are calculated from the distances measured on a chromatogram

Since the spots cannot travel further than the solvent that transports them, R_F values are always in the range of 0 to 1. The distance travelled by a spot can vary depending on factors such as the composition of the solvent, temperature, pH and type of paper used (in paper chromatography). R_F values are reproducible and therefore can be used to identify substances by comparing experimental values to accepted values, provided that the conditions are the same.

Worked example 4

Consider the thin layer chromatogram shown in figure 57. Calculate the retardation factor for the red and blue spots.

Given that the stationary phase is silica and the solvent is a hydrocarbon, determine which component is more polar.

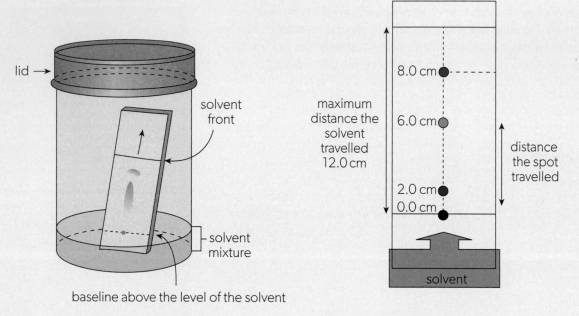

▲ Figure 57 The apparatus for thin layer chromatography (left) and the resulting chromatogram (right)

Solution

Figure 58 shows that the mixture contains three components, which appear as distinct spots at different distances from the baseline: at 8.0 cm (red), 6.0 cm (green) and 2.0 cm (blue). The distance between the baseline and the solvent front is 12.0 cm.

You can calculate the retardation factors (R_F) for the red and blue spots (top and bottom) as follows:

$$R_F(\text{red}) = \frac{\text{distance travelled by the red spot}}{\text{distance travelled by the solvent}}$$

$$= \frac{8.0 \text{ cm}}{12.0 \text{ cm}}$$

$$= 0.67 \text{ (2 significant figures)}$$

$$R_F(\text{blue}) = \frac{\text{distance travelled by the blue spot}}{\text{distance travelled by the solvent}}$$

$$= \frac{2.0 \text{ cm}}{12.0 \text{ cm}}$$

$$= 0.17 \text{ (2 significant figures)}$$

$R_F(\text{red})$ is larger than $R_F(\text{blue})$. This means that the component(s) of the red spot are more soluble in the mobile phase (non-polar hydrocarbon solvent) and the blue spot is composed of substances that have a greater affinity for the stationary phase. The stationary phase (silica) is polar, so the blue spot contains substances that are more polar than the substances in the red spot.

Chromatograms composed of clear dots are frequently shown in books. If you do a chromatography experiment, you will often obtain a chromatogram in which each colour blends into the next. If you want a clearer separation between substances in the mobile phase, you can try using different solvents.

▶ Figure 58 This experimental chromatogram has less clear separation between substances than the previous examples we looked at

Thin-layer chromatography of plant pigments

The pigments found in plants can be separated by chromatography, depending on their relative affinities to the mobile and stationary phases. The separation can be optimized by changing the phases. In this investigation, you will compare the chromatograms obtained when the mobile and stationary phases are changed.

Relevant skills

- Tool 1: Measuring length
- Tool 1: Paper or thin layer chromatography
- Tool 3: Record uncertainties in measurements
- Tool 3: Propagate uncertainties in processed data
- Inquiry 3: Discuss the impact of uncertainties on the conclusions

Safety

- Wear eye protection.
- Flammable solvents. Keep away from flames and other sources of ignition.
- Organic solvents are toxic to the environment.
- Work in a well-ventilated space and dispense solvents in a fume cupboard.

Materials

- Pestle and mortar
- Chromatography spotter
- Chromatography paper
- TLC plate
- Pencil
- Ruler
- Chromatography tank or beaker
- Lid or aluminium foil for the chromatography tank
- Propanone
- Fresh green leaves
- Chromatography solvents (9:1 petroleum ether:propanone mixture)

Method
Part 1: Preparation of the chromatography plate

1. Grind the leaves with a few cm^3 of propanone until you have obtained a concentrated pigment extract.

2. Handle the TLC plate carefully by the edges. Draw a line in pencil, across the plate, 1 to 2 cm from the bottom.

3. Spot the plant pigment extract onto the starting line, taking care to create a small, concentrated spot of pigment. Allow it to dry.

Part 2: Chromatography

4. Place a few cm^3 of the chromatography solvent into the tank. Place the lid on top, to ensure that the atmosphere in the chamber becomes saturated with the solvent vapours. You can place a paper towel soaked in the solvent vertically inside the tank to speed up the process.

5. Then, lower the TLC plate into the solvent, taking care to keep the solvent below the pigment spot on the starting line. Ensure the plate does not touch the sides of the beaker. If you are using paper, you must make sure that it is upright in the tank by attaching it to the lid or to a splint placed across the top of the tank.

6. Remove the plate from the tank once the solvent is near the top. Quickly draw a line in pencil over the solvent front before it evaporates. Draw circles around the pigment spots that are visible in the chromatogram. Allow the plate to dry.

7. Measure the distance travelled by the solvent. Measure the distance travelled by each pigment from the base line to the centre of the spot.

8. Determine a reasonable uncertainty for the distances you have measured.

9. Calculate the R_F value of each pigment, and the uncertainty of each.

10. Compare your results to those of other members of your class. Compare your results to literature values for the R_F values, if available. Search online for information that will help you identify each of the pigments.

11. Comment on the precision and, if possible, accuracy, of your class's results. Discuss the impact of measurement uncertainty on the validity of your conclusions.

Part 3: Changing the phases

12. Consider how the results would differ if you used a different solvent system, or a different stationary phase. Then, try it out.

13. Clear up as instructed by your teacher, taking care not to dispose of the solvents down the sink.

Practice questions

36. Calculate the retardation factor, R_F, for the green spot at 6.0 cm in figure 58.

37. Suggest why the baseline in paper chromatograms is drawn in pencil.

38. Chromatographic analysis of a mixture of dyes (A) and three individual pigments (B, C and D) produced the chromatogram on the right.

    Identify and explain which pigments are present in A.

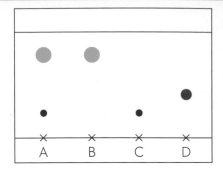

Resonance (*Structure 2.2.11*)

Sometimes, a molecule cannot be described by a single Lewis formula. Instead, there are two or more possible **resonance structures** that collectively represent the molecule. Resonance often happens when there is more than one position for a double or triple bond in a molecule. For example, two Lewis formulas can be drawn for ozone, O_3 (figure 59). Note that a double-headed arrow is used to show alternative resonance structures.

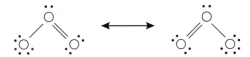

▲ Figure 59 Ozone has two possible resonance structures

These structures might suggest that the two oxygen–oxygen bonds in ozone are different: one single and one double. In reality, the ozone molecule is a hybrid of these two resonance structures. Bond length and strength data (table 13) show that the two bonds in ozone are:

• identical to each other

• intermediate between a single O–O bond and a double O=O bond in terms of bond strength and length

• have a bond order of 1.5, because there are 3 bonding electron pairs distributed across two domains.

Bond	Length / 10^{-12} m	Average bond enthalpy / kJ mol^{-1}	Bond order
O–O single bond	148	144	1
O⸗O bond in ozone	128	362	1.5
O=O double bond	121	498	2

▲ Table 13 Bond data for oxygen–oxygen bonds

Global impact of science

Diatomic oxygen, O_2, and ozone, O_3, are oxygen allotropes that absorb different wavelengths of electromagnetic radiation corresponding to the ultraviolet (UV) section of the spectrum. The double bond in O_2, with bond order 2, is stronger and therefore has a higher bond enthalpy than the oxygen–oxygen bond in O_3, with bond order 1.5. Breaking the oxygen–oxygen double bond requires higher energy.

The dissociation of the double oxygen–oxygen bond in O_2 can be brought about by electromagnetic radiation with wavelengths of 240 nm and below. The bonds in ozone can be broken by wavelengths of 330 nm and below. These wavelengths correspond to the UV-C and UV-B regions of the electromagnetic spectrum, respectively.

The presence of O_3 in the stratosphere prevents nearly all solar UV-C and UV-B radiation from reaching the Earth's surface (figure 60).

UV-C and UV-B radiation can cause DNA mutations, resulting in skin cancer (melanomas) and cataracts. Ozone layer depletion caused by chlorofluorocarbons (CFCs) in the atmosphere means that more UV radiation can reach the Earth's surface. International collaboration involving the 1987 Montreal Protocol and the banning of ozone-depleting substances has slowed down the rate of depletion. Evidence in recent years suggests the concentration of ozone in the stratosphere has increased, resulting in the recovery of the ozone layer (figure 61).

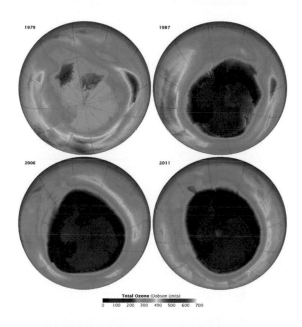

▲ Figure 61 Satellite images of a hole in the ozone layer over Antarctica from 1979 to 2011. The largest ozone hole to date was recorded in 2006. Since then, data suggests that the hole has been getting smaller again

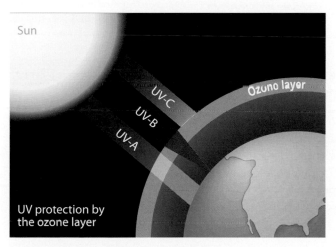

▲ Figure 60 The ozone layer in the stratosphere absorbs UV-C radiation and most UV-B radiation

◀ Figure 62 Former US president Barack Obama presents the Presidential Medal of Freedom to Mexican chemist and environmental scientist Mario Molina. Molina helped to uncover the link between CFCs and ozone layer depletion

Delocalization

Resonance structures can be described as a single structure using the concept of **delocalization**. In covalent bonds, two atoms share a pair of electrons, which are localized between the two atoms. Delocalization occurs when electrons are shared by more than two atoms in a molecule or ion, as opposed to being localized between a pair of atoms.

Consider the ion NO_2^-. It is represented by two resonance structures (figure 63a). The nitrogen–oxygen bond has bond order 1.5, and its length and strength are between those of a single N–O bond and a double N=O bond. One pair of electrons is delocalized across the two N–O bonding domains. For this reason, the ion can be represented with a structure containing a dashed line that shows the delocalization (figure 63b).

Practice questions

39. Draw the structure of ozone that shows the delocalized electron pair.

40. Draw the delocalized structure of the carbonate ion, CO_3^{2-}, and explain why the C–O bond order in this ion is 1.33.

41. Compare the C–O bond lengths and strengths in the carbonate ion to those in carbon dioxide and methanol, CH_3OH.

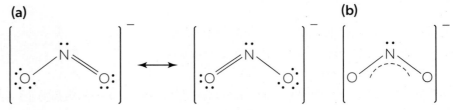

(a)   **(b)**

▲ Figure 63 (a) The ion NO_2^- can be represented by two Lewis formulas (b) An alternative representation of the ion shows the delocalized electron pair

Note that the structure in Figure 63b is not a Lewis formula because it does not specify locations of bonding and non-bonding electron pairs.

Resonance does not involve the interconversion between the various resonance structures through rotation or bond movement. Instead, resonance structures collectively describe a single molecule that cannot be represented by a single Lewis formula due to the presence of electron delocalization. Therefore, the actual molecule is somewhere "in-between" the various resonance structures, which is best represented by the single delocalized structure.

Linking question

Why are oxygen and ozone dissociated by different wavelengths of light? (Structure 1.3)

Benzene and resonance (*Structure 2.2.12*)

Benzene, C_6H_6, is an important example of a molecule that demonstrates resonance.

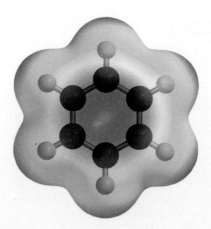

▲ Figure 64 The ball-and-stick structure and electron cloud of benzene, C_6H_6

The Lewis formula for benzene contains a six-membered ring of carbon atoms connected through alternating single and double bonds. Two resonance structures are possible (figure 65).

◀ Figure 65 Resonance structures of benzene

Some of the electrons in the carbon–carbon bonds in benzene are delocalized around the ring. For this reason, these delocalized electrons are often represented with a circle:

In skeletal formulas, straight lines represent bonds. The ends and vertices (corners) of the bonds represent carbon atoms, and hydrogen atoms are not drawn. Detailed instructions for drawing skeletal formulas of molecules are given in *Structure 3.2*.

Benzene is an aromatic hydrocarbon. The term "**aromatic**" is used to describe cyclic molecules that are planar and stabilized due to the presence of delocalized electrons. Benzene is the simplest aromatic hydrocarbon, and structures similar to it can be found in a large range of substances (figure 66).

Activity

1. One of the resonance structures of 1,2-dichlorobenzene is shown below. Draw the other resonance structure.

   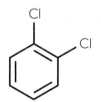

2. Draw a structure of 1,2-dichlorobenzene that contains a circle to represent the delocalized electrons.

3. Research the structures of the following molecules and identify the benzene rings in each:

TOK

Skeletal formulas are an example of "chemistry shorthand" that are used extensively by chemists. To what extent are chemical symbols, formulas and equations like a language? What is gained or lost using chemical notation?

(a)

curcumin

(b)

Kevlar

(c)

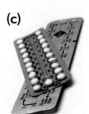

estradiol

▲ Figure 66 (a) Curcumin, found in the yellow food spice turmeric (b) Kevlar is used for making personal armour (c) Synthetic estradiol is used as medication in menopausal hormone therapy

▲ **Figure 67** The delocalized electrons in the p orbitals of benzene form a ring of electron density above and below the plane of the molecule

In ordinary conversation, the term "aromatic" can be used to describe something as having a pleasant smell. In chemistry, however, this term is used to indicate a benzene derivative or the presence of a benzene ring in a compound. However, some aromatic compounds are non-benzenoids. For example, azulene—a component of some blue pigments found in nature—contains alternating single and double bonds in two planar rings that are joined together:

Can you think of other examples of chemical terms that have different meanings in everyday life?

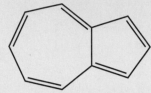

In benzene, one electron in each of the six carbon atoms occupies a p orbital. The six p orbitals overlap, forming a ring of electron density above and below the plane of the molecule. The electrons in this ring are delocalized (figure 67). The carbon atoms are sp^2 hybridized (see **Structure 2.2.16**).

Physical evidence for the structure of benzene

Once it was established that C_6H_6 was the molecular formula for benzene, the next step was to deduce its structure (figure 68). The 1:1 ratio of carbon to hydrogen suggested the presence of multiple bonds. Eventually a cyclic arrangement of carbon atoms with alternating single and double bonds was proposed (figure 69). August Kekulé was credited with the structure, which is said to have come to him in a dream. The structure proposed was a lop-sided hexagon because of the differing carbon-carbon bond lengths.

▲ **Figure 68** Some of the 19th century suggestions for the structure of benzene

▶ **Figure 69** (a) August Kekulé on an East German postage stamp (b) Kekulé's suggestion for the structure of benzene

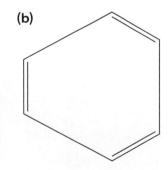

X-ray diffraction patterns later showed a regular hexagonal arrangement of carbon atoms in benzene where all carbon–carbon bonds had the same length, suggesting they were all equivalent (figure 70).

The carbon–carbon bond length in benzene is intermediate between that of a single and a double carbon–carbon bond (table 14). Bond enthalpy data shows that they are also of intermediate strength.

Bond	Bond length / 10^{-12} m	Average bond enthalpy / kJ mol^{-1}
C–C single bond	154	346
C⚌C in benzene	140	507
C=C double bond	134	614

▲ Table 14 Bond length and strength data for the carbon–carbon bond in benzene compared to single and double carbon–carbon bonds

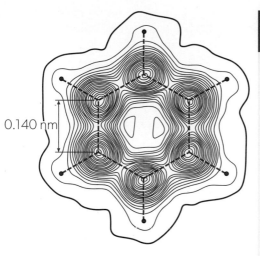

0.140 nm

▲ Figure 70 A technique called X-ray diffraction shows a contour map of the electron density in an individual benzene molecule. Each carbon–carbon bond length is 0.140 nm

Chemical evidence for the structure of benzene

The chemical behaviour of benzene differs from what would normally be expected from alkenes. Like alkenes, benzene is unsaturated, but unlike alkenes, it does not readily undergo addition reactions. Alkenes undergo addition reactions, where the double bond breaks and the carbon atoms on either end form bonds with new species. In benzene, addition would disrupt the stabilizing effect of the delocalized electron ring, which would be energetically unfavourable. Instead, benzene tends to undergo substitution reactions.

Addition and substitution reactions are discussed in *Reactivity 3.4*.

Resonance energy of benzene

Resonance increases the stability of the benzene molecule. This is quantified by a value called **resonance energy**. For benzene it is 152 kJ mol^{-1}. This value is established by comparing the thermochemistry data for benzene and those for cyclic six-membered alkenes.

For example, cyclohexene contains one double bond. When we add hydrogen to this double bond, the change in energy is –120 kJ mol^{-1}.

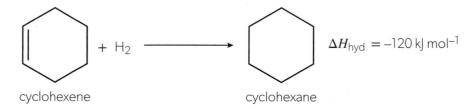

cyclohexene   cyclohexane

$\Delta H_{hyd} = -120$ kJ mol^{-1}

This reaction is called a hydrogenation reaction, and the change in energy is referred to as the **enthalpy of hydrogenation**, ΔH_{hyd}. 1,3-cyclohexadiene is a similar molecule with two double bonds, so its enthalpy of hydrogenation is nearly double: –232 kJ mol^{-1}.

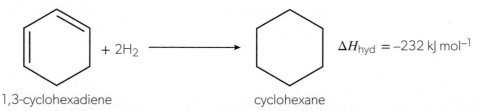

1,3-cyclohexadiene   cyclohexane

$\Delta H_{hyd} = -232$ kJ mol^{-1}

Changes in energy across a reaction are known as enthalpy changes. Enthalpy changes are discussed in *Reactivity 1.1*. Naming cyclic compounds is covered in *Structure 3.2*.

The next molecule in the series matches Kekulé's suggested structure and is the theoretical molecule, 1,3,5-cyclohexatriene. 1,3,5-cyclohexatriene contains three double bonds and its enthalpy of hydrogenation is predicted to be $-120 \times 3 = -360\,kJ\,mol^{-1}$ (figure 71). However, the actual value for benzene is $-208\,kJ\,mol^{-1}$. This is due to the added stability provided by resonance in the benzene ring. The difference between these enthalpy values is $152\,kJ\,mol^{-1}$, which is the resonance energy.

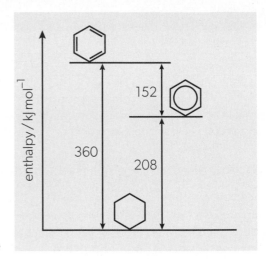

▶ **Figure 71** The resonance energy of benzene ($152\,kJ\,mol^{-1}$) compared to the theoretical compound 1,3,5-cyclohexatriene

Hydrogenation of benzene is less **exothermic** than expected. This is because the reaction has an additional energetic cost to break the delocalized electron rings.

1,2-disubstituted benzene compounds

Consider the Kekulé structure of benzene, containing alternating double and single bonds. When chlorine, Cl_2, reacts with benzene to form 1,2-dichlorobenzene, you would expect that two possible **isomers** can be formed: one with the two chlorine atoms at either end of a carbon–carbon single bond, whereas in the other isomer, the chlorines would be at either end of a carbon–carbon double bond (figure 72a). However, because all electrons are evenly distributed, these two isomers represent the same molecule (figure 72b).

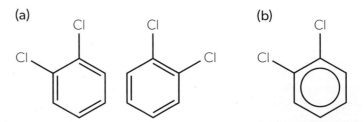

▲ **Figure 72** (a) The two theoretical isomers of 1,2-dichlorobenzene for the Kekulé structure of benzene (b) The actual structure of 1,2-dichlorobenzene has only one isomer because the carbon atoms are all chemically equivalent

Practice questions

42. Describe two pieces of evidence—one physical and one chemical—for the structure of benzene.

43. How many delocalized electrons does each of the carbon atoms in a benzene ring contribute to the delocalized electron cloud?

44. State the bond angle in benzene and the molecular geometry around each of the carbon atoms.

 Linking questions

How does the resonance energy in benzene explain its relative unreactivity? (Reactivity 2.1, 2.2)

What are the structural features of benzene that favour it undergoing electrophilic substitution reactions? (Reactivity 3.4)

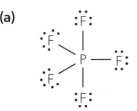

(a)

(b)

▲ **Figure 73** Two species with expanded octets: PF_5 and ICl_4^-, with five and six pairs of electrons around the central atom, respectively

Expanded octets (*Structure 2.2.13*)

Lewis formulas and VSEPR were discussed in the SL section of this chapter. At HL you should also be familiar with molecules containing atoms with more than eight valence electrons. This is known as an **expanded octet**. The central atoms in these molecules can have five or six pairs of electrons, amounting to 10 or 12 electrons, respectively. Examples of such species include phosphorus pentafluoride, PF_5, and the tetrachloroiodide ion, ICl_4^-, shown in figure 73.

Drawing Lewis formulas for expanded octets

You can follow the same steps used for drawing Lewis formulas at SL for molecules containing expanded octets. However, in this case, we omit the final step—we do not check that the central atom has a full octet.

Worked example 5

Draw the Lewis formulas of each of the following species with expanded octets:

a. sulfur hexafluoride, SF_6 b. triiodide ion, I_3^- c. xenon trioxide, XeO_3

Solution

Follow the steps above for each molecule:

	a sulfur hexafluoride, SF_6	b triiodide ion, I_3^-	c xenon trioxide, XeO_3
Step 1 Count valence electrons	Sulfur: 6 Fluorine: $7 \times 6 = 42$ Total: $6 + 42 = 48$	Iodine: $7 \times 3 = 21$ This polyatomic ion has a -1 charge meaning that it has an additional electron. Total: $21 + 1 = 22$	Xenon: 8 Oxygen: $6 \times 3 = 18$ Total: $8 + 18 = 26$
Step 2 Calculate the number of electron pairs	$\dfrac{48}{2} = 24$ pairs	$\dfrac{22}{2} = 11$ pairs	$\dfrac{26}{2} = 13$ pairs

	a sulfur hexafluoride, SF_6	b triiodide ion, I_3^-	c xenon trioxide, XeO_3
Step 3 Arrange the atoms	F F F S F F F	I I I	O Xe O O
Step 4 Draw the single bonds	 6 pairs used so far…	I—I—I 2 pairs used so far…	 3 pairs used so far…
Step 5 Put non-bonding pairs on peripheral atoms	 The fluorine atoms now have full octets. 24 pairs used.	 The peripheral iodine atoms now have noble gas configurations and therefore do not need any more electrons 8 pairs used so far…	 The oxygen atoms now have full octets. 12 pairs used so far…
Step 6 Put any remaining electron pairs on the central atom	We have used all 24 available electron pairs, so the structure above is final. There are 6 pairs of electrons around the central sulfur atom. It has an expanded octet.	 The remaining three pairs of electrons are assigned to the central iodine atom. This species is an ion, so square brackets and the charge are needed. We can see that the central iodine atom has an expanded octet, with 5 pairs of electrons around it.	 We assign the final electron pair to the xenon atom. All 13 available electron pairs have been used.

In the structure deduced for XeO_3, each atom has eight electrons and none have an expanded octet. However, this structure is incorrect, and an exception to the method we use for deducing Lewis formulas.

Later in this chapter, we will discuss the concept of **formal charge**: this will help us deduce the true structure of XeO_3 and explain why it is considered to have an expanded octet.

Practice questions

45. Draw the Lewis formula of each of the following:
    - a. PCl_5
    - b. BrF_3
    - c. IF_5
    - d. ICl_2^-
    - e. ICl_2^+
    - f. SOF_4

Molecular geometry of expanded octets

In the SL section of this chapter, we discussed the geometries that arise when atoms have two, three or four electron domains. Several additional geometries occur for expanded octets containing five or six electron pairs.

Five domains: trigonal bipyramidal geometry

If there are five electron domains, the electrons in those domains repel each other and therefore adopt positions at 120° and 90° from each other. This electron domain geometry is called **trigonal bipyramidal**. The five domains form a shape consisting of two pyramids that share the same triangular base.

The electron domains in a trigonal bipyramidal molecule fall into two categories:

1. **Equatorial**: the bonds forming the triangular base of the pyramids, at 120° from each other.

2. **Axial**: the bonds that form the axis of the molecule, at 90° from the plane of the triangular base.

A trigonal bipyramidal electron domain geometry gives rise to four possible molecular geometries, depending on the presence of non-bonding domains:

*   When all five domains are bonding domains, the molecule has **trigonal bipyramidal** geometry.

*   When four of the five domains are bonding domains, and one is a non-bonding domain, the molecule has **seesaw** geometry.

*   When three of the five domains are bonding domains, and two are non-bonding domains, the molecule has **T-shaped** geometry.

*   When only two of the five domains are bonding domains, and three are non-bonding domains, the molecular has **linear** geometry.

Examples of each type are shown in figure 75.

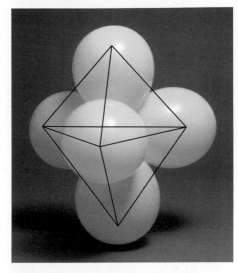

▲ **Figure 74** A balloon model of the trigonal bipyramidal electron domain geometry

▼ **Figure 75** The four types of trigonal bipyramidal electron domain geometry: (a) trigonal bipyramidal (b) seesaw (c) T-shaped (d) linear

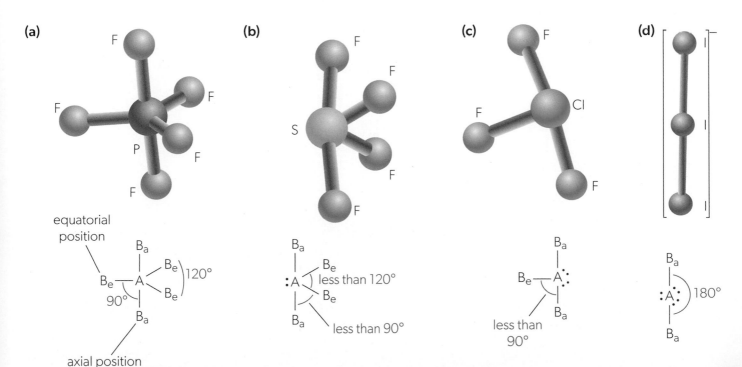

Practice question

46. Consider the seesaw and T-shaped molecular geometries. Why do you think the non-bonding domains are located in equatorial positions, not in axial positions?

▲ **Figure 76** A balloon model of the octahedral electron domain geometry

Like other geometries, the 120° and 90° bond angles in a trigonal bipyramidal shape are distorted by the presence of non-bonding pairs, due to their greater repulsion.

Six domains: octahedral geometry

If there are six domains, the electron pairs in those domains adopt positions at 90° from each other. This electron domain geometry is called **octahedral**. This is because the domains form an octahedron, which is an eight-sided polyhedron consisting of eight identical equilateral triangles.

The electron domains in an octahedral molecule are all at 90° to each other, so there is no need to distinguish between axial and equatorial domains.

An octahedral electron domain geometry gives rise to three common molecular geometries, depending on the presence of non-bonding domains:

- When all six domains are bonding domains, the molecule has **octahedral** geometry.

- When five of the six domains are bonding domains, and one is a non-bonding domain, the molecule has **square pyramidal** geometry.

- When four of the six domains are bonding domains, and two are non-bonding domains, the molecule has **square planar** geometry.

Examples of each type are shown in figure 77.

The 90° bond angles in octahedral molecular geometries are distorted by the presence of non-bonding pairs, due to their greater repulsion.

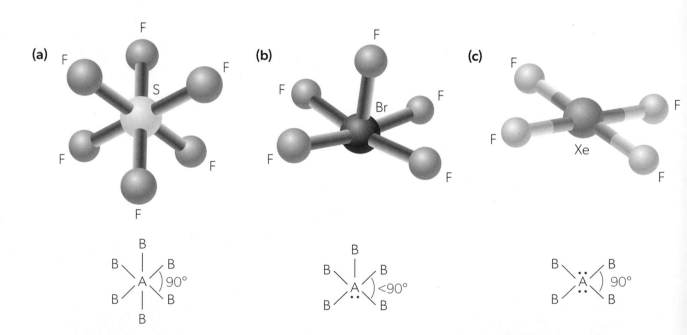

▲ **Figure 77** The three types of octahedral electron domain geometry: (a) octahedral (b) square pyramidal (c) square planar

Electron domain geometry	Number of domains	Number of bonding domains	Number of non-bonding domains	Molecular geometry	Bond angle	Example
trigonal bipyramidal	5	5	0	trigonal bipyramidal	90°, 120°	PF_5
		4	1	seesaw	<90°, <120°	SF_4
		3	2	T-shaped	<90°	ClF_3
		2	3	linear	180°	I_3^-
octahedral	6	6	0	octahedral	90°	SF_6
		5	1	square pyramidal	<90°	BrF_5
		4	2	square planar	90°	XeF_4

▲ Table 15 Summary of molecular geometries with five and six electron domains

Activity

1. Identify the electron domain geometry and molecular geometry of a species with the following bonding and non-bonding domains:
   a. six bonding domains
   b. four bonding domains and one non-bonding domain
   c. five bonding domains
   d. two bonding domains and three non-bonding domains
   e. four bonding domains and two non-bonding domains

2. Identify the number of bonding and non-bonding domains around atoms with the following molecular geometries:
   a. trigonal bipyramidal
   b. T-shaped
   c. octahedral
   d. square planar
   e. seesaw
   f. square pyramidal
   g. linear (there are two possible answers here)

3. Deduce the electron domain geometry and molecular geometry of each of the following:
   a. PCl_5
   b. BrF_3
   c. IF_5
   d. ICl_2^-
   e. ICl_2^+
   f. SOF_4
   g. SiF_6^{2-}
   h. IF_4^+
   j. SCl_4

4. Predict the bond angles in molecules **a** to **j** above.

Linking question

How does the ability of some atoms to expand their octet relate to their position in the periodic table? (Structure 3.1)

Formal charge (*Structure 2.2.14*)

Formal charge is the charge an atom would have if all the bonding electrons in the molecule were shared equally, and if its non-bonding electrons were not shared at all. We can assign formal charge to atoms in a molecule to select the best Lewis formula when several are possible. Formal charge is calculated by looking at the electrons an atom has before and after bonding:

Formal charge = number of electrons before bonding – number of electrons after bonding

$$FC = VE - (NBE + \tfrac{1}{2}BE)$$

Where:

FC = formal charge

VE = number of valence electrons of the atom before bonding

NBE = number of non-bonding electrons assigned to the atom in the Lewis formula

BE = number of bonding electrons assigned to the atom in the Lewis formula

The sum of the formal charges of all atoms in a molecule or ion should be equal to the overall charge of that molecule or ion.

Worked example 6

Show that the overall charge of nitrogen trichloride, NCl_3, is zero by calculating the formal charges of the atoms in the molecule. The Lewis formula of NCl_3 is shown below.

Solution

The formal charge of the nitrogen atom can be calculated as follows:

VE = 5 (because nitrogen is in group 15)

NBE = 2 (because in the Lewis formula the nitrogen atom has two non-bonding electrons)

BE = 6 (because in the Lewis formula the nitrogen atom has three bonds, which are equivalent to six bonding electrons)

$FC(N) = 5 - (2 + 0.5 \times 6) = 0$

Therefore, the formal charge of the nitrogen atom in this Lewis formula is zero.

The same process can be repeated to determine the formal charge of each chlorine atom:

VE = 7 (because chlorine is in group 17)

NBE = 6 (because in the Lewis formula each chlorine atom has six non-bonding electrons)

BE = 2 (because in the Lewis formula each chlorine atom has one bond, therefore two bonding electrons)

$FC(Cl) = 7 - (6 + 0.5 \times 2) = 0$

Therefore, the formal charge of each chlorine atom in this Lewis formula is zero.

The sum of the formal charges is zero, which is also the overall charge of the molecule.

The following conditions are generally favourable:

1. formal charges close to zero

2. a difference in formal charges in the molecule close to zero

3. negative formal charge(s) assigned to the more electronegative atom(s).

When presented with alternative Lewis formulas, we look for the one that satisfies these requirements most.

Worked example 7

Assign formal charges to the atoms in the sulfate ion, SO_4^{2-}, to deduce its Lewis formula.

Solution

We start by drawing the Lewis formula of SO_4^{2-} in a way that follows the octet rule:

Now we can calculate the formal charges of the sulfur and oxygen atoms based on this Lewis formula:

$$FC(S) = 6 - (0 + 0.5 \times 8) = +2$$

$$FC(O) = 6 - (6 + 0.5 \times 2) = -1 \text{ (all the oxygen atoms are equivalent in this Lewis formula and therefore each have the same formal charge of } -1)$$

The formal charges in this Lewis formula are not close to zero, and the difference between them is 3: these conditions are not favourable.

We know that sulfur is prone to forming expanded octets, so the following Lewis formula is proposed:

Calculating the formal charges of the sulfur and oxygen atoms in this Lewis formula gives:

$$FC(S) = 6 - (0 + 0.5 \times 12) = 0$$

$$FC(O \text{ singly bonded}) = 6 - (6 + 0.5 \times 2) = -1$$

$$FC(O \text{ doubly bonded}) = 6 - (4 + 0.5 \times 4) = 0$$

The more electronegative oxygen has the negative charge, same as that in the previously suggested structure. However, the formal charge values in this second Lewis formula are closer to zero and have a range of 1: these conditions are more favourable. Therefore, the second Lewis formula is preferred.

In both cases, the sums of the formal charges are −2, which is equal to the charge of the polyatomic ion.

Practice questions

47. Determine the formal charge of every atom in each of the molecules below. You will need to draw the Lewis formulas first.

   a. H_2O
   b. CO_2
   c. OH^-
   d. NO_3^-
   e. SF_6
   f. BH_3

Practice questions

48. Based on formal charge considerations, determine which Lewis formula is preferred:

    a. $:\ddot{N}\!=\!\!=\!\!N\!=\!\!=\!\ddot{O}:$ $:\ddot{N}\!=\!\!=\!\!O\!=\!\!=\!\ddot{N}:$

    b. 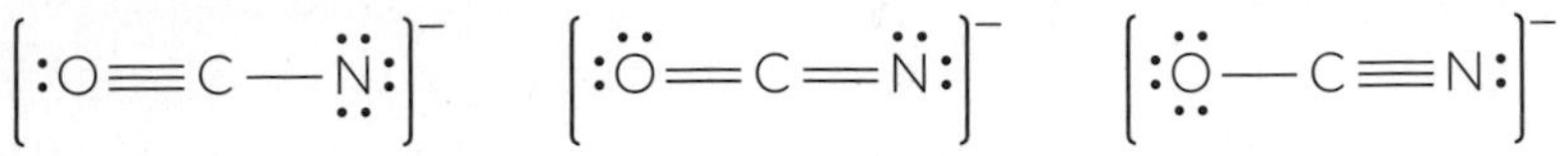 $\left[:O\!\equiv\!C\!-\!\ddot{\ddot{N}}:\right]^{-}$ $\left[:\ddot{\ddot{O}}\!=\!\!=\!C\!=\!\!=\!\ddot{N}:\right]^{-}$ $\left[:\ddot{\ddot{O}}\!-\!C\!\equiv\!N:\right]^{-}$

49. Determine the formal charge of each atom in the XeO_3 molecule shown below and suggest an alternative Lewis formula in which the formal charges are all zero.

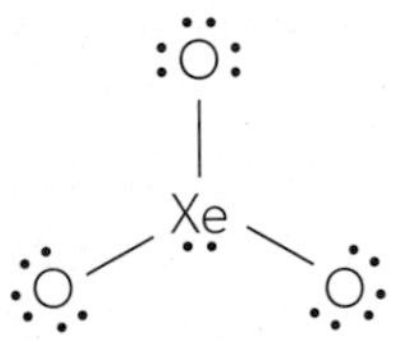

50. Draw two alternative Lewis formulas for ClO_3^-. Determine, using formal charge, which of the two is favourable.

Sigma bonds (σ) and pi bonds (π) (*Structure 2.2.15*)

Earlier in this topic, we defined covalent bonding as the electrostatic attraction between two nuclei and a shared pair of electrons. In this section, we will delve deeper into how pairs of electrons occupying atomic orbitals on neighbouring atoms are shared using **valence bond theory**. This is the idea that valence electrons form bonds when orbitals in neighbouring atoms overlap with each other, allowing the electrons to pair their **spins**.

We know that bond strength increases with greater bond order. Consider the average bond enthalpy for a single C–C bond ($346\ kJ\ mol^{-1}$) and contrast it with that of a double C=C bond ($614\ kJ\ mol^{-1}$). The average bond enthalpy for the double bond is significantly higher but **not double** that of the single bond. This is because the two bonds in the double bond are not the same: one is a **sigma bond (σ)** and the other is a **pi bond (π)**. Pi bonds are generally weaker than sigma bonds. This is why the strength of the double bond is not exactly double the strength of the single bond.

In a triple carbon–carbon bond, one of the three bonds is a sigma bond and the other two are pi bonds (figure 78).

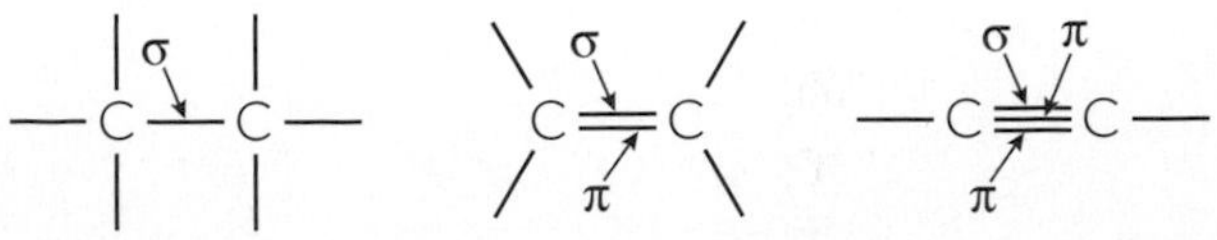

▲ Figure 78 Single bonds are sigma bonds. Double bonds contain one sigma and one pi bond. Triple bonds contain one sigma and two pi bonds

Linking question

What are the different assumptions made in the calculation of formal charge and of oxidation states for atoms in a species? (Structure 3.1, Reactivity 3.2)

Sigma bonds (σ)

Imagine two hydrogen atoms approaching each other. They each have one unpaired electron, in the 1s orbital. Eventually their 1s orbitals will overlap. When they do so, the electrons will pair up forming a covalent bond.

Imagine an invisible line between the nuclei of the two bonding atoms. This line is known as the **bond axis** or **internuclear axis.** The two 1s orbitals will overlap along this axis to form a sigma bond. This means that there is a region of high electron density along the bond axis. Figure 79 shows that sigma bonds can form between pairs of orbitals of different types.

<u>S</u>ingle bonds are always <u>s</u>igma bonds.

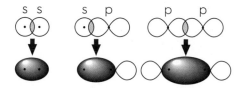

▲ Figure 79 Sigma bonds are formed when orbitals overlap along the bond axis

---

ATL **Thinking skills**

Sigma bonds get their name from the Greek letter **σ**, which is where the letter *s* came from in the Latin alphabet. When you look directly along the bond axis, the sigma bond looks spherical, like an s orbital.

Pi bonds get their name from the Greek letter **π**, which corresponds to the letter **p**. When you look along the bond axis, the pi bond looks like a p orbital, with lobes above and below the axis.

What other things in science are named after things they resemble?

The shapes of s and p orbitals are discussed in *Structure 1.3*.

Pi bonds (π)

If p orbitals are present in two neighbouring atoms, they can overlap sideways, above and below the bond axis. This type of overlap forms a pi bond (figure 80). In pi bonds, the electron density is concentrated at opposite sides of the bond axis.

One pi bond consists of *two* lobes at opposite sides of the bond axis, as shown in figure 80. A pi bond contains two electrons, which occupy both lobes.

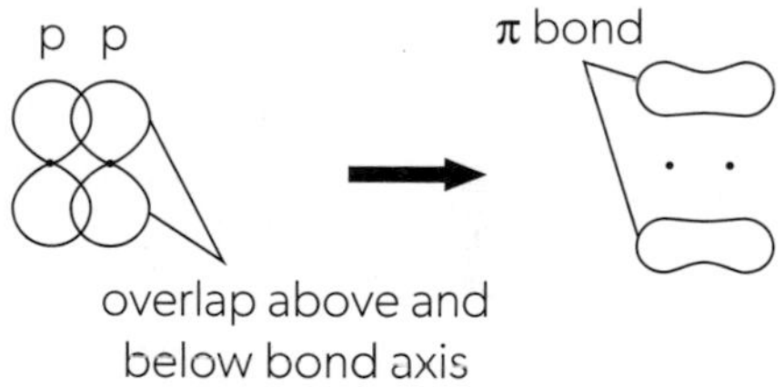

p   p   π bond

overlap above and below bond axis

▲ Figure 80 Pi bonds are formed when p orbitals overlap above and below the bond axis

---

Worked example 8

Determine the number of sigma and pi bonds in the molecule $(NC)_2C=CH_2$. The structural formula is given below.

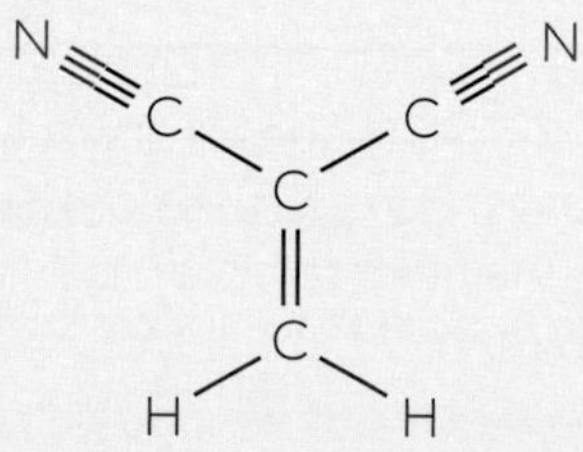

Solution

First, count the sigma bonds in the structural formula. Remember that single bonds are sigma bonds and each of the multiple bonds contains one sigma bond. Therefore, there are 7 sigma bonds.

Then, count the pi bonds. Each triple bond contains two pi bonds. Each double bond contains one pi bond. There are 5 pi bonds in total.

Worked example 9

Determine the number of sigma and pi bonds in the phosphate anion, PO_4^{3-}. The Lewis formula is given below.

Solution

The three single bonds and one of the bonds in the double bond are all sigma bonds, giving a total of 4 sigma bonds. There is one pi bond: the second bond in the double bond.

Practice questions

51. Compare and contrast sigma bonds (σ) and pi bonds (π).
52. Explain why two overlapping s orbitals cannot form a pi bond.

Activity

Look up the Lewis formula of each of the following molecules. Determine their number of sigma and pi bonds.

a. ethanoic acid, CH_3COOH
b. propyne, CH_3CCH
c. ethanenitrile, CH_3CN
d. carbonate ion, CO_3^{2-}
e. ammonium ion, NH_4^+
f. phosphorus trichloride, PCl_3
g. azide ion, N_3^-
h. methanoate ion, $HCOO^-$

Theories

While valence bond theory helps to explain molecular geometry, it fails to give information on electron energies or properties such as magnetism. Many discussions of bonding amongst chemists are based on a more sophisticated theory, **molecular orbital theory**. This theory assumes that electrons are spread across the entire molecule rather than being localized in specific bonds.

Sometimes a phenomenon is best described using both theories. For example, the aromatic stability of benzene can be explained by two models:

- Valence bond theory describes the strong, hexagonal sigma bond framework of benzene.

- Molecular orbital theory shows that delocalization gives benzene a distinct energetic advantage.

Molecular orbital theory draws heavily from quantum mechanics. Can you think of any other examples where we use the other sciences to enhance our understanding of chemistry?

Hybridization (*Structure 2.2.16*)

By now you should know that carbon is **tetravalent**, meaning it can form four covalent bonds. If these bonds are single bonds, they are usually arranged in a tetrahedral structure around the carbon atom at approximately 109.5° angles from one another. For example, methane, CH_4, has a tetrahedral structure.

However, the ground state electron configuration of carbon, $1s^2 2s^2 2p^2$, contradicts these observations. Carbon contains two unpaired 2p electrons (figure 81a) and therefore should form two bonds (one involving each unpaired electron), not four. Furthermore, the two occupied 2p orbitals are at 90° from one another, not 109.5°. This means that the atomic orbitals must undergo certain changes when they form bonds.

Hybridization is the concept of mixing atomic orbitals to form new **hybrid orbitals** for bonding. An example of this process is shown in figure 81 for the carbon atom. There are two steps:

1. **Promotion**: A 2s electron is promoted to one of the 2p orbitals.

2. **Hybridization**: The singly occupied 2s and 2p atomic orbitals are hybridized, meaning they combine and give rise to orbitals of new shapes. The resulting orbitals are called hybrid orbitals and they all have the same energy.

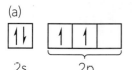

(a)

2s   2p

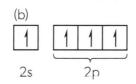

(b)

2s   2p

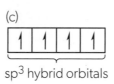

(c)

sp^3 hybrid orbitals

▲ Figure 81 (a) The ground state electron configuration of carbon
(b) One of the 2s electrons is promoted to a 2p orbital
(c) The four atomic orbitals are hybridized to form four sp^3 hybrid orbitals of identical energy

Overall, hybridization is energetically favourable. The promotion step absorbs energy, but the energy released by the subsequent bond formation outweighs this. Promotion does not require much energy because it relieves the 2s electron of the repulsion it experiences when paired.

sp^3 hybrid orbitals

The number of resulting hybrid orbitals is equal to the number of atomic orbitals combined to make them. If four atomic orbitals (one 2s and three 2p) are combined in carbon, four equivalent **sp^3** hybrid orbitals are produced. Each of these hybrid orbitals is a mixture composed of one part 2s and three parts 2p, and therefore has 25% s character and 75% p character (figure 82).

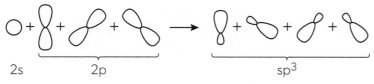

2s   2p sp^3

▲ Figure 82 One 2s and three 2p atomic orbitals combine to form four equivalent sp^3 hybrid orbitals

The four sp^3 hybrid orbitals arrange themselves tetrahedrally, leading to the 109.5° bond angles. In carbon, these orbitals are each occupied by one electron and therefore can form four sigma bonds. For example, in methane, CH_4, each of the sp^3 hybrid orbitals overlaps with a 1s atomic orbital on a hydrogen atom, forming four covalent bonds (figure 83).

You can deduce the electron configuration of an element from its atomic number. This is covered in *Structure 1.3*.

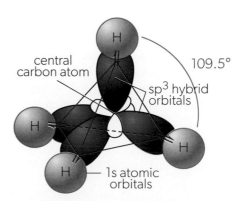

central carbon atom

109.5°

sp^3 hybrid orbitals

1s atomic orbitals

▲ Figure 83 In methane, each carbon sp^3 orbital overlaps with a hydrogen 1s orbital. The geometry around the carbon atom is tetrahedral

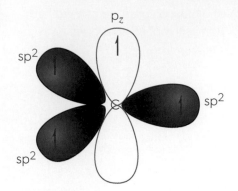

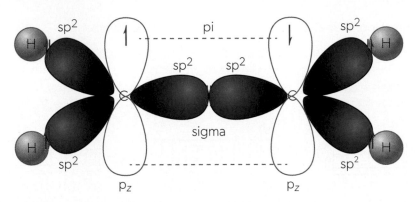

sp² hybrid orbitals

Carbon can also become **sp²** hybridized. The combination of one 2s and two 2p atomic orbitals produces three sp² hybrid orbitals. These sp² hybrid orbitals arrange themselves in a trigonal planar fashion, at 120° from one another (figure 84).

Each hybridized orbital contains one electron, so they can form sigma bonds. The remaining unhybridized p_z orbital can then go on to form a pi bond with a parallel p_z orbital on a different atom (figure 85).

▲ Figure 84 Each of the three sp² hybrid orbitals (blue) contains one electron. The unhybridized p_z atomic orbital (white) also contains one electron

▲ Figure 85 In ethene, C_2H_4, the two carbon atoms each have three sp² hybrid orbitals (blue) which form sigma bonds with each other and two hydrogen atoms. The remaining unhybridized p_z orbitals (white) overlap side-by-side and form a pi bond, resulting in a carbon–carbon double bond

sp hybrid orbitals

The combination of one 2s and one 2p atomic orbital leads to the formation of two **sp** hybrid orbitals. They adopt a linear arrangement, with 180° between them (figure 86).

The hybrid orbitals can form sigma bonds. The remaining unhybridized p_y and p_z orbitals can form two pi bonds with parallel p orbitals on a neighbouring atom (figure 87).

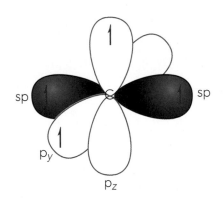

▲ Figure 86 Each of the two sp hybrid orbitals (blue) contains one electron. The two unhybridized p_y and p_z atomic orbitals (white) also contain one electron each

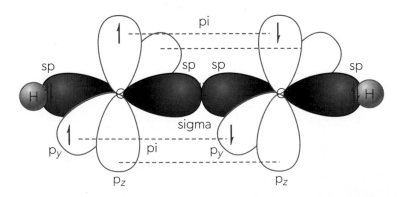

▲ Figure 87 In ethyne, C_2H_2, the one sp–sp overlap and the two 1s–sp overlaps form three sigma bonds. The four unhybridized p orbitals form two pi bonds, resulting in a carbon–carbon triple bond

Hybridization in other atoms

Hybridization also occurs in atoms other than carbon. The number of electrons in the p orbitals will differ, but the general principles are the same. For example, consider the oxygen atom in water, H_2O, which is sp^3 hybridized:

- The oxygen atom's ground state electron configuration is $1s^2\, 2s^2\, 2p^4$.

- We can distinguish between the three 2p orbitals to show that one of them contains a pair of electrons and the other two are singly occupied: $1s^2\, 2s^2\, 2p_x^2\, 2p_y^1\, 2p_z^1$.

- The 2s and 2p orbitals hybridize to form four equivalent sp^3 hybrid orbitals containing six electrons in total (figure 88). Two of the sp^3 hybrid orbitals contain a pair of electrons each and therefore do not form bonds: they constitute the two lone pairs on oxygen that you are familiar with. The remaining two hybrid orbitals are singly occupied and go on to form sigma bonds with the s orbitals on hydrogen atoms.

Since the four hybrid orbitals are tetrahedrally arranged, the bond angles are close to 109.5°. However, two of the hybrid orbitals are already full, so the oxygen atom has two lone pairs. This corresponds to bent molecular geometry. The fact that the bond angle is not exactly 109.5° but rather 104.5° suggests that the hybridization is close to but not exactly sp^3.

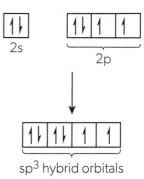

▲ Figure 88 The 2s and 2p atomic orbitals in oxygen are hybridized to form four sp^3 hybrid orbitals

Practice questions

53. Draw orbital box diagrams to show the promotion and hybridization process for the oxygen atom in water, described above.

54. Describe the changes undergone by the atomic orbitals of nitrogen when it bonds to hydrogen to form ammonia, NH_3.

Hybridization and geometry

Hybridization and sigma bonding are closely linked to electron domain geometry. The number of hybrid orbitals formed by an atom is equal to the number of its electron domains.

You will remember from the discussion of VSEPR that double and triple bonds are treated as single additional electron domains when determining molecular shape. This is because pi bonds do not involve hybridized orbitals, so they do not have a large effect on the geometry of a molecule.

Hybridization	Number of hybrid orbitals	Number of electron domains	Electron domain geometry	Number of non-bonding domains	Molecular geometry
sp^3	4	4	tetrahedral	0	tetrahedral
				1	trigonal pyramidal
				2	bent
sp^2	3	3	trigonal planar	0	trigonal planar
				1	bent
sp	2	2	linear	0	linear

▲ Table 16 Hybridization, electron domain geometry and molecular geometry

Practice questions

55. State the hybridization of:
    a. an atom with tetrahedral electron domain geometry
    b. the carbon atom in ethene, C_2H_4
    c. the oxygen atom in oxygen difluoride, OF_2
    d. the nitrogen atom in molecular nitrogen, N_2
    e. an atom with trigonal pyramidal molecular geometry
    f. an atom with bent molecular geometry (two answers are possible here).

56. Consider the following species: CF_4, HCN, N_2H_2. For each of the carbon and nitrogen atoms in these species, deduce the following:
    a. number of electron domains
    b. electron domain geometry
    c. hybridization.

57. The Lewis formula of methyl propanoate is shown below. Deduce the hybridization, number of electron domains, electron domain geometry and molecular geometry for atoms A, B and C.

Hybridization and delocalization

Consider the ethanoate ion, CH_3COO^-. It forms when ethanoic acid loses a hydrogen ion. By doing so, the carbon–oxygen bond lengths change: the bond orders change from 2 and 1 to 1.5 each. The electrons in the double bond become delocalized across the two carbon–oxygen domains. This is shown in figure 89.

(a)
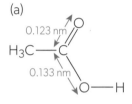

In the C=O bond of ethanoic acid, CH_3COOH, the carbon and oxygen are both sp^2 hybridized. The unhybridized 2p orbital electron in each atom forms the pi bond between them. The other oxygen atom, in –OH, has sp^3 hybridization.

When the –OH hydrogen is lost to form ethanoate, the remaining oxygen atom adopts an sp^2 hybridization. As a result, the oxygen atoms and the carbon atom each have three electron domains around them. Three electron domains correspond to sp^2 hybridization and one unhybridized 2p orbital each. The orbitals overlap allowing the p electrons in them to be delocalized as shown in figure 90.

(b)

▲ Figure 89 When ethanoic acid loses a hydrogen ion, the C–O bonds become equivalent

 Activity

Draw the two resonance structures of the ethanoate ion.

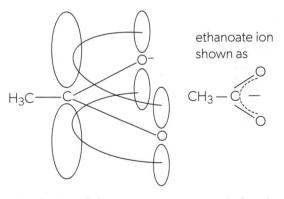

▲ Figure 90 The delocalization of electrons due to resonance in the ethanoate anion

End-of-topic questions

Topic review

1. Using your knowledge from the *Structure 2.2* topic, answer the guiding question as fully as possible:

   What determines the covalent nature and properties of a substance?

2. Compare and contrast ionic bonding, covalent bonding and intermolecular forces.

3. Compare and contrast molecular and covalent network substances.

Exam-style questions

Multiple-choice questions

4. Which of the following species is molecular?

   A. Na_2O

   B. KBr

   C. NH_4NO_3

   D. N_2O_4

5. Which molecule has the shortest carbon–oxygen bond?

   A. CH_3CH_2OH

   B. $(CH_3)_2O$

   C. $(CH_3)_2CO$

   D. CO

6. What is the electron domain geometry, the molecular geometry and the Cl–P–Cl bond angle for the molecule of phosphorus trichloride, PCl_3?

	Electron domain geometry	Molecular geometry	Cl–P–Cl bond angle / °
A.	tetrahedral	tetrahedral	109.5
B.	tetrahedral	trigonal pyramidal	109.5
C.	tetrahedral	trigonal pyramidal	100.3
D.	trigonal pyramidal	trigonal pyramidal	100.3

7. The hydronium ion is formed when a water molecule, H_2O, reacts with a hydrogen ion, H^+. Which type of bond is formed in this reaction?

   A. coordination bond

   B. ionic bond

   C. hydrogen bond

   D. intermolecular force

8. The electronegativities, χ, of four elements are:

Element	H	C	O	Cl
χ	2.2	2.6	3.4	3.2

   Which bond is the most polar?

   A. C–H

   B. O–H

   C. H–Cl

   D. C–O

9. Which of the following allotropes of carbon is molecular?

   A. graphite

   B. graphene

   C. buckminsterfullerene

   D. diamond

10. What are the intermolecular forces present between molecules of CH_3F?

    A. London (dispersion) forces

    B. London (dispersion) forces and dipole–dipole forces

    C. London (dispersion) forces, dipole–dipole forces and hydrogen bonding

    D. hydrogen bonding

11. What are the formal charges on P and O in the Lewis formula of the phosphate oxoanion?

    A. P is −1 and O is 0

    B. P is +5 and O is −2

    C. P is 0 and O is 0 and −1

    D. Both are −3

    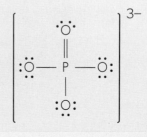

12. What is the electron domain geometry of the sulfite oxoanion, $[SO_3]^{2-}$?

    A. trigonal planar

    B. trigonal pyramidal

    C. tetrahedral

    D. bent (V-shaped)

13. What is the molecular geometry of BrF_5?

    A. octahedral

    B. square planar

    C. T-shaped

    D. square pyramidal

14. What is the molecular geometry of $[PF_6]^-$?

    A. trigonal planar

    B. trigonal bipyramidal

    C. square pyramidal

    D. octahedral

15. Which of the following molecules is non-polar?

    A. SF_4

    B. ClF_3

    C. $BrCl_5$

    D. SeF_6

16. Which of the combinations of atomic orbitals shown below left results in a sigma bond?

    A. I and II only

    B. I and III only

    C. II and III only

    D. I, II and III

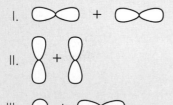

17. What is the hybridization of the oxygen atom in ethanol, CH_3CH_2OH?

    A. sp

    B. sp^2

    C. sp^3

    D. It is not hybridized.

18. What is the hybridization of the carbon atom in hydrogen cyanide, HCN?

    A. sp

    B. sp^2

    C. sp^3

    D. It is not hybridized.

19. How many sigma and pi bonds are present in a molecule of hydrogen cyanide, HCN?

	sigma	pi
A.	1	3
B.	2	1
C.	2	2
D.	3	1

Extended-response questions

20. Describe the meaning of the term **covalent bond.** [2]

21. The phosphonium ion, PH_4^+, is formed when a hydrogen ion, H^+, reacts with a molecule of phosphine, PH_3. For each species, phosphine and phosphonium:

    a. Draw the Lewis formula. Identify any coordination bonds [3]

    b. Deduce the electron domain geometry and molecular geometry. [2]

    c. Suggest the bond angle. [2]

    d. Deduce whether it is polar or non-polar. Explain your reasoning. [2]

22. Methane, CH_4, ammonia, NH_3, and water, H_2O, all have tetrahedral electron domain geometry with bond angles of 109.5°, 107° and 104.5°, respectively. Explain these differences in bond angle values. [2]

23. Diamond and graphite are allotropes of carbon. Describe and explain the electrical conductivity of these two materials. [3]

24. Household vinegar is made from aqueous ethanoic acid, CH_3COOH.

    a. Draw the Lewis formula of ethanoic acid. [2]

    b. Draw a diagram to represent the intermolecular interaction between a molecule of ethanoic acid and a water molecule. [2]

    c. Ethanoic acid molecules can form dimers, particularly when gaseous or dissolved in a non-polar solvent. Dimerization means that two ethanoic acid molecules can associate as shown below.

    i. Calculate the molar mass of an ethanoic acid dimer. [1]

    ii. Suggest why ethanoic acid is more likely to form dimers when dissolved in non-polar solvents (such as hexane) than in polar solvents (such as water). [2]

    d. Ethanoic acid, CH_3COOH, dissolves readily in water. The aqueous solubility of hexanoic acid, $CH_3CH_2CH_2CH_2CH_2COOH$, is very low. Explain this difference in solubility. [2]

    e. State the number of pi and sigma bonds in a molecule of ethanoic acid. [1]

    f. Deduce the hybridization of the carbon and oxygen atoms in ethanoic acid. [2]

25. The following substances all have similar molar masses.

    i ethanoic acid, CH_3COOH

    ii propan-1-ol, $CH_3CH_2CH_2OH$

    iii methoxyethane, $CH_3CH_2OCH_3$

    iv butane, $CH_3CH_2CH_2CH_3$

    a. Use your knowledge of intermolecular forces and structure to list these four substances in order of increasing boiling point. Explain your reasoning. [4]

    b. Explain why, when comparing the strength of different intermolecular forces, it is helpful to compare substances that have similar molar masses. [1]

26. Explain why the boiling point of the group 17 elements increases down the group. [2]

27. Water is an excellent solvent.

    a. Oxygen dissolved in water bodies supports the presence of aquatic organisms. Describe the type of intermolecular forces that occur between oxygen molecules, O_2, and water molecules, H_2O. [1]

    b. Ionic compounds often dissolve in water but not in organic solvents. Explain why ionic compounds do not readily dissolve in hydrocarbon solvents. [2]

28. Morphine and diamorphine are strong painkillers belonging to a group of substances known as opiates. The esterification of morphine produces diamorphine (heroin), as shown below:

    a. Identify and explain which of these two opiates has a higher aqueous solubility. [3]

    b. The potency of opiates depends on their ability to cross from the blood into the brain. The blood–brain barrier is composed of lipids which are non-polar. Identify and explain which of these two opiates dissolves more readily in non-polar environments. [1]

morphine

diamorphine (heroin)

AHL

29. A paper chromatogram for a mixture of amino acids is obtained using a non-polar solvent.

   a. Calculate the retardation factor, R_F, for the spot labelled X in the chromatogram. [2]

   b. Deduce which of the three components is likely to have the lowest polarity. [1]

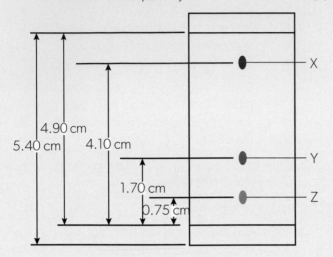

30. The diagram below is a chromatogram of an extract from a supermarket curry sauce (labelled S). Four reference samples of food colourings have also been run on the chromatogram. Outline three conclusions that can be derived from the information in the chromatogram. [3]

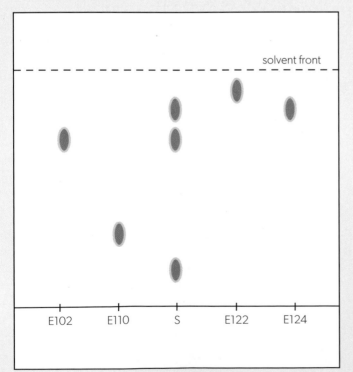

31. The carbonate ion contains delocalized electrons.

   a. Draw the three resonance structures of the carbonate ion, CO_3^{2-}. [2]

   b. State the bond order of the carbon–oxygen bond in the carbonate ion. [1]

32. Benzene, C_6H_6, contains delocalized electrons and is often represented as follows.

   a. Explain, with reference to hybridization, why the delocalized electrons in benzene form a ring above and below the plane of the molecule. [2]

   b. Outline how bond length data confirms the structure of benzene. [1]

   c. Describe one piece of chemical evidence that further supports the structure of benzene shown above. [2]

33. Sulfur trioxide, SO_3, is an acidic gas. If present in the atmosphere, sulfur trioxide can dissolve in rainwater leading to acid rain.

   a. Draw two Lewis formulas for SO_3, one that follows the octet rule, and one where the sulfur atom has an expanded octet. [3]

   b. Determine the formal charges of each of the atoms in the Lewis formulas you drew in (a). [2]

   c. Deduce, using the concept of formal charge, which of the two Lewis formulas is preferred. Explain your reasoning. [1]

AHL

34. Ethene, C_2H_4, belongs to a group of substances known as the alkenes.

   a. Draw the Lewis formula for ethene. [1]

   b. Deduce the molecular geometry of each of the carbon atoms in ethene. [1]

   c. Suggest values for the following bond angles in ethene:

       i.   HCH

       ii. HCC [2]

   d. State the number of sigma and pi bonds in a molecule of ethene. [1]

   e. Deduce the hybridization of the carbon atoms in ethene. [1]

   f. Explain why molecular rotation is restricted around the carbon–carbon bond in ethene. [2]

   g. Identify the main type of intermolecular force present between ethene molecules. Explain your reasoning. [2]

35. State the hybridization of the carbon atoms in diamond and in graphite. Using these data, explain the electrical conductivity of each of the two materials. [4]

36. The condensed structural formula of phenylamine is $C_6H_5NH_2$. A molecular model of phenylamine is shown below.

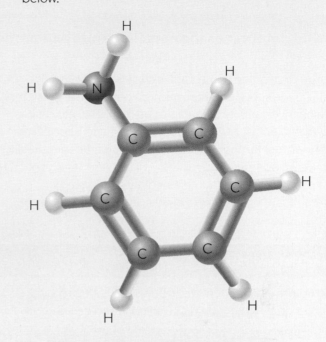

   a. Deduce the electron domain geometry and molecular geometry of the carbon and nitrogen atoms in phenylamine. [2]

   b. State the hybridization of the carbon and nitrogen atoms in phenylamine. [2]

   c. A theoretical study of the electron structure of phenylamine suggests that the H–N–H bond angle in phenylamine should be 112.79°, which is very close to the experimental value. Discuss what you may conclude about the molecular geometry around the nitrogen in the –NH_2 group in the structure of phenylamine, and deduce the hybridization state of nitrogen on this basis. [3]

37. Carbon atoms can form single, double or triple covalent bonds with other carbon atoms. Suggest why two carbon atoms are unlikely to form quadruple bonds with each other. [2]

What determines the metallic nature and properties of an element?

A large proportion of the Earth's crust is composed of metallic elements, and humans have been using metals for millennia. The ability of metals to be moulded into shapes, drawn into wires, and conduct electricity and heat makes them very useful and versatile. These properties are a result of metallic bonding.

Some metals are found as elementary substances in nature, but many are found in ores, in their oxidized state.

The reduction of iron(II) from its ore is achieved with the addition of carbon in a blast furnace. Higher reactivity metals are reduced by **electrolysis** of their molten compounds.

Metal extraction processes require large amounts of energy, posing environmental risks. However, many metals can be recycled, provided that metallic waste is collected and treated safely and correctly.

Understandings

Structure 2.3.1 — A metallic bond is the electrostatic attraction between a lattice of cations and delocalized electrons.

Structure 2.3.2 — The strength of a metallic bond depends on the charge of the ions and the radius of the metal ion.

Structure 2.3.3 — Transition elements have delocalized d-electrons.

AHL

Metallic structures (*Structure 2.3.1*)

Nearly 100 of the 118 known elements are metals. Bonding in metals can be described as the electrostatic attraction between metal cations and delocalized electrons. The degree of electron delocalization in pure elements is inversely related to their electronegativity, which generally increases across periods and decreases down groups. Therefore, metallic properties of elements demonstrate the opposite trend: they increase down groups and decrease across periods. For example, the properties of period 3 elements, from sodium to argon, change gradually from metallic to non-metallic (table 1).

▶ Figure 1 Ancient copper awl, a piercing tool, discovered in the Jordan Valley. It is estimated to be 6,000 years old, and one of the oldest metal artefacts known

	Na	Mg	Al	Si	P	S	Cl	Ar
Metal, metalloid, or non-metal?	metal	metal	metal	metalloid	non-metal	non-metal	non-metal	non-metal
Structure	metallic	metallic	metallic	covalent network	molecular covalent	molecular covalent	molecular covalent	monatomic
Electrical conductivity	high	high	high	semi-conductor	low	low	low	low
Electronegativity	0.9	1.3	1.6	1.9	2.2	2.6	3.2	N/A

▲ Table 1 Properties of period 3 elements

ATL Thinking skills

Chemistry often involves thinking about how microscopic behaviour affects observable features of matter and materials. The connection between structure and physical properties is a central idea in this chapter. As you read, make a conscious effort to connect each of the observable characteristics of metals to the way their particles behave. When you have finished, compare your notes to a peer's and see if there are any similarities and/or differences.

The **metallic–covalent bonding continuum** is shown in figure 2. On the left, there are elements with low electronegativity values that tend to form metallic bonds. On the right, there are highly electronegative non-metals that form covalent bonds. Metalloids are somewhere in the middle: they have both metallic and non-metallic properties.

The metallic bonding model

When metallic atoms bond to other metallic atoms, their valence electrons become delocalized. Delocalized electrons are not attached to individual ions, but rather move freely within the lattice of closely packed cations. This creates a negatively charged "sea" of electrons that surrounds the cations.

Metallic bonding results from the electrostatic attraction between metal cations and the sea of delocalized electrons around them (figure 3).

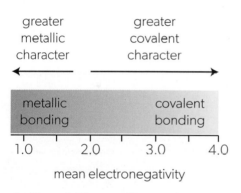

▲ Figure 2 The metallic–covalent bonding continuum

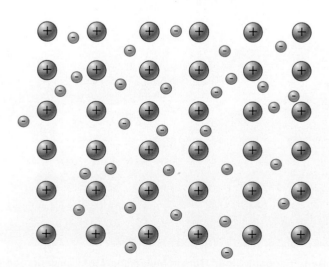

▲ Figure 3 Structure of a metal showing an array of positive ions (cations) surrounded by a "sea" of delocalized electrons

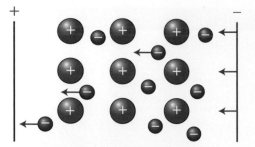

▲ Figure 4 When a potential difference is applied, the delocalized electrons are donated by the negative terminal and move towards the positive terminal

The properties of metallic structures

The presence of delocalized electrons gives rise to many characteristic properties of metals. Metals are good **electrical conductors, thermal conductors**, and they are **malleable** and **ductile**.

Substances can conduct electricity when mobile charged particles are present. Consider a sample of metal. It contains a lattice of cations surrounded by free-moving, negatively charged electrons. The delocalized electrons move randomly throughout the metallic structure. When a potential difference is applied, there is a net movement of electrons away from the negative terminal and towards the positive terminal (figure 4).

TOK

Metals are lustrous and reflect light well, which is why they are used to make mirrors. Reflective surfaces have historically sparked human fascination and curiosity. In some cultures, mirrors were believed to have magical properties. They have also been perceived as valuable, status-signalling objects. Mirrors have appeared in literature, both as instruments of clarity and misrepresentation. Nowadays, we know the delocalized electrons in metals reflect incident light, and most household mirrors are made by applying a layer of silver metal to a glass surface. These are examples of how different areas of knowledge might approach the same topic.

Are some types of knowledge more open to interpretation than others?

▲ Figure 5 Anish Kapoor's Cloud Gate in Millennium Park, Chicago, reflects the people, buildings and sky around it on its curved, stainless steel surface

One of the reasons metals are used to make cookware is their thermal conductivity, meaning that they conduct heat well. The thermal energy of a substance is related to the **kinetic energy** of its particles. When a substance is heated, its particles gain kinetic energy, so their vibrations and movement become more vigorous.

When a metal is heated, the vibrations of its cations increase in magnitude. These vibrations are easily passed along to other cations because of their closely packed arrangement. The vibrating cations also transfer energy to the surrounding electrons through collisions. Since the electrons in metals are mobile, they can pass this increased kinetic energy to other parts of the lattice. This explains why metals are good conductors of heat.

As thermal energy in metals increases, ions in the lattice vibrate more, so there are more collisions between electrons and ions. Some kinetic energy is converted to heat with each collision. These collisions are the cause of **electrical resistance** in metals, which increases with temperature (figure 6). Conversely, a decrease in temperature reduces the frequency of collisions, so the electrons move in a more direct path. In this case, the resistance decreases.

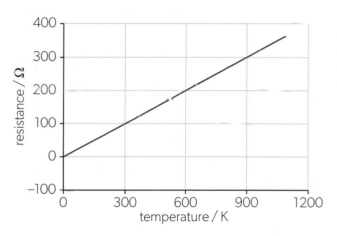

◀ Figure 6 Resistance increases linearly with temperature for most metals

Superconductors are materials that offer no resistance to electric current below a certain "critical" temperature. At very low temperatures many metals and some other materials exhibit this property.

Malleability is the ability of a solid to be pressed or pounded into different shapes without breaking. This useful property is the reason why metallic objects have such a wide array of shapes, ranging from sewing needles to aeroplanes to zippers to anchors.

Metallic bonding is non-directional. When a force is applied to a metallic structure, the layers of cations can slide past each other without breaking the electrostatic attraction to the surrounding delocalized electrons. The cations can therefore be rearranged, allowing the metal to take on a new shape (figure 7). Aluminium foil, often used to wrap food, demonstrates that the metal aluminium is readily malleable.

A related property, ductility, is the ability of a solid to be stretched into wires. Ductility and electrical conductivity are a very useful combination of properties that allows electrical wires to be produced at the industrial scale.

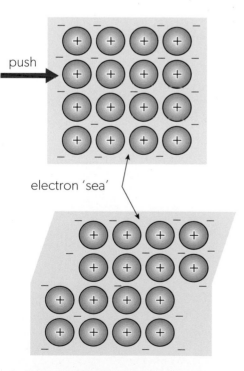

push

electron 'sea'

▲ Figure 7 Metallic bonding remains intact even after a metal is hammered into a sheet or other shape

To summarize, metals are typically:

- good electrical conductors because they contain mobile delocalized electrons

- good thermal conductors because they contain mobile delocalized electrons and closely packed cations

- malleable and ductile, because the layers of cations can slide past each other without breaking the metallic bonds.

▲ Figure 8 Liquid iron metal from a blast furnace

▲ Figure 9 Aluminium cans in a recycling plant. Like other metals, aluminium retains its properties after being recycled

 Database investigation of the properties of metals

In this skills task, you will use a database to explore the properties of metals.

Relevant skills

- Tool 2: Identify and extract data from databases
- Inquiry 1: Formulate research questions
- Inquiry 2: Collect and record sufficient relevant quantitative data

Instructions

Using at least one database of your choice, identify, extract and analyse data that will allow you to investigate the trend relating to the properties of metals. This may involve looking at electrical conductivity, melting point or thermal conductivity of a number of metals across a period or down a group. (If you are an AHL student, you may wish to explore a property of the transition elements.)

State a research question that is addressed by your data collection and processing.

 Before starting this task, create a plan of the steps you think it will require and estimate the time needed to complete each step. Then, do the task and keep a log of the actual time you spent completing the various steps.

When you have finished, compare the two sets of steps and times. How do they compare? How can these observations help you plan extended tasks in the future?

Practice questions

1. State the property (or properties) of metals that makes them suitable for manufacturing:
   a. kitchen pans
   b. electric power cables
   c. radiators
   d. cars
   e. artificial hip joints
   f. guitar strings.
2. Draw a labelled diagram to explain how bonding occurs in metals.
3. Draw labelled diagrams to explain why metals are:
   a. malleable
   b. good electrical conductors
   c. good thermal conductors.

 Linking questions

What experimental data demonstrate the physical properties of metals, and trends in these properties, in the periodic table? (Tool 1, Inquiry 2, Structure 3.1)

What trends in reactivity of metals can be predicted from the periodic table? (Reactivity 3.2)

The strength of metallic bonds (*Structure 2.3.2*)

The stronger the attractions between the cations and delocalized electrons, the stronger the metallic bond.

Metallic bond strength *decreases* with *greater ionic radius*, and *increases* with *greater ionic charge*.

A third factor, related to ionic charge, is the electron density of the "sea" of delocalized electrons.

Melting point and boiling point data trends can be used to compare the strength of metallic bonding in different metals. Consider the melting and boiling points of the group 1 metals lithium, sodium and potassium (table 2).

Element	Charge of the cation	Ionic radius / 10^{-12} m	Melting point / °C	Boiling point / °C
lithium	1+	76	181	1342
sodium	1+	102	98	883
potassium	1+	138	63	759

◀ Table 2 Melting and boiling point data for group 1 metals

The melting and boiling points decrease down the group as the ionic radius increases. A greater ionic radius means a greater average distance between the cations and the delocalized electrons. This weakens the electrostatic forces of attraction between them.

Now compare sodium against two other period 3 metals, magnesium and aluminium.

▶ Table 3 Melting and boiling point data for period 3 metals

Element	Charge of the cation	Ionic radius / 10^{-12} m	Melting point / °C	Boiling point / °C
sodium	1+	102	98	883
magnesium	2+	72	650	1090
aluminium	3+	54	660	2519

The melting and boiling points of sodium, magnesium and aluminium increase across the period. The reasons for this are:

1. **Decrease in ionic radius**: this reduces the distance between the cations and electrons, increasing the electrostatic attraction between them

2. **Increase in ionic charge**: this increases the strength of the electrostatic attraction between the cations and delocalized electrons

3. **Greater number of delocalized electrons per ion**: Each aluminium atom contributes three valence electrons to the sea of electrons. Sodium atoms have only one valence electron that becomes delocalized.

▲ Figure 10 Metallic bonding in sodium is relatively weak. As a result, sodium is a soft metal that can easily be cut with a knife

🧪 Using trends in data to predict properties

In this skills task, you will examine the trends in group 1 melting points to predict the properties of unknown elements.

Relevant skills

- Tool 3: Construct and interpret charts and graphs
- Tool 3: Extrapolate graphs
- Inquiry 2: Carry out relevant and accurate data processing

Instructions

1. Examine the melting point and boiling point data in table 2. Plot a graph of these data that will help you analyse the melting and boiling point trends down group 1.

2. Extrapolate your graph to predict the melting and boiling points of the other metals in group 1 (rubidium and caesium).

3. Look the values up in the chemistry data booklet and compare them to your predicted values. Calculate the percentage errors.

4. Briefly evaluate the validity of the extrapolation.

5. Draw a second graph of these data, including Rb and Cs this time.

6. Reflect on how plotting data that covers a larger range affects the quality of your second graph.

Practice questions

4. The melting points of potassium and calcium are 63 °C and 842 °C, respectively. State three reasons why the melting point of calcium is higher than that of potassium.

5. List these metals in order of increasing melting point.
    A. Be, Ca, Sr, Ba
    B. Be, Ba, Ca, Sr
    C. Ca, Sr, Ba, Be
    D. Ba, Sr, Ca, Be

6. Which of the following explains why aluminium has a higher melting point than magnesium?
    I.   Al contains more delocalized electrons per ion than magnesium
    II.   The ionic radius of Al^{3+} is smaller than that of Mg^{2+}.
    III. The charge of an aluminium ion is greater than that of a magnesium ion.
        A. I only
        B. II only
        C. II and III only
        D. I, II and III

Linking question

What are the features of metallic bonding that make it possible for metals to form alloys? (Structure 2.4)

Bonding in transition elements (*Structure 2.3.3*)

Transition elements are found in the d-block in the periodic table, and they have many properties typically associated with metals: hardness, strength and high density. They exhibit variable oxidation states and, frequently, catalytic properties.

Transition elements are defined as those that have atoms, or give rise to ions, with incomplete d sublevels. Most of the d-block elements are transition elements, but there are some exceptions. For example, zinc is not a transition element because its 3d sublevel is full, both when it is an atom and when it forms the only stable ion Zn^{2+}.

Observations

Scientists are observers, looking at Earth and all other parts of the universe, to obtain data about natural phenomena. The name of the transition element chromium alludes to the colourful compounds it forms. Chromium was discovered by French chemist Nicolas-Louis Vauquelin, who observed and analysed a lead ore sample containing chromium from Siberia. After precipitating out the lead, Vauquelin noted the different colours and named this new element after the Greek word for colour, *chromos.*

The position of the d-block and transition elements in the periodic table are discussed in *Structure 3.1.*

AHL

▲ Figure 11 Car catalytic converter made from platinum, rhodium and palladium. Transition metals often have catalytic properties due to their variable oxidation states

In transition elements, the electrons in the d sublevel become delocalized, as well as the electrons of the outer level. The greater electron density strengthens the electrostatic forces of attraction between the cations and the electron sea that surrounds them. This results in higher melting points for transition elements compared to group 1 and group 2 metals.

The large number of delocalized electrons in transition elements also allows for good electrical conductivity. This is because there are more delocalized electrons that can move along the metal sample when a potential difference is applied.

The chemical properties of transition elements are discussed in *Reactivity 3.4*.

▶ Figure 12 The glowing filament in this incandescent light bulb contains tungsten. Tungsten's extraordinarily high melting point makes it suitable for high temperature applications. It is also exceptionally strong and dense. In fact, its name derives from the Swedish for heavy stone: "tung sten"

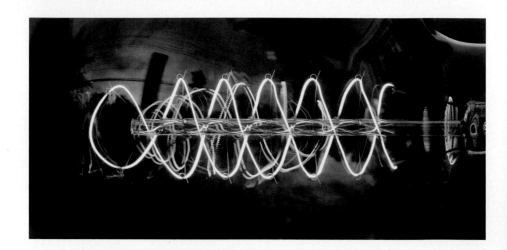

Practice questions

7. Which of the following properties of transition elements is a result of the presence of delocalized d electrons?

   A. coloured compounds
   B. high density
   C. good electrical conductivity
   D. catalytic properties

8. Which of the following statements is **incorrect**?

   A. Transition elements generally have high melting points and boiling points due the presence of unpaired d electrons.
   B. Many transition element compounds are coloured because their ions often contain incomplete d sublevels.
   C. Transition elements are good electrical conductors due to the presence of delocalized d electrons.
   D. Transition elements often have multiple oxidation states.

9. Explain, with reference to the electron configurations of Zn and Zn^{2+}, why zinc is not a transition element.

Linking question

Why is the trend in melting points of metals across a period less evident across the d-block? (Structure 3.1)

ATL Thinking skills

As you know, there are various linking questions in this book that aim to interconnect the course content and encourage the development of a networked understanding of chemistry. Write a linking question of your own, that explores the connection between the content in this section and at least one other part of the DP chemistry course. Share it with a partner and compare answers.

End-of-topic questions

Topic review

1. Using your knowledge from the *Structure 2.3* topic, answer the guiding question as fully as possible:

   What determines the metallic nature and properties of an element?

2. Compare and contrast ionic, metallic and covalent bonding.

3. Compare and contrast ionic, metallic, covalent network, and molecular structures.

Exam-style questions

Multiple-choice questions

4. Which statement best describes metallic bonding?

   A. Electrostatic attractions between a lattice of positive ions and delocalized electrons.

   B. Electrostatic attractions between a lattice of negative ions and delocalized protons.

   C. Electrostatic attractions between protons and electrons.

   D. Electrostatic attractions between oppositely charged ions.

5. Which substance has a metallic structure?

	Melting point / °C	Solubility in water	Electrical conductivity when molten	Electrical conductivity when solid
A	36	high	low	low
B	186	insoluble	low	low
C	1083	high	high	low
D	1710	insoluble	high	high

Extended-response questions

6. Metals have many useful properties including malleability and electrical conductivity.

Element	sodium, Na	magnesium, Mg	aluminium, Al	copper, Cu
Electrical conductivity / $\times 10^7 \, S \, m^{-1}$	2.1	2.3	3.8	5.9

a. Explain why metals are malleable. Include a labelled diagram in your answer. [3]

b. The electrical conductivity of several elements is shown below:

    i. Describe and explain the trend in the electrical conductivity of sodium, magnesium and aluminium. [2]

    ii. Explain why the electrical conductivity of metals decreases with increasing temperature. [2]

    iii. Suggest why the electrical conductivity of copper is significantly higher than that of the other three elements shown in the table. [1]

7. Boiling point data for the period 3 elements is shown below.

Element	Sodium, Na	Magnesium, Mg	Aluminium, Al	Silicon, Si	Phosphorus, P_4	Sulfur, S_8	Chlorine, Cl_2	Argon, Ar
Boiling point / °C	883	1 090	2 519	3 265	281	445	−34	−186

a. Identify the structure of each of the elements in period 3. [1]

b. Plot a graph of boiling point vs atomic number. [1]

c. Explain the trend in boiling point across the period. [3]

d. Describe and explain the trend in electrical conductivity across the period. [3]

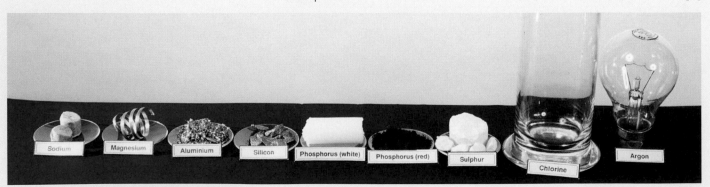

▲ Figure 13 The period 3 elements

8. Explain why melting point and boiling point are good indicators of metallic bond strength. [1]

AHL

Structure 2.4 | From models to materials

What role do bonding and structure have in the design of materials?

So far in *Structure 2*, we have discussed how materials can be classified according to their bonding: ionic, covalent and metallic. Materials can also be classified according to their use, origin, and properties.

History has characterized civilizations by the materials they use: Stone Age, Bronze Age, and Iron Age. Uses for materials were developed based on observations of their properties, before explanations for those properties had been proposed.

From metals to nanotechnology, research into the uses of materials is sometimes deliberate, and sometimes their unique properties are discovered by chance.

Materials science is the study of the structure and properties of materials and their application, and the development of new materials. It draws heavily from chemistry, physics and engineering.

Another way to classify materials involves grouping them into three types: metals, polymers and ceramics. Two of these categories are considered in *Structure 2.4*: metals (specifically, alloys) and polymers. Your knowledge of structure and bonding will help you understand the nature of alloys and polymers. Alloys are predominantly metallic structures, while polymers are covalent compounds. Polymers have molecular structure, although their molecules are very large.

Understandings

Structure 2.4.1 — Bonding is best described as a continuum between the ionic, covalent and metallic models, and can be represented by a bonding triangle.

Structure 2.4.2 — The position of a compound in the bonding triangle is determined by the relative contributions of the three bonding types to the overall bond.

Structure 2.4.3 — Alloys are mixtures of a metal and other metals or non-metals. They have enhanced properties.

Structure 2.4.4 — Polymers are large molecules, or macromolecules, made from repeating subunits called monomers.

Structure 2.4.5 — Addition polymers form by the breaking of a double bond in each monomer.

Structure 2.4.6 — Condensation polymers form by the reaction between functional groups in each monomer with the release of a small molecule.

AHL

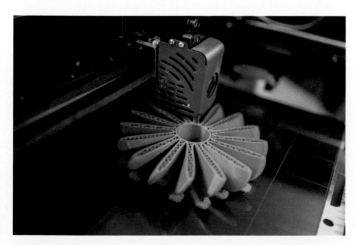

▲ Figure 1 A 3D printer creating a model from a polymer material, and traditional Javanese gamelan metallophones

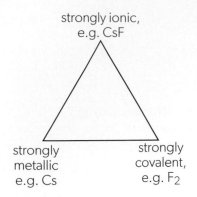

▲ Figure 2 The van Arkel-Ketelaar triangular bonding diagram

The bonding continuum (*Structure 2.4.1*)

Some materials cannot be categorized easily into substances with metallic, ionic and covalent bonding. We saw this in **Structure 2.1**, with compounds such as aluminium chloride, $AlCl_3$, having a hybrid of covalent and ionic bonding. We also discussed metalloids in **Structure 2.3**, which exhibit covalent and metallic properties. Therefore, bonding is best described as a continuum between these models.

Dutch chemists Anton Eduard van Arkel and Jan Ketelaar constructed a triangular diagram to represent the continuum of the three bonding types. The three corners represent "pure" metallic, ionic and covalent bonding (figure 2). The sides of the triangle represent the intermediates between one bonding type and another. For example, an Al–F bond can be positioned roughly halfway along the ionic–covalent continuum.

 Patterns and trends

We use classification in science to organize objects, information and concepts. This can help us to recognize trends and patterns as well as anomalies. Materials can be classified according to different criteria such as use, origin, properties, structure, and bonding. What are the advantages and disadvantages of each system?

The properties of materials with different bonding types

The **solubility** and **volatility** of a substance can be explained in terms of bond type, structure and, if applicable, intermolecular forces. The **electrical conductivity** of a substance depends on the presence of mobile charged particles. Table 1 summarizes these and other properties of substances with different bonding types.

Property	Bonding type			
	Metallic	**Ionic**	**Molecular covalent**	**Covalent network**
melting point and boiling point	varies	high	low	high
volatility	low	low	high	low
water soluble	no	varies	varies	no
electrical conductor	yes	no when solid yes when liquid/ aqueous	no	no (except for graphite and graphene)
thermal conductor	yes	no	no	varies
brittle	no	yes	no, except some polymers	yes
susceptible to corrosion	yes	varies	varies	no

▲ Table 1 Typical properties of metallic, covalent and ionic substances. This table is meant as a general guide and there are exceptions where substances do not exhibit typical behaviour

Brittleness is the opposite of malleability. Ionic crystals are brittle while metals are malleable. Brittle substances snap and break when subjected to a force because they cannot be deformed easily. The atoms or ions within the substance are unable to slide past each other.

Elastic materials will change shape when subjected to a force, and return to their original shape after the force is removed. In a metal spring, the metallic bonding is responsible for the elasticity. In contrast, rubber is elastic due to its long polymer chains being able to uncoil and coil up again. You can observe this property in an elastic band. The opposite property is **plasticity**, which means that the material retains its deformed shape even after the external force is removed. Modelling clay is an example of a material with plasticity.

The term **corrosion** is often used to describe the oxidation of a metal, such as iron, in the presence of oxygen and water. Corrosion more generally refers to a chemical reaction between a material and its surrounding environment, which damages the material in some way.

Many materials do not behave in a manner typical of its bonding type. For example, aluminium has a low density while mercury is volatile, despite the fact that both substances are metals. Graphite and graphene are covalent network solids that conduct electricity. Molecular covalent substances have variable properties because their molecular size and polarity vary widely. Waxes are soft molecular covalent substances, whereas some molecular covalent plastics are hard. In some cases, more than one bonding type can contribute to the properties of a substance. This helps to explain the variations of properties observed within a given bonding category.

▲ Figure 3 Rusting is a type of corrosion. Iron and steel rust in the presence of water and oxygen

Worked example 1

Explain the trends in melting points for silver halides and potassium halides, given that the bonds in silver halides have greater covalent character than those in potassium halides. Use the data given in table 2 and table 3.

	Melting point (°C)	
AgF	435	Melting point generally increases
AgCl	455	
AgBr	432	
AgI	558	

▲ Table 2 Melting points of the silver halides

	Melting point (°C)	
KF	859	Melting point decreases
KCl	773	
KBr	734	
KI	681	

▲ Table 3 Melting points of the potassium halides

Solution

The partial covalent character of silver halides means that their melting points depend on the strength of van der Waals forces. The increasing number of electrons going down the group from AgF to AgI increases the strength of the London (dispersion) forces, causing the melting point to increase (table 2).

The bonding in potassium halides is predominantly ionic, with minimal covalent character. Therefore, the strength of the bonds depends on how well the cations and anions interact with one another. From KF to KI, the size of the anion increases, so the electrostatic attraction between oppositely charged ions decreases and therefore the melting point also decreases (table 3).

You can try to locate these compounds in the triangular bonding diagram in figure 2. The silver halides are all in the covalent region of the diagram, near the ionic section. In contrast, the potassium halides are all firmly located in the ionic region of the triangle.

 Data-based questions

Plot a graph of the data given in tables 2 and 3. Compare your graph with the tables and suggest which is a better method for illustrating the data.

Use of triangular bonding diagrams (*Structure 2.4.2*)

Over time, chemists have further developed van Arkel and Ketelaar's original work. Eventually, electronegativity was incorporated in a way that merges the two continua in figures 4a and 4b.

You first encountered figure 4a in *Structure 2.2*, and figure 4b in *Structure 2.3*.

The resulting triangular bonding diagram (or "bonding triangle") is shown in figure 4c.

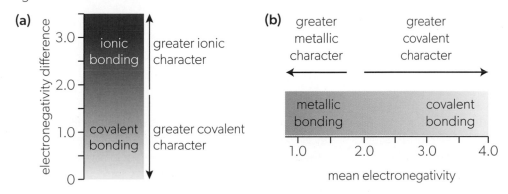

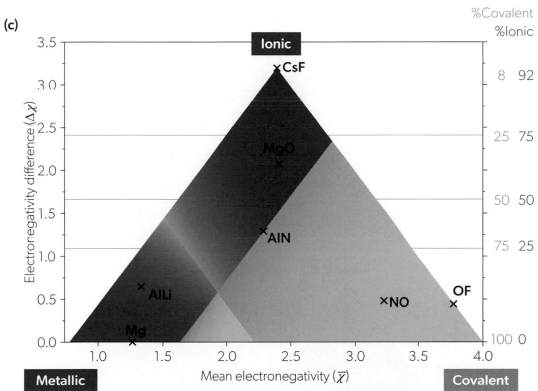

▲ Figure 4 (a) The ionic–covalent bonding continuum (b) The metallic–covalent bonding continuum (c) The resulting triangular bonding diagram. This diagram incorporates different bonding categories, the mean electronegativity and the difference in electronegativity. Examples of several substances are included

The position of a substance in the bonding triangle is determined by the relative contributions of the three bonding types to the overall bond. To locate it, we need to know the electronegativity values of the two elements involved in the bond. We then need to work out two parameters:

- difference in electronegativities, $\Delta\chi$, which is the ionic–covalent parameter

- mean electronegativity, $\overline{\chi}$, which is the metallic–covalent parameter.

We do not need to consider molecular formula or bond order. As a result, two or more different bonds may have identical parameters. For example, the double carbon–oxygen bond in carbon dioxide is in the same place in the bonding diagram as the triple carbon–oxygen bond in carbon monoxide.

Worked example 2

Determine the position of the following substances in the triangular bonding diagram:

a. carbon tetrafluoride, CF_4

b. pure silicon, Si

c. barium iodide, BaI_2

Solution

a. carbon tetrafluoride, CF_4

For this example, we need to look at the bond between C and F in CF_4. Carbon has an electronegativity of 2.6 and fluorine is 4.0. From this, we can calculate the electronegativity difference, $\Delta\chi$:

$$\Delta\chi = 4.0 - 2.6$$
$$= 1.4$$

Then, calculate the mean electronegativity, $\bar{\chi}$:

$$\bar{\chi} = \frac{2.6 + 4.0}{2}$$
$$= 3.3$$

With these parameters, we can locate the carbon–fluorine bond in the upper part of the covalent region of the triangle (figure 5). This means that the bond between C and F is polar covalent.

b. pure silicon, Si

In this example, we consider the bond between two Si atoms. In elemental substances like this, the atoms are all the same type and thus have the same electronegativity. Therefore, the electronegative difference between two Si atoms is zero, and the mean electronegativity is equal to the Pauling electronegativity of one Si atom (1.9).

This places silicon in the region between metallic and covalent in the bonding triangle (figure 5).

c. barium iodide, BaI_2

The two atoms in the bond are barium ($\chi = 0.9$), and iodine ($\chi = 2.7$).

$$\Delta\chi = 2.7 - 0.9 = 1.8$$
$$\bar{\chi} = \frac{2.7 + 0.9}{2} = 1.8$$

With these parameters, we can locate the bond in the ionic region of the diagram (figure 5).

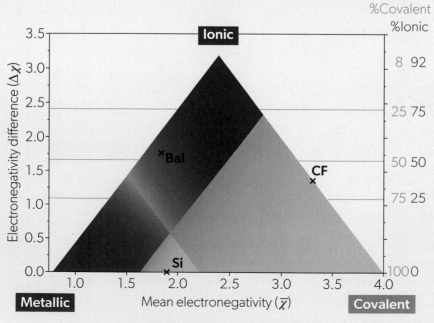

▲ Figure 5 The location of certain bonds in the triangular bonding diagram

Practice questions

1. Using electronegativity values, deduce the main type of bonding in each of the following substances by determining their position in the bonding triangle.
   a. fluorine, F_2
   b. caesium fluoride, CsF
   c. lithium oxide, Li_2O
   d. brass, composed of Cu and Zn
   e. aluminium chloride, $AlCl_3$
   f. silicon dioxide, SiO_2

2. Qualitatively describe the relative contributions of the three different types of bonding in each of the substances in the previous question.

3. Explain why nitrogen monoxide, NO, and nitrogen dioxide, NO_2, are located at the same point in the bonding diagram.

The properties of a substance can be predicted by its position in the bonding triangle. We have already discussed the typical properties of pure ionic (*Structure 2.1*), covalent (*Structure 2.2*), and metallic (*Structure 2.3*) substances. Now we will look at a selection of substances with significant contributions from more than one type of bonding.

Silicon: metallic and covalent character

Silicon is positioned between the metallic and covalent regions in the bonding triangle. It is a metalloid, which means that it exhibits both covalent and metallic properties. Silicon is lustrous (shiny) like a metal. On the other hand, it forms a covalent network structure in which atoms share electrons. Like covalent substances, silicon is brittle and forms a (weakly) acidic oxide. In terms of electrical conductivity, it is an intermediate between conductors (like metals) and insulators (like most covalent substances) and is classified as a **semiconductor**. Semiconductors are generally poor electrical conductors, but their conductivity increases when heated, illuminated, or in the presence of certain impurities.

◀ Figure 6 A silicon wafer containing hundreds of tiny microchips for use in electronics.

Lattice enthalpy is discussed in *Structure 2.1*, and in *Reactivity 1.2* at additional higher level.

Magnesium iodide: ionic and covalent character

In magnesium iodide, $\Delta\chi = 1.4$ and $\bar{\chi} = 2.0$. The percentage ionic and covalent character are both about 50%. As a result of this, magnesium iodide has an unusually high lattice enthalpy.

Data-based questions

You can test the assumptions of a model by comparing predicted and experimental values. For example, the theoretical lattice enthalpy for an ionic compound assumes completely ionic behaviour. However, if the bonds between ions are partly covalent, the experimental lattice enthalpy tends to be significantly greater than the theoretical prediction. This often happens with larger ions, particularly those with lower electronegativity values, because their electron clouds are more polarisable. This allows a certain degree of electron sharing (covalent bonding) in addition to ionic bonding.

Table 4 shows the difference between experimental and theoretical values of lattice enthalpy for sodium chloride and silver chloride.

Compound	Experimental lattice enthalpy / kJ mol^{-1}	Theoretical lattice enthalpy / kJ mol^{-1}
sodium chloride, NaCl	790	750
silver chloride, AgCl	918	734

▲ **Table 4** Comparison of experimental and theoretical lattice enthalpy values for two compounds

Based on the differences between theoretical and experimental lattice enthalpy values, assess and compare the ionic and covalent character of bonding in each compound.

Aluminium chloride: ionic and covalent character

Another compound which is classified as ionic but has a significant covalent contribution is aluminium chloride. Unlike magnesium iodide, aluminium chloride has an unusually low melting point, even though it appears in the same region of the bonding triangle as magnesium iodide. This shows that the bonding triangle cannot always be used to predict the properties of a substance. This is particularly true if the bonding in that substance has a mixed character.

In aluminium chloride, $\Delta\chi = 1.6$ and $\bar{\chi} = 2.4$. Therefore, aluminium chloride is right at the boundary between ionic and polar covalent in the bonding triangle. As a result, aluminium chloride exhibits both covalent and ionic properties. $AlCl_3$ forms ionic lattices when solid. At high pressure, it melts into Al_2Cl_6 dimers (figure 7) at 190 °C, which is very low for an ionic compound. Also, unlike a typical ionic compound, aluminium chloride is soluble in non-polar solvents such as trichloromethane.

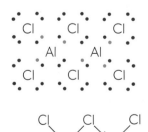

▲ **Figure 7** Two ways of representing bonding in the aluminium chloride dimer, Al_2Cl_6

Activity

Use your knowledge of the behaviour of compounds containing a hydrogen–oxygen bond to predict where in the bonding triangle you expect to find this bond. Then calculate the $\Delta\chi$ and $\bar{\chi}$ values to see if your predictions were close.

 Linking question

Why do composites like reinforced concretes, which are made from ionic and covalently bonded components and steel bars, have unique properties? (Structure 2.1, 2.2, 2.3)

Practice questions

4. Mean electronegativity and difference in electronegativity values for different bonds are shown below. Outline the properties you would expect a compound with these bonds to have.
   a. $\bar{\chi} = 2.9$ and $\Delta\chi = 1.4$
   b. $\bar{\chi} = 1.9$ and $\Delta\chi = 0.3$
   c. $\bar{\chi} = 2.3$ and $\Delta\chi = 1.9$

Alloys (*Structure 2.4.3*)

Pure metallic elements can be mixed with other elements, metallic or non-metallic, to form **alloys**. Different atoms or ions can be held within the structure, while still maintaining the delocalized sea of electrons throughout. Alloys therefore retain many metallic characteristics such as electrical conductivity and lustre. However, the properties of alloys are different to those of pure metals. Properties such as hardness, corrosion resistance and melting point, can be enhanced by alloying a metal with other elements.

Alloys are **mixtures**. The ratio of the components in an alloy can vary without changing the identity of the substance. For example, the proportion of carbon in steel ranges from traces to about 2%. In addition, the components of alloys retain many of their original properties. All these characteristics are typical of mixtures.

In *Structure 2.3*, we discussed how metallic structures contain a lattice of cations in a sea of delocalized electrons, and the fact that bonding between the cations and electrons is non-directional. This accounts for the malleability of metals.

Properties of alloys

In pure metals, all the cations in the metallic lattice are the same size. When a force is applied, the layers of cations slide past each other easily.

While malleability is a useful property of metals, sometimes we require a metallic structure to be stronger. Alloying involves the addition of atoms or ions with a different radius to the cations of the pure metal, which disrupts the regular structure of the lattice. When an alloy is struck with a force, the layers of cations do not slide past each other as easily. Therefore, alloys are usually stronger than pure metals.

iron

alloy

◀ Figure 8 In pure metals, the ions are packed in a regular pattern, whereas the addition of another element to form an alloy disrupts this arrangement, reducing the malleability

Examples of alloys

NaK is a sodium–potassium alloy that has a lower melting point than its constituent elements, so it is a liquid at room temperature. Because of this, NaK is used as a nuclear reactor coolant, which is non-volatile and can be pumped as any other liquid.

Memory metals are alloys that return to their original shape upon heating. They are used to make objects that are prone to being deformed through use, such as spectacle frames.

Iron is an abundant metal in the Earth's crust. Pure iron can be deformed relatively easily. This limitation can be overcome by alloying it with carbon to create **steel**. Steels are harder and stronger than pure iron, making them ideal for a variety of uses ranging from construction to tools. There are many varieties of steel, each containing different amounts of carbon and other elements.

Like iron, steel **rusts** in the presence of water and elemental oxygen, O_2. Rusting is problematic because it transforms iron into hydrated iron(III) oxide. This is an ionic compound, so it does not have the valuable properties of metals. Rust also flakes off easily, lowering the volume of metal and exposing the iron underneath to further corrosion.

The degradation of objects made from iron and steel has economic implications. Large steel-containing structures (for instance, ships and bridges) need to be protected from rusting to preserve their physical integrity. Rust protection methods include barrier methods (such as painting and oiling) and sacrificial methods (such as **galvanising**).

Stainless steels are iron alloys that contain at least 11% chromium. The chromium reacts with oxygen in the air or water to form a thin layer of chromium oxide on the surface of the steel. This chromium oxide layer prevents rusting, and therefore stainless steels have useful applications in cooking and medical-grade equipment.

(a)

(b)

▲ Figure 9 (a) The Eiffel Tower in Paris, France, is made of iron. It is painted every few years to protect it from rusting. The painting process itself lasts several months (b) Adding chromium to steel creates stainless steel, a rust-resistant alloy suitable for kitchen equipment

How does the carbon content affect the hardness of steel?

Steels that are made up of 0.3% to 0.6% carbon by mass are known as medium steels. The hardness of medium-steel samples with varying carbon contents were tested, giving the results shown in figure 10.

Hardness values (units: HV) were determined using a method known as the Vickers hardness test. Harder materials have greater Vickers hardness values.

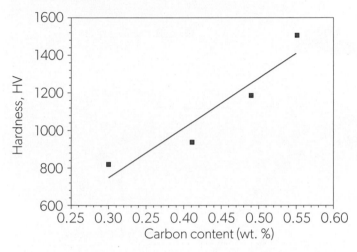

▲ Figure 10 Graph showing the hardness of carbon-containing steels. Source of data: Calik, A., Akin, D., Sahin, O. and Ucar, N. (2010) "Effect of Carbon Content on the Mechanical Properties of Medium Carbon Steels", Z. *Naturforsch*. 65a, 468–472

Relevant skills

- Inquiry 1: Demonstrate independent thinking, initiative or insight
- Inquiry 1: Formulate research questions and hypotheses
- Inquiry 2: Interpret graphs
- Inquiry 3: Interpret processed data and analysis to draw and justify conclusions

Instructions

1. Research how the Vickers hardness test is performed.

2. Based on the information given, compose a possible research question that would be answered by the data shown above. Make sure you include the:

   - variables
   - range of the independent variable
   - methodology
   - specific system being studied (in this case, medium carbon steels).

3. Describe and explain the trend shown in the graph.

4. Try to think of at least one alternative interpretation.

5. What do you think might happen to the hardness in steels with carbon content greater than 0.55%? Outline your ideas.

6. Outline an experiment that could be done to explore your answers to questions 4 and 5 above.

Bronze is an alloy of copper and tin. It is harder than pure copper and resistant to corrosion. Before steel became common, bronze was used in shipbuilding, tools and various household artefacts including coins.

Brass is a highly malleable alloy of copper and zinc. It is used to make musical instruments due to its acoustic properties. The copper in brass has **antimicrobial** properties, which is why it is often used to make door handles in public buildings such as hospitals.

◄ Figure 11 Old bronze coins from China

▲ Figure 12 An 19th century theodolite made of brass. This instrument was used for measuring angles

(ATL) Research skills and social skills

The life cycle of a product involves the process and materials involved in its production, usage and disposal. The substances present undergo various transformations throughout (figure 13). Structure, as well as chemical and physical properties, determines many of the processes involved in a product life cycle.

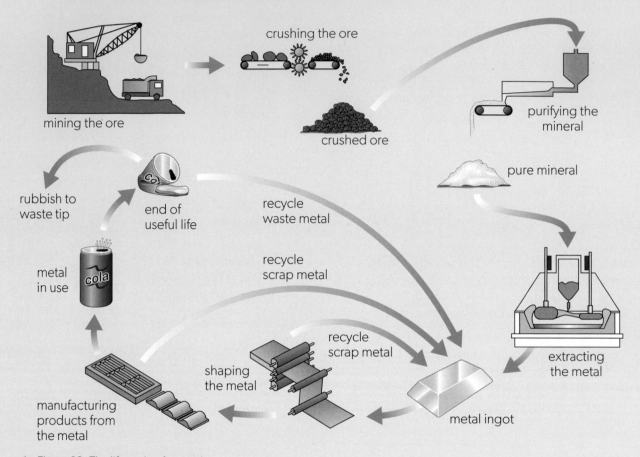

▲ **Figure 13** The life cycle of a metal

In this task, you will work with a group to research the life cycle of a particular product. You will also reflect on the collaboration skills involved. Work with one or two other people.

1. Collaboratively, choose one of the following substances and products to investigate:
    - aluminium in an aircraft
    - iron in a ship
    - carbon in a plastic bottle.

2. Briefly reflect on how you collectively reached a choice.

3. Plan to research the various stages of the product's life cycle:
    a. extraction of raw materials
    b. manufacturing
    c. use
    d. disposal.

4. Consider
    - if and how you will divide up the work within your team
    - how you and your team will share your findings with each other.

5. Carry out your research.

6. Prepare a visual representation of your research, depicting the life cycle of your product. Before you start, take a moment to reflect on how to take everyone's opinions into account when deciding how to represent your work.

7. Share your work with your class.

8. Individually, reflect on your role in your group, and identify one IB learner profile attribute that you developed during this task.

Practice questions

5. Explain why alloys such as steel are harder than pure metals.
6. Suggest why the electrical conductivity of an alloy is often lower than that of a pure metal.

Linking question

Why are alloys more correctly described as mixtures rather than as compounds? (Structure 1.1)

Polymers (*Structure 2.4.4*)

Polymers are a class of covalently bonded materials characterized by their low thermal and electrical conductivity, and low density compared to other types of materials. Polymers are **macromolecules**, which are very large molecules. Plastics and synthetic textiles such as nylon are examples of **synthetic polymers. Natural polymers** include fibrous materials like lignin and silk, as well as many of the biological molecules involved in life processes, such as DNA, starch and proteins.

Long, covalently bonded polymer chains are made when small molecules called **monomers** join end-to-end (figure 15), like beads strung together to create a necklace. This process is called **polymerization**. There are two types of polymerization: **addition** and **condensation**.

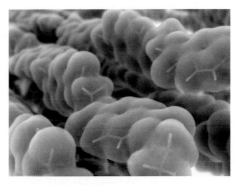

▲ Figure 14 A digital model of polyethene terephthalate (PET), a synthetic polymer

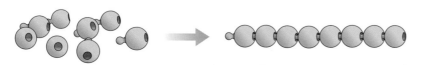

▲ Figure 15 In polymerization, monomers join up to make a chain called a polymer. Here the small beads represent monomers and the resulting chain, the polymer

Repeating units

A **repeating unit** is a group of atoms that appears repeatedly along a polymer chain. The exact lengths of polymers can vary, so the best way to describe the structure of a polymer is with the structural formula of the repeating unit. In the formula, the repeating unit is shown enclosed in large brackets with bonds drawn across the brackets, indicating that the polymer chain continues on either side. The subscript n after the closing bracket represents the number of repeating units in the polymer. A typical polymer contains hundreds or thousands of repeating units, so n represents a very large number.

Worked example 3

A section of the polypropene molecule is shown below.
Draw the repeating unit of polypropene.

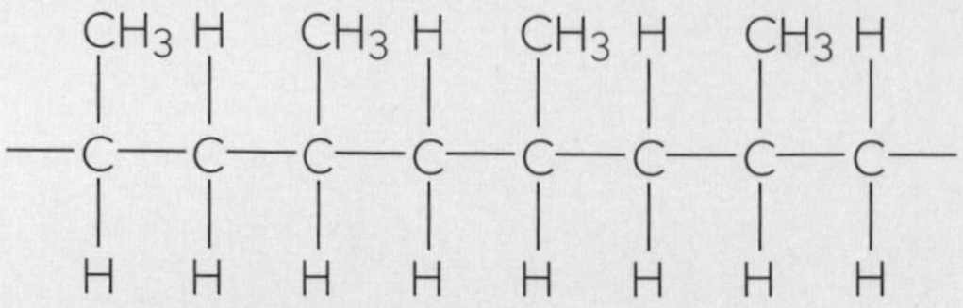

Solution

Looking at the section, you can see that there is a repeating pattern every two carbon atoms along the polymer chain.

Therefore the repeating unit is

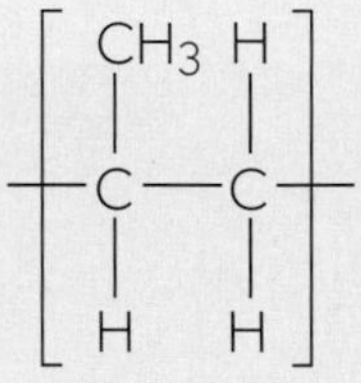

When drawing a repeating unit, make sure that the connecting bonds on both sides cross the brackets. If you need to represent the whole polymer chain, do not forget to include the subscript n to signify that the polymer is composed of many repeating units:

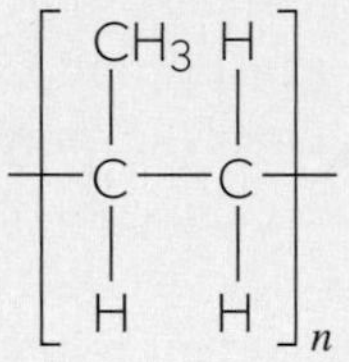

Practice questions

7. Draw the repeating unit for each of the polymer sections shown.

    a.

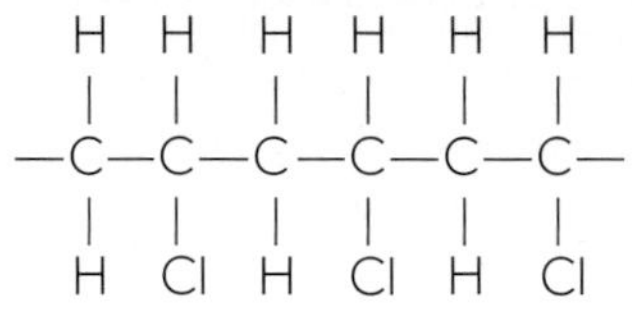

    b.

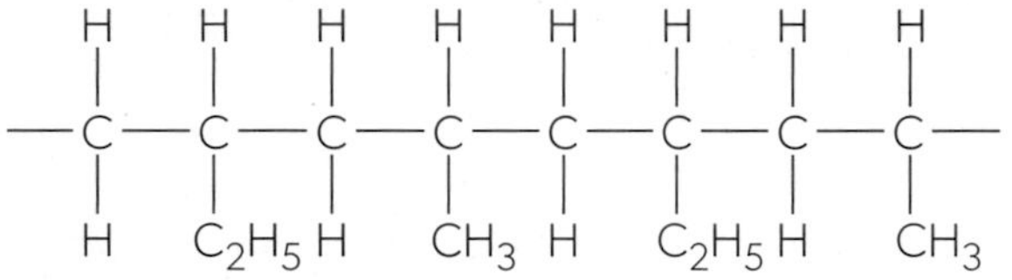

    c.

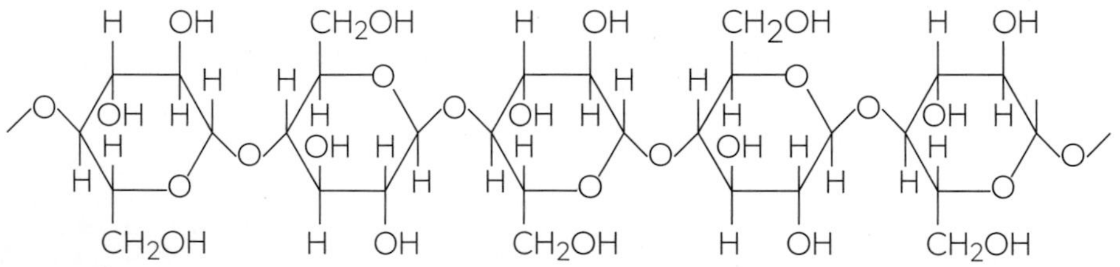

8. The structure of poly(phenylethene), also known as polystyrene, is shown below.

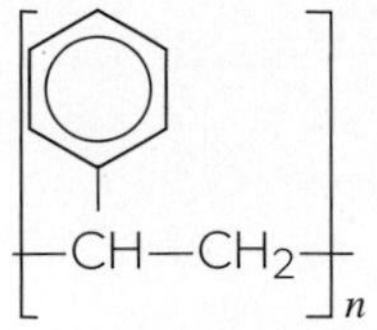

    Draw a section of the polymer showing three repeating units.

9. The structure of nylon-6,6 is shown below. Draw a section of the polymer showing two repeating units.

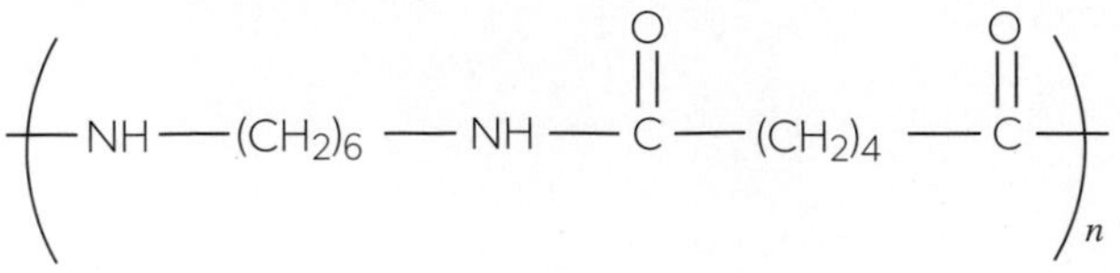

The structure of the repeating unit originates from the monomer used to make the polymer. As a result, some polymer names are monomer name in brackets, with the prefix **poly-**. However, many natural polymers have unique names that do not refer to the monomer. Similarly, many synthetic polymers names are referred to by their commercial brand names. Can you think of any examples?

Polymers can be classified according to their source (natural or synthetic), or type of polymerization reaction undergone by the monomers (addition or condensation).

Some examples are shown in table 5.

Name	Repeating unit	Examples of uses	Natural or synthetic?	Type of polymer
cis-1,4-poly(isoprene) (natural rubber)	CH₂ CH₂ / C=C / H₃C H	balloons, elastic bands, car tyres (vulcanized rubber)	natural	addition
polyethene	H H / C–C / H H	carrier bags, packaging, cable insulation, toys	synthetic	addition
cellulose	(ring structure with OH, O, HO, OH)	cell walls, paper, cellophane	natural	condensation
Kevlar	(aromatic structure with N, H, C=O)	sports equipment, personal armour	synthetic	condensation

▲ Table 5 Examples of polymers

◀ Figure 16 Kevlar was first synthesized by Polish-American chemist Stephanie Kwolek. Over 3,000 saved lives have been directly linked to the use of Kevlar in bulletproof vests and anti-stab clothing. Kwolek remarked, "I don't think there's anything like saving someone's life to bring you satisfaction and happiness."

Properties of polymers

In general, polymers are strong, because of their macromolecular nature. Like most covalent materials, they do not contain moving charged particles, so their thermal and electrical conductivities are also low.

Though molecular, polymers have relatively high melting points and boiling points. This is due to the number of intermolecular forces that occur along the length of the polymer molecules. The long polymer strands can also wind around each other and become entangled, requiring energy to disentangle and separate them.

The specific properties of a polymer depend on its chain length, the type of forces holding the chains together, the type of monomers and the way these monomers are arranged.

Many biological molecules, including proteins, carbohydrates and nucleic acids, are natural polymers. Some provide structure to living organisms, while others support life processes. Starch is a natural polymer that acts as an energy store in plants. Others include collagen, which gives our skin structure and elasticity, and DNA, which houses an organism's genetic information.

Cellulose is a carbohydrate and a natural linear polymer of glucose. Plants make glucose through photosynthesis. Cellulose is found in cell walls. It supports fibrous structures such as cotton, as well as stalks and tree trunks. You can see the repeating unit of cellulose in table 5.

The presence of –OH groups (hydroxyl groups) in cellulose means that the polymer chains can form hydrogen bonds with one another. The resulting fibres are uniquely strong and insoluble in water as well as many common solvents. The strength of cellulose makes it an important material for the structural integrity of plants. Humans have taken advantage of these properties by using cellulose to make paper, textiles, and other polymer materials like cellophane.

Starch is the main energy store in plants and is also a glucose polymer, like cellulose. However it has different properties: it is not as strong, so the chains can be broken down easily and digested by humans. The main difference lies in the way the monomers are connected: the polymer chains in starch have an irregular branched structure. As a result, the polymer chains in starch cannot pack as tightly as in cellulose and therefore form fewer hydrogen bonds with one another.

Synthetic polymers are humanmade. Many of them are petroleum-based, which means that they are derived from crude oil. Examples include PET bottles, nylon textiles, non-stick Teflon coating on pans and synthetic rubber tyres. Semi-synthetic polymers are derived from renewable biological materials produced by plants and bacteria. These polymers provide a sustainable alternative to crude oil-based polymers, and some have important biomedical applications.

▲ Figure 17 Cotton, made up of cellulose, is a natural polymer

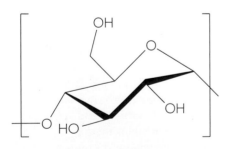

▲ Figure 18 The repeating unit of starch

Practice questions

10. Explain why polymers have relatively high melting points.

11. Explain why polymers are generally electrical insulators.

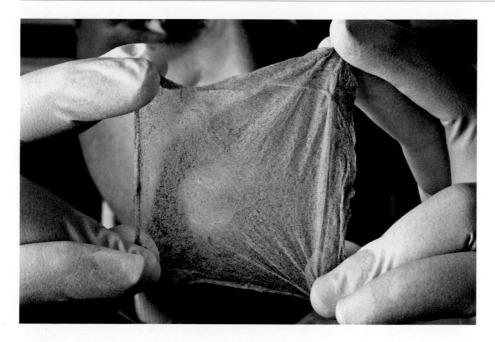

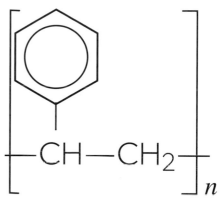

◀ Figure 19 This artificial tissue was created from a mixture of living cells and semi-synthetic polymer. Tissues like this can be used in organ transplantation

Plastics are synthetic polymers composed mainly of carbon and hydrogen. They are often chemically unreactive, although they can catch fire and sustain combustion.

Poly(phenylethene), commonly known as **polystyrene**, is a widely used plastic found in packing peanuts and food containers.

The structure of polystyrene is shown in figure 20. It is a non-polar hydrocarbon and so it is insoluble in water. London dispersion forces (LDFs) hold the polymer chains together. These LDFs can be overcome by heating, so polystyrene can be melted down and reshaped easily. On cooling, new LDFs can form between polystyrene molecules, so the polymer solidifies again.

▲ Figure 20 The structure of polystyrene

Polystyrene chains cannot rotate easily due to the bulky phenyl groups (C_6H_5), making it brittle. Polystyrene is a good thermal insulator, particularly in its expanded form, so it is used to make coolers and insulating panels for the construction industry.

London dispersion forces (LDFs), and the other intermolecular forces are discussed in *Structure 2.2*.

One of the reasons plastics are useful is that, due to their strong covalent bonds, they are chemically inert and therefore durable. As a result, many plastics do not typically **biodegrade.** Biodegradation involves the breakdown of materials, by microorganisms, into small molecules. Polymers rarely contain reactive functional groups that could act as bacterial binding sites. In some plastics, the polymer chains are covalently bonded together, which is known as **crosslinking**. A dense covalent network formed by crosslinking provides very few entry points for microorganisms.

 Activity

Compare and contrast two of the polymers discussed in this topic.

Investigating hydrogels

Smart materials undergo a change in properties in response to surrounding conditions (such as pH, temperature or magnetic fields). Hydrogels are smart materials used in a range of industrial and biomedical applications. They are made up of crosslinked polymer networks that can absorb and hold water, causing the gel to swell. In this task, you will investigate the response of a hydrogel to variations in the surrounding conditions.

The hydrogel chains contain numerous carboxyl groups. In acidic media, these –COOH groups form hydrogen bonds with one another, causing the hydrogel to contract (figure 21). In basic environments, they become ionized, forming carboxylate ions, –COO⁻. The negative charges on neighbouring carboxylate groups repel each other, making the hydrogel swell.

Relevant skills

- Inquiry 1: Justify the range and quantity of measurements
- Inquiry 2: Interpret qualitative and quantitative data

Safety

Wear eye protection.

Materials

- Commercially available soft contact lenses (manufactured from poly(2-hydroxyethyl methacrylate-co-methacrylic acid))
- Contact lens buffer solution
- Sodium chloride
- Weak acid solution (for example, household vinegar)
- Weak base solution (for example, sodium hydrogencarbonate 0.06 mol dm⁻³)
- Distilled water
- Standard laboratory equipment including petri dishes, beakers, pipettes and balances.
- Ruler

Instructions

Part 1: The effect of pH variations on the hydrogel

1. Prepare three clear shallow dishes, such as petri dishes, each containing one of the following:
   - contact lens buffer solution
   - weak acid solution
   - weak base solution
2. Place the dishes on a white surface.
3. Place a contact lens in the first dish and measure its diameter. (If it is difficult to see, it might be inside out. Reverse it and try again.)
4. Place contact lenses in the other two dishes and note any changes in diameter.
5. Swap the contact lenses in the acidic and basic solutions and observe what happens. Are the changes reversible?

Part 2: The effect of salinity variations on the hydrogel

6. Repeat the Part 1 experiment, using these solutions instead:
   - contact lens buffer solution
   - distilled water
   - saturated sodium chloride solution

You can reuse the Part 1 lenses, provided they are not damaged.

7. Describe your results.
8. Propose a hypothesis to explain the observed results of the Part 2 experiment.
9. Devise an experiment that you could do to test your hypothesis. Identify the independent, dependent and control variables. What measurements would you make? Justify the range and quantity of these measurements.

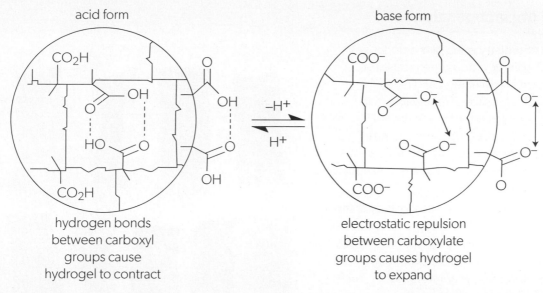

acid form

base form

hydrogen bonds between carboxyl groups cause hydrogel to contract

electrostatic repulsion between carboxylate groups causes hydrogel to expand

▲ Figure 21 Interconversion between –COOH groups and –COO⁻ groups associated with the pH adjustments

Polymers and the environment

After reduction and reuse, recycling is the third approach for mitigating the environmental impact of plastics. Plastic recycling poses several challenges. The process is energy-intensive: plastic waste must be sorted and cleaned beforehand. It then needs to be broken up and melted, which prevents some plastic types from being recycled because they release atmospheric pollutants on heating. Unlike glass and aluminium, where the quality is largely unaffected by recycling, plastics often deteriorate when they are recycled.

As a result, most plastic waste is not recycled and is instead incinerated or disposed of in landfills. Much of discarded plastic ends up in the oceans, where it persists and poses a risk to wildlife.

The challenge, and an area of growing interest, is designing plastic materials that are durable enough to be used for a certain period of time but biodegradable in the longer term. Work is also being done to find ways to make existing petroleum-based plastics more appetizing feedstock for microorganisms.

Biodegradable plastics facilitate bacterial action by having increased surface area, shorter polymer chains, or functional groups that are attractive to bacteria or light-sensitive. For example, plant-based hydro-biodegradable plastics contain carbohydrates that are broken down in a process called hydrolysis. Care must be taken to dispose of them in conditions that provide enough oxygen for bacterial action. In low oxygen environments found in landfills, decomposition can slow down or even produce environmentally harmful products, such as the greenhouse gas methane.

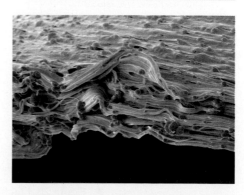

▲ Figure 22 Discarded plastics can end up in the ocean. The cold and dark ocean environment slows down plastic degradation

Greenhouse gases are discussed at AHL in *Structure 3.2.*

▲ Figure 23 A scanning electron micrograph of starch granules embedded in a biodegradable plastic carrier bag. The starch absorbs water and swells up, providing greater surface area for bacterial action

Global impact of science

The 12 **green chemistry** principles emphasize the benefits of non-hazardous chemicals and solvents, efficient use of energy and reactants, reduction of waste, choice of renewable materials, use of biodegradable materials and prevention of accidents. The philosophy of green chemistry has been adopted by many educational and commercial organisations and eventually passed into national and international laws.

Plant-based biodegradable plastics, or **bioplastics**, are aligned with two principles of **green chemistry**: use of renewable feedstocks and designing materials to be biodegradable. Most plastics are petroleum-based, so their feedstock is non-renewable. The raw materials for bioplastics are renewable, such as maize. Even though plastic biodegradation releases carbon dioxide, plant-based polymers are sometimes considered carbon-neutral because the same quantity of carbon dioxide is removed from the atmosphere by the biosynthesis of monomers in plants.

To evaluate the environmental impact of a product, you have to consider many factors. For example, growing maize for plant-based plastics uses land that could otherwise be used to grow other crops, which can potentially cause food shortages. Additionally, farming maize may require fertilizers or pesticides, which have their own associated environmental implications.

▲ Figure 24 These bioplastic samples were made from household food waste such as banana peel, onion skins and even pineapple crowns

Practice questions

12. Outline two challenges associated with recycling plastic waste.

13. Outline two structural features of biodegradable plastics.

Microplastics are very small fragments of plastic, broken down over time to sizes under 5 mm. This process is not the same as biodegradation because the plastics remain chemically unchanged. Because of their size, they can enter the food chain and find their way into unexpected places. In 2022, microplastics of PET, polystyrene and polyethene were discovered in human blood for the first time. Their health impact of microplastics is still unknown.

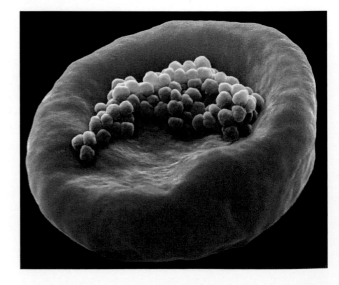

▶ Figure 25 Electron micrograph showing microplastic polyethene particles, known as microbeads, on a single red blood cell. Microbeads are sometimes added to health and beauty products, but this use is now banned in many countries

Addition polymers (*Structure 2.4.5*)

During polymerization, monomers combine together to form a polymer. One way this can happen is **addition polymerization**, which involves monomers that contain a double bond. The double bond breaks and the two electrons form a covalent bond with another monomer. Because of the change in bond order, and due to their extraordinary length, polymer molecules have quite different physical and chemical properties than their corresponding monomers.

Ethene, C_2H_4, is the simplest alkene. It polymerizes to form polyethene. Consider two ethene monomers:

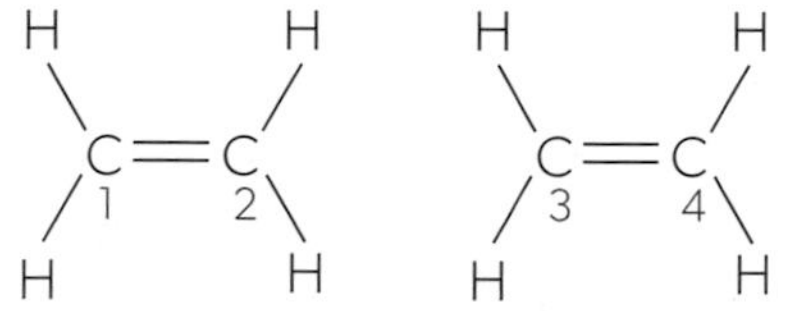

The electrons in their double bonds open up...

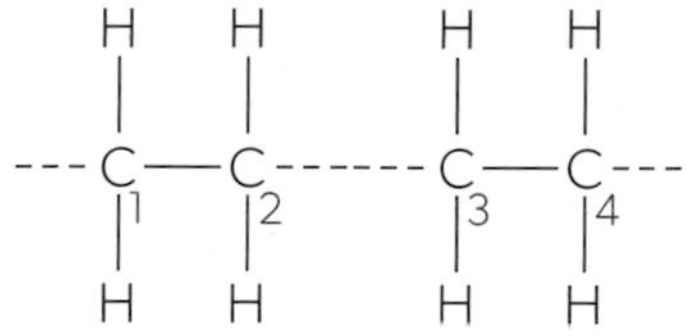

... and these electrons are shared between carbon-2 and carbon-3 to form a single bond, joining the monomers together:

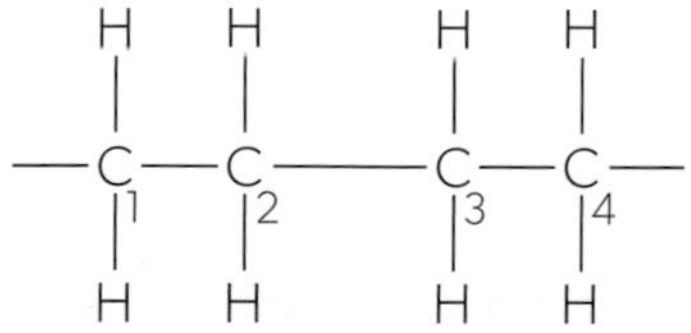

Now imagine a similar process that involves a very large number of ethene monomers:

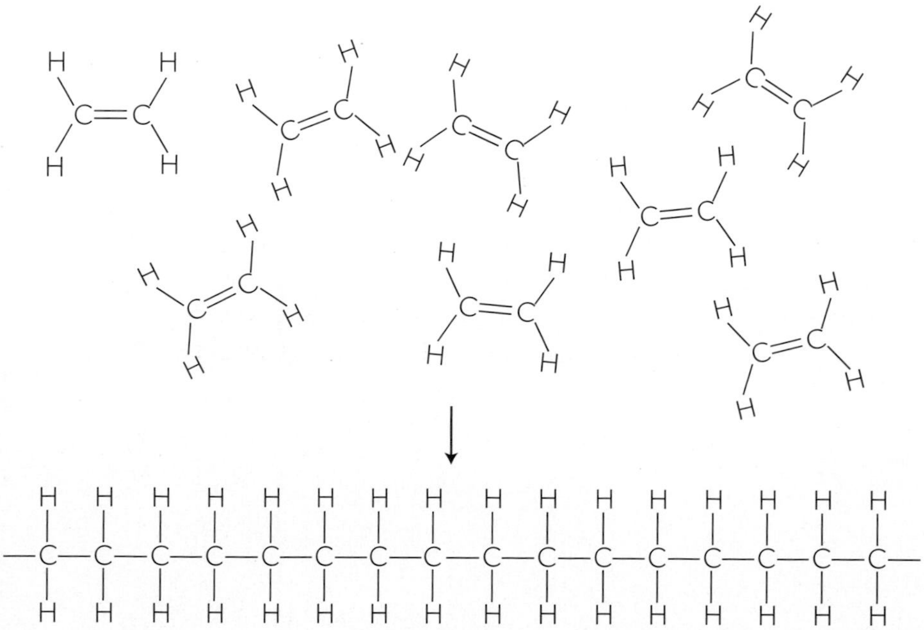

We can simplify the equation for the addition polymerization reaction using **n** to convey that there are a very large number of monomers:

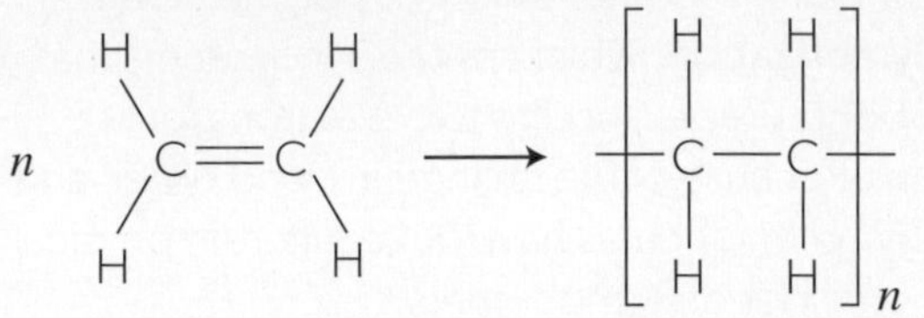

We can also use condensed structural formulas:

$$n \quad H_2C = CH_2 \longrightarrow -(CH_2CH_2)_n$$

Any monomer that contains a carbon–carbon double bond can undergo addition polymerization. Polyvinyl chloride, PVC, is a polymer made from chloroethene molecules. Its IUPAC name is poly(chloroethene). Every other carbon in the resulting polymer has a chlorine substituent. The carbon–chlorine bond is polar, which results in dipole–dipole forces between the polymer chains, unlike in non-polar polyethene where only London dispersion forces are present. PVC is therefore the stronger of the two polymers.

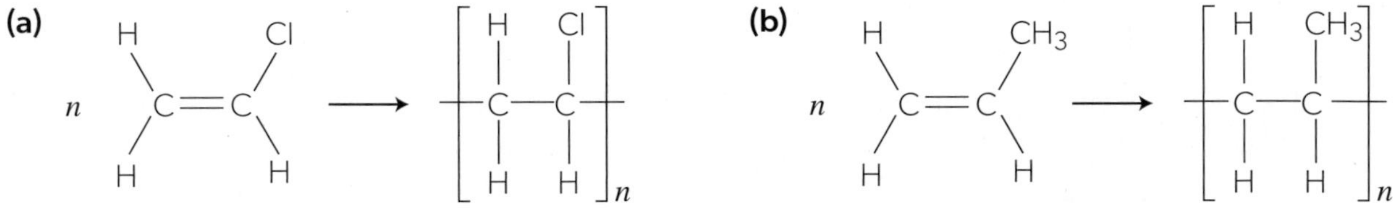

▲ Figure 26 Addition polymerization reactions to form (a) poly(chloroethene) and (b) polypropene

TOK

Structure 2.4 illustrates the difference between **science** and **technology**, and the changing relationship between these two disciplines over time. Science is concerned with observing and understanding the physical and natural world. The focus in technology is the practical application of scientific knowledge.

In the past, technological developments preceded scientific knowledge. People used and developed materials for centuries before understanding the underlying reasons behind the materials' observable properties. Empirical observation of early technologies contributed to scientific understanding of how and why they worked. Nowadays, the scientific research often occurs first. Desirable properties for a particular application are identified, and then the required materials are engineered using scientific knowledge.

What is the role of empirical observation in science and technology?

Practice questions

14. For each of the following monomers:
    i.   draw a section of the polymer containing three repeating units
    ii. deduce the structure of the repeating unit of the polymer.
    a.   tetrafluoroethene

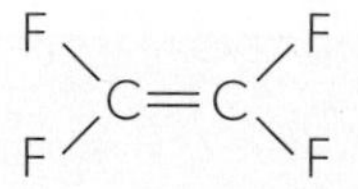

    F   F
     \   /
      C = C
     /   \
    F   F

    b.   acrylonitrile (prop-2-enenitrile)

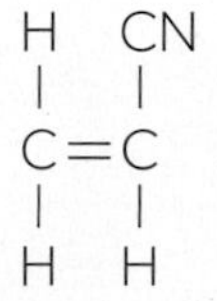

    H   CN
    | |
    C = C
    | |
    H   H

    c.   but-2-ene

         H   H   H   H
         | | | |
    H — C — C = C — C — H
         | |
         H   H

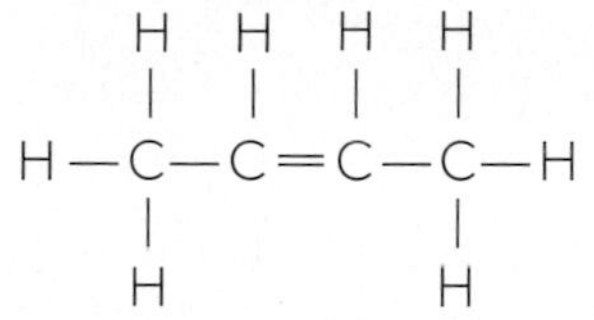

15. For each of the polymer sections shown:
    i.   deduce the structure of the repeating unit
    ii. draw the structure of the monomer.

    a.
         CH₃   CH₃   CH₃   CH₃
          | | | |
    ··· — CH₂— CH — CH₂ — CH— CH₂ — CH— CH₂— CH — ···

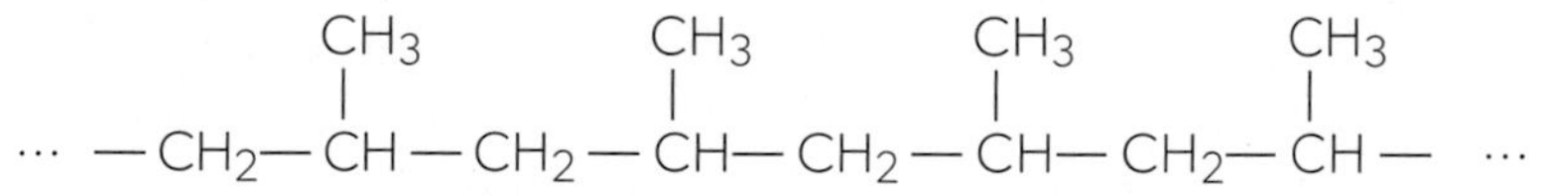

    b.
         O   OCH₃   O   OCH₃
          \\   /   \\   /
           C   H   CH₃ H   C
       H | H | | | | |
       | | | | | | | |
    —— C — C — C — C — C — C ——
       | | | | | |
       H   CH₃ H | H   CH₃
                   C
                  // \
                 O   OCH₃

 Linking questions

What functional groups in molecules can enable them to act as monomers for addition reactions? (Structure 3.2)

Why is the atom economy 100% for an addition polymerization reaction? (Reactivity 2.1)

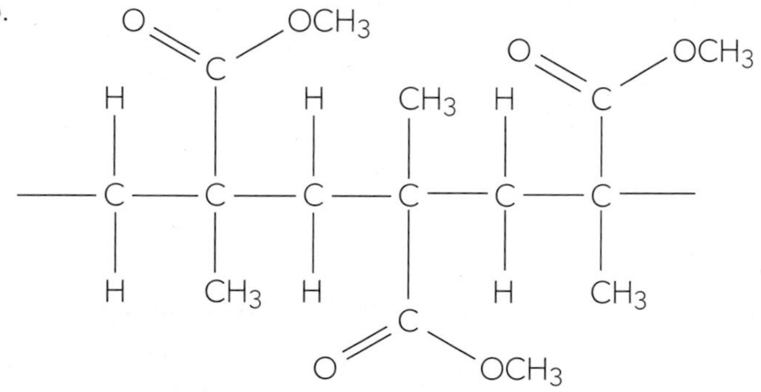

16. Compare and contrast ethene and polyethene in terms of their structure, bonding, melting point and electrical conductivity.

Condensation polymers (*Structure 2.4.6*)

Condensation polymers, which include nylon and cellulose, are produced in a reaction between monomers that have reactive functional groups on either end. For every bond formed between two monomers, a small molecule is released (for instance, water or HCl).

▶ Figure 27 Slacklining is a practice in balance that typically uses nylon webbing tensioned between two anchor points. Nylon is a condensation polymer

Esters are formed in a condensation reaction between an alcohol and a carboxylic acid (figure 28). In this reaction, the hydroxyl and carboxyl functional groups form an ester linkage, and a water molecule is released. This condensation reaction is known as **esterification**.

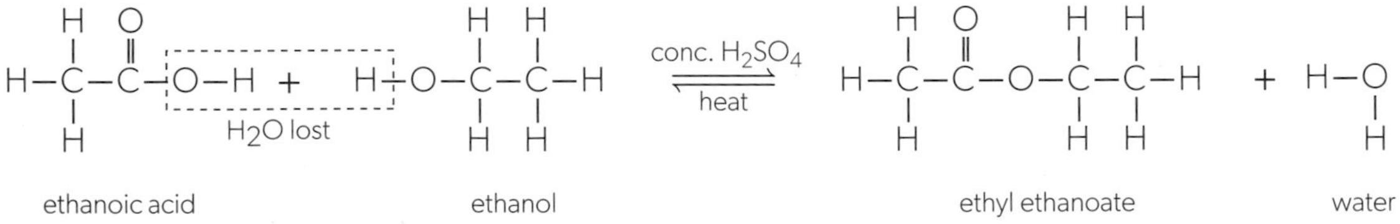

ethanoic acid   ethanol   ethyl ethanoate   water

▲ Figure 28 Ethyl ethanoate, an ester, is formed in the condensation reaction between ethanoic acid and ethanol

The structure and properties of esters are discussed further in *Structure 3.2*.

The reactions used to make condensation polymers are similar to esterification: two functional groups react to form a linkage, and a small molecule is released. However, in polymerization, the reacting monomers have **two** reactive functional groups, one at each end of the molecule, which allows a polymer chain to form. We will look at two examples of how this can happen.

Condensation between two different monomers

One monomer with the same functional group at either end can react with another monomer with a different functional group at either end. This creates a polymer of alternating monomer structure: for example, if monomer A reacts with monomer B, a polymer with the pattern ABABABAB is formed. The general reaction between a dicarboxylic acid monomer and difunctional alcohol (diol) monomer to form an ester linkage is shown in figure 29.

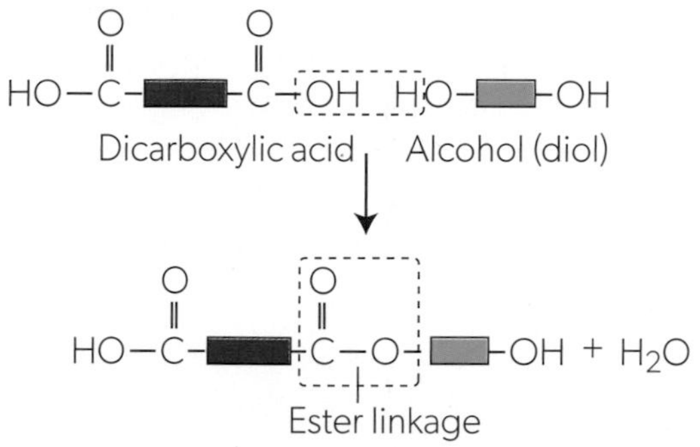

◀ Figure 29 A condensation reaction between two monomers to form an ester

The product molecule in figure 29 has two reactive ends that can subsequently react with more monomers (figure 30). This forms a long polymer chain with an alternating pattern based on each monomer structure, with each monomer joined by an ester linkage. The resulting polymer product is called a **polyester**.

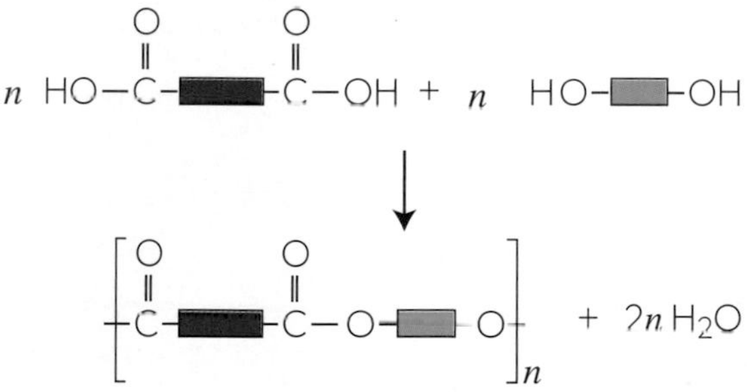

◀ Figure 30 A generalized condensation reaction to make a polyester, where n is the number of monomer pairs involved and the length of the resulting polymer chain

The polymer nylon-6,6 is produced in the condensation reaction between hexanedioic acid and hexane-1,6-diamine (figure 31). The product is known as a **polyamide** due to the multiple amide linkages formed during the condensation process.

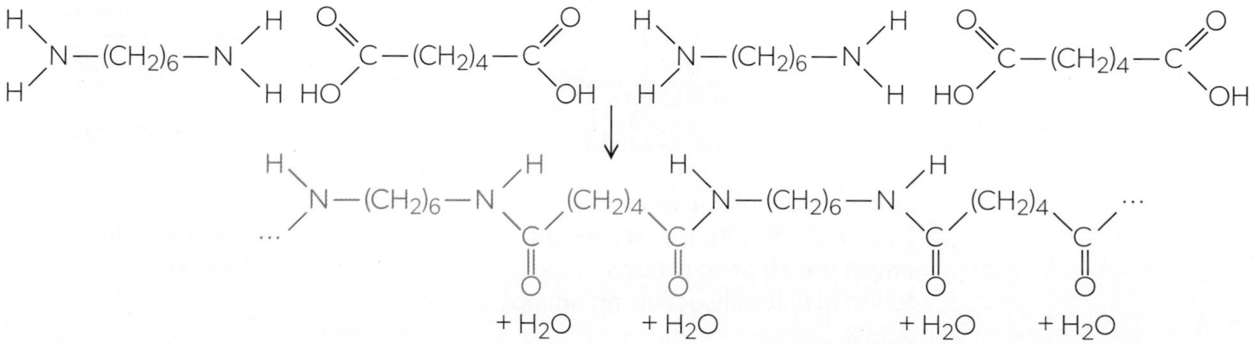

▲ Figure 31 The formation of nylon-6,6. The repeating unit is shown in green

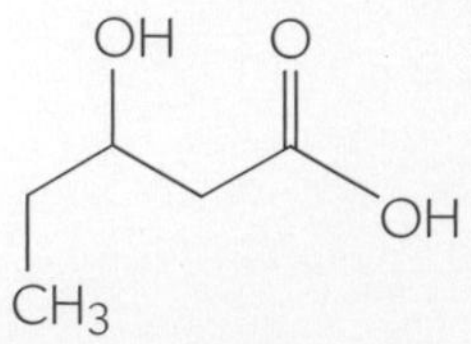

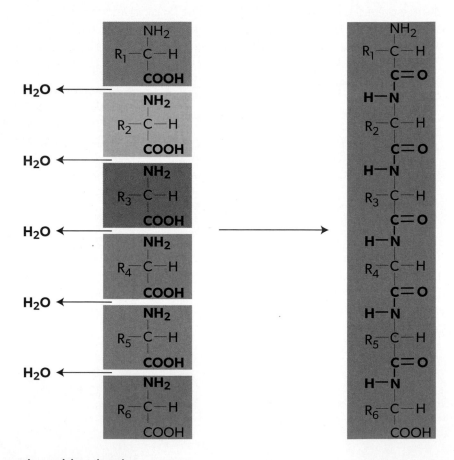

▲ Figure 32 The structure of the 3-hydroxypentanoic acid monomer

Condensation of the same monomer

If there is a different functional group on each end of a monomer, then it can polymerize with itself by condensation. For example, 3-hydroxypentanoic acid (figure 32) contains both a hydroxyl group and a carboxyl group. Multiple monomers of 3-hydroxypentanoic acid can react with one another to form a polymer.

Proteins are biopolymers that play a central role in metabolic processes, transport and sensory functions, structural integrity, and virtually all other molecular aspects of life. Proteins are **polypeptides**, which are polymers made up of monomers called 2-amino acids. There are 20 types of 2-amino acids used by living organisms as building blocks of proteins. 2-amino acids have an –NH_2 group and a –COOH group, so they can form amide linkages with each other through condensation polymerization. In a polypeptide, the amide link is also known as a **peptide bond**:

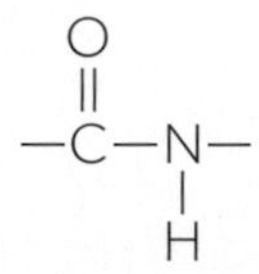

▶ Figure 33 The formation of a polypeptide chain

amino acid molecules
(where R_1, R_2, R_3 etc. represent any of the 20 or so groups found in naturally occurring amino acids)

polypeptide (part of)

Activity

Figure 34 shows the 3D structure of lysozyme, the natural antibacterial enzyme contained in saliva, milk, mucus, tears and egg whites. The carbon atoms are shown as white, the oxygen atoms are red and the nitrogen atoms are blue. Can you find any amide linkages?

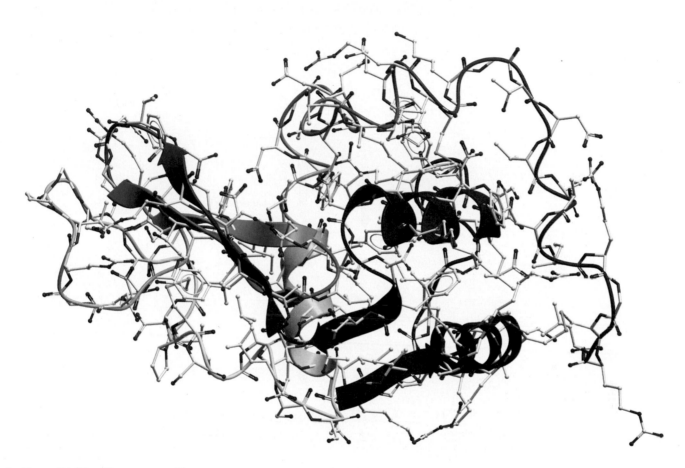

▲ Figure 34 The 3D structure of lysozyme

Practice questions

17. Draw the repeating unit of a condensation polymer made from 3-hydroxypentanoic acid (figure 32).

18. Outline two differences between addition and condensation polymerization.

19. Identify the type of polymer made and the small molecule released when monomers containing these functional groups form polymers:

   a. A hydroxyl group, –OH, and a carboxyl group, –COOH

   b. A carboxyl group, –COOH, and an amino group, –NH$_2$

   c. Two hydroxyl groups, –OH

   d. An acyl chloride group, –COCl, and an amino group, NH$_2$

Hydrolysis

The reverse of condensation reactions is called **hydrolysis**, where the linkage formed by condensation is split up by a molecule of water. All biological macromolecules form by condensation reactions and break down by hydrolysis. Figure 35 shows how monosaccharides are polymerized via condensation and formed again by the reverse hydrolysis reaction. Note the two reactive ends of monosaccharides are hydroxyl groups, which form polymers with a glycosidic linkage. Polysaccharides make up carbohydrates, which are used as energy sources and energy reserves in living organisms.

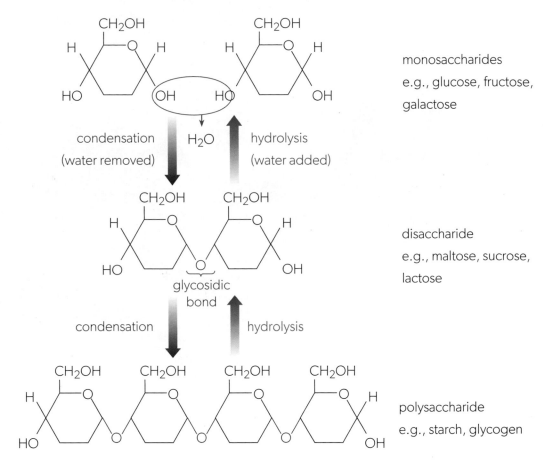

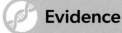

▲ Figure 35 Condensation and hydrolysis of monosaccharides, the polymers that make up carbohydrates. Some —OH groups are omitted for clarity

 Evidence

Water is often referred to as the 'molecule of life' due to its vital role in biological reactions. For this reason, NASA scientists have been searching for water on Mars, as the presence of water is thought to be required for life forms to exist.

Linking question

What functional groups in molecules can enable them to act as monomers for condensation reactions? (Structure 3.2)

End-of-topic questions

Topic review

1. Using your knowledge from the *Structure 2.4* topic, answer the guiding question as fully as possible:

   What role do bonding and structure have in the design of materials?

Exam-style questions

Multiple-choice questions

2. Calculate the mean electronegativity, $\bar{\chi}$, and electronegativity difference, $\Delta\chi$, for the bonds found in lead(II) bromide, $PbBr_2$.

   χ (Pb) = 1.8 and χ (Br) = 3.0.

	Mean electronegativity, $\bar{\chi}$	Electronegativity difference, $\Delta\chi$
A	4.2	2.6
B	2.6	4.2
C	2.4	1.2
D	1.2	2.4

3. Which of the following substances has properties of both ionic and covalent compounds? Refer to the triangular bonding diagram and the electronegativity values below.

Element	Na	K	Cs	F	Zn	S	Ag	Br
Electronegativity	0.9	0.8	0.8	4.0	1.6	2.6	1.9	3.0

   A. NaK, a sodium-potassium alloy

   B. Caesium fluoride, CsF

   C. Zinc sulfide, ZnS

   D. Bromine, Br_2

4. Which of the following mixtures is an alloy?
   A. sodium chloride and water
   B. iron and vanadium
   C. carbon fibre and a polymer
   D. magnesium and chlorine

5. Which of these substances does **not** have the ability to polymerize?
   A. ethane, CH_3CH_3
   B. tetrafluoroethene, CF_2CF_2
   C. propene, CH_2CHCH_3
   D. propenenitrile, CH_2CHCN

6. Which of the following pairs of molecules will **not** react to form a polymer?

   A.
   and

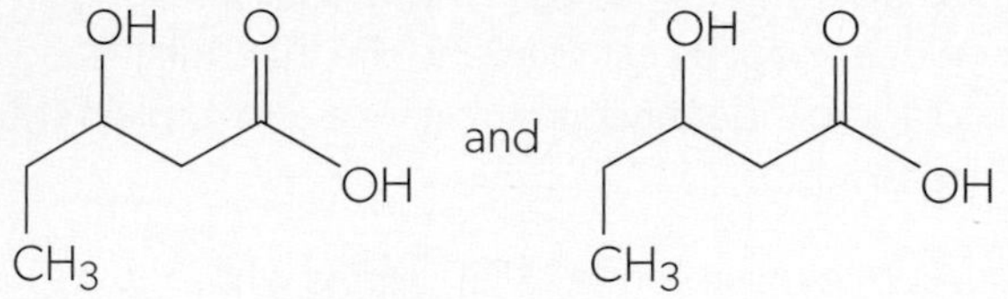

   B.
   and

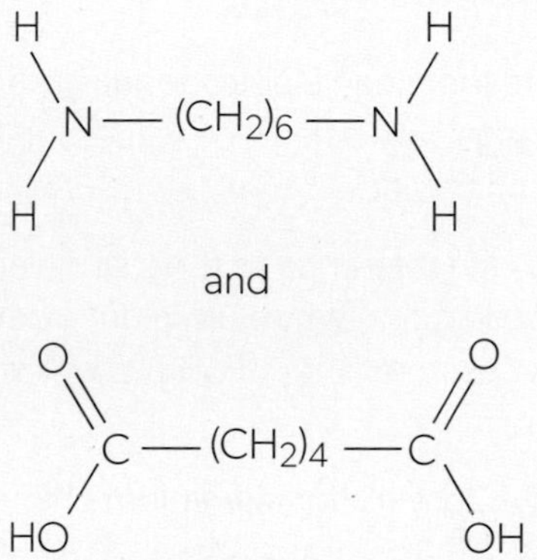

   C.
   and

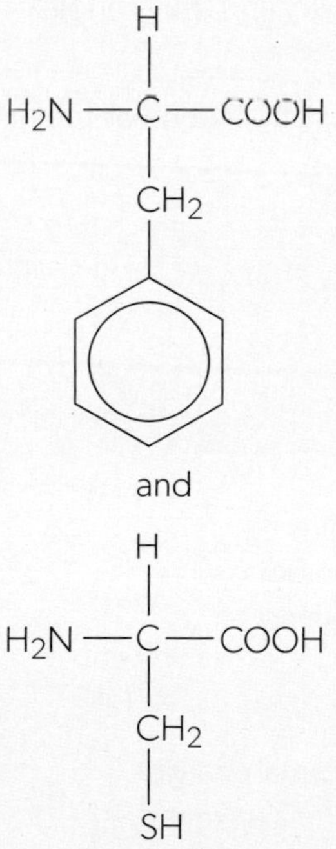

   D.
   and

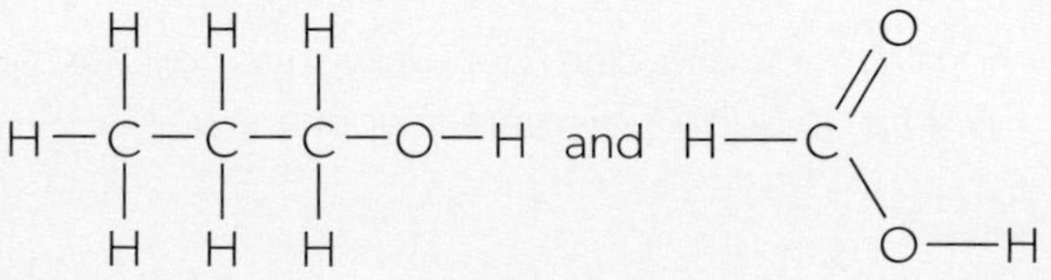

Extended-response questions

7. Tin(II) chloride has a melting point of 247 °C while lead(II) chloride has a melting point of 500 °C. Both are used in the production of aurene glass, an iridescent artisan glassware. One of these two substances exists as discrete molecules in the vapour phase. Using electronegativity tables and the triangular bonding diagram, identify that substance and explain your reasoning. [3]

8. Aluminium chloride, $AlCl_3$, forms white crystals with a low melting point and boiling point.

   a. i. Calculate the mean electronegativity and electronegativity difference values for aluminium chloride. [2]

      ii. Explain, with reference to the triangular bonding diagram, why aluminium chloride displays characteristics of both covalent and ionic substances. [1]

   b. Draw the Lewis formula for aluminium chloride. [2]

   c. Using your knowledge of VSEPR theory, deduce the molecular geometry and bond angle of aluminium chloride. [2]

   d. Aluminium chloride is often found as the dimer Al_2Cl_6.

      i. The dimer contains two coordination bonds. Outline the meaning of the term *coordination bond*. [1]

      ii. Determine the formal charges of the atoms in the aluminium chloride dimer above. [2]

9. Explain why alloys are generally stronger than pure metals. Include a diagram in your answer. [3]

10. Chloroethene can react to form a polymer known as poly(chloroethene). The structure of chloroethene is shown below.

   a. Identify the structural feature of chloroethene that allows it to form a polymer. [1]

   b. State the name of this type of polymerization. [1]

   c. Draw the repeating unit of the polymer. [2]

   The mean molecular mass of the polymer chains in a sample of poly(chloroethene) is found to be $69,000 \, g \, mol^{-1}$.

   d. Deduce the mean number of repeating units per polymer chain. State your answer to two significant figures. [2]

   e. The melting point of chloroethene is −154 °C, whereas the melting point of poly(chloroethene) is around 210 °C. Explain this difference in melting points. [2]

11. Polylactic acid (PLA) is a bioplastic made from renewable feedstocks such as sugarcane and maize. It is broken down by microorganisms under high temperature and high humidity conditions.

   2-hydroxypropanoic (lactic) acid

   polylactic acid (PLA)

   a. Deduce why PLA is more susceptible to bacterial action than petroleum-based plastics such as poly(propene). [2]

   b. Outline two advantages and two disadvantages of bioplastics such as PLA. [4]

AHL

12. Kevlar is a high tensile-strength polymer first synthesized in 1964 by chemist Stephanie Kwolek. The structure of Kevlar is shown below.

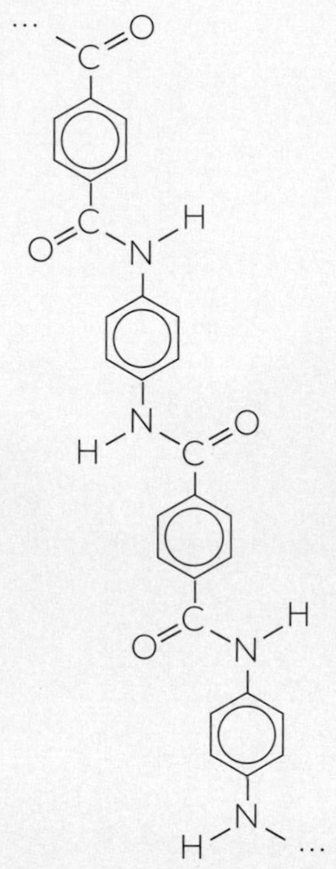

a. Identify the amide linkage in Kevlar. [1]

b. Draw the structural formulas of the monomers from which Kevlar is made. [2]

c. State the formula of the other compound produced when Kevlar monomers polymerize and state the name of this type of polymerization. [2]

d. Kevlar chains are attracted to one another through hydrogen bonding. Identify which features of Kevlar allow hydrogen bonding to occur. [1]

e. Draw a diagram showing hydrogen bonds between two adjacent Kevlar polymer chains. [2]

AHL

13. The repeating units of polymers A and B are shown below.

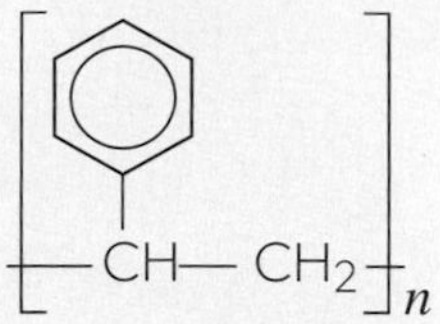

polymer A

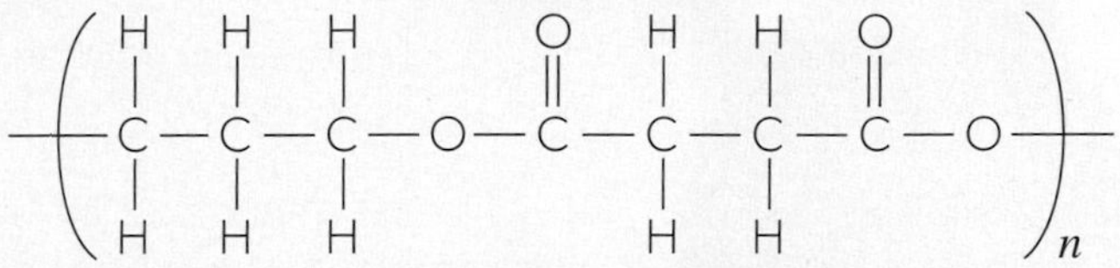

polymer B

a. For each polymer, state whether it is an addition or condensation polymer. [1]

b. For each polymer, deduce and draw the structural formulas of its monomer(s). [3]

14. Two condensation monomers are shown below:

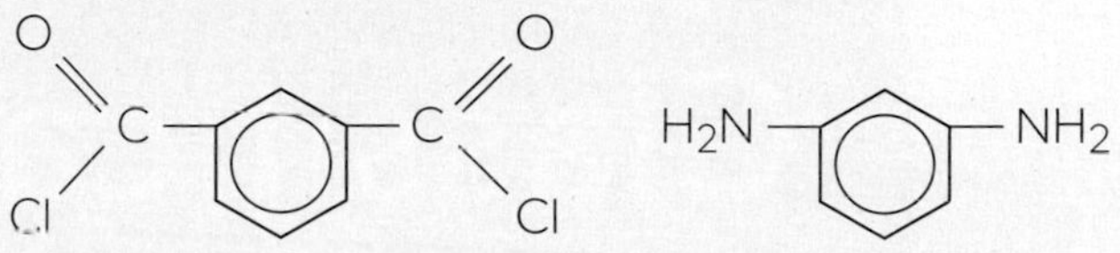

a. Draw the repeating unit of the resulting polymer. [2]

b. Deduce the formula of the inorganic product formed in the polymerization process. [1]

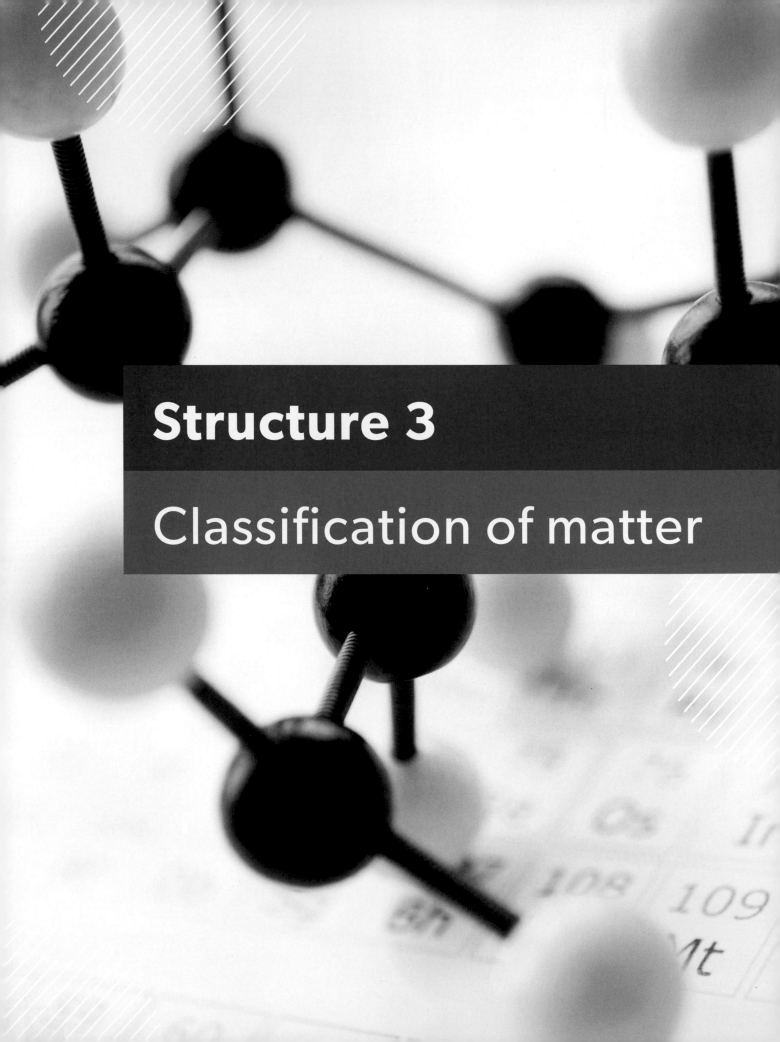

Structure 3

Classification of matter

The periodic table: Classification of elements

How does the periodic table help us to predict patterns and trends in the properties of elements?

One of the ways we can classify matter is based on what we use it for. For example, motor oils can be classified according to their viscosity, because they are used for different purposes depending on how thick they are. So, how can we classify the elements?

The increase and decrease of certain properties closely correlates with the position of elements across the periods or down the groups of the periodic table. These patterns and trends in the periodic table can be described visually with graphs.

Understandings

Structure 3.1.1—The periodic table consists of periods, groups and blocks.

Structure 3.1.2—The period number shows the outer energy level that is occupied by electrons. Elements in a group have a common number of valence electrons.

Structure 3.1.3—Periodicity refers to trends in properties of elements across a period and down a group.

Structure 3.1.4—Trends in properties of elements down a group include the increasing metallic character of group 1 elements and decreasing non-metallic character of group 17 elements.

Structure 3.1.5—Metallic and non-metallic properties show a continuum. This includes the trend from basic metal oxides through amphoteric to acidic non-metal oxides.

Structure 3.1.6—The oxidation state is a number assigned to an atom to show the number of electrons transferred in forming a bond. It is the charge that atom would have if the compound were composed of ions.

Structure 3.1.7—Discontinuities occur in the trend of increasing first ionization energy across a period.

Structure 3.1.8—Transition elements have incomplete d-sublevels that give them characteristic properties.

Structure 3.1.9—The formation of variable oxidation states in transition elements can be explained by the fact that their successive ionization energies are close in value.

Structure 3.1.10—Transition element complexes are coloured due to the absorption of light when an electron is promoted between the orbitals in the split d-sublevels. The colour absorbed is complementary to the colour observed.

Periods, groups and blocks in the periodic table (*Structure 3.1.1* and *Structure 3.1.2*)

The **periodic table** consists of horizontal **periods** and vertical **groups**. The periodic table is also divided into four **blocks: s, p, d** and **f**. Elements are organized into these blocks according to the arrangement of their outermost valence electrons. For example, elements in groups 1 and 2 have their outermost valence electrons in the s sublevel, so these are in the s-block. The blocks are shown in figure 1 on the next page.

You can find a detailed version of the periodic table at the back of this book.

We can also classify elements in the periodic table as **metals**, **non-metals** and **metalloids**. Metals generally have three or fewer valence electrons, which are delocalized and thus contribute to metallic bonding (*Structure 2.3*). Non-metals usually have four or more valence electrons which are not delocalized.

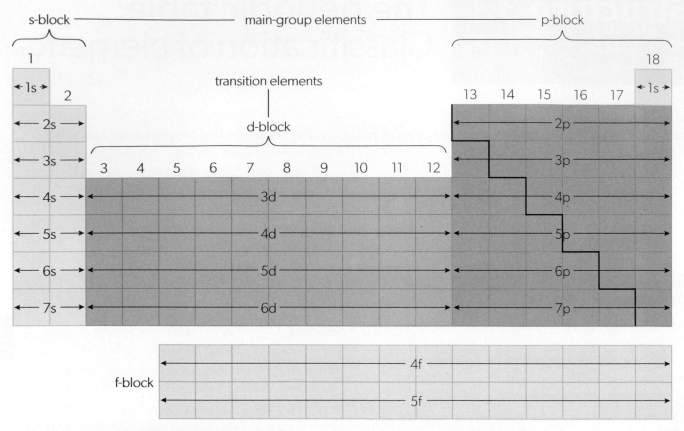

▲ Figure 1 The s, p, d and f blocks of the periodic table

Group number(s)	Group name
1	alkali metals
3–11	transition elements
17	halogens
18	noble gases

▲ Table 1 Names of groups in the periodic table

Metalloids show hybrid metallic and non-metallic behaviour. They can form both ionic and covalent bonds. Elements considered to be metalloids are usually directly either side of the red zig-zag in figure 1. Metals are to the left of the metalloids, and non-metals are on the right.

Groups in the periodic table are numbered from 1 to 18. Elements in some groups are given a collective name, which are summarized in table 1.

Periods are numbered from 1 to 7. The period number corresponds to the **principal quantum number**, n, of the outermost electron sublevels for elements in this period. For example, elements in period 2 have $n = 2$, and their valence electrons are found in the 2s or 2p sublevels.

Elements in the same group all have the same number of valence electrons. For groups 1 and 2, the group number is equal to the number of valence electrons. For example, elements in group 1, the **alkali metals**, all have one electron in their valence sublevel. For groups 13 to 18, the last digit of the group number is equal to the number of valence electrons. For example, elements in group 15 all have five valence electrons.

You can use this information to deduce the electron configurations of elements. This is straightforward for periods 2 and 3. Consider phosphorus: being in period 3 means that $n = 3$ for the valence sublevels, and being in group 15 means that it has five electrons in these sublevels. You can use the symbol of the previous noble gas to represent the inner electrons to give the condensed electron configuration:

$$[Ne]\ 3s^2\ 3p^3$$

With period 4, we need to take the 3d sublevel into account, which fills after the 4s sublevel and before the 4p sublevel. The group number is equal to the total number of electrons occupying the 4s, 3d and 4p sublevels. Consider bromine: being in group 17 means that 17 electrons fill the outer sublevels. Two go into 4s, ten go into 3d and the remaining five go into 4p, so the electron configuration of bromine is $[Ar] 4s^2 3d^{10} 4p^5$.

> Remember that s, p, d and f sublevels all have a different maximum number of electrons. Detailed guidance on how to fill sublevels with electrons is given in *Structure 1.3*.

Patterns and trends

Imagine trying to learn chemistry without the periodic table. How many concepts are connected to this universal way of organizing the elements? How often do you refer to it during a typical chemistry lesson?

Classifying things and exploring their similarities, differences and trends helps us to better understand them. Without knowing the order and patterns depicted in the periodic table, remembering and understanding the properties of all the known elements would be very difficult. By the mid to late 19th century, chemists had long been classifying elements according to their properties. Some had also put them in order of atomic mass and found certain patterns. Russian chemist Dmitri Mendeleev developed an early version of the periodic system based on atomic masses of elements, which eventually led to the periodic table as we know it today.

Organizing the elements like this not only helped scientists understand the properties of the elements known at the time, but the presence of gaps in Mendeleev's table also signalled the possible existence of other undiscovered elements. Gallium, for example, was not known to Mendeleev, but he had assigned an element with atomic mass of around 68 (near the scribbles at the centre of figure 2) and predicted some of its properties. When gallium was discovered, many of these properties were close to Mendeleev's predictions.

Classification of the elements is not always clear-cut. For example, you may have questioned the location of hydrogen in the periodic table. Does it belong to group 1,

group 17 or a different location altogether? A second example is the ongoing discussion about which elements belong to group 3: is it Sc, Y, La and Ac, or Sc, Y, Lu and Lr? Where in the periodic table do you think hydrogen should be? Do you think it matters? What is the purpose of classification?

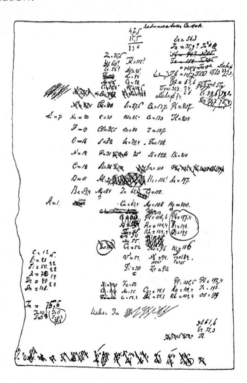

▲ **Figure 2** Some of Mendeleev's notes leading up to his 1869 periodic system

Activity

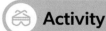

For the element positions described in **a**, **b** and **c**, use the periodic table at the back of this book to deduce the following information:

 i. name of the element
 ii. name of the group
 iii. principal quantum number
 iv. full electron configuration
 v. condensed electron configuration

a. period 2, group 16   b. period 3, group 2   c. period 4, group 8

Linking question

How has the organization of elements in the periodic table facilitated the discovery of new elements? (Structure 1.2)

Periodicity: Trends in the periodic table (*Structure 3.1.3*)

Elements are organized in the periodic table according to their electron structure. As a result, you can observe trends in the periodic table, which helps you to make predictions about the chemical reactivity of elements. The study of trends in the periodic table is known as **periodicity**. Trends are observed for various properties going across periods and down groups, such as **atomic radius**, **ionic radius**, **ionization energy**, **electron affinity** and **electronegativity**. You can graph these trends using the values for each property from sections 9 and 10 of the data booklet.

Atomic radius

Figure 3 shows the atomic radii of atoms with increasing atomic number.

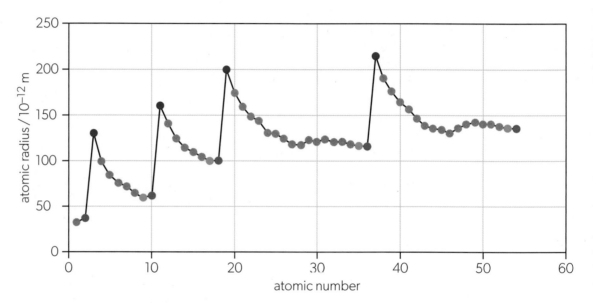

▲ Figure 3 Plot of the atomic radius against atomic number. The alkali metals have data points highlighted in red, the noble gases in blue, and the halogens in green

From the start to the end of each period (from red to blue), the atomic radius decreases. This is because each successive element has an additional proton, and so a higher nuclear charge. The increased nuclear charge going across a period means that there is greater attraction between the valence electrons and the nucleus, causing the radius to become smaller with each additional proton. As the electrons are added to the same energy level going across the period, there is no increase in shielding by the inner electrons.

The atomic radius generally increases going down a group in the periodic table. This is shown in figure 3 for the alkali metals (red dots). Although the number of protons increases, and therefore the nuclear charge also increases, the valence electrons are added to a higher energy level. This results in the outermost electrons being shielded from the attractive electrostatic force of the nucleus by a greater number of inner levels of electrons.

The diminished electrostatic force experienced by a shielded valence electron is known as the **effective nuclear charge, Z_{eff}.**

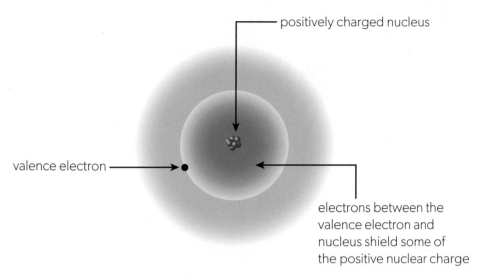

▲ Figure 4 The concept of effective nuclear charge based on electron shielding

Ionic radius

Ions with a positive charge are called cations. To form a cation, an atom loses one or more of its valence electrons, resulting in more protons than electrons. Therefore, the effective nuclear charge is greater for cations, and the ionic radius is smaller than the atomic radius. This effect increases with the charge magnitude, so larger positive charges lead to smaller radii.

Groups 1, 2 and 13 tend to form cations, with the charge equal to the number of valence electrons. Therefore, these ions have noble gas electron configurations. Similar to atomic radius, ionic radius decreases going from group 1 to group 2, and from group 2 to group 13, as the nuclear charge increases. For example, Na^+ is larger than Mg^{2+}, which is in turn larger than Al^{3+}.

Ions with a negative charge are called anions. To form an anion, an atom gains electrons to fill its valence sublevel, resulting in more electrons than protons. The force of attraction between the nucleus and these additional electrons is smaller, and the ionic radius is greater than the atomic radius. The greater the magnitude of negative charge, the larger the ionic radius.

Groups 15, 16 and 17 tend to form anions with the charge equal to the number of valence electrons minus eight. Ionic radius decreases from group 15 to group 17, as the nuclear charge increases. For example, N^{3-} is larger than O^{2-}, which in turn is larger than F^-.

However, going down a group, ionic radius increases for both cations and anions due to the increasing number of electron energy levels and increased shielding of the nucleus. This is similar to the trend for atomic radius.

Figure 5 shows atomic and ionic radii for cations and anions.

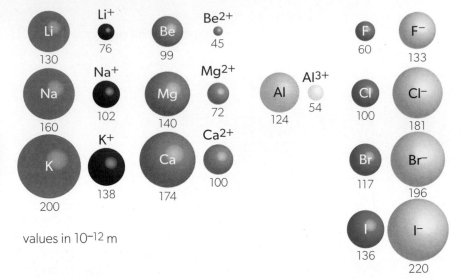

values in 10^{-12} m

 Data-based question

Use the values from the data booklet to plot a graph of the ionic radii for the period 2 and period 3 elements against their atomic number. Try to explain the trends in your graph.

▲ Figure 5 Radii of cations are smaller than those of the parent atoms. Radii of anions are larger than those of the parent atoms

Isoelectronic species have the same electron configuration. For example, Mg^{2+}, Na^+, Ne, F^- and O^{2-} all have the configuration $1s^2 2s^2 2p^6$. Because they are formed from different elements, with a different number of protons, their size varies depending on the ratio of protons to electrons (figure 6).

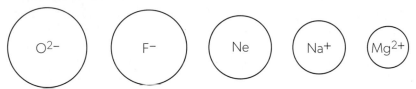

▲ Figure 6 All of these species have the same electron configuration, $1s^2 2s^2 2p^6$, but different sizes

Ionization energy

As discussed in *Structure 1.3 (AHL)*, ionization energy (*IE*) is the minimum energy required to eject an electron out of a neutral atom or molecule in its ground state.

$$X(g) + IE \rightarrow X^+(g) + e^-$$

The first ionization energy (IE_1) generally decreases down the groups of the periodic table. This is because the number of energy levels increases and the shielding effect becomes greater, so less energy is required to ionize atoms.

Going across a period, the number of protons in the nucleus increases, so the outermost electrons are held closer to the nucleus by the **increased nuclear charge**. At the same time, the shielding effect remains nearly constant because the number of inner electrons does not change. Therefore, more energy is required to remove outermost electrons, so ionization energy increases across a period.

There are some discontinuities in the trend of increasing ionization energy across a period, which are discussed in *Structure 3.1.7 (AHL)*.

The general trend of decreasing ionization energy down a group and increasing across a period is shown in figure 7.

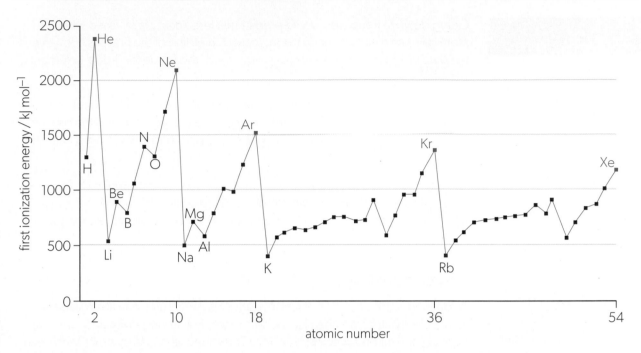

▲ Figure 7 Plot of first ionization energy against atomic number for the elements from hydrogen to xenon

Electron affinity

Electron affinity (EA) is the energy released when an additional electron is attached to a neutral atom or molecule.

$$X(g) + e^- \rightarrow X^-(g) + EA$$

The more favourable this reaction is for a given element, the higher the electron affinity. The trend for first electron affinity (EA) in the periodic table going across a period is similar to that of ionization energy: the nuclear charge increases and so more energy is released when an electron is added. The general trend of increasing electron affinity across a period is shown in figure 8.

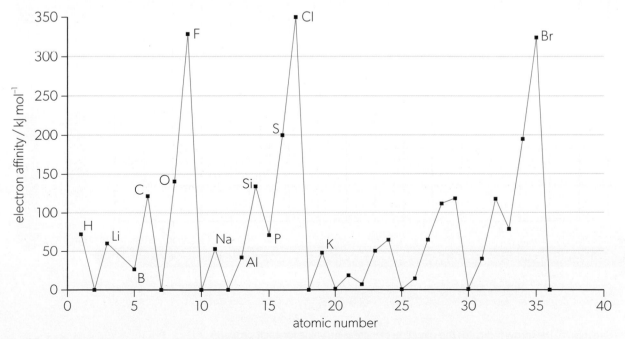

▲ Figure 8 Plot of first electron affinity against atomic number for the elements from hydrogen to krypton

US chemist Linus Pauling introduced the definition of electronegativity described here. Pauling was also the first person to win Nobel Prizes in two different categories. In 1954, he was awarded the Nobel Prize in Chemistry for his work on chemical bonding and structure. Eight years later, he won the Nobel Peace Prize for his opposition to weapons of mass destruction.

What role can scientists play in the promotion of peace in the world today?

What are the ethical responsibilities of scientists?

Going down a group, you would predict that less energy is released when an atom gains an electron due to the increased shielding of the nucleus. However, the data in figure 8 does not show a clear trend: this is true for group 1 elements, but not for the other groups.

For some elements, adding an electron to an atom results in a less stable electron configuration, which will reduce the energy released. This explains the drop in electron affinity between silicon and phosphorus, as phosphorus has a half-filled 3p sublevel.

We do not always have accurate data for an element's electron affinity. This is usually the case in elements with full or half-filled sublevels, such as beryllium, nitrogen, magnesium, manganese, zinc and the noble gases, where the process of adding an electron is not favourable.

Electronegativity

As defined in *Structure 2.2*, electronegativity is a measure of an atom's ability to attract electrons from a chemical bond to itself. Electronegativity decreases down a group, and increases across a period. This is because smaller atoms with nearly complete valence shells will attract electrons more easily than larger atoms with fewer valence electrons.

Figure 9 shows a summary of the periodic trends discussed.

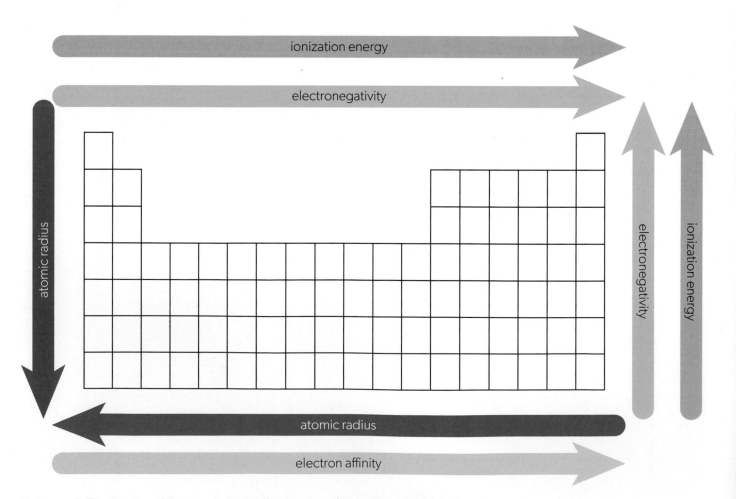

▲ Figure 9 The direction of the arrow indicates the direction of increasing value for each property

Activity

Identify the element with the **greatest** value for each property. The first column has been done for you.

Property	Strontium or iodine	Mg or Ca	C or Si	Li or F	Ba or Cl	K or Br	C or Pb	Cs or F
atomic radius	Sr							
ionic radius	I							
electron affinity	I							
ionization energy	I							
electronegativity	I							

Periodic trends in atomic volume

In this task, you will follow some of the work done by Julius Lothar Meyer, who explored the periodicity of the known chemical elements in the 18th century.

Relevant skills

- Tool 2: Use spreadsheets to manipulate data

- Tool 2: Represent data in graphical form

- Tool 3: Use and interpret scientific notation

- Tool 3: Apply and use SI prefixes and units

- Inquiry 2: Identify, describe and explain patterns, trends and relationships

Method

1. In a spreadsheet, collect atomic mass and atomic radius data for the elements from hydrogen to strontium (see example of table format below).

2. Calculate the volume of a single hydrogen atom, using the equation for the volume of a sphere.

3. Calculate the volume of a single atom of the other elements by copying and pasting the formula down the column.

4. Calculate the atomic volume of hydrogen. Atomic volume is the volume taken up by one mole of an element, in $m^3\ mol^{-1}$. Again, copy and paste the formula down the column to obtain data for the other elements.

5. Plot a graph of atomic volume vs atomic mass.

6. Discuss the features of the graph that illustrate the phenomenon of periodicity.

7. At the time, chemists described the pattern of similar properties every eighth element as the law of octaves. To what extent does the pattern in your graph support the law of octaves?

8. Are there any elements that do not fit the general trend?

	A	B	C	D	E	F	G
1	Element	Atomic number	Atomic mass	Atomic radius $/\times 10^{-12}\ m$	Atomic radius/m	Volume of a single atom $/m^3$	Atomic volume $/m^3\ mol^{-1}$
2	H		1				
3	He		2				
4	Li		3				
5	Be		4				

TOK

You have now learnt about various interconnected concepts in DP chemistry that can be used to construct explanations. It can be interesting to see how far back you can keep asking 'why'. For example:

The C—F bond is highly polar.

Why?

Because the electrons in the bond are not distributed equally between the two atoms.

Why?

Because the difference in electronegativity is large.

Why?

Because fluorine is more electronegative than carbon.

Why?

Because fluorine has a stronger tendency to attract a bonded pair of electrons towards itself than carbon.

Why?

Because fluorine has a smaller atomic radius than carbon.

Why?

Because fluorine atoms have a higher nuclear charge than carbon atoms, and both types of atoms have the same number of energy levels.

Why?

Because fluorine atoms contain nine protons each, whereas carbon atoms contain six. Fluorine and carbon atoms have two electrons in the first energy level, and their valence electrons are in the second energy level.

Why?

Because the first energy level holds up to two electrons.

Why?

Because the first energy level only has one orbital.

And so on. How did you acquire the knowledge that you use to build explanations such as this? What makes a good explanation?

Think about the "explain" questions in your IB examinations. How can you know how detailed your explanations should be?

Metallic character and periodicity (*Structure 3.1.4*)

Metallic character is not a quantitative property, that is, we do not assign a value to it. However, the metallic character of an element is closely linked to its ionization energy. Elements with high metallic character have delocalized electrons in their structure. Therefore, elements with lower ionization energies are more likely to be metallic.

For example, the group 1 elements, known as the alkali metals, have decreasing ionization energy and increasing metallic character going down the group. This means that they are also more reactive going down the group, as they have a greater tendency to donate their single valence electron in chemical reactions.

The group 17 elements are known as the halogens. We describe the halogens in terms of their non-metallic character—this property is the opposite to metallic character and is linked to electronegativity. The non-metallic character and electronegativity of the halogens decrease going down the group. Halogens higher up the group are therefore more reactive, as they have a greater tendency to accept an electron in chemical reactions and fill their valence sublevel.

We can illustrate these trends in the alkali metals and the halogens by looking at some typical chemical reactions.

Reactions of alkali metals with water

A **Brønsted–Lowry acid** is a substance that donates protons, H^+. A **Brønsted–Lowry base** is a substance that accepts protons, H^+. The alkali metals all form aqueous Brønsted–Lowry bases when reacted with water. The general equation for the reaction of a group 1 metal with water is as follows:

$$M(s) + H_2O(l) \rightarrow \frac{1}{2}H_2(g) + MOH(aq)$$

For example, lithium reacts with water as follows:

$$Li(s) + H_2O(l) \rightarrow \frac{1}{2}H_2(g) + LiOH(aq)$$

Ionization energy decreases and the metallic character of the alkali metals increases going down the group. As a result, reactivity increases and the reactions of alkali metals with water occur faster and more vigorously going down the group.

Reactions of halide ions with halogens

Elemental halogens react with other species to gain electrons, and to achieve a complete octet, becoming anions. The electronegativity and non-metallic character of the halogens decrease going down the group, and hence their reactivity also decreases going down the group.

We can illustrate the reactivity of the halogens by looking at the reactions between elemental halogens and the anions of other halogens. When elemental fluorine reacts with a chloride ion, elemental fluorine gains an electron, forming a fluoride ion and elemental chlorine as follows:

$$\frac{1}{2}F_2 + Cl^- \rightarrow F^- + \frac{1}{2}Cl_2$$

The reverse reaction will not happen, however: chlorine cannot oxidize fluoride ions. This is because fluorine is more electronegative than chlorine, so the reaction is not favourable.

The greater the electronegativity difference between the two reacting species, the more readily the reaction occurs. Hence, F_2 will react faster with I^- than with Cl^-. This is also the case for the reactions between halogens and alkali metals: F_2 reacts faster with K than with Li. The equation for the former reaction is shown below:

$$\frac{1}{2}F_2 + K \rightarrow K^+ + F^-$$

Brønsted–Lowry acids and bases, and the distinction between alkaline and basic substances are covered in *Reactivity 3.1*.

▲ Figure 10 Elemental sodium metal reacts vigorously with water. This effect increases going down group 1

In this reaction, chloride loses electrons (undergoes **oxidation**) and fluorine gains electrons (undergoes **reduction**). Oxidation and reduction are discussed in *Reactivity 3.2*.

📚 Activity

Complete the table to show which reactions can occur between halogen species. The first row has been done for you.

	F⁻	Cl⁻	Br⁻	I⁻
F_2	No reaction	$F_2 + 2Cl^- \rightarrow 2F^- + Cl_2$	$F_2 + 2Br^- \rightarrow 2F^- + Br_2$	$F_2 + 2I^- \rightarrow 2F^- + I_2$
Cl_2		No reaction		
Br_2			No reaction	
I_2				No reaction

Practice questions

1. Write the word and symbol equations for the reaction between potassium and water.

2. List the alkali metals from the most reactive to the least reactive, in terms of their reactions with water.

3. Identify the most reactive pair of reactants in **a** to **d**.
   a. $Li + I_2$ or $Li + F_2$
   b. $Ba + I_2$ or $Ba + Cl_2$
   c. $K + Cl_2$ or $Li + Cl_2$
   d. $F_2 + KCl$ or $F_2 + KI$

 Linking question

Why are simulations and online reactions often used in exploring the trends in chemical reactivity of group 1 and group 17 elements? (Inquiry 2, Tool 2)

ATL Thinking skills

The modern periodic table is a very useful way of organizing the elements. Conduct an online search for alternative representations of the periodic table. An example is shown in figure 11.

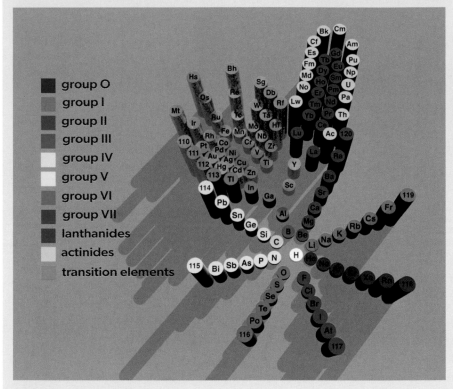

group O
group I
group II
group III
group IV
group V
group VI
group VII
lanthanides
actinides
transition elements

▲ **Figure 11** This periodic table has a "bicycle wheel" structure, with each group being a spoke of the wheel

Select two or three alternative periodic tables and consider:
- What features of the elements are being highlighted in each case?
- To what extent are they essentially the same as the standard periodic table?
- What are the advantages and disadvantages of the different periodic tables?

Metal oxides and non-metal oxides (*Structure 3.1.5*)

Another way to describe acids and bases is through their ability to accept or donate a pair of electrons: a Lewis acid can accept an electron pair, and a Lewis base can donate an electron pair. Many metal oxides are Lewis bases. They react with water to form hydroxides, also bases, by donating an electron pair to hydrogen in water. Reactions of alkali metal oxides with water have the general equation:

$$M_2O(s) + H_2O(l) \rightarrow 2MOH(aq)$$

Here are some examples:

$$Na_2O(s) \quad + \quad H_2O(l) \quad \longrightarrow \quad 2NaOH(aq)$$
$$\text{sodium oxide} \qquad\qquad\qquad\qquad\qquad \text{sodium hydroxide}$$

$$Li_2O(s) \quad + \quad H_2O(l) \quad \longrightarrow \quad 2LiOH(aq)$$
$$\text{lithium oxide} \qquad\qquad\qquad\qquad\qquad \text{lithium hydroxide}$$

Group 2 oxides are also Lewis bases. They also react with water to form hydroxides, with the following general equation:

$$MO(s) + H_2O(l) \rightarrow M(OH)_2(aq)$$

Here are some examples:

$$MgO(s) \quad + \quad H_2O(l) \quad \longrightarrow \quad Mg(OH)_2(s)$$
$$\text{magnesium oxide} \qquad\qquad\qquad\qquad\qquad \text{magnesium hydroxide}$$

$$BaO(s) \quad + \quad H_2O(l) \quad \longrightarrow \quad Ba(OH)_2(aq)$$
$$\text{barium oxide} \qquad\qquad\qquad\qquad\qquad \text{barium hydroxide}$$

Non-metallic oxides are Lewis acids. They react with water to form other acids by accepting an electron pair from oxygen in water. Here are some examples:

$$CO_2(g) \quad + \quad H_2O(l) \quad \longrightarrow \quad H_2CO_3(aq)$$
$$\text{carbon dioxide} \qquad\qquad\qquad\qquad\qquad \text{carbonic acid}$$

$$SO_2(g) \quad + \quad H_2O(l) \quad \longrightarrow \quad H_2SO_3(aq)$$
$$\text{sulfur dioxide} \qquad\qquad\qquad\qquad\qquad \text{sulfurous acid}$$

$$SO_3(l) \quad + \quad H_2O(l) \quad \longrightarrow \quad H_2SO_4(aq)$$
$$\text{sulfur trioxide} \qquad\qquad\qquad\qquad\qquad \text{sulfuric acid}$$

$$P_4O_{10}(s) \quad + \quad 6H_2O(l) \quad \longrightarrow \quad 4H_3PO_4(aq)$$
$$\text{phosphorus(V) oxide} \qquad\qquad\qquad\qquad\qquad \text{phosphoric acid}$$

Going across any period, the oxides of the elements become less basic and more acidic. For example, $Na_2O(s)$ forms a stronger base than $MgO(s)$ in its reaction with water and $SO_3(l)$ forms a stronger acid than $P_4O_{10}(s)$.

A chemical species that behaves as both a Lewis acid and a Lewis base is termed **amphoteric**. For example, aluminium oxide, $Al_2O_3(s)$, is amphoteric: it can accept an electron pair or donate an electron pair depending on what it is reacting with.

For example, when reacting with aqueous sodium hydroxide, a base, aluminium oxide acts as a Lewis acid:

$$Al_2O_3(s) \quad + \quad 2NaOH(aq) \quad + \quad 3H_2O(l) \quad \rightarrow \quad 2Na[Al(OH)_4](aq)$$

| aluminium oxide | sodium hydroxide | | sodium aluminate |

When reacting with hydrochloric acid, aluminium oxide acts as a Lewis base:

$$Al_2O_3(s) \quad + \quad 6HCl(aq) \quad \rightarrow \quad 2AlCl_3(aq) \quad + \quad 3H_2O(l)$$

| aluminium oxide | hydrochloric acid | aluminium chloride |

Both of these reactions are neutralization reactions: an acid and a base react with each other to form a salt.

The acid–base properties of some period 3 elements are shown in table 2.

Formula of oxide	$Na_2O(s)$	$MgO(s)$	$Al_2O_3(s)$	$SiO_2(s)$	$P_4O_{10}(s)$	$SO_3(l)$ and $SO_2(g)$
Acid or base?	basic	basic	amphoteric	acidic	acidic	acidic

▲ Table 2 Trend in the acid–base properties of the oxides of some period 3 elements

At the standard level of DP chemistry, you do not need to know the definition of Lewis acids and bases. Lewis acids and bases are discussed in more detail in *Reactivity 3.4 (AHL)*.

Acid rain and ocean acidification

Pure water at 298 K has a pH of 7.0, which corresponds to a neutral solution: it is neither acidic nor basic. Rainwater is naturally acidic due to the presence of dissolved carbon dioxide, which forms weak carbonic acid:

$$CO_2(g) + H_2O(l) \rightarrow H_2CO_3(aq)$$

A typical pH value for rainwater is 5.6. We know that other oxides, such as nitrogen oxides, NO_x and sulfur dioxide, SO_2, are more acidic than CO_2 because nitrogen and sulfur are further to the right in the periodic table. If these gases dissolve in rainwater, the ensuing rain is more acidic than normal, and is known as **acid rain**, which has a pH of less than 5.6.

Sulfur dioxide and nitrogen oxides are produced naturally by volcanic eruptions and decomposing vegetation. These pollutants can also be released into the atmosphere by industrial processes, such as the combustion of fossil fuels with high levels of sulfur impurities.

You have seen the reaction of SO_2 with water that produces sulfurous acid. The reaction between water and nitrogen dioxide produces nitric acid and nitrous acid:

$$2NO_2(g) \quad + \quad H_2O(l) \quad \rightarrow \quad HNO_3(aq) \quad + \quad HNO_2(aq)$$

| nitrogen dioxide | | | nitric acid | nitrous acid |

Ultimately, nitrous acid and sulfurous acid are oxidized by atmospheric oxygen to nitric acid and sulfuric acid, respectively:

$$2HNO_2(aq) + O_2(g) \rightarrow 2HNO_3(aq)$$
$$2H_2SO_3(aq) + O_2(g) \rightarrow 2H_2SO_4(aq)$$

Oceans absorb a large proportion of the carbon dioxide released into the atmosphere. As a result, carbonic acid is formed in the ocean. Increased ocean acidity can affect the ability of coral reefs and shellfish to calcify their skeletons.

Linking question

How do differences in bonding explain the differences in the properties of metal and non-metal oxides? (Structure 2.1 and Structure 2.2)

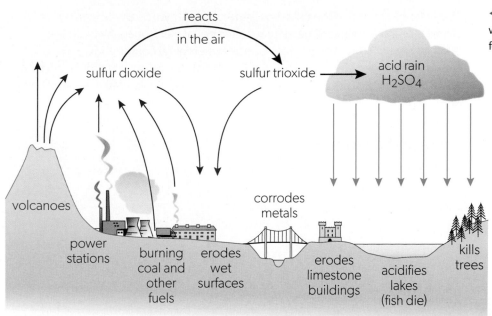

Oxidation states (*Structure 3.1.6*)

The bonding triangle introduced in *Structure 2.4* shows that bonding is often not purely ionic or covalent, but rather occurs on a continuum with unequal sharing of electrons. The concept of **oxidation state**, or **oxidation number**, is used to represent the charge on an atom in a compound if it were composed of ions. It also describes the number of electrons shared or transferred when forming a bond.

Water, for example, is a predominantly covalent molecule: H_2O. If it were an ionic compound, then oxygen would gain two electrons and each hydrogen atom would lose one, as oxygen is more electronegative than hydrogen. Therefore, oxygen is assigned an oxidation state of -2 and each hydrogen has an oxidation state of $+1$.

▲ Figure 13 The oxidation states of elements in water

Deducing oxidation states in compounds

In compounds, the more electronegative atoms are assigned negative oxidations states and the less electronegative atoms are assigned positive oxidation states. There are a number of useful rules for deducing the oxidation states of atoms:

Rule 1 The oxidation state of any free element is zero.

Rule 2 The sum of the oxidation states of all atoms in a neutral compound is zero.

Rule 3 The sum of the oxidation states of all atoms in a polyatomic ion is equal to the charge of the ion.

Rule 4 The oxidation state of fluorine is -1 in all compounds.

Rule 5 The oxidation state of the group 1 metals is always $+1$ and the oxidation state of the group 2 metals is always $+2$.

Rule 6 The oxidation state of oxygen is -2 except in OF_2, where fluorine is more electronegative, so oxygen has an oxidation state of $+2$, and in peroxides such as H_2O_2, where it is -1.

Rule 7 The oxidation state of hydrogen is $+1$ when it is combined with more electronegative elements (most non-metals), and -1 when it is combined with less electronegative elements (metals). Hydrogen in a -1 oxidation state is called a hydride ion.

The first rule states that the oxidation state of an element is always zero. This is because atoms of the same element have equal electronegativity, so the electrons are shared evenly across bonds in elemental compounds. For example, consider elemental oxygen, O_2: the molecule is symmetrical, and the electronegativity of the atoms is identical. Therefore, the oxidation state of each atom is zero.

Worked example 1

Deduce the oxidation state of

a. Mn in $KMnO_4$
b. the atoms in H_2, Cu, Zn, Xe, and Au
c. P in the phosphate ion, PO_4^{3-}
d. all the elements in $LiAlF_4$

Solution

a. According to rule 2, as $KMnO_4$ is a neutral compound, the sum of the oxidation states of its atoms is equal to zero:

(oxidation state of K) + (oxidation state of Mn) + (oxidation state of O × 4) = 0

We multiply the oxidation state value for oxygen by four, as there are four oxygen atoms in the compound.

You can assign oxidation states to O and K, and then find the value for Mn that would make the sum of the oxidation states zero.

According to rule 5, the oxidation state of potassium is +1. According to rule 6, the oxidation state of oxygen is −2. Substituting these values into the above equation above gives:

$1 + $ (oxidation state of Mn) $+ (-2 \times 4) = 0$

oxidation state of Mn $= +7$

b. These are all elemental species, so each atom has an oxidation state of 0 (rule 1).

c. According to rule 3, as PO_4^{3-} is a polyatomic ion, the sum of the oxidation states of its atoms is equal to the charge on the ion:

(oxidation state of P) + (oxidation state of O × 4) = −3

According to rule 6, the oxidation state of oxygen is −2. Substituting this value into the above equation above gives:

(oxidation state of P) $+ (-2 \times 4) = -3$

oxidation state of P $= +5$

d. Using rule 2: (oxidation state of Li) + (oxidation state of Al) + (oxidation state of F × 4) = 0

According to rule 5, the oxidation state of lithium is +1. According to rule 4, the oxidation state of fluorine is −1. Substituting these values into the above equation above gives:

$1 + $ (oxidation state of Al) $+ (-1 \times 4) = 0$

oxidation state of Al $= +3$

Practice questions

4. Assign oxidation states to each atom in the following compounds:
    a. $K_2Cr_2O_7$
    b. CH_4
    c. CO_2
    d. $HClO_4$
    e. OF_2
    f. SO_3^{2-}
    g. NaH
    h. Fe_2O_3
    j. CH_3OH
    k. CH_2O
    l. Na_2O_2

Naming of oxyanions

Oxyanions are polyatomic anions that include oxygen atoms. When naming oxyanions, you should include the oxidation state of the non-oxygen atom as a Roman numeral at the end of the name. The Roman numerals are enclosed in brackets with no space between the ion name and the bracket.

For example, consider the anion MnO_4^-. The anion is named manganate(VII), as Mn has an oxidation state of +7. The neutral compound $KMnO_4$ is called potassium manganate(VII). This compound is also known under its traditional name, potassium permanganate.

While the Roman numerals are often used to refer to the oxidation state of a transition metal, the non-oxygen atom in an oxyanion can also be a non-metal. $KClO_4$, for example, is called potassium chlorate(VII).

In practice, the Roman numeral is omitted for some common oxyanions. These are shown in table 3.

Ionic formula	Common name	Systematic name
NO_2^-	nitrite	nitrate(III)
NO_3^-	nitrate	nitrate(V)
SO_3^{2-}	sulfite	sulfate(IV)
SO_4^{2-}	sulfate	sulfate(VI)

▲ Table 3 The common names and systematic names of some oxyanions. Both are acceptable in examinations

Practice questions

5. Write the chemical formulas for the following compounds:
   a. hydrogen nitrate
   b. sodium phosphate(V)
   c. magnesium bromate(III)
   d. potassium sulfite

 Linking question

How can oxidation states be used to analyse redox reactions? (Reactivity 3.2)

(ATL) Communication skills

Scientists have used symbols to identify different elements for a long time (figure 14). Nowadays, we use the symbols in the periodic table, and the nomenclature conventions encouraged by the International Union of Pure and Applied Chemistry (IUPAC). Your study of chemistry will involve using symbols as they are stated in the data booklet, writing chemical formulas correctly (including subscripts and brackets where necessary) and naming substances according to IUPAC guidance.

Test your chemical communication skills by finding the error in each of the following and writing the correct answer:

1. HCL
2. NE
3. O3

4. MgOH2
5. cuprum(II) oxide

▲ Figure 14 The symbols used by alchemists for copper and iron, respectively

Discontinuities in ionization energy trends (*Structure 3.1.7*)

Earlier in this topic, you saw that there was a trend of increasing ionization energy going across a period. You may have noticed two discontinuities from group 2 to group 13, and from group 15 to group 16 (figure 15).

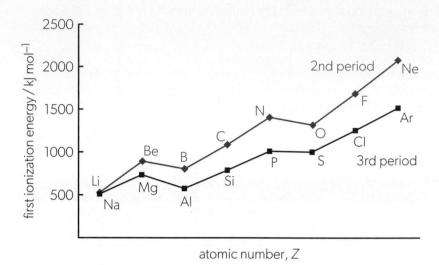

▲ Figure 15 The ionization energies of the second and third period elements

The first discontinuity, between beryllium and boron, can be explained by looking at their electron configurations. The electron configuration of boron is $1s^2 2s^2$ and the electron configuration of beryllium is $1s^2 2s^2 2p^1$. Despite the greater nuclear charge in boron, it is easier to remove an electron from this element as the valence electron is in a higher energy sublevel, which is shielded from the nucleus by the 2s sublevel. This drop in ionization energy therefore provides evidence for the existence of s and p sublevels.

The second discontinuity, between nitrogen and oxygen, can be explained by electron repulsion in orbitals. The electron configuration of nitrogen is $1s^2 2s^2 2p^3$ while for oxygen it is $1s^2 2s^2 2p^4$. In nitrogen, the three electrons in the 2p orbitals do not come into close proximity. The paired electrons in oxygen occupy the same region of space and have increased repulsion. The higher-energy paired electron is therefore easier to remove. This explains the drop in ionization energy between the two groups.

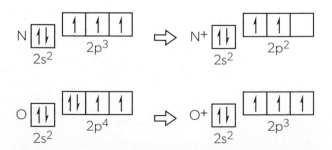

▲ Figure 16 A half-filled p sublevel is stable due to decreased electron repulsion, as none of the electrons are paired

Properties of the transition elements (*Structure 3.1.8*)

A transition element is defined as an element that has a partially filled d-sublevel or can form a stable cation with an incomplete d-sublevel. Having an incomplete d-sublevel gives the transition elements some characteristic properties.

For example, the d electrons have high energy and are delocalized throughout the metallic structure. The attraction between the delocalized electrons and the metal ions in transition elements is strong, so these elements have high melting points. Most transition elements have unpaired d electrons, and as a result they are paramagnetic. This is because the spin of an unpaired d electron creates a magnetic moment, which can align with an external magnetic field.

Zinc is not considered a transition element despite being in the d-block. This is because it does not form cations with an incomplete d-sublevel. Its electron configuration is $[Ar]4s^2 3d^{10}$, so it readily loses its 4s electrons to form 2+ ions, but the complete d-sublevel remains intact. All of the d electrons in elemental zinc are also paired in its neutral state, which means that zinc is not magnetic.

A **catalyst** is a species that reduces the activation energy required for a reaction to occur, while not being used up in that reaction. **Heterogeneous catalysts** are a type of catalyst that exist in a separate phase to the reactants. Transition elements are often used as heterogeneous catalysts. Gas molecules adsorb onto a solid surface of the transition element, and the molecules undergo chemical changes. The products then desorb (figure 17).

Platinum, palladium and rhodium are transition elements used as heterogeneous catalysts in catalytic converters. Catalytic converters adsorb carbon monoxide and oxygen molecules in the exhaust of vehicle, and desorb carbon dioxide:

$$2CO + O_2 \rightarrow 2CO_2$$

> Catalysts and activation energy are discussed further in *Reactivity 2.2*.

> 🔗 **Linking question**
>
> What are the arguments for and against including scandium as a transition element? (Structure 2.3)

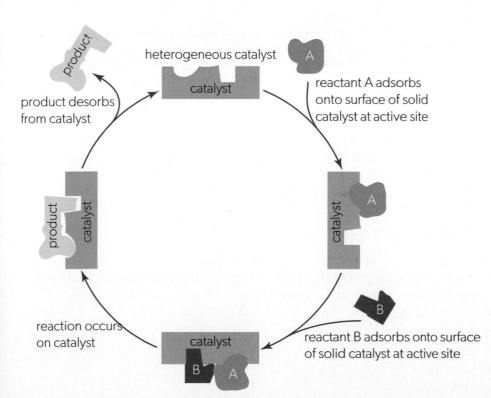

product desorbs from catalyst

heterogeneous catalyst

reactant A adsorbs onto surface of solid catalyst at active site

reaction occurs on catalyst

reactant B adsorbs onto surface of solid catalyst at active site

◀ Figure 17 **The action of heterogeneous catalysts. Transition elements are often good heterogeneous catalysts**

This reduces the emission of the products of incomplete combustion created when burning fuel. Platinum is also used as a heterogeneous catalyst in hydrogen fuel cells. In this process, $H_2(g)$ and $O_2(g)$ adsorb onto the catalyst surface and $H_2O(l)$ desorbs.

Transition elements are also able to form compounds known as **complex ions**, some of which have distinctive colours. The chemistry behind this phenomenon is detailed later in this topic.

Variable oxidation states in transition elements (*Structure 3.1.9*)

Deducing electron configurations of transition elements

As with any other element, if you know the position of a transition element on the periodic table, you can determine its electron configuration.

Worked example 2

Determine the condensed electron configuration of iron, Fe.

Solution

The condensed electron configuration starts with the previous noble gas in square brackets, which in this case is argon: [Ar]

Then, identify the valence sublevels. Iron is in period 4, so its valence sublevels are 4s and 3d. Remember, 3d comes after 4s because it is higher in energy.

Then, determine the number of valence electrons. This can be deduced from the group number. Iron is in group 8, so it has eight valence electrons. Two go into the 4s sublevel, and the remaining six go into the 3d sublevel.

Therefore, the condensed electron configuration of iron is $[Ar]\,4s^2\,3d^6$. If you need the full electron configuration, expand the electron configuration of argon: $1s^2\,2s^2\,2p^6\,3s^2\,3p^6\,4s^2\,3d^6$.

The two exceptions to the method described in the worked example are chromium and copper. In both cases, a 4s electron is promoted into the 3d sublevel to achieve a more stable electron configuration. Therefore, chromium is $[Ar]\,4s^1\,3d^5$ and copper is $[Ar]\,4s^1\,3d^{10}$.

Variable oxidation states

Another property of the transition elements is their ability to form a variety of stable ions in different oxidation states. These are known as **variable oxidation states.** Some of the most common oxidation states of transition elements are given in figure 18.

The common oxidation states of the transition elements are given in the data booklet.

	Sc	Ti	V	Cr	Mn	Fe	Co	Ni	Cu
					7				
				6	6	6			
oxidation			5	5	5	5	5		
states		4	4	4	4	4	4	4	
that occur									
in	3	3	3	3	3	3	3	3	3
compounds	2	2	2	2	2	2	2	2	2
									1
	1	1	1	1	1	1	1		

▲ Figure 18 Common oxidation states of the period 4 transition elements, scandium to zinc. The most common oxidation states are shown in a larger font

From figure 18, you can see that the most common oxidation states of manganese are +2, +4 and +7. This means that it forms the following oxides: manganese(II) oxide, manganese(IV) oxide, and manganese(VII) oxide.

The formation of variable oxidation states in transition elements can be explained by the fact that their successive ionization energies are close in value. The successive ionization energies of chromium and copper are shown in figure 19, along with their orbital filling diagrams.

> Methods for drawing orbital filling diagrams are described in *Structure 1.3*.

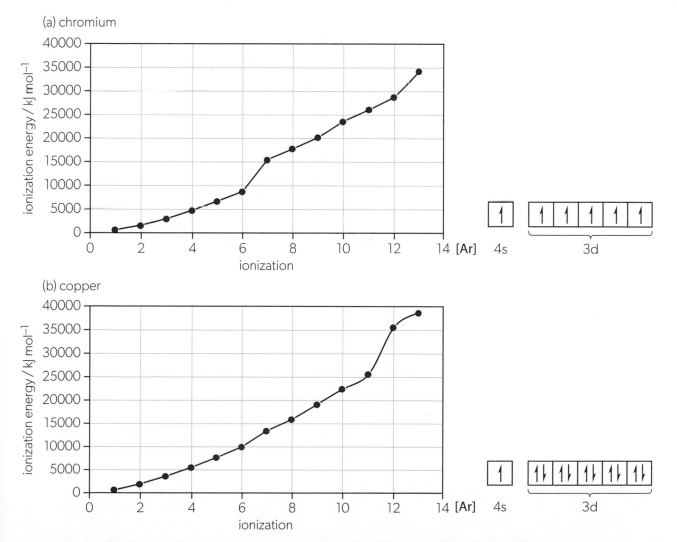

▲ Figure 19 The first 13 successive ionization energies and orbital filling diagrams for (a) chromium (b) copper

Activity

Find data for the first 13 ionization energies of the other period 4 transition elements and plot a graph for each element. Using your graphs, look for connections between the electron configurations, the difference in successive ionization energies, and the common ions for each transition element.

Chromium forms ions with oxidation states ranging from +1 to +6, losing the $4s^1$ electron first and then the d electrons. The most common ions are Cr^{3+}, which has an electron configuration of $[Ar]\,3d^3$, and CrO_4^{2-} and $Cr_2O_7^{2-}$, in which chromium has an electron configuration of $[Ar]$. There is a very small increase in energy between losing the first electron through to the sixth electron, so all six oxidation states are possible.

The increments between successive ionization energies for copper are larger than those for chromium, so it forms fewer stable ions. The common ions are Cu^+, which has an electron configuration of $[Ar]\,3d^{10}$ and Cu^{2+}, which has an electron configuration of $[Ar]\,3d^9$. Cu(III) and Cu(IV) exist in some compounds, such as in superconductors, but higher oxidation states are not stable.

Scandium has the electron configuration $[Ar]\,4s^2\,3d^1$. Sc^{3+} was the only known ion up until the 1920s, and it was not considered a transition element, as a transition element has to be able to form ions with partially filled d-sublevels. However, scandium can also exist in the +2 oxidation state, having an electron configuration of $[Ar]\,3d^1$. An example of a scandium(II) compound is $CsScCl_3$.

Practice questions

6. Manganese and iron both can exist in +2 and +3 oxidation states. The condensed orbital filling diagrams of these elements and their ions are given here:

atom/ion		electron structure 3d					4s
Mn	[Ar]	↑	↑	↑	↑	↑	↑↓
Mn^{2+}	[Ar]	↑	↑	↑	↑	↑	
Mn^{3+}	[Ar]	↑	↑	↑	↑		
Fe	[Ar]	↑↓	↑	↑	↑	↑	↑↓
Fe^{2+}	[Ar]	↑↓	↑	↑	↑	↑	
Fe^{3+}	[Ar]	↑	↑	↑	↑	↑	

Explain why Mn^{2+} is more common than Mn^{3+} and why Fe^{2+} can be easily oxidized to Fe^{3+}.

7. Vanadium commonly forms compounds where it has a +5 oxidation state.

   a. Determine the condensed electron configuration of elemental vanadium.

   b. Explain why the most common oxidation state of vanadium is +5.

Transition element complexes (*Structure 3.1.10*)

Transition elements form coloured compounds known as **complexes** or complex ions when bonded with **ligands**. Ligands are molecules or ions with a lone pair of electrons that can be donated to a transition element cation. This results in a coordination bond formed between the ligands and the central cation (figure 20).

When ligands form coordination bonds with transition element cations, the orbitals in the d-sublevel are split into two sets with different energies (figure 21). Normally, all five d orbitals have the same energy: they are **degenerate**. The energy gap between the split orbitals corresponds to a wavelength within the visible light region of the electromagnetic spectrum.

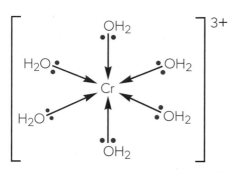

▲ Figure 20 The complex ion $[Cr(H_2O)_6]^{3+}$ has six coordination bonds from the lone pairs on the H_2O molecules to the central Cr^{3+} ion. It is violet in colour

> The formation of complex ions is detailed in *Reactivity 3.4*.

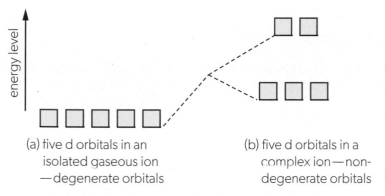

(a) five d orbitals in an isolated gaseous ion —degenerate orbitals

(b) five d orbitals in a complex ion—non-degenerate orbitals

▲ Figure 21 The effect of ligands on the d orbitals of the transition element ion

The difference in energy between the split orbitals allows electrons in the lower orbitals to be promoted into the higher orbitals. This happens when the complex absorbs light with a wavelength corresponding to the energy gap between the non-degenerate orbitals. The colour observed is the complementary colour to the colour absorbed. Complementary colours are opposites on the colour wheel given in figure 22. The colour wheel is also shown in the data booklet.

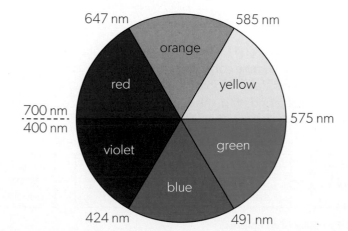

▲ Figure 22 The colour wheel

For example, $[Cr(H_2O)_6]^{3+}$ absorbs yellow light when electrons are promoted between the split d orbitals. As a result, the complex appears violet, as this is the opposite colour on the colour wheel.

The larger the splitting of the d orbitals, the greater the energy required to promote an electron. Therefore, for a larger splitting, the light needs to have a shorter wavelength and higher frequency. Remember the relationships between wavelength, frequency and energy:

speed of light $(3.00 \times 10^8 \, m \, s^{-1})$ = frequency (in s^{-1}) × wavelength (in m)

$$c \qquad = \qquad f \qquad \times \qquad \lambda$$

energy (in J) = Planck constant $(6.63 \times 10^{-34} \, J \, s)$ × frequency (in s^{-1})

$$E \qquad = \qquad h \qquad \times \qquad f$$

Worked example 3

The colour of $[CuCl_4]^{2-}$ ions is equivalent to visible light with a wavelength of 647 nm. Calculate the energy gap of the split d orbitals in $[CuCl_4]^{2-}$.

Solution

If the colour of the complex ion corresponds to the wavelength 647 nm, the wavelength of light absorbed is opposite on the colour wheel (figure 22): 491 nm.

To determine the frequency of the light, use $c = f \times \lambda$. Remember to convert the wavelength to metres.

$$3 \times 10^8 \, m \, s^{-1} = 491 \times 10^{-9} \, m \times f$$

$$f = 6.11 \times 10^{14} \, s^{-1}$$

Then use $E = h \times f$ to determine the energy gap between the orbitals:

$$E = 6.63 \times 10^{-34} \, J \, s \times 6.11 \times 10^{14} \, s^{-1}$$

$$E = 4.05 \times 10^{-19} \, J$$

The identities of the ligand and the transition element cation are two factors that affect the colour of the complex observed. The same ligand will cause different splitting in the d orbitals depending on the central metal cation. Ligands that form stronger coordination bonds with the central ion will produce a greater energy gap between the split d orbitals (figure 23).

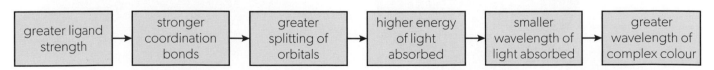

▲ Figure 23 The relationship between ligand strength and the colour of complexes

The third factor affecting the colour observed is the charge on the central transitional element cation. A Cr^{3+} complex will have a different splitting of d orbitals than a Cr^{2+} complex with the same ligands, and they will therefore be of different colours.

Practice question

8. A complex $[Cu(H_2O)_6]^{2+}$(aq) changes colour when excess Cl^-(aq) is added. Chloride ions are considered to be weaker ligands than water. Explain the colour change observed.

 Colorimetric analysis of a solution of unknown concentration

In this task, you will determine the unknown concentration of a transition metal in solution by interpolating a calibration curve. You will need a colorimeter, or a mobile phone app that can measure RGB values.

Relevant skills

- Tool 1: Use and apply colorimetry
- Tool 1: Prepare a standard solution
- Tool 1: Carry out dilutions
- Tool 3: Construct and interpret graphs
- Tool 3: Interpolate graphs
- Tool 3: Draw and interpret uncertainty bars
- Inquiry 1: Calibrate measuring apparatus, including sensors
- Inquiry 3: Discuss the impact of uncertainties on the conclusions
- Inquiry 1: Formulate research questions

Possible transition metal solutions include those containing manganate(VII) ions, MnO_4^-, hexaaquacopper(II) complex ions, $[Cu(H_2O)_6]^{2+}$, and pentaaquathiocyanatoiron(III) complex ions, $[Fe(H_2O)_5(SCN)]^{2+}$. Knowledge of complex ions is not necessary to use this important analytical technique, so this task is also suitable for SL students.

A simplified version of this task involves determining the concentration of a complex in solution prepared by your teacher or technician. You can also use it to determine the concentration of a particular coloured component in everyday substances.

Method

1. Review the colorimetry section of the *Skills* chapter.
2. Read through the procedure below and make a list of the equipment you will need.

Part 1: Calibration curve preparation

3. Prepare a standard solution of the ion you will be analysing, and perform a serial dilution to obtain five solutions of different concentrations.
4. Depending on the colour of the solution, determine a suitable wavelength of incident light.
5. Calibrate the colorimeter.
6. Determine the absorbances of the five standard solutions.
7. Plot a calibration curve that shows absorbance vs concentration.
8. Check for errors by verifying that your line of best fit goes through the origin.
9. Propagate uncertainties and include error bars on your graph. Draw lines to determine the maximum and minimum gradients.

Part 2: Analysis of solution of unknown concentration

10. Obtain a small sample of the solution of unknown concentration. Determine the absorbance of the analysed solution.
11. Use your calibration curve to determine the concentration of the analysed ion in the solution.
12. Determine the absolute, relative and percentage uncertainty of the concentration.

Part 3: Developing a research question

13. Brainstorm possible research questions that could involve using this analytical technique.

 Linking question

What is the nature of the reaction between transition element ions and ligands in forming complex ions? (Reactivity 3.4)

How can colorimetry or spectrophotometry be used to calculate the concentration of a solution of coloured ions? (Tool 1, Inquiry 2)

End of topic questions

Topic review

1. Using your knowledge from the *Structure 3.1* topic, answer the guiding question as fully as possible:

    How does the periodic table help us to predict patterns and trends in the properties of elements?

Exam-style questions

Multiple-choice questions

2. Which of the following does **not** apply to zinc?

    A. metal

    B. d-block element

    C. transition element

    D. element

3. Which of the following oxides forms an acidic solution when added to water?

    I. SiO_2

    II. SO_2

    III. SrO

    A. II only

    B. III only

    C. I and II only

    D. II and III only

4. Which of the following trends does **not** explain why electronegativity increases across periods?

    A. increasing number of valence electrons

    B. atomic radius decreases

    C. nuclear charge increases

    D. constant shielding

5. Which of the following shows a second ionization energy?

    A. $X(g) \rightarrow X^{2+}(g) + 2e^-$

    B. $X^-(g) + e^- \rightarrow X^{2-}(g)$

    C. $X^{2+}(g) \rightarrow X^{3+}(g) + e^-$

    D. $X^+(g) \rightarrow X^{2+}(g) + e^-$

6. Which of the following represents the total amount of energy, in kJ, required to form one mole of oxide ions from one mole of oxygen atoms?

    (1st *EA* of oxygen = -141 kJ mol^{-1}; 2nd *EA* of oxygen = 753 kJ mol^{-1}; N_A = 6.02 x 10^{23} mol^{-1})

    A. -141

    B. $-141 \times 2 \times 6.02 \times 10^{23}$

    C. $(-141 + 753) \times 6.02 \times 10^{23}$

    D. $-141 + 753$

7. Which pair of elements would react most vigorously?

    A. Cs and F_2

    B. Li and I_2

    C. Cs and I_2

    D. Li and F_2

8. Which statement about the halogens is correct?

    A. Cl_2 is a weaker reducing agent than F_2

    B. I_2 will oxidize Br^-

    C. I_2 has a lower tendency to be reduced than Br_2

    D. Cl_2 will reduce I^-

9. Deduce the oxidation state of nitrogen in sodium nitrate.

    A. +1   B. +2   C. +5   D. +6

10. Which of the following is **not** a typical property of transition elements?

    A. catalytic properties

    B. variable oxidation states

    C. low electrical conductivity

    D. complex ion formation

11. Which of the following statements explains why transition metal compounds are coloured?

    A. Electrons absorb energy when they move to a higher-energy d-orbital

    B. Electrons release energy when they drop down to a lower-energy level

    C. Transition elements have variable oxidation states

    D. Electrons are delocalized in transition elements

12. What is the oxidation state of cobalt in the complex ion $[CoCl_4]^{2-}$?

    A. -2 B. 0   C. +2   D. +4

AHL

13. What is the ligand in the complex $[NH_4]_2[Fe(H_2O)_6][SO_4]_2$?

    A. Fe^{2+}

    B. $[SO_4]^{2-}$

    C. H_2O

    D. $[NH_4]^+$

14. $[Ni(NH_3)_6]^x$ is a complex ion with nickel in the +2 oxidation state. What is the overall charge, x, of the complex?

    A. 0

    B. 1+

    C. 2+

    D. 3+

Extended-response questions

15. This question is about the halogens. The halogens are in group 17 of the periodic table.

    a. Explain why the atomic radius of the halogens increases down the group. [2]

    b. Explain why fluorine has a greater electronegativity than oxygen. [2]

    c. Explain why the chloride ion has a greater radius than the chlorine atom. [2]

    d. When chlorine gas is bubbled through aqueous potassium bromide, an orange-brown liquid is produced.

        i.   Write a balanced equation, including state symbols, for the reaction described above. [2]

        ii. Explain why this reaction occurs. [2]

16. The alkali metals react vigorously with water.

    a. State the group number of the alkali metals. [1]

    b. Explain why the ionization energy of the alkali metals decreases down the group. [2]

    c. Phenolphthalein indicator is added to a large container full of water. Then, a small piece of sodium is added to the water.

        i.   Describe and explain the change in the colour of the indicator. [1]

        ii. Suggest a value for the pH of the solution in the container once the reaction has ended. [1]

    d. Write a balanced equation, including state symbols, for the reaction of rubidium with water. [2]

17. This question is about acid deposition.

    a. Explain, using relevant chemical equation(s), why rain is naturally acidic. [2]

    b. Explain why burning coal leads to the production of acid rain. [2]

    c. Figure 12 on page 243 shows some of the consequences of acid deposition.

        i.   List the five consequences of acid deposition shown in figure 12. [1]

        ii. Choose **one** of the consequences of acid rain. Suggest how it can be offset. [2]

18. a. Sketch a graph showing the first 10 ionization energies of copper. [2]

    b. State the abbreviated electron configuration of copper. [1]

    c. State the full electron configuration of a copper(II) ion. [1]

19. Explain why aqueous copper(II) nitrate forms a coloured solution, whereas aqueous zinc nitrate is colourless. [2]

AHL

20. When excess hydrochloric acid is added to aqueous cobalt(II) ions, a ligand exchange reaction takes place, causing the colour of the solution to change from pink to blue:

$$[Co(H_2O)_6]^{2+} + 4Cl^- \rightleftharpoons [CoCl_4]^{2-} + 6H_2O$$

    pink   blue

    a. The pink $[Co(H_2O)_6]^{2+}$ complex ion absorbs light with a wavelength of approximately 540 nm. Calculate the energy difference between the split d orbitals of cobalt in this complex ion. [2]

    b. Explain how the colour change in this reaction shows that chloride ions are weaker ligands than water molecules. [2]

21. Explain, by referring to successive ionization energies, why titanium can exist in variable oxidation states, but calcium only occurs in the +2 oxidation state. [2]

Structure 3.2 Functional groups: Classification of organic compounds

How does the classification of organic molecules help us to predict their properties?

Classification is a common human approach utilized in many fields of science. Just as biology uses scientific taxonomy to classify organisms on the basis of shared characteristics, chemists utilize a unique system of nomenclature to group and name compounds that share important features of structure and reactivity.

Understandings

Structure 3.2.1—Organic compounds can be represented by different types of formulas. These include empirical, molecular, structural (full and condensed), stereochemical and skeletal.

Structure 3.2.2—Functional groups give characteristic physical and chemical properties to a compound. Organic compounds are divided into classes according to the functional groups present in their molecules.

Structure 3.2.3—A homologous series is a family of compounds in which successive members differ by a common structural unit, typically CH_2. Each homologous series can be described by a general formula.

Structure 3.2.4—Successive members of a homologous series show a trend in physical properties.

Structure 3.2.5—"IUPAC nomenclature" refers to a set of rules used by the International Union of Pure and Applied Chemistry to apply systematic names to organic and inorganic compounds.

Structure 3.2.6—Structural isomers are molecules that have the same molecular formula but different connectivities.

Structure 3.2.7—Stereoisomers have the same constitution (atom identities, connectivities and bond multiplicities) but different spatial arrangements of atoms.

Structure 3.2.8—Mass spectrometry (MS) of organic compounds can cause fragmentation of molecules.

Structure 3.2.9—Infrared (IR) spectra can be used to identify the type of bond present in a molecule.

Structure 3.2.10—Proton nuclear magnetic resonance spectroscopy (^{1}H NMR) gives information on the different chemical environments of hydrogen atoms in a molecule.

Structure 3.2.11—Individual signals can be split into clusters of peaks.

Structure 3.2.12—Data from different techniques are often combined in structural analysis.

Formulas of organic compounds (*Structure 3.2.1*)

Introduction to organic chemistry

Organic chemistry is the field of chemistry where carbon-based compounds are studied. Carbon atoms have four valence electrons, so they can form four bonds to other atoms. Carbon atoms can undergo **catenation**, which is the process by which many identical atoms are joined together by covalent bonds. This produces straight-chain, branched, or cyclic structures. This special feature of carbon is the reason why organic chemistry is such a wide and varied field of study. Organic compounds include fuels, paints, dyes, alcohol, plastics, industrial solvents, drugs and medicines, foods, pesticides and fertilizers. Most biological molecules, such as DNA, contain carbon atoms and are also considered to be organic.

Types of formulas

You can represent the structure of an organic compound in several ways by using different types of formulas.

An **empirical formula** represents the simplest ratio of atoms present in a molecule. The **molecular formula** describes the actual number of atoms present in the molecule. Both these types of formula offer you little or no information about the possible structures of larger, more complex molecules. For example, a molecule of the carbohydrate glucose has the molecular formula $C_6H_{12}O_6$ and an empirical formula of CH_2O. From this information, the structure of glucose cannot be deduced.

> Empirical and molecular formulas were introduced in *Structure 1.4*.

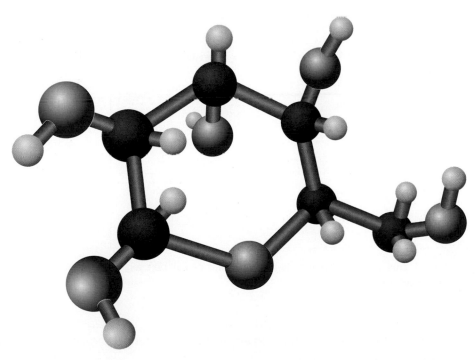

▲ **Figure 1** Glucose, $C_6H_{12}O_6$, is a simple six-carbon sugar with hydroxyl functional groups. The structure of the ring is not reflected by the molecular formula

Structural formulas, unlike molecular formulas, describe the structure of a compound. There are three types of structural formula: full, condensed and skeletal.

- **Full structural formulas** or **displayed formulas** are two-dimensional representations showing all the atoms and bonds, and their positions relative to one another in a compound.

- In a **condensed structural formula**, all the atoms and their relative positions are shown, but some or all the bonds are omitted.

- A **skeletal formula** is the most basic representation of the structural formula, where the carbon and hydrogen atoms are not shown but the end of each line and each vertex represents a carbon atom. The atoms present in functional groups are also included as shown in table 1.

Name	Full structural formula	Condensed structural formula	Skeletal formula
propane	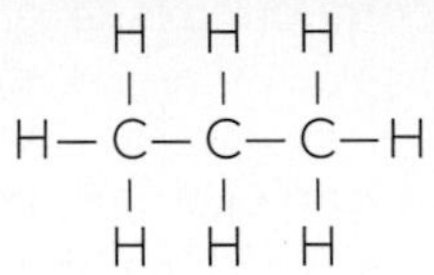	$CH_3CH_2CH_3$	
propan-2-ol	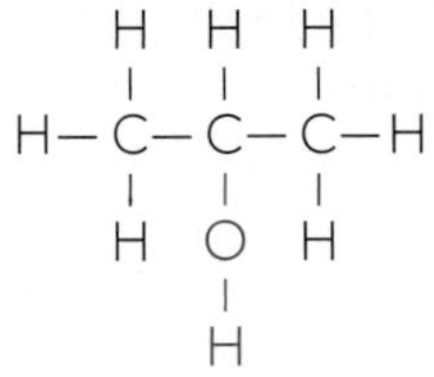	$CH_3CH(OH)CH_3$	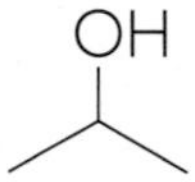
propanal	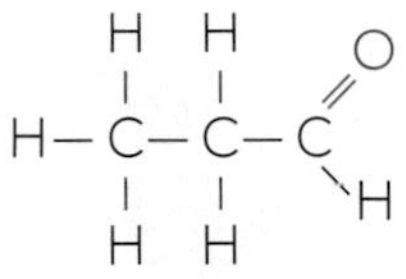	CH_3CH_2CHO	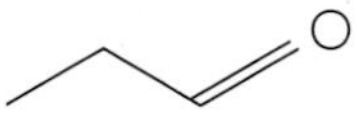
propanone	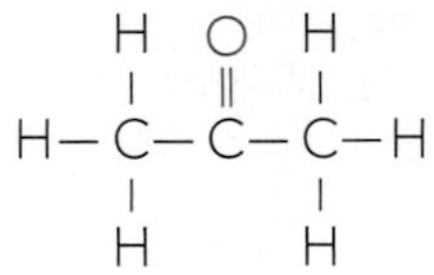	$CH_3C(O)CH_3$	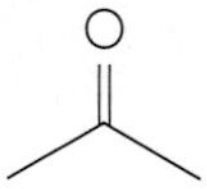
propene	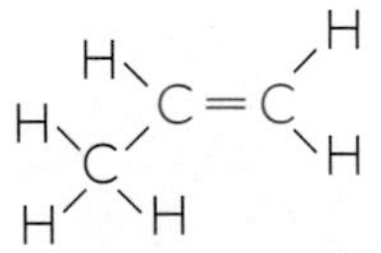	$CH_3CH=CH_2$ or CH_3CHCH_2	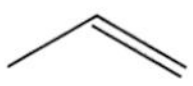

▲ Table 1 Full, condensed and skeletal formulas can all be used to represent structures of organic compounds

In structural formulas, a covalent bond between two atoms is represented by a single line that describes two bonding electrons. For a double bond, two lines are used, and for a triple bond, three lines are used.

Constructing 3D models of organic compounds is an excellent interactive technique, enhancing your ability to visualize of molecules and mutual orientation of individual atoms. Models can enhance understanding of a variety of concepts, from the naming of organic molecules and visualizing stereoisomers (including optical isomers) to complex reaction mechanisms.

Practice questions

1. Draw the full structural formula and skeletal formula for each of the following organic compounds.

    a. hexane, $CH_3CH_2CH_2CH_2CH_2CH_3$

    b. butan-1-ol, $CH_3CH_2CH_2CH_2OH$

    c. chloroethene, $H_2C=CHCl$

2. Which of these molecular formulas are also empirical formulas?

    I. C_2H_6O

    II. $C_2H_4O_2$

    III. C_5H_{12}

    A. I and II only

    B. I and III only

    C. II and III only

    D. I, II and III

3. Which of the following compounds has the same empirical formula and molecular formula?

	Empirical formula	Molar mass / $g\,mol^{-1}$
A	CO_2H	90
B	CH_3O	62
C	C_2H_4O	88
D	C_4H_8O	72

 Linking question

What is unique about carbon that enables it to form more compounds than the sum of all the other elements' compounds? (Structure 2.2)

 What are the advantages and disadvantages of different depictions of an organic compound (structural formula, stereochemical formula, skeletal formula, 3D models, etc.)? (Structure 2.2)

Models

Models are useful when direct observation is difficult. Two-dimensional representations of molecules (such as structural formulas) do not represent some of the more complex features that become clear in three-dimensional physical and computer models. You can explore some of the different types of 3D models in the *Skills* chapter.

Functional groups (*Structure 3.2.2*)

A **functional group** is an atom, or a group of atoms, that gives organic compounds their physical and chemical characteristics.

There are tens of millions of organic compounds, and this number is constantly rising as new compounds are synthesized in pharmaceutical companies, industrial laboratories and research universities. **Natural compounds** found in plants and animals are synthesized by living organisms.

▲ Figure 2 Digitalin is a cardioactive steroid found naturally in the foxglove plant. It is used to treat congestive heart failure and irregular heartbeats or atrial fibrillation. It features hydroxyl, alkoxy and ester functional groups

All organic compounds are divided into **classes of organic compounds** dependent on the specific functional group found in their molecules. Compounds of the same class have similar chemical properties. Some functional groups and their corresponding classes are given in table 2. R groups are used in table 2 to symbolize any carbon-containing groups. They are useful for writing general formulas for classes of organic compounds. Carbon-containing groups with no other functionalities are called **alkyl groups**.

| Functional group | | Suffix | Class | Example |
Name	Formula			
halogeno (fluoro, chloro, bromo or iodo)	R–X	—	halogenoalkanes	chloromethane $H_3C{-}Cl$
hydroxyl	R–OH	-ol	alcohols	ethanol 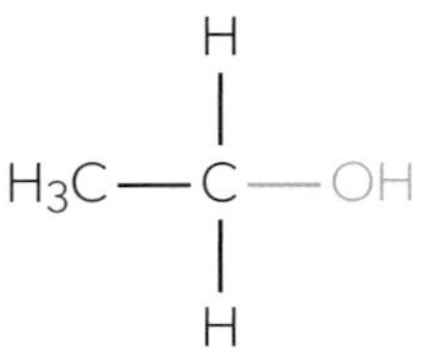 $H_3C{-}\overset{\overset{H}{\mid}}{\underset{\underset{H}{\mid}}{C}}{-}OH$
carbonyl	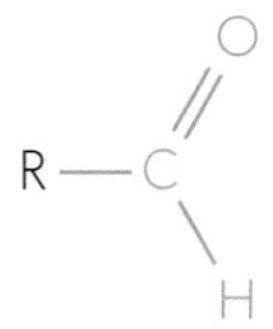 $R{-}\overset{\overset{O}{\parallel}}{C}{-}H$	-al	aldehydes	ethanal 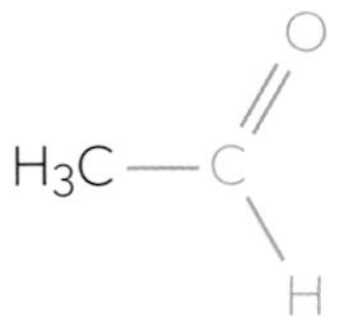 $H_3C{-}\overset{\overset{O}{\parallel}}{C}{-}H$
carbonyl	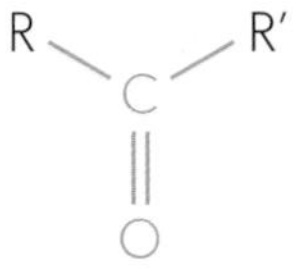 $\underset{\underset{O}{\parallel}}{\overset{R\quad R'}{C}}$	-one	ketones	propanone 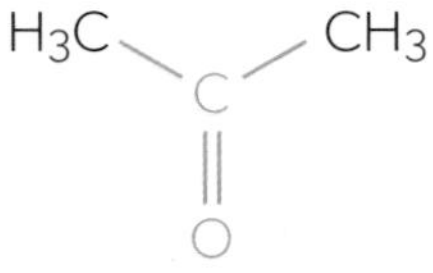 $\underset{\underset{O}{\parallel}}{\overset{H_3C\quad CH_3}{C}}$
carboxyl	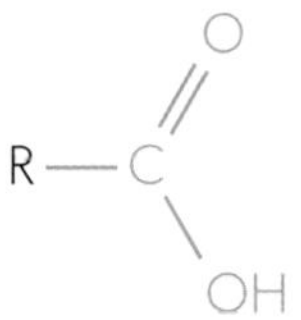 $R{-}\overset{\overset{O}{\parallel}}{C}{-}OH$	-oic acid	carboxylic acids	ethanoic acid 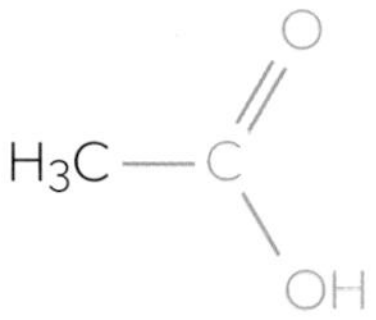 $H_3C{-}\overset{\overset{O}{\parallel}}{C}{-}OH$
alkoxy	R–O–R'	—	ethers	methoxymethane $H_3C{-}O{-}CH_3$
amino	R–NH$_2$	-amine	primary amines	ethanamine 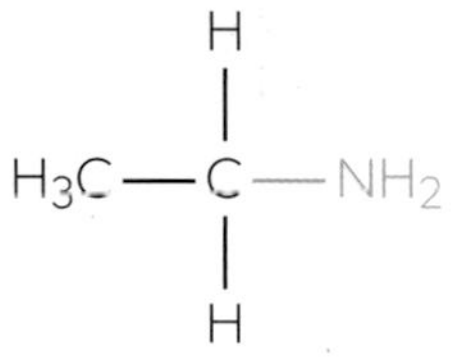 $H_3C{-}\overset{\overset{H}{\mid}}{\underset{\underset{H}{\mid}}{C}}{-}NH_2$
amido	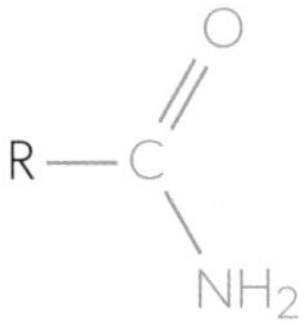 $R{-}\overset{\overset{O}{\parallel}}{C}{-}NH_2$	-amide	amides	ethanamide 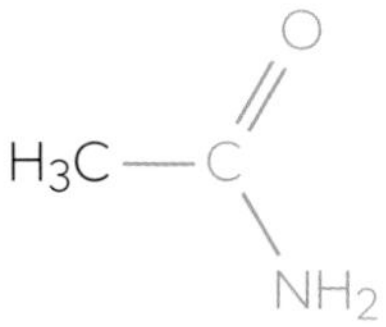 $H_3C{-}\overset{\overset{O}{\parallel}}{C}{-}NH_2$
ester	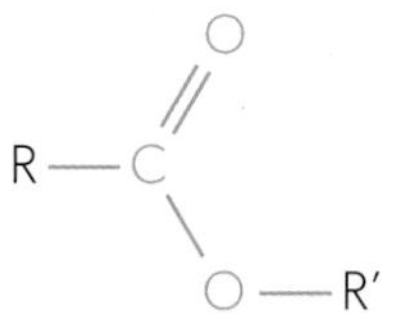 $R{-}\overset{\overset{O}{\parallel}}{C}{-}O{-}R'$	-oate	esters	methyl ethanoate 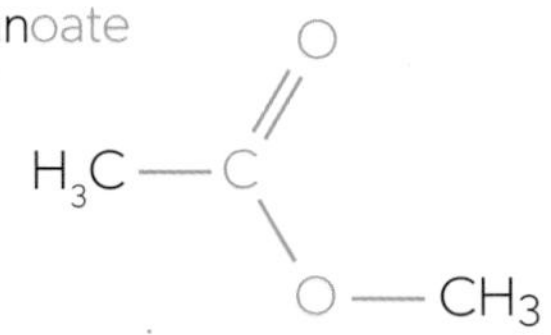 $H_3C{-}\overset{\overset{O}{\parallel}}{C}{-}O{-}CH_3$
phenyl	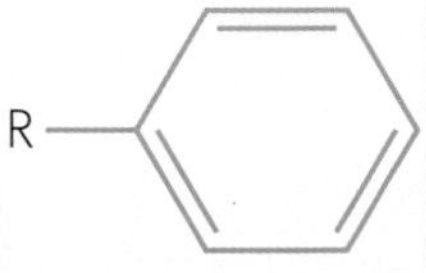 R–⬡	—	aromatics	methylbenzene H₃C–⬡ 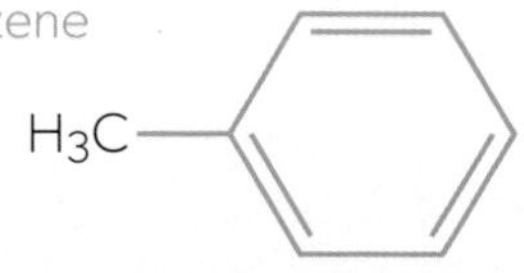

▲ Table 2 Summary of classes of organic compounds and their functional groups

ATL Self-management skills

How can you develop familiarity with the various functional groups, their names, and structures? You may wish to create a flashcard deck to help you to build and maintain your fluency. As you will see in *Reactivity* 3, recognizing the various functional groups will help you to predict the chemical behaviour of various substances.

Saturated and unsaturated hydrocarbons

Organic compounds can also be classified according to whether they are saturated or unsaturated. In a saturated compound, all the carbon–carbon bonds are single. The simplest example of a saturated hydrocarbon is methane, CH_4, a member of the **alkane** family. The primary carbon chain in an unsaturated hydrocarbon must contain one or more double or triple carbon–carbon bonds. Hydrocarbons with carbon–carbon double bonds are called **alkenes**. The simplest alkene is ethene, $H_2C{=}CH_2$. Alkanes and alkenes are both **aliphatic**. This means their molecules contain no aromatic rings.

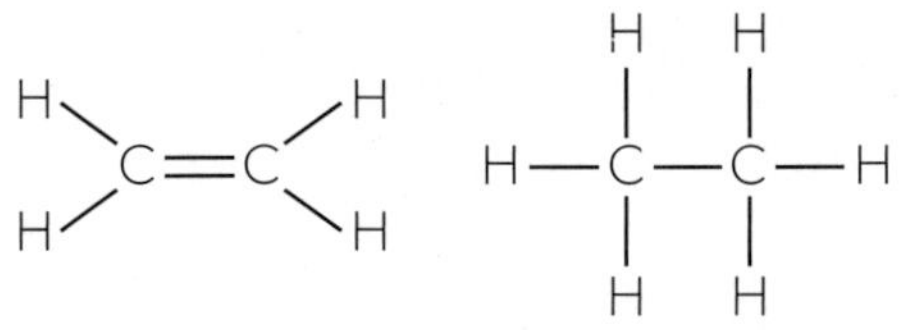

▲ Figure 3 Ethene (left) with a carbon–carbon double bond, is unsaturated, whereas ethane (right) is saturated

TOK

Terms such as *natural*, *organic* and *chemical* have precise meanings in chemistry, which can differ from their meanings in everyday language. Consider *organic*. This word takes on a variety of meanings depending on its context: organic food, organic chemistry and organic household waste all use the word *organic* to signify a different thing.

What is the role of context in the choice and interpretation of language?

Linking questions

AHL What is the nature of the reaction that occurs when two amino acids form a dipeptide? (Structure 2.4)

How can functional group reactivity be used to determine a reaction pathway between compounds, e.g., converting ethene into ethanoic acid? (Reactivity 3.2, 3.4)

Homologous series (*Structure 3.2.3*)

A **homologous series** is a family of compounds that can be grouped together based on similarities in their structure and reactivity. A homologous series has the same general formula, which varies from one member to another by one CH_2 (methylene) group. If you know the general formula for each homologous series, you can identify which group a particular compound belongs to without drawing its structure.

The **alkane** series has the general formula C_nH_{2n+2}. Alkanes are **hydrocarbons** that contain carbon and hydrogen atoms only. Table 3 shows the alkane homologous series. You should be able to identify how the structural formula of each successive member differs from the previous by a single CH_2 group.

Name	Molecular formula	Condensed structural formula	Full structural formula
methane	CH_4	CH_4	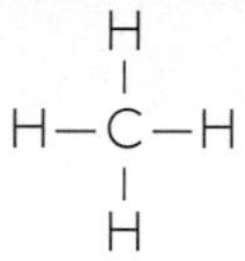 $\begin{array}{c} H \\ \mid \\ H-C-H \\ \mid \\ H \end{array}$
ethane	C_2H_6	CH_3CH_3	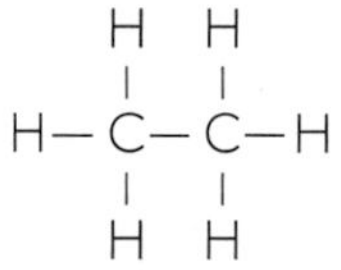 $\begin{array}{cc} H & H \\ \mid & \mid \\ H-C-C-H \\ \mid & \mid \\ H & H \end{array}$
propane	C_3H_8	$CH_3CH_2CH_3$	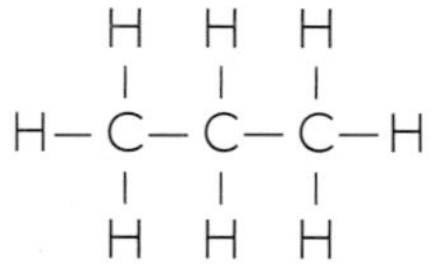 $\begin{array}{ccc} H & H & H \\ \mid & \mid & \mid \\ H-C-C-C-H \\ \mid & \mid & \mid \\ H & H & H \end{array}$
butane	C_4H_{10}	$CH_3CH_2CH_2CH_3$	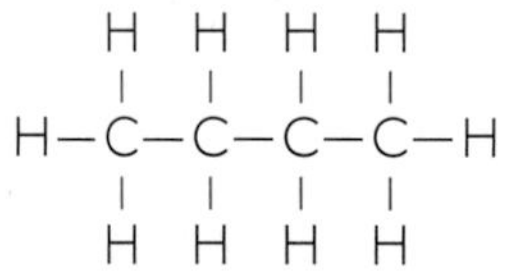 $\begin{array}{cccc} H & H & H & H \\ \mid & \mid & \mid & \mid \\ H-C-C-C-C-H \\ \mid & \mid & \mid & \mid \\ H & H & H & H \end{array}$
pentane	C_5H_{12}	$CH_3CH_2CH_2CH_2CH_3$	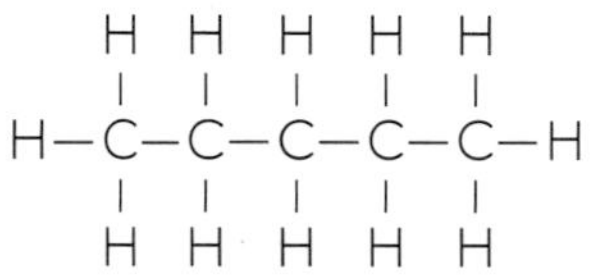 $\begin{array}{ccccc} H & H & H & H & H \\ \mid & \mid & \mid & \mid & \mid \\ H-C-C-C-C-C-H \\ \mid & \mid & \mid & \mid & \mid \\ H & H & H & H & H \end{array}$
hexane	C_6H_{14}	$CH_3CH_2CH_2CH_2CH_2CH_3$	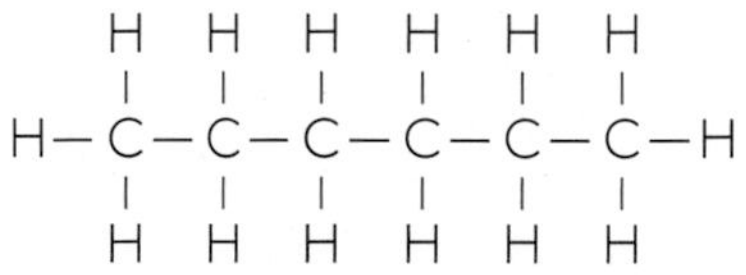 $\begin{array}{cccccc} H & H & H & H & H & H \\ \mid & \mid & \mid & \mid & \mid & \mid \\ H-C-C-C-C-C-C-H \\ \mid & \mid & \mid & \mid & \mid & \mid \\ H & H & H & H & H & H \end{array}$

▲ Table 3 The homologous series of alkanes

You can synthesize halogenoalkanes from alkanes by a process known as radical substitution. This form of substitution introduces a highly polar carbon–halogen bond into what was a relatively inert alkane molecule. Radical substitution is detailed in *Reactivity 3.3*.

The **alkenes** are a homologous series of unsaturated hydrocarbons that contain a carbon–carbon double bond. The alkene series has the general formula of C_nH_{2n}. The **alkyne** family have carbon–carbon triple bonds, with the general formula C_nH_{2n-2}. **Halogenoalkanes** are alkanes where one of the hydrogen atoms has been substituted with a halogen atom. They have the general formula $C_nH_{2n+1}X$, where X is the halogen atom.

Homologous series	alkenes	alkynes	halogenoalkanes
General formula	C_nH_{2n}	C_nH_{2n-2}	$C_nH_{2n+1}X$
C_2	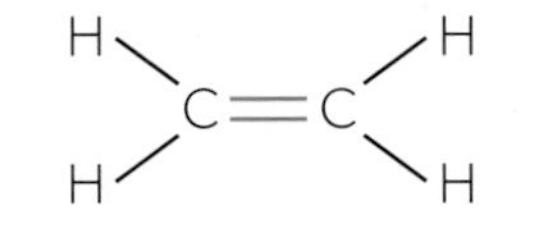 ethene	 ethyne	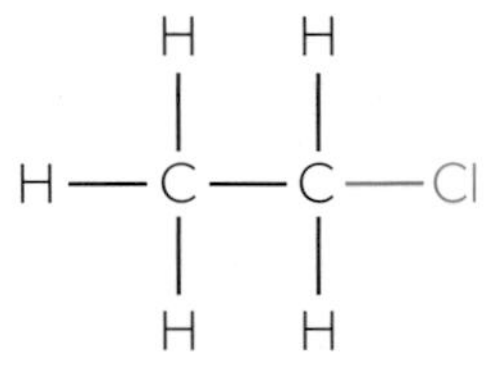 chloroethane
C_3	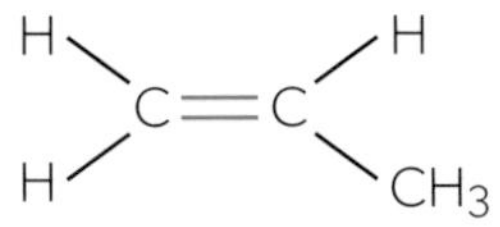 propene	 propyne	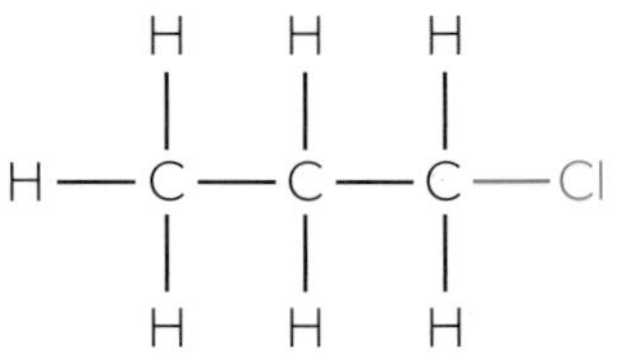 1-chloropropane
C_4	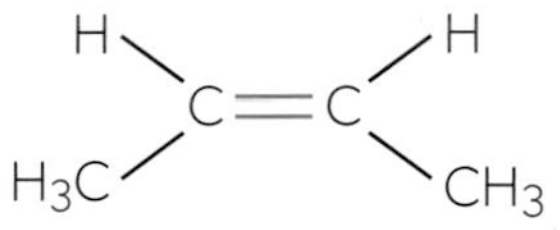 but-2-ene	 but-2-yne	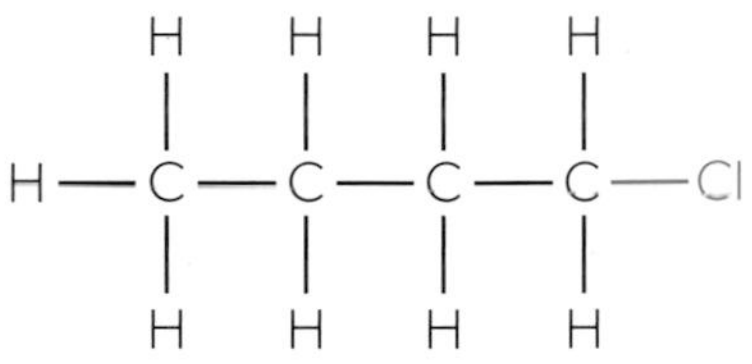 1-chlorobutane

▲ Table 4 The homologous series of straight-chain alkenes, alkynes and halogenoalkanes

Members of the homologous series of alcohols all contain the hydroxyl functional group (–OH). They have a general formula of $C_nH_{2n+1}OH$. Alcohols can undergo combustion reactions (*Reactivity 1.3*) and oxidation reactions (*Reactivity 3.2*). Another homologous series, the aldehydes, have a general formula of $C_nH_{2n}O$. Aldehydes contain the carbonyl functional group (C=O) in the terminal position of the carbon chain. Ketones have the same general formula and functional group as aldehydes, but the carbonyl group is found anywhere on the carbon chain except the ends. These homologous series are summarized in table 5.

Homologous series	alcohols	aldehydes	ketones
General formula	$C_nH_{2n+1}OH$	$C_nH_{2n}O$	$C_nH_{2n}O$
C_2	ethanol	ethanal	no C_2 ketones
C_3	propan-1-ol	propanal	propanone
C_4	butan-1-ol	butanal	butanone
C_5	pentan-1-ol	pentanal	pentan-3-one

▲ Table 5 The homologous series of alcohols, aldehydes and ketones

Carboxylic acids have the general formula $C_nH_{2n}O_2$, and have a carboxyl group. Methanoic acid, HCOOH, where $n = 1$, is the simplest carboxylic acid of this homologous family.

Ethers have the general formula ROR′, where R and R′ represent alkyl groups. The –OR group is called an alkoxy group. The simplest alkoxy group has one carbon atom: –OCH$_3$. This is known as a methoxy group. Here are some examples of ethers:

$H_3C - O - CH_3$ $H_3C - O - CH_2CH_3$

methoxymethane   methoxyethane

Members of the amine homologous series all have the general formula $C_nH_{2n+3}N$. There are three types of amines: primary amines, secondary amines and tertiary amines. In primary amines, the nitrogen atom is bonded to only one carbon atom. In secondary amines, the nitrogen is bonded to two carbon atoms, and in tertiary amines, it is bonded to three carbon atoms. This is summarized in table 6.

Esters have the general formula RCO_2R'. They are derived from carboxylic acids and have a variety of applications ranging from flavouring agents and medications to solvents and explosives.

Amides are also derived from carboxylic acids and have a general formula of $C_nH_{2n+1}NO$. They contain the amido functional group, $-C(O)NH_2$.

Type of amine	Number of carbon atoms bonded to nitrogen	Example
primary	1	propanamine
secondary	2	dimethylamine
tertiary	3	trimethylamine

 Table 6 Primary, secondary and tertiary amines

Linking question

How useful are 3D models (real or virtual) to visualize the invisible? (Tool 2)

Practice questions

4. Which functional group is circled in this molecule?

    A. hydroxyl

    B. carboxyl

    C. carbonyl

    D. ether

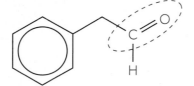

5. Which functional groups are present in this molecule?

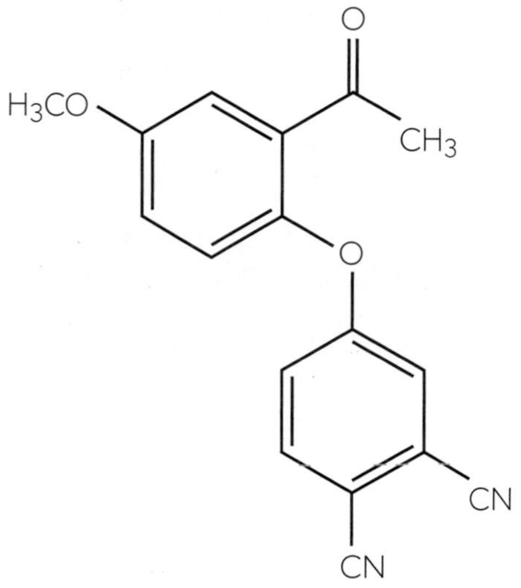

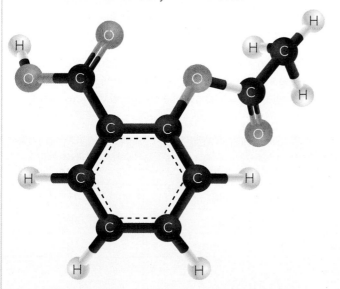

    A. carbonyl, ether, phenyl

    B. carbonyl, ester, phenyl

    C. carboxyl, ether, amino

    D. carboxyl, ester, amino

6. Which of the following functional groups are present in the aspirin molecule?

    I. ether   II. carboxyl   III. ester

    A. I and II only   B. I and III only

    C. II and III only   D. I, II and III

7. Which of the following functional groups is present in paracetamol?

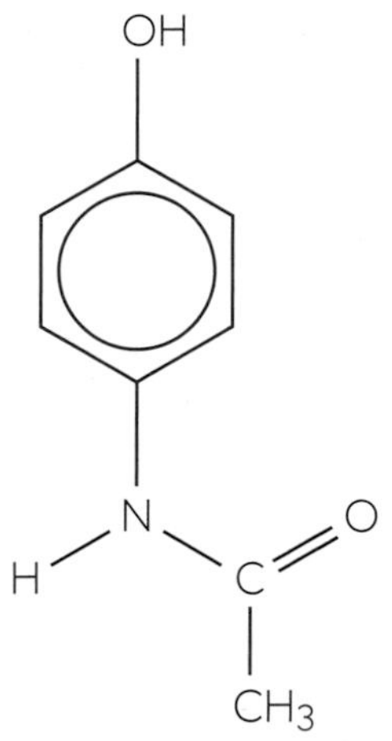

    A. carboxyl   B. amino

    C. alkene   D. hydroxyl

8. Galantamine is a drug used to treat symptoms of Alzheimer's disease. Its structure is shown below.

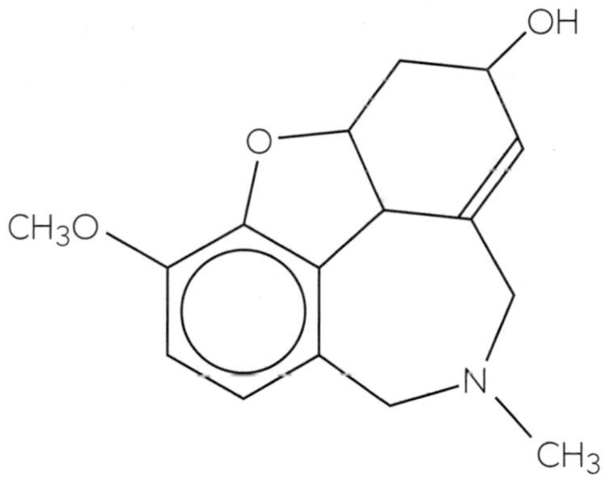

    Which of the following functional groups are present in this molecule?

    A. amido, hydroxyl, ester

    B. amino, hydroxide, ether

    C. amino, hydroxyl, ether

    D. amido, hydroxide, ester

Research skills

Our atmosphere is mostly made up of nitrogen, N_2, and oxygen, O_2. Organic compounds, which may have been the precursors of biomolecules, are thought to have formed billions of years ago when the composition of the atmosphere was very different. An attempt was made to recreate these conditions in the mid-20th century, known as the Miller–Urey experiment.

Research two sources on the Miller–Urey experiment (or another theory of the origin of biomolecules), and summarize your findings in your own words, drawing as many connections as possible to your knowledge of chemistry. Fully document the information sources, and evaluate their reliability.

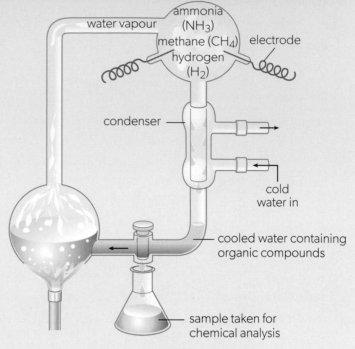

▲ Figure 4 The apparatus for the Miller–Urey experiment

Physical trends in homologous series (*Structure 3.2.4*)

Members of the same homologous series have similar chemical properties, as they have the same functional group. The functional group is responsible for the types of characteristic reactions the compound undergoes.

For a given homologous series, physical properties such as melting and boiling points change gradually with each successive member. This is because the length of the carbon chain increases by CH_2 each time, altering the intermolecular forces and therefore the physical properties.

The boiling points and melting points for the straight-chain alkanes are shown in table 7. These values increase as the number of carbon atoms increases. As a consequence, boiling point and melting point increase with increasing molar mass.

Name	Formula	Boiling point/°C	Melting point/°C
methane	CH_4	−161	−183
ethane	C_2H_6	−89	−172
propane	C_3H_8	−42	−188
butane	C_4H_{10}	−0.5	−135
pentane	C_5H_{12}	36	−130
hexane	C_6H_{14}	69	−95

▲ Table 7 The melting and boiling points of the straight-chain alkanes

The close correlation between carbon chain length and boiling point enables you to predict the properties of neighbouring members of the homologous series. The trend in melting points is not as smooth as that of boiling points and would have less certainty in terms of the prediction of properties.

You can use simple distillation apparatus and a temperature probe to measure the boiling points for successive members of a homologous series.

You learned about distillation techniques in *Structure 1.1*.

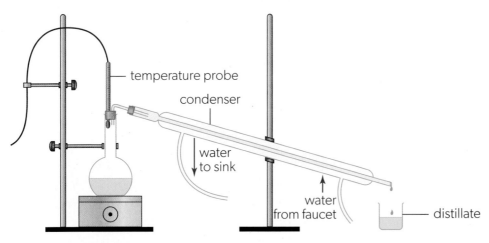

▲ Figure 5 Distillation apparatus incorporating a temperature probe

The trend in increasing boiling point results from stronger London (dispersion) forces as the carbon chain increases in length. London (dispersion) forces are attractive forces between non-polar molecules (for example, hydrogen H$_2$). They also exist between polar molecules (for example, HCl). Every molecule will experience London (dispersion) forces, whether it is non-polar or polar.

London (dispersion) forces and other intermolecular forces were covered in *Structure 2.2*. Which of the homologous series discussed so far experience hydrogen bonding?

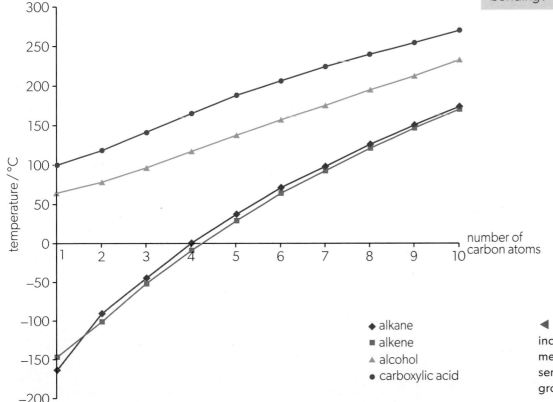

◀ Figure 6 The boiling point increases with each successive member of a homologous series as the carbon chain grows in length

Constructing a simple graphical model

In this task you will construct and test a simple graphical model of the boiling points of a homologous series of your choice.

Relevant skills

- Inquiry 1: State and explain predictions using scientific understanding
- Tool 2: Identify and extract data from databases
- Tool 2: Use spreadsheets to manipulate data
- Tool 2: Represent data in graphical form
- Tool 3: Sketch graphs to qualitatively describe trends
- Tool 3: Apply the coefficient of determination (R^2)
- Tool 3: Construct graphs and draw lines or curves of best fit
- Tool 3: Extrapolate graphs
- Tool 3: Calculate and interpret percentage error
- Inquiry 2: Describe and explain patterns, trends and relationships

Instructions

Part 1: Prediction

1. List the first six members of a homologous series of your choice.
2. Predict the relationship between boiling point and carbon chain length. Justify your prediction, referring to your knowledge of intermolecular forces and functional groups.
3. Sketch a graph that qualitatively describes the relationship you have predicted.

Part 2: Data collection

4. Identify an online database that contains boiling point data for a variety of substances (for instance, NIST Chemistry Web Book).
5. Collect data on the boiling point for the six organic compounds you have selected and enter the data into a spreadsheet.
6. Organize the data: enter the name, formula, and number of carbon atoms, and convert all temperatures into a consistent unit.

Part 3: Graph

7. Create a scatter graph of the data.
8. Using software, draw a line or curve of best fit that describes the data well.
9. Determine the R^2 value and the equation of the line or curve.

Part 4: Test the model

You have now created a simple graphical model of the relationship between boiling point and carbon chain length. You can test your model by using it to make predictions and then checking the accuracy of these predictions.

10. Use the equation of the line or curve of best fit to predict the boiling point of the 7th, 10th, and if possible, the 20th members of the homologous series.
11. Look up their actual boiling point values.
12. Calculate the percentage errors of each of the predictions.

Part 5: Interpret the results

13. Compare and contrast the percentage errors for the three predictions. Consider the following questions:
    - How do their accuracies compare? Why might this be?
    - Does the accuracy of the predictions change for longer carbon chain lengths?
    - This is an example of extrapolation because you predicted a value that was outside the range of the data. How might the accuracy change if you had interpolated the data instead?
14. Describe and explain the trend shown in the graph you plotted.
15. If uncertainties were reported along with the boiling point data, what impact did these have on your analysis?
16. Briefly evaluate your model, giving examples of its specific strengths and weaknesses.
17. How might the accuracy of this simple model be improved?

Extension

Choose a different homologous series and repeat the task. Compare and contrast the two sets of results.

The different boiling points of members of homologous series mean that different members will condense at different temperatures. In the petrochemical industry, this principle is used as a separation technique. For example, crude oil is a mixture of hydrocarbons that vary in the length of their carbon chain. **Fractional distillation** is a physical separation process where crude oil is vaporized and then passed through a column. As the vapour rises through the column, it cools. Therefore, compounds will condense at different levels in the column depending on the length of their carbon chain. Compounds with a similar boiling point, and therefore similar chain length, will condense at the same level. Long-chain compounds will condense further down the column, as they have a higher boiling point. Shorter-chain compounds have a lower boiling point, so will condense further up the column, where it is cooler.

The density and viscosity of members of a homologous series also increase with carbon chain length.

Linking question

What is the influence of the carbon chain length, branching and the nature of the functional groups on intermolecular forces? (*Structure 2.2*)

Naming organic compounds (*Structure 3.2.5*)

With the existence of millions of different chemical substances, names help us to differentiate them. The name of an organic compound needs to provide enough information to identify the class of the compound, the length of its longest carbon chain, and any **substituents** and functional groups present. A substituent is any part of an organic compound that is not part of the longest carbon chain. For example, any alkyl group in an alkane compound that is not part of the main chain is considered a substituent. An alkane with alkyl substituents is known as a **branched-chain alkane**.

The International Union of Pure and Applied Chemistry (IUPAC) has developed a universal chemical naming system for both organic and inorganic compounds: **IUPAC nomenclature**. IUPAC nomenclature rules helps chemists to avoid confusion, eliminate language barriers, and freely communicate knowledge. An overview of the IUPAC naming system for organic compounds is shown in figure 7:

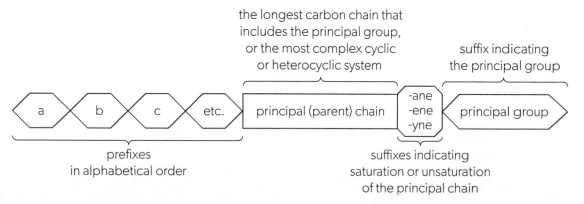

▲ Figure 7 Outline of the nomenclature of organic compounds

Naming alkanes

Alkanes are saturated organic molecules that only contain carbon–carbon single bonds. You can use the following steps to deduce the IUPAC name of an alkane:

1   Examine the structure of the compound and determine the longest continuous carbon chain. This provides the **root name** for the alkane depending on the number of carbon atoms (table 8). If an alkane is a straight-chain molecule, its name is the same as its root name. All alkanes have the same suffix: -ane.

Number of carbon atoms	Name
1	methane
2	ethane
3	propane
4	butane
5	pentane
6	hexane

▲ Table 8 IUPAC root names for the alkane series

2.   If alkyl substituents are present in the compound, the name of a substituent is determined by the number of carbon atoms, with the suffix changing from "-ane" to "-yl". Some examples are shown in table 9.

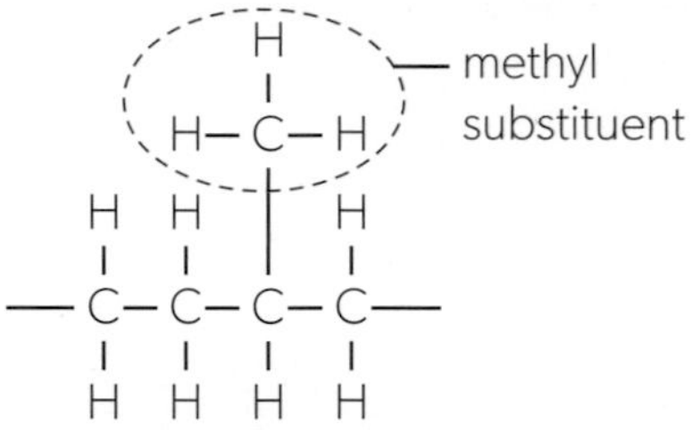

Number of carbon atoms in alkyl group	Substituent name	Condensed formula
1	methyl	$-CH_3$
2	ethyl	$-CH_2CH_3$
3	propyl	$-CH_2CH_2CH_3$
4	butyl	$-CH_2CH_2CH_2CH_3$

▲ Table 9 IUPAC names for the alkyl substituents

3.   You can specify the position of a substituent by assigning numbers, known as **locants**, to the main carbon chain. The carbon chain is numbered such that the substituent is given the lowest value locant.

In the example below, numbering from left to right results in the methyl substituent being on carbon 2. Numbering from right to left would incorrectly have the substituent on carbon 5.

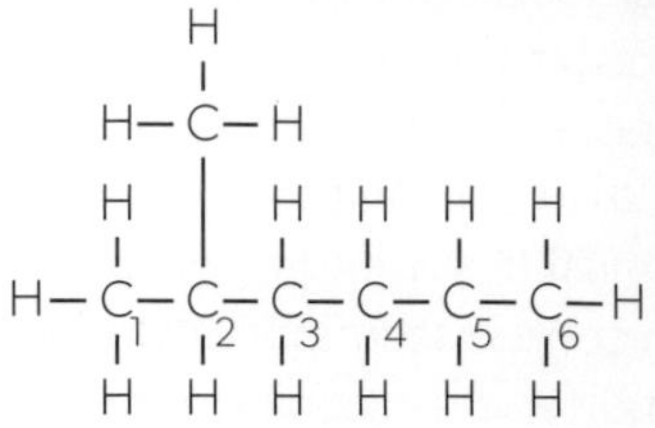

4.   The name of the alkyl substituent goes before the root name as a prefix, and a hyphen is used to separate the locant number and the prefix. Therefore, the above compound is called 2-methylhexane.

5.   If there are multiple different substituents, they are arranged in alphabetical order prior to the root name. For example, the below compound is called 3-ethyl-2-methylhexane.

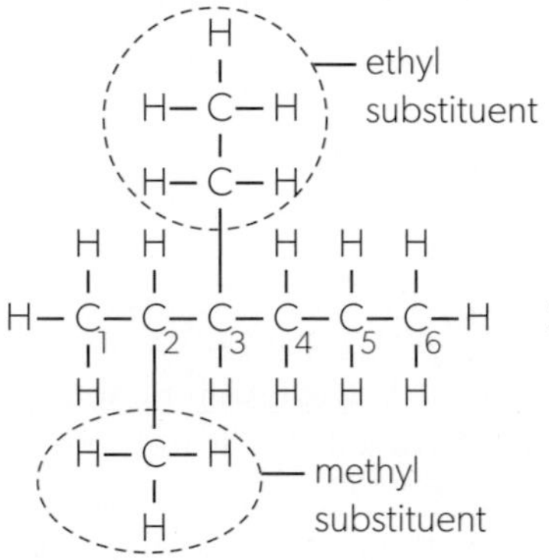

6.   If there are multiple identical substituents, the number of substituents is indicated by a multiplier before the substituent name, as summarized in table 10.

Number of identical substituents	Multiplier name
2	di
3	tri
4	tetra
5	penta

▲ Table 10 Numerical multipliers in the IUPAC nomenclature

The locants of each substituent are separated by a comma. For example, the below compound is called 2,3-dimethylhexane.

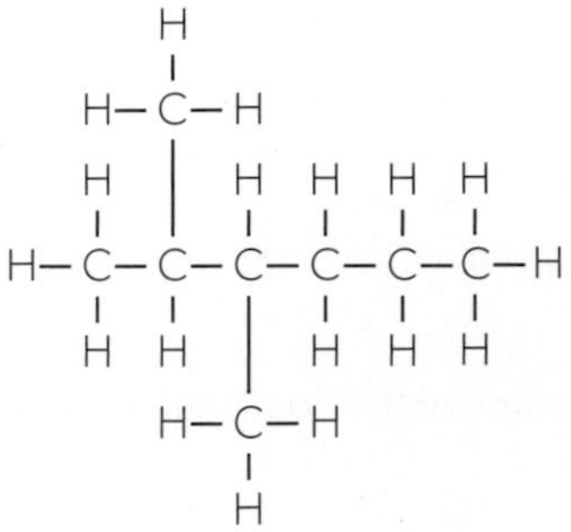

Structural isomers are compounds that have the same molecular formula but different structural formulas. Different isomers have unique physical and chemical properties. In the following worked example, we will look at the structural isomers of the hydrocarbon hexane, and practise using the rules above to name these isomers.

Worked example 1

Three different structural isomers of hexane, C_6H_{12}, are shown below. Using the IUPAC nomenclature, name these isomers.

a.

```
    H  H  H  H  H
    |  |  |  |  |
H − C− C− C− C− C− H
    |  |  |  |  |
    H  H  H  |  H
             |
          H− C− H
             |
             H
```

b.

```
             H
             |
          H− C− H
             |
       H  |  H  H
       |  |  |  |
    H− C− C− C− C− H
       |  |  |  |
       H  H  |  H
             |
          H− C− H
             |
             H
```

c.

```
             H
             |
          H− C− H
             |
       H  |  H  H
       |  |  |  |
    H− C− C− C− C− H
       |  |  |  |
       H  |  H  H
          |
       H− C− H
          |
          H
```

Solution

a. First, count the number of carbon atoms in the longest straight carbon chain. There are five carbon atoms, so the parent chain is **pentane**.

Then, number the carbon chain so that the substituent has the lowest possible locant. The **methyl** ($-CH_3$) substituent is bonded to carbon 2.

The IUPAC name for this compound is 2-methylpentane.

b. There are four carbon atoms in the longest straight carbon chain, and so the parent chain is **butane**.

In this case, the direction of numbering of the carbon chain is irrelevant as the compound is symmetrical, so both approaches would give the same locants for the substituents.

There are two **methyl** ($-CH_3$) substituents, one bonded to carbon 2 and one to carbon 3.

Since the two substituents are identical, a numerical multiplier is required to indicate the number of these substituents. In this case, *di* indicates two substituents.

The IUPAC name for this compound is 2,3-dimethylbutane. Remember that commas are used between numbers, and hyphens are used between a number and a letter.

c. Like b, the parent chain is **butane**. Two **methyl** ($-CH_3$) substituents are bonded to carbon 2. The IUPAC name for this compound is therefore 2,2-dimethylbutane.

Practice questions

9. Applying IUPAC nomenclature rules, state the name of each of the following compounds:

a. $CH_3 - CH_2 - CH_2 - CH - CH_2 - CH_3$
    with CH_3 below the fourth carbon

b. $CH_3 - CH_2 - C - CH_2 - CH - CH_3$
    with CH_3 above the third carbon, CH_3 below the third carbon, and CH_3 below the fifth carbon

c. $CH_3 - CH_2 - CH - CH - CH_3$
    with $CH_2 - CH_3$ above the fourth carbon, and $CH_3 - CH - CH_3$ below the third carbon

10. Draw the structural formulas of the following molecules:

a. 2,2,3-trimethylpentane

b. 5-ethyl-4,4-dimethyloctane

Naming alkenes

Alkenes are unsaturated organic molecules that contain a carbon–carbon double bond. The IUPAC nomenclature rules for the naming of alkene molecules are the same as those for the alkane series. All alkene names end with the suffix -ene, instead of -ane used for alkanes.

The following compound has both a carbon–carbon double bond, and an alkyl side chain.

$$H_3C - \underset{1}{CH} = \underset{2}{CH} - \underset{3}{CH} - \underset{4}{\underset{|}{CH}} - \underset{5}{CH_2} - \underset{6}{CH_3}$$
$$CH_3$$

When you number the longest carbon chain, the position of the carbon–carbon double bond must have the lowest locant, taking priority over any substituents present in the molecule. Numbering from right to left would incorrectly place the carbon–carbon double bond on carbon-4. Therefore, numbering occurs from left to right in this example.

With the methyl (–CH₃) substituent at locant 4, the name of this molecule is as follows:

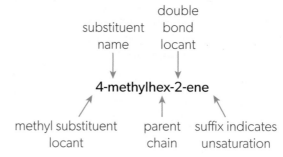

Worked example 2

Using the IUPAC rules, name the following alkenes.

a.

b.

Solution

a. There are four carbon atoms in the longest straight carbon chain, so the parent chain is **butane**.

The molecule contains a carbon–carbon double bond, so the suffix changes from -ane to **-ene**.

Numbering the carbon chain from right to left gives the unsaturated carbon–carbon double bond the lowest possible locant. The double bond follows carbon-1.

The IUPAC name for this compound is but-1-ene.

b. There are six carbon atoms in the longest straight carbon chain, so the parent chain is **hexane**.

The molecule contains a carbon-carbon double bond, so the suffix changes from -ane to **-ene**.

The double bond follows carbon-2 in the longest carbon chain. Two **methyl** (–CH₃) substituents are bonded to carbon atoms number 4 and 5.

The IUPAC name for this molecule is 4,5-dimethylhex-2-ene.

Naming halogenoalkanes

When naming halogenoalkanes, the position of the halogen substituent is identified by the locant on the parent chain. Table 11 lists the prefixes used for each halogeno group.

For derivatives of methane and ethane featuring only one halogeno group, all locants for the substituent are equivalent, so they are omitted. Consider the following compound:

Halogeno group	Substituent name
–F	fluoro
–Cl	chloro
–Br	bromo
–I	iodo

▲ Table 11 Names of halogeno substituents

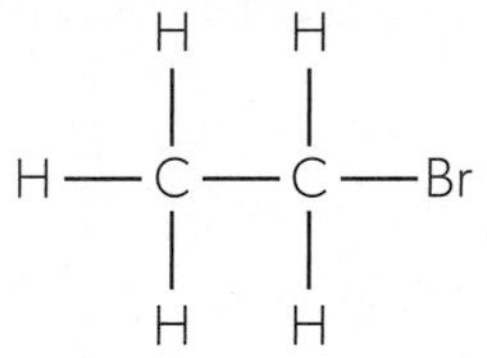

Moving the bromine atom to other positions will form the same molecule. The IUPAC name for this molecule is bromoethane, not 1-bromoethane.

For a halogenoalkane with more than one halogeno group, list the halogens in alphabetical order. For example, the following compound is named 2-chloro-1,1,1-trifluoroethane.

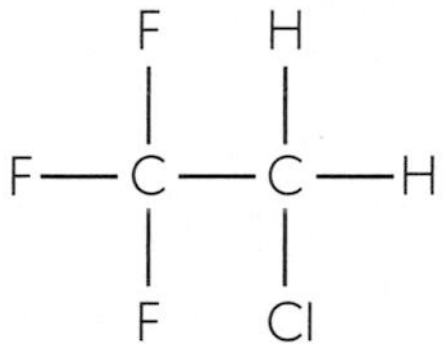

If a halogeno compound also features a carbon–carbon double bond, the double bond takes priority over the halogeno group in terms of assigning locants. So the following molecule is 3-chlorobut-1-ene, not 2-chlorobut-3-ene.

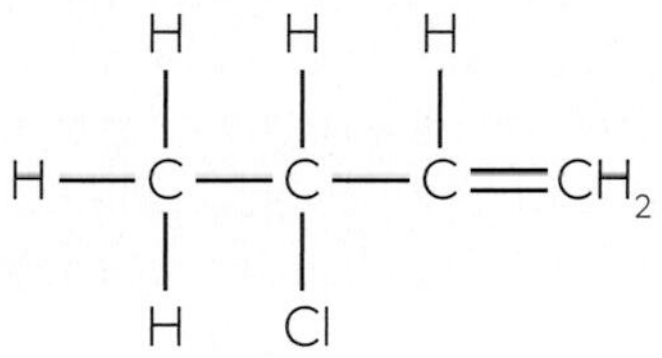

Practice questions

11. Apply the IUPAC nomenclature rules to deduce the names of the following compounds.

a.

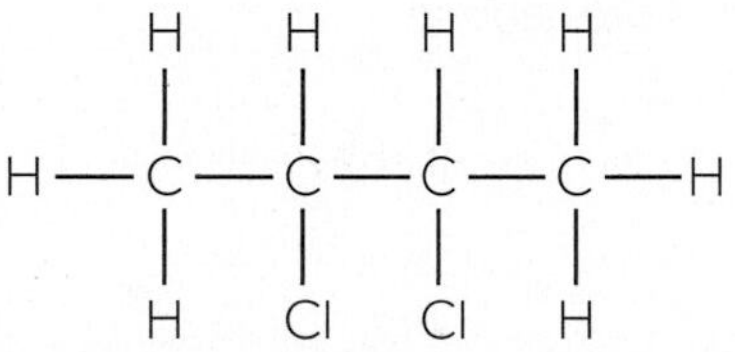

b.

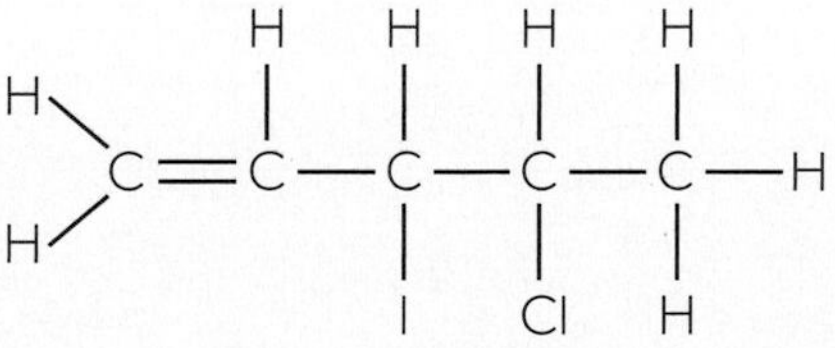

c.

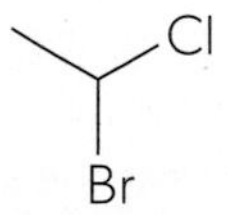

12. Deduce the structural formulas for the following compounds.

a. 2,3,4-tribromohexane

b. 1,3-dibromo-2-chlorobutane

c. 2-fluorobut-2-ene

Naming alcohols

All alcohols contain the hydroxyl (–OH) functional group. When naming alcohols, you need to change the parent chain suffix from -*ane* to -*anol*. The carbon chain is numbered such that the hydroxyl group has the lowest possible locant, taking priority over other substituents or carbon–carbon double bonds.

The simplest molecule in this homologous series is methanol. The first two members of this homologous series do not require locants. This is because the hydroxyl group in any position results in the same molecule, as seen in methanol and ethanol below.

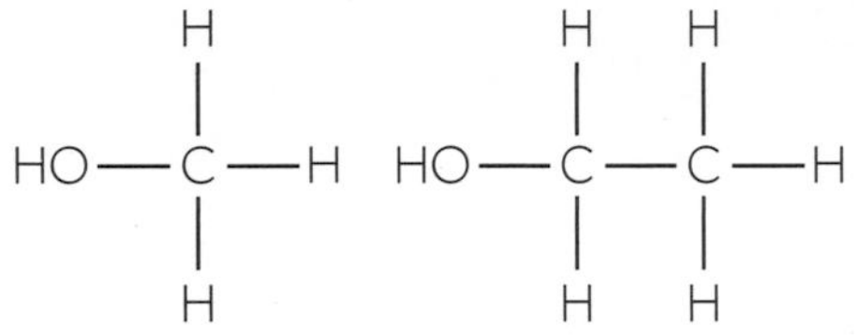

methanol   ethanol

Alcohol compounds with three or more carbon atoms require locants to specify the position of the hydroxyl group. The locant comes before the -ol suffix, for example:

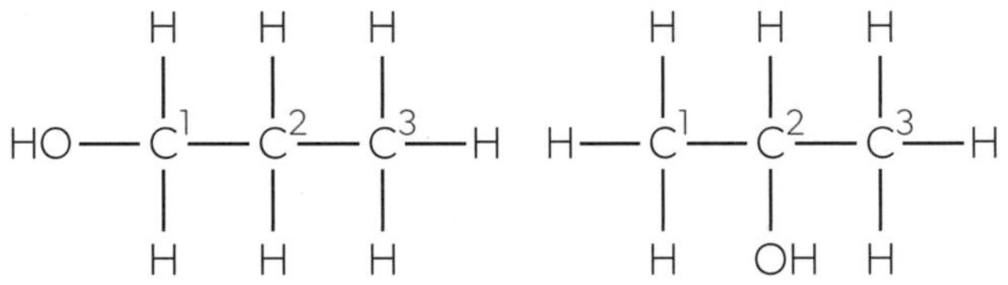

propan-1-ol   propan-2-ol

Naming compounds with carbonyl groups

The carbonyl group (C=O) is the functional group for two classes of compounds: aldehydes and ketones. In aldehydes, the carbonyl group is always positioned at the end of the carbon chain. This is known as the **terminal** position. There is no need to number the carbonyl group in aldehydes, as it always occurs at the end of the chain. When the aldehyde includes a substituent, the numbering of carbon atoms begins at the carbon attached to the carbonyl group. For example, aldehyde names have the suffix -al, so ethane becomes ethanal.

When the carbonyl functional group is not located at the terminal position in the primary carbon chain, the compound is a ketone. In ketones, the carbon chain is numbered such that the carbonyl group has the lowest possible locant, taking priority over other substituents or carbon–carbon double bonds. Ketone names have the suffix -one, so propane becomes propanone.

Worked example 3

Using the IUPAC rules, name the following aldehydes.

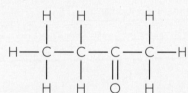

a.

b.

c.

Solution

a.   The parent chain comes from the C₁ alkane, **methane**.

The molecule contains a carbonyl group in the terminal position, so the suffix -ane becomes -al.

There are no substituents, so the IUPAC name is methanal.

b.   The parent chain comes from the C₅ alkane, **pentane** and the suffix becomes -al.

The carbon numbering starts from the carbonyl group, and therefore the methyl substituent is attached to carbon-4.

The IUPAC name is 4-methylpentanal.

c.   This is an example of an aldehyde featuring a halogeno substituent. The parent chain comes from the C₄ alkane, **butane**, and the suffix becomes -al.

The carbon numbering starts from the carbonyl group, and therefore the fluoro substituent is attached to carbon-3.

The IUPAC name is 3-fluorobutanal.

Worked example 4

Using the IUPAC rules, name the following ketones.

a.

b.

c.

Solution

a.   The parent chain comes from the C₃ alkane, **propane**. The molecule contains a carbonyl group that is not in the terminal position, so the suffix becomes -one.

The IUPAC name is propanone. The locant is not required because propanone is the only three-carbon ketone that can exist; if the carbonyl group were at carbon-1 or carbon-3, it would be an aldehyde. Therefore, all the information required to deduce the structure is in the name 'propanone'.

b.   The parent chain comes from the C₄ alkane, **butane**, and the suffix becomes -one. The locant is not required, as this is the only four-carbon ketone that can exist. So the final IUPAC name will be butanone.

c.   The parent chain comes from the C₅ alkane, **pentane**, and the suffix becomes -one. The carbon atoms are numbered from right to left such that the carbonyl group has the lowest locant. The carbonyl group is at carbon-2, and the methyl group is at carbon-4. Therefore, the IUPAC name is 4-methylpentan-2-one.

Naming carboxylic acids

Carboxylic acids are a class of compound featuring a carboxyl group (–COOH). Carboxylic acids have the suffix -oic acid. So a carboxylic acid with an ethane parent chain is called ethanoic acid.

In carboxylic acids, carbon chains are numbered starting with the functional group, including the carbon from the carboxyl group in the chain. Like in aldehydes, there is no need to assign locants to carboxyl groups, as they are always in the terminal position. Here are two examples of carboxylic acids:

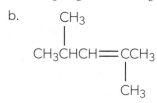

3-bromopropanoic acid   4-methylpentanoic acid

Practice questions

13. Draw the structures of the following molecules based on the IUPAC name given:

   a. hexan-3-one
   b. 2-chloropropan-1-ol
   c. 3,3-dimethylpentane
   d. 5-ethyl-2-methylhept-3-ene
   e. 2-methylpentan-2-ol

14. Apply IUPAC nomenclature rules to name the following molecules:

   a. $CH_3CH_2CH=CHCH_2CH_2CH_3$

   b.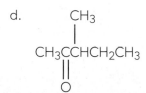

   c.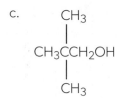

   d.

   e.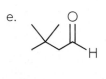

▲ **Figure 8** Methanoic acid, commonly known as formic acid, occurs naturally in some ants

Science as a shared endeavour

Scientists collaborate across the world, and agreed-upon terminology facilitates unambiguous communication. The naming rules you have learnt in this chapter enable you to communicate chemical compounds using both structures and names. If you look at the ingredients listed on the label of a bottle of a personal hygiene product such as shampoo, you will now be able to recognize some of the naming stems and suffixes.

The standardization of organic nomenclature encourages the use of names that follow IUPAC guidelines, but certain non-systematic names are still used widely. Many of these allude to the uses and characteristics of substances. For example, methanoic acid is commonly known as formic acid, after the Latin for ant: *formica*. If you speak Italian, French, Portuguese or Spanish, you will recognize this as the root of *formica*, *fourmi*, *formiga* and *hormiga*, respectively.

In some cases, IUPAC names are far too long to be useful. For example, the IUPAC name for the mild painkiller ibuprofen is (*RS*)-2-(4-(2-methylpropyl) phenyl)propanoic acid.

Structural isomers (*Structure 3.2.6*)

Structural or **constitutional isomers** are compounds that have the same molecular formula but different connectivity of the atoms. There are several types of structural isomers.

Chain isomers

Chain isomers are structural isomers with different lengths of carbon chains. For example, butane and methylpropane both have the molecular formula C_4H_{10}. Butane is a straight-chain molecule because it has no substituents on the carbon chain. However, methylpropane is a branched-chain molecule, because it has an alkyl substituent.

Isomer	butane	methylpropane
Molecular formula	C_4H_{10}	
Full structural formula	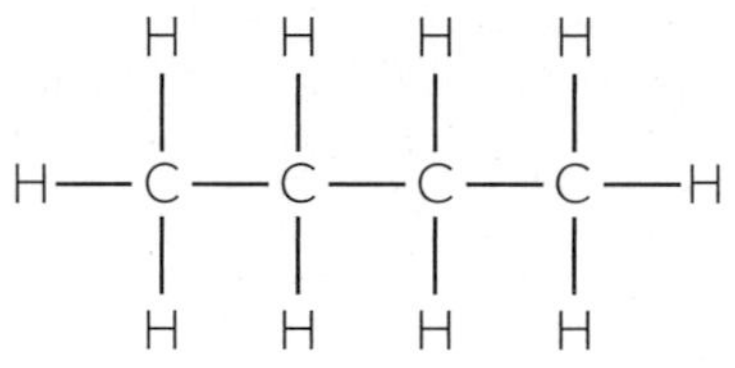	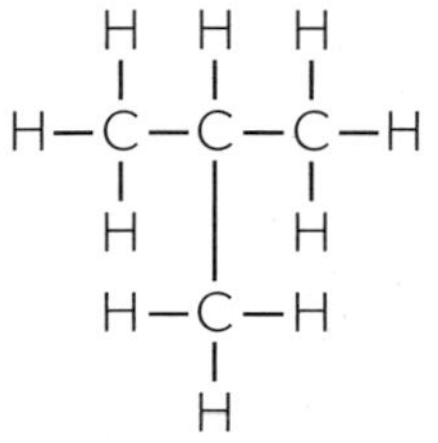
Condensed structural formula	$CH_3CH_2CH_2CH_3$	$CH(CH_3)_3$
Skeletal formula		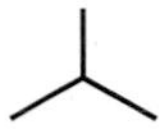

▲ Table 12 Butane and methylpropane are chain isomers

Positional isomers

Positional isomers are structural isomers where the position of the functional group changes. For example, bromopentane has three different positional isomers, all with the molecular formula $C_5H_{11}Br$.

Isomer	1-bromopentane	2-bromopentane	3-bromopentane
Molecular formula	$C_5H_{11}Br$		
Full structural formula	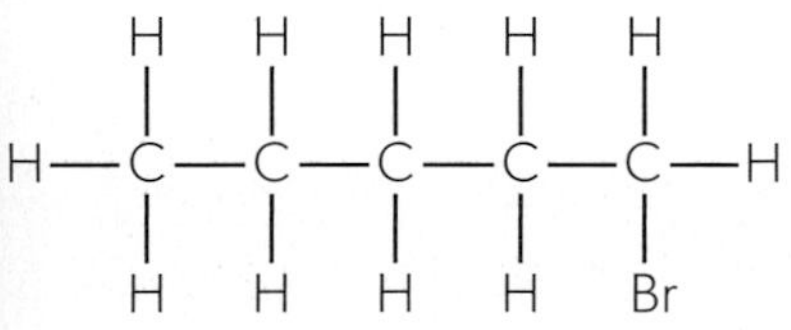	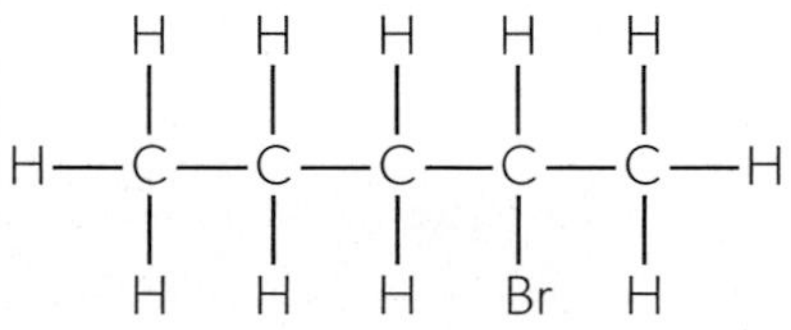	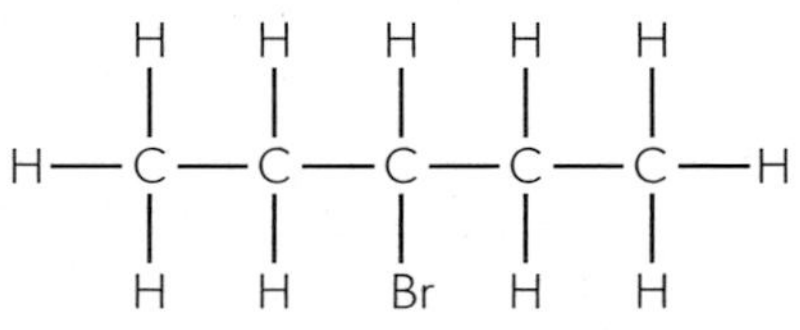
Skeletal formula	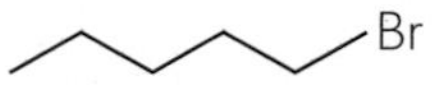	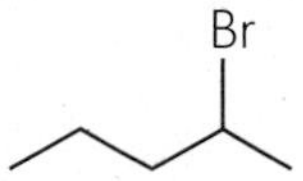	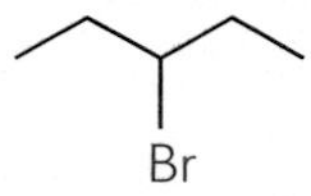

▲ Table 13 Positional isomers of bromopentane

Worked example 5

Draw the structural formulas of all possible structural isomers of a compound with the molecular formula C_4H_9F. Name the isomers using IUPAC nomenclature.

Solution

First, draw the longest possible straight carbon chain. For this molecule, it is a four-carbon chain.

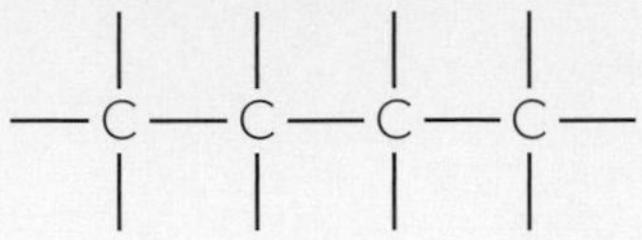

$$-C-C-C-C-$$

Next, number the carbon chain and place the fluorine substituent on the carbon chain in as many positions as possible, then add the hydrogen atoms. Remember that the numbering of the carbon chain can occur in either direction.

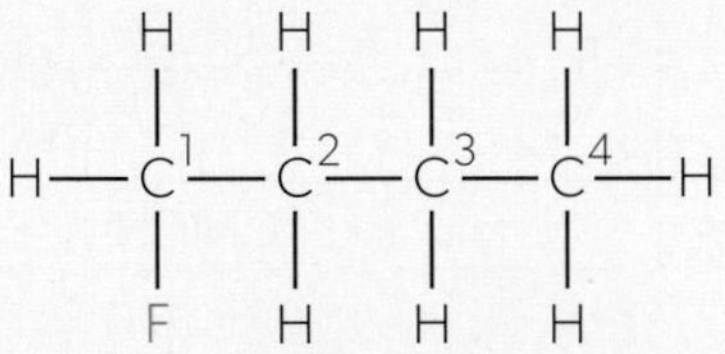

1-fluorobutane

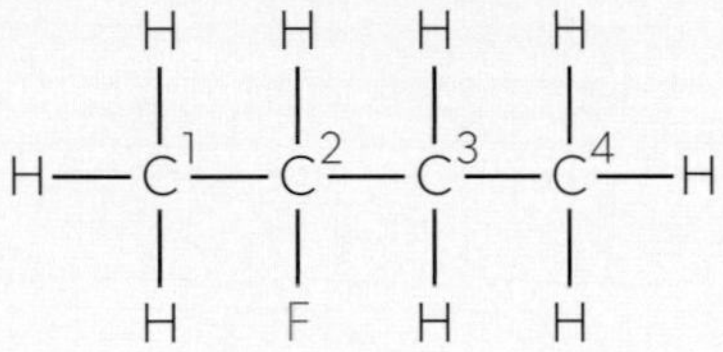

2-fluorobutane

3-fluorobutane and 4-fluorobutane are not possible isomers: 3-fluorobutane is equivalent to 2-fluorobutane, and 4-fluorobutane is equivalent to 1-fluorobutane, and the lower locants are favoured.

Having identified and named all the butane isomers, you need to look at possible chain isomers. Shorten the longest chain by one carbon, creating a propane chain.

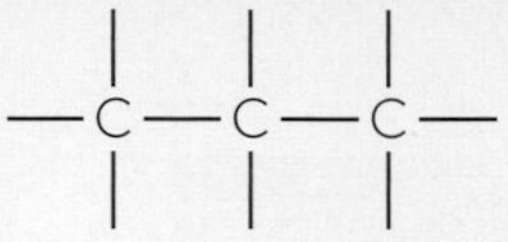

$$-C-C-C-$$

To share the same molecular formula as the isomers above, you need to add a methyl group to the propane chain. For this molecule, the only possible position is on the central carbon in the chain, as adding the methyl group anywhere else will create a four-carbon butane chain.

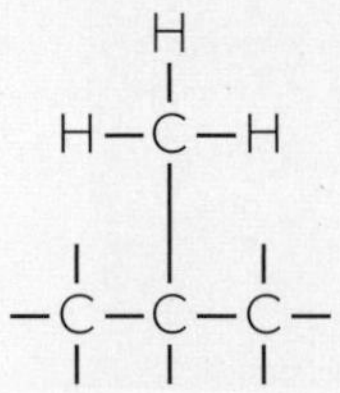

Then, position the fluorine substituent on the carbon chain in as many ways as possible to create unique molecules.

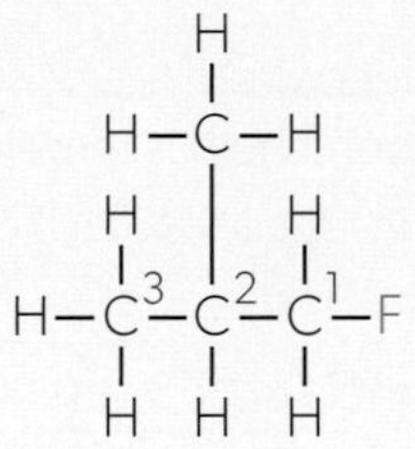

1-fluoro-2-methylpropane 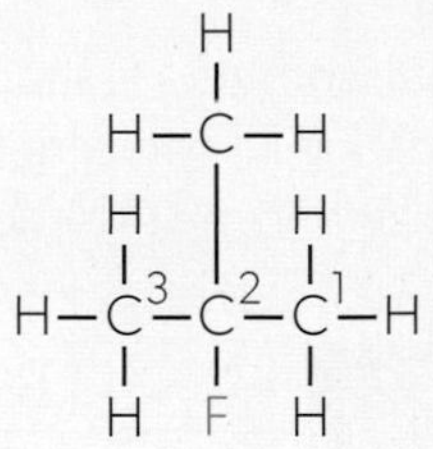   2-fluoro-2-methylpropane

---

🔖 Activity

Draw and name all the structural isomers of chloropentane, $C_5H_{11}Cl$.

Primary, secondary and tertiary compounds

Changing the position of the functional group on the parent carbon chain affects the chemical properties of compounds. You can classify compounds such as alcohols, halogenoalkanes and amines as **primary**, **secondary** and **tertiary**, depending on the position of the functional group.

In halogenoalkanes, a halogeno group is bonded to a carbon atom. In order to determine whether the compound is primary, secondary or tertiary, we count the number of carbon atoms bonded to this carbon atom. A **primary (1°)** carbon atom is bonded to one other carbon atom, a **secondary (2°)** carbon atom is bonded to two other carbon atoms, and a **tertiary (3°)** carbon atom is bonded to three other carbon atoms (figure 9).

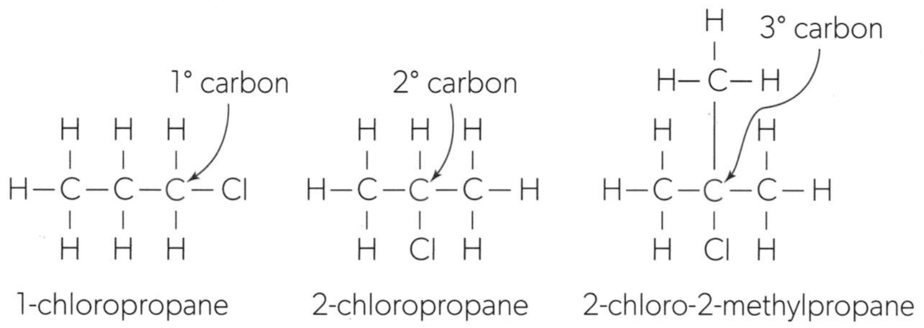
1° carbon
2° carbon
3° carbon
1-chloropropane
2-chloropropane
2-chloro-2-methylpropane

▲ **Figure 9** Primary (1°), secondary (2°) and tertiary (3°) halogenoalkanes

The reaction mechanisms and conditions used to form primary, secondary and tertiary halogenoalkanes are discussed in *Reactivity 3.3*.

Alcohols are classified in the same way as halogenoalkanes: the number of carbon atoms attached to the carbon atom next to the functional group indicates the class of the alcohol (figure 10).

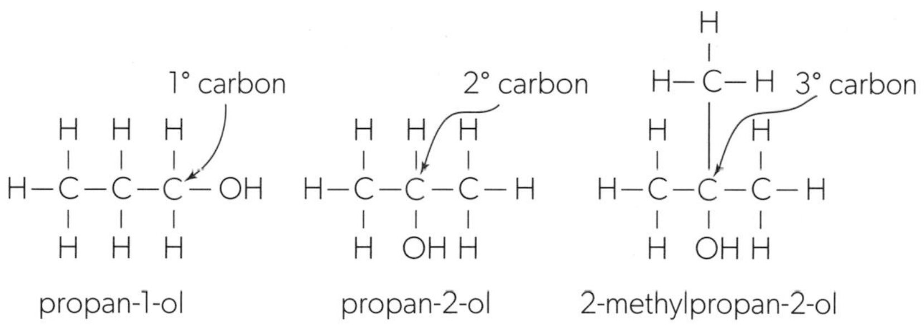
1° carbon
2° carbon
3° carbon
propan-1-ol
propan-2-ol
2-methylpropan-2-ol

▲ **Figure 10** Primary (1°), secondary (2°) and tertiary (3°) alcohols

Primary, secondary and tertiary alcohols form different products in oxidation reactions. This is discussed in *Reactivity 3.2*.

Classification of amines as primary, secondary or tertiary is slightly different to the method used with alcohols and halogenoalkanes. Amine classification depends on the number of alkyl groups bonded directly to the *nitrogen* atom of the functional group, unlike halogenoalkanes and alcohols that consider the carbon atom next to the functional group.

When naming amines, the parent name loses the suffix -e, which is replaced by "amine". When naming secondary and tertiary amines, N is used to signify that the substituents are bonded to the nitrogen atom. In figure 11, the substituents are methyl groups.

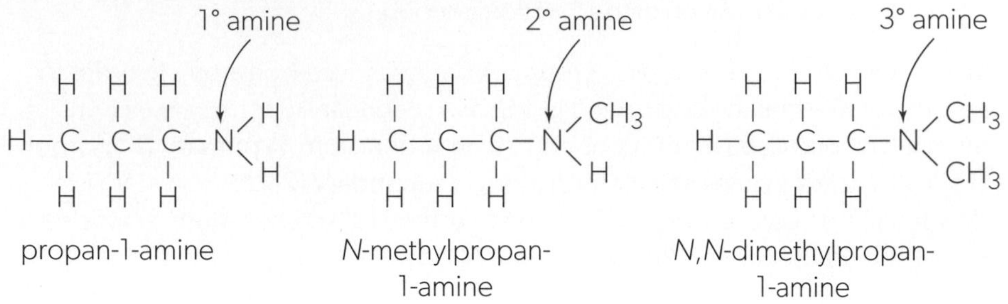

▲ Figure 11 Primary (1°), secondary (2°) and tertiary (3°) amines

Practice questions

15. Propan-2-ol is a useful organic solvent.

    a. Draw the skeletal formula of propan-2-ol.

    b. Classify propan-2-ol as a primary, secondary or tertiary alcohol, giving a reason.

16. Deduce which of the following is a secondary alcohol.

A.

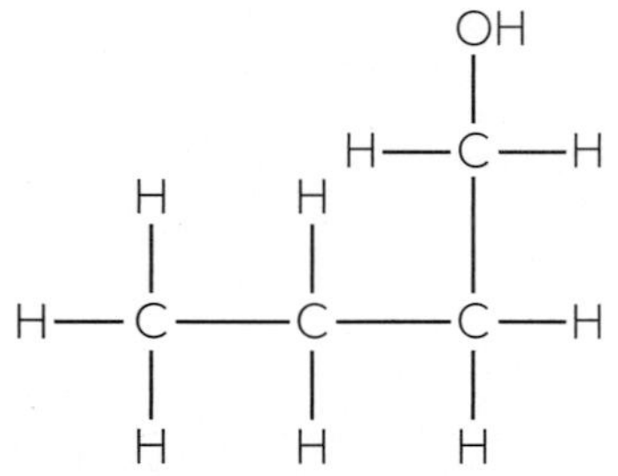

B.

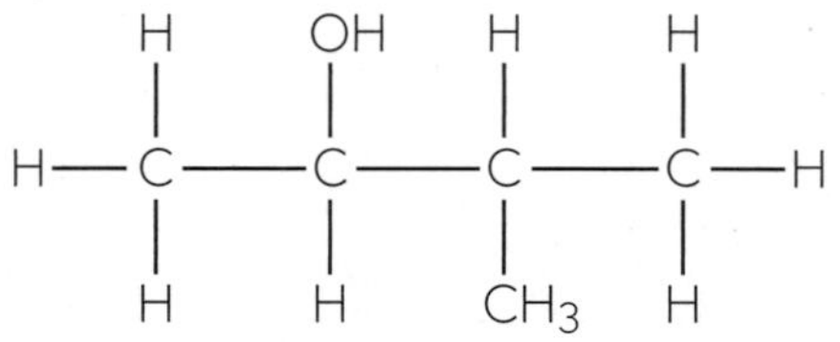

C.

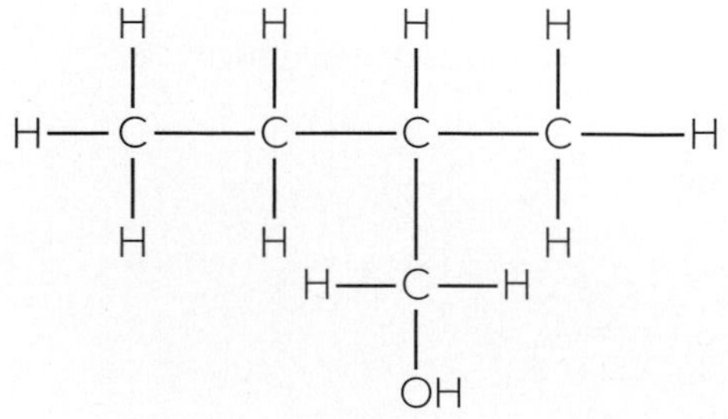

D.

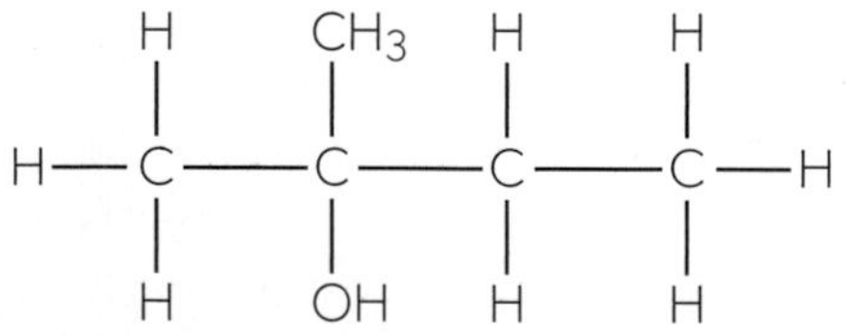

Functional group isomers

Functional group isomers are structural isomers with the atoms arranged differently such that they have different functional groups.

Compounds with hydroxyl and alkoxy functional groups can be structural isomers of each other. For example, consider the molecular formula C_2H_6O. Functional group isomers of this molecule include ethanol, an alcohol with a hydroxyl group, and methoxymethane, an ether with an alkoxy group.

ethanol   methoxymethane

Aldehydes and ketones can be functional group isomers of each other, as they both have carbonyl groups. Propanone, the simplest ketone, is a functional group isomer of propanal. Both isomers have the molecular formula C_3H_6O.

propanal   propanone

Compounds with ester and carboxyl groups can also be functional group isomers. The simplest ester, methyl methanoate $C_2H_4O_2$, is an isomer of the carboxylic acid, ethanoic acid.

methyl methanoate   ethanoic acid

Molecular models

Propan-1-ol, propan-2-ol and methoxyethane all have the molecular formula C_3H_8O. In this task, you will reflect on the information obtained from different types of formulas and models of these substances.

Relevant skills

- Tool 1: Physical and digital molecular modelling

Instructions

1. State the full structural formula, condensed structural formula and skeletal formula for each of the three isomers.

2. Using a molecular model kit, construct models of the three isomers.

3. Create digital models of the three isomers using online molecular visualization software (for example, MolView). Explore the different visualization possibilities, including: ball and stick model, space-filling model, molecular electrostatic potential (MEP) surface.

4. Compare and contrast the strengths and uses of each of the various formulas and molecular models.

5. How does the molecular structure of these three compounds relate to their physical properties and reactivity?

Linking question

How does the fact that there are only three isomers of dibromobenzene support the current model of benzene's structure? (*Structure 2.2*)

ATL Thinking skills

In the *Molecular models* task, you gathered information about three compounds, including: types of formulas, types of models, discussion of strengths and properties. Reflect on how this information could be organized. Would you opt for a table, Venn diagram, concept map, or anything different? Do you have a preferred method? Why does effective organization of information support learning?

Stereoisomers (*Structure 3.2.7*)

Another type of isomerism exhibited by some organic compounds is **stereoisomerism**. A pair of **stereoisomers** have the same molecular formula, connectivities and bond multiplicity. However, they have different spatial arrangements of the atoms. Stereoisomers can be subdivided into two major classes: **conformational isomers** and **configurational isomers** (figure 12).

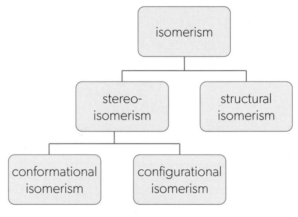

▲ Figure 12 Types of isomerism

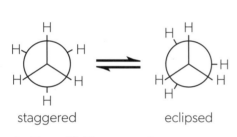

staggered   eclipsed

▲ Figure 13 The two conformers of ethane, viewed along the carbon–carbon bond

If a molecule contains two groups bonded together by a single bond, these groups can rotate about the single bond to form different conformational isomers. For example, in ethane, CH_3CH_3, the two CH_3 groups can rotate about the carbon–carbon bond resulting in two possible orientations, known as conformational isomers or **conformers** (figure 13).

This process does not involve breaking any bonds, and it is virtually impossible to separate the individual conformers.

Configurational isomers can only be interconverted by breaking and reforming bonds. Configurational isomers can be subdivided into **cis–trans isomers** and **optical isomers** (figure 14).

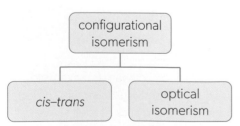

▲ Figure 14 Types of configurational isomerism

Cis–trans isomers

In aliphatic alkenes, groups bonded to the unsaturated carbon atoms are not able to rotate about the carbon–carbon double bond. If these two groups are different, two isomers can exist: a *cis* isomer where two identical substituents are on the same side of the carbon–carbon double bond, and a *trans* isomer where two identical substituents are on opposite sides of the double bond. In Latin, *cis* means "this side of" and *trans* means "the other side of".

For example, the symmetrical molecule of but-2-ene has *cis–trans* configurational isomers about the double bond (figure 15). The horizontal axis of the double bond is the **reference plane**: it is used to determine whether an isomer is *cis* or *trans*, depending on which groups are either side of the plane.

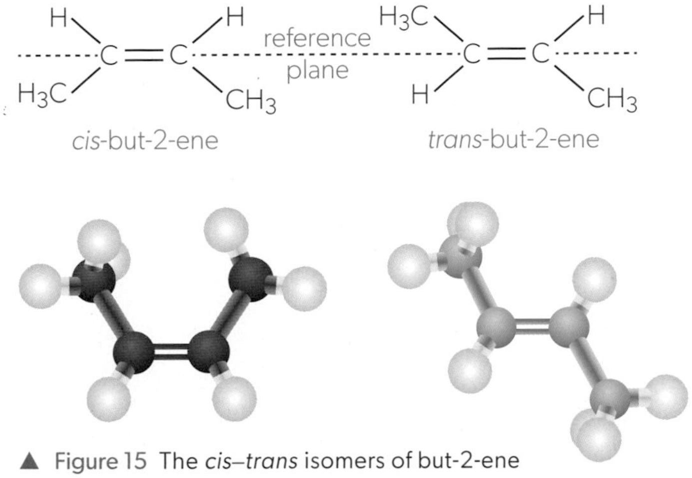

▲ Figure 15 The *cis–trans* isomers of but-2-ene

In figure 15, the *cis*-isomer of but-2-ene has the two methyl substituents on the same side of the reference plane, while the *trans*-isomer has the two methyl substituents on opposite sides of the reference plane.

For an alkene to exhibit *cis–trans* isomerism, the two groups attached to each unsaturated carbon atom have to be different. For example, propene, $H_2C=CHCH_3$, will not experience *cis–trans* isomerism because one of the carbon atoms has two identical groups (hydrogen atoms), so the potential isomers will also be identical (figure 16).

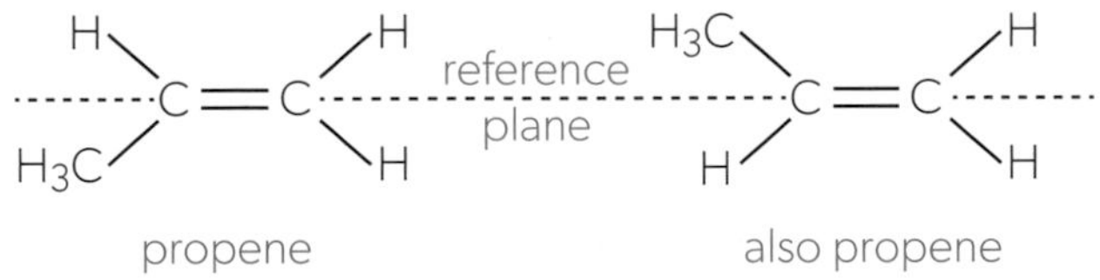

▲ Figure 16 Propene does not exhibit *cis–trans* isomerism

Cis–trans isomerism also occurs in disubstituted cycloalkane molecules. In this case, the reference plane is the flat face of the ring structure. For example, the compound 1,2-dimethylcyclobutane has *cis–trans* configurational isomers (figure 17).

The *cis*-isomer has both methyl substituents on the same side of the plane of the ring. The *trans*-isomer has methyl substituents on opposite sides of the plane of the ring. Disubstituted three-carbon rings also experience *cis–trans* isomerism (figure 18).

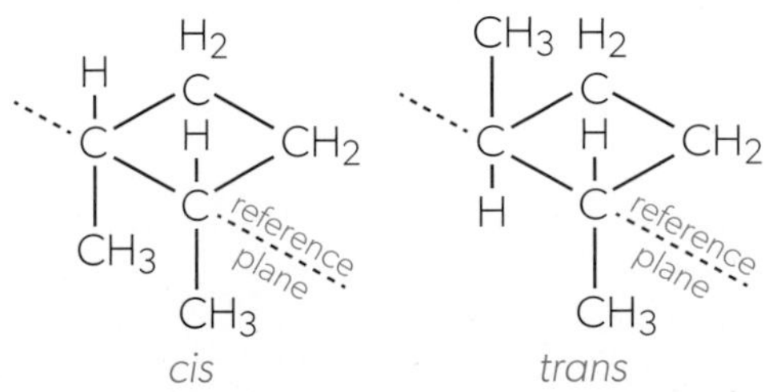

▲ Figure 17 The *cis–trans* isomers of 1,2-dimethylcyclobutane

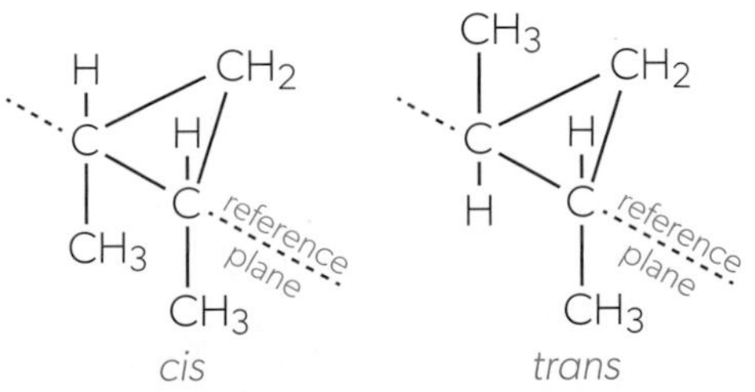

▲ Figure 18 The *cis–trans* isomers of 1,2-dimethylcyclopropane

Practice questions

17. Deduce and draw the *cis-* and *trans*-isomers of hex-2-ene.

18. For each of the following molecules, state whether they exhibit *cis–trans* isomerism. For the molecules that do, deduce whether they are in the *cis* or *trans* configuration. Name each molecule using IUPAC rules.

a.

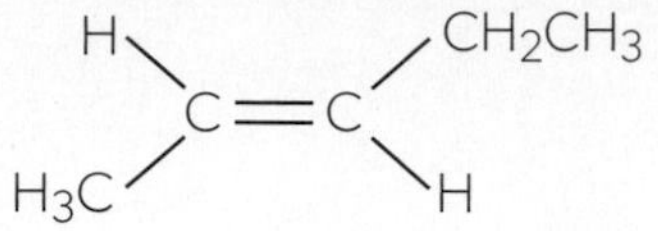

$$H \diagdown C = C \diagup CH_2CH_3$$
$$H_3C \diagup \qquad \diagdown H$$

b.

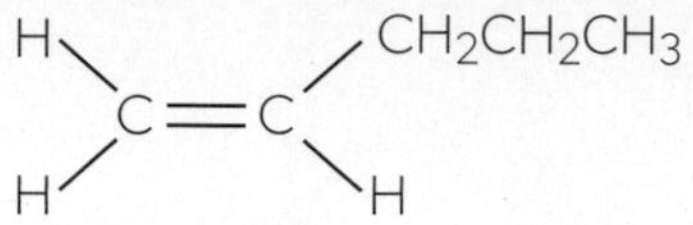

$$H \diagdown C = C \diagup CH_2CH_2CH_3$$
$$H \diagup \qquad \diagdown H$$

c.

$$H_3CH_2C \diagdown C = C \diagup CH_2CH_3$$
$$H \diagup \qquad \diagdown H$$

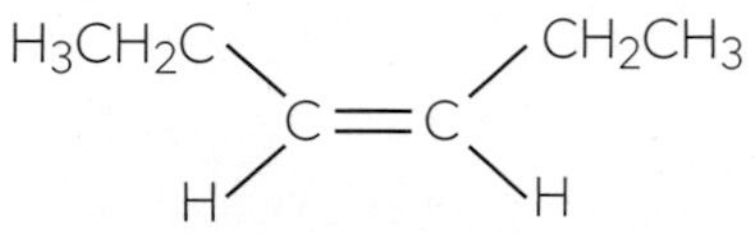

d.

$$H_3C \diagdown C = C \diagup H$$
$$H_3C \diagup \qquad \diagdown H$$

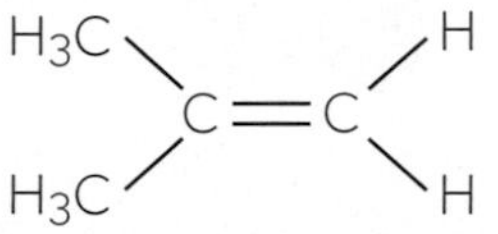

19. Draw the structural formula of the following compounds, and then deduce if they can form *cis–trans* isomers.

a. 2-bromoprop-1-ene

b. pent-2-ene

c. 1-bromo-2-chloroethene

d. 1,1-dibromobut-1-ene

e. 1-ethyl-2-methylcyclopropane

f. 2,3-dimethylpent-2-ene

Optical isomers

A **chiral** carbon is defined as a carbon atom bonded to four different atoms or groups of atoms. It is also known as a **stereocentre** or **asymmetric centre**. Molecules with one or more chiral carbon atoms exhibit a type of configurational isomerism called **optical isomerism**. A pair of optical isomers are called **enantiomers**.

Chiral molecules have the ability to rotate plane-polarized light; this is known as **optical activity**.

Penicillamine is a drug used in the treatment of rheumatoid arthritis. It has the molecular formula $C_5H_{11}NO_2S$. Its structural formula is shown in figure 19.

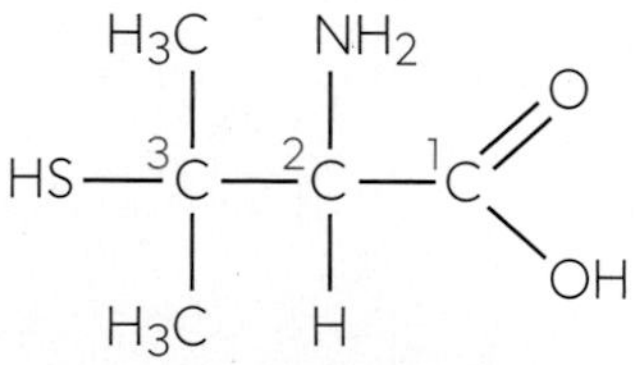

Carbon-1 is only bonded to three atoms, so it is not chiral, or **achiral**. Carbon-2 and carbon-3 each have four bonded atoms or groups. However, two of the four groups bonded to carbon-3 are identical (the two methyl groups), so carbon-3 is achiral. Carbon-2 is bonded to four different atoms or groups of atoms, and it is therefore chiral. The presence of this chiral carbon atom means that penicillamine is optically active and can exist as a pair of two different enantiomers. Enantiomers are non-superimposable mirror images of each other, and they have no plane of symmetry.

$$HS \overset{H_3C}{\underset{H_3C}{\overset{|}{\underset{|}{\overset{3}{C}}}}} \overset{NH_2}{\underset{H}{\overset{|}{\underset{|}{\overset{2}{C}}}}} \overset{1}{C} \overset{O}{\underset{OH}{\diagup}}$$

▲ Figure 19 Penicillamine has the IUPAC name 2-amino-3-methyl-3-sulfanylbutanoic acid

Structural formulas, like the one in figure 19, provide information on how the atoms and groups present in a molecule are connected. However, these formulas do not allow you to differentiate between enantiomers in a molecule with a chiral centre. As a result, enantiomers are drawn as three-dimensional structures known as **stereochemical formulas** (figure 20), where specific symbols are used to designate the direction of a bond (table 14).

Name	Symbol	Directionality
line bond	C—	aligned with the plane of paper
wedge bond	C◀	coming out of the plane (towards the viewer)
dash bond	C⠒⠒⠒	going behind the plane (away from the viewer)

▲ Table 14 Summary of directional bonds used in stereochemical formulas

You will recall from *Structure 2.2* that a central atom with four regions of electron density adopts a tetrahedral arrangement with bond angles of 109.5°. For the tetrahedral arrangement around a chiral carbon atom, two of the bonds are usually drawn as line bonds, one as a wedge bond, and one as a dash bond. Wedge and dash bonds are tapered, meaning that they start off narrow from the carbon atom and widen.

To draw the enantiomer of a stereochemical formula, you need to draw its mirror image. For example, the stereochemical formulas of the penicillamine enantiomers are shown in figure 21.

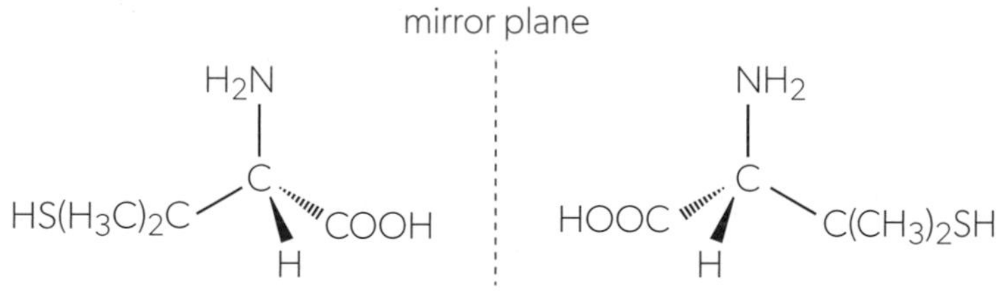

mirror plane

H_2N NH_2

$HS(H_3C)_2C$ —C⠒⠒⠒ COOH   HOOC ⠒⠒⠒C— $C(CH_3)_2SH$

H   H

▲ Figure 21 Enantiomers of penicillamine

Enantiomers, like those in figure 21, are non-superimposable. This means that if you were to create 3D models of them using a molecular model kit, it would be impossible to align the models such that the functional groups of the same type point in the same direction. However, if you were to hold a mirror to one of the enantiomers, the mirror image would show the structure of the other enantiomer.

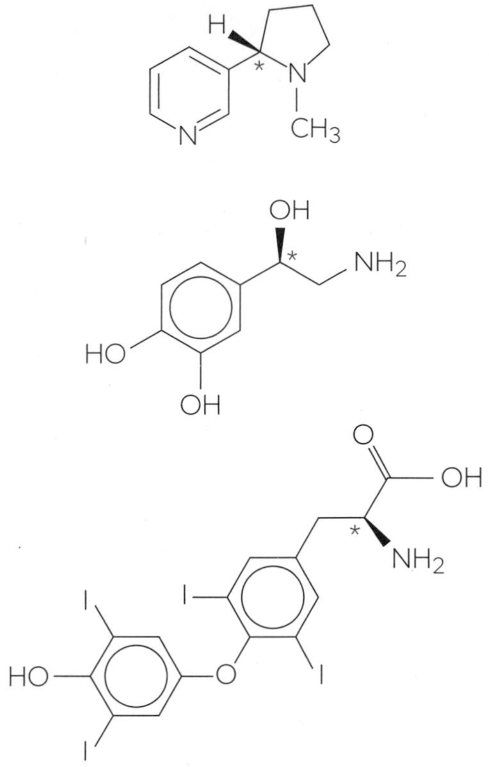

▲ Figure 20 All of these molecules have a chiral carbon atom, designated by *. Nicotine (top) is naturally synthesized by the tobacco plant, norepinephrine (middle) is a neurotransmitter, and thyroxine (bottom) is a hormone from the thyroid gland

◀ Figure 22 Molecular models of the enantiomers of 2,3-dihydroxypropanal, $HOCH_2CH(OH)CHO$. Can you identify the chiral centres?

Worked example 6

Draw the stereochemical formulas of the two enantiomers of 2-aminopropanoic acid showing a tetrahedral arrangement around the chiral carbon atom.

Solution

First, draw the structural formula to deduce which carbon atom is chiral.

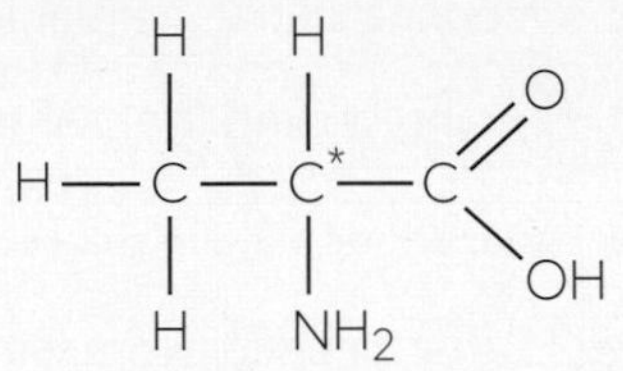

The chiral carbon atom is marked with an asterisk. It is attached to four different atoms or groups of atoms: a hydrogen atom and methyl, amino and carboxyl groups.

To draw the stereochemical formula, arrange the atoms and groups around the chiral carbon atom such that one is projected out of the plane of the paper (wedge bond), one is projected behind the plane of the paper (dash bond), and two atoms or groups of atoms are in the plane of the paper (line bonds). This gives you your first enantiomer.

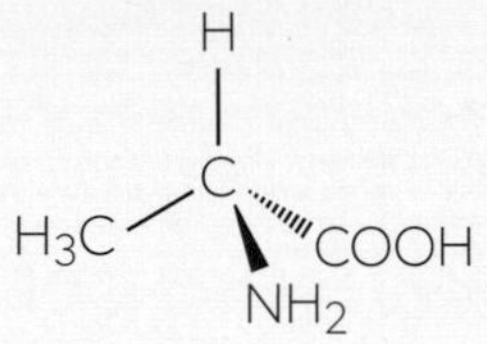

It does not matter which bond is assigned to which group for this first enantiomer, as the bonds in the tetrahedral structure are all equivalent. The wedges and dashes are just used to give you a sense of the 3D structure.

Finally, to draw the other enantiomer, draw a dashed vertical line to represent a mirror and draw the reflected image of your existing enantiomer.

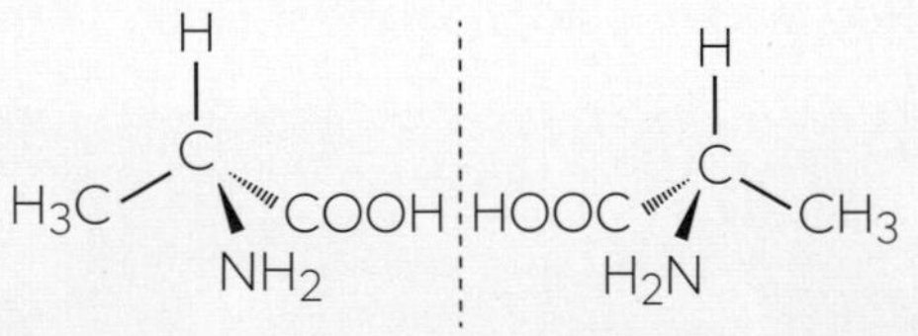

Make sure that the connectivities between the central atom and the groups are correct. For example, the methyl group has been reversed (H_3C) in the left-hand image to correctly show a carbon to carbon bond.

Earlier, we discussed the *cis–trans* isomerism of a cycloalkane, 1,2-dimethylcyclobutane. The *trans*-isomer of this molecule also exhibits optical isomerism. This is because its mirror images are non-superimposable (figure 23).

Different enantiomers of the same compound have different odours. Our perceived taste can also be affected by a specific enantiomer present in foods. The different chemical properties of enantiomers are identical in non-chiral environments, but become different in the presence of other chiral compounds, such as biological molecules in the human body.

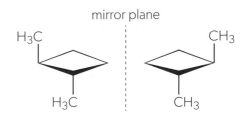

▲ Figure 23 The optical isomerism of *trans*-1,2-dimethylcyclobutane

A pair of enantiomers under the same conditions will rotate plane-polarized light by the same angle, but in opposite directions (figure 24). One enantiomer rotates the plane of polarization clockwise and is designated the (+) enantiomer. The other enantiomer rotates the plane of polarization anti-clockwise and is designated the (–) enantiomer.

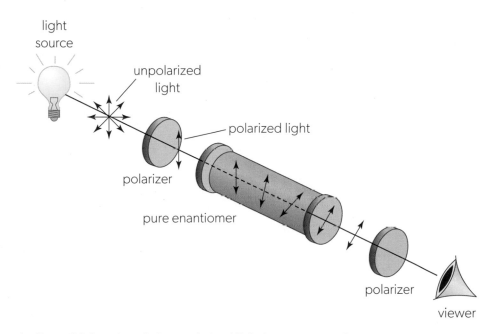

▲ Figure 24 Rotation of plane-polarized light by a pure enantiomer

A 50:50 mixture of two enantiomers is called a **racemic mixture** (or **racemate**), and it does not rotate plane-polarized light.

Activity

For each of the following molecules, complete the tasks (a)-(c):

    i. butan-2-ol, $C_4H_{10}O$

    ii. 2-hydroxypropanoic acid, $C_3H_6O_3$

a. Identify and explain which carbon atom is chiral.

b. Draw stereochemical formulas showing the tetrahedral arrangement around the chiral carbon atom.

c. Construct 3D models for the pair of enantiomers.

AHL

Rotation of plane-polarized light by optically active compounds

Optically active compounds rotate plane-polarized light. Here you will explore this phenomenon using simple polarizing filters, and consider ways to develop the investigation further.

Relevant skills

- Inquiry 1: Demonstrate independent thinking and creativity in the design of an investigation
- Inquiry 1: Identify control variables
- Inquiry 1: Design and explain a valid methodology
- Inquiry 2: Identify and record relevant qualitative observations

Safety

Wear eye protection.

Your teacher will provide you with further safety precautions, depending on the identity of the substances being analysed.

Materials

- Two polarizing filters
- Light source, for example, a mobile phone flashlight or torch
- Clear glass tube with a clear cover at one end, for example, a measuring cylinder
- Clamp and stand
- Samples of optically active substances, such as sucrose, glucose or fructose.

Instructions

1. Prepare aqueous solutions of the substance(s) you wish to test. Set them aside.
2. Place the light source on a flat surface, pointing up. Place a polarizing filter over the light source.
3. Pour distilled water into the tube and clamp it in place above the source of plane-polarized light.
4. Record the height of water in the tube.

5. Place the second polarizing filter above the tube.
6. Dim the lights and look at the light source through the second polarizing filter.

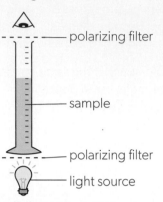

7. Rotate the second polarizing filter and observe the changes to the light intensity.
8. Repeat the experiment, this time placing one of your prepared solutions in the tube. Explore the effect of this solution on the rotation of plane-polarized light.

Questions

1. List some of the variables that you must keep constant in order to explore the effect of different compounds on the rotation of plane-polarized light.
2. Identify and explain the changes you would make to this method in order to do one or more of the following:
   - measure the angle of rotation
   - investigate the effect of concentration on the rotation of plane-polarized light
   - investigate the effect of path length on the rotation of plane-polarized light
   - investigate the effect of the composition of a mixture on the rotation of plane-polarized light.
3. Show your ideas to your teacher, and if you have time, try them out.

Mass spectrometry (MS) (*Structure 3.2.8*)

The operational details of the mass spectrometer are discussed in *Structure 1.2*.

Mass spectrometry (MS) is an analytical technique that can be used to break up organic compounds into fragments, some of which will be ions. This is known as **fragmentation**. The molecular masses of these fragments will correspond to the masses of certain functional groups or parts of the carbon chain. As a result, MS can be used to obtain information about the functional groups and structural fragments present in an organic compound.

The molecular masses of the different fragments are recorded in a graph called a **mass spectrum**. The different peaks within a mass spectrum are known collectively as a **fragmentation pattern**. The peak with the highest mass to charge ratio (m/z) in a mass spectrum corresponds to the parent compound. This is known as the **molecular ion peak**.

Worked example 7

The mass spectrum for propan-1-ol is shown in figure 25.

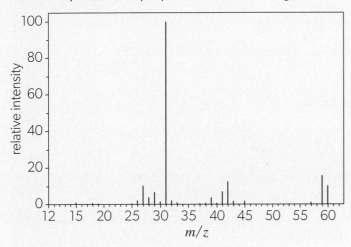

▲ Figure 25 Mass spectrum of propan-1-ol

Use the mass spectrum to show that the fragmentation pattern corresponds to the structure of propan-1-ol.

Solution

If you know the name of the analysed compound, begin by drawing its structural formula. This will help you to figure out the possible fragmentation patterns. When you have more than one option, draw the structural formulas of all the expected molecules. In this case, there is only one possible structural formula (propan-1-ol):

$$H-\underset{\underset{H}{|}}{\overset{\overset{H}{|}}{C}}-\underset{\underset{H}{|}}{\overset{\overset{H}{|}}{C}}-\underset{\underset{H}{|}}{\overset{\overset{H}{|}}{C}}-OH$$

Then, examine the spectrum and identify the peaks and their m/z values. Use the m/z values to propose structures for each of the fragments:

m/z	Molecular fragment	Explanation
60	$[CH_3CH_2CH_2OH]^+$	Molecular ion (the original molecule that has lost one electron)
31	$[CH_2OH]^+$	Loss of CH_3CH_2 from the m/z 60
29	$[CH_3CH_2]^+$	Loss of CH_2OH from the m/z 60

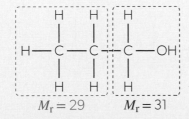

$M_r = 29$ $M_r = 31$

Data on specific MS fragments are provided in section 21 of the data booklet. Here are some examples:

- A peak with $m/z = 15$ corresponds to a CH_3 fragment

- A peak with $m/z = 17$ corresponds to an OH fragment

- A peak with $m/z = 29$ corresponds to a CHO or CH_2CH_3 fragment

- A peak with $m/z = 31$ corresponds to a CH_3O or CH_2OH fragment

- A peak with $m/z = 45$ corresponds to a COOH fragment

Practice questions

20. A straight-chain ketone with the molecular formula $C_5H_{10}O$ has two structural isomers.

   a. Deduce the structural formulas of the two possible isomers. Mass spectra **A** and **B** of the two isomers are given.

   b. Explain which spectrum is produced by each compound using section 22 of the data booklet.

A

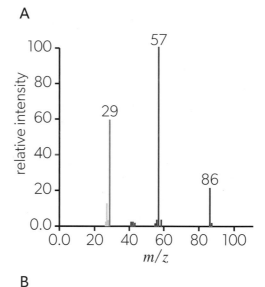

B

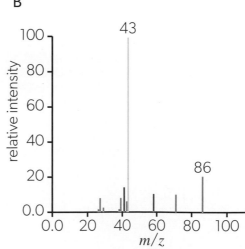

Infrared (IR) spectroscopy (*Structure 3.2.9*)

Infrared (IR) spectroscopy is an analytical technique that can be used to identify the types of bonds present in an organic compound, and therefore determine its functional groups. IR spectroscopy exposes a sample of an organic compound to IR radiation. Molecules absorb IR radiation, causing certain bonds to vibrate. IR spectroscopy is a type of **vibrational spectroscopy**.

There are two basic types of vibration that are possible in molecules: stretching/compression and bending. The type of movement observed depends on whether the molecule is diatomic or polyatomic. For example, the diatomic molecule of hydrogen chloride can only undergo stretching and compression (figure 26).

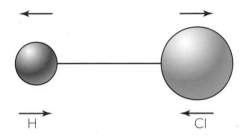

▲ **Figure 26 Stretching and compression in HCl molecule**

When a bond in a molecule absorbs IR radiation, the frequency of the resultant vibration depends on the bond enthalpy and the masses of the atoms in the bond. The higher the bond enthalpy, the stronger the bond. This will in turn result in a shorter wavelength, and higher frequency of the vibration. The larger the masses of the atoms, the lower the frequency of the vibration.

Hydrogen iodide contains a large iodine atom and has a relatively low bond enthalpy, so its vibration occurs at a lower frequency when compared to the other hydrogen halide molecules (table 15).

Molecule	Bond enthalpy / kJ mol^{-1}	Frequency of vibration / 10^{11} s^{-1}	Wavenumber / cm^{-1}
H–Cl	431	8.66	2886
H–Br	366	7.68	2559
H–I	298	6.69	2230

▲ **Table 15 Bond enthalpy and wavenumbers of halogen halide molecules**

In IR spectra, the absorption is usually described in terms of its **wavenumber**, $\bar{\nu}$, shown in the fourth column of table 14. The wavenumber is the reciprocal of the wavelength: $\bar{\nu} = \dfrac{1}{\lambda}$.

As described in *Structure 1.3*, frequency (f) is related to wavelength (λ) via the relationship $f = \dfrac{c}{\lambda}$, where c is the speed of light. This means that frequency is related to the wavenumber as follows: $\bar{\nu} = c \times f$.

In polyatomic molecules, the stretching can be symmetric or asymmetric. Polyatomic molecules can also undergo bending, a vibration where the bond angle changes. Both water and carbon dioxide molecules stretch and bend when they absorb IR radiation (figure 27), so they can be analysed by IR spectroscopy. In other words, such molecules are IR-active.

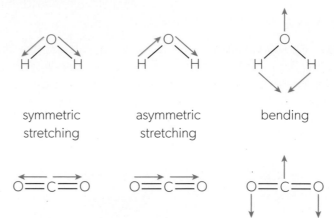

symmetric stretching

asymmetric stretching

bending

▶ Figure 27 Modes of vibration of water and carbon dioxide molecules. All three modes of vibration are IR active

For a compound to be IR active, the molecule must experience a change in dipole moment during the vibration; otherwise, no peak will be observed in the IR spectrum. Diatomic homonuclear molecules that do not have a permanent dipole, such as hydrogen, H_2, oxygen, O_2, and chlorine, Cl_2, are IR inactive.

Diatomic heteronuclear molecules, such as hydrogen fluoride, HF, are only capable of performing the stretching and compression vibrations. Because there is a change in the dipole moment, such molecules are IR active.

Plotting the wavenumbers of the molecular vibrations against their intensity gives a graph known as an **IR spectrum**. Certain wavenumbers are characteristic of specific functional groups, which are listed in section 20 of the data booklet. We can therefore analyse the IR spectrum to determine the types of functional group present in a molecule. For example, the IR spectrum of butanoic acid is shown in figure 28.

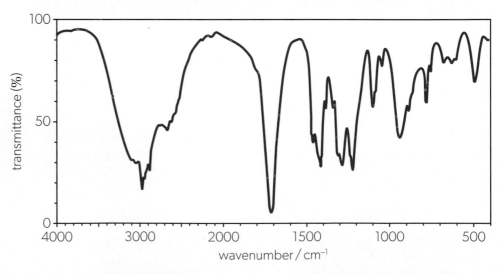

▲ Figure 28 IR spectrum of butanoic acid

There is a strong, very broad peak in the range 2500–3000 cm^{-1} characteristic of the O–H bond in carboxylic acids. The strong peak in the range 1700–1750 cm^{-1} is characteristic of the C=O bond.

IR spectroscopy only provides information about the functional groups present in a molecule, so this technique alone cannot be used to determine the structure of an unknown molecule. Therefore, information gathered from IR spectroscopy must be used in conjunction with information derived from other analytical techniques, such as mass spectroscopy, **nuclear magnetic resonance (NMR) spectroscopy** or **combustion analysis**.

Worked example 8

An organic compound containing only carbon, hydrogen and oxygen is composed of 62.02% carbon and 10.43% hydrogen by mass.

a. Determine the empirical formula of the compound, showing your working.

b. The infrared spectrum of the compound is shown below. Deduce the functional group present in the molecule.

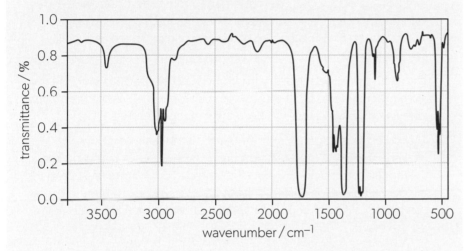

c. The mass spectrum of the compound is shown below. Deduce the relative molecular mass of the compound, and hence determine the molecular formula.

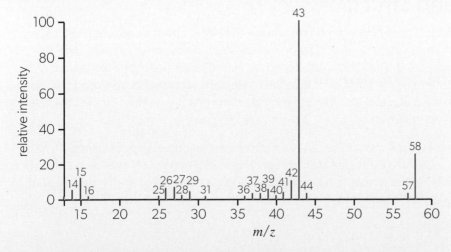

d. Through experimental analysis, it was determined that the compound did not contain a terminal carbonyl group. Deduce the structural formula of the compound.

Solution

a. The method for determining an empirical formula is shown below:

	C	H	O
Step 1: Divide the percentage by the relative atomic mass	$\dfrac{62.02}{12.01} = 5.16$	$\dfrac{10.43}{1.01} = 10.33$	$\dfrac{27.55}{16.00} = 1.72$
Step 2: Divide each result by the lowest value (1.72) to determine the whole number ratio, rounding if necessary	$\dfrac{5.16}{1.72} = 3.00$	$\dfrac{10.33}{1.72} = 6.01$	$\dfrac{1.72}{1.72} = 1.00$

The empirical formula is therefore C_3H_6O.

b. A strong peak in the range of 1700–1750 cm^{-1} shows the presence of a carbonyl group, C=O. Possible organic molecules include aldehydes, ketones, carboxylic acids and esters, but the question has asked you to deduce the functional group of the compound, so the answer is carbonyl/C=O.

c. The relative molecular mass of the compound is represented by the molecular ion peak: 58. If you add up the atomic masses of the empirical formula, this is also equal to 58, so the molecular formula is C_3H_6O.

d. The molecule is not an ester or carboxylic acid, as there is only one oxygen atom in the molecular formula. As the carbonyl group is not in the terminal position of the molecule, the carbonyl group must be in the middle of the carbon chain, so the molecule is a ketone. Therefore, the compound is propanone, and its structural formula is CH_3COCH_3.

Proton nuclear magnetic resonance (^{1}H NMR) spectroscopy (*Structure 3.2.10* and *Structure 3.2.11*)

Proton nuclear magnetic resonance (^{1}H NMR) spectroscopy is perhaps the most useful analytical technique for structural determination of organic molecules. The spectra produced by ^{1}H NMR spectroscopy help you to determine the number of different chemical environments of hydrogen atoms in the molecule. An atom's chemical environment is defined by the other atoms attached to it.

^{1}H NMR spectroscopy is an example of **spin resonance spectroscopy**. The nucleus of a hydrogen atom ^{1}H contains a single positively charged proton. This nucleus has two possible spin orientations, as described by the **nuclear spin quantum number (m_I)**, with possible values of $m_I = +\frac{1}{2}$ or $m_I = -\frac{1}{2}$.

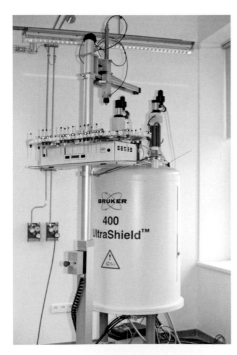

▲ Figure 29 A nuclear magnetic resonance (NMR) spectrometer

The hydrogen nucleus behaves as a spinning charged particle that can act as a small magnet. Similar to the electron spin, the two possible spin states of a ^1H nucleus are degenerate: they exist at the same energy level. When a strong magnetic field is applied to the sample containing ^1H nuclei, the two spin states become non-equivalent and split into two different energy levels (figure 30).

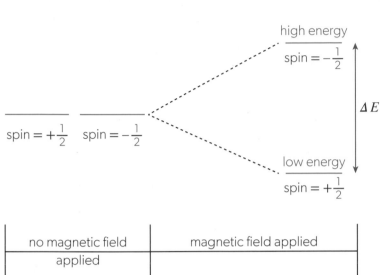

the magnetic field generated by the higher energy spin state is opposite to the applied external magnetic field

high energy

$$spin = -\frac{1}{2}$$

ΔE

$$spin = +\frac{1}{2} \quad spin = -\frac{1}{2}$$

low energy

$$spin = +\frac{1}{2}$$

the magnetic field generated by the lower energy spin state is aligned with the applied external magnetic field

| no magnetic field applied | magnetic field applied |

▲ Figure 30 The two spin states of the hydrogen nucleus are different in energy when a magnetic field is applied

$$H_3C - \underset{\underset{CH_3}{|}}{\overset{\overset{CH_3}{|}}{Si}} - CH_3$$

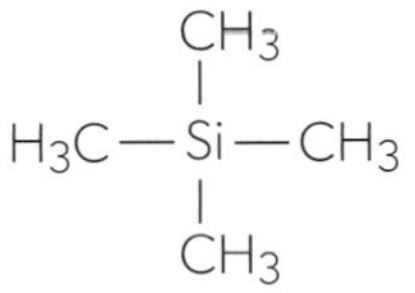

▲ Figure 31 TMS is the universally accepted reference standard used in ^1H NMR spectroscopy

The energy difference, ΔE, between the spin states increases along with the external magnetic field strength. For the nucleus to be promoted from the lower energy state to the higher energy state, it must be exposed to radio waves of energy equal to the difference between the two spin states. The exact energy required for this process depends on the chemical environment of hydrogen atoms in the analysed molecule. So, hydrogen atoms in different chemical environments absorb electromagnetic radiation of different frequencies.

The frequencies of radiation absorbed by different hydrogen environments are displayed in a ^1H **NMR spectrum** as **signals**. These signals are given in terms of their **chemical shift**, δ, expressed in units of parts per million (ppm). Chemical shift is measured using a relative scale, where the signal produced by a tetramethylsilane (TMS) standard is assigned a chemical shift of 0 ppm. TMS is a symmetrical molecule with four equivalent methyl groups (figure 31).

A summary of chemical shift values for different hydrogen environments is given in section 21 of the data booklet.

Low-resolution ^1H NMR

When examining a low resolution ^1H NMR spectrum, the area under each signal is proportional to the number of ^1H nuclei generating this signal. The area under a signal is small and difficult to measure, so an **integration trace** is added to the spectrum to indicate the relative number of ^1H nuclei represented by each signal.

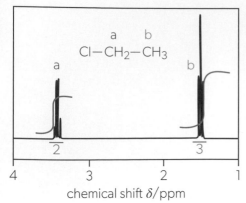

▲ Figure 32 The ¹H NMR spectrum of chloroethane, C_2H_5Cl

Consider the ¹H NMR spectrum of chloroethane, shown in figure 32. There are two signals in the spectrum indicating two hydrogen environments, one with an integration trace of 2, and one with an integration trace of 3. These signals therefore correspond to the CH_2 and CH_3 groups in chloroethane, respectively.

Figure 33 shows the ¹H NMR spectrum of 2-bromopropane. There are three groups with hydrogen atoms in the molecule, but only two signals in the spectrum. This is because the molecule is symmetrical, so the two methyl groups are identical and therefore form a single hydrogen environment. The signal with an integration trace of 6 corresponds to these methyl groups, as they have six hydrogen atoms. The other signal corresponds to the CH group.

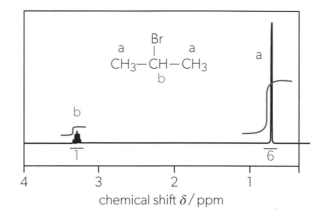

▲ Figure 33 The ¹H NMR spectrum of 2-bromopropane, C_3H_7Br

Sometimes integration traces are given as a ratio. For instance, imagine a ¹H NMR spectrum with two signals where the area under one signal is equal to 1, and the area under the other is equal to 2. These correspond to the relative number of hydrogen atoms in each environment, so the actual numbers of hydrogen atoms in different environments could be 2 and 4, or 3 and 6, or any other values in a 1:2 ratio.

TOK

The interactions between particles and different types of electromagnetic radiation underpin many of the spectroscopic methods we have looked at. These interactions result in nuclear spin changes (NMR) or bond stretching and bending (IR). When electromagnetic radiation of a certain wavelength encounters a molecule, the radiation is scattered.

Occasionally, the wavelengths of incident and scattered radiation are not the same. This phenomenon is known as "Raman scattering" after Sir Chandrasekhara Venkata Raman who was awarded the 1930 Nobel Physics Prize for his work in this area.

The Raman scattering effect formed the basis of Raman spectroscopy, which is nowadays used to determine structural features of various materials. Evidence for Raman scattering was very difficult to obtain at first. Eventually, with ongoing experiments, Raman and his team arrived at conclusive evidence for the existence of the effect.

How much evidence is required in order to support a robust conclusion?

▲ Figure 34 C. V. Raman, pictured here, discovered the Raman effect with his coworker, Sir Kariamanikkam Srinivasa Krishnan

Worked example 9

Sketch the ^{1}H NMR spectra for the following molecules:

a. propanone, $CH_3C(O)CH_3$
b. ethanoic acid, CH_3COOH

Solution

a. Propanone, $CH_3C(O)CH_3$, is a symmetrical ketone. The structural formula shows a carbonyl functional group and two methyl groups. The chemical environment for all the hydrogen atoms is identical:

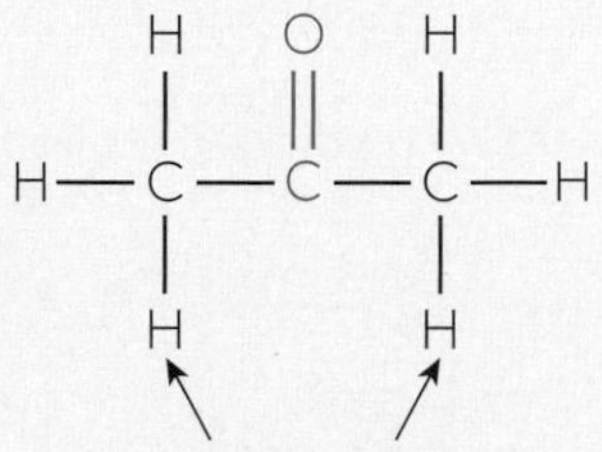

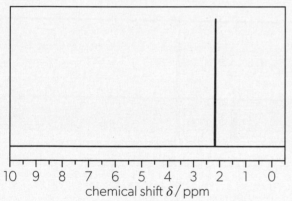

identical chemical
environments

The equivalent protons in propanone will produce one signal in the range of 2.2–2.7 ppm (data booklet, section 21). Integration traces are not useful here because there is only one hydrogen environment. A sketch of the ^{1}H NMR spectrum is shown below:

b. Ethanoic acid, CH_3COOH, has a total of four protons in two different chemical environments.

The proton of the carboxyl group will produce a signal in the region of 9.0–13.0 ppm, and the methyl group will produce a signal in the region of 2.0–2.5 ppm. The areas under the signals should reflect the ratio of protons in the two environments, which is 1:3. A sketch of the ^{1}H NMR spectrum is shown below:

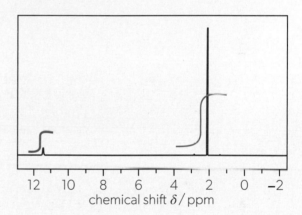

Practice questions

21. The structural formulas of two organic compounds are shown below.

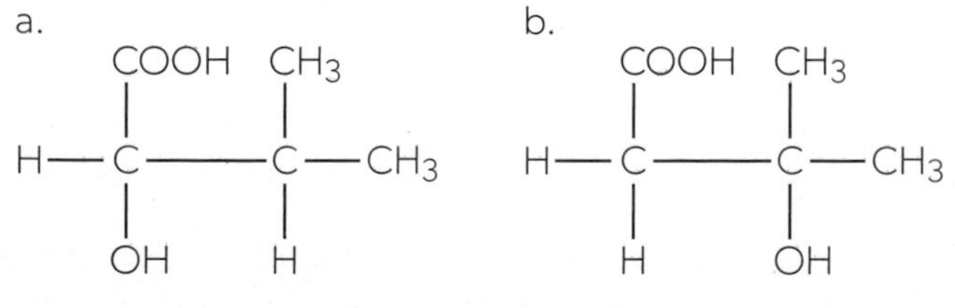

a.

$$COOH \quad CH_3$$
$$H-C-----C-CH_3$$
$$OH \quad\quad H$$

b.

$$COOH \quad CH_3$$
$$H-C-----C-CH_3$$
$$H \quad\quad OH$$

Deduce the number of signals and the ratio of areas under the signals in the ^{1}H NMR spectra of these compounds.

High-resolution ¹H NMR

Advances in technology have led to improvements in the accuracy of analytical instruments like ¹H NMR. There are distinct differences between low resolution and high resolution ¹H NMR spectra and the information that you can extract from them.

We have already demonstrated that low resolution ¹H NMR provides us with information about the chemical environment of hydrogen atoms in a molecule. The number of signals corresponds to the number of different chemical environments, while the integration trace shows the number of hydrogen atoms associated with each signal. In a high resolution ¹H NMR spectrum, individual signals are split into clusters of peaks. The splitting pattern of a signal provides us with information about the number of hydrogen atoms attached to neighbouring atoms in a molecule.

Consider 1,1,2-trichloroethane, which has two hydrogen environments: H_a and H_b:

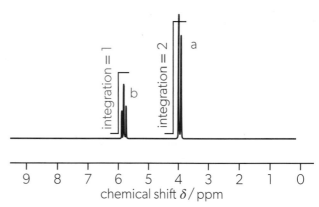

There are two signals in the spectrum of 1,1,2-trichloroethane (figure 35). The signal for the H_a atoms is split into a cluster of two peaks and is referred to as a **doublet**. The signal for the H_b atom is split into a cluster of three peaks and is referred to as a **triplet**.

▲ Figure 35 The high resolution ¹H NMR spectrum of 1,1,2-trichloroethane

The different splitting patterns of the signals for the hydrogen environments are caused by the spin–spin interactions of neighbouring hydrogen nuclei. As discussed previously, each proton has nuclear spin that can either be aligned with or against the external magnetic field. If hydrogen atoms in a molecule are in close proximity, the magnetic fields generated by their nuclei can affect each other, resulting in NMR signal splitting. This is known as **spin–spin coupling**.

The number of peaks in a cluster for an individual signal, known as **multiplicity**, can be calculated by using the $N + 1$ rule:

If the proton H_x has N protons attached to its nearest neighbours, the NMR signal of H_x will be split into $N + 1$ peaks.

Examples are given in table 16. The ratio of the heights of the peaks in a cluster, the **peak ratio**, is deduced using Pascal's triangle.

Number of neighbouring hydrogen atoms (N)	Number of peaks in cluster ($N + 1$)	Splitting pattern	Peak ratio
0	1	singlet (s)	1
1	2	doublet (d)	1:1
2	3	triplet (t)	1:2:1
3	4	quartet (q)	1:3:3:1

▲ Table 16 Splitting patterns due to neighbouring hydrogen atoms in ^{1}H NMR spectra. Splitting patterns of five and more peaks are often called "multiplets"

For example, in 1,1,2-trichloroethane, the spins of the nuclei in the H_a environment will couple with the spins in the adjacent H_b environment. The H_b environment has one proton, and therefore the H_a signal will be split into a doublet (d) as per the $N + 1$ rule. The heights of the two peaks will be equal. The H_a environment has two protons, and therefore the H_b signal will be split into a triplet (t), and the peak ratio will be 1:2:1. The magnetically equivalent H_a nuclei do not couple with each other.

TOK

The splitting patterns observed in high-resolution ^{1}H NMR spectra correspond to the numbers in a mathematical pattern known as Pascal's triangle (figure 36). In each row of the triangle, adjacent numbers are added together to form the new row. Evidence exists that this pattern was known by ancient civilizations in different parts of the world including China and Persia.

What is the role of mathematics in scientific knowledge-building?

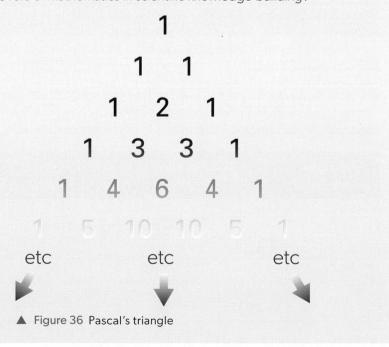

etc   etc   etc

▲ Figure 36 Pascal's triangle

Worked example 10

Show that the splitting patterns in the signals of the ^{1}H NMR spectrum in figure 37 correspond to the structure of butanone, $H_3CC(O)CH_2CH_3$.

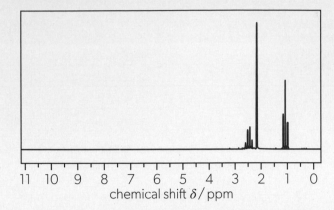

▲ Figure 37 High resolution ^{1}H NMR spectrum of butanone

Solution

First, draw the structure of butanone and deduce the number of hydrogen environments.

$$H_3C - \overset{\overset{\displaystyle O}{\|}}{C} - CH_2 - CH_3$$

A   B   C

There are three different hydrogen environments, which correspond to the three signals in the spectrum.

Hydrogen environment A is not adjacent to any other hydrogen environments, so $N = 0$. According to the $N + 1$ rule, this environment corresponds to the singlet in the spectrum.

Environment B is adjacent to environment C, which has three hydrogen atoms. Therefore, environment B corresponds to the quartet. The signal corresponding to environment C is split into a triplet by the two hydrogen atoms in environment B.

The peak ratio in each cluster corresponds to that predicted by Pascal's triangle. The chemical shifts of the signals for B and C also closely match the predicted values from the data booklet. This is summarized in table 17:

Hydrogen environment	$N + 1$ rule	Splitting pattern	Peak ratio	Predicted δ / ppm	Actual δ / ppm
A	$N = 0$ $0 + 1 = 1$	singlet (s)	1	2.2–2.7	2.1
B	$N = 3$ $3 + 1 = 4$	quartet (q)	1:3:3:1	2.2–2.7	2.4
C	$N = 2$ $2 + 1 = 3$	triplet (t)	1:2:1	0.9–1.0	1.1

▲ Table 17 You can display data like this when matching the structural details of a compound to signals in a ^{1}H NMR spectrum

Combining analytical techniques (*Structure 3.2.12*)

To determine the structure of a molecule, you may be required to interpret a wide variety of data. For example, different methods are needed to deduce the types of functional group present and the arrangement of the carbon atoms in a molecule. It is the combination of data that will enable you to deduce the unknown structure.

This data may include:

* combustion analysis data (to calculate empirical formula)

* mass spectroscopy (MS) data

* the infrared (IR) spectrum

* the ^1H NMR spectrum

* information from the *data booklet*

* description of reactions that the compound can undergo (covered in the *Reactivity* section of DP chemistry).

 Predict spectra from molecular models

In the skills task on page 283, you were asked to create molecular models of propan-1-ol, propan-2-ol and methoxyethane. Now you will examine these molecular models and consider the spectroscopic data of these substances.

Skills

* Tool 2: Identify and extract data from databases

Instructions

1. Create molecular models of propan-1-ol, propan-2-ol and methoxyethane using a molecular model kit, if you have not done so already.

2. Examine the molecular models and determine the number of different hydrogen environments in each molecule. Then, predict the ^1H NMR spectrum of each (consider the number of signals, area under each signal, splitting patterns and chemical shifts).

3. Predict the major IR spectrum signals expected for each compound.

4. For each compound, predict some of the peaks found in its mass spectrum.

5. Search for the spectra in a suitable database and see how the actual spectra compare to your predictions. Examples of databases include the AIST Spectral Database for Organic Compounds.

Worked example 11

Ethane reacts with chlorine to form product X. Combustion analysis was used to determine that product X has the following composition by mass: carbon 24.27%, hydrogen 4.08%, chlorine 71.65%.

a. Determine the empirical formula of product X.

b. The mass and ^{1}H NMR spectra of product X are shown below. Deduce, giving your reasons, its structural formula, and hence the name of the compound.

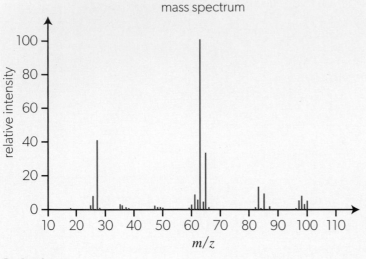

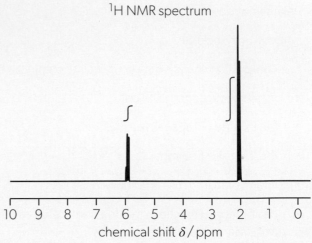

Solution

a. Divide the percentages by the relative atomic mass for each element.

carbon: $\dfrac{24.27}{12.01} = 2.021$ hydrogen: $\dfrac{4.08}{1.01} = 4.04$

chlorine: $\dfrac{71.65}{35.45} = 2.021$

The empirical formula is therefore $C_2H_4Cl_2$.

b. When analysing mass spectra of compounds containing chlorine, we need to consider that chlorine has two abundant isotopes of different relative atomic mass: chlorine-35 and chlorine-37. Therefore, the fragment ions of product X will contain a mixture of chlorine-35 and chlorine-37 atoms (table 18).

From the mass spectrum, we can deduce that the empirical formula is identical to the molecular formula because the average of the molecular ion peak m/z values is equal to the relative molecular mass of the empirical formula (99). We can also deduce that the following structural elements occur in the molecule: C_2H_4Cl and C_2H_4. Therefore, the structural formula of product X is likely to be CH_3CHCl_2. We can confirm this using ^{1}H NMR spectroscopy.

The ^{1}H NMR spectrum has two signals, indicating two different hydrogen environments.

The integration trace shows a ratio of 3:1, indicating there are three times as many hydrogen atoms in one environment than in the other.

The signal at 2.0 ppm is a doublet (H_A), indicating a neighbouring hydrogen environment with one proton. At 5.8 ppm there is a quartet (H_B), indicating a neighbouring hydrogen environment with three protons. We can therefore conclude that product X is 1,1-dichloroethane:

m/z	Molecular fragment	Reason
100	$[C_2H_4{}^{35}Cl^{37}Cl]^+$	molecular ion containing chlorine-37 and chlorine-35 isotopes
98	$[C_2H_4{}^{35}Cl_2]^+$	molecular ion containing two chlorine-35 isotopes
65	$[C_2H_4{}^{37}Cl]^+$	loss of chlorine-35 atom from molecular ion peak (m/z 100)
63	$[C_2H_4{}^{35}Cl]^+$	loss of chlorine-35 atom from molecular ion peak (m/z 98)
27	$[C_2H_3]^+$	loss of HCl and Cl

▲ Table 18 Fragment ions of product X

$$\begin{array}{ccc} H_A & & Cl \\ | & & | \\ H_A - C & - & C - H_B \\ | & & | \\ H_A & & Cl \end{array}$$

End of topic questions

Topic review

1. Using your knowledge from the *Structure 3.2* topic, answer the guiding question as fully as possible:

   How does the classification of organic molecules help us to predict their properties?

Exam-style questions

Multiple-choice questions

2. What is the name of the following compound, applying IUPAC rules?

   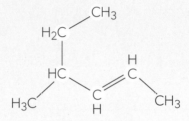

   A. 4-methylhex-2-ene

   B. 4-ethylpent-2-ene

   C. 2-ethylpent-3-ene

   D. 3-methylhex-4-ene

3. Which of the following is in the same homologous series as CH_3OCH_3?

   A. CH_3COCH_3

   B. CH_3COOCH_3

   C. $CH_3CH_2CH_2OH$

   D. $CH_3CH_2CH_2OCH_3$

4. Which of the following analytical techniques would show the difference between propan-2-ol, $CH_3CH(OH)CH_3$, and propanal, CH_3CH_2CHO?

   I. mass

   II. infrared

   III. ¹H NMR

   A. I and II only

   B. I and III only

   C. II and III only

   D. I, II and III

5. Which substance has the following ¹H NMR spectrum?

   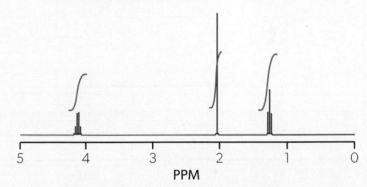

   PPM

   A. propane

   B. propanal

   C. ethyl ethanoate

   D. butanoic acid

Extended-response questions

6. a. State, giving a reason, if but-1-ene exhibits *cis–trans* isomerism. [1]

   b. The reaction which occurs between but-1-ene and hydrogen iodide at room temperature produces the halogenoalkane, 2-iodobutane. State, giving a reason, if the product of this reaction exhibits stereoisomerism. [1]

7. Organic chemistry can be used to synthesize a variety of products. Combustion analysis of an unknown organic compound indicated that it contained only carbon, hydrogen and oxygen.

a. Deduce two features of this molecule that can be obtained from the mass spectrum, namely m/z 58 and m/z 43. Use section 22 of the data booklet. [2]

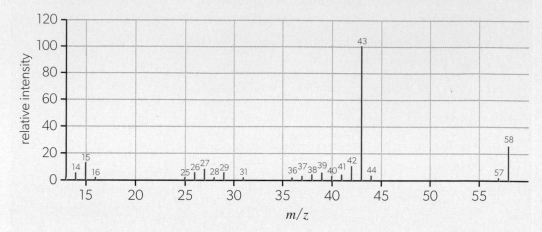

b. Identify the bond responsible for the absorption at **A** in the infrared spectrum. Use section 20 of the data booklet. [1]

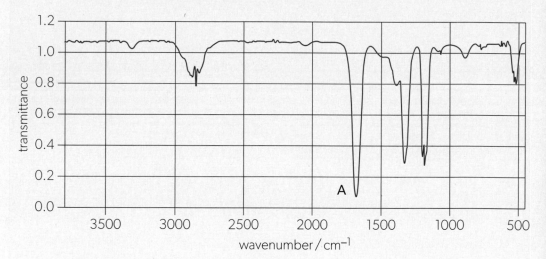

c. Deduce information from the ^{1}H NMR and then identify the unknown compound using the previous information, the ^{1}H NMR spectrum and section 21 of the data booklet. [3]

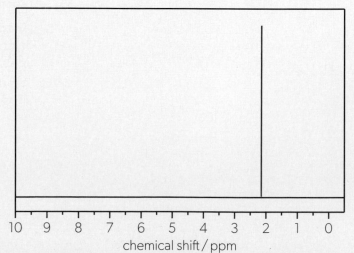

8. a. Draw the stereoisomers of butan-2-ol using wedge-dash type
        representations.   [1] **AHL**

    b. Outline how two enantiomers can be distinguished using a polarimeter. [2]

9. Compound **A** is in equilibrium with compound **B**.

    a. Identify the functional groups present in each molecule.   [2]

    b. Compound **A** and **B** are isomers. Draw two other structural isomers
        with the same formula.   [2]

    c. The IR spectrum of one of the compounds is shown:

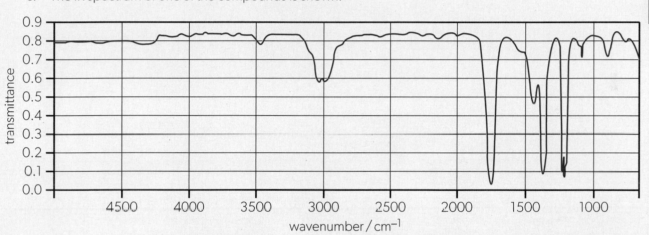

        Deduce, giving a reason, the compound producing this spectrum.   [1]

10. The mass spectrum of chlorine is shown.

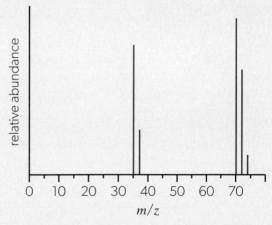

    a. Outline the reason for the two peaks at $m/z = 35$ and 37.   [1]

    b. Explain the presence and relative abundance of the peak at $m/z = 74$. [2]

11. a. Draw the structural formula of ethoxyethane.   [1]

    b. Deduce the number of signals and chemical shifts with splitting patterns in the
        ^{1}H NMR spectrum of ethoxyethane. Use section 21 of the data booklet. [3]

Tools for Chemistry

Introduction

The purpose of this section is to provide you with a toolkit containing key experimental, technological and mathematical skills that support scientific inquiry in DP chemistry. You should refer to this section frequently throughout the course to develop fluency with these skills and practise applying them in a variety of contexts. Such contexts might include skills tasks, practical work, lab-based questions, the collaborative sciences project, and your scientific investigation for internal assessment (IA).

▲ Figure 1 Al-Rāzī (c. 860–930) was an Islamic Persian scholar, physician and alchemist

Tool 1: Experimental techniques

Safety and respect of self, others and environment

Being principled is one of the learner profile attributes. This means that you act with integrity and honesty, with a strong sense of fairness and justice, and that you take responsibility for your actions and their consequences. In conducting experimental work, being principled involves lab safety and academic integrity.

Safety

Two of the twelve green chemistry principles relate to prevention: **accident prevention** and **waste prevention**.

Accident prevention is an integral and important part of labwork. It includes wearing protective gear, conducting risk assessments, putting the correct safety measures in place, and eliminating risks by opting for safer alternatives if possible. Safety considerations must protect the experimenter, the environment, as well as anybody else using the laboratory space.

Familiar personal protective equipment (PPE) includes safety goggles and lab coats for the people present during the experiment. The use of PPE is commonplace in schools and laboratories these days, but it has not always been the case. When looking at historical photographs or images of science experiments, see if you can identify any safety risks and discuss suitable control measures.

A risk assessment needs to be completed prior to carrying out experiments. Assessing risk involves identification of hazards, ways of minimizing risk, and what to do if something goes wrong. The main parts of a risk assessment are:

- Hazards: substances or activities that have the potential to cause harm (for example, a hot crucible).

- Risks: evaluating the severity and likelihood of harm, including who could get harmed and how (continuing with our example, the person doing the experiment could burn themselves if they touch the crucible. This is likely to happen if they need to measure the mass of the crucible and its contents).

- Control measures: what will be done to minimize the risk and, if possible, remove the hazards (the experimenter should wait 5 minutes for the crucible to cool before placing it on the balance, and pick it up using tongs. The balance should be located near the experimenter's lab bench).

Safety is not only relevant during the experiment, but also after its completion. Chemists need to think about the products of their experiments and identify suitable disposal procedures. The disposal of chemicals needs to be done in accordance with local rules and policies in order to protect the environment, as well as people involved in the disposal process. If possible, left-over reagents should be recovered for future use.

▲ Figure 2 A selection of hazard symbols that you are likely to see on reagent bottles in a chemistry lab. Do you know what each of them means?

ATL Research skills

Research the 12 principles of green chemistry and try to complete the blanks in figure 3.

▲ **Figure 3** This comic strip shows two people discussing green chemistry. Think of examples to fill in the blanks

Experiments should involve using the smallest possible amounts of reagents. Doing so further minimizes the environmental and economic impact of practical work.

Academic integrity

Respecting others includes respecting their ideas and giving them credit by citing information sources. You must clearly distinguish between your own words and those of others using quotation marks followed by an appropriate citation. In-text citations are also required when you paraphrase another person's ideas. Full references should then be included in a section at the end of the document (usually titled bibliography, works cited or references). The IB does not mandate a particular referencing style, but it is expected that the minimum information included are the name of the author, date of publication, title of source, page numbers and URL, if applicable.

Practice questions

1. State which hazard symbols are you likely to find on a bottle containing:

   a. concentrated nitric acid

   b. 0.10 mol dm^{-3} sulfuric acid

   c. solid copper(II) sulfate

   d. solid sodium metal

   e. cyclohexane.

2. Find out what referencing system is used at your school, then write a full reference for this book.

3. What are some of the hazards associated with the two reagents in figure 4?

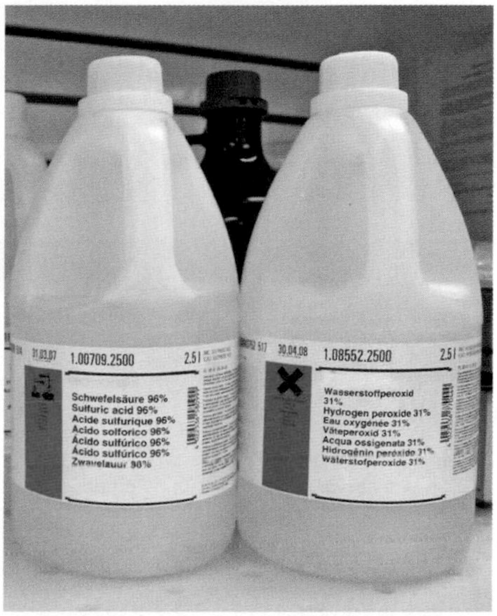

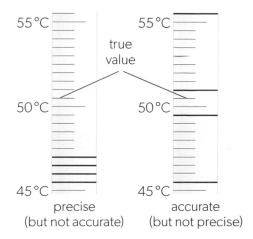

▲ Figure 4 Bottles containing concentrated solutions of sulfuric acid and hydrogen peroxide

Measuring variables

Handling data is a vital skill of scientific research. Qualitative data include all non-numerical information not obtained from measurement, such as colour changes or bubbling. Quantitative data are numerical information obtained from measurement. Measurement is a core aspect of science and is important to scientists because it provides quantitative data. One of the factors affecting data quality is the type of instrument used to obtain these data, as well as whether it was used correctly or not.

Accuracy and precision

Measured values (and processed values) can be evaluated in terms of their precision and accuracy (figure 5). Repeating measurements under the same conditions allows us to assess their **precision**: precise values are close to each other. Therefore, if a high-precision instrument is used to determine a particular measurement several times, the measured values will occur over a small range. **Accuracy** refers to how close a measured or processed value is to the true value.

55°C — 55°C
true value
50°C — 50°C
45°C — 45°C
precise (but not accurate) — accurate (but not precise)

▲ Figure 5 Precise values are close to each other. The mean of accurate values is close to the true value

311

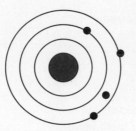   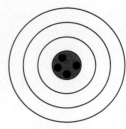

low precision and low accuracy | low precision and high accuracy | high precision and low accuracy | high precision and high accuracy

▲ **Figure 6** The difference between accuracy and precision. The bullseye represents the true value

Data-based question

After performing a calorimetry experiment, a group of students determined the enthalpy of combustion of ethanol. Their results are shown below.

Trial	1	2	3	Mean
ΔH / kJ mol^{-1} ± 5 kJ mol^{-1}	−705	−707	−701	−704

Ethanol's standard enthalpy of combustion is reported in the data booklet as −1 367 kJ mol^{-1}.

How would you describe these results in terms of accuracy and precision?

A. high accuracy, low precision

B. low accuracy, high precision

C. high precision, high accuracy

D. low precision, low accuracy

ATL Thinking skills

In a calorimetry experiment, for example, the assumption that the specific heat capacity of a dilute solution is equal to that of water does not invalidate the analysis because dilute solutions are mostly composed of water (*Reactivity 1.1*). Can you think of other reasons why this is usually a valid assumption? Can you think of any situations in which this assumption would not be valid?

Validity and reliability

The quality of measurements, and indeed of entire pieces of research, is often assessed in terms of validity and reliability.

Validity is concerned with whether something measures what it is meant to measure. Measuring pH with an improperly calibrated pH probe will result in an invalid measurement. A methodology or analysis is valid when it addresses the research question appropriately.

Reliability is connected to whether doing the same investigation again produces the same results. Reliability can be assessed in terms of **repeatability**: the agreement of results when trials are repeated using identical methods, carried out under the same conditions and by the same experimenter.

Repeating trials during an experiment is good practice. When you do a school practical, you can compare the results of different trials to assess their repeatability. Conducting several trials allows you to calculate a mean, assess precision and identify any clear outliers. The more repeats, the more reliable the results, but doing three trials is usually a good rule of thumb.

In some cases (certain statistical tests for example) greater numbers of replicates are required. In others (such as titrations), measurements should be repeated until several **concordant** values are obtained. Concordance means that the values agree with each other, and in titrations this generally means a range of less than 0.1 cm³ (although this value is not set in stone, and it is best to specify it when writing up lab investigations).

Reproducibility is the consistency of results obtained by following the same method and conditions but carried out by different people, instruments and reagent sources. Comparing the outcomes of an experiment to those obtained by different research teams gives scientists insight into the reproducibility of their results. One of the reasons scientists include detailed experimental methods when reporting findings is to allow other scientists to reproduce these findings.

Next, we will consider several key types of measurements in chemistry: mass, volume, time, temperature, length, pH, electric current and electric potential difference.

Mass

Different balances have different levels of precision. In a chemistry lab, the greater the precision of a balance, the greater the number of decimal places it will measure mass to. When writing down a value, do not leave out any trailing zeroes. For example, a mass recorded to four decimal places (e.g. 0.5000 g) was measured with a more precise balance than a mass recorded to only one decimal place (e.g. 0.5 g).

Gravimetric analyses involve the measurement of mass changes during a chemical reaction. For example, the reaction mixture that produces a gas will decrease in mass over time because the mass of the escaped gas will be lost. Another example is the mass increase that happens when magnesium burns in air. The mass of the reaction mixture before and after the reaction provides information on how much gas was produced or absorbed. It is worth noting that not all containers, however similar they may seem, have the same mass. Remember to record the mass of the empty container when conducting a gravimetric analysis.

Practice questions

4. Explain why measurements taken with a pH probe are more precise than pH values determined using pH paper.

5. Can experimental outcomes be reliable but not valid?

▲ **Figure 7** A digital balance. Why is there a number on the display even though there is nothing on top of the balance?

🔵 Data-based question

A student wishing to determine the enthalpy of hydration of copper(II) sulfate collected the following data. A mistake was made when the mass was being measured. Identify the mistake and suggest what might have caused it:

	Mass / g ± 0.01 g
Mass of empty crucible	13.55
Mass of crucible and hydrated copper(II) sulfate	11.96
Mass of crucible and anhydrous copper(II) sulfate	11.60

Practice questions

6. Which of the following mass measurements was obtained using the most precise balance?

    A. 0.0007 g

    B. 2.0 g

    C. 2.030 g

    D. 100.10 g

7. Is the feather shown in the photo really massless?

Volume

The suitability of a particular type of volume-measuring equipment depends on its intended function and the desired level of precision. Measuring cylinders designed to measure approximate volume are not very precise.

Volumetric analyses involve the measurement of volume changes during a chemical reaction. For volumetric analyses, a pipette, burette and/or volumetric flasks are needed. Beakers and conical flasks give an approximate measurement of the volume and are designed for holding liquids, not measuring volume.

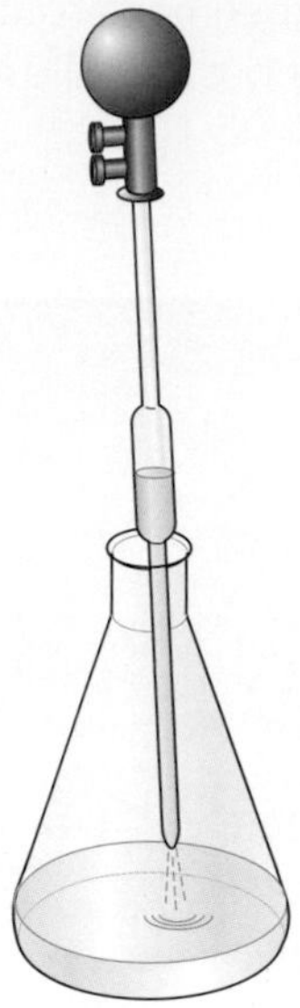

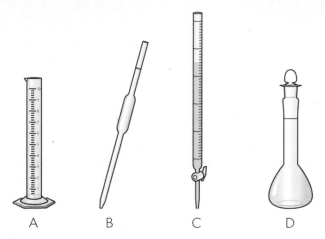

▲ **Figure 8** A liquid being released from a volumetric pipette into a conical flask

▲ **Figure 9** Various volume-measurement devices. A – measuring cylinder, B – volumetric pipette, C – burette, D – volumetric flask

Many liquids, including aqueous solutions, form a U-shaped **meniscus** inside a narrow vessel, such as a burette or the neck of a volumetric flask. Volume measurements should be read from the bottom of the meniscus (figure 10).

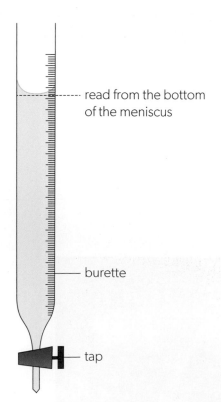

read from the bottom of the meniscus

burette

tap

▲ **Figure 10** Water and aqueous solutions form a U-shaped meniscus

▲ **Figure 11** Mercury forms an upside-down meniscus, where the centre bulges upwards above the sides

Practice questions

8. Which of the following are appropriate for precise volume measurements?

    • volumetric pipette

    • volumetric flask

    • burette

    • beaker

    • conical flask

    • measuring cylinder

9. What glass container would you use to:

    a. determine the volume of liquid added during a titration

    b. prepare a standard solution from a solid reagent

    c. carry the acid you need for a pH experiment from the reagent bottle to your lab bench

    d. approximately measure 25 cm³ of liquid?

10. What is the volume of liquid contained in the measuring cylinder in figure 12?

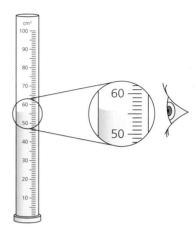

▲ **Figure 12** The meniscus should be read at eye-level

Time

The SI unit for time is the second (s). When measuring the rate of a chemical reaction, for example, you might use a stopwatch to record how long it takes, in seconds, for a certain volume of gas to be produced. The uncertainty associated with time measurements usually depends on the context in which time is being measured. The uncertainty of the stopwatch itself is often far smaller than the uncertainty arising from the experimenter's reaction time.

For example, if you are recording how long it takes for a clock reaction to produce a change in colour, there will be a short delay between the instant the colour changes and your finger stopping the stopwatch (figure 13). The stopwatch display might indicate a time of 16.45 seconds, which might suggest an uncertainty of ± 0.01 s indicated by the two-decimal-places display (figure 13). A more realistic uncertainty, given your reaction time, would be around 0.5 to 1 second.

▲ **Figure 13** In the "iodine clock" reaction, the clear and colourless reaction mixture suddenly turns blue-black after a certain period of time

> (ATL) **Thinking skills**
>
> Some experiments take place over long periods of time, such as hours, days, or even weeks. In these cases, is it always necessary to record times to the nearest second?

Temperature

You can use digital or analogue thermometers to measure temperature. The required accuracy and precision of a temperature measurement is highly dependent on the context of the experiment.

Continuous monitoring of temperature can be achieved by connecting a temperature probe to a data logger as shown in figure 14.

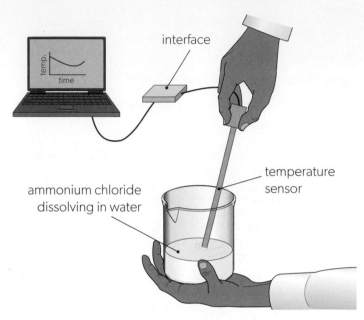

interface

temperature sensor

ammonium chloride dissolving in water

▲ **Figure 14** Temperature sensors and data loggers can be used to monitor the temperature changes over time

Length

Significant figures are discussed in detail in *Tool 3: Mathematics*.

Length measurements usually involve rulers although other devices, such as calipers, can also be used. As with other analogue devices, reading the scale usually involves estimating the last digit in the measurement, depending on its location between two scale divisions. Scale resolution will therefore have an impact on the number of significant figures in the measurement (figure 15).

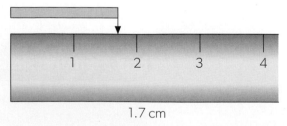

1.7 cm

This ruler gives a reading to two significant figures

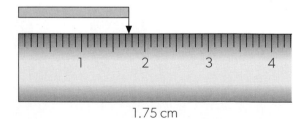

1.75 cm

This ruler gives a reading to three significant figures

▲ **Figure 15** The relationship between scale resolution and significant figures

Practice questions

11. A student conducted three trials of an iodine clock reaction. The reagent concentrations and temperatures were the same in all three trials. The results obtained, in minutes, seconds and milliseconds (min:s.ms) are shown below.

    a. State the times to the nearest second.

    b. Suggest a suitable uncertainty for the time measurements, in seconds.

    c. Calculate the mean time, in seconds.

| Trial | Time taken for the colour change to occur | | |
	/ min:s.ms ± 1 ms	/ s	Mean / s
1	01:15.120		
2	01:10.400		
3	01:12.985		

12. Identify and explain the limitations of each of the following temperature measurement scenarios and propose an alternative method.

    a. A student heats $5 \, cm^3$ of water in a beaker and checks the temperature by placing a thermometer in the water, with the bulb touching the bottom of the beaker.

    b. A student wishing to perform a reaction at 2 °C cools one of the reagents down to this temperature in a beaker, removes the beaker from the ice bath, and adds the second reagent.

    c. In a study of an enzyme action at 40 °C, a student places a test tube containing a sample of the enzyme solution into a 40 °C water bath. The student then immediately adds the substrate to the enzyme solution.

13. What is the length of the orange bar?

14. Figure 16 shows a device called a caliper. What chemistry-related measurements might you obtain using a caliper?

▲ Figure 16 A vernier caliper

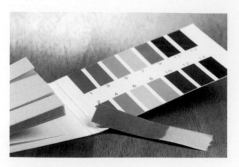

▲ Figure 17 pH paper turns different colours depending on the pH of the substance being tested

pH of solution

The pH of a solution is related to the concentration of hydrogen ions in that solution. pH can be estimated with indicators, such as universal indicator, or the anthocyanins extracted from red cabbage. These indicators vary in colour across the entire pH scale. The colour shown is compared to a key, allowing a pH value to be estimated (figure 17).

The pH of a solution can be more reliably determined with a properly calibrated pH probe (figure 18).

▲ Figure 18 A pH probe reading compared to the colour of the universal indicator

Using a natural acid–base indicator

In this task, you will use the juice of red cabbage as a natural acid–base indicator for determining the pH of household products.

Red cabbage juice is rich in **anthocyanins** (HA), which are amphiprotic organic compounds. In acidic solutions with low pH, anthocyanins accept protons and form cations, H_2A^+. In basic solutions with high pH, anthocyanins lose protons and form anions, A^-:

$$H_2A^+ \underset{+H^+}{\overset{-H^+}{\rightleftharpoons}} HA \underset{+H^+}{\overset{-H^+}{\rightleftharpoons}} A^-$$

cationic form   neutral form   anionic form
(red)   (blue)   (green to yellow)

All three forms (cationic, neutral and anionic) of anthocyanins absorb visible light at different wavelengths (*Structure 1.3*), producing the characteristic colour changes of red cabbage juice.

Relevant skills
- Tool 1: Measuring variables
- Inquiry 1: Designing

Safety
- Wear gloves and eye protection
- Do not use products with a corrosive safety warning

Materials
- red cabbage
- two liquid household products, such as vinegar or liquid detergent
- two solid household products, such as baking soda or antacid tablets
- five clear glasses
- cloth or filter paper
- standard laboratory glassware

Method
1. Prepare the indicator by blending several leaves of red cabbage with a glass of water and filter the resulting mixture through cloth or paper.

2. To test one of your liquid household products, pour 20–30 cm³ of the filtered cabbage juice into a clear glass and add 2–3 cm³ of the analysed solution.

3. To test one of your solid household products, crush the solid and dissolve it in water, before mixing with the cabbage juice.

4. Compare the colours of your solutions with those shown in figure 19 to determine their approximate pH.

▲ Figure 19 Red cabbage juice at pH 1 (left) to 10 (right)

Practice questions

15. What are the pH values of each of the substances tested in figures 17 and 18? What might a suitable uncertainty be in each case?

16. Outline one advantage and one disadvantage of

   a. pH indicators

   b. pH probes.

Electric current

Electric current is the rate of flow of electrical charge. It is measured in amperes (A), using an ammeter. To measure current, the ammeter must be connected in series in the circuit. The simplest type of circuit is one where all the components are connected one after another, following a single path, with no branches. Current is the same throughout a series circuit, so the ammeter can be placed anywhere in the circuit.

Figure 20 shows how to connect solid and liquid samples in a series circuit powered by a battery pack. Inserting a solid into a circuit is usually straightforward: you simply connect the leads to either end of the solid. Inserting a conductive liquid or solution (collectively known as electrolytes) requires electrodes made of inert materials, such as graphite or platinum. Care must be taken to ensure that the electrodes do not touch each other.

Current depends on factors such as electrode surface area, distance between the electrodes, voltage applied and the materials involved. These factors all affect the opposition to the flow of charge within the circuit, known as resistance.

Electrical conductivity, often given in microsiemens per centimetre, $\mu S\,cm^{-1}$, is a related property, which is the ability to conduct charge and which can be measured directly with a probe. Electrical conductivity measurements can be used to monitor the progress of a reaction in which ions are consumed or produced because conductivity of a solution depends on the concentration and identity of mobile ions present.

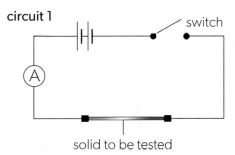

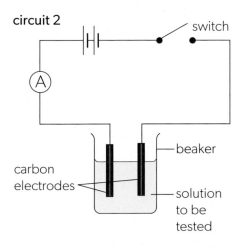

▲ Figure 20 Ammeters are connected in series. Circuit 1 shows how to connect a solid sample while circuit 2 shows how to connect a solution

▲ **Figure 21** Voltmeters are connected in parallel to a component to measure its potential difference (in this case, the potential difference of the lamp is being measured)

Electric potential difference

Potential difference is the energy difference per unit charge between two points in a circuit. It is expressed in volts (V) and measured using a voltmeter. Voltmeters measure the potential difference across circuit components and are therefore connected in parallel to the component being measured (figure 21).

In chemistry, you will most likely encounter potential difference measurements in the context of **voltaic cells**. The potential difference between the anode and cathode is known as cell potential. Measuring cell potential involves connecting the voltmeter to the two half-cells and completing the circuit with a salt bridge as shown in figure 22.

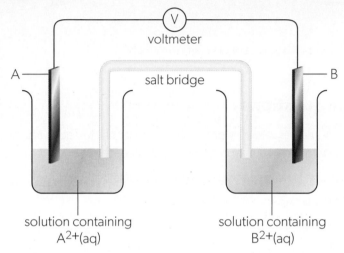

▲ **Figure 22** The potential difference across the two half-cells is measured by connecting a voltmeter to the electrodes. A and B represent metals

Another term you might come across is electromotive force (emf), which is the maximum value of potential difference across a cell when no current is flowing. This is the reason why high-resistance voltmeters are used to measure cell potential. Their high resistance means that virtually no current flows through them, hence the cell potential measured is close to the emf.

Practice questions

17. Outline how current passes through solid materials, such as a copper wire or graphite electrode.

18. Outline how current passes through electrolytes, such as molten lead(II) bromide or aqueous sodium chloride.

Practice questions

19. Explain why the following mixtures undergo a change in conductivity as the reaction proceeds.

    a. $HCl(aq) + NaOH(aq) \rightarrow NaCl(aq) + H_2O(l)$

    b. $AgNO_3(aq) + NaBr(aq) \rightarrow AgBr(s) + NaNO_3(aq)$

20. In this section you have considered how to measure key types of variables: mass, volume, time, temperature, pH, current and potential difference. For each variable, think of an experiment you have done in the past which involved measuring that variable, and state the instrument used.

Applying experimental techniques

In this section, we will discuss experimental techniques for preparing, isolating and analysing chemical species and phenomena.

Preparation

Preparation of a standard solution

Preparation and dilution of standard solutions requires measuring masses and volumes with high precision. This is achieved by using an analytical balance and **volumetric glassware**, such as burettes, pipettes and volumetric flasks (figure 23).

A standard solution is usually prepared as follows:

1. The solid (or sometimes liquid) solute is weighed in a clean and dry beaker using an analytical balance.

2. A small volume of deionized water is added to the beaker, and the mixture is stirred with a glass rod until the solid dissolves completely.

3. The solution is transferred to a volumetric flask using a glass funnel.

4. The beaker, glass rod and funnel are rinsed three times with deionized water, each time adding this water into the volumetric flask to ensure that the solute is transferred completely.

5. Deionized water is added to the volumetric flask until its level reaches the **graduation mark** on the flask.

6. The flask is stoppered and turned over at least ten times to ensure that the solution is mixed thoroughly.

7. The solution is transferred to a reagent bottle with a label showing the formula or name of the substance, its concentration, the date and initials of the person who prepared the solution, as well as any hazard labels, if relevant.

▲ Figure 23 Volumetric flask and pipette

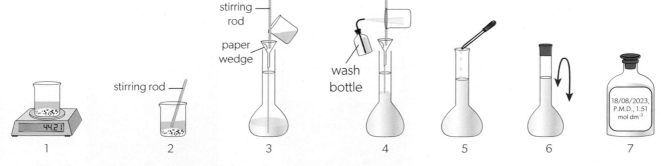

▲ Figure 24 Preparation of a standard solution

Measuring cylinders and beakers with graduation marks should never be used for preparing or diluting standard solutions, as they do not allow measurement of volumes with sufficient precision. Similarly, the balance must show the mass of a sample with at least four significant figures.

Practice questions

21. These steps outline how to prepare a standard solution. List them in the correct order.

    A. Add deionized water up to the mark on the volumetric flask.

    B. Dissolve the solute in a small amount of deionized water in a beaker.

    C. Transfer the solution to a labelled reagent bottle for storage.

    D. Pour the solution into a volumetric flask.

    E. Weigh the desired mass of the solute.

    F. Rinse the beaker, funnel and stirring rod with deionized water over the volumetric flask.

    G. Stopper and invert the flask several times.

22. A student prepares $100\,cm^3$ of a $0.50\,mol\,dm^{-3}$ solution of copper(II) sulfate by carrying out the following steps. Identify the mistakes.

    A. A sample of $7.98\,g$ of hydrated copper(II) sulfate was weighed and transferred into a volumetric flask using a funnel.

    B. Tap water was added to the volumetric flask until it was close to the mark on the neck of the flask. At this point water was added dropwise. Too much water was accidentally added, so the meniscus went above the mark. Excess water was carefully removed until the meniscus was back at the mark.

    C. The flask was stoppered and inverted twice.

    D. A label was placed on the volumetric flask for storage.

23. Sodium hydroxide is hygroscopic, meaning that it absorbs air moisture. Suggest the implications of this property for the preparation of standard solutions of sodium hydroxide.

Preparation of solutions by dilution

To dilute a standard solution, a certain volume of the solution is measured using a burette or volumetric pipette and transferred to a volumetric flask. Deionized water is added to the graduation mark, and the stoppered flask is turned over several times to achieve a uniform composition of the final solution.

bottle A   bottle B

▲ **Figure 25** The solution in bottle B can be prepared by performing a simple dilution of the stock solution in bottle A

Worked example 1

A teacher wishes to prepare $0.500\,dm^3$ of $0.400\,mol\,dm^{-3}$ HCl for a class practical. A $11.0\,mol\,dm^{-3}$ HCl stock solution is available. Determine the volume of stock solution required to prepare the desired solution.

Solution

We know that:

$$c_1 \times V_1 = c_2 \times V_2$$

where 1 indicates the stock solution and 2 indicates the desired solution.

Substituting the known values gives:

$$(11.0\,mol\,dm^{-3}) \times V_1 = (0.400\,mol\,dm^{-3}) \times (0.500\,dm^3)$$

Solving for V_1 gives:

$$V_1 = \frac{0.400\,mol\,dm^{-3} \times 0.500\,dm^3}{11.0\,mol\,dm^{-3}}$$

$$V_1 = 0.01818182...\,dm^3$$

Convert to cm^3, which are more easily measured in a lab, and round to three significant figures:

$$V_1 = 18.1818182...\,cm^3 \approx 18.2\,cm^3\ (3\ sf)$$

Serial dilutions are used to accurately prepare a series of solutions of increasingly lower concentration. Solutions for calibration curves are often prepared using this method, as extremely low concentrations are often required. Figure 26 shows how to carry out a serial dilution where the solution in each successive tube is ten times more dilute than that in the previous one. The solution in each test tube is prepared by transferring $1\,cm^3$ of the preceding solution and adding $9\,cm^3$ of deionized water. This procedure is repeated until the desired final concentration is obtained.

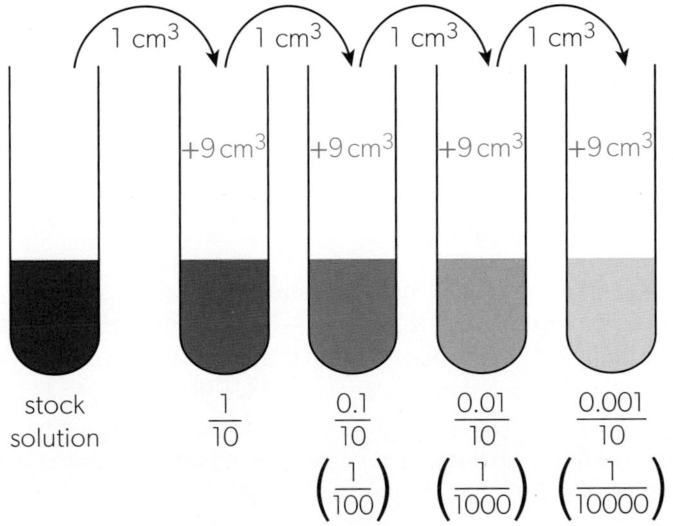

Calibration curves are introduced in *Structure 1.4*.

Practice questions

24. A technician wishes to prepare the following solutions by diluting a $5.00\,mol\,dm^{-3}$ stock HCl solution. What volume of the stock solution is needed in each case?

    a. $750\,cm^3$ of $1.00\,mol\,dm^{-3}$ HCl (aq)

    b. $3.00\,dm^3$ of $0.150\,mol\,dm^{-3}$ HCl (aq)

25. Suppose that the concentration of the stock solution in figure 26 is $0.50\,mol\,dm^{-3}$. Determine the concentration, in $mol\,dm^{-3}$, of each of the diluted solutions.

▲ **Figure 26** A serial dilution

Reflux and distillation

Reflux and distillation are easily confused because they involve similar equipment (figure 27). However, these methods are used for very different purposes. Reflux is a method used to heat reaction mixtures, whereas distillation is a separation method. A piece of equipment known as a **condenser** is used in both cases. The condenser is essentially a tube within another tube. The outer tube is cooled by a continuous flow of cold water. As a result, the temperature of the gaseous substances in the inner tube decreases, causing them to condense. You will also notice that both techniques use anti-bumping granules to ensure smooth boiling. Next, we will discuss the differences between reflux and distillation.

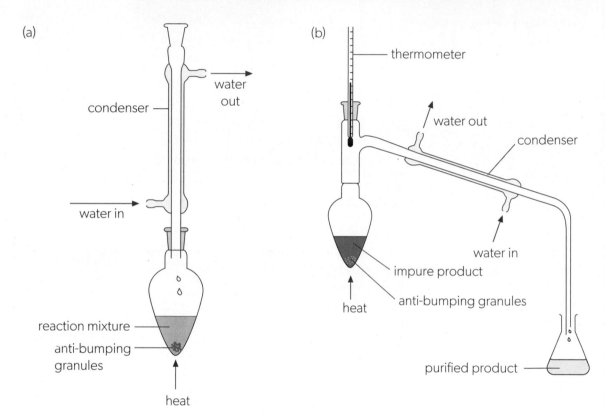

▲ **Figure 27** (a) Reflux apparatus (b) simple distillation apparatus

Reflux is used to minimize the loss of volatile substances during continuous heating of a reaction mixture. A condenser is placed above the reaction flask, so that any vapours rising from the reaction mixture condense and drip back down into the flask instead of escaping.

Distillation is used to separate the components of a mixture according to their boiling points. The mixture is heated until the boiling point of the more volatile component is reached. The substance with the lower boiling point rises into the still head, condenses in the condenser, and is collected at the bottom of the condenser. Remember that temperature stays constant during state changes, so the temperature shown on the thermometer will remain constant at the boiling point of whichever substance is being distilled off.

Figure 27(b) above depicts **simple distillation**, which separates a volatile liquid from a non-volatile solid (or two liquids with very different boiling points). For example, an aqueous solution of copper(II) sulfate can be separated into copper(II) sulfate and water using this method.

Fractional distillation is a related technique, used to separate mixtures of two or more miscible and volatile liquids with similar boiling points (figure 28), such as an ethanol–water mixture. A fractionating column is fitted to the flask. As the mixture is heated, the substances in the mixture continually vaporize, rise into the column, condense on the glass beads and drip back down into the distillation flask. The cycle repeats many times. The vapours contain a greater proportion of the lower-boiling-point component, which eventually reaches the top of the column and enters the condenser. The receiving flask is sometimes placed in an ice-water bath, particularly if the distillate is volatile. A fresh receiver is used to collect each distillate.

Practice questions

26. Compare and contrast reflux and simple distillation.

27. Compare and contrast simple and fractional distillation.

28. Primary alcohols such as ethanol are oxidized by acidified potassium dichromate, first to aldehydes. Eventually, and if there is enough oxidizing agent, the aldehydes are further oxidized to carboxylic acids. Identify and outline which technique, distillation or reflux, you would use if you were looking to obtain (a) an aldehyde and (b) a carboxylic acid.

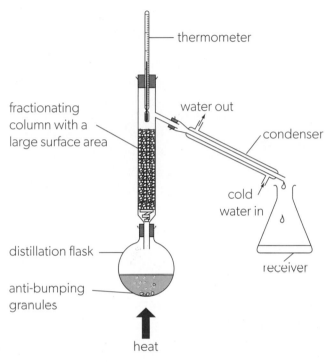

▲ Figure 28 Fractional distillation apparatus

Industrial-scale fractional distillation is used to separate mixtures such as crude oil (figure 29) and liquid air.

◄ Figure 29 An industrial fractional distillation facility, used to separate the components of crude oil

Isolation

Drying to a constant mass

When a solid is obtained in the lab, it may contain trace amounts of water, even if it looks dry. Solid samples are often **dried to a constant mass** to ensure that they do not contain any water (or other volatile impurities). Drying to a constant mass involves:

1. heating the sample

2. allowing it to cool for a few minutes

3. weighing it

4. repeating the process until two consecutive equal masses are obtained.

A suitable heating method should be chosen to ensure that the desired solid does not decompose. Common heating equipment includes Bunsen burners, hotplates and drying ovens.

Practice question

29. You can determine the water of hydration of a hydrated salt, such as magnesium sulfate, by heating the salt to remove the water and measuring the mass before and after heating:

$$MgSO_4 \bullet xH_2O(s) \xrightarrow{\text{heat}} xH_2O(g) + MgSO_4(s)$$

Explain why, when determining the water of hydration of magnesium sulfate, the magnesium sulfate should be heated to a constant mass.

Separation of mixtures

The method used to separate a mixture depends on the properties of the mixture components. For example, if one of the components is magnetic, it can be removed using a magnet. Mixtures of solids and liquids can be separated using filtration, crystallization or simple distillation. Mixtures of liquids can be separated with fractional distillation or using a piece of equipment known as a separating funnel.

Filtration separates particles according to their sizes by passing them through a medium, such as filter paper, which contains tiny holes or pores. Filtration is often used to separate insoluble solids from liquids. Liquid or aqueous particles are usually much smaller than solid particles, so they easily pass through the holes in the filter paper. The solid left behind in the filter paper is known as the **residue** and the liquid that passes through is the **filtrate**.

In **gravity filtration** (figure 30), as its name suggests, the filtrate passes through the paper due to gravity.

▲ **Figure 30** Gravity filtration

This process can be very slow, so it is sped up by **fluting** the filter paper. Fluting involves folding the filter paper into an accordion shape (figure 31) before placing it inside the funnel.

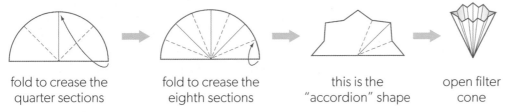

fold to crease the quarter sections → fold to crease the eighth sections → this is the "accordion" shape → open filter cone

▲ Figure 31 Figure showing the steps for making a fluted cone from circular filter paper

Vacuum filtration is quicker than gravity filtration, and it is often used when the residue is the desired component of the mixture. The mixture is loaded onto a sheet of moist filter paper laid flat on the perforated plate inside a Buchner funnel. Suction from a vacuum pump draws the liquid through the filter paper and into a receiving flask underneath (figure 32). Care must be taken not to overfill the funnel to prevent the solid from slipping under the filter paper. The residue is usually rinsed three or four times with the solvent to remove soluble impurities.

A solid can be **crystallized** out of a solution by removing the solvent. Most of the solvent is evaporated off by heating it over boiling water until the solution becomes saturated. The solution can be checked for saturation by spotting a drop of it onto a cold tile and watching for the formation of crystals as it cools. The bulk of the remaining solution is left to cool slowly to allow crystals to form. Finally, any remaining solvent is removed from the crystals by filtration (if necessary), and the crystals are allowed to dry to a constant mass on a watch glass.

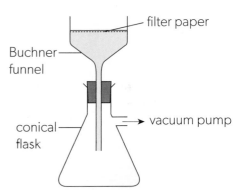

▲ Figure 32 Vacuum filtration apparatus

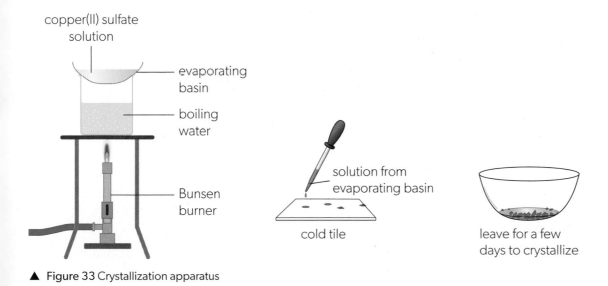

▲ Figure 33 Crystallization apparatus

A solvent can also be removed from a solute through simple distillation, or rotary evaporation, which is a type of distillation performed at low pressures to reduce the solvent's boiling point.

Practice questions

30. Figure 34 shows the preparation and isolation of salt crystals from the reaction between an acid and an insoluble base. Copper(II) sulfate crystals can be made this way, from the reaction between sulfuric acid and copper(II) oxide.

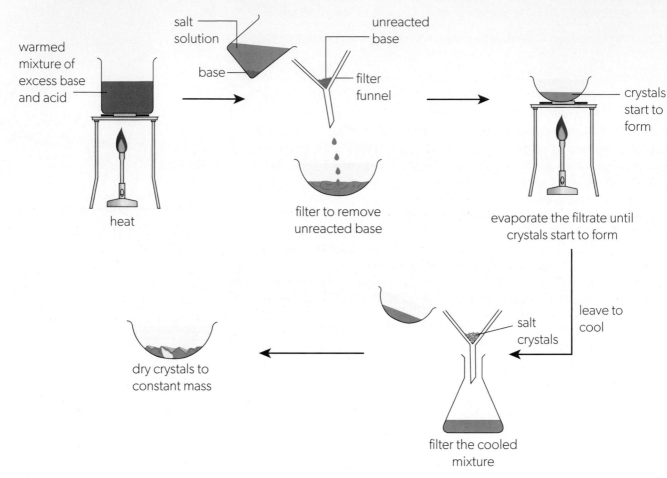

▲ **Figure 34** Summary of the preparation and isolation of an inorganic salt by reacting an insoluble base with an acid

a. Identify what is removed in the first filtration step.

b. Sometimes solvents are evaporated by heating the evaporating dish directly as shown. Discuss the advantages and disadvantages of doing so, compared with heating the evaporating dish over a hot water bath as discussed earlier in the chapter (figure 33).

c. Suggest why the solution is not evaporated to dryness.

Miscible liquids can be separated by **fractional distillation.** Two immiscible liquids (such as oil and water) can be separated by placing them in a **separating funnel** (figure 35). After ensuring that the stopcock is closed, the mixture is poured into the funnel. Soon after, the liquids form two distinct layers. The denser, lower layer is then drained into an underlying beaker by opening the stopcock, slowing down as the interface between the two layers approaches the narrower neck at the bottom end of the funnel.

Separating funnels are also used to selectively extract solutes from one solvent into another.

Recrystallization

Recrystallization is a purification process based on selective solubility that can be used to isolate a desired solid from a mixture of solids. Recrystallization is often used to purify organic synthesis products. This isolation technique requires a solvent that selectively dissolves different components of the mixture depending on the temperature.

The mixture of solids typically contains three types of components, each with different solubilities in the identified solvent:

*   impurities that are insoluble in the solvent at all temperatures

*   impurities that are soluble in the solvent at all temperatures

*   the desired solid that we wish to isolate, which must be soluble in the solvent when hot, but insoluble when cold.

The recrystallization process is detailed in figure 36.

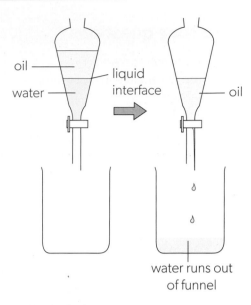

▲ **Figure 35** Separating an oil–water mixture with a separating funnel

Miscibility of liquids, molecular polarity and intermolecular forces are all discussed in *Structure 2.2*.

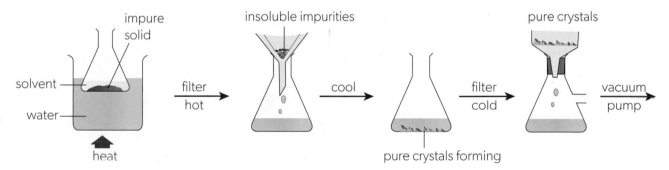

▲ **Figure 36** Recrystallization process

1. First, the impure solid is dissolved in hot solvent. This causes the desired solid to dissolve, along with any soluble impurities.

2. Any insoluble impurities are removed by hot filtration.

3. The filtrate is allowed to cool slowly. Crystals of the desired solid form while the soluble impurities remain in solution.

4. Cold vacuum filtration follows, removing the soluble impurities and isolating the crystals which are then dried.

In a more complex form of recrystallization, two solvents are used. The process starts off as usual, but after the hot filtration step, a second solvent is added that dissolves the remaining impurities but not the desired solid. This causes the desired solid to precipitate out.

Practice questions

31. What method would you use to separate the following mixtures?

    a. a suspension containing solid barium sulfate and an aqueous sodium chloride solution

    b. an ethanol–water mixture

    c. a sodium chloride solution

    d. a mixture of water and cyclohexane

    e. aspirin, from a mixture of soluble and insoluble impurities

Analysis

Melting point determination

The purity of a solid can be assessed by measuring its melting point. Organic solids frequently melt at low to moderate temperatures. A small sample of the solid is placed in a capillary tube. The sample is heated, observed closely, noting down the temperature(s) at which it starts and finishes melting. Pure substances have sharp melting points that agree with published values. If impurities are present, the melting point is usually lowered and the solid melts over a range of temperatures.

Figure 37 shows how a Thiele tube can be used in a melting point determination. The capillary tube containing the sample is attached to a thermometer and lowered into an oil bath. The oil bath is then heated, slowing down the heating rate as the expected melting point is approached. Two temperatures are recorded: first, when the tiny yet visible droplets appear, and second, when the entire sample has just melted.

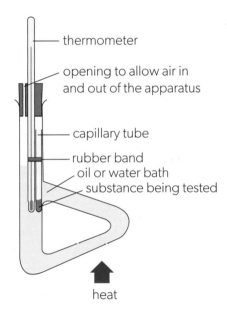

▲ **Figure 37** Melting point determination with a Thiele tube

Electronic melting point apparatus may contain a built-in magnifying glass that facilitates observation of the sample.

🕐 Data-based question

The following melting point data were collected after performing certain organic syntheses. Outline what can be concluded about the purity of the products.

Experiment	Experimental melting point of product / °C ± 1 °C	Theoretical melting point of product / °C
Synthesis of aspirin	122–129	136
Synthesis of methyl 3-nitrobenzoate	76–77	78

Chromatography

Chromatography is a collective term for a group of methods in which a mixture is analysed by the separation of its components according to their relative affinities to the mobile and stationary phases. The two types of chromatography you are likely to encounter as a DP chemistry student are paper chromatography (figure 38) and thin-layer chromatography (TLC) (figure 39).

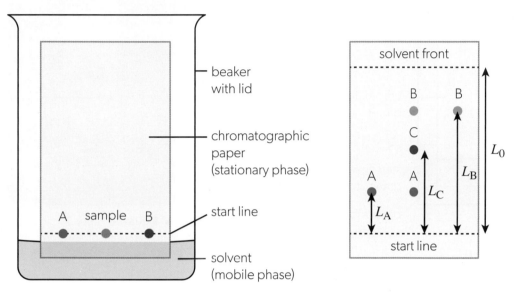

▲ **Figure 38** The set-up for paper chromatography, and the resulting chromatogram

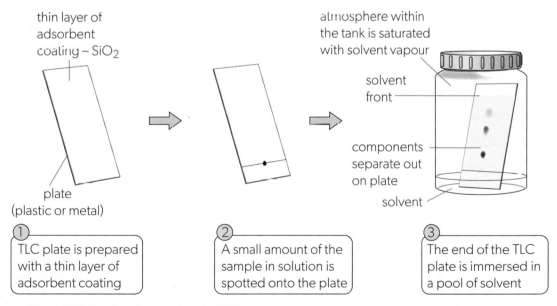

▲ **Figure 39** Thin-layer chromatography (TLC)

In figure 38, the lengths marked L_A, L_B, L_C and L_0 indicate the distances travelled by components A, B, C and the solvent, respectively. These lengths can be used to calculate R_F values.

The calculation of R_F values was discussed in *Structure 2.2*. Review the content of *Structure 2.2* and answer the practice questions on the next page.

Practice questions

32. Figure 40 shows a chromatography experiment involving leaf pigments.

    a. Outline why a lid has been placed on the container.

    b. Explain which component in the mixture has the greatest affinity for the solvent.

33. A student investigating the components of a black food colouring sets up a chromatography experiment as shown below:

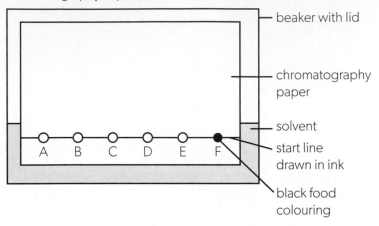

Describe and explain two errors in the set-up shown.

34. Complex mixtures can be separated using a technique known as two-dimensional chromatography. The chromatogram is run in one solvent, rotated 90°, and run again in a second solvent (figure 41).

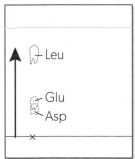 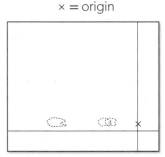 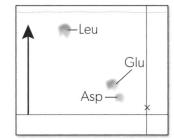

× = origin

run the chromatogram in solvent 1 (butanol / ethanoic acid / water mixture)

turn the paper 90°

run the chromatogram in solvent 2 (phenol / water) then add locating agent

▲ **Figure 41** Two-dimensional chromatography of an amino acid mixture

Study figure 41 and answer the questions below:

a. State which amino acid has the greatest affinity for solvent 1.

b. Amino acids all contain amino ($-NH_2$) and carboxyl ($-COOH$) groups, which are polar and form hydrogen bonds. Between these two functional groups is a carbon atom, to which a substituent known as an R group is attached. The identity of the R group is different for different amino acids. The structures of aspartic acid (Asp), glutamic acid (Glu) and leucine (Leu) are shown below. With reference to their structural features, suggest why these amino acids have different affinities for the two solvent systems.

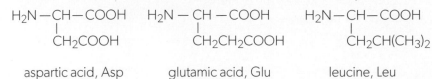

$$H_2N-CH-COOH \qquad H_2N-CH-COOH \qquad H_2N-CH-COOH$$
$$\quad\quad | \qquad\qquad\qquad\quad | \qquad\qquad\qquad\quad |$$
$$\quad CH_2COOH \qquad\quad CH_2CH_2COOH \qquad\quad CH_2CH(CH_3)_2$$

aspartic acid, Asp   glutamic acid, Glu   leucine, Leu

▲ **Figure 40** Paper chromatogram of leaf pigments

Practice questions

35. Remember that the R_F value is calculated as follows:

$$R_F = \frac{\text{distance travelled by the spot}}{\text{distance travelled by the solvent}}$$

Refer to the chromatogram below and:

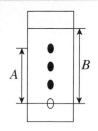

a.   calculate the R_F value of the top spot. Assume that length $A = 3.95\,\text{cm}$ and $B = 5.02\,\text{cm}$.

b.   estimate the R_F value of the middle and bottom black spots.

Calorimetry

We can measure the amount of heat released (or absorbed) during a chemical reaction using a technique known as **calorimetry**, which is discussed in *Reactivity 1.1*. The basic principle of calorimetry is that the reaction's enthalpy change (which we cannot measure directly) causes a measurable change in the temperature of a known mass of water. It is assumed that no energy is lost to the surroundings, although in practice heat loss is a large source of error in calorimetry experiments. Heat loss can be minimized by insulation.

Figure 42 illustrates a calorimetry experiment involving the reaction between magnesium and dilute hydrochloric acid. Note that it is carried out in an insulated container, such as a polystyrene cup, to minimize heat loss to the surroundings.

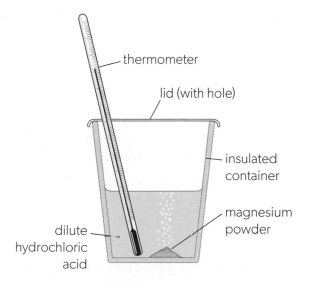

▲ **Figure 42** Calorimetry of the reaction between magnesium and dilute hydrochloric acid

Practice questions

Review *Reactivity 1.1* and answer these questions.

36. The experimental set-up in figure 42 is used to determine the enthalpy of the reaction between magnesium and dilute hydrochloric acid.

a.   Identify two measures shown in the diagram that will lower the heat lost to the surroundings during the reaction.

b.   Suggest a modification that would further minimize heat loss.

37. A calorimetry experiment is performed in an insulated container using 5.0 g of zinc and excess hydrochloric acid to determine the enthalpy of reaction. The temperature rise was 11.0 °C and the mass of the acid solution was 200 g.

a.   Write a balanced equation for this reaction, including state symbols.

b.   State and explain whether the reaction is exothermic or endothermic.

c.   Calculate the amount of heat released, in kJ.

d.   Determine the enthalpy change for this reaction, in kJ mol⁻¹.

e.   Identify two assumptions made in the calculation above.

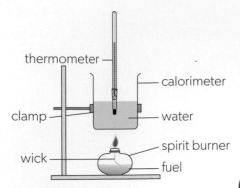

▲ **Figure 43** Determining the combustion enthalpy of a fuel

Practice question

38. A student performed the experiment shown in figure 43. They took the following steps:

   - poured 20 cm³ of water into the calorimeter

   - weighed and lit the spirit burner without the cap

   - measured the temperature of the water for one minute

   - extinguished the spirit burner, keeping the cap off

   - replenished the water in the calorimeter

   - weighed the spirit burner a second time to find the mass of fuel consumed.

   Suggest two practical measures that would improve the reliability of the experiment.

Enthalpies of combustion can be determined using the apparatus shown in figure 43. A measured mass of the fuel is combusted in a spirit burner below a copper calorimeter containing a known mass of water. In this case, the calorimeter is made of a thermally conductive material to facilitate the energy transfer from the combustion reaction to the water. Heat from the combustion reaction is also transferred to the calorimeter, and this should be considered when processing the results. Heat is also transferred to the surrounding air, but this is difficult to quantify and should therefore be minimized.

🥧 Data-based question

A student carried out the experiment shown in figure 43 to determine the enthalpy of combustion of butan-1-ol, C_4H_9OH. The following results were obtained:

Mass of copper calorimeter / g ± 0.01 g	24.03
Mass of copper calorimeter and water / g ± 0.01 g	99.92
Temperature of water before combustion / °C ± 0.1 °C	21.7
Mass of butan-1-ol spirit burner before combustion / g ± 0.01 g	75.47
Temperature of water after combustion / °C ± 0.1 °C	60.1
Mass of butan-1-ol spirit burner after combustion / g ± 0.01 g	74.38

Use these data to calculate the experimental enthalpy of combustion. Note that the specific heat capacity of water is 4.18 kJ kg⁻¹ K⁻¹ and the specific heat capacity of copper is 0.385 kJ kg⁻¹ K⁻¹.

Titration

The unknown concentration of a substance in solution can be determined by reacting it with a standard solution of known concentration and volume in a **titration** (figure 44).

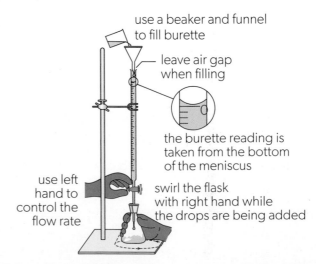

▲ **Figure 44** Titration apparatus. Note: left-handed chemists usually swirl the flask with their left hand and control the flow rate with their right

The **equivalence point**, the point at which the reagents are present in stoichiometric amounts, must be easily identifiable, and this is often achieved by observing a colour change. In acid–base titrations, and many redox titrations, an indicator is added to the reaction mixture. This indicator changes colour at or near the equivalence point. Some redox titrations are said to be self-indicating because the change in oxidation state of at least one of the reagents is accompanied by a colour change.

Normally the burette holds the standard solution, and the flask below contains the solution of unknown concentration (or **analyte**). The titration is performed as follows:

▲ Figure 45 A student performing a redox titration. The solution in the burette is potassium permanganate, $KMnO_4$, which is dark purple and obscures the meniscus. In this case the top of the meniscus should be read

- **Rinse the burette:** A small amount of standard solution is added to the burette, which is then tilted and rotated to ensure that its entire inner surface comes into contact with the solution. The solution is then drained from the burette through the tap and the process is repeated twice more.

- **Fill the burette:** With a funnel, the burette is filled with standard solution to the zero mark or just below. The funnel is removed, bubbles are removed from the tip by draining some of the titrant, and the burette volume is noted by taking a reading at the bottom of the meniscus.

- **Rinse the volumetric pipette:** The volumetric pipette is rinsed with a small amount of analyte three times.

- **Prepare the conical flask:** Using the volumetric pipette, a known volume of analyte is transferred into a clean conical flask. A few drops of indicator are added, and the flask is placed under the burette. You may choose to place it on a white paper or tile to be able to better distinguish the colour change.

- **Rough titration:** The standard solution is quickly delivered from the burette into the conical flask, which is continuously swirled, until the end point is reached (when the indicator changes colour). The purpose of this step is to give you a rough idea of how much standard solution is required to reach the end point. The final burette volume is noted, and the burette is refilled if necessary. The contents of the conical flask are discarded in a suitable waste container. The flask is rinsed with distilled water and it is placed back below the burette. The results will not be affected if it is slightly wet, as long as it is clean.

- **Accurate titrations:** The titration is repeated. This time, the standard solution is added swiftly, with swirling, and the addition is slowed down within a few cm^3 of the end point. Dropwise addition of the standard solution as the end point is approached will allow an exact (to the nearest drop) volume of titrant to be established. The volume of titrant is recorded, and the accurate titration is repeated until at least two concordant results are obtained (typically within 0.1 cm^3 of each other).

Titrations do not always need indicators. The end point can be determined from changes in certain properties of the solution, such as electrical conductivity or temperature, which can be measured over the course of the titration. These titrations are known as conductometric and thermometric titrations, respectively.

Practice questions

39. a. Explain why the burette and pipette need to be rinsed with the solutions of reagents, instead of pure water.

    b. Explain why a rough titration is needed.

    c. Explain why it does not matter if the conical flask holding the analyte is wet.

    d. Identify at least three mistakes to avoid when performing a titration.

    e. A group of students know that the concentration of a hydrochloric acid in their lab is approximately $0.2\,mol\,dm^{-3}$. They find the exact concentration of HCl(aq) by titrating it against a solution of sodium carbonate, Na_2CO_3(aq). Sodium carbonate is a primary standard: it is a stable solid that is available in a high-purity form and does not absorb water or carbon dioxide from the atmosphere.

        i.   Write a balanced equation for the reaction between hydrochloric acid and sodium carbonate.

        ii. The students found that $0.300\,g$ of sodium carbonate was neutralized exactly by $23.83\,cm^3$ of the hydrochloric acid solution. Determine the concentration, in $mol\,dm^{-3}$, of HCl(aq) in the solution from the students' results.

Construction of electrochemical cells

In DP chemistry you need to know how to construct two types of electrochemical cells: electrolytic and voltaic. They both involve electrodes, electrolytes, and a complete circuit, allowing electrons to be gained by chemical species at one electrode (reduction) and lost at the other (oxidation).

Electrolytic cells convert electrical energy into chemical energy (*Reactivity 3.2*). Electrolysis is a non-spontaneous process and therefore it requires a continuous energy input, usually in the form of a cell or battery pack. The electrodes are usually made of an inert material, such as graphite or platinum. Some applications, such as electroplating, involve active electrodes. These are electrodes that are themselves oxidized or reduced, rather than simply acting as a site for oxidation or reduction to occur.

The silver anode in figure 46 is an example of an active electrode.

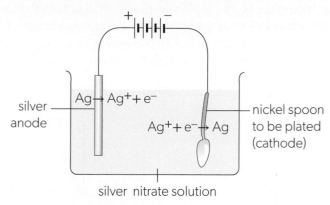

silver anode — Ag → Ag⁺ + e⁻

Ag⁺ + e⁻ → Ag

nickel spoon to be plated (cathode)

silver nitrate solution

▲ **Figure 46** An electrolytic cell used to electroplate an object placed at the cathode

Electrode placement depends on the nature of the products released at the electrodes. Gaseous products can be collected in inverted tubes. Figure 47 shows a Hofmann voltameter, in which slightly acidified water is being electrolysed, producing hydrogen gas and oxygen gas. The electrodes are located at the bottom of the tube to allow the gas bubbles to rise for collection at the top.

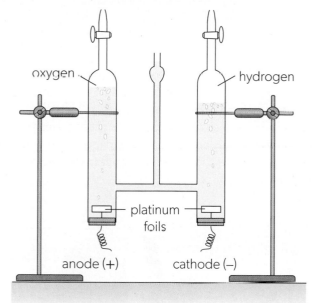

oxygen

hydrogen

platinum foils

anode (+)   cathode (−)

▲ **Figure 47** A Hofmann voltameter is used for electrolysing water. Can you explain the relative volumes of the gases collected at the top of the tubes? How could the identity of the gases be confirmed? Note that the power supply has been omitted from the diagram

If electrolysis leads to the formation of a solid product, it will usually plate the electrode and therefore be readily observable. The quantification of solids produced in electrolysis requires measuring the difference in mass of the electrode, as well as collecting any solid particles or flakes that might fall to the bottom of the electrolytic cell.

Quantitative electrolytic investigations require careful control of many variables, including electrode surface area, temperature, time and potential difference applied. The electrodes should be thoroughly cleaned and dried before the experiment.

Electrolytic and voltaic cells are discussed in *Reactivity 3.2*.

In voltaic cells, chemical energy is spontaneously converted into electrical energy. When two half-cells are connected in a circuit, a potential difference arises, causing current to flow as oxidation occurs in one half-cell, and reduction occurs in the other (see *Reactivity 3.2*). The potential difference can be measured by connecting a high-resistance voltmeter to the electrodes (figure 48).

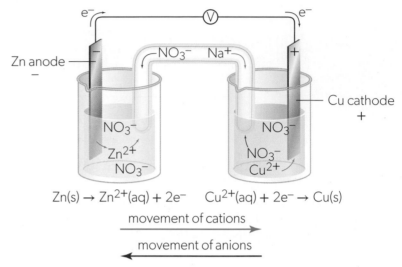

$$Zn(s) \rightarrow Zn^{2+}(aq) + 2e^- \qquad Cu^{2+}(aq) + 2e^- \rightarrow Cu(s)$$

▲ **Figure 48** A voltaic cell. Electrons flow in the external circuit from anode to cathode. In the salt bridge, cations and anions flow towards the cathode and anode, respectively

A salt bridge connects the two half-cells to complete the circuit and prevent the build-up of charge. It is a source of mobile inert ions. A salt bridge can be constructed by filling a U-tube with agar mixed with an inert electrolyte (e.g. potassium chloride or sodium sulfate) and plugging it at either end with cotton wool. A simpler alternative is using a strip of filter paper soaked in a saturated solution of the inert electrolyte and dipping its ends in the two half-cells (figure 49).

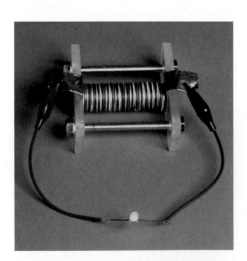

▲ **Figure 50** A voltaic pile powering an LED. It is a series of alternating copper coins, paper soaked in vinegar and zinc washers, similar to the first battery constructed by Alessandro Volta in the 18th century

▲ **Figure 49** The zinc and copper half-cells in this voltaic cell are connected by a paper salt bridge

Electrode identity, electrolyte concentration and temperature are among the factors affecting cell potentials. Other factors include electrode surface area, placement and how long the cell has been running (because this affects the electrolyte concentrations). As with electrolytic cells, the electrodes should be thoroughly cleaned and dried before connecting them in the circuit.

Practice question

40. Compare and contrast electrolytic and voltaic cells.

Colorimetry and spectrophotometry

Spectrophotometry is an analytical technique that is based on a sample's interaction with light of a certain wavelength. Spectrophotometry uses a range of UV, visible or IR radiation. **Colorimetry** is a similar technique that is limited to wavelengths of visible light.

In both spectrophotometry and colorimetry, light is passed through the sample, then detected (figure 51). The incident and transmitted intensities of light are compared to determine the **absorbance**. The relationship between absorbance and concentration is quantifiable, and it is often presented in the form of a **calibration curve**. The calibration curve can be used to determine the concentration of a solution by measuring its absorbance of light of a particular wavelength.

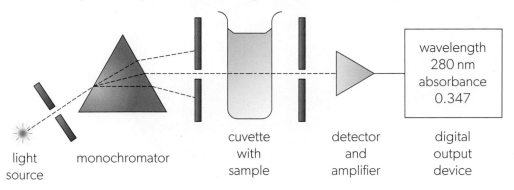

cuvette with sample   detector and amplifier   digital output device

light source   monochromator

wavelength 280 nm absorbance 0.347

▲ **Figure 51** A single-beam UV-vis spectrophotometer

Mathematically, the absorbance (A) is calculated from light intensity as follows:

$$A = \log\left(\frac{I_0}{I_t}\right)$$

where I_0 is the intensity of the incident light and I_t is the intensity of the transmitted light.

Absorbance is also proportional to the concentration of the solute (c):

$$A = \varepsilon c l$$

where l is the cuvette (sample container) length, in cm, and ε is a constant (known as the molar extinction coefficient) that depends on the solvent nature and the temperature of the solution. If the same cuvette and experimental conditions are used, the product of ε and l also becomes a constant. Therefore, the concentration can be determined by comparing an absorbance measurement to a calibration curve of absorbance vs concentration (figure 52). Note that this relationship becomes non-linear at high concentrations, so solutions with low concentrations should be used for precise measurements.

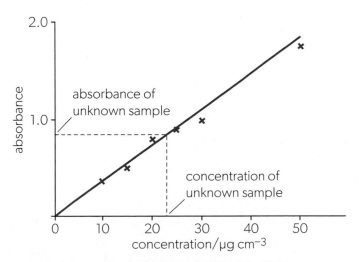

▲ **Figure 52** Use of a calibration curve to determine the concentration of an unknown sample

While spectrophotometers often measure absorbance over a range of wavelengths, a simple colorimeter can only determine the sample's absorbance at a specific wavelength of visible light. The colorimeter's wavelength setting should correspond to a wavelength that is strongly absorbed by the sample. The light absorbed by a coloured solution is **complementary** to the colour observed. Complementary colours are placed opposite each other in the colour wheel (figure 53). For example, a blue solution absorbs orange light corresponding to wavelengths between 585 and 647 nm. A colorimetry experiment involving this substance should therefore use a wavelength that falls within this range.

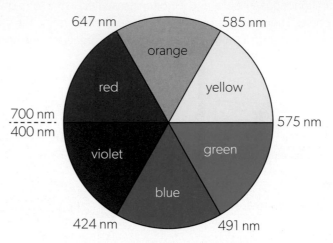

▲ **Figure 53** The colour wheel

Practice questions

41. A student constructed a calibration curve by recording the absorbance at 490 nm of solutions containing varying concentrations of $[FeSCN]^{2+}$.

    a. Suggest the colour of the $[FeSCN]^{2+}$ solution.

    b. The calibration curve gave a linear relationship with the following equation:

    $$A = 14\,000 \times c$$

    where A is absorbance at 490 nm and c is the concentration of $[FeSCN]^{2+}$, in $mol\,dm^{-3}$.

    Determine the concentration of a sample of $[FeSCN]^{2+}(aq)$ with an absorbance of 0.225.

⊕ Data-based question

The absorbance of a protein solution of unknown concentration was found to be 0.285. Using the calibration curve below, determine the concentration of the protein in this solution.

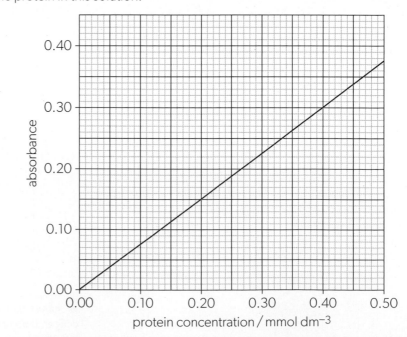

Molecular models

Certain features of molecules, such as molecular geometry, are not apparent from Lewis or structural formulas. Models help us visualize the three-dimensional structure of a molecule. In addition, models can be rotated (manually or digitally) to observe different structural features and molecular geometry, as well as allowing us to predict how molecules might interact with one another.

Models can be built from molecular model kits or constructed using digital modelling software. While model kits are usually ball-and-stick models, digital models can be toggled between this and other types of visualizations, including space-filling models (often based on van der Waals radii) and molecular electrostatic potential (MEP) surfaces. Digital models can easily be saved as files for future reference or communication.

▲ Figure 54 A molecular model of ice

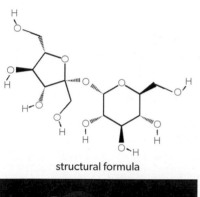

structural formula

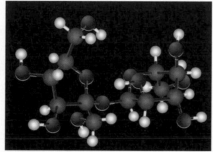

ball-and-stick model

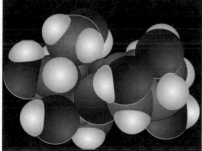

space-filling model

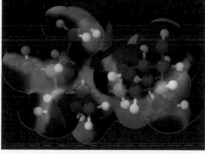

MEP surface

▲ Figure 55 Various representations of sucrose

Practice questions

42. Discuss the advantages and disadvantages of digital molecular models compared to physical models in the following scenarios:

    a. school science labs

    b. scientific research.

The construction of digital models of very large molecules, such as proteins and other biomolecules, has revolutionized modern drug development. The interactions between potential drug candidate molecules and their biological targets are now explored using software. This process is known as computer-aided drug design. If a promising group of molecules is identified computationally (*in silico*), they can then be synthesized and investigated in a laboratory (*in vitro*). Drug development is extremely resource-intensive; technology and computing power can help to streamline the process by quickly screening vast numbers of different molecules and targets.

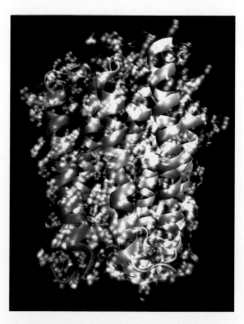

▲ Figure 56 Molecular model of the protein rhodopsin, which is involved in vision

341

Tool 2: Technology

Sensors

Sensors are digital devices that can be used to measure physical properties and transmit the results electronically. Common sensors in school chemistry labs include temperature probes, pH probes, light sensors and conductivity probes. Digital sensors and other devices, such as colorimeters, are convenient because they:

- provide a practical and easy way to measure certain properties that would otherwise require bulky equipment or time-consuming analyses (for example, light sensors or dissolved oxygen sensors)

- often produce more precise results than the alternative equipment or traditional laboratory techniques (for example, pH sensors as compared to acid–base indicators)

- can be connected to data loggers that collect and save large amounts of data at regular intervals.

However, sensors often require calibration before use. If you are going to use one, find out how to calibrate it by checking the manual first. Colorimeters and pH probes are examples of digital devices that require frequent calibration.

Smartphones also contain sensors, and along with certain apps they can make useful measurements in a chemistry lab. US chemists Thomas Kuntzleman and Erik Jacobson devised a way to create a simple yet effective colorimeter using a cardboard cuvette-holding chamber, a smartphone red green blue (RGB) app and coloured construction paper that was published in the Journal of Chemical Education. In the case of a blue copper(II) sulfate solution, where the complementary colour red is absorbed, the incident and transmitted R (red) values were used as I_0 and I_t, respectively, to calculate absorbance (figure 57).

Because phones are items that people frequently touch and carry outside the laboratory, care must be taken to keep them well away from hazards. A 3D-printed cuvette holder could be designed for sturdier support of the analysed sample.

▲ **Figure 57** The view through the smartphone RGB analyser on a blank cuvette (top) and a 0.50 mol dm⁻³ copper(II) sulfate solution (bottom). The phone application records the average R, G, and B values of the pixels within the circle. Given the R values for the blank (190) and the analysed solution (53), an absorbance of 0.554 was obtained for the copper(II) sulfate sample

🍳 Activity

1. Make a list of the sensors available in your school.

2. If available, download an RGB app onto a smartphone or tablet. Investigate how the following affect the RGB values registered by the app. Remember to get approval from your teacher before trying any of these out.

   a. Colour of a solution (you could experiment with distilled water and different food colouring dyes, for example).

   b. Concentration of a coloured solution of your choice.

   c. Colour of the background placed behind the sample.

   d. Path length (i.e. the length of sample the light travels through before reaching the detector).

◀ **Figure 58** Modern-day scientists researching a frozen lake in Antarctica. What might they be investigating? Find out about some of the current chemistry research being done in the Antarctic

Spreadsheets

Spreadsheets are powerful data manipulation and analysis tools that you can use for entering and processing large amounts of data. Operators, formulas and functions tell the software to perform certain operations. Even with a small selection of simple operators, spreadsheets will help you to process experimental data quickly and efficiently.

Figure 59 shows two screenshots of an Excel spreadsheet. The values in the last two columns shown in the first screenshot have been calculated automatically using the formulas shown in the same columns of the second screenshot. Note that each formula in Excel begins with an equal sign (=) followed by cell references (e.g. E7) and operators, such as multiplication (*) and division (/). Brackets are used to change the order of operations. Note that all processed values should be strictly numerical because most operators in Excel do not recognize numbers that are entered into the same cell as letters.

(a)

	A	B	C	D	E	F	G
1	Primary alcohol	Mass of water / g	Change in temperature / K	Change in spirit burner mass / g	Molar mass of alcohol / g mol-1	Amount of alcohol / mol	Enthalpy change / kJ mol-1
2	methanol	78.8	38.1	1.26	32.04	0.039325843	−319. 1166302
3	ethanol	80	38.8	1.22	46.07	0.026481441	−489.9552052
4	butan-1-ol	79.9	38.4	1.09	74.12	0.014705882	−872.0937984
5	pentan-1-ol	80.3	41	1.12	88.15	0.012705615	−1083.128486

(b)

	A	B	C	D	E	F	G
1	Primary alcohol	Mass of water / g	Change in temperature / K	Change in spirit burner mass / g	Molar mass of alcohol / g mol-1	Amount of alcohol / mol	Enthalpy change / kJ mol-1
2	methanol	78.8	38.1	1.26	32.04	=D2/E2	=−(B2*4.18*C2)/(1000*F2)
3	ethanol	80	38.8 .	1.22	46.07	=D3/E3	=−(B3*4.18*C3)/(1000*F3)
4	butan-1-ol	79.9	38.4	1.09	74.12	=D4/E4	=−(B4*4.18*C4)/(1000*F4)
5	pentan-1-ol	80.3	41	1.12	88.15	=D5/E5	=−(B5*4.18*C5)/(1000*F5)

▲ **Figure 59** An Excel spreadsheet used to calculate the enthalpy of combustion of primary alcohols from experimental data. The values and formulas are shown in the upper and lower images, respectively

While symbols such as * and / and ^ are used in spreadsheets (and some calculators), you must not use this notation outside the spreadsheet, for example, when explaining your calculations as part of a lab report. Common spreadsheet operators and functions are summarized in table 1.

Operator or function	Action	Example
+	addition	=A5+B5 Finds the sum of values in cells A5 and B5.
−	subtraction	=A5−B5 Subtracts the value in cell B5 from that in A5.
*	multiplication	=A5*B5 Finds the product of values in cells A5 and B5
/	division	=A5/B5 Divides the value in cell A5 by that in B5.
^	power	=A5^2 Finds the square of the value in cell A5.
=EXP()	exponent	=EXP(A5) Raises e (base of natural logarithm, 2.718282…) to the power of the value in cell A5. This function is inverse of =LN()
=AVERAGE()	mean (average)	=AVERAGE(A5:A8) Finds the mean (average) of the values in cells A5 to A8.
=MAX()	maximum	=MAX(A5:A8) Finds the maximum value in cells A5 to A8.
=MIN()	minimum	=MIN(A5:A8) Finds the minimum value in cells A5 to A8.
=SUM()	sum	=SUM(A5:A8) Adds together the values in cells A5 to A8.
=LOG()	logarithm (base 10)	=LOG(A5) Computes the log (base 10) of the value in cell A5.
=LN()	natural logarithm	=LN(A5) Computes the natural log (base e) of the value in cell A5. This function is inverse of =EXP().

▲ Table 1 Common spreadsheet functions

More advanced spreadsheet functions that you might want to explore next could involve sorting data, conditional formatting, inserting scroll bars, vertical lookup, recording macros, etc.

You can easily plot various types of graphs in spreadsheets. Remember to include all the features of good-quality graphs (see the Graphing section in Tool 3 for details).

Most spreadsheets and graphing software packages can be used to plot a simple scatter plot, line of best fit and error bars. Depending on what program you use, some functionality may be limited, particularly when trying to plot different error bars on each point, drawing curves of best fit, or omitting certain points from the line of best fit. In such cases, you might be able to find a workaround, or print the graph and draw these on by hand.

Practice questions

43. State the formulas you would enter into a spreadsheet to calculate the following:

    a. $\dfrac{0.400}{5 \times 4.18}$

    b. The mean of 42, 32, 45, 46 and 48.

    c. $-\log(0.0034)$

    d. $10^{-12.5}$

    e. the range of the values in cells A5 to A8.

44. Describe the operation performed by each of the following spreadsheet formulas:

    a. =MAX(C6:C10)

    b. =SUM(D5,E7,G3)

    c. =−(A3*B3*4.18)/(C4/46.07).

45. An investigation into the effect of an impurity on the freezing point of water gave the following results.

Mass of impurity / g ± 1 g	Freezing point / °C ± 0.5 °C			
	1	2	3	Mean
0	0.8	0.7	0.7	
10	−1.6	−1.7	−1.4	
15	−1.0	−1.2	−1.1	
30	−3.2	3.1	−3.3	
45	−5.3	−5.5	−5.3	

    a. Enter these values into a spreadsheet and calculate the mean freezing points.

    b. Use the spreadsheet to plot a graph of these results.

Databases

Databases are digital repositories of large amounts of information, all organized to facilitate retrieval and continuous updates. In chemistry you might use databases to gather information on:

- Elements and their properties (e.g. WebElements, created by Mark Winter at the University of Sheffield, UK)

- Physical properties (e.g. Chemspider, created by the Royal Society of Chemistry, UK)

- Energetics data (e.g. NIST Chemistry WebBook, edited by Linstrom and Mallard at the National Institute of Standards and Technology, USA)

- Spectral data (e.g. SDBSWeb, created by the National Institute of Advanced Industrial Science and Technology, Japan)

- Molecular geometry data (e.g. CoolMolecules, created by Hanson *et al.* at St. Olaf College and Washington State University, USA).

You may use databases to look up two or three entries at a time, but sometimes you will conduct more complex database investigations involving large amounts of data. These tips will help you when extracting data from databases as part of extended investigations:

- Spend some time exploring the database to get a sense of the types of data that are available.

- Make a note of the search terms and search parameters you can use.

- Organize the data you collect into a spreadsheet as you go along.

- Many databases cite the original information source, so you should note this as well as the reference to the database itself.

ATL Communication skills

Choose one of the databases mentioned in this section. Spend 15 minutes looking around the database. Come up with three independent and three dependent variables that you could explore with data extracted from the database.

Share your ideas with your class. Tell them:

- which database you looked at and what types of information it contains

- what you found easy and difficult when using the database

- your list of three independent and three dependent variables.

Listen to your peers' database ideas and identify one that you would like to look at further.

Activity

1. Using a database of your choice, investigate the melting points of:

    - group 1 elements

    - group 17 elements

    - period 3 elements.

    Explain any trends you observe. Refer to your knowledge of bonding and physical properties (*Structure 2*).

2. Explore the electrical conductivity of the elements in a periodic table group or period of your choice by identifying and extracting data from a database of your choice.

    (Note: you might only find resistivity values. A material's resistance to electrical flow is its resistivity. The relationship between resistivity and electrical conductivity is electrical conductivity $= \dfrac{1}{\text{resistivity}}$).

Modelling

You have already come across models in your study of chemistry, for instance the ideal gas model, or atomic models. Scatter graphs help us model the relationship between two variables. As described in the NOS section of the subject guide, "models are simplifications of complex systems". With the aid of technology, scientists can model the structure and reactivity of intricate chemical phenomena. The best way to develop a model is to start with a simple relationship or concept and build up from there. Here we will look at two examples of modelling tasks that are facilitated by technology: spreadsheet modelling and molecular mechanics.

Spreadsheet modelling

Spreadsheets can be set up to transform input variables into output variables in the context of a given system, such as equilibria. You can explore the effect of external disturbances on equilibrium concentrations, for example. US chemist Charles Marzzacco proposed an introductory spreadsheet model of a simple coin-flipping equilibrium:

$$\text{heads (H)} \underset{k_r}{\overset{k_f}{\rightleftharpoons}} \text{tails (T)}$$

The input parameters are the initial concentrations of H and T, and the forward and reverse rate constants, k_f and k_r. These parameters then determine the H and T concentrations over time, from which a concentration vs time graph is generated. We can now explore the effects of changing the initial concentrations and the rate constants on the equilibrium concentrations and the equilibrium constant, K.

The resulting spreadsheet values and formulas are shown in figure 60. Here we have added scrollbars to the spreadsheet proposed by Marzzacco, as well as a calculation of the equilibrium constant, K, from the forward and reverse rate constants k_f and k_r, where $K = \dfrac{k_f}{k_r}$.

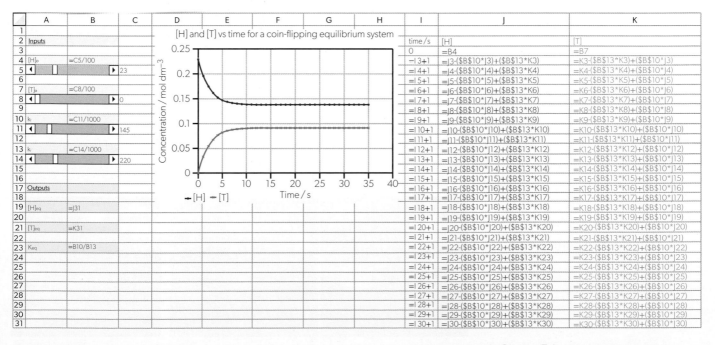

▲ Figure 60 Spreadsheet model of a simple equilibrium system. Source of data: C. Marzzacco, *J. Chem. Ed.*, 1993, 70(12), p.993

Activity

Replicate the spreadsheet in figure 60. (If you are not able to include scroll bars, leave them out and simply change the input parameters manually.)

a. Adjust the input parameters and explore the effect of changing the following parameters on the equilibrium:

    i. initial concentrations

    ii. rate constants.

Remember to check what happens when you input both intermediate values and the extremes (very low and very high values).

b. Explore the effect of a sudden addition of H to the equilibrium mixture at 15 seconds by manually entering a number into cell J18 that is larger than the number currently in that cell.

c. Suggest how you could explore the effect of removing H or T.

d. Determine some of the limitations of this model.

e. Extend the spreadsheet to investigate:

    i. an equilibrium system composed of three species, for example, $A + B \rightleftharpoons C$.

    ii. two equilibria related by a common species, for example, $A \rightleftharpoons B \rightleftharpoons C$.

Molecular mechanics

The development of molecular modelling has profoundly impacted our understanding of chemistry by greatly extending the scope of what chemists can "see" and study. Here we will introduce molecular mechanics, which is a field of molecular modelling that treats atoms as balls joined by springs, in line with classical mechanics. A more complex area of molecular modelling, which is beyond the scope of the DP chemistry course, involves the use of quantum mechanical principles to study the distribution of electron density in reacting species.

Figure 61 shows molecular models of water, methanol and methoxymethane, which can be analysed to explore the effect of substituents bonded to the oxygen on the R–O–R′ bond angle. The bond angle data derived from each model are shown in table 2.

Molecule	R–O–R′ bond angle
water	104.0°
methanol	107.1°
methoxymethane	111.5°

▲ Table 2 Bond angles in molecular models of water, methanol and methoxymethane (R and R′ represent either H or CH_3 substituents)

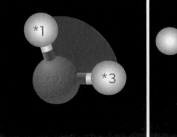
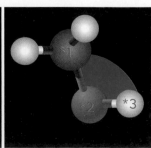
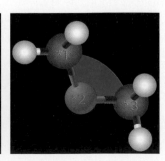

▲ Figure 61 Molecular models of water, methanol and methoxymethane

Activity

a. Outline the effect of replacing the H atoms with methyl groups on the bond angle.

b. Using a molecular editor of your choice, replicate the molecules shown in figure 61.

    i. Compare the bond angle values in your models with those given in table 2.

    ii. Suggest possible extensions to this activity.

Figure 62 shows a model of two water molecules approaching one another. For each distance, the potential energy between the two water molecules was calculated. This is summarized in table 3.

 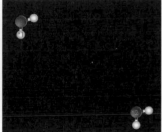  

▲ Figure 62 A molecular model of two water molecules at distances of 15.5, 11.2, 5.07 and 1.78 Å apart. In the last case, the molecules are close enough for a hydrogen bond to form

Distance between molecules / Å	Potential energy / kJ mol⁻¹
15.5	−0.15
11.2	−0.45
5.07	−3.88
1.78	−27.66

▲ Table 3 Distances between two water molecules and the corresponding potential energy values

To begin with, the hydrogen atom in one molecule is located 15.5 Å (or 15.5×10^{-10} m) from the oxygen atom in the other molecule, and the potential energy between the two molecules is −0.15 kJ mol⁻¹. As the molecules approach each other, the distance and energy both decrease until a hydrogen bond is formed. The model predicts that the hydrogen bond will have a length of 1.78 Å (or 1.78×10^{-10} m) and an energy of −27.7 kJ mol⁻¹. The validity of the model can be verified by comparing these predicted values with experimental data.

Activity

a. Build a model of the water molecule using a molecular editor of your choice, such as Avogadro.

b. Explore the tools and visualizations available in that editor.

c. Add a second water molecule.

d. Using the molecular editor, calculate the energy of the interaction between the two water molecules.

e. Find out what happens when you add more water molecules and run an energy optimization.

f. Build a different molecule and explore its interactions with other molecules.

Tool 3: Mathematics

The purpose of this section is to help you ensure, understand and evaluate the reliability of your quantitative measurements and analyses as well as good communication. Mathematical skills, units, significant figures, uncertainties and graphs are relevant to nearly every topic in DP chemistry. If you take advantage of every opportunity to reinforce these concepts, they will become habitual.

Units

According to the International Bureau of Weights and Measures (*Bureau International des Poids et Mesures*, BIPM), **SI units** should be used, and these conventions followed:

- A space between a numerical value and its unit.

- Decimal markers should be preceded by a number, even if this number is zero.

- Different units should be separated by a space.

Practice questions

46. Which of the following are SI units?

   - metre, m

   - calorie, cal

   - ounce, oz

   - foot, ft

   - cubic centimetre, cm^3

   - degree Fahrenheit, °F

   - millimetre of mercury, mmHg

   - pascal, Pa

   - kelvin, K

   - metre per second, m/s

47. What is the SI unit for each of the following quantities?

   a. time

   b. energy

   c. volume

   d. amount of substance

   e. pressure

48. Identify and correct the mistake in each of the following:

   a. time for reaction to happen = 16 sec

   b. temperature = 304 °K

   c. enthalpy of reaction = -77.5 kJmol^{-1}

   d. mass = 0.33g

   e. melting point = 277 °F

   f. density = 0.78 g / dm^{-3}

   g. volume = 500 cc

   h. amount = .5 mol

   i. amount = 3.0 g

   j. $R = 8.31$ J/(Kmol)

Uncertainties

Measurements (and the values derived from them) are always inexact. Repeating a measurement under the same conditions and with the same instrument produces randomly varying values. Factors influencing this random variation, known as **random error**, include instrument imprecision, fluctuations during readings and human reaction time. Random error is described by the **uncertainty**. Measurements should therefore be recorded along with their associated uncertainty and unit. For example, the temperature is given as 36.6 °C ± 0.1 °C or (36.6 ± 0.1) °C in figure 63. This tells us that the measured value is somewhere in the range of 36.5 °C to 36.7 °C.

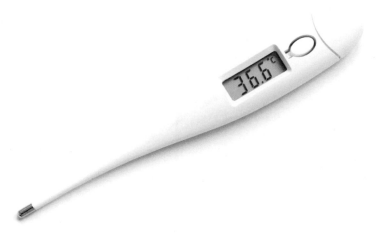

▲ **Figure 63** The temperature shown here is 36.6 °C + 0.1 °C or (36.6 ± 0.1) °C

Instruments with greater precision give rise to measurements that have lower uncertainties. For instance, while the uncertainty of a measuring cylinder may be around ± 0.5 cm³, the uncertainty of a burette is typically ± 0.05 cm³. The measurements obtained using the burette are therefore more precise.

Measured values should be recorded with the correct level of precision and the uncertainty must be stated. The determination of the uncertainty of a particular measurement usually involves one of the following:

*   estimating the uncertainty based on the display or scale on the instrument you are using

*   uncertainty stated by the manufacturer of the measuring instrument

*   estimating the range over which a value fluctuates.

Estimating the uncertainty of an instrument

The way to estimate an instrument's uncertainty depends on whether it is digital or analogue (has a scale).

The uncertainty of a digital instrument is its **least count**, that is, the lowest value above zero that the instrument can register. This is usually the reading of "1" in the lowest decimal place on the display. Consider the mass shown in figure 64. The least count is 0.1 g, so the uncertainty is ± 0.1 g and the mass of the powder is recorded as 0.7 g ± 0.1 g.

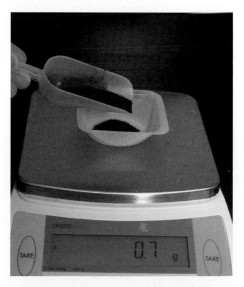

▲ **Figure 64** The uncertainty of the measurement is ± 0.1 g

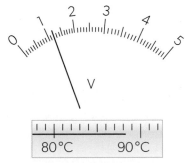

Analogue instruments have scales marked on them. The uncertainty is estimated as half the smallest scale division. The thermometer in figure 65 has scale divisions at every 1 °C. The uncertainty is half the smallest scale division, ± 0.5 °C. Therefore, the temperature should be recorded as 37.0 °C ± 0.5 °C. Remember that the last digit in the temperature value is estimated. In this case, it is zero, which means that the reading is on level with the 37 °C tick mark on the scale.

▲ **Figure 65** The temperature is 37.0 °C ± 0.5 °C

Practice questions

49. State the uncertainty of the measuring cylinder shown below. The scale has tick marks at every 1 cm³.

50. State the measurement, including units and uncertainties, shown in the three images below.

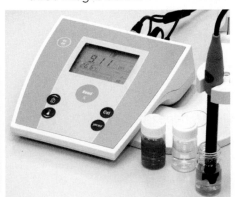

▲ **Figure 66** The uncertainty of this burette is stated in the label: ± 0.05 ml (or ± 0.05 cm³)

Uncertainty given by the manufacturer

Sometimes the uncertainty is stated on a label somewhere on the instrument (figure 66).

Values that fluctuate

Sometimes the measured value fluctuates over time, for instance in the case of certain conductivity probe displays. In these cases, you can try to estimate the uncertainty based on the variation in values you observe.

For example, the rate of reaction between marble chips (calcium carbonate, $CaCO_3$) and hydrochloric acid can be explored by measuring the rate of carbon dioxide gas production using the apparatus in figure 67. However, determining the volume of gas at specific points of time is difficult because the bubbles interfere with the meniscus. In this case the measurement uncertainty is greater than half the smallest scale division on the measuring cylinder. To better estimate the volume measurements and their associated uncertainties, you could try slowing down the reaction. Alternatively, you could make a video recording during the reaction and subsequently pause it at different points to examine the scale. Another option is to use a gas syringe instead of the measuring cylinder.

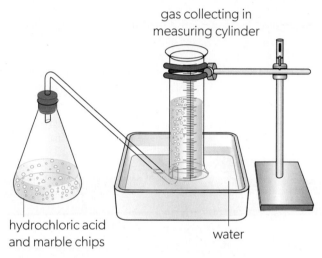

▲ **Figure 67** In this set-up for measuring the rate of a gas-producing reaction, the bubbles interfere with the volume measurements and increase the uncertainty

Further sources of uncertainty

Some sources of uncertainty can be difficult to quantify. Examples include human reaction time or change in the appearance of the reaction mixture. These should be noted by the experimenter. For example, the reaction between sodium thiosulfate and hydrochloric acid produces a precipitate, making the reaction mixture progressively opaque. The kinetics of this reaction can be studied by measuring how long it takes for a picture placed below the mixture to be obscured by the precipitate (figure 68).

The stopwatch reading is 42 seconds, and the uncertainty of the instrument is $\pm 1\,s$ (the least count). However, the actual uncertainty of the time measurement is probably much higher, as it must include the human reaction time and the uncertainty associated with the experimenter's perception of the moment when the cross becomes no longer visible.

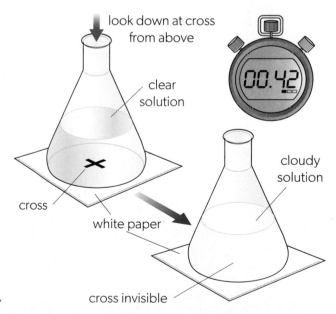

▲ **Figure 68** Studying the kinetics of the reaction between hydrochloric acid and sodium thiosulfate involves measuring the time taken for the reaction mixture to obscure a cross on a piece of paper under the flask

TOK

Mathematical techniques help scientists communicate measurements and their uncertainties, make predictions, process and analyse data from experiments. What is the role of mathematics in the production of scientific knowledge? How does this compare to the role of mathematics in other areas of knowledge?

In the case of the reaction between sodium thiosulfate and hydrochloric acid (figure 68), the random error can be estimated by repeating measurements and determining the range over which they vary. Suppose that the measurement is repeated five times under the same conditions and the following data are obtained for the time taken for the cross to become invisible:

42 s   39 s   43 s   44 s   45 s

The mean time is 42.6 s $\approx$ 43 s. The uncertainty of a set of replicate values can be estimated in two ways: one based on the range of readings, and the other on the mean value.

1. **Estimating the uncertainty by halving the range of readings**

We can estimate the uncertainty by halving the range of the readings. Here, the range is 45 s − 39 s = 6 s. Half the range is $\frac{6}{2}$ = 3 s. Therefore, the time for the cross to become invisible is 43 s $\pm$ 3 s.

2. **Estimating the uncertainty by finding the furthest reading from the mean value**

The values range from 39 s to 45 s. The mean value is 2 s away from the top of the range (45 s − 43 s = 2 s) and 4 s away from the bottom of the range (43 s − 39 s = 4 s). The larger of the two is used as an estimate of the uncertainty. Therefore, the mean time taken for the cross to be obscured under these conditions can be recorded as 43 s $\pm$ 4 s.

Data-based question

A student measured the volume of hydrogen produced in the reaction between magnesium and dilute sulfuric acid after 2 minutes, producing the following results.

Volume of H_2 (g) produced after 2 minutes / $cm^3 \pm 0.5$ cm^3		
47.7	43.5	44.0

a. Calculate the mean volume.

b. Estimate the uncertainty by halving the range of readings.

c. Estimate the uncertainty by finding the furthest reading from the mean value.

d. Which uncertainty value would you report with the data? Why?

Measurement

Measurements are limited in precision and accuracy. Some values in science are exact and have no uncertainty, for instance, the seven defining constants in the SI system. These constants are shown below:

Defining constant	Symbol	Numerical value	Unit
hyperfine transition frequency of Cs–133	Δv_{Cs}	9 192 631 770	Hz
speed of light in vacuum	c	299 792 458	$m\,s^{-1}$
Planck constant	h	6.626×10^{-34}	J s
elementary charge	e	$1.602\,176\,634 \times 10^{-19}$	C
Boltzmann constant	k	$1.380\,649 \times 10^{-23}$	$J\,K^{-1}$
Avogadro constant	N_A	$6.022\,140\,76 \times 10^{23}$	mol^{-1}
luminous efficacy	K_{cd}	683	$lm\,W^{-1}$

▲ Table 4 The seven defining constants of the SI

As we have seen, there are several ways of determining measurement uncertainties. While the equipment uncertainty (least count or half the smallest scale division) is a common way to do this, you might need to account for additional uncertainties. Effective communication of data collected during experiments includes noting how you estimated the uncertainty of the different data you measured. (Was it the least count? Was it given by the manufacturer? Did you estimate it based on fluctuations in the values?) Doing so will help you (and your reader) understand the impact of measurement uncertainty on your conclusions.

Expressing uncertainties

Uncertainties can be expressed in different formats: absolute, relative or percentage. To illustrate this, consider the following volume measurement: $15.0\,cm^3 \pm 0.2\,cm^3$. Let the measured value be x, in this case $15.0\,cm^3$, and the associated uncertainty, $u(x)$, which is $\pm 0.2\,cm^3$.

- **Absolute uncertainty:** The uncertainty associated with a given value. It has the same units as the value it is associated with. In the example above, $\pm 0.2\,cm^3$ is the absolute uncertainty.

- **Percentage uncertainty:** The absolute uncertainty expressed as a percentage of the value it is associated with.

$$\text{percentage uncertainty} = \frac{\text{absolute uncertainty of } x}{x} \times 100\%$$

$$= \frac{u(x)}{x} \times 100\%$$

Using the example above:

$$\text{percentage uncertainty} = \frac{0.2\,cm^3}{15.0\,cm^3} \times 100\%$$

$$= 1.3\% \text{ (2 sf)}$$

- **Relative (or fractional) uncertainty:** The ratio comparing the magnitude of the absolute uncertainty to the magnitude of the associated value:

$$\text{relative uncertainty} = \frac{\text{absolute uncertainty of } x}{x}$$

$$= \frac{u(x)}{x}$$

Using the example above:

$$\text{relative uncertainty} = \frac{0.2 \text{ cm}^3}{15.0 \text{ cm}^3}$$

$$= 0.013 \text{ (2 sf)}$$

Note that the relative uncertainty is the percentage uncertainty expressed as a decimal fraction.

Uncertainties are often stated to one significant figure. Sometimes two significant figures are acceptable, particularly if the uncertainty is very small. Either convention is acceptable, as long as you use it consistently.

Uncertainty is an estimate of precision, so the precision of a measured value must correspond to that of its uncertainty. In other words, the value itself should have the same number of decimal places as the uncertainty associated with it. For example, a mass given as $10.3 \text{ g} \pm 0.01 \text{ g}$ is inconsistent in terms of precision because the uncertainty suggests that the balance gives readings to 2 decimal places, but the mass is given only to 1 decimal place. Perhaps the uncertainty was incorrectly written down and it is in fact $\pm 0.1 \text{ g}$, giving a correct mass $10.3 \text{ g} \pm 0.1 \text{ g}$. Alternatively, the experimenter might have failed to write down a trailing zero after the 3 (which should not be omitted because it is significant), so the correct mass is $10.30 \text{ g} \pm 0.01 \text{ g}$.

Activity

Copy and complete the table:

Measured value	Absolute uncertainty	Percentage uncertainty	Relative uncertainty
19.96 cm³	± 0.04 cm³		
1.08 V		± 2%	
78.5 °C			± 0.01

Decimal places and significant figures

Any measurement involves uncertainty, so the result of the measurement always has a limited number of **significant figures** (sf). The length of a small object measured with a ruler typically has no more than three sf as shown in figure 69.

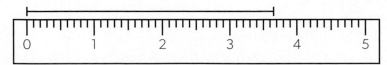

▲ **Figure 69** Measurement uncertainty when using a ruler

In figure 69, the measured length is 3.67 cm. The first two figures, 3 and 6, are certain, as the length is greater than 3.6 but less than 3.7 cm. The last figure, 7, is uncertain, as the actual length could be 3.66 or 3.68 cm. There is absolutely no way of getting the fourth figure using this ruler, as we are not even sure about the third figure.

As measurements are never exact, neither are the results of the calculations involving these measurements. There are two rules of thumb, which depend on the type of calculation being carried out:

1. Adding and subtracting measured values: the answer should be recorded to the least number of decimal places present in the values used.

2. Multiplying and dividing measured values: the answer should be recorded to the least number of significant figures present in the values used.

Worked example 2

At the beginning of a titration, the initial burette reading was 1.03 cm^3. At the end point, the final burette reading was 24.13 cm^3. Calculate the volume of the titrant used.

Solution

Volume used = final volume − initial volume

$$= 24.13\,cm^3 - 1.03\,cm^3$$

$$= 23.10\,cm^3\,(2\,dp)$$

Both values used have two decimal places. They are subtracted and therefore the answer should also be stated to two decimal places.

Worked example 3

Calculate the amount of magnesium oxide, MgO, in mol, in a 0.500 g sample of pure magnesium oxide.

Solution

Molar mass of MgO = M(MgO) = 24.31 + 16.00 = 40.31 g mol^{-1}

$$n = \frac{m}{M}$$

$$n(MgO) = \frac{0.500\,g}{40.31\,g\,mol^{-1}}$$

$$= 0.012404...\,mol \approx 0.0124\,mol\,(3\,sf)$$

This operation involves a multiplication, so the answer should be rounded to the least number of significant figures in the values used. The answer is therefore rounded to three significant figures.

Many calculations require several steps. Rounding too early could lead to incorrect answers due to rounding errors. Round the final answer to the required number of decimal places or significant figures, but not the answers to intermediate steps. Always carry two or three extra significant figures through intermediate calculations.

Propagating uncertainties

Processed data are obtained by manipulating raw measurements and therefore also have uncertainties associated with them. The overall uncertainty of a calculated result can be estimated by **propagating** the uncertainty introduced by each of the measurements. The way this is done depends on the type of calculation done with the raw data and it can be categorized into three types:

1. Addition and subtraction

2. Multiplication and division

3. Exponents (AHL only)

At this level, you will rarely need to consider the uncertainty of values or constants that are not given to you with an uncertainty. Examples include the speed of light, molar mass values and specific heat capacities. For most purposes you can assume that values such as these are exact and have no associated uncertainty.

In this section, we will discuss only simplified methods of uncertainty propagation that can be used in typical laboratory experiments with a small number of variables. More advanced statistical methods of uncertainty propagation will not be assessed in this course.

In a calculation involving addition or subtraction, we propagate uncertainties by adding the absolute uncertainties of the values that are being added or subtracted.

Worked example 4

The mass of an ethanol spirit burner is measured before and after combustion giving 70.971 g and 70.350 g, respectively. The uncertainty for both mass values is ±0.001 g. Calculate the mass of ethanol combusted and the associated uncertainty.

Solution

First, calculate the change in mass:

mass of ethanol combusted = initial mass – final mass

$m = 70.971\,g - 70.350\,g$

$m = 0.621\,g$

Then propagate the uncertainties. You are subtracting the mass values, so you must add the absolute uncertainties:

$u(m) = \pm 0.001\,g + (\pm 0.001\,g)$

$u(m) = \pm 0.002\,g$

Finally, write the overall result:

$m = 0.621\,g \pm 0.002\,g$

Worked example 5

A student uses a volumetric pipette to transfer $25.00\,cm^3 \pm 0.05\,cm^3$ of dilute acid into a conical flask and then adds a further $10.0\,cm^3 \pm 0.5\,cm^3$ of the acid solution using a measuring cylinder. Calculate the total volume of acid added to the flask, along with its associated uncertainty.

Solution

First, calculate the total volume added:

total volume of acid = first volume + second volume

$V = 25.00\,cm^3 + 10.0\,cm^3$

$V = 35.0\,cm^3$ (1 dp)

Note that the result is rounded to one decimal place to be consistent with the second volume, which has the least number of decimal places in the raw data.

You are adding the volume values and therefore to propagate the uncertainties you must add the absolute uncertainties:

$u(V) = \pm 0.05\,cm^3 + (\pm 0.5\,cm^3)$

$= \pm 0.55\,cm^3$

$\approx \pm 0.6\,cm^3$ (1 sf)

This gives the overall result:

$V = 35.0\,cm^3 \pm 0.6\,cm^3$

The large uncertainty of the measuring cylinder largely eclipses the high precision of the volumetric pipette.

Propagating uncertainties in calculations involving multiplication or division requires you to add the percentage uncertainties of the values that are being multiplied or divided. Since percentage uncertainties are just the relative uncertainty in decimal form, you can also think of this as adding the relative uncertainties and expressing them as percentages at the end of the calculation.

Worked example 6

A sucrose solution is prepared by dissolving $10.35\,g \pm 0.02\,g$ of solid sucrose in water to produce $100.00\,cm^3 \pm 0.10\,cm^3$ of solution. Calculate the concentration of sucrose, in $g\,dm^{-3}$, and the associated uncertainty.

Solution

First calculate the mass concentration of sucrose:

$c = \dfrac{m}{V}$

$= \dfrac{10.35\,g}{100.00 \times 10^{-3}\,dm^3}$

$= 103.5\,g\,dm^{-3}$

Note that the result is given to four significant figures, as this is the least number of significant figures in the raw data.

The calculation involves a division, therefore to propagate the uncertainties you must compute the percentage uncertainties and then add them together:

$u(c) = \left(\dfrac{\pm 0.02\,g}{10.35\,g} + \dfrac{\pm 0.10\,cm^3}{100.00\,cm^3} \right) \times 100\%$

$= \pm 0.29324...\% \approx \pm 0.3\%$ (1 sf)

This gives the overall result:

$c = 103.5\,g\,dm^{-3} \pm 0.3\%$

This percentage uncertainty can be converted to an absolute uncertainty:

$u(c) = \dfrac{0.3}{100} \times 103.5\,g\,dm^{-3}$

$= 0.3105\,g\,dm^{-3}$

$\approx 0.3\,g\,dm^{-3}$ (1 sf)

$c = 103.5\,g\,dm^{-3} \pm 0.3\,g\,dm^{-3}$

Worked example 7

The temperature of a 50.00 g ± 0.01 g sample of water is 23.0 °C ± 0.1 °C. The sample is then heated with an ethanol spirit burner until it reaches 33.0 °C ± 0.1 °C.

Calculate the heat absorbed by the water, in J, and its associated uncertainty. The specific heat capacity of water is 4.18 J g^{-1} K^{-1}.

Solution

First, find the temperature difference, ΔT, and use this to determine the heat absorbed, Q:

$$\Delta T = T_{final} - T_{initial}$$

$$= 33.0\,°C - 23.0\,°C$$

$$= 10.0\,°C$$

$$= 10.0\,K$$

$$Q = mc\Delta T$$

$$= 50.00\,g \times 4.18\,J\,g^{-1}\,K^{-1} \times 10.0\,K$$

$$= 2090\,J\,(3\,sf)$$

Note that the result is rounded to three significant figures to be consistent with the least number of significant figures in the values used in the calculation.

The first calculation involves a subtraction, so to propagate the uncertainties in temperature you add them together:

$$u(T) = \pm0.1\,°C + (\pm0.1\,°C)$$

$$= \pm0.2\,°C$$

$$= \pm0.2\,K$$

Then, in $Q = mc\Delta T$, we are multiplying values and therefore we must add the percentage uncertainties:

$$u(Q) = \left(\frac{\pm0.01\,g}{50.00\,g} + \frac{\pm0.2\,K}{10.0\,K}\right) \times 100\%$$

$$= \pm2.02\%$$

$$\approx \pm2\%\,(1\,sf)$$

Note the specific heat capacity value is assumed to be exact (because it was not measured in this experiment) and hence is assumed to have no uncertainty. The overall result is as follows:

$$Q = 2090\,J \pm 2\%$$

This percentage uncertainty can be converted to an absolute uncertainty:

$$u(Q) = \frac{2}{100} \times 2090\,J$$

$$= 42.8\,J$$

$$\approx 40\,J\,(1\,sf)$$

$$Q = 2090\,J \pm 40\,J$$

Propagating uncertainties in calculations involving exponents requires percentage uncertainties. The percentage uncertainty of the raw data is multiplied by the value of the exponent. For instance, if the raw data is cubed (raised to the third power), then propagating its uncertainty involves multiplying the percentage uncertainty by three.

Worked example 8

The rate equation for the decomposition of hydrogen iodide is found to be:

$$\text{rate} = k[\text{HI}]^2$$

where k is the rate constant and [HI] is the concentration of hydrogen iodide. Calculate the rate of the decomposition of hydrogen iodide when $[\text{HI}] = 0.60\,\text{mol}\,\text{dm}^{-3} \pm 0.03\,\text{mol}\,\text{dm}^{-3}$. The value of k at this temperature is $1.58\,\text{dm}^3\,\text{mol}^{-1}\,\text{s}^{-1}$.

Solution

$$\text{rate} = k[\text{HI}]^2$$

$$= (1.58\,\text{dm}^3\,\text{mol}^{-1}\,\text{s}^{-1}) \times (0.60\,\text{mol}\,\text{dm}^{-3})^2$$

$$= 0.5688\,\text{mol}\,\text{dm}^{-3}\,\text{s}^{-1}$$

$$\approx 0.57\,\text{mol}\,\text{dm}^{-3}\,\text{s}^{-1}\ (2\,\text{sf})$$

The result is rounded to two significant figures to be consistent with the least number of significant figures in the values used.

The calculation involves an exponent therefore we must first compute the percentage uncertainty:

$$u([\text{HI}]) = \frac{\pm 0.03\,\text{mol}\,\text{dm}^{-3}}{0.60\,\text{mol}\,\text{dm}^{-3}} \times 100\%$$

$$= \pm 5\%$$

Then, multiply it by the value of the exponent, which is 2 in this case:

$$u(\text{rate}) = \pm 5\% \times 2 = \pm 10\%$$

This gives the overall result:

$$\text{rate} = 0.57\,\text{mol}\,\text{dm}^{-3}\,\text{s}^{-1} \pm 10\%$$

We can also express this as an absolute uncertainty:

$$u(\text{rate}) = \frac{10}{100} \times 0.57\,\text{mol}\,\text{dm}^{-3}\,\text{s}^{-1}$$

$$= 0.057\,\text{mol}\,\text{dm}^{-3}\,\text{s}^{-1}$$

$$\approx 0.06\,\text{mol}\,\text{dm}^{-3}\,\text{s}^{-1}\ (1\,\text{sf})$$

$$\text{rate} = 0.57\,\text{mol}\,\text{dm}^{-3}\,\text{s}^{-1} \pm 0.06\,\text{mol}\,\text{dm}^{-3}\,\text{s}^{-1}$$

ATL Communication skills

What is good mathematical communication? Why is it important? Why are you encouraged to show your working out for all your calculations?

You will notice that propagation increases the overall uncertainty. As raw data are processed, the range of values they can cover becomes greater.

Uncertainties and means

Trials are often repeated to check for repeatability and minimize random error. Uncertainties do not need to be propagated when calculating the mean of a set of values.

For example, consider the following set of results from a titration:

	Titration trial			
	Rough	**1**	**2**	**3**
Initial burette reading / cm³ ± 0.03 cm³	0.00	0.00	11.00	0.10
Final burette reading / cm³ ± 0.03 cm³	11.25	11.00	21.95	11.05
Titrant volume / cm³ ± 0.06 cm³	11.25	11.00	10.95	10.95

Subtracting the final volume from the initial volume gives the titrant volume, hence the volume uncertainty is $2 \times \pm 0.03 = \pm 0.06 \text{ cm}^3$.

The mean of the titrations, omitting the rough trial, is as follows:

$$\text{mean volume} = \frac{11.00 \text{ cm}^3 + 10.95 \text{ cm}^3 + 10.95 \text{ cm}^3}{3}$$

$$= 10.966... \text{ cm}^3$$

$$\approx 10.97 \text{ cm}^3 \pm 0.06 \text{ cm}^3$$

The answer is rounded to two decimal places because the measured values are all given to two decimal places. Note that the uncertainty of the mean volume is the same as the uncertainty of each of the three trials.

🕑 Data-based question

A student performed an experiment to find the density of butan-1-ol. The results are given below.

Volume of butan-1-ol / cm³ ± 0.5 cm³	Mass of measuring cylinder / g ± 0.01 g	Mass of butan-1-ol and measuring cylinder / g ± 0.01 g
8.1	42.82	49.38

a. Determine the mass of the butan-1-ol sample.

b. Calculate the density of the butan-1-ol sample.

c. Calculate the uncertainty of the density. Express your answer as (i) an absolute uncertainty and (ii) a relative uncertainty.

Graphs and tables

Graphs show how changes in an independent variable affect a dependent variable. Graphical techniques can be used to examine the nature of the relationship between the variables and to predict unknown quantities.

Sketching graphs

Sketched graphs have labelled but unscaled axes, and they are used to show qualitative trends, such as variables that are proportional or inversely proportional to each other. Examples of sketched graphs are shown in figure 70.

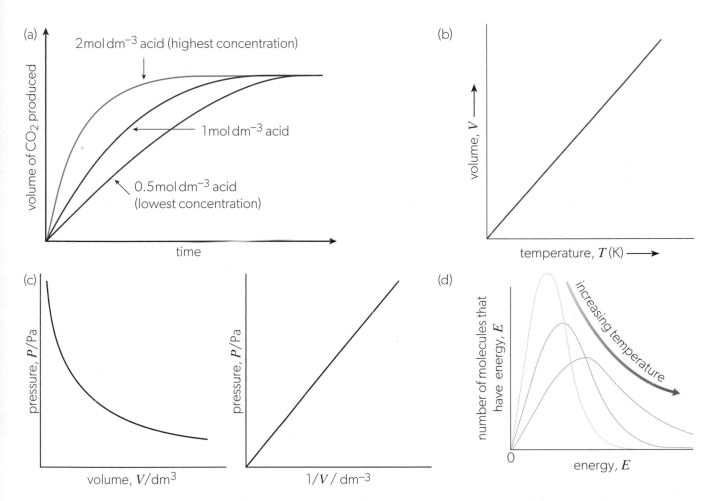

▲ **Figure 70** Examples of sketched graphs: (a) curves showing the effect of concentration on the rate of the reaction between an acid and a metal carbonate (b) graph showing that gas volume is directly proportional to temperature (c) two graphs showing that pressure is inversely proportional to volume (d) Maxwell–Boltzmann distributions of molecular energies for a fixed mass of gas at three different temperatures

---

📚 Activity

Sketch graphs to show:

a. concentration of reactants and products over time in an equilibrium system

b. the melting point of the elements across period 3

c. how pH changes when a strong acid is added to a strong base.

Data, particularly quantitative data, can be organized into tables and then processed into charts. There are different types of charts, including graphs.

Tables

Quantitative data can be presented as a table. It is customary to include the following features in data tables:

- **Descriptive title to aid communication**. If more than one table is being presented, each should be numbered as well.

- **Independent variable in the leftmost column**, listed in order of increasing value.

- **Dependent variable(s) in the columns to the right of the independent variable**. List the results from repeating trials next to each other, and the mean in the rightmost column.

- **Descriptive column headings.**

- **Units and uncertainties in column headings.**

- **Consistent precision in each column.**

- **Clear identification of values of importance** (for example, concordant values in a titration, or anomalous results).

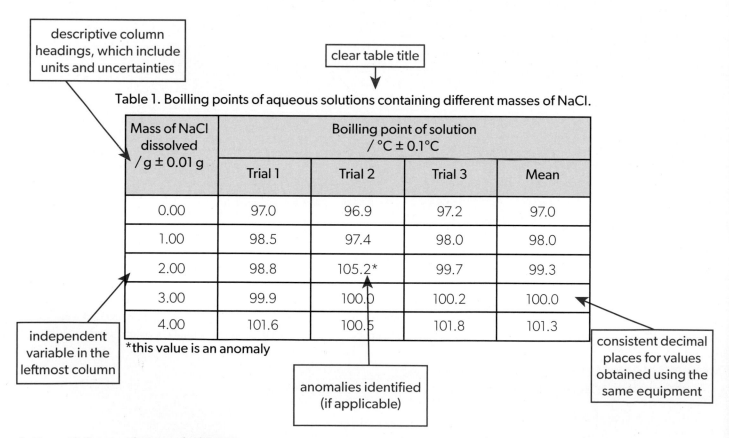

descriptive column headings, which include units and uncertainties

clear table title

Table 1. Boiling points of aqueous solutions containing different masses of NaCl.

Mass of NaCl dissolved /g ± 0.01 g	Boiling point of solution / °C ± 0.1°C			
	Trial 1	Trial 2	Trial 3	Mean
0.00	97.0	96.9	97.2	97.0
1.00	98.5	97.4	98.0	98.0
2.00	98.8	105.2*	99.7	99.3
3.00	99.9	100.0	100.2	100.0
4.00	101.6	100.5	101.8	101.3

*this value is an anomaly

independent variable in the leftmost column

anomalies identified (if applicable)

consistent decimal places for values obtained using the same equipment

▲ Figure 71 Common features of a data table

Control variables should be recorded but they are not usually included in tables because they should not change throughout the experiment. Tables can sometimes include a space for qualitative data, where applicable. Otherwise, qualitative data can be included directly above or below the table.

Sometimes you will need to think of a way to effectively record additional information in your table (such as variables that need to be monitored or citations for data that you have found in databases). You can find examples of how to do this in scientific journal articles.

 Activity

A student did an experiment looking at how the potential difference of a voltaic cell changes with different combinations of half-cells and wrote down the following data during the practical. Organize the data into a suitable table.

> zinc – copper: 0.80 V, 0.78 V, 0.74 V
>
> copper – silver: 0.50 V, 0.53 V, 0.51 V
>
> zinc – silver: 1.30 V, 1.25 V, 1.27 V
>
> magnesium – copper: 2.05 V, 2.00 V, 1.34 V
>
> magnesium – zinc: 1.54 V, 1.43 V, 1.23 V

Charts and graphs

Charts and graphs represent all the data simultaneously, allowing us to explore trends and patterns in the data. There are different types of charts and graphs, each used to present information in a way that aids interpretation. Examples include:

- **Pie charts** allow us to compare parts of a whole, usually expressed as percentages.

- **Bar charts** are used to group data into distinct categories known as categorical variables, such as type of intermolecular forces, or identity of a substituent atom.

- **Histograms** involve grouping data according to a range in the value of the quantitative variable such as length (e.g. 0–1 cm, 1–2 cm, 2–3 cm).

- **Line graphs** also show quantitative data, but they contain points joined by straight lines. Line graphs are particularly useful for continuous data, which means that they can be organized along a numbered line and have any value within a continuous range.

- **Scatter graphs** represent the relationship between two variables. The data points are plotted and then a line or curve of best fit (also known as a trendline) is added to show the relationship between the two variables. Unlike line graphs, the points on scatter graphs are not necessarily joined up because what matters is the overall relationship they show.

(1)

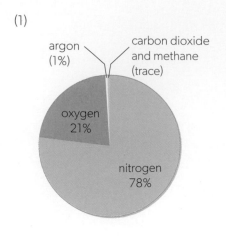

(2)

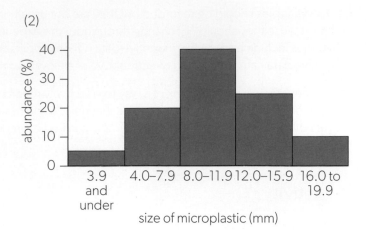

(3)

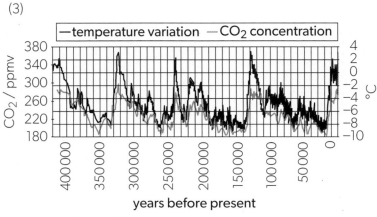

ppmv = parts per million by volume

(4)

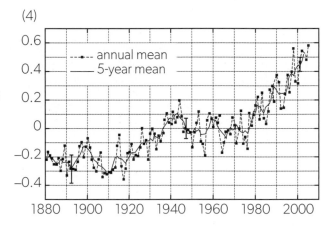

(5)

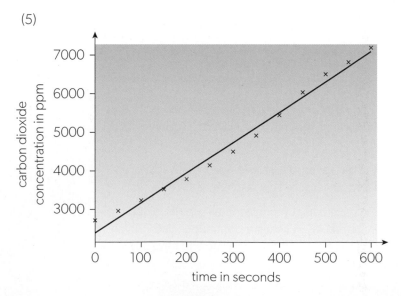

▲ **Figure 72** Different types of charts and graphs. (1) Pie chart showing the present composition of the Earth's atmosphere (2) Histogram showing the distribution in the size of microplastics in a sample (3) Line graph showing global temperature variation and atmospheric CO_2 concentration obtained through isotopic analysis of ice cores (4) Line graph showing global mean temperatures in recent years (5) Scatter graph showing the linear increase of carbon dioxide concentration over time.

Note that the line of best fit does not go through all the points because it represents the overall relationship between the two variables

Plotting graphs

All graphs, regardless of whether they are drawn by hand or plotted using technology should have the following elements:

- Descriptive title to aid communication. If more than one graph is being presented, each should be numbered as well.

- Independent variable on the x-axis, dependent variable on the y-axis

- Axis scales that cover a suitable range and contain equally spaced marks

- Axis labels, including units and uncertainties

- Accurately plotted points.

Data sometimes contains **outliers**, which are points that do not fit the trend of the line or curve. Sometimes scientists use advanced statistical tests to determine outliers. At this level, you can identify outliers by eye. It is important to consider, when discussing your results, possible underlying causes of an outlier.

Logarithmic scales can be used when plotting data that cover a very large range. Doing so compacts the information in the graph and allows us to easily identify significant changes in the data (figure 73).

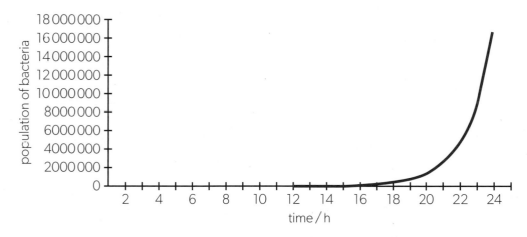

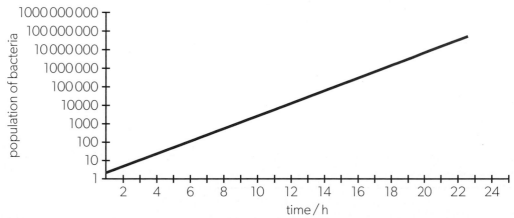

▲ **Figure 73** The scales on both axes on the first graph are linear. Plotting the same data with a logarithmic scale on the y-axis in second graph gives a straight line, confirming that the trend in the data is exponential

Data-based questions

1. Identify the outlier in the graph below:

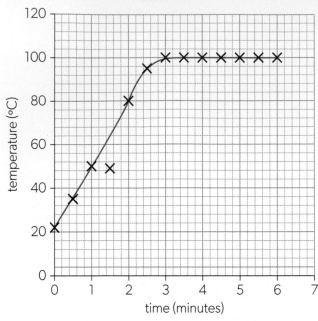

2. The graph below shows the logarithm of the successive ionization energies for potassium.

   Explain why a logarithmic scale on the y-axis is useful in this case.

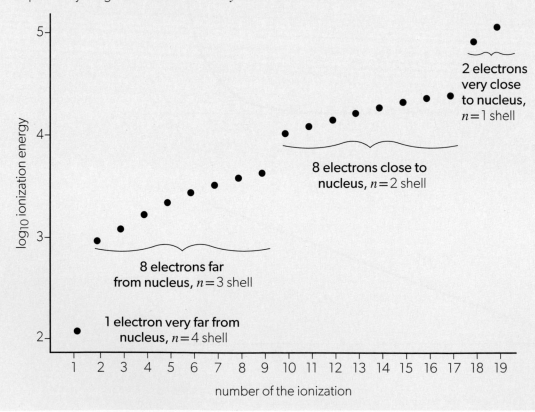

Error bars

Error bars (also known as uncertainty bars) are graphical representations of uncertainties, effectively showing that a data point represents a range of values. Vertical error bars show the uncertainty of the y-axis variable, whereas horizontal error bars show the uncertainty of the x-axis variable.

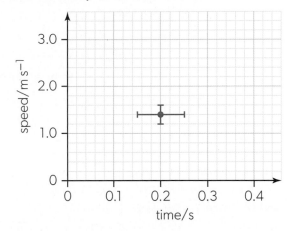

◀ **Figure 74** Error bars show the uncertainty of a point on a graph. The x and y values for this point are, time $= 0.2\,s \pm 0.05\,s$, and speed $= 1.4\,m\,s^{-1} \pm 0.2\,m\,s^{-1}$

Graph-plotting software often has an option to include error bars when the absolute or percentage uncertainty is specified. However, if there is a different uncertainty associated with each point, the software might not accommodate this. In this case you can either draw the error bars for each point by hand or have the program draw error bars corresponding to the largest uncertainty in the data set (and add a note to this effect). You should determine which option is best in the context of your analysis.

Lines or curves of best fit

A **line** or **curve of best fit** represents the general trend in a graph and can be used for further analysis of the relationship between the variables. The line or curve of best fit does not have to pass through every single point, but it must have roughly the same number of points on either side of it. If there are any obvious outliers amongst your data, you might need to ignore them when drawing the line or curve of best fit. This does not mean, however, that they should be excluded from the analysis. It is good practice to circle and label outliers on graphs, as well as discussing their possible sources and significance in the analysis.

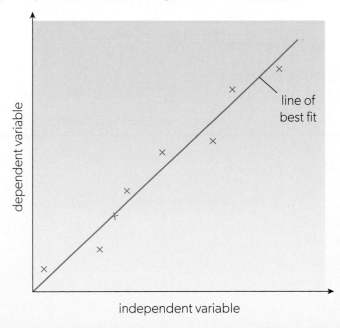

◀ **Figure 75** A generalized graph showing a line of best fit

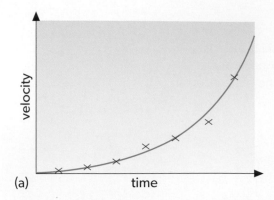

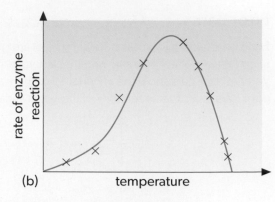

▲ **Figure 76** Depending on the nature of the data, curves of best fit can be simple (a) or complex (b)

Technology can be useful when drawing lines and curves of best fit, but you must consider whether computer-generated lines or curves of best fit make sense in the context of the data. For example, software may skew the line due to an outlier, or it may show a straight line where a curve would be more suitable. Lines and curves of best fit do not always have to pass through the origin.

If points on a graph have error bars, then the curve or line of best fit should pass through as many of the error bars as possible. If the error bars of a point are outside the line of best fit, it may indicate that the point is an outlier, or possibly that the uncertainty is greater than expected.

Data-based questions

A reaction rate investigation produced the following data:

Concentration of reactant A / mol dm^{-3} ± 0.0002	Rate of reaction / mol dm^{-3} s^{-1} ± 5%
0.0010	0.015
0.0020	0.031
0.0030	0.027
0.0040	0.065
0.0050	0.089

a. Plot a graph of the results.

b. Draw error bars on your graph.

c. Draw a line of best fit.

d. Identify the outlier and suggest what might have caused it.

Interpretation of features of graphs

The relationship between variables can be further analysed by interpreting the features of a graph: correlation, form, gradient, intercept, maxima and minima, and area. The main thing to remember is that the features of a graph can often represent an aspect of the idea being studied. We will also discuss the use of a coefficient of determination (R^2), interpolation and extrapolation.

The first thing to consider when interpreting a graph is the **correlation** it shows (if any). If the dependent (y) variable increases when the independent variable (x) increases, then the variables are positively correlated. A negative correlation describes a relationship where an increase of the independent variable causes the dependent variable to decrease. If the points are randomly scattered in the graph, the correlation is weak or absent (figure 77).

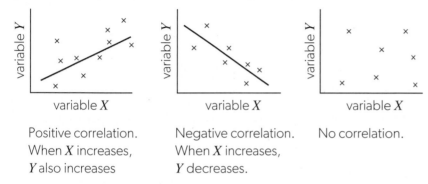

Positive correlation. When X increases, Y also increases

Negative correlation. When X increases, Y decreases.

No correlation.

▲ **Figure 77** Sketched graphs showing positive, negative and absent correlation

The relationships shown in graphs can be linear or non-linear. A linear relationship corresponds to a straight line in the graph that is described by the equation $y = mx + c$, where m is the gradient (slope) of the line and c is the y-intercept. **Direct proportionality** is a special type of linear relationship, in which the two variables are in a constant ratio. When one variable increases, the other increases by the same rate. For example, the volume of a gas is directly proportional to its temperature, provided that the temperature is in kelvin (figure 78). If two variables are directly proportional, the straight-line graph must go through the origin.

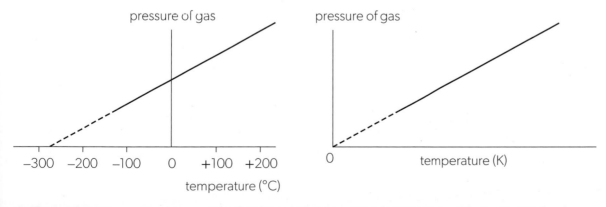

▲ **Figure 78** Pressure and temperature of an ideal gas are linearly related, but the relationship is only directly proportional when the temperature is in kelvin

Variables are inversely proportional when one increases and the other decreases by the same rate. Unlike direct proportionality, graphs showing inverse proportionality are not linear. Plotting one of the variables in its reciprocal form, however, gives a straight-line graph. The relationship between pressure and volume of an ideal gas is a good example of inverse proportionality (figure 79).

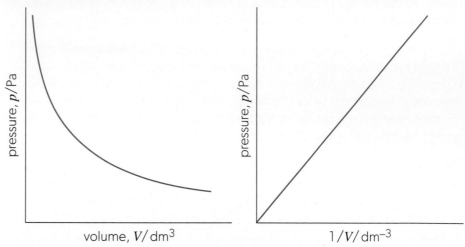

▲ **Figure 79** Pressure and volume of an ideal gas are inversely proportional to each other: plotting p vs $\frac{1}{V}$ gives a straight-line graph

Non-linear is a term that includes all relationships that are not straight lines. Sometimes, non-linear relationships can be described by functions, such as exponential, logarithmic and quadratic functions.

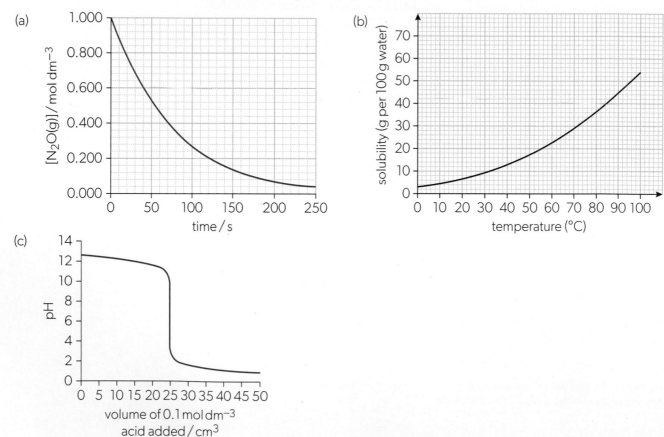

▲ **Figure 80** Graphs showing non-linear relationships: (a) the decomposition of nitrous oxide, (b) a solubility curve and (c) a pH curve for the addition of a strong acid to a strong base

Gradient (or slope)

The **gradient** (or **slope**) of a line is a measure of its steepness, but it also has a physical meaning: it is the rate of change of one variable with respect to the other variable. To find the gradient of a line, you need to choose two points on that line, ideally well separated from each other. Gradients can be positive or negative, depending on whether the graph shows a positive or a negative correlation, respectively.

$$m = \frac{\Delta y}{\Delta x} = \frac{y_2 - y_1}{x_2 - x_1}$$

For example, a graph of mass of a substance (along the y-axis, in g) vs its volume (along the x-axis, in cm^3) gives a straight line. The gradient of the line has units $g\,cm^{-3}$, which gives us the density of that substance.

The uncertainty for the gradient can be determined by drawing the steepest and most gradual lines of best fit possible within the constraints of the error bars, giving values for maximum and minimum gradients. Half the difference between these two lines can be used to estimate the uncertainty of the gradient.

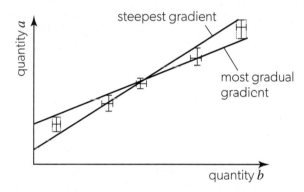

▲ **Figure 81** Maximum and minimum gradients on a graph can be determined from the most and least steep lines of best fit

Straight-line graphs have constant gradients while curves have changing gradients. For instance, data for product volume vs time will produce a curve, because the rate decreases during the reaction (figure 82).

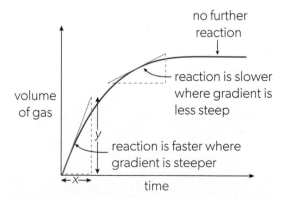

▲ **Figure 82** The rate decreases over the course of a reaction, so the gradient decreases over time

We can determine the gradient at a specific point on the curve by drawing a **tangent line** and working out its gradient. In figure 82, two tangent lines are shown in green. The one at $t = 0$ is clearly steeper because the rate is higher at that point in time. Figure 83 shows how to calculate the gradient of a tangent line.

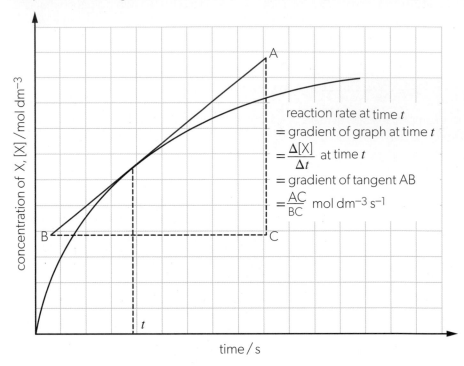

reaction rate at time t

$=$ gradient of graph at time t

$= \dfrac{\Delta[X]}{\Delta t}$ at time t

$=$ gradient of tangent AB

$= \dfrac{AC}{BC}$ mol dm^{-3} s^{-1}

▶ **Figure 83** The rate of reaction at time t can be determined by calculating the gradient of the tangent line at that point

Intercepts

The intercepts are the points where the line or curve crosses the axes. Intercepts can be found by extending the line until it meets the axis (this is called extrapolation). You can also use the equation of the line in the form $y = mx + c$ (or $y = mx + b$) where c (or b) represents the y-intercept.

As with gradient, the significance of the intercepts depends on the context of the data being plotted. In some cases, you would expect the y-intercept to be 0, for instance when plotting the concentration of a product *vs* time as a part of a kinetics investigation. In other cases, the intercept can be used to derive a numerical quantity (figure 84).

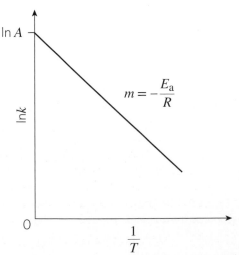

▶ **Figure 84** The y-intercept of a graph showing $\ln k$ *vs* $\dfrac{1}{T}$ is $\ln A$, where A is the Arrhenius factor

Similar to gradient, the uncertainty of the y-intercept can be determined by drawing the most and least steep lines of best fit possible within the constraints of the error bars. This gives maximum and minimum values for the y-intercept, from which you can find the uncertainty (figure 85).

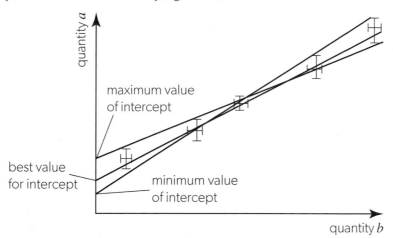

▲ **Figure 85** Maximum and minimum values for the y-intercept can be determined from the most and least steep lines of best fit

Maxima, minima and areas on graphs

Further interpretation of graphs may involve inspecting maximum values, minimum values or the area under a curve. The significance of each of these depends on the variables being represented in the graph.

The absorbance spectrum of a solution of an unknown substance (figure 86) shows a maximum at around 520 nm, which corresponds to the green region of the visible spectrum. The colour of the unknown solution must be red, as it is complementary to green.

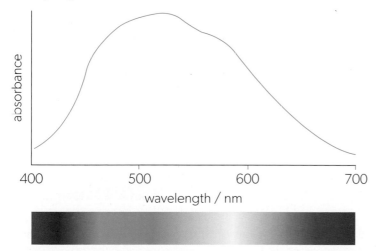

▲ **Figure 86** The maximum absorbance shown in this graph helps us to determine the colour of the solution

The changes in the concentrations of ozone, O_3, and chlorine monoxide radicals, $ClO\bullet$, 18 km high in the atmosphere across different latitudes are shown in figure 87. The minimum in ozone concentration occurs at the same point as a maximum in $ClO\bullet$ concentration, suggesting an association between low ozone levels and high $ClO\bullet$ levels.

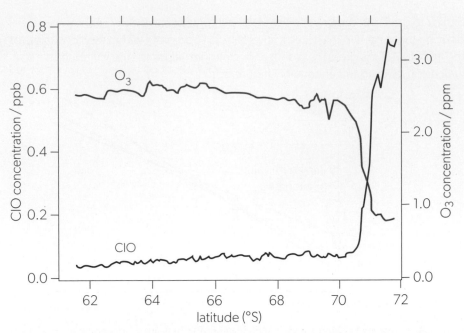

▲ **Figure 87** The minimum ozone, O_3, concentration corresponds to a maximum chlorine monoxide radical, CIO•, concentration

The area under a curve (or the area under a graph) is the area enclosed by the curve and x-axis. It represents an accumulation of the y-axis variable over the course of a certain x-axis variable change.

The area under the curve is useful in Maxwell–Boltzmann distributions, where it represents the number of particles in the sample. Figure 88 shows the Maxwell–Boltzmann distributions for the same species at two different temperatures.

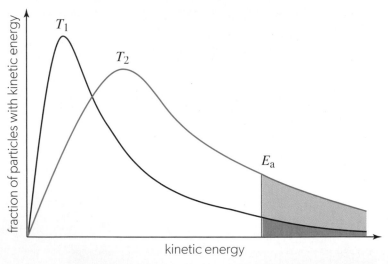

▲ **Figure 88** The area under the curve in a Maxwell-Boltzmann distribution represents the number of particles, which can be used to compare the proportion of particles that have sufficient energy to react at two temperatures T_1 (red) and T_2 (green), where $T_2 > T_1$

The particles that have sufficient energy to react are represented by the shaded regions. Their kinetic energy is equal to or greater than the activation energy, E_a. At the higher temperature, T_2, the shaded area is larger (red + green) and therefore the number of particles with sufficient energy to react is higher than at T_1, at which the shaded area is smaller (red only).

Data-based questions

1. Students investigating the effect of temperature on air volume plotted the following graph of their results:

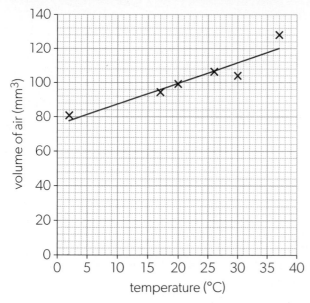

a. State whether the correlation between volume and temperature is positive or negative.

b. Does the graph show direct proportionality? Why or why not?

c. Determine the gradient of the line of best fit. State your answer to two significant figures and include the units.

d. Describe briefly the physical significance of the gradient.

2. The following data were recorded using six standard solutions in order to determine the concentration of a sample of aqueous copper(II) sulfate using a technique called atomic emission spectroscopy.

Concentration, c / mol dm^{-3}	Absorbance, A
0.1002	0.130
0.2008	0.270
0.2819	0.380
0.4000	0.540
0.5082	0.685
0.6000	0.810
Unknown	0.460

a. Plot a suitable graph of these results and include a line of best fit.

b. Determine the concentration, in mol dm^{-3}, of copper(II) sulfate.

c. Deduce the equation of the line of best fit, stating the gradient and y-intercept both to two significant figures.

d. Discuss whether this graph shows a direct proportionality between absorbance and concentration.

3. The graph (right) shows the volume of hydrogen gas collected over time for two experiments. A tangent line has been drawn at 25 seconds for experiment 1.

   a. Calculate the instantaneous rate of reaction for experiment 1 at $t = 25$ s.

   b. Describe and explain whether the instantaneous rate of reaction for experiment 2 at $t = 25$ s is greater or less than that of experiment 1.

   c. Describe and explain whether the overall rate of reaction in experiment 2 is greater or less than that in experiment 1.

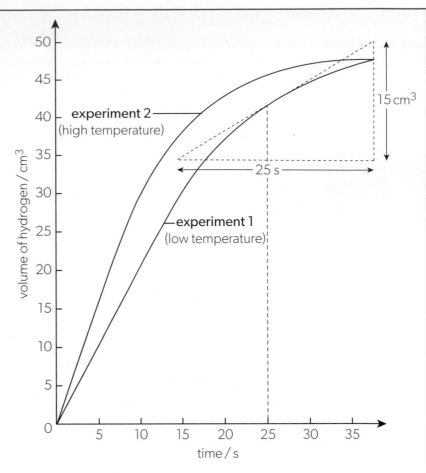

4. The graph shows how the concentration of dinitrogen pentoxide, N_2O_5, changes over time. A tangent has been drawn at $t = 0$.

   a. Calculate the initial rate of reaction. Give your answer to two significant figures and state the units.

   b. Calculate the average rate of reaction in the first 30 seconds.

   c. Explain why your answers to (a) and (b) above are different.

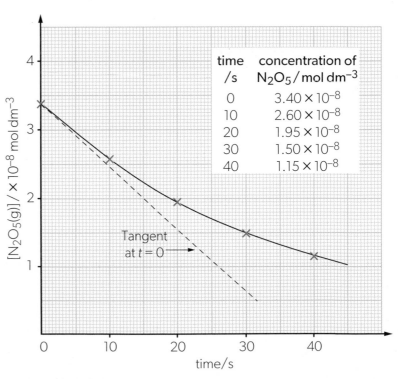

time /s	concentration of N_2O_5 / mol dm^{-3}
0	3.40×10^{-8}
10	2.60×10^{-8}
20	1.95×10^{-8}
30	1.50×10^{-8}
40	1.15×10^{-8}

Coefficient of determination, R^2

We have seen that all measurements, and data derived from them, have an associated uncertainty. As a result of this, the data points in a graph are not likely to fall directly on the line of best fit, but rather be scattered on either side of it. One of the tools we can use to assess the predictive power of the relationship between the variables in a graph is R^2, the **coefficient of determination**.

In DP chemistry, you will use technology to find R^2 values, and you will be expected to interpret them. However, you will not be required to calculate R^2 yourself. The value of R^2 lies between 0 and 1. The closer it is to 1, the more closely the data points match the line of best fit (figure 89) and the more successful the model is at predicting the changes in the y-axis variable. R^2 is only suitable for linear models.

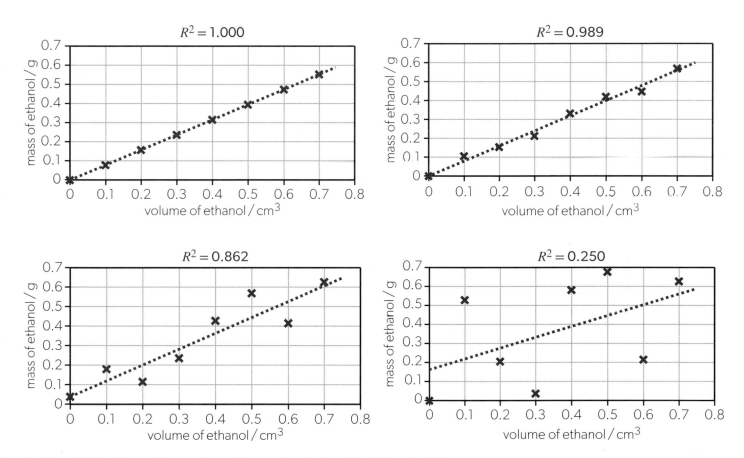

▲ **Figure 89** The R^2 value is 1 when the points are all on the line of best fit (graph 1). The points in graphs 2 to 4 become increasingly more spread out, causing the R^2 to decrease

Interpolation and extrapolation

Lines and curves of best fit can be interpolated or extrapolated. Interpolation involves predicting values within the range of the experimental data, whereas extrapolation involves predicting values outside the range of these data (figure 90).

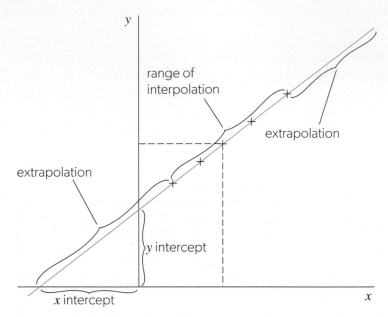

▲ **Figure 90** Interpolation and extrapolation using a straight line

Consider the absorbance versus concentration graph shown in figure 91. The points on the line were obtained by measuring the absorbance of six solutions of known concentration.

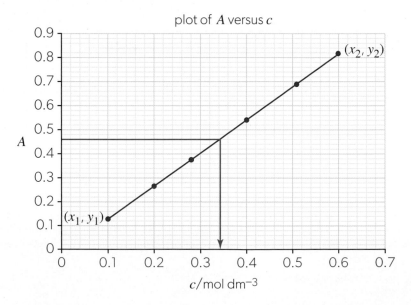

▲ **Figure 91** Interpolation involves finding values within the range of the experimental data

The absorbance of a sample of unknown concentration is found to be 0.460. Its concentration can then be found by interpolation, as shown, giving 0.34 mol dm^{-3}. If known, the equation of the line could also be used to find the concentration, by entering 0.460 in for y and solving for x.

In a calorimetry experiment, the temperature of the reaction mixture containing copper(II) sulfate solution and zinc metal is monitored over time and the results are plotted on a graph (figure 92). The maximum temperature value is needed to calculate ΔT and subsequently the amount of heat, Q (and later, the enthalpy change, ΔH). The highest temperature measured during the experiment is lowered due to heat loss to the surroundings.

Extrapolating the cooling curve (the part of the graph that shows a decreasing temperature over time) backwards to the time at which the zinc was added to the copper(II) sulfate solution, allows us to predict a corrected value of T_{max}. Doing so assumes a constant rate of heat loss during the reaction.

Extrapolation always involves the assumption that the relationship between variables stays the same beyond the range of data. In practice, this is not always the case. For example, the relationship between absorbance and concentration in figure 91 is linear, but the linearity is lost at higher concentrations.

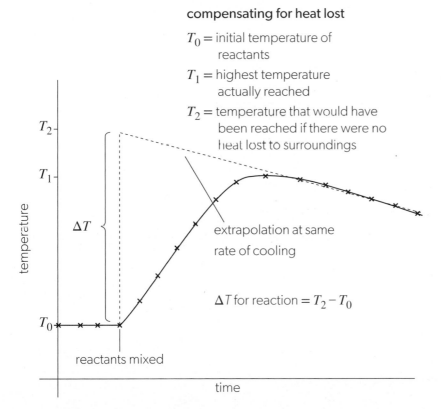

compensating for heat lost

$T_0 =$ initial temperature of reactants

$T_1 =$ highest temperature actually reached

$T_2 =$ temperature that would have been reached if there were no heat lost to surroundings

extrapolation at same rate of cooling

ΔT for reaction $= T_2 - T_0$

reactants mixed

time

▲ **Figure 92** Extrapolation of temperature data during a zinc metal and copper(II) sulfate reaction to determine a corrected value of T_2, the maximum temperature

Data-based questions

1. A student performed a calorimetry experiment to determine the enthalpy change for the reaction between zinc metal and copper(II) sulfate. A graph of the student's results is shown below.

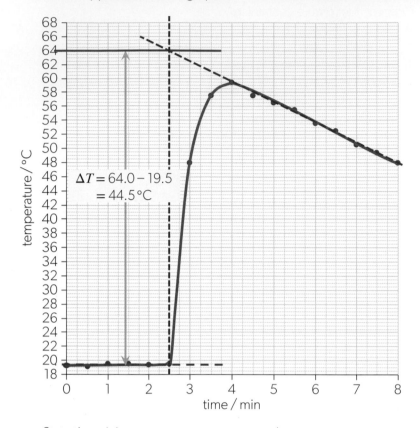

    a. State the minimum temperature measured.

    b. State, giving a reason, whether the reaction is exothermic or endothermic.

    c. Explain why the student extrapolated the rate of cooling back to $t = 2.5$ min.

    d. Outline what other information you would need to calculate the heat, Q, and enthalpy change, ΔH, for this reaction.

2. A student collected temperature data every minute while titrating hydrochloric acid, HCl(aq), with 0.600 mol dm^{-3} sodium hydroxide, NaOH(aq). Data for the volume of sodium hydroxide added and temperature are shown in the table on the right.

    a. Plot a graph of these data.

    b. Find the maximum temperature by extrapolation of the heating and cooling curves.

    c. The volume of hydrochloric acid was 25.00 cm³. Determine the concentration of the acid.

    d. Calculate the enthalpy change of neutralization, $\Delta H^{\ominus}$, for this reaction, in kJ mol^{-1}.

Volume of NaOH (aq) / cm³	Temperature / °C
0.00	22.4
5.00	23.1
10.00	23.9
15.00	24.6
20.00	25.4
25.00	25.6
30.00	25.5
35.00	25.4
40.00	25.3
45.00	25.2
50.00	25.1

Sources of experimental error

Experimental errors result from the fact that measurements are never exact: they always have uncertainties associated with them. Experimental errors can be systematic or random.

The word "error" in everyday language can mean "mistake". In science, "error" has a very different meaning. One of the features of a thorough scientific research is the consideration of sources of experimental errors, their impact on the outcome of an investigation, and possible ways to minimize them.

Random and systematic error

Random errors arise due to natural variations in measured values, leading to measurements that are equally likely to be too high as too low. Random errors can be minimized, but not eliminated, by repeating measurements. Large random errors reduce precision because the values are spread over a greater range. The uncertainty provides an estimate of random error.

Sources of random errors include readings that fluctuate over time, uncertainty associated with reading a scale, and random variations in the measurements made with a particular instrument.

Systematic errors affect results in the same direction (either making the results too high or too low). For example, consistently taking a burette reading from above the meniscus will result in readings that are too high (figure 93).

Systematic errors reduce accuracy and can be minimized, but not eliminated, by making changes to the experimental methodology. If a particular systematic error can be quantified, the values can then be corrected for it.

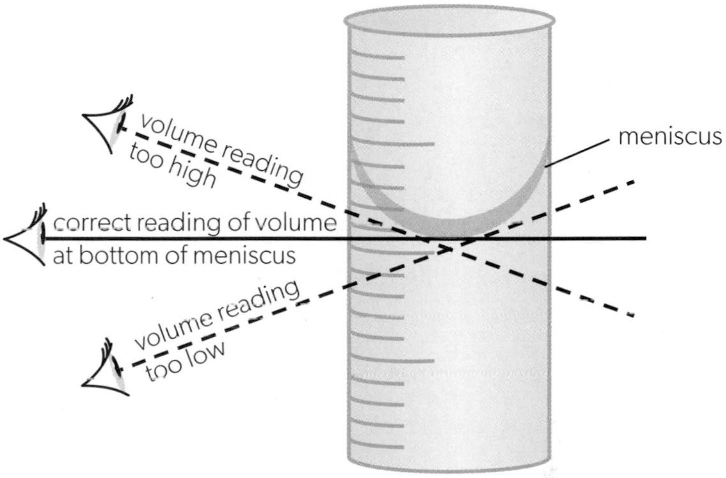

Sources of systematic errors include a miscalibrated temperature probe, presence of background radiation, leaks in gas collection apparatus, and heat loss to the surroundings during calorimetry experiments.

▲ Figure 93 Examples of systematic error when taking burette readings

Errors and processed results

Estimates of random errors, obtained by uncertainty propagation, are useful in the analysis of errors in experimental results. Comparisons between percentage uncertainties, and, where available, theoretical values and percentage errors can give a basis for evaluating the precision and accuracy of a processed result.

Percentage error (not to be confused with percentage uncertainty) is a way of quantifying how far an experimental value is from the theoretical value:

$$\text{Percentage error} = \left| \frac{\text{experimental value} - \text{theoretical value}}{\text{theoretical value}} \right| \times 100\%$$

Note that the theoretical value is sometimes called the accepted value or literature value.

For example, the theoretical value for the activation energy of a reaction is reported in the literature as $65 \, kJ \, mol^{-1} \pm 1\%$. Students A and B determine this activation energy experimentally and obtain the data shown in table 5. Student A finds the activation energy to be $60 \, kJ \, mol^{-1} \pm 10\%$. The $\pm 10\%$ part is an estimate of the random error, indicating that the result is in the range of 54–$66 \, kJ \, mol^{-1}$. We can see that the theoretical value lies within the range of the experimental value, suggesting that the two are aligned. However, the high percentage uncertainty (10%) indicates low precision, particularly when compared to that of the theoretical value (1%).

	Experimental E_a value / $kJ \, mol^{-1}$	Uncertainty / %	Absolute uncertainty / $\pm kJ \, mol^{-1}$	Range / $kJ \, mol^{-1}$	Percentage error / %
student A	60	10	6	54–66	7.7
student B	50	4	2	48–52	23

▲ Table 5 Student results for the experimentally determined activation energy of a reaction

Now consider the result obtained by student B for the same reaction: $50 \, kJ \, mol^{-1} \pm 4\%$. The random error is smaller than that obtained by student A, indicating greater precision (though still not as precise as the theoretical value). The result is in the range of 48–$52 \, kJ \, mol^{-1}$. This time, the theoretical value does not lie in this range, therefore the result is inaccurate. In addition, since the uncertainty is 4% and the percentage error is 23%, only up to 4% of the 23% error can be attributed to random errors. This suggests that systematic errors contributed significantly to the difference between student B's result and the theoretical value.

Practice questions

51. Classify each of the following errors as systematic or random:

    a. heat loss to the surroundings in an investigation involving calorimetry

    b. use of a miscalibrated colorimeter to determine the absorbance of a sample

    c. taking a burette reading by positioning the viewer slightly below the meniscus rather than positioning the eyes on a level with the meniscus

    d. use of a pH probe that reports values to the nearest whole number.

52. A student calculated the enthalpy change for the combustion of butan-1-ol and obtained $-1430 \, kJ \, mol^{-1}$. The theoretical value is $-2676 \, kJ \, mol^{-1}$. Calculate the percentage error in the student's result.

53. Two students investigated the enthalpy of neutralization of a strong acid with a strong base. Their results were:

    student C: $-35 \, kJ \, mol^{-1} \pm 4 \, kJ \, mol^{-1}$
    student D: $-50 \, kJ \, mol^{-1} \pm 10 \, kJ \, mol^{-1}$

    The theoretical value for the enthalpy of neutralization is $-57 \, kJ \, mol^{-1}$.

    a. Calculate the percentage errors of the students' results.

    b. Comment on the relative impact of random errors on each of the results.

Errors and graphs

Graphs can also provide information regarding systematic and random error (figure 94). Random errors cause data points to deviate from the line of best fit. Systematic errors cause all the data to shift in the same direction (either up or down). On graphs, this means that the points are systematically located higher (or lower) than expected. This is especially noticeable in the y-intercept: a y-intercept that is not where you would expect it to be suggests the presence of a systematic error.

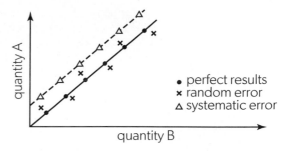

• perfect results
× random error
△ systematic error

▲ **Figure 94** The presence of systematic and random errors can be inferred from graphical data

Practice questions

54. The combustion of primary alcohols methanol to octan-1-ol was determined experimentally (table 6).

Number of carbon atoms in primary alcohol	Experimental enthalpy of combustion / kJ mol⁻¹ ± 40 kJ mol⁻¹
1	−230
2	−400
3	−680
4	−814
5	−990
6	−1300
7	−1440
8	−1700

▲ Table 6 Experimental enthalpy of combustion of the primary alcohols

a. Plot a graph of the data in table 6, including error bars and a line or curve of best fit.

b. Using your graph, and published enthalpy of combustion data, comment on the impact of systematic error on the results.

c. Using your graph, comment on the impact of random error on the results.

55. A student investigated the effect of concentration of sodium hydrogencarbonate on the rate of its reaction with citric acid. The reaction produces carbon dioxide, so the student timed how long it took to collect 25 cm³ of this gas. The student's data collected is shown in table 7.

Concentration of sodium hydrogencarbonate / mol dm⁻³ ± 5 %	Time taken to produce 25 cm³ of carbon dioxide / s ± 1 s		
	1	2	3
0.25	210	277	166
0.50	195	135	118
1.00	66	93	53
1.50	21	27	35
2.00	15	25	15

▲ Table 7 Time taken to produce 25 cm³ of carbon dioxide in the reaction of sodium hydrogencarbonate and citric acid at different concentrations of sodium hydrogencarbonate

a. Calculate the mean time for each of the sodium hydrogencarbonate concentrations.

b. Determine an estimate of the time uncertainty by halving the range of each set of three trials.

c. Plot a graph to show the relationship between concentration of sodium hydrogencarbonate and time, including error bars and a line or curve of best fit.

d. Using your graph, comment on the impact of random error on the analysis.

e. Calculate the rate by finding $\frac{1}{T}$ for each concentration.

f. Plot a graph to show the relationship between concentration of sodium hydrogencarbonate and rate.

g. Using your graph, comment on the impact of systematic and random error on the analysis.

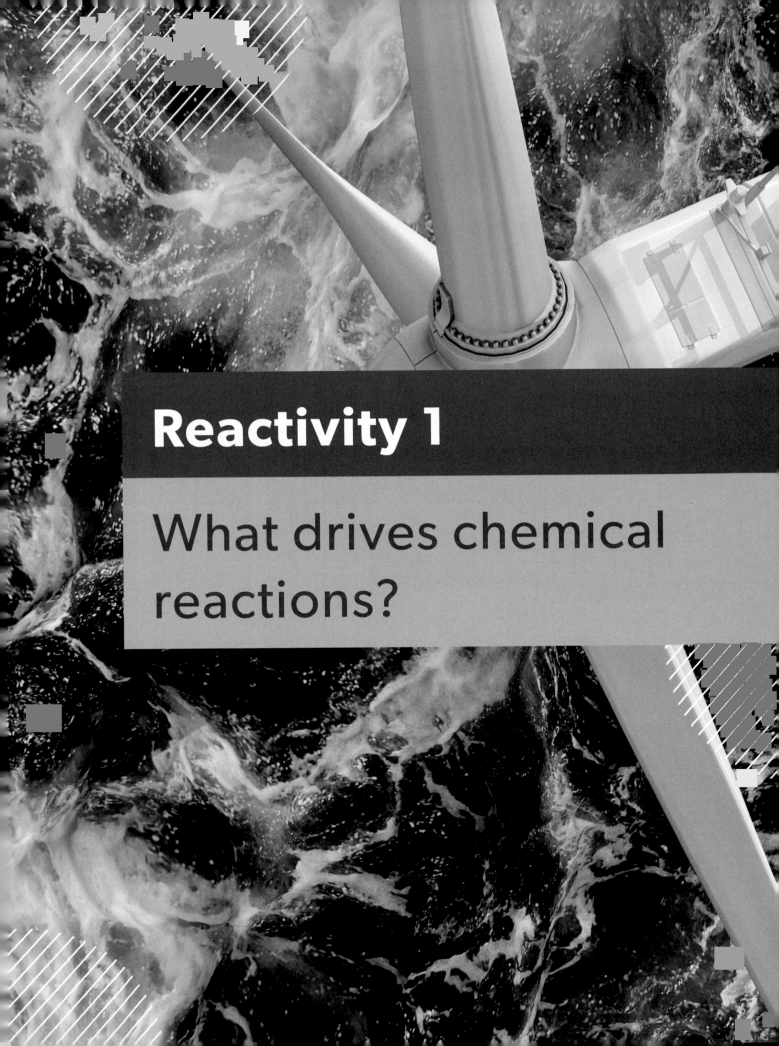

Reactivity 1

What drives chemical reactions?

What can be deduced from the temperature change that accompanies chemical or physical change?

Chemistry involves the study of chemical reactions and physical changes of state of the elements and their compounds. Conservation of energy is a fundamental principle of science, which is examined through observation and experimentation. The use of models, empirical or experimental data, the language of mathematics and scientific terminology, all contribute to our understanding of energy changes associated with chemical reactions.

An understanding of the relationships that exist between chemistry and energy involves understanding how energy is transferred between a chemical system and the surroundings. This information can in turn be used to develop an understanding of the relative stability of reactants and products, leading to better control over the progress of the reaction being studied.

Understandings

Reactivity 1.1.1—Chemical reactions involve a transfer of energy between the system and the surroundings, while total energy is conserved.

Reactivity 1.1.2—Reactions are described as endothermic or exothermic, depending on the direction of energy transfer between the system and the surroundings.

Reactivity 1.1.3—The relative stability of reactants and products determines whether reactions are endothermic or exothermic.

Reactivity 1.1.4—The standard enthalpy change for a chemical reaction, $\Delta H^{\ominus}$, refers to the heat transferred at constant pressure under standard conditions and states. It can be determined from the change in temperature of a pure substance.

Energy transfer in chemical reactions (*Reactivity 1.1.1*)

In a chemical reaction, total energy is conserved. Chemical potential energy is stored in the chemical bonds of **reactants** and **products**, while the temperature of the reaction mixture is a function of the kinetic energy of the atoms, ions and molecules present.

All chemical reactions involve energy changes. Energy may be released into the **surroundings** from the **reaction system** or it may be absorbed by the reaction system from the surroundings. Most commonly, the energy is transferred in the form of heat, but it may also be in the form of electricity, sound or light.

In an **open system**, the transfer of matter and energy is possible across its boundary (for example, matter can be added to a beaker, and energy can be transferred through its sides). A **closed system** allows no transfer of matter, though energy may be transferred across the boundary. In an **isolated system**, matter and energy can neither enter nor exit the system.

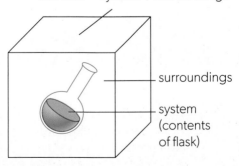

universe = system + surroundings

surroundings

system (contents of flask)

▲ Figure 1 The universe is the combination of the system and its surroundings

▲ **Figure 2** In each of the above scenarios, energy is transferred. In hot springs, energy is transferred as heat, and in fireworks, as heat, sound and light. Both scenarios are open systems

 ## Models

Most industrial processes take place in open or closed systems. The loss of heat during an industrial process not only affects the efficiency of the chemical reaction, but also contributes to a loss of useful energy, an increase in thermal pollution. Thermography can be used to model heat flow and loss from structures in chemical industries as a heat map, where red is hot and purple is cold.

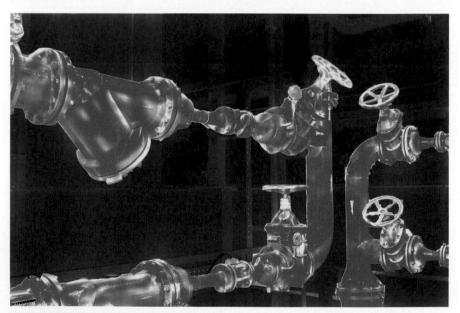

▲ **Figure 3** Thermograph of industrial engineering system

What are the advantages of modelling heat distribution and transfer? How can chemical engineers use the data collected to improve the efficiency of industrial processes? How does this help our environment?

What is the difference between heat and temperature?

Temperature, T, is an example of a **state function**. For a state function, any change in value is independent of the pathway between the initial and final measurements.

For example, if you take the temperature of the water in a swimming pool early in the morning (the initial value) and then again in the afternoon (final value), it will not tell you the complete story of any temperature fluctuations that may have occurred throughout the day. The calculation of the temperature change is simple:

$$\Delta T = T_{final} - T_{initial}$$

▲ **Figure 4** If you record the temperature of a pool at the beginning and the end of a day, it will not give you an indication of the heating and cooling that has occurred throughout the day, only the overall temperature change

Other examples of state functions include volume, enthalpy and pressure.

Heat, q, is a form of energy that is transferred from a warmer body to a cooler body, as a result of the **temperature gradient**. Heat is sometimes referred to as **thermal energy**. It can be transferred by the processes of conduction, convection and radiation.

Heat has the ability to do **work**. When heat is transferred to an object, the result is an increase in the average **kinetic energy** of its particles. This causes an increase in temperature and potentially a phase change, for example, a change of state from liquid to gas.

At **absolute zero**, 0 K (-273.15 °C), all motion of the particles theoretically stops and the **entropy**, S, of a system reaches its minimum possible value. The absolute temperature (in kelvin) is proportional to the average kinetic energy of the particles of matter. As the temperature increases, the kinetic energy of the particles also increases.

Entropy is defined and explored in *Reactivity 1.4*.

(ATL) Communication skills

Your communication skills will develop incrementally throughout the entire chemistry programme. Communication skills include consistent and accurate application of scientific terminology. Your use of the terms "heat" and "temperature" in explanations will demonstrate your understanding of these concepts.

For example, consider the simple reaction between magnesium metal and hydrochloric acid.

$$Mg(s) + 2HCl(aq) \rightarrow MgCl_2(aq) + H_2(g)$$

In this reaction, heat is released from the reaction system into the surroundings, and the temperature of the aqueous solution rises. When you think about heat, you are considering the transfer of thermal energy in the system. When you refer to the temperature of a system, you are describing the average kinetic energy of the particles within that system.

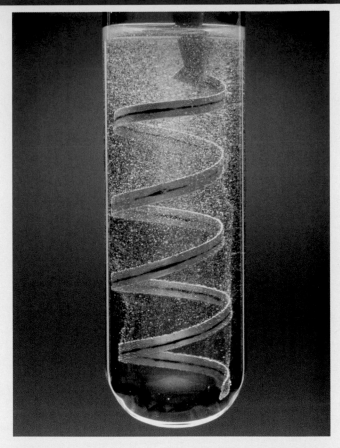

▲ **Figure 5** Magnesium ribbon reacting with hydrochloric acid

Thermochemistry is the study of heat changes that occur during chemical reactions. Heat changes are often described in terms of **enthalpy**. At constant pressure, the **enthalpy change**, ΔH, is defined as the heat transferred from a closed system to the surroundings during a chemical reaction. The terms "enthalpy change" and "heat of reaction" are commonly used when describing the thermodynamics of a reaction. The most common unit of enthalpy change is kJ.

 Activity

Imagine a glass of water containing ice cubes sitting in the summer sun. It will undergo a change in enthalpy.

- Is the glass of water an open, closed or isolated system?

- Identify the system and the surroundings.

- Explain the movement of energy in the form of heat, between the system and the surroundings.

- What would you observe on the outside of the glass? Explain this observation in terms of a change of state and transfer of energy.

Exothermic and endothermic reactions (*Reactivity 1.1.2*)

A chemical reaction in which heat is transferred from the system to the surroundings is defined as an **exothermic** reaction. In contrast, chemical reactions in which heat is absorbed into the system from the surroundings are defined as **endothermic** reactions.

When a chemical reaction takes place, the atoms of the reactants are rearranged to create new products. Chemical bonds in the reactants are broken, and new chemical bonds are made to form products. Energy is absorbed by the reaction system to break the chemical bonds, and therefore bond breaking is an endothermic process. This energy is termed the **bond dissociation energy** and it can be quantified for each type of bond. Energy is released into the surroundings when new chemical bonds are made, and therefore bond making is an **exothermic** process. The **transfer of energy** between the surroundings and the system is an important part of your understanding of the energy changes in a reaction.

Exothermic reactions have a negative enthalpy change, and endothermic reactions have a positive enthalpy change. The sign of the enthalpy change is defined from the perspective of the system and not the surroundings. For example, in an exothermic reaction, heat is being lost by the system and so the enthalpy change is negative. For an endothermic reaction, heat is absorbed by the system, so the enthalpy change is positive.

To determine whether a reaction is endothermic or exothermic, we can use a calorimeter. A calorimeter is any apparatus used to measure the amount of heat being exchanged between the system and the surroundings. In the school laboratory, experiments focus on the change in temperature, ΔT, of the reaction solvent, which in most cases is water.

Observations

In the laboratory, observations can be made using human senses, or with the aid of instruments such as data-logging equipment. The application of digital technology to collect data is one of the essential skills in the study of chemistry.

Energy profiles (*Reactivity 1.1.3*)

Energy profiles are a visual representation of the enthalpy change during a reaction. From an energy profile, you can determine the enthalpy of the reactants and the products, the **activation energy** (E_a), and the enthalpy change for the reaction.

Many chemical reactions are exothermic. In these reactions, energy is released from the system to the surroundings. The reactants of this reaction are at a higher energy level and considered to be lower in stability. Products for exothermic reactions are at a lower energy level and considered to be more energetically favourable.

Activation energy, E_a, is the minimum energy required for the reaction to take place. You will study activation energy in *Reactivity 2.2*.

Consider the reaction between zinc and copper(II) sulfate solution. It is a single displacement reaction involving the displacement of the copper(II) ion by zinc:

$$Zn(s) + CuSO_4(aq) \rightarrow Cu(s) + ZnSO_4(aq)$$

Measured quantities of copper(II) sulfate solution and zinc are mixed in a calorimeter. The mixture is stirred, and the change in temperature of the solution is measured using a thermometer or data-logging equipment. In this reaction, heat is generated by the reaction system. As a result, the temperature of the solution increases, and the heat is transferred to the surroundings. The reaction is therefore exothermic.

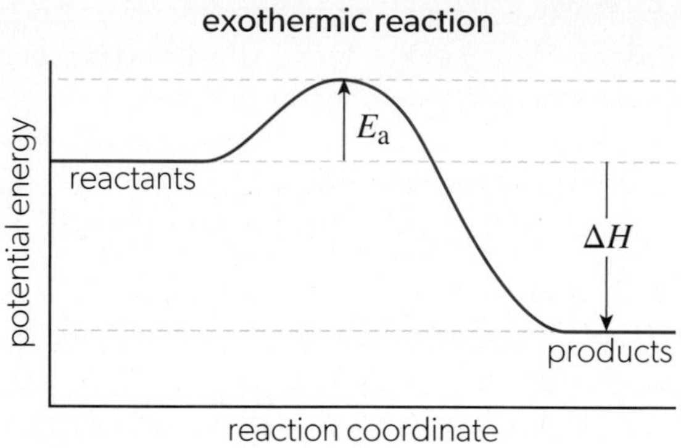

exothermic reaction

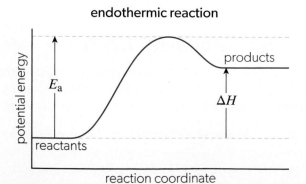

▲ **Figure 6** Using a thermometer or a temperature probe, you would observe an increase in the temperature of the reaction mixture in an exothermic reaction. The enthalpy of the products is lower than that of the reactants. You would describe the products as being energetically more stable than the reactants

The graph in figure 6 is an example of an **energy profile**.

If you consider an endothermic reaction, the products of the reaction are at a higher energy level and therefore are less stable than the reactants.

Ammonium nitrate, NH_4NO_3, is an important component of fertilizers. When the solid dissolves in water to form aqueous ammonium and nitrate ions, the temperature of the solution decreases.

$$NH_4NO_3(s) \rightarrow NH_4^+(aq) + NO_3^-(aq)$$

This heat is absorbed by the reaction system from the surroundings. The apparatus containing the reaction mixture will feel cold to touch. This is an example of an endothermic reaction (figure 7).

endothermic reaction

▲ **Figure 7** Using a thermometer or a temperature probe, you would observe a decrease in the temperature of the reaction mixture in an endothermic reaction. The enthalpy of the products is greater than that of the reactants. The products are described as being energetically less stable than the reactants

Global impact of science

Developments in science may have ethical, environmental, political, social, cultural and economic consequences, which must be considered during decision making. The pursuit of science may have unintended consequences. German chemist Fritz Haber was awarded the Nobel Prize in Chemistry in 1918 for developing a method to chemically extract nitrogen from the air by reacting it with hydrogen. Haber's discovery allowed for the large-scale production of fertilizers that began during the green revolution and continues today. However, his process also provided Germany with a source of ammonia that was used for the production of explosives during the First World War. The global impact of science is evident in Haber's research.

ATL Communication skills

Communication skills cover a wide range of skills and forms of communication. Your ability to effectively communicate verbally and in written form most often comes to mind when you are thinking about improving communication skills. However, communication also involves your ability to read and write different forms of texts intended for different audiences. In science, you need to be able to write formal laboratory reports using specific terminology and accepted writing styles. Another form of communication you would utilize in writing reports and answering examination questions, is your ability to sketch graphs and extract data and meaningful information from graphs. Can you read and analyse the energy profiles that represent exothermic and endothermic reactions? Try to accurately sketch these diagrams, including all of the components.

Practice question

1. Barium hydroxide, $Ba(OH)_2$, reacts with ammonium chloride, NH_4Cl:

$Ba(OH)_2(s) + 2NH_4Cl(s) \rightarrow BaCl_2(aq) + 2NH_3(g) + 2H_2O(l)$ $\Delta H = +164 \text{ kJ mol}^{-1}$

Which of the following is correct for this reaction?

	Temperature	Enthalpy	Stability
A	increases	products have lower enthalpy than the reactants	products are less stable than the reactants
B	decreases	products have lower enthalpy than the reactants	products are more stable than the reactants
C	decreases	products have higher enthalpy than the reactants	products are less stable than the reactants
D	increases	products have higher enthalpy than the reactants	products are more stable than the reactants

Standard enthalpy change, $\Delta H^{\ominus}$ (Reactivity 1.1.4)

The **standard enthalpy change for a reaction**, $\Delta H^{\ominus}$, refers to the heat transferred at constant pressure under standard conditions and states. It can be determined from the change in temperature of a pure substance. The units of $\Delta H^{\ominus}$ are kJ mol^{-1}.

To calculate $\Delta H^{\ominus}$ for a reaction, you therefore need to determine the amount of heat released or absorbed in the course of that reaction. This can be done by measuring the change in temperature of a pure substance to or from which this heat is transferred. When calculating the amount of heat lost or gained by a pure substance such as water, you need to know the **specific heat capacity**, c, of that substance.

The specific heat capacity of a pure substance is defined as the amount of heat needed to raise the temperature of 1 kg of that substance by 1 °C or 1 K. For example, the specific heat capacity of ethanol is 2.44 kJ kg^{-1} K^{-1}, so it takes 2.44 kJ to raise the temperature of 1 kg of ethanol by 1 K. The lower the specific heat capacity of a given substance, the higher the rise in temperature when the same amount of heat is transferred to the sample.

Specific heat capacity is an **intensive property** that does not vary in magnitude with the size of the system being described. For example, a 10 cm³ sample of copper has the same specific heat capacity as a 1 ton block of copper. When you heat up a pure substance, the rise in temperature depends on:

- the identity of that substance
- the mass of that substance
- the amount of heat supplied.

Standard temperature and pressure (STP) conditions are denoted by the symbol $\ominus$. STP is a temperature of 273.15 K and a pressure of 100 kPa. Standard ambient temperature and pressure (SATP) refer to more practical reaction conditions of 298.15 K and 100 kPa. STP and SATP conditions are given in the section 2 of the data booklet.

Substance	Specific heat capacity / kJ kg^{-1} K^{-1}
water	4.18
ethanol	2.44
copper	0.385

▲ Table 1 The specific heat capacities of water, ethanol and copper

Practice questions

2. Using table 1, calculate how much energy is required to raise the temperature of the following by 1 K.
   a. 1 kg of water
   b. 1000 kg of copper
3. When equal masses of two different substances, X and Y, absorb the same amount of energy, their temperatures rise by 5 °C and 10 °C, respectively. Which of the following is correct?
   a. The specific heat capacity of X is twice that of Y.
   b. The specific heat capacity of X is half that of Y.
   c. The specific heat capacity of X is one fifth that of Y.
   d. The specific heat capacity of X is the same as that of Y.
4. Using table 1, state which of the following statements is correct.
   a. More heat is needed to increase the temperature of 50 g of water by 50 °C than 50 g of ethanol by 50 °C.
   b. If the same heat is supplied to equal masses of ethanol and water, the temperature of the water will increase more.
   c. If equal masses of water at 20 °C and ethanol at 50 °C are mixed together, the final temperature will be 35 °C.
   d. If equal masses of water and ethanol at 50 °C cool down to room temperature, ethanol gives out more heat.

Specific heat capacity, c, is used to calculate the amount of heat, Q, absorbed by a pure substance using the relationship:

$$Q = mc\Delta T$$

where m is mass of the pure substance in kg and ΔT is the change in temperature of that substance in K.

Heat, Q, is related to enthalpy change, ΔH, by the following equation:

$$\Delta H = -\frac{Q}{n}$$

where n is the number of moles of the **limiting reactant**. In a reaction, the limiting reactant is the reacting substance with the least stoichiometric amount present, which therefore limits how much product can be formed. In contrast, the other reacting substances are said to be in **excess**.

Measurement

Performing reactions in a polystyrene coffee cup to measure the enthalpy change is a convenient experimental procedure. This method introduces systematic errors that can be analysed and the effect of their directionality assessed.

Systematic errors are a consequence of the experimental procedure. Their effect on empirical data is constant and always in the same direction. With the coffee-cup calorimeter, the measured change in enthalpy for a reaction will always be lower in magnitude than the actual value, as some heat will be transferred between the contents and the surroundings in every experiment.

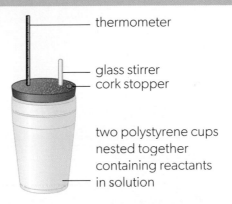

thermometer

glass stirrer
cork stopper

two polystyrene cups
nested together
containing reactants
in solution

▲ **Figure 8** A coffee-cup calorimeter

Worked example 1

1. When a 1.15 g sample of anhydrous lithium chloride, LiCl, was added to 25.0 g of water in a coffee-cup calorimeter, a temperature rise of 3.80 K was recorded. Calculate the enthalpy change of dissolution for 1 mol of lithium chloride. Assume that the heat capacity of lithium chloride itself is negligible.

2. 180.0 J of heat is transferred to a 100.0 g sample of iron, resulting in a temperature rise from 22.0 °C to 26.0 °C. Calculate the specific heat capacity of iron.

Solution

1. $Q = mc\Delta T$

    $= 0.025 \text{ kg} \times 4.18 \text{ kJ kg}^{-1} \text{ K}^{-1} \times 3.80 \text{ K}$

    $= 0.397 \text{ kJ}$

    Now you need to determine the energy gained for 1 mol of LiCl.

    $n(\text{LiCl}) = \dfrac{1.15 \text{ g}}{42.39 \text{ g mol}^{-1}} = 0.0271 \text{ mol}$

    $\Delta H = -\dfrac{Q}{n} = \dfrac{-0.397 \text{ kJ}}{0.0271 \text{ mol}} = -14.6 \text{ kJ mol}^{-1}$

2. First, determine the change in temperature, ΔT:

    $\Delta T = (299 - 295) \text{ K} = 4 \text{ K}.$

    Substitute the values into $Q = mc\Delta T$:

    $0.180 \text{ kJ} = 0.100 \text{ kg} \times c \times 4 \text{ K}$

    Make c the subject of the equation and solve:

    $c = \dfrac{0.180 \text{ kJ}}{0.100 \text{ kg} \times 4 \text{ K}} = 0.450 \text{ kJ kg}^{-1} \text{ K}^{-1}$

Practice questions

5. Calculate the energy absorbed by water when the temperature of 30 g of water is raised by 30 °C. The specific heat capacity of water is 4.18 J g^{-1} K^{-1}.

6. 0.675 kJ of heat is transferred to 125 g of copper metal. Copper metal has a specific heat capacity of 385 J kg^{-1} K^{-1}. Calculate the change in temperature of the copper metal.

Investigation to find the enthalpy change for a reaction

In this skills task, we will look at the method used to calculate the enthalpy change for the exothermic metal displacement reaction between zinc and copper(II) sulfate:

$$Zn(s) + CuSO_4(aq) \rightarrow Cu(s) + ZnSO_4(aq)$$

Relevant skills

- Tool 1: Measuring variables

- Tool 1: Applying techniques

- Tool 2: Applying technology to process data

- Tool 3: Processing uncertainties

- Tool 3: Graphing

- Inquiry 1: Controlling variables

- Inquiry 2: Processing data

- Inquiry 3: Evaluating

Materials

- electronic balance

- coffee-cup calorimeter

- measuring cylinder

- thermometer or temperature probe

- 1.0 mol dm^{-3} copper(II) sulfate solution

- zinc powder

Method

1. Using an electronic balance, accurately measure the mass of 25 cm^3 of 1.0 mol dm^{-3} CuSO$_4$ solution. Transfer the solution to the coffee-cup calorimeter.

2. Using a thermometer or a temperature probe, record the temperature of the solution every 30 seconds for up to three minutes, or until a constant temperature is achieved.

3. At three minutes, introduce between 1.3 g and 1.4 g of zinc powder, record the exact mass of zinc and commence stirring.

4. Continue to take temperature readings for up to five minutes after the maximum temperature has been reached.

5. Produce a temperature versus time graph to determine the change in temperature.

6. Use your value of ΔT to calculate the heat released, Q, and the enthalpy change for the reaction, ΔH.

Assumptions and errors

A number of assumptions are made when using this method:

- The heat released from the reaction is completely transferred to the water.

- The coffee cup acts as an insulator against heat loss to the surroundings. However, the coffee cup also has a heat capacity and heat is transferred to it from the water, but it is much smaller than that of water.

- The maximum temperature reached is an accurate representation of the heat evolved during the reaction.

- The specific heat capacity of an aqueous solution is the same as that of water.

Loss of heat from the system to the surroundings is the main source of error in this experiment and one that is difficult to quantify. The change in temperature, ΔT, calculated from a graph will include a systematic error. This loss of heat means that the maximum temperature recorded will be lower than the theoretical value, making the calculated value of Q lower than the actual value. The effect of errors on the result of subsequent calculations is important in considering improvements in experimental procedures.

→

An accepted method of calculating the maximum temperature to compensate for systematic errors in data is to look at the cooling section of the curve after the reaction is complete, and extrapolate this back to the moment when zinc is introduced at 3 minutes, as shown in figure 9. A more accurate value for ΔT can then be determined.

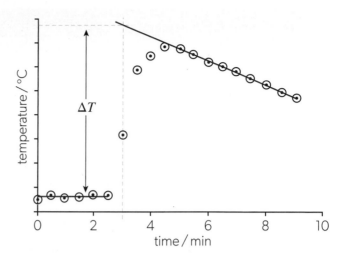

▲ Figure 9 Example of a temperature vs time graph for a calorimetry experiment

Worked example 2

A coffee-cup calorimeter was used to measure the temperature change for the reaction between zinc powder and a 1.0 mol dm^{-3} solution of copper(II) sulfate. The following results were recorded:

Mass of copper(II) sulfate solution / g	28.8
Mass of zinc / g	1.37
ΔT / °C	39.0

Determine the amount of heat released and the enthalpy change for this reaction.

Solution

First, use $Q = mc\Delta T$ to determine the amount of heat released:

$$Q = 0.0288 \text{ kg} \times 4.18 \text{ kJ kg}^{-1} \text{ K}^{-1} \times 39.0 \text{ K}$$

$$= 4.69 \text{ kJ}$$

Then, determine the limiting reactant for the reaction.

$$\text{Number of moles of zinc, } n(\text{Zn}) = \frac{m}{M_r}$$

$$= \frac{1.37 \text{ g}}{65.38 \text{ g mol}^{-1}}$$

$$= 0.0210 \text{ mol}$$

Number of moles of copper(II) sulfate, $n(\text{CuSO}_4) = c \times v$

$$= 1.00 \text{ mol dm}^{-3} \times 0.0288 \text{ dm}^3$$

$$= 0.0288 \text{ mol}$$

Zinc is present in a smaller amount, so it is the limiting reactant. You can calculate the enthalpy change of reaction from $\Delta H = -\dfrac{Q}{n}$:

$$\Delta H = -\frac{4.69 \text{ kJ}}{0.0210 \text{ mol}} = -223 \text{ kJ mol}^{-1}$$

TOK

In theory of knowledge, there are 12 concepts in focus. These are: evidence, certainty, truth, interpretation, power, justification, explanation, objectivity, perspective, culture, values and responsibility. Scientists perform experiments and process the raw data to enable us to draw conclusions. We compare experimental and theoretical values. What concepts do we utilize when justifying our conclusions? How do we use evidence? Are our judgments subjective or objective? When analysing and appraising experimental limitations, how do assumptions have an impact on our perceptions?

Combustion of primary alcohols

You can determine the enthalpy change of combustion of common alcohols in a school laboratory. After repeating the experiment several times with a homologous series of alcohols, you can subsequently analyse this data and identify patterns.

Relevant skills

*   Tool 1: Recognise and address the relevant safety, ethical or environmental issues in an investigation
*   Tool 1: Measuring temperature and mass
*   Tool 1: Calorimetry
*   Inquiry 1: Appreciate when and how to insulate against heat loss or gain
*   Inquiry 2: Identify and record relevant qualitative observations and sufficient relevant quantitative data

Materials

*   five spirit burners, each containing one of the following alcohols: methanol, ethanol, propan-1-ol, butan-1-ol and pentan-1-ol
*   electronic balance
*   beaker or metal calorimeter
*   tripod
*   temperature probe or thermometer

Safety

Alcohols should be handled and disposed of with care because they are generally flammable, hazardous and volatile.

Instructions

1. Using suitable sources, identify the hazards and complete a risk assessment for this experiment. In your risk assessment, you should:
    *   identify the **hazards**
    *   assess the level of **risk**
    *   determine relevant **control measures**
    *   identify suitable **disposal methods** aligned with your school's health and safety policies.

2. Determine the initial mass of the spirit burners using an electronic balance.

3. Accurately determine the mass of 30 cm³ of water contained in a 250 cm³ beaker or metal calorimeter.

4. Using either a temperature probe or a thermometer, determine and record the initial temperature of the water.

5. Ignite a spirit burner under the beaker or calorimeter and allow the alcohol to burn to heat the water. The period over which it burns can be set in one of two different ways:

    a. allow each alcohol to burn until a temperature change of 30 °C is reached

    b. allow each alcohol to burn for a period of two minutes.

6. Determine the final mass of each spirit burner immediately after the flame is extinguished. Take extra care because the burner will be hot.

7. Use your values of ΔT of the water and Δm of the burner to calculate the amount of heat released, Q, and the enthalpy change of combustion, ΔH, for each alcohol.

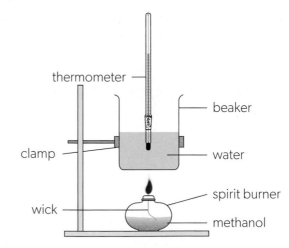

▲ Figure 10 A typical arrangement of experimental apparatus for an enthalpy of combustion determination

ATL Research skills

Cite your sources fully, according to your school's citing and referencing system.

Worked example 3

A metal calorimeter was used to measure the amount of heat released by the combustion of methanol. The following results were recorded:

Mass of water / g	31.2
Change in mass of methanol / g	0.348
ΔT / °C	30.0

Determine the amount of heat released and the enthalpy change of combustion for methanol.

Solution

First, use $Q = mc\Delta T$ to determine the amount of heat released:

$$Q = 0.0312 \text{ kg} \times 4.18 \text{ kJ kg}^{-1} \text{ K}^{-1} \times 30.0 \text{ K}$$

$$= 3.91 \text{ kJ}$$

Methanol reacts with oxygen in a combustion reaction. Methanol is the limiting reactant for this reaction because oxygen is present in excess in the air.

Number of moles of methanol, $n(CH_3OH) = \dfrac{m}{M_r}$

$$= \frac{0.348 \text{ g}}{32.05 \text{ g mol}^{-1}}$$

$$= 0.0109 \text{ mol}$$

You can calculate the enthalpy change of reaction from $\Delta H = -\dfrac{Q}{n}$:

$$\Delta H = -\frac{3.91 \text{ kJ}}{0.0109 \text{ mol}} = -359 \text{ kJ mol}^{-1}$$

Thermochemistry experiments provide a useful set of raw data, and involve experimental procedures that can be evaluated for random and systematic errors. The identification of the systematic errors and examination of their directionality are essential aspects of the analysis of experimental results. Calorimetry experiments typically give a smaller change in temperature than is predicted from theoretical values. This is the result of heat loss from the system, which is difficult to measure. Scientists usually make the assumption that the heat lost to the environment is negligible. TOK helps us to understand our judgments of discrepancies between experimental and theoretical values.

Thermometric titration

The neutralization reaction between an acid and a base is exothermic. In this skills task, you will determine the unknown concentration of hydrochloric acid by measuring the change in temperature while sodium hydroxide is added to the acid. The temperature will reach a maximum when the acid and base are mixed together in stoichiometric amounts.

Relevant skills

- Tool 1: Calorimetry and acid–base titration

- Tool 2: Use sensors

- Tool 3: Calculate and interpret percentage error

- Tool 3: Understand the significance of uncertainties in raw and processed data

- Tool 3: Propagate uncertainties and state them to an appropriate level of precision

- Tool 3: Extrapolate graphs

- Inquiry 1: Appreciate when and how to insulate against heat loss or gain

- Inquiry 3: Identify and discuss sources of systematic and random error

Safety

- Wear eye protection.

- Sodium hydroxide solution is corrosive.

- Hydrochloric acid is corrosive.

Materials

- two 250 cm³ polystyrene cups

- thermometer or temperature probe

- graduated pipette and filler

- burette

- ~50.0 cm³ sodium hydroxide solution of known concentration

- 30.0 cm³ hydrochloric acid of unknown concentration

Method

1. Read through the safety, materials and method. Use this information, and relevant safety data, to complete a risk assessment for this practical work and show it to your teacher.

2. Review the titration, percentage error and uncertainties sections in the *Skills* chapter.

3. Rinse and fill the burette with sodium hydroxide solution. Record its concentration.

4. Add 25 cm³ of acid solution to the cup and place it under the burette. Nest it inside a second cup, for additional thermal insulation. For safety, these cups should be placed inside a beaker to avoid tipping over.

5. Position the temperature probe in the acid and record the initial temperature of the acid in the cup.

6. Add a small volume (~5 cm³) of sodium hydroxide solution to the acid, stirring gently. Record the highest temperature reached with this addition.

7. Continue adding small volumes of sodium hydroxide solution and recording the temperature until the temperature decreases over several consecutive readings.

8. Clear up as instructed by your teacher.

Questions

1. Plot a graph showing temperature vs volume of sodium hydroxide solution added.

2. Extrapolate the two sections of the graph to find the maximum temperature reached during the titration.

3. Determine the concentration of the acid, along with absolute and percentage uncertainties. Make sure you state all values to an appropriate level of precision.

4. Calculate the percentage error of your experimental acid concentration.

5. Determine the enthalpy of neutralization, along with absolute and percentage uncertainties. State all values to an appropriate level of precision.

6. Calculate the percentage error of your experimental enthalpy of neutralization.

7. Comment on the relative impacts of systematic and random error on the values obtained for the acid concentration and enthalpy of neutralization.

8. Suggest and explain two improvements that could be made to this method.

Measurement

Experimental enthalpy values can be assessed in terms of their accuracy and precision. Random errors in measurement lead to imprecision, whereas systematic errors cause inaccuracy. What are some of the sources of random and systematic errors in an enthalpy of neutralization experiment? To what extent are these errors quantifiable?

ATL Thinking skills

Calorimetry experiments conducted in research laboratories utilize the same principles as the calorimetry experiments described in this chapter. The instrument used is called a **bomb calorimeter** (figure 11). A sample is burned inside a chamber (called a "bomb"), and the resulting temperature change of the surrounding water is measured.

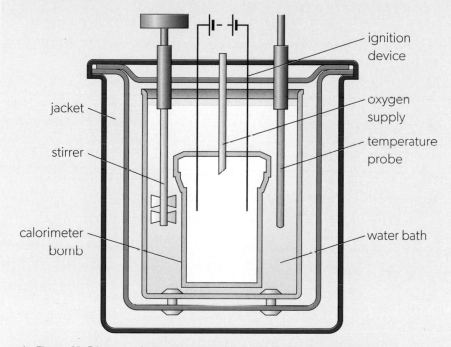

▲ **Figure 11** Diagram of a bomb calorimeter used in research laboratories

1. Study the diagram carefully and list all the features that are labelled.

2. Deduce the purpose of each feature.

3. Consider why the measurements obtained with a bomb calorimeter are highly accurate and precise.

4. What properties of water make it suitable for calorimetry experiments?

End of topic questions

Topic review

1. Using your knowledge from the *Reactivity 1.1* topic, answer the guiding question as fully as possible:

   What can be deduced from the temperature change that accompanies chemical or physical change?

Exam-style questions

Multiple-choice questions

2. Which is correct for the following reaction?

   $$2Al(s) + 6HCl(aq) \rightarrow 2AlCl_3(aq) + 3H_2(g)$$
   $$\Delta H = -1049 \text{ kJ mol}^{-1}$$

   A   Reactants are less stable than products and the reaction is endothermic.

   B   Reactants are more stable than products and the reaction is endothermic.

   C   Reactants are more stable than products and the reaction is exothermic.

   D   Reactants are less stable than products and the reaction is exothermic.

3. Which statement is correct?

   A   In an exothermic reaction, the products have more energy than the reactants.

   B   In an exothermic reversible reaction, the activation energy of the forward reaction is greater than that of the reverse reaction.

   C   In an endothermic reaction, the products are more stable than the reactants.

   D   In an endothermic reversible reaction, the activation energy of the forward reaction is greater than that of the reverse reaction.

4. Which statement is correct for this reaction?

   $$Fe_2O_3(s) + 3CO(g) \rightarrow 2Fe(s) + 3CO_2(g)$$
   $$\Delta H = -26.6 \text{ kJ mol}^{-1}$$

   A   13.3 kJ are released for every mole of Fe produced.

   B   26.6 kJ are absorbed for every mole of Fe produced.

   C   53.2 kJ are released for every mole of Fe produced.

   D   26.6 kJ are released for every mole of Fe produced.

5. In which reaction do the reactants have a lower energy than the products?

   A $CH_4(g) + 2O_2(g) \rightarrow CO_2(g) + 2H_2O(g)$

   B $HBr(g) \rightarrow H(g) + Br(g)$

   C $Na^+(g) + Cl^-(g) \rightarrow NaCl(s)$

   D $NaOH(aq) + HCl(aq) \rightarrow NaCl(aq) + H_2O(l)$

Extended-response questions

6. Nitrogen dioxide and carbon monoxide react according to the following equation:

   $NO_2(g) + CO(g) \rightarrow NO(g) + CO_2(g)$
   $\Delta H = -226 \text{ kJ mol}^{-1}$

   a. Calculate the enthalpy change for the reverse reaction. [1]

   b. State the equation for the reaction of NO_2 in the atmosphere to produce acid deposition. [1]

7. Powdered zinc was reacted with 25.00 cm³ of 1.000 mol dm⁻³ copper(II) sulfate solution in an insulated beaker. Temperature was plotted against time.

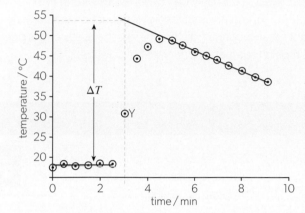

   a. Estimate the time at which the powdered zinc was placed in the beaker. [1]

   b. State what point **Y** on the graph represents. [1]

   The maximum temperature used to calculate the enthalpy change of reaction was chosen at a point on the extrapolated (red) line.

   c. State the maximum temperature that should be used, and outline **one** assumption made in choosing this temperature on the extrapolated line. [1]

   d. To determine the enthalpy of reaction, the experiment was carried out five times. The same volume and concentration of copper(II) sulfate was used but the mass of zinc was different each time. Suggest, with a reason, if zinc or copper(II) sulfate should be in excess for each trial. [1]

   The formula $q = mc\Delta T$ was used to calculate the amount of energy released.

   The values used in the calculation were $m = 25.00$ g and $c = 4.18 \text{ J g}^{-1} \text{K}^{-1}$.

   e. State an assumption made when using these values for m and c. [1]

   f. Predict, giving a reason, how the final enthalpy of reaction calculated from this experiment would compare with the theoretical value. [1]

8. A potato chip (crisp) was ignited, and the flame was used to heat a test tube containing water.

Mass of water/g	7.8
Mass of chip/g	1.2
Initial temperature/°C	21.3
Final temperature/°C	22.6

   a. Calculate the heat required, in kJ, to raise the temperature of the water, using data in the table above and from section 2 of the data booklet. [1]

   b. Determine the enthalpy of combustion of the potato chip, in kJ g⁻¹. [1]

How does application of the law of conservation of energy help us to predict energy changes during reactions?

The utilization of energy is central to our lives. Agricultural, industrial, and domestic activities all consume vast amounts of energy daily. **Thermodynamics** is the study of energy and its interconversions. The first law of thermodynamics states that energy can be converted from one form to another and that the total amount of energy for a given system remains constant.

This law is often called the **law of conservation of energy**. According to this law, energy can be neither created nor destroyed; it can only be converted from one form to another. This means that for a given reaction we can account for and quantify all the energy changes. This is one of the most fundamental principles of science.

Understandings

Reactivity 1.2.1—Bond-breaking absorbs and bond-forming releases energy.

Reactivity 1.2.2—Hess's law states that the enthalpy change for a reaction is independent of the pathway between the initial and final states.

Reactivity 1.2.3—Standard enthalpy changes of combustion, $\Delta H_c^{\ominus}$, and formation, $\Delta H_f^{\ominus}$, data are used in thermodynamic calculations.

Reactivity 1.2.4—An application of Hess's law uses enthalpy of formation data or enthalpy of combustion data to calculate the enthalpy change of a reaction

Reactivity 1.2.5—A Born–Haber cycle is an application of Hess's law, used to show energy changes in the formation of an ionic compound.

AHL

Bond-breaking and bond-forming (*Reactivity 1.2.1*)

When a chemical reaction takes place, the atoms of the reactants are rearranged to create products. Chemical bonds in the reactants are broken and new chemical bonds are made to form the products. Energy is required to break the chemical bonds: bond-breaking is an **endothermic** process. Energy is released when new chemical bonds form: bond-making is an **exothermic** process. This transfer of energy between the system and the surroundings is an essential part of understanding the energy changes involved in a chemical reaction (*Reactivity 1.1*).

Laws and theories

The **law of conservation of energy** states that energy can be neither created nor destroyed; it can only be converted between different forms. Laws allow predictions to be made but, unlike theories, laws do not have explanatory power. What other laws have you come across in your study of chemistry?

Bond enthalpy

Imagine a simple hydrogen molecule, H_2. The breaking of the hydrogen molecule into individual hydrogen atoms requires energy. The **bond enthalpy** (*BE*) is defined as the energy required to break one mole of bonds by **homolytic fission** in one mole of gaseous covalent molecules under standard conditions. The process of homolytic fission distributes the electrons from the bond equally between the two new species. This results in the formation of radicals, indicated by the • symbol.

For example, the bond enthalpy of the H–H bond is equal to the enthalpy change of the following reaction:

$$H_2(g) \rightarrow 2H\bullet(g) \qquad \Delta H = +436 \text{ kJ mol}^{-1}$$

Bond-breaking is an endothermic process, and therefore it has a **positive enthalpy value**.

Selected *BE* values are provided in table 1 (on the next page) and section 12 of the data booklet. Bond enthalpy values are derived from experimental data on the breaking of the same bond in a wide variety of compounds. For example, the C–H bond enthalpy will vary through the alkane series because the chemical environment of the individual bonds changes.

If a molecule of methane underwent a series of steps in which one hydrogen atom was removed at a time, the bond enthalpy for each removal would be different. This is because the chemical environment changes upon the removal of successive hydrogen atoms.

$$CH_4(g) \rightarrow \bullet CH_3(g) + \bullet H(g) \qquad BE_1(\text{C–H}) = +439 \text{ kJ mol}^{-1}$$

$$\bullet CH_3(g) \rightarrow \bullet CH_2(g) + \bullet H(g) \qquad BE_2(\text{C–H}) = +462 \text{ kJ mol}^{-1}$$

$$\bullet CH_2(g) \rightarrow \bullet CH(g) + \bullet H(g) \qquad BE_3(\text{C–H}) = +424 \text{ kJ mol}^{-1}$$

$$\bullet CH(g) \rightarrow \bullet C(g) + \bullet H(g) \qquad BE_4(\text{C–H}) = +338 \text{ kJ mol}^{-1}$$

Can you recognize the differences in the products of each equation and explain how the chemical environment changes for each successive step?

Bond enthalpies are **average values** and are therefore only an approximation. The average bond enthalpy value of the C–H bond is $+414 \text{ kJ mol}^{-1}$.

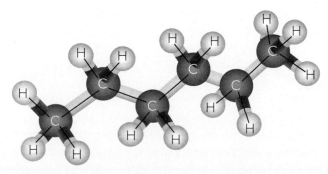

▲ **Figure 1** When you examine the structure of this hexane molecule, can you see how the chemical environment of the C–H bonds differs throughout the molecule?

Bond	Average bond enthalpy / kJ mol⁻¹	Bond	Average bond enthalpy / kJ mol⁻¹
H–H	436	C=O	804
O–O	144	C=N	615
O=O	498	C≡N	890
O–H	463	N–N	158
C–H	414	N=N	470
C–C	346	N≡N	945
C=C	614	Cl–Cl	242
C≡C	839	Br–Br	193
C–O	358	I–I	151

▲ Table 1 Average bond enthalpies at 298 K

TOK

Measured energy changes in chemical reactions can be explained by the model of bonds broken and bonds formed. These explanations are based on empirical data and further data can be obtained to modify and refine the model. What role do models play in the acquisition of knowledge in the natural sciences? How do we determine whether a model is useful? How are models used in other areas of knowledge?

You can use bond enthalpy data to calculate the enthalpy change of reaction, ΔH:

$$\Delta H = \Sigma(BE \text{ of bonds broken}) - \Sigma(BE \text{ of bonds formed})$$

The enthalpy change of a reaction calculated from bond enthalpies will differ from its actual value, as bond enthalpies are average values. Additionally, bond enthalpy data do not take intermolecular forces into account, which is particularly important when substances in a reaction are solid or liquid.

When performing bond enthalpy calculations, it is useful to draw out the full structural formulas of the reactants and products involved in the reaction. From this, you can determine the type and number of each chemical bond present.

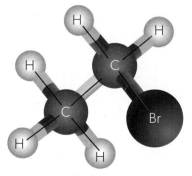

▲ Figure 2 A molecule of bromoethane, C_2H_5Br, has one C–C bond, one C–Br bond and five C–H bonds

Models

The model of bond breaking and formation is used to represent the energy changes taking place within a reaction system. We can assess the extent to which this model is in agreement with empirical data, by comparing the theoretical values generated from bond energy data with experimental evidence. Using empirical data to replace or modify proposed models is a central methodology in science.

Worked example 1

Using bond enthalpy data, determine the enthalpy change of reaction for the following process:

$$C_2H_4(g) + HBr(g) \rightarrow C_2H_5Br(g)$$

Solution

Start by drawing the full structural formulas of all of the reactants and products:

In the reactants, there are four C–H bonds, one C=C bond and one H–Br bond. In the product, there are five C–H bonds, one C–C bond, and one C–Br bond. Therefore, the enthalpy change for the reaction is as follows:

$$\Delta H = \Sigma(BE \text{ of bonds broken}) - \Sigma(BE \text{ of bonds formed})$$

$$= [4BE(\text{C–H}) + BE(\text{C=C}) + BE(\text{H–Br})] - [5BE(\text{C–H}) + BE(\text{C–C}) + BE(\text{C–Br})]$$

Then, substitute in the *BE* values from the data booklet:

$$= [(4 \times 414) + 614 + 366] - [(5 \times 414) + 346 + 285]$$

$$= 2636 - 2701$$

$$= -65 \text{ kJ mol}^{-1}$$

Practice questions

1. The chloralkali process is an industrial process used to make chlorine gas and sodium hydroxide. Much of the chlorine gas is combined with hydrogen gas to produce hydrogen chloride:
    $$Cl_2(g) + H_2(g) \rightarrow 2HCl(g)$$
    The enthalpy change for this reaction is -185 kJ mol^{-1}. Using section 12 of the data booklet, calculate the bond enthalpy, in kJ mol^{-1}, for the H–Cl bond.

2. Enthalpy changes depend on the number and type of bonds broken and formed. Hydrogen gas can be formed industrially by the reaction of natural gas with steam:
    $$CH_4(g) + H_2O(g) \rightarrow 3H_2(g) + CO(g)$$
    Determine the enthalpy change, ΔH, for this reaction, in kJ mol^{-1}, using section 12 of the data booklet.

3. Methane and chlorine react to produce chloromethane and hydrogen chloride.
    a. Write the balanced chemical equation for this reaction.
    b. Using bond enthalpy values from the data booklet, determine the enthalpy change for this reaction.
    c. Deduce and explain whether this reaction is exothermic or endothermic.
    d. State which are more energetically stable, the reactants or the products.

Linking question

How would you expect bond enthalpy data to relate to bond length and polarity? (*Structure 2.2*)

How does the strength of a carbon–halogen bond affect the rate of a nucleophilic substitution reaction? (*Reactivity 3.4*)

Hess's law (*Reactivity 1.2.2*)

If you have travelled to New York City, Shanghai, Paris, Seoul, London, Madrid or Delhi, you have experienced the subways that criss-cross these enormous cities and transport millions of people every day. In any transport network there is more than one way to travel between point A and B. If you are an adventurous traveller, half the fun is often working out the fastest route.

▲ **Figure 3** The extensive metro system in Seoul provides many alternative pathways for navigating the urban sprawl

The same idea can be true for chemistry. A chemical equation usually shows the net reaction – it can be a summary of a number of different reactions, which, when added together, result in an overall reaction.

For example, the overall reaction for the oxidation of elemental sulfur, $S(s)$, to sulfur trioxide, $SO_3(l)$, is as follows:

$$S(s) + \frac{3}{2}O_2(g) \rightarrow SO_3(l)$$

However, this process is difficult to carry out in one step. Typically, sulfur is oxidized first to sulfur dioxide, $SO_2(g)$, and then sulfur dioxide is oxidized further to sulfur trioxide:

Step 1: $S(s) + O_2(g) \rightarrow SO_2(g)$

Step 2: $SO_2(g) + \frac{1}{2}O_2(g) \rightarrow SO_3(l)$

If you add up the enthalpy changes that occur in each of these individual steps, the total would be the same as the enthalpy change of the overall reaction. This is the basis of Hess's law:

Regardless of the route by which a chemical reaction proceeds, the enthalpy change will always be the same, as long as the initial and final states of the system are the same.

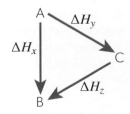

▲ **Figure 4** Hess's law

Hess's law is an application of the conservation of energy law. Figure 4 illustrates Hess's law: the enthalpy change for reaction A → B (ΔH_x) is equal to the sum of the enthalpy changes for the reactions A → C and C → B ($\Delta H_y + \Delta H_z$).

Hess's law can be applied to find the unknown enthalpy change for a given reaction by combining other reactions with known enthalpy changes.

Worked example 2

The reaction for the formation of methanol is shown below:

$$C(s) + 2H_2(g) + \frac{1}{2}O_2(g) \rightarrow CH_3OH(l)$$

Use Hess's law and equations 1–3 to calculate the enthalpy change for the reaction above.

1	$CH_3OH(l) + \frac{3}{2}O_2(g) \rightarrow CO_2(g) + 2H_2O(l)$	$\Delta H = -726\,kJ\,mol^{-1}$
2	$C(s) + O_2(g) \rightarrow CO_2(g)$	$\Delta H = -394\,kJ\,mol^{-1}$
3	$H_2(g) + \frac{1}{2}O_2(g) \rightarrow H_2O(l)$	$\Delta H = -286\,kJ\,mol^{-1}$

Solution

There are two methods for using Hess's law for a question like this: the **summation of equations method**, and the **enthalpy cycle diagram method**.

Summation of equations method

First, inspect the formation of methanol reaction scheme. In your calculation, you need to use enthalpy values corresponding to reactions that involve carbon, hydrogen and oxygen as reactants, and methanol as a product.

Start with the reactants. Carbon is a reactant in equation 2, so this equation can be used as written:

$$C(s) + O_2(g) \rightarrow CO_2(g) \qquad \Delta H = -394\,kJ\,mol^{-1}$$

You require two moles of hydrogen as a reactant. Therefore, equation 3 can be used in the direction as written, but with doubled stoichiometric coefficients. This means that the enthalpy value must also be doubled:

$$2H_2(g) + O_2(g) \rightarrow 2H_2O(l) \qquad \Delta H = -572\,kJ\,mol^{-1}$$

We also need half a mole of oxygen. Oxygen is present as a reactant in two of the above equations, so ignore this for now.

For the product methanol, you need to use equation 1. However, the equation must be reversed to make methanol a product, not a reactant. When reversing the chemical equation, the sign of the enthalpy change must also be changed:

$$CO_2(g) + 2H_2O(l) \rightarrow CH_3OH(l) + \frac{3}{2}O_2(g) \qquad \Delta H = +726\,kJ\,mol^{-1}$$

Now add the three equations together, cancelling the species common to both sides of the reaction, and add the enthalpy values:

$C(s) + O_2(g)$	$\rightarrow$	$CO_2(g)$	$\Delta H = -394\,kJ\,mol^{-1}$
$2H_2(g) + O_2(g)$	$\rightarrow$	$2H_2O(l)$	$\Delta H = -572\,kJ\,mol^{-1}$
$CO_2(g) + 2H_2O(l)$	$\rightarrow$	$CH_3OH(l) + \frac{3}{2}O_2(g)$	$\Delta H = +726\,kJ\,mol^{-1}$

Total $\quad C + 2O_2 \frac{1}{2}O_2 + 2H_2 + CO_2 + 2H_2O \rightarrow CO_2 + 2H_2O + CH_3OH + \frac{3}{2}O_2 \quad \Delta H = -240\,kJ\,mol^{-1}$

Cancelling the common species gives the overall equation $C + 2H_2 + \frac{1}{2}O_2 \rightarrow CH_3OH$, which is identical to the reaction in the question. Therefore, the total enthalpy change for the reaction is equal to $-240\,kJ\,mol^{-1}$. Note that here and below the states are omitted to save space.

Enthalpy cycle diagram method

You can represent the calculations needed to determine the enthalpy change of reaction in the form of an enthalpy cycle diagram.

First, write the equation you are trying to find the enthalpy change for: $C + 2H_2 + \frac{1}{2}O_2 \rightarrow CH_3OH$. Below this, write the other species from equations 1, 2 and 3 not included in the original reaction: $CO_2 + 2H_2O$. Oxygen, O_2, is usually not included in enthalpy cycle diagrams.

Then, draw two arrows from the reactants to the species at the bottom to represent equations 2 and 3. Draw an arrow from the products to the species at the bottom to represent equation 1.

Finally, label the arrows with the enthalpy change values. As with the summation of equations method, the enthalpy change for equation 3 needs to be doubled to give the value for two moles of hydrogen. The resulting enthalpy cycle diagram is shown in figure 5.

Then, calculate the sum of the enthalpy change values, following the alternative pathway from the reactants to the products, via the CO_2 and $2H_2O$ intermediates. If the pathway you are following is in the opposite direction of an arrow (in this case, the last arrow), reverse the sign of the enthalpy change. This calculation is summarized below:

$$(-394) + (-286 \times 2) + (+726) = -240 \text{ kJ mol}^{-1}$$

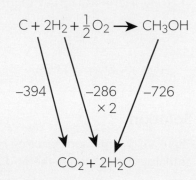

▲ **Figure 5** Enthalpy cycle diagram for the formation of methanol

---

(ATL) Communication skills

In the previous worked example, you have seen two methods for representing and solving Hess's law calculations: the summation of equations method and the enthalpy cycle diagram method. In what situations would you use an enthalpy cycle? In what situations would you use the summation of equations method? What are the strengths and limitations of each method? To what extent are the methods different if they represent the same concept?

Whichever method you prefer, make sure you show your reasoning clearly in assessments. Often candidates make arithmetical errors when calculating the enthalpy of reaction. It is advisable to clearly show your full working rather than simply recording the final answer. This gives the examiner the opportunity to assign partial marks where applicable.

Practice questions

4. Calculate the enthalpy change for the following reaction, using the summation of equations method, and equations 1–3 below:

$$3H_2(g) + 2C(s) + \frac{1}{2}O_2(g) \rightarrow C_2H_5OH(l)$$

1	$C_2H_5OH(l) + 3O_2(g) \rightarrow 2CO_2(g) + 3H_2O(l)$	$\Delta H = -1367 \text{ kJ mol}^{-1}$
2	$C(s) + O_2(g) \rightarrow CO_2(g)$	$\Delta H = -394 \text{ kJ mol}^{-1}$
3	$H_2(g) + \frac{1}{2}O_2(g) \rightarrow H_2O(l)$	$\Delta H = -286 \text{ kJ mol}^{-1}$

5. Determine the enthalpy change for the following reaction, using the enthalpy cycle diagram method and equations 1–3 below:

$$4NH_3(g) + 5O_2(g) \rightarrow 4NO(g) + 6H_2O(l)$$

1	$N_2(g) + O_2(g) \rightarrow 2NO(g)$	$\Delta H = +66 \text{ kJ mol}^{-1}$
2	$N_2(g) + 3H_2(g) \rightarrow 2NH_3(g)$	$\Delta H = -92 \text{ kJ mol}^{-1}$
3	$2H_2(g) + O_2(g) \rightarrow 2H_2O(l)$	$\Delta H = -572 \text{ kJ mol}^{-1}$

Using Hess's law to find enthalpy change

When heated, potassium hydrogencarbonate decomposes to produce potassium carbonate, carbon dioxide and water. The enthalpy change for this decomposition is difficult to determine directly. You will determine the enthalpy change for two other reactions, and then apply Hess's law to find the enthalpy change for the decomposition of potassium hydrogencarbonate.

Relevant skills

- Tool 1: Calorimetry
- Tool 3: Record, propagate and express uncertainties
- Inquiry 2: Carry out relevant and accurate data processing
- Inquiry 3: Evaluate the implications of methodological weaknesses, limitations and assumptions on the conclusions

Safety

- Wear eye protection.
- Hydrochloric acid is an irritant.
- Anhydrous potassium carbonate is an irritant.
- The reaction between carbonates and acids effervesces. Cover the reaction vessel with a lid to prevent loss of reagents and spillage.

Materials

- $2 \, mol \, dm^{-3}$ hydrochloric acid
- anhydrous potassium carbonate (solid)
- potassium hydrogencarbonate (solid)

Method

1. Read through the method and write a list of equipment required for the practical. Show it to your teacher for approval before starting.
2. Place approximately 3 g of potassium carbonate into a weighing boat. Record the exact mass.
3. Use a measuring cylinder to collect $30 \, cm^3$ of hydrochloric acid and transfer it into a coffee-cup calorimeter. Record the temperature of the acid.
4. Add the potassium carbonate to the acid solution and stir the mixture, monitoring the temperature. Record the maximum temperature reached.
5. Dispose of the reaction mixture appropriately.

6. Repeat the procedure, using potassium hydrogencarbonate instead of potassium carbonate. (Note that this time, record the lowest temperature reached because the reaction is endothermic.)

Analysis
Part 1: Known enthalpy changes of reaction

1. Write a balanced equation, including state symbols, for the reaction between:
   a. potassium carbonate and hydrochloric acid
   b. potassium hydrogencarbonate and hydrochloric acid
2. Show that the acid is in excess in both cases.
3. Process your experimental data to determine the enthalpy change for the reaction between:
   a. potassium carbonate and hydrochloric acid
   b. potassium hydrogencarbonate and hydrochloric acid

Part 2: Applying Hess's law

4. Write a balanced equation, including state symbols, for the decomposition of potassium hydrogencarbonate to produce potassium carbonate and other products.
5. Draw an enthalpy cycle connecting the three reactions:
   - potassium carbonate and hydrochloric acid
   - potassium hydrogencarbonate and hydrochloric acid
   - thermal decomposition of potassium hydrogencarbonate
6. Apply Hess's law, and your processed data from part 1, to determine the enthalpy change for the thermal decomposition of potassium hydrogencarbonate.
7. Propagate uncertainties to determine the absolute uncertainty of the enthalpy values you have obtained.
8. Look up the theoretical value for the enthalpy change for the decomposition of potassium hydrogencarbonate online. Use this value to determine the percentage error of your result.
9. Identify two assumptions made throughout this investigation and discuss their validity.
10. Identify and explain two major sources of systematic error in this investigation.

Standard enthalpy changes of combustion, $\Delta H_c^{\ominus}$, and formation, $\Delta H_f^{\ominus}$ (*Reactivity 1.2.3 and Reactivity 1.2.4*)

When describing a standard enthalpy change, there are a number of steps that must be completed.

1. Determine the type of reaction.

2. Write an equation to describe the reaction.

3. Include state symbols in the equation.

The **standard enthalpy change of combustion**, $\Delta H_c^{\ominus}$, is the enthalpy change that occurs when one mole of a substance in its standard state is burned completely in oxygen. The standard state of a pure substance is the form that it takes under standard conditions: 25.00 °C (298.15 K), which is taken as being room temperature, and a pressure of 1.00×10^5 Pa.

When a **hydrocarbon** undergoes combustion, it reacts with oxygen to produce carbon dioxide, CO_2, and water, H_2O. Butane, C_4H_{10}, is a highly flammable gas and a component of liquefied petroleum gas (LPG). The equation for its combustion is as follows:

$$C_4H_{10}(g) + \frac{13}{2}O_2(g) \rightarrow 4CO_2(g) + 5H_2O(l)$$

The standard enthalpy of combustion for butane, $\Delta H_c^{\ominus}$, is -2878 kJ mol^{-1}. This value, and the standard enthalpies of combustion for other substances, are given in section 14 of the data booklet, along with their standard states.

The chemical equation for the combustion of butane can also be written with the enthalpy of combustion value included in the equation. The negative enthalpy change indicates an **exothermic** reaction, so energy is released into the surroundings. Therefore, the value is included on the product side:

$$C_4H_{10}(g) + \frac{13}{2}O_2(g) \rightarrow 4CO_2(g) + 5H_2O(l) + 2878 \text{ kJ mol}^{-1}$$

Exothermic and endothermic reactions are introduced in *Reactivity 1.1*.

 Activity

Write equations to describe the standard enthalpy change of combustion for the following compounds, and state the enthalpy change values by referring to section 14 of the data booklet. Ensure that you include state symbols and that the equations are balanced:

a. octane, C_8H_{18}

b. cyclohexanol, $C_6H_{12}O$

c. methanoic acid, CH_2O_2

d. glucose, $C_6H_{12}O_6$

e. chloroethane, C_2H_5Cl (hint: a strong, corrosive acid is one of the products)

Global impact of science

Much of the world's energy supply is produced by the combustion of hydrocarbons. Increasingly, the scientific community is researching alternative energy sources to reduce harmful emissions of greenhouse gases. This will enable governments to reach emission targets set by international agreements, such as COP21 in Paris and COP26 in Glasgow.

The combustion of ammonia is being investigated as a possible carbon-neutral fuel source. However, one problem with this new technology is the high level of NO_x emissions. Industrial processes are rarely totally "green": there are environmental costs associated with every method of energy production. How do we measure the environmental impact of different fuel sources, and which factors are most important for a fuel to be sustainable? Who should make decisions about what fuel sources are used to produce energy?

The **standard enthalpy change of formation**, $\Delta H_f^\ominus$, of a substance is the energy change that occurs when one mole of a substance is formed from its constituent elements in their standard states. Most elements are solids in their standard state, but there are 11 elements that are gases under standard conditions, for example argon, Ar, and fluorine, F_2. Two elements are liquid under standard conditions: mercury, Hg, and bromine, Br_2.

The constituent elements of butane are carbon, C, and hydrogen, H_2. Therefore, the formation of butane is described by the following equation:

$$4C(s) + 5H_2(g) \rightarrow C_4H_{10}(g)$$

The enthalpy change of formation of butane, $\Delta H_f^\ominus$, is $-126\,kJ\,mol^{-1}$. This value, and other values for the standard enthalpy change of formation for many common compounds, can be found in section 13 of the data booklet. The standard states of these compounds are given in the same section.

Activity

Write equations to describe the standard enthalpy change of formation for the following compounds, and state the enthalpy change values by referring to section 13 of the data booklet:

a. propane, C_3H_8

b. chloromethane, CH_3Cl

c. ethanol, C_2H_6O

d. benzoic acid, C_6H_5COOH

e. carbon monoxide, CO

f. methylamine, CH_3NH_2

You will notice that when writing chemical equations for standard enthalpy changes of combustion and formation, fractional stoichiometric coefficients may be necessary to balance the equation. This is one of the few cases when it is correct to use fractions to balance the final chemical equation.

AHL

Under standard conditions, the element carbon exists as several allotropes, for example, diamond, graphite and buckminsterfullerene. Allotropes of carbon have different $\Delta H_f^{\ominus}$ values. If an element has several allotropes, the normal convention is that the most stable allotrope is the standard state. The enthalpy of formation is $0\,kJ\,mol^{-1}$ for all elemental substances in their standard states. Graphite is taken as the standard for the carbon allotropes, and so it has a standard enthalpy of formation of $0\,kJ\,mol^{-1}$.

The allotropes of carbon and their structures are discussed in *Structure 2.2*.

 Linking question

Would you expect allotropes of an element, such as diamond and graphite, to have different $\Delta H_f^{\ominus}$ values? (Structure 2.2)

ATL Thinking skills

Reflect on the rationale for averaging bond enthalpy values. Evaluate the strengths and limitations of this approach.

Applying Hess's law to enthalpy changes of combustion

As discussed earlier in this topic, enthalpy is a state function, so the enthalpy change of a reaction is the difference between the initial and final values of the enthalpy of the system. The pathway between this initial and final state of a chemical reaction does not affect the overall enthalpy change.

The overall standard enthalpy change for a reaction, $\Delta H_r^{\ominus}$, can be calculated using enthalpy of combustion data. The difference between the sum of the enthalpy changes of combustion of the reactants and the sum of the enthalpy changes of combustion of the products is equal to the overall enthalpy change of the reaction:

$$\Delta H_r^{\ominus} = \Sigma(\Delta H_c^{\ominus}\ \text{reactants}) - \Sigma(\Delta H_c^{\ominus}\ \text{products})$$

This is a direct application of Hess's law.

Practice question

6. Using Hess's law and enthalpy of combustion data, the enthalpy change of formation of pentane is determined to be $-177\,kJ\,mol^{-1}$, but the value given in the data booklet is $-173\,kJ\,mol^{-1}$. Suggest why the values are different.

Applying Hess's law to enthalpy changes of formation

The standard enthalpy change for a reaction can also be calculated using enthalpy of formation data. The difference between the sum of the enthalpy changes of formation of the products and the sum of the enthalpy changes of formation of the reactants is equal to the overall enthalpy change of the reaction:

$$\Delta H_r^{\ominus} = \sum(\Delta H_f^{\ominus}\ \text{products}) - \sum(\Delta H_f^{\ominus}\ \text{reactants})$$

Worked example 3

Calculate the standard enthalpy change of formation of pentane, C_5H_{12}, using the enthalpy of combustion data from section 14 of the data booklet.

Solution

Step 1: Write a balanced chemical equation for the formation of 1 mol of pentane: $5C(s) + 6H_2(g) \rightarrow C_5H_{12}(l)$

Step 2: Write equations for the combustion of carbon, hydrogen and pentane and find their standard enthalpy change values from section 14 of the data booklet.

1	$C(s) + O_2(g) \rightarrow CO_2(g)$	$\Delta H^\ominus = -394\,kJ\,mol^{-1}$
2	$H_2(g) + \frac{1}{2}O_2(g) \rightarrow H_2O(l)$	$\Delta H^\ominus = -286\,kJ\,mol^{-1}$
3	$C_5H_{12}(l) + 8O_2(g) \rightarrow 5CO_2(g) + 6H_2O(l)$	$\Delta H^\ominus = -3509\,kJ\,mol^{-1}$

Step 3: Use these equations with either the summation of equations method or enthalpy cycle diagram method to determine the enthalpy change for the reaction. These methods were introduced earlier in this topic.

Summation of equations method

Inspect the formation of pentane reaction scheme. In your calculation, you need to use enthalpy values corresponding to the combustion reactions that involve carbon, hydrogen and pentane. Carbon is a reactant in equation 1, but five moles are required for the overall equation. Therefore you should multiply the enthalpy of combustion of carbon by five:

$$5C(s) + 5O_2(g) \rightarrow 5CO_2(g) \quad \Delta H = -1970\,kJ\,mol^{-1}$$

You also require six moles of hydrogen as a reactant. Equation 2 should be used, with the enthalpy change multiplied by six:

$$6H_2(g) + 3O_2(g) \rightarrow 6H_2O(l) \quad \Delta H_c^\ominus = -1716\,kJ\,mol^{-1}$$

For the product pentane, you should use equation 1, but the equation must be reversed to make pentane a product, not a reactant. As a result, the enthalpy change sign changes from negative to positive:

$$5CO_2(g) + 6H_2O(l) \rightarrow C_5H_{12}(l) + 8O_2(g) \quad \Delta H^\ominus = +3509\,kJ\,mol^{-1}$$

Now add the three equations together, cancelling the species common to both sides of the reaction, and add the enthalpy values:

$5C(s) + 5O_2(g)$	$\rightarrow$	$5CO_2(g)$	$\Delta H_c^\ominus = -1970\,kJ\,mol^{-1}$
$6H_2(g) + 3O_2(g)$	$\rightarrow$	$6H_2O(l)$	$\Delta H_c^\ominus = -1716\,kJ\,mol^{-1}$
$5CO_2(g) + 6H_2O(l)$	$\rightarrow$	$C_5H_{12}(l) + 8O_2(g)$	$\Delta H^\ominus = +3509\,kJ\,mol^{-1}$
Total $5C + 5\cancel{O_2} + 6H_2 + 3\cancel{O_2} + 5\cancel{CO_2} + 6\cancel{H_2O}$	$\rightarrow$	$5\cancel{CO_2} + 6\cancel{H_2O} + C_5H_{12} + 8\cancel{O_2}$	$\Delta H^\ominus = -177\,kJ\,mol^{-1}$

Cancelling the common species gives the overall equation $5C(s) + 6H_2(g) \rightarrow C_5H_{12}(l)$. Therefore, the total enthalpy change for the reaction is $-177\,kJ\,mol^{-1}$.

Enthalpy cycle diagram method

The enthalpy cycle diagram is on the right. Oxygen, $O_2(g)$, is not included in enthalpy cycle diagrams.

Calculate the sum of the enthalpy change values, following the arrows from the reactants to the products, via the $CO_2(g)$ and $H_2O(l)$ intermediates. If you are going against the direction of an arrow, reverse the sign of the enthalpy change. This calculation is summarized below:

$$\Delta H^\ominus = 5 \times (-394) + 6 \times (-286) + 3509 = -177\,kJ\,mol^{-1}$$

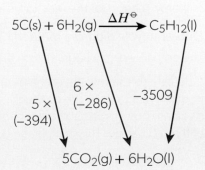

Worked example 4

Calculate the enthalpy of combustion of pentane, $C_5H_{12}(l)$, using the enthalpy of formation data from section 13 of the data booklet.

Solution

Step 1: Write a balanced chemical equation for the combustion of 1 mol of pentane:

$$C_5H_{12}(l) + 8O_2(g) \rightarrow 5CO_2(g) + 6H_2O(l)$$

Step 2: Write equations for the formation of pentane, carbon dioxide, and water from their elements and find their standard enthalpy change values from section 12 of the data booklet.

1	$5C(s) + 6H_2(g) \rightarrow C_5H_{12}(l)$	$\Delta H_f^{\ominus} = -173\,kJ\,mol^{-1}$
2	$C(s) + O_2(g) \rightarrow CO_2(g)$	$\Delta H_f^{\ominus} = -394\,kJ\,mol^{-1}$
3	$H_2(g) + \frac{1}{2}O_2(g) \rightarrow H_2O(l)$	$\Delta H_f^{\ominus} = -286\,kJ\,mol^{-1}$

Step 3: Use these equations with either the summation of equations method or enthalpy cycle diagram method to determine the enthalpy change for the reaction.

Summation of equations method

In your calculation, you need to use enthalpy values corresponding to the formation reactions that involve pentane, carbon dioxide and water. Pentane is a reactant in the overall equation, but a product in equation 1. Therefore, equation 1 must be reversed. As a result, the enthalpy change sign changes from negative to positive:

$$C_5H_{12}(l) \rightarrow 5C(s) + 6H_2(g) \qquad \Delta H^{\ominus} = +173\,kJ\,mol^{-1}$$

Carbon dioxide is a product in an equation 2, but five moles are required for the overall equation. Therefore, you should multiply the enthalpy of formation of carbon dioxide by five:

$$5C(s) + 5O_2(g) \rightarrow 5CO_2(g) \qquad \Delta H_f^{\ominus} = -1970\,kJ\,mol^{-1}$$

You require six moles of water as a product. Equation 3 should be used, with the enthalpy change multiplied by six:

$$6H_2(g) + 3O_2(g) \rightarrow 6H_2O(l) \qquad \Delta H_f^{\ominus} = -1716\,kJ\,mol^{-1}$$

Now add the three equations together, cancelling the species common to both sides of the reaction, and add the enthalpy values:

	$C_5H_{12}(l)$	$\rightarrow$	$5C(s) + 6H_2(g)$	$\Delta H^{\ominus} = +173\,kJ\,mol^{-1}$
	$5C(s) + 5O_2(g)$	$\rightarrow$	$5CO_2(g)$	$\Delta H_f^{\ominus} = -1970\,kJ\,mol^{-1}$
	$6H_2(g) + 3O_2(g)$	$\rightarrow$	$6H_2O(l)$	$\Delta H_f^{\ominus} = -1716\,kJ\,mol^{-1}$
Total	$C_5H_{12} + 5C + 5O_2 + 6H_2 + 3O_2$	$\rightarrow$	$5C + 6H_2 + 5CO_2 + 6H_2O$	$\Delta H^{\ominus} = -3513\,kJ\,mol^{-1}$

Cancelling the common species gives the overall equation $C_5H_{12}(l) + 8O_2(g) \rightarrow 5CO_2(g) + 6H_2O(l)$. Therefore, the total enthalpy change for the reaction is $-3513\,kJ\,mol^{-1}$.

Enthalpy cycle diagram method

The enthalpy cycle diagram is on the right.

Calculate the sum of the enthalpy change values, following the arrows from the reactants to the products, via the C(s) and $H_2(g)$ intermediates. If you are going against the direction of an arrow, reverse the sign of the enthalpy change. This calculation is summarized below:

$$\Delta H^{\ominus} = (+173) + 5 \times (-394) + 6 \times (-286) = -3513\,kJ\,mol^{-1}$$

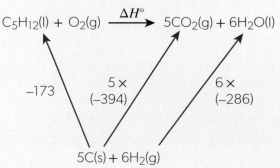

Practice questions

7. Calculate the enthalpy change of formation of propanone, $CH_3COCH_3(l)$, using enthalpy change of combustion data from section 14 of the data booklet.

   Use the summation of equations method and the enthalpy cycle diagram method to support your answer.

8. Hydrogen gas can be made using the following reaction:

   $$CH_4(g) + H_2O(g) \rightarrow 3H_2(g) + CO(g)$$

   Section 13 of the data booklet lists the standard enthalpies of formation, $\Delta H_f^\ominus$, for some of the species in the reaction above.

   a. Outline why no value is listed for $H_2(g)$.

   b. Determine the value of $\Delta H^\ominus$, in kJ mol^{-1}, for the reaction using the standard enthalpy change of formation values.

   c. Outline why the value of enthalpy change of this reaction calculated from bond enthalpies is less accurate.

9. Consider the following enthalpy cycle diagram:

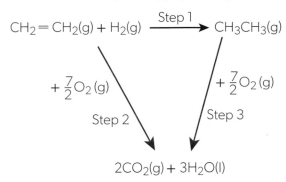

   a. Calculate the standard enthalpy change, $\Delta H^\ominus$, of step 2 using section 14 of the data booklet.

   b. Determine the standard enthalpy change, $\Delta H^\ominus$, of step 1.

   c. Suggest one reason why the calculated value of $\Delta H^\ominus$ using Hess's law in part **b** can be considered accurate and one reason why it can be considered approximate.

Comparing ΔH values

The standard enthalpy of combustion of propan-1-ol is -2021 kJ mol^{-1}. In this task, you will compare this value to those obtained with some of the methods described in this topic.

Relevant skills

- Tool 3: Use arithmetic and algebraic calculations to solve problems

- Tool 3: Calculate and interpret percentage error

- Inquiry 3: Compare the outcomes of an investigation to the accepted scientific context

- Inquiry 3: Identify and discuss sources and impacts of systematic and random errors

Method

1. Determine the enthalpy of combustion of propan-1-ol by each of the following methods:

   a. Refer to average bond enthalpy values.

   b. Conduct a calorimetry experiment. This could be real (as described in *Reactivity 1.1*), or simulated, if you have access to a suitable simulation software. Record the relevant measurement uncertainties and propagate them.

   c. Refer to standard enthalpies of formation.

2. Determine the percentage error of each of the values obtained in methods **a**, **b** and **c**.

3. Discuss and explain the differences between the standard enthalpy of combustion of propan-1-ol and each of your calculated values obtained by methods **a**, **b** and **c**.

Born–Haber cycles (*Reactivity 1.2.5*)

Born–Haber cycles are another application of Hess's law. These cycles combine the enthalpy changes associated with several steps involved in the formation of an ionic compound. The steps in the cycle include ionization energy, enthalpy of atomization, electron affinity, lattice enthalpy, and enthalpy of formation.

Ionization energy, IE, is the standard enthalpy change that occurs on the removal of one mole of electrons from one mole of atoms or positively charged ions in the gaseous state. For metal ions with multiple valence electrons, the first, second, and sometimes further ionization energies are defined:

$$IE_1: \quad M(g) \rightarrow M^+(g) + e^- \qquad\qquad IE > 0$$

$$IE_2: \quad M^+(g) \rightarrow M^{2+}(g) + e^- \qquad\qquad IE > 0$$

The process of ionization is endothermic. The values of first ionization energy can be found in section 9 of the data booklet.

The **enthalpy of atomization**, $\Delta H_{at}^{\ominus}$, is the standard enthalpy change that occurs when one mole of gaseous atoms of an element is formed. This process is endothermic. For a solid monatomic species, the value of $\Delta H_{at}^{\ominus}$ is equal to the enthalpy of sublimation:

$$M(s) \rightarrow M(g) \qquad\qquad \Delta H_{at}^{\ominus} > 0$$

For a gaseous diatomic species, $\Delta H_{at}^{\ominus}$ is equal to half of the bond enthalpy (section 12 of the data booklet):

$$\frac{1}{2}X_2(g) \rightarrow X(g) \qquad\qquad \Delta H_{at}^{\ominus} > 0$$

The **electron affinity**, EA, is the standard enthalpy change on the addition of one mole of electrons to one mole of atoms in the gaseous state:

$$X(g) + e^- \rightarrow X^-(g) \qquad\qquad EA < 0$$

Electron affinity of non-metals is typically negative, but there are exceptions, such as the electron affinity of helium or nitrogen. Values of first and second electron affinity can be found in section 9 of the data booklet.

The **lattice enthalpy**, $\Delta H_{lattice}^{\ominus}$, is defined as the standard enthalpy change that occurs when gaseous ions are formed from one mole of structural units of a solid ionic lattice:

$$MX(s) \rightarrow M^+(g) + X^-(g) \qquad \Delta H_{lattice}^{\ominus} > 0$$

The process is endothermic. Experimental values of lattice enthalpy at 298.15 K can be found in section 16 of the data booklet.

Lattice enthalpies are often quoted as negative values that represent the exothermic formation of the solid lattice from gaseous ions. However, in DP chemistry, the lattice enthalpy always refers to the endothermic formation of gaseous ions from the solid lattice.

We have already defined the standard enthalpy of formation, $\Delta H_f^{\ominus}$, of a substance as the enthalpy change that occurs when one mole of a substance is formed under standard conditions from its constituent elements in their standard states. Standard enthalpy of formation of an ionic substance, such as NaCl(s), can be represented by a single equation:

$$Na(s) + \frac{1}{2}Cl_2(g) \rightarrow NaCl(s) \qquad \Delta H_f^{\ominus} = -411\,kJ\,mol^{-1}$$

(ATL) Self-management skills

The correct use of subject-specific terminology is essential to your ability to demonstrate your knowledge and understanding of new concepts in chemistry.

You will encounter several key terms in this section: lattice enthalpy, standard enthalpy of formation, standard state, enthalpy of atomization, ionization energy and electron affinity. Make a glossary that has an entry for each term, detailing the following:

Term	Symbol	Definition	Equation
lattice enthalpy	$\Delta H_{lattice}^{\ominus}$	Standard enthalpy change that occurs on the formation of gaseous ions from one mole of a solid lattice	$MX(s) \rightarrow M^+(g) + X^-(g)$

You can continue to add terms to your chemistry glossary over time.

Using the Born–Haber cycle

The lattice enthalpy, enthalpy of atomization, ionization energy, electron affinity and enthalpy of formation can be combined to construct a Born–Haber cycle. A generalized Born–Haber cycle is shown in figure 6.

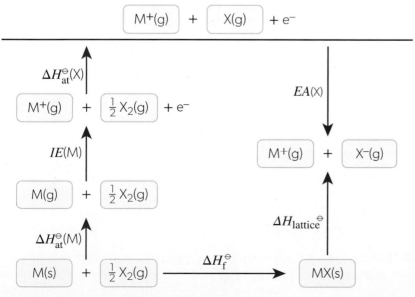

▲ Figure 6 A generalized Born–Haber cycle

The unknown value of enthalpy change for any step in a Born–Haber cycle can be determined by following the opposite pathway in the cycle. For example, if you know the values of enthalpy of atomization, ionization energies, electron affinities and lattice enthalpy, you can find the standard enthalpy of formation of an ionic compound. The enthalpy of formation is equal to the sum of the enthalpy changes for each of the other steps. Much like with the enthalpy cycle diagrams you constructed before, you should reverse the sign of any enthalpy changes where the pathway goes against the direction of the arrow in the cycle.

Worked example 5

Determine the lattice enthalpy of potassium bromide using the Born–Haber cycle in figure 7.

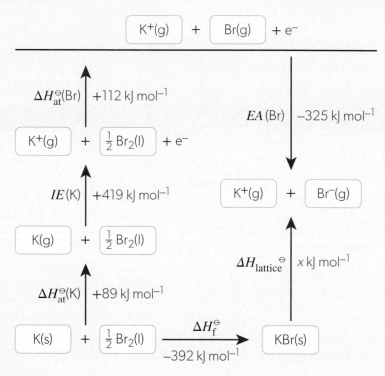

▲ Figure 7 Born–Haber cycle to calculate the lattice enthalpy of potassium bromide

Solution

The lattice enthalpy is the enthalpy change for the reaction $KBr(s) \rightarrow K^+(g) + Br^-(g)$. In figure 7, this step is shown from the bottom right of the cycle, to the middle right, labelled $\Delta H_{lattice}^{\ominus}$. To find the lattice enthalpy, follow the alternative pathway, going from $KBr(s)$ clockwise around to $K^+(g) + Br^-(g)$. Add the values of the enthalpy change for each step:

$$\Delta H_{lattice}^{\ominus}(KBr) = -\Delta H_f^{\ominus} + \Delta H_{at}^{\ominus}(K) + IE(K) + \Delta H_{at}^{\ominus}(Br) + EA(Br)$$

$$= -(-392) + (+89) + (+419) + (+112) + (-325)$$

$$= +687 \text{ kJ mol}^{-1}$$

Note that you followed the reverse of the enthalpy of formation reaction arrow, so the sign was reversed: +392 instead of −392. Remember to express your answer for the lattice enthalpy of an ionic compound as a positive value, because it is an endothermic process.

Practice question

10. Interpret and utilize the values from the Born–Haber cycle in figure 8 to calculate the enthalpy of formation for the ionic compound magnesium oxide, MgO. MgO has doubly charged ions.

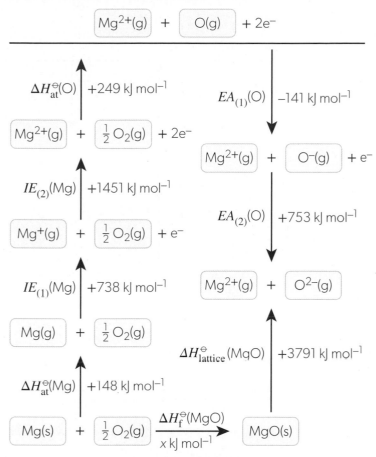

▲ **Figure 8** Born–Haber cycle to calculate the enthalpy of formation for magnesium oxide

▲ **Figure 9** The lattice structure of crystalline magnesium oxide: grey = Mg^{2+}, red = O^{2-}

 Linking question

What are the factors that influence the strength of lattice enthalpy in an ionic compound? (*Structure 2.1*)

Factors affecting the value of lattice enthalpy are detailed in *Structure 2.1*.

End of topic questions

Topic review

1. Using your knowledge from the *Reactivity 1.2* topic, answer the guiding question as fully as possible:

   How does application of the law of conservation of energy help us to predict energy changes during reactions?

Exam-style questions

Multiple-choice questions

2. Which equation represents the N–H bond enthalpy in NH_3?

   A $NH_3(g) \rightarrow N(g) + 3H(g)$

   B $\frac{1}{3}NH_3(g) \rightarrow \frac{1}{3}N(g) + H(g)$

   C $NH_3(g) \rightarrow \frac{1}{2}N_2(g) + \frac{3}{2}H_2(g)$

   D $NH_3(g) \rightarrow \bullet NH_2(g) + \bullet H(g)$

3. Bond enthalpies can be used to calculate the enthalpy change of the following reaction, in $kJ\,mol^{-1}$.

   $$N_2(g) + 3H_2(g) \rightleftharpoons 2NH_3(g)$$

   Referring to section 12 of the data booklet, deduce which of the following represents the calculation of the enthalpy change of reaction from bond enthalpies.

   A $(6 \times 391) - [(3 \times 436) + 945]$

   B $(3 \times 391) - (436 + 945)$

   C $-[(3 \times 436) + 945] + (3 \times 391)$

   D $-(6 \times 391) + [(3 \times 436) + 945]$

4. Which equation represents lattice enthalpy?

   A $NaCl(g) \rightarrow Na^+(g) + Cl^-(g)$

   B $NaCl(s) \rightarrow Na^+(g) + Cl^-(g)$

   C $NaCl(s) \rightarrow Na^+(aq) + Cl^-(aq)$

   D $NaCl(s) \rightarrow Na^+(s) + Cl^-(s)$

5. Using the Born–Haber cycle below, determine which calculation represents the lattice enthalpy of strontium chloride, $SrCl_2$, in $kJ\,mol^{-1}$.

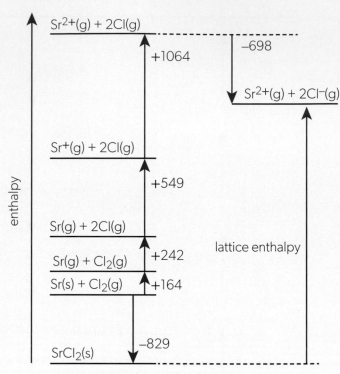

   A $-(-829) + 164 + 242 + 549 + 1064 - (-698)$

   B $-829 + 164 + 242 + 549 + 1064 - 698$

   C $-(-829) + 164 + 242 + 549 + 1064 - 698$

   D $-829 + 164 + 242 + 549 + 1064 - (-698)$

AHL

AHL

AHL

6. The Born–Haber cycle for potassium oxide, K_2O, is shown below:

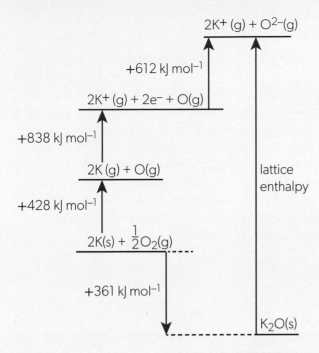

Which calculation represents the lattice enthalpy in $kJ\,mol^{-1}$?

A $-361 + 428 + 838 + 612$

B $-(-361) + 428 + 838 + 612$

C $-361 + 428 + 838 - 612$

D $-(-361) + 428 + 838 - 612$

Extended-response questions

7. The following equation describes the industrial production of ammonia using the Haber process.

$$N_2(g) + 3H_2(g) \rightleftharpoons 2NH_3(g)$$

a. Calculate the enthalpy change of reaction for this process using bond enthalpy data. [3]

b. State and explain whether the reaction is exothermic or endothermic. [2]

8. Ammonia reacts with oxygen to form nitrogen and steam.

a. Write a balanced chemical equation for this reaction, showing the states of all species. [1]

b. Draw full structural formulas for all reactants and products to determine the nature of the covalent bonds present. [4]

c. Using bond enthalpy values from section 12 of the data booklet, determine the enthalpy change for this reaction. [3]

9. Octane, C_8H_{18}, is a component of liquid gasoline. When gasoline is burned as fuel, octane is combusted in oxygen via the following reaction:

$$C_8H_{18}(l) + \frac{25}{2}O_2(g) \rightarrow 8CO_2(g) + 9H_2O(g)$$

a. Using bond enthalpy data, calculate the enthalpy change for this reaction. [3]

b. The enthalpy change for this reaction was determined experimentally to be $-5470\,kJ\,mol^{-1}$. Give **two** reasons why the enthalpy change value you obtained from bond enthalpy data differs from the experimental value. [2]

Reactivity 1.3   Energy from fuels

When we burn fossil fuels, chemical energy is converted into thermal energy. Thermal energy can be used to generate electrical energy to power devices in our homes and in industry. The use of fossil fuels is being increasingly questioned as advances in science and technology reveal the damage that their use is having on our planet. In recent years, the cost of alternative renewable energy sources has decreased and renewable energy technology has advanced.

This, coupled with scientific evidence of the rate of change in global warming, has increased the urgency for governments to commit to net-zero carbon emissions by 2050.

The challenges of using chemical energy from non-renewable, carbon-based fuels, such as crude oil, natural gas and coal, which provide 80% of our energy, are clearly understood by the scientific community. Many industrialized nations depend on the use of fossil fuels to sustain their fast-developing economies. This is a major problem that has to be overcome if the use of renewable, clean sources to meet our global energy needs is to exceed the use of fossil fuels.

Understandings

Reactivity 1.3.1—Reactive metals, non-metals and organic compounds undergo combustion reactions when heated in oxygen.

Reactivity 1.3.2—Incomplete combustion of organic compounds, especially hydrocarbons, leads to the production of carbon monoxide and carbon.

Reactivity 1.3.3—Fossil fuels include coal, crude oil and natural gas, which have different advantages and disadvantages.

Reactivity 1.3.4—Biofuels are produced from the biological fixation of carbon over a short period of time through photosynthesis.

Reactivity 1.3.5—A fuel cell can be used to convert chemical energy from a fuel directly to electrical energy.

Combustion (*Reactivity 1.3.1* and *Reactivity 1.3.2*)

In combustion reactions, substances are burned in oxygen. Metals, non-metals and hydrocarbons all react with oxygen when combusted, producing different products.

Combustion of metals

Oxidation and reduction are defined in terms of electron transfer in *Reactivity 3.2*.

The combustion of reactive metals in the presence of oxygen results in the **oxidation** of the metal, the **reduction** of oxygen, and the formation of an ionic compound. This type of reaction is therefore known as a **redox** reaction. Oxidation can be defined as a gain of oxygen or a loss of electrons. Reduction is defined as a loss of oxygen or a gain of electrons.

The general equation for the reaction of a metal with oxygen is as follows:

metal + oxygen → metal oxide

Lithium burns in oxygen to release heat and produce lithium oxide:

$$4Li(s) + O_2(g) \rightarrow 2Li_2O(s)$$

Magnesium readily reacts with oxygen, producing a brilliant white light and magnesium oxide:

$$2Mg(s) + O_2(g) \rightarrow 2MgO(s)$$

In both of these reactions, the reactive metals are being oxidized to form metal ions. The half-equations for each of these metals reveal the loss of electrons:

$$Li \rightarrow Li^+ + e^-$$

$$Mg \rightarrow Mg^{2+} + 2e^-$$

The oxygen atoms are reduced to form O^{2-} ions in each reaction:

$$O_2 + 4e^- \rightarrow 2O^{2-}$$

Most metal oxides are ionic compounds. Rules for naming ionic compounds are outlined in *Structure 2.1*.

Combustion of non-metals

Non-metals are also oxidized when combusted in oxygen, forming non-metal oxides:

$$\text{non-metal} + \text{oxygen} \rightarrow \text{non-metal oxide}$$

Sulfur, a non-metal, can be found as impurities in fossil fuels, such as coal and crude oil. Coal may contain up to 3% of sulfur. The combustion of sulfur-containing compounds in oxygen predominantly produces sulfur dioxide, SO_2:

$$S(s) + O_2(g) \rightarrow SO_2(g)$$

Sulfur dioxide can then further react with oxygen in the atmosphere to produce sulfur trioxide:

$$2SO_2(g) + O_2(g) \rightleftharpoons 2SO_3(g)$$

Sulfur trioxide reacts with water in the atmosphere, to form sulfuric acid:

$$SO_3(g) + H_2O(l) \rightarrow H_2SO_4(aq)$$

In industry, sulfur dioxide is produced in vast quantities as feedstock for the synthesis of sulfuric acid. The majority of the sulfuric acid is then used in the production of fertilizers, along with paper, paints, textiles and a wide variety of other products.

 Activity

Research the applications of sulfur dioxide, and evaluate the environmental impact of its industrial production. Research other non-metal oxides and their applications and environmental impacts.

Complete combustion of organic compounds

Hydrocarbons are organic compounds composed only of carbon and hydrogen atoms. Alkanes are the simplest hydrocarbons, which are present in fossil fuels, such as crude oil and natural gas. They are relatively inert. This is because they have a low bond polarity, strong covalent carbon–carbon bonds (bond enthalpy = 346 kJ mol^{-1}) and strong carbon–hydrogen bonds (bond enthalpy = 414 kJ mol^{-1}). However, alkanes do participate in some reactions, including combustion.

▲ **Figure 1** Liquefied petroleum gas contains liquid propane, and is used as a fuel for gas barbecues

The chemical equation for the combustion of alkanes shows the general formula of the homologous series of alkanes: C_nH_{2n+2}. Homologous series of organic compounds were introduced in *Structure 3.2*.

Alkanes are commonly used as fuels, releasing large amounts of energy in combustion reactions. For combustion to occur, a fuel must be volatile. Volatility is the tendency of a substance to change state from liquid to gas. As the length of the carbon chain increases in the alkane series, the boiling point also increases, and therefore volatility decreases. As a result, short-chain alkanes tend to be used as fuels. Liquefied petroleum gas (LPG) consists predominantly of compressed propane, C_3H_8, while petrol (gasoline) is a mixture of hydrocarbons from butane, C_4H_{10}, to dodecane, $C_{12}H_{26}$.

Alkanes undergo **complete combustion** in the presence of excess oxygen. This reaction is exothermic, and it produces carbon dioxide and water. The general equation for the complete combustion of alkanes is shown below:

$$\text{alkane} + \text{oxygen} \rightarrow \text{carbon dioxide} + \text{water}$$

$$C_nH_{2n+2} + \frac{3n+1}{2}O_2 \rightarrow nCO_2 + (n+1)H_2O$$

In this, and following combustion reactions, the states of all species depend on the reaction conditions. For example, at high temperatures, water can be produced as steam, $H_2O(g)$, while at lower temperatures it will condense into liquid, $H_2O(l)$.

Petrol, also known as gasoline, is a mixture of hydrocarbons obtained from oil, with octane present in the highest proportion. The reaction for the combustion of octane is shown below:

$$C_8H_{18}(l) + \frac{25}{2}O_2(g) \rightarrow 8CO_2(g) + 9H_2O(l) \qquad \Delta H_c^{\ominus} = -5470 \, \text{kJ mol}^{-1}$$

Falsification

Up until the late 18th century, combustion was explained in terms of the phlogiston theory. A substance known as "phlogiston" was thought to be present in all combustible materials. Burning was believed to release phlogiston into the air. The combustion of carbon would lead to a mass loss, which was interpreted as the loss of phlogiston. The fact that some substances, such as metals, gain mass when combusted was a puzzling observation—further hypotheses were developed to justify it. Examination of the mass losses and gains when different elements burnt in air, along with the discovery of oxygen, led to the falsification of the phlogiston theory. It was superseded by the theory of combustion that you have learnt about and that explains combustion in terms of oxidation.

The phlogiston theory may seem unreasonable or even peculiar to someone who knows what we know now. If we consider the knowledge base at the time, however, phlogiston was not an implausible concept. Scientific claims, hypotheses or theories can be falsified. This can lead to the development of new theories, which must then be corroborated by experimental evidence in order to eventually become accepted (although theories can never be proven with certainty). Scientists must remain open-minded with respect to new evidence.

Think of a claim, hypothesis or theory you have encountered in your study of science. What type of evidence could challenge it? How much evidence would be needed to falsify it?

Alcohols are another class of organic compounds. They have a wide range of applications, such as fuels, solvents and antiseptics. Like alkanes, alcohols undergo complete combustion reactions releasing carbon dioxide and water.

The general equation for the complete combustion of alcohols is shown below:

alcohol + oxygen → carbon dioxide + water

$$C_nH_{2n+1}OH + \frac{3n}{2}O_2 \rightarrow nCO_2 + (n+1)H_2O$$

Ethanol is used in medicine for its antiseptic properties. It is also used as a fuel for vehicle engines. It is known as a **biofuel** because it can be produced by plants, a renewable resource, as opposed to fossil fuels, the supply of which is finite. The reaction for the complete combustion of ethanol is shown below:

$$C_2H_5OH(l) + 3O_2(g) \rightarrow 2CO_2(g) + 3H_2O(l) \qquad \Delta H_c^{\ominus} = -1367 \text{ kJ mol}^{-1}$$

This reaction is strongly exothermic. The use of ethanol as a renewable biofuel is increasing. The scientific community is working to resolve the problems associated with the production of biofuels, such as the high cost of production, the impact of using farmland on food supply, and the lower amount of energy produced per unit mass or volume of the fuel.

Global impact of science

Carbon dioxide has a significant environmental impact, as it contributes to global warming. This is the major reason why governments around the world promote clean energy, and legislate for limiting the production of new cars that use fossil fuels. Ethanol is seen as a more environmentally friendly fuel, as it comes from a renewable source.

How can the scientific community, industry and governments work together to reduce the use of fossil fuels? Who has the greatest responsibility to reduce our fossil fuel use?

▲ **Figure 2** Ethanol-fuelled public transport is becoming more common as governments look for ways to reduce reliance on fossil fuels

Practice questions

1. Deduce balanced equations for the **complete** combustion of:
   a. methane
   b. propane
   c. pentane
   d. hexane

2. Deduce balanced equations for the **complete** combustion of:
   a. propanol
   b. pentanol
   c. heptanol

Incomplete combustion of organic compounds

If there is a limited supply of a species participating in a chemical reaction, this species is known as the **limiting reactant**. Combustion reactions usually occur in a system where atmospheric air and oxygen are in excess, so the limiting reactant is the organic compound (fuel). However, when this is not the case, such as in a car engine, oxygen becomes the limiting reactant, which causes **incomplete combustion** of the fuel.

In the incomplete combustion of hydrocarbons, carbon monoxide, CO(g), and/or elemental carbon, C(s), can be produced. Carbon monoxide is a poisonous gas that irreversibly binds to hemoglobin in the red blood cells, reducing the oxygen-carrying capacity of the blood (figure 3). Elemental carbon produced in incomplete combustion reactions is often referred to as soot.

▶ Figure 3 The carbon monoxide molecule irreversibly binds to the large hemoglobin molecule

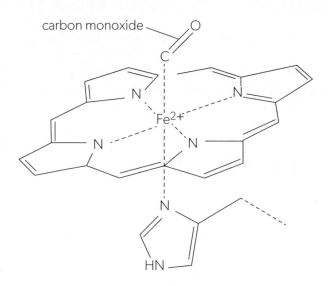

carbon monoxide

The general equation for the incomplete combustion of an alkane to form carbon monoxide is as follows:

alkane + oxygen → carbon monoxide + water

$$C_nH_{2n+2} + \frac{2n+1}{2}O_2 \rightarrow nCO + (n+1)H_2O$$

The general equation for the incomplete combustion of an alkane to form carbon (soot) is as follows:

alkane + oxygen → carbon + water

$$C_nH_{2n+2} + \frac{n+1}{2}O_2 \rightarrow nC + (n+1)H_2O$$

The incomplete combustion reactions of methane are shown below:

$$CH_4(g) + \frac{3}{2}O_2(g) \rightarrow CO(g) + 2H_2O(l)$$

$$CH_4(g) + O_2(g) \rightarrow C(s) + 2H_2O(l)$$

These incomplete combustion reactions can occur simultaneously with complete combustion reactions in different ratios. Incomplete combustion reactions are less exothermic than the corresponding complete combustion reactions.

Incomplete combustion can be observed in the laboratory, with the appearance of black soot on the bottom of glassware that has been heated over a Bunsen or spirit burner flame.

Practice questions

3. Deduce balanced equations for the **incomplete** combustion of the following alcohols, where carbon monoxide is one of the products:
    a. propanol
    b. pentanol
    c. heptanol

Fossil fuels (*Reactivity 1.3.3*)

Fossil fuels include crude oil, coal and natural gas. Crude oil, or petroleum, is a non-renewable resource, and its use as a fuel is deeply embedded in the global society. Crude oil is a natural mixture of hydrocarbons, organic compounds containing nitrogen, sulfur and oxygen, and a wide variety of other elements.

Coal, petroleum and natural gas are the main fuels used to generate electricity in power stations and the internal combustion engines of cars. As the global demand for energy increases, so does the consumption of fossil fuels. Globally, governments are making plans and legislating to limit the consumption of fossil fuels and promote the use of renewable energy. However, the transition to clean energy will take many decades.

One consequence of the use of fossil fuels is the release of large quantities of carbon dioxide, a product of the combustion reaction, into the atmosphere. Carbon dioxide is a **greenhouse gas**, which means that it traps heat energy inside the Earth's atmosphere. This is known as the **greenhouse effect** (figure 4).

Linking questions

Why is high activation energy often considered to be a useful property of a fuel? (*Reactivity 2.2*)
Which species are the oxidizing and reducing agents in a combustion reaction? (*Reactivity 3.2*)
What might be observed when a fuel such as methane is burned in a limited supply of oxygen? (*Inquiry 2*)
How does limiting the supply of oxygen in combustion affect the products and increase health risks? (*Reactivity 2.1*)

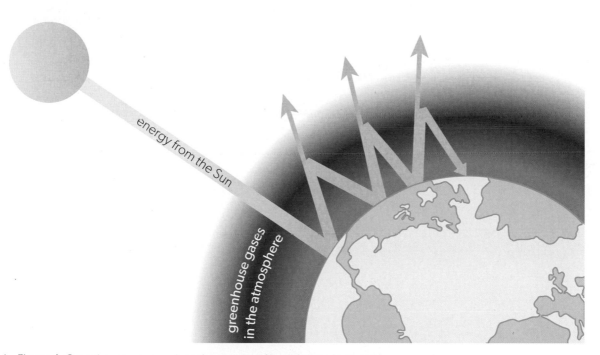

energy from the Sun

greenhouse gases in the atmosphere

▲ Figure 4 Greenhouse gases reduce the amount of heat radiated into space

Nitrogen, N_2, and oxygen, O_2, make up over 99% of atmospheric air. Neither nitrogen or oxygen have the ability to absorb infrared radiation that enters the Earth's atmosphere from the Sun. By comparison, carbon dioxide, CO_2, constitutes approximately 0.04% of the atmosphere. Despite the small proportion of carbon dioxide, the increase in the concentration of this greenhouse gas is causing significant damage to our environment. A carbon dioxide molecule can absorb infrared radiation resulting in the vibration of bonds within the molecule. After absorbing the infrared radiation and undergoing vibration, the molecule will emit infrared radiation back into the atmosphere. Some of this radiation will be directed towards the Earth's surface, which will increase the global temperature.

The IR activity of carbon dioxide is discussed in *Structure 3.2*.

Other greenhouse gases include methane, nitrous oxide, water vapour and fluorinated substances such as hydrofluorocarbons.

As a result of the greenhouse effect, average temperatures around the world are increasing, which is known as global warming. There is widespread agreement in the scientific community that the main cause of global warming is the increase in the levels of greenhouse gases: in particular, carbon dioxide (figure 5), and to a lesser extent, methane.

▶ **Figure 5** Annual mean atmospheric CO_2 levels recorded at Mauna Loa Observatory

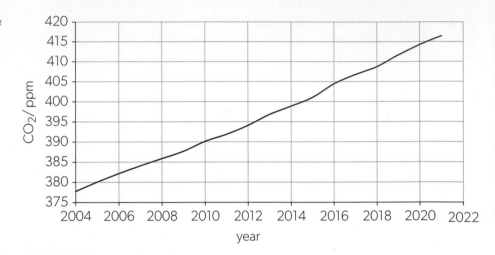

The combustion of different fuels releases different quantities of greenhouse gases (figure 6).

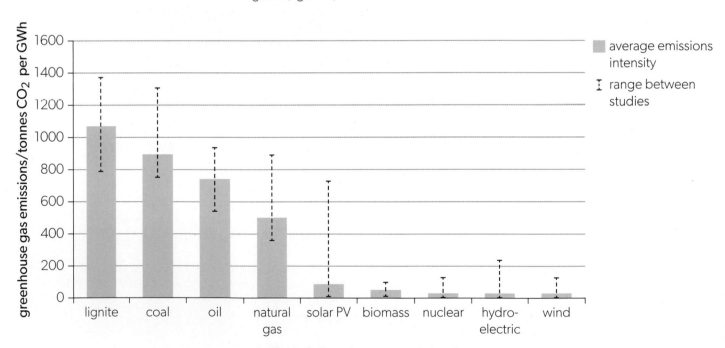

▲ **Figure 6** Greenhouse gas emissions by energy source

Figure 7 shows the historic patterns of atmospheric carbon dioxide concentration during the last three glacial cycles, constructed by the analysis of ice-core samples. The current CO_2 levels are much higher than historical peaks, suggesting that human-made emissions are largely responsible for the additional quantities of carbon dioxide in the atmosphere.

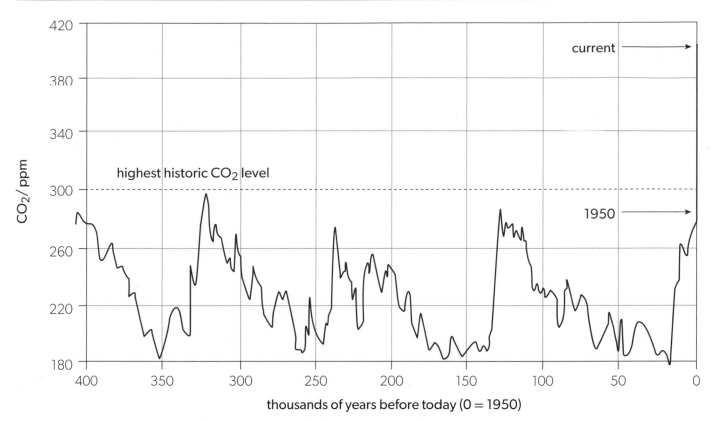

▲ Figure 7 Historic CO_2 levels during the last three glacial cycles, as reconstructed from ice cores

TOK

One of the optional themes in TOK is knowledge and indigenous societies. The global population of indigenous people is estimated to be over 450 million, living all over the world. Indigenous communities have a long-standing relationship with, and a highly developed understanding of, the natural environment and biodiversity. Many are particularly vulnerable to the adverse effects of climate change, despite the fact that their contribution to greenhouse gas emissions is exceptionally low. International organizations such as the United Nations (UN) have not only called for awareness and action to protect indigenous peoples, but also recognized that their participation in decision-making processes is of value to all: "indigenous peoples' knowledge should be considered an important element within the international debate regarding adaptation to climate change".

▲ Figure 8 Participants at the United Nations Permanent Forum for Indigenous Issues

What values and assumptions underpin the use of the term "indigenous" knowledge?

Many different sources of fuels are used in everyday life. The choice of fuels depends on the economic development of nations and the natural resources available. Each fuel has a different **specific energy**: the amount of heat energy released per mass of the fuel. Wood, a traditional means of generating energy for cooking and heating, has the lowest specific energy of all common fuels.

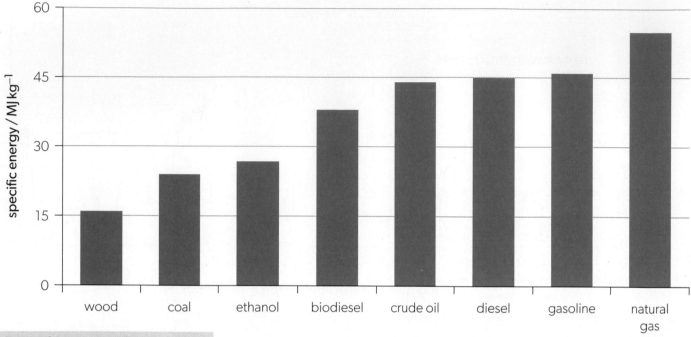

▲ Figure 9 The specific energy of different fuels

Practice questions

4. Propane gas C_3H_8 is a hydrocarbon that is classified as a fossil fuel.

   a. Write the balanced chemical equation for the complete combustion of propane and identify the greenhouse gas that is produced.

   b. Enthalpy of combustion of propane is $-2219\,kJ\,mol^{-1}$. Calculate the energy released in burning 1 gram (the specific energy) of propane.

   c. Calculate the mass of carbon dioxide produced per gram of propane burned.

   d. Calculate the carbon footprint, in tonnes of carbon dioxide, from burning 1000 kg (1 tonne) of propane gas. Identify some of the environmental issues caused by the release of this greenhouse gas that was previously locked within the Earth.

Common fuels vary in composition. The chain length of hydrocarbons present in these fuels also varies. In general, the longer the hydrocarbon chain, the greater the tendency of the fuel to undergo incomplete combustion. As discussed, incomplete combustion results in the release of poisonous carbon monoxide and/or elemental carbon (soot). But it also produces a smaller amount of heat energy per unit mass of the fuel when compared to the complete combustion of the same hydrocarbon. Larger hydrocarbons have a reduced volatility due to stronger **London (dispersion) forces (LDFs)**. This affects the way the hydrocarbon molecules interact with the oxygen molecules and the type of combustion that occurs.

 Linking questions

Why do larger hydrocarbons have a greater tendency to undergo incomplete combustion? (*Structure 3.2*)

 Why is carbon dioxide described as a greenhouse gas? (*Structure 3.2*)

 What are some of the environmental, economic, ethical and social implications of burning fossil fuels? (*Reactivity 3.2*)

Air pollution database task

The concepts in this topic are directly and indirectly related to air pollution, including carbon monoxide (released during the incomplete combustion of hydrocarbons), NO_x and SO_x (released in the combustion of substances containing nitrogen and sulfur, which is also discussed in *Structure 3.1*) and greenhouse gases (methane, carbon dioxide and CFCs). In this task you will explore a factor affecting the environmental impact of an atmospheric pollutant of your choice.

Relevant skills

- Tool 2: Identify and extract data from databases
- Inquiry 1: Develop investigations that involve databases
- Inquiry 1: Justify the range and quantity of measurements
- Inquiry 2: Carry out relevant and accurate data processing
- Inquiry 2: Assess reliability and validity
- Inquiry 3: Relate the outcomes of an investigation to the stated research question
- Inquiry 3: Discuss the impact of uncertainties on the conclusions
- Inquiry 3: Evaluate the implications of methodological weaknesses, limitations and assumptions on conclusions

Method

1. Access an air quality database online. Possible sources include:
   - World Health Organization's air pollution data portal
   - World Air Quality Index project
   - Air pollution data shared by the European Environment Agency
   - Your local ministry of the environment

2. Browse the database, carry out some background research, and select one or two pollutants to focus on.

3. Decide how much data to collect and the range of data required to address your aim.

4. Formulate a focused research question that includes the variables, methodology and range of data.

5. Collect data in a well-organized table, including the data sources.

6. Estimate the uncertainty of the data, stating clearly how you have done so and any assumptions you have made.

7. Transform your data by processing and graphing. Include an explanation of how you have processed the data.

8. Discuss and interpret the processed data and their uncertainties.

9. Assess the validity and reliability of the outcome of your investigation.

10. Evaluate the outcome of your investigation: how confident are you in your conclusion?

11. Evaluate the methodology: discuss the strengths and limitations of the data, assumptions and analysis.

12. Outline possible extensions to your investigation. Try to relate one of these proposed extensions to another diploma subject, or TOK.

Biofuels (*Reactivity 1.3.4*)

With the global population continuing to increase at approximately 1% per year, our understanding of how to efficiently use our finite resources is of vital importance. Alternatively, renewable energy resources can be used instead of fossil fuels. Sustainable energy is a UN initiative, with the goal of doubling the share of renewable energy contributed to global energy production by 2030. The cost of renewable energy has decreased over the decades, and now represents a viable option for governments, businesses and individuals.

Renewable energy resources include those that depend on the heat or motion of the Earth (geothermal, wind, and tidal power) or the Sun's radiation (solar power and biomass). Non-renewable energy technology generally uses fossil fuels, such as coal and natural gas, which are finite in their supply.

Biofuels are renewable resources, produced from organic compounds, which in turn are generated from carbon dioxide during biological processes. The production of organic compounds from carbon dioxide is known as **biological carbon fixation**. For example, green plants use photosynthesis to absorb carbon dioxide from the atmosphere and transform it into glucose, which can be converted into ethanol, a biofuel, by fermentation.

In the process of photosynthesis, radiant energy in the form of sunlight is converted by plants into chemical energy. Plants contain chlorophyll molecules, which are capable of absorbing light energy. This light energy is used for photosynthesis, a complex series of reactions that results in the conversion of carbon dioxide and water into glucose, $C_6H_{12}O_6$, and oxygen:

$$6CO_2(g) + 6H_2O(l) \rightarrow C_6H_{12}O_6(aq) + 6O_2(g)$$

Glucose stores chemical energy in its bonds. Photosynthesis is an example of biological fixation of carbon.

 ## Global impact of science

Clean, renewable energy has become a global priority, as there is a growing acceptance that the planet is undergoing rapid climate change leading to global warming, more frequent extreme weather events, and often irreversible changes in habitats. The ability of scientists to effectively harness energy and minimize the use of non-renewable energy sources in industry is of paramount importance today.

Scientists are developing new types of catalysts that absorb light energy and transfer this energy to chemical reactions.

▲ **Figure 10** A light-powered catalyst at the Massachusetts Institute of Technology, USA

What would be the benefits of utilizing light energy to promote chemical reactions, in the same way as plants using sunlight in the process of photosynthesis?

The fermentation of glucose produces ethanol, C_2H_5OH, a biofuel:

$$C_6H_{12}O_6(aq) \rightarrow 2C_2H_5OH(aq) + 2CO_2(g)$$

Carbon dioxide, a greenhouse gas, is also produced in the fermentation process. However, this is offset by the absorption of a greater amount of carbon dioxide in the process of photosynthesis.

The industrial production of biofuels in many countries has economic and environmental implications. Brazil has undertaken large-scale production of ethanol from sugarcane for decades. An increased demand for renewable biofuels has both advantages and disadvantages (table 1).

Advantages	Disadvantages
Renewable resource	Use of agricultural land, water resources, fertilizers and pesticides for growing crops
Reduced greenhouse emissions	Diversion of food production to the production of biofuels
Sustainable resource	Monocultures can result in a reduction of biodiversity
Wide range of plant materials and waste can be used for biofuel production	Possible deforestation as the demand for biofuels increases
Economic security with a reduced dependence on imported oil supplies	High cost of production

▲ **Table 1** Advantages and disadvantages of the use of biofuels

Practice questions

5. a. Write the balanced chemical equation for the fermentation of glucose.

   b. Explain why ethanol-based fuels are said to have a lower carbon footprint than petroleum-based fuels, even though they both release similar amounts of carbon dioxide upon combustion.

   c. Present the arguments for both the advantages and disadvantages of the production and use of biofuels, if you were asked to address a United Nations Climate Change conference.

Source analysis

In this task, you will research information about biofuel from a variety of sources, and evaluate the quality of the information.

Relevant skills

- Inquiry 1: Consult a variety of sources

Method

1. Consult a variety of sources related to biofuels. Try to cover at least three different types of sources, for example:
   - Government or international organization website
   - Academic journal article
   - Secondary data from a database or publicly available data set
   - Newspaper or newspaper article
   - Textbook
   - Video
   - Online image search

2. Each of your sources will provide information about biofuels. Summarize this information.

3. Construct a table assessing the following aspects of each source you have chosen: citation, source type, accuracy, bias, credibility, relevance, advantages, and disadvantages.

Data-based questions

Data summarizing the global biofuel production are given in figure 11.

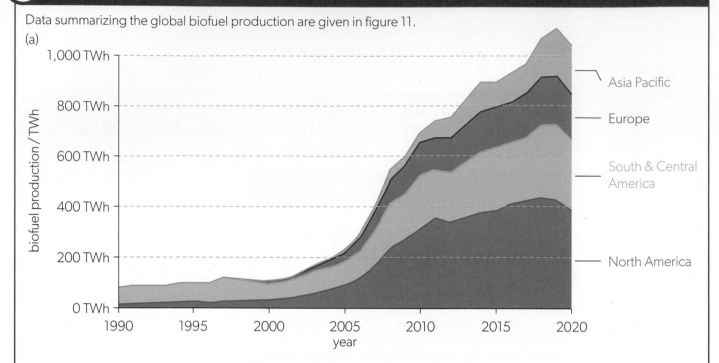

(a)

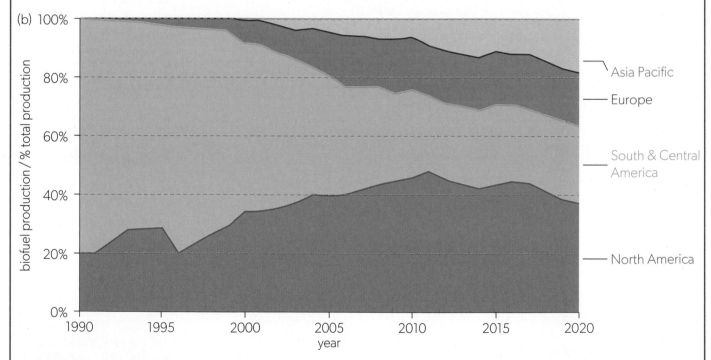

(b)

▲ Figure 11 Biofuel production by world region (a) in terawatt-hours (b) as a percentage of total production.
(Source of data: Statistical Review of World Energy, BP, 2021)

1. Describe three conclusions that can be drawn from the graphs in figure 11.

2. Formulate two questions, stemming from the information in the graphs in figure 11.

3. The two graphs were produced from the same data, but they represent these data in slightly different ways. Compare and contrast the purpose of the two graphs.

Fuel cells (*Reactivity 1.3.5*)

In *Reactivity 3.2*, you will learn that a **primary (voltaic) cell** is an electrochemical cell that converts chemical energy from spontaneous redox reactions into electrical energy. These reactions are mostly irreversible, and the cells of this type are mainly used for low-current applications, such as hand torches, long-life fire alarms, calculators and wall clocks. However, primary cells often contain materials that are toxic to humans and the environment if they are not disposed of properly. Primary cells typically contain zinc metal, manganese dioxide and potassium hydroxide.

In a **secondary cell**, the chemical reactions are reversible, so the battery can be recharged. Therefore, secondary cells are more environmentally friendly than primary cells. The lead–acid battery is a type of rechargeable battery, found in the majority of gasoline-powered vehicles. Electric vehicles are powered by lithium-ion batteries, another type of secondary cell.

While secondary cells have advantages over traditional primary cells, there are still problems associated with these types of batteries, such as overheating, limited lifespan and environmental concerns in terms of their disposal.

A **fuel cell** is used to convert chemical energy of a fuel directly into electrical energy. The first fuel cell was invented in 1838 by Welsh scientist, William Groves. The difference between a fuel cell and a voltaic cell is that in a fuel cell, the fuel is continuously supplied from an external source, whereas a voltaic cell contains finite amounts of reactants locked within the cell. Therefore, a fuel cell can produce electricity indefinitely, while a voltaic cell stops working when all reactants within the cell are consumed.

Hydrogen fuel cell

The **hydrogen fuel cell** is an electrochemical cell that uses hydrogen and oxygen gases as fuel. The redox reaction between hydrogen and oxygen produces water, electricity and heat.

The following steps occur in a hydrogen fuel cell:

1. The **proton exchange membrane (PEM)** selectively allows hydrogen ions to diffuse between the **cathode** and **anode** but prevents the passage of other ions, molecules and electrons between these **electrodes**.

2. Hydrogen gas is oxidized at the anode on the surface of a platinum-based catalyst. The half-equation for the reaction at the anode is as follows:

    $$H_2(g) \rightarrow 2H^+(aq) + 2e^-$$

3. The electrons cannot move through the PEM, and therefore have to leave the cell through an external circuit, producing the electrical output of the cell.

4. The protons formed at the anode move across the PEM to the cathode, where they combine with the oxygen gas and electrons. Oxygen gas is reduced to form water as a waste product. The half-equation for the reaction at the cathode is as follows:

    $$O_2(g) + 4e^- + 4H^+(aq) \rightarrow 2H_2O(l)$$

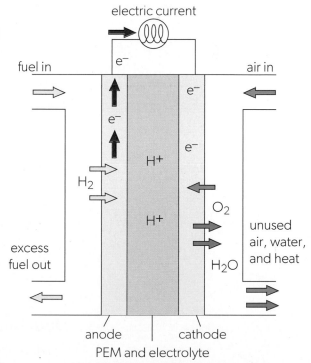

▲ **Figure 12** The main components of a commercial hydrogen cell include an electrolyte, a proton exchange membrane (PEM) and the electrodes

The overall redox reaction is:

$$2H_2(g) + O_2(g) \rightarrow 2H_2O(l)$$

Hydrogen fuel cells are considered to be a clean energy source because their only product is water. However, hydrogen gas, $H_2(g)$, is not abundant on Earth and needs to be produced by the following methods, one of which has environmental implications.

1. Electrolysis of water

The electrolysis of water is the process of using electricity to split water into oxygen and hydrogen gas. This is the reverse process of the reaction that occurs in a hydrogen fuel cell.

$$2H_2O(l) \rightarrow 2H_2(g) + O_2(g)$$

The electrolysis of water can be powered by renewable energy sources, such as solar energy. This environmentally friendly source of hydrogen gas only makes up a small proportion of the global supply.

2. Steam reforming of hydrocarbons

Hydrocarbons, such as natural gas, gasified coal and biomass, diesel and other liquid fuels, undergo steam reforming resulting in the production of toxic carbon monoxide gas and hydrogen gas. The general equation for the steam reforming of alkanes is as follows:

$$C_nH_{2n+2}(g) + nH_2O(g) \rightleftharpoons nCO(g) + (2n + 1)H_2(g)$$

Carbon monoxide then reacts with steam to form the greenhouse gas carbon dioxide:

$$CO(g) + H_2O(g) \rightleftharpoons CO_2(g) + H_2(g)$$

The annual production of hydrogen from the various sources is responsible for carbon dioxide emissions greater than the total annual carbon dioxide emissions of the United Kingdom. The scientific community continues to search for new green sources of hydrogen gas.

 Experiments

Alternatives to fossil fuels, ranging from hydrogen fuel cells to biofuels, have arisen from extensive experimentation by numerous scientists over time. Creativity and imagination play a role in experimental design, interpretation and conclusions. Once considered an electrochemical curiosity, fuel cells are now seen as a promising source of renewable energy.

Direct-methanol fuel cell

In a **direct-methanol fuel cell (DMFC)**, the source of the hydrogen ions, H^+, is methanol rather than hydrogen gas. Methanol is a liquid fuel that can be produced from renewable resources through fermentation. Methanol is cleaner than hydrogen because the method of its production has less impact on the environment in terms of formation of greenhouse gases. Methanol also has a far greater energy density (energy per unit volume) than hydrogen gas.

DMFCs use the electrochemical reactions below:

anode half-equation: $CH_3OH(aq) + H_2O(l) \rightarrow CO_2(g) + 6H^+(aq) + 6e^-$

cathode half-equation: $\frac{3}{2}O_2(g) + 6H^+(aq) + 6e^- \rightarrow 3H_2O(l)$

overall reaction: $CH_3OH(aq) + \frac{3}{2}O_2(g) \rightarrow CO_2(g) + 2H_2O(l)$

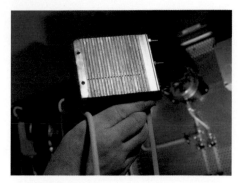

▲ Figure 13 A portable direct-methanol fuel cell (DMFC), which can be used to power laptops and video cameras

The electrochemical reaction produces carbon dioxide, a greenhouse gas. The reaction at the anode requires a catalyst that contains expensive precious metals, usually ruthenium and palladium. Another disadvantage is the toxicity of methanol.

Methanol can also be used to supply hydrogen gas for hydrogen fuel cells. Steam reforming of methanol at 250 °C produces $CO_2(g)$ and $H_2(g)$, along with a small amount of $CO(g)$.

A distinct advantage of DMFCs is their high energy density (figure 14). The slope of the graph gives the energy per unit volume, showing that the methanol fuel cell has a much higher energy density than the lithium-ion battery.

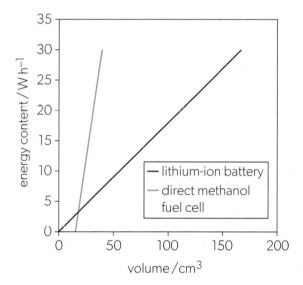

▲ Figure 14 Comparison of the energy density for the lithium-ion battery and the direct-methanol fuel cell

Practice questions

6. a. Outline the differences between a primary and secondary cell.

    b. What are the advantages a hydrogen fuel cell has over the traditional lead-acid battery.

    c. Identify and explain the function of the feature of a hydrogen fuel cell that involves the passage of ions.

    d. A direct methanol fuel cell (DMFC) converts chemical energy to electrical energy.

       i. Deduce the anode and cathode half equations and the overall chemical equation for this electrochemical cell.

       ii. Outline one advantage and one disadvantage of the methanol cell (DMFC) compared to a hydrogen fuel cell.

When comparing fuels, the energy density (energy released per unit volume) and specific energy (energy released per unit mass) can give quite different pictures (table 2).

Fuel source	Energy density / MJ dm^{-3}	Specific energy / MJ kg^{-1}
compressed hydrogen	1.9	120
methanol	16	20
liquefied natural gas	21	50
liquid propane	27	46
gasoline	32	46

▲ Table 2 Comparing fuels in terms of energy density and specific energy

The specific energy of hydrogen is more than double that of any other fuel in table 2. In hydrogen, H_2, 2.02 g contains 1 mol of fuel, compared with 32.05 g for 1 mol of methanol, CH_3OH, or approximately 110 g for 1 mol of gasoline. Therefore, it could be easy to believe that hydrogen is the best fuel choice.

However, fuels need to be stored and transported. The molar volume of a gas at room temperature and 1 atm pressure is approximately 24 dm^3. One mole of gaseous hydrogen under these conditions would require a 24 dm^3 storage tank, which adds to the weight if the device is to be portable, such as in a car. One mole of methanol would occupy 40.4 cm^3, and the same 24 dm^3 storage tank could hold over 545 mol of methanol fuel. Even when compressed, the hydrogen gas occupies a much larger volume, and regulators and compressors further increase the weight of the storage system. Gasoline offers the highest energy density but has associated environmental problems.

(ATL) Communication skills

The general public needs to have a good understanding of certain scientific issues in order to make informed choices. This presents scientists with the challenge of accurately conveying complex information and specific terminology in a succinct and accessible way. For example, this chapter contains information about fuels that is communicated in different types of tables, diagrams, charts and graphs. What are the advantages and disadvantages of different data communication methods?

Design an infographic to communicate the science behind an issue of your choice to a particular audience (also of your choice). Possible issues related to this chapter include renewable energy, biofuels, fuel cells and incomplete combustion. Possible audiences include younger pupils at your school, your family or your local community. You may even wish to discuss this with your creativity, activity, service (CAS) coordinator. Depending on the nature of the issue and the audience, this task could be connected to a CAS experience.

Linking question

What are the main differences between a fuel cell and a primary (voltaic) cell? (Reactivity 3.2)

End of topic questions

Topic review

1. Using your knowledge from the *Reactivity 1.3* topic, answer the guiding question as fully as possible:

   What are the challenges of using chemical energy to address our energy needs?

Exam-style questions

Multiple-choice questions

2. Which of the following statements is false?

   A. The complete combustion of a hydrocarbon forms carbon dioxide and water.

   B. All combustion reactions are endothermic reactions.

   C. The incomplete combustion of hydrocarbons and alcohols form the same products.

   D. Complete combustion reactions require an excess of oxygen gas as a reactant.

3. Which equation(s) represent the incomplete combustion of methane?

   I. $CH_4(g) + 2O_2(g) \rightarrow CO_2(g) + 2H_2O(g)$

   II. $CH_4(g) + 1\frac{1}{2}O_2(g) \rightarrow CO(g) + 2H_2O(g)$

   III. $CH_4(g) + O_2(g) \rightarrow C(s) + 2H_2O(g)$

   A. I, II and III   C. III only

   B. II and III only   D. II only

Extended-response questions

4. a. Write the chemical equation for the incomplete combustion of methane. [1]

   b. Explain how one of the products of incomplete combustion can lead to increased health risks. [2]

   c. Calculate the enthalpy change for the incomplete combustion of methane, in $kJ\,mol^{-1}$, using the bond enthalpy data given below. [2]

Bond	Average bond enthalpy / $kJ\,mol^{-1}$
C—H	414
O—H	463
O=O	498
C≡O	1077

5. Ethanol is a biofuel that can be mixed with gasoline.

   a. Write the equation for the complete combustion of ethanol. [1]

   b. Outline the evidence that relates global warming to increasing concentrations of greenhouse gases in the atmosphere. [3]

   c. Explain, including a suitable equation, why biofuels are considered to be carbon neutral. [2]

6. A hydrogen fuel cell uses pure hydrogen gas as the fuel and a proton exchange membrane as the electrolyte. A diagram of the hydrogen fuel cell is shown below.

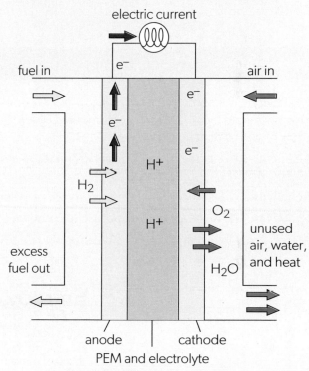

   a. Suggest an advantage of the hydrogen fuel cell over the lead–acid battery for use in cars. [1]

   b. Outline **one** advantage and **one** disadvantage of the methanol cell (DMFC) compared with a hydrogen fuel cell. [2]

7. Even though fuel cells, primary cells and rechargeable cells have similar fundamental characteristics, there are important differences between them.

   a. Suggest one common feature that fuel cells, primary cells and rechargeable cells share. [1]

   b. Outline the difference between primary and rechargeable cells. [1]

   c. Identify **one** factor that affects the voltage of a cell and **a different factor** that affects the current it can deliver. [2]

Reactivity 1.4 Entropy and spontaneity (AHL)

Systems that are approaching equilibrium, or that are in equilibrium, are common. Understanding the nature of chemical equilibrium enables chemists to control the direction and rate of chemical change, maximizing the yield of chemical reactions and minimizing the formation of waste and by-products. There are two key features of a chemical reaction that determine the direction of chemical change for a given set of conditions:

1. The enthalpy change (whether the reaction is exothermic or endothermic)
2. The entropy change (whether the reaction increases or decreases the disorder of the system).

Understanding these features allows chemists to predict how temperature and other factors, such as the states of reactants and products, affect the direction of a particular chemical change.

Understandings

Reactivity 1.4.1—Entropy, S, is a measure of the dispersal or distribution of matter and/or energy in a system. The more ways the energy can be distributed, the higher the entropy. Under the same conditions, the entropy of a gas is greater than that of a liquid, which in turn is greater than that of a solid.

Reactivity 1.4.2—Change in Gibbs energy, ΔG, relates the energy that can be obtained from a chemical reaction to the change in enthalpy, ΔH, change in entropy, ΔS, and absolute temperature, T.

Reactivity 1.4.3—At constant pressure, a change is spontaneous if the change in Gibbs energy, ΔG, is negative.

Reactivity 1.4.4—As a reaction approaches equilibrium, ΔG becomes less negative and finally reaches zero.

Entropy, S (Reactivity 1.4.1)

Chemists need to understand the conditions under which chemical reactions will proceed, so that they can modify and control chemical systems to achieve desired outcomes. A reaction is said to be **spontaneous** when it moves towards either completion or **equilibrium** under a given set of conditions, without external intervention. This intervention may be in the form of a change in temperature, pressure or concentration of a reactant. Reactions that are spontaneous can occur at different rates and may be either **endothermic** or **exothermic**. Reactions that do not take place under a given set of conditions are said to be **non-spontaneous**.

Chemical equilibrium is achieved when the forward and backward reactions in a reversible process occur at the same rate. This is detailed in *Reactivity 2.3*.

Exothermic reactions are usually spontaneous, but there are many exceptions to this rule. To better understand the spontaneity of chemical changes, we need to examine a number of different aspects of a chemical reaction.

Entropy, S, is a measure of the dispersal or distribution of the total available energy or matter in a system. If energy and matter are localized in one place within a chemical system, the entropy of the system is low. Conversely, if energy and matter are randomly distributed throughout a system, the entropy of the system is high. Entropy is often said to be a measure of the disorder of a system.

TOK

Entropy is difficult to conceptualize, so it is popularly defined in terms of disorder or randomness. However, this is generally considered an oversimplification that could lead to misconceptions. Have you come across the notion of entropy-as-disorder? Have you found it useful? To what extent do the benefits of an explanation that is not entirely correct outweigh its shortcomings?

Entropy can be defined thermodynamically in terms of dispersal of matter and energy (as it is done in DP chemistry). An alternative statistical definition of entropy looks at the number of possible ways, known as microstates, in which molecular energy can be distributed. The relationship between entropy and the number of microstates was proposed by Austrian physicist Ludwig Boltzmann, and the formula describing this relationship is written on his tombstone in Vienna, Austria (figure 1).

The two perspectives of entropy, thermodynamic and statistical, are complementary.

▲ **Figure 1** Ludwig Boltzmann's tombstone, with the equation $S = k \log W$. S is entropy, and W is the number of microstates. Do you know what k is?

Spontaneous reactions lead to an increase in the total entropy of the system and its surroundings. If you understand the freedom of movement of particles in a system, you can quantify the total entropy change during a reaction. This allows you to predict the direction of that reaction.

Entropy and physical changes

Imagine the condensation that appears on the outside of a glass containing iced water on a hot day (figure 2). There is a temperature difference between the system (iced water and the glass) and the surroundings (everything outside the system). This temperature difference results in thermal energy being transferred from the surrounding atmosphere to the glass and its contents, until an equilibrium is reached. With this thermal energy transfer, the entropy of the water/ice mixture will increase, while the entropy of the surroundings will decrease.

When a substance changes state from solid to liquid, or from liquid to gas, energy is absorbed by the particles of that substance. Therefore, the kinetic energy of the particles increases. This increased kinetic energy gives the particles more freedom of movement and more ways of distributing the energy. Therefore, the entropy of a gas is greater than that of a liquid, which in turn is greater than that of a solid, under the same conditions (figure 3). For example, in figure 2, the condensed water (liquid state) on the surface of the glass is lower in entropy than the water vapour (gaseous state) in the atmosphere.

▲ **Figure 2** Changes in entropy are associated with every chemical and physical process

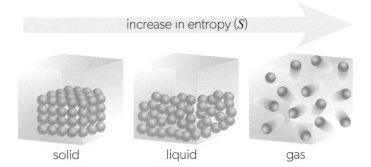

increase in entropy (S)

solid   liquid   gas

▲ **Figure 3** Entropy increases from the solid phase to the liquid phase, and from the liquid phase to the gaseous phase

The total entropy change that occurs during a reaction is the sum of the entropy changes of the reaction system and the surroundings:

$$\Delta S_{total} = \Delta S_{system} + \Delta S_{surroundings}$$

The **second law of thermodynamics** says that chemical reactions are spontaneous if they result in an overall increase in the total entropy value. When the total entropy remains constant, the system is at equilibrium. If the total entropy change is found to be negative, we say that the reaction is non-spontaneous. In this case, the spontaneous reaction would proceed in the opposite direction to the way in which it was written. For a reversible reaction, if the forward reaction has a negative entropy change, the reverse reaction has a positive entropy change.

Entropy change	Spontaneity of reaction
$\Delta S_{total} > 0$	spontaneous (forward reaction is spontaneous)
$\Delta S_{total} = 0$	equilibrium (neither forward nor backward reaction is favoured)
$\Delta S_{total} < 0$	non-spontaneous (reverse reaction is spontaneous)

▲ Table 1 The second law of thermodynamics allows us to predict the direction of a spontaneous reaction

Theories vs laws

Scientific laws, like theories, can be used to make predictions, but, unlike theories, laws do not have explanatory power. In this chapter, you are learning about the second law of thermodynamics. Think of another law that you have come across in chemistry. What predictions can you make from it?

You can make predictions about whether you expect a chemical reaction to have a positive or negative entropy change based on the degree of disorder in the products and the reactants. Begin by examining the states of matter of all reactants and products—remember that entropy increases from a solid to a liquid to a gas. The coefficients used to balance the equation must also be considered when predicting changes in entropy.

Worked example 1

Predict whether the following reactions will have a positive or negative entropy change, ΔS.

a. $NH_4Cl(s) \rightarrow NH_3(g) + HCl(g)$

b. $2C_2H_6(g) + 7O_2(g) \rightarrow 4CO_2(g) + 6H_2O(l)$

Solution

a. In this decomposition reaction, 1 mol of solid ammonium chloride changes into 1 mol of ammonia gas and 1 mol of hydrogen chloride gas.

- A change in state from a solid to a gas indicates an increase in the entropy of the system.
- The coefficients balancing the equation indicate an increase in disorder, with 1 mol of solid changing into 2 mol of gases.

This reaction will therefore have a positive entropy change, as there is an increase in the disorder of the system.

b. In this combustion reaction, 2 mol of ethane and 7 mol of oxygen are converted into 4 mol of carbon dioxide and 6 mol of water.

- A change in state from gas to liquid indicates a decrease in the entropy of the system.
- The total number of moles of products (10) is greater than that of the reactants (9). However, the number of moles of gaseous products (4) is lower than that of the gaseous reactants (9). Since gases have much higher entropy values than liquids or solids, we can ignore condensed phases and pay attention only to gases. The decrease in the number of moles of gases, from 10 to 4, means that the overall order of the system increases.

This reaction will therefore have a negative entropy change, as there is a decrease in the disorder of the system.

Practice questions

1. Predict and explain whether the following reactions will have a positive or negative entropy change, ΔS.

   a. $NH_4NO_3(s) \rightarrow N_2O(g) + 2H_2O(l)$

   b. $N_2(g) + 3H_2(g) \rightarrow 2NH_3(g)$

   c. $N_2O_4(g) \rightarrow 2NO_2(g)$

   d. $CaCO_3(s) \rightarrow CaO(s) + CO_2(g)$

Calculating entropy changes

Entropy is a state function, so a change in entropy across a reaction can be determined by the difference between the final entropy and the initial entropy of the reacting species. Therefore, the **standard entropy change**, $\Delta S^\ominus$, of a system can be calculated from **standard entropy values**, $S^\ominus$, of the reactants and the products.

$$\Delta S^\ominus = \sum S^\ominus(\text{products}) - \sum S^\ominus(\text{reactants})$$

The $\ominus$ symbol indicates standard conditions. Standard entropy values of some substances are given in section 13 of the data booklet. The units for standard entropy values are $J\,K^{-1}\,mol^{-1}$.

When performing entropy change calculations, the following points need to be considered:

- Remember that values for entropy are specific for different states of matter, for example, $S^{\ominus}(H_2O(g)) = 189\,J\,K^{-1}\,mol^{-1}$ while $S^{\ominus}(H_2O(l)) = 70\,J\,K^{-1}\,mol^{-1}$.

- The coefficients used to balance the equation must be applied to standard entropy values when calculating the entropy change for a reaction.

- Examine the chemical reaction and predict whether you expect the reaction to have a positive or negative entropy change, based on the degree of disorder in the products and reactants. This prediction can be used to check your final calculation.

Worked example 2

Calculate the standard entropy change for the following reactions using the standard entropy values given in the table.

a. $H_2(g) + Cl_2(g) \rightarrow 2HCl(g)$

b. $H_2(g) + \dfrac{1}{2}O_2(g) \rightarrow H_2O(l)$

c. $N_2O_4(g) \rightarrow 2NO_2(g)$

d. $NH_4Cl(s) \rightarrow NH_3(g) + HCl(g)$

Substance	$S^{\ominus}/J\,K^{-1}\,mol^{-1}$
$H_2(g)$	131
$Cl_2(g)$	223
$O_2(g)$	205
$N_2O_4(g)$	304
$NO_2(g)$	240
$NH_4Cl(s)$	95
$NH_3(g)$	193
$HCl(g)$	187
$H_2O(l)$	70

Solution

a. $\Delta S^{\ominus} = \sum S^{\ominus}(products) - \sum S^{\ominus}(reactants)$

$\quad = [2S^{\ominus}(HCl(g))] - [S^{\ominus}(H_2(g)) + S^{\ominus}(Cl_2(g))]$

$\quad = (2 \times 187) - (131 + 223)$

$\quad = +20\,J\,K^{-1}\,mol^{-1}$

The entropy change is small and positive, indicating a small increase in disorder. The value is small because there is no change in the number of moles of gas from reactants to products.

b. $\Delta S^{\ominus} = [S^{\ominus}(H_2O(l))] - [S^{\ominus}(H_2(g)) + \dfrac{1}{2}S^{\ominus}(O_2(g))]$

$\quad = (70) - (131 + \dfrac{205}{2})$

$\quad = -164\,J\,K^{-1}\,mol^{-1}$

The large negative entropy change associated with this reaction indicates a large decrease in disorder (greater order), with 1.5 mol of gases changing into 1 mol of a liquid.

c. $\Delta S^{\ominus} = [2S^{\ominus}(NO_2(g))] - [S^{\ominus}(N_2O_4(g))]$

$\quad = (2 \times 240) - (304)$

$\quad = +176\,J\,K^{-1}\,mol^{-1}$

This reaction has a large positive entropy change that reflects an increase in disorder, from 1 mol of gas on the reactant side to 2 mol of gases on the product side.

d. $\Delta S^{\ominus} = [S^{\ominus}(NH_3(g)) + S^{\ominus}(HCl(g))] - [S^{\ominus}(NH_4Cl(s))]$

$\quad = (193 + 187) - (95)$

$\quad = +285\,J\,K^{-1}\,mol^{-1}$

Transforming 1 mol of a solid into 2 mol of gases results in a large increase in disorder, hence the large positive entropy change.

Practice questions

2. Which reaction causes the greatest increase in entropy of the system?

    A. $HCl(g) + NH_3(g) \rightarrow NH_4Cl(s)$

    B. $(NH_4)_2Cr_2O_7(s) \rightarrow Cr_2O_3(s) + N_2(g) + 4H_2O(g)$

    C. $CaCO_3(s) \rightarrow CaO(s) + CO_2(g)$

    D. $I_2(g) \rightarrow I_2(s)$

3. Which is correct for the reaction $H_2O(g) \rightarrow H_2O(l)$?

    A. Enthalpy increases and entropy increases.

    B. Enthalpy decreases and entropy increases.

    C. Enthalpy increases and entropy decreases.

    D. Enthalpy decreases and entropy decreases.

4. Determine the standard entropy change, in $J\,K^{-1}\,mol^{-1}$, for the decomposition of dinitrogen monoxide:

$$N_2O(g) \rightarrow N_2(g) + \frac{1}{2}O_2(g)$$

Substance	$S^{\ominus}\,/\,J\,K^{-1}\,mol^{-1}$
$N_2O(g)$	220
$N_2(g)$	192
$O_2(g)$	205

 Linking question

Why is the entropy of a perfect crystal at 0 K predicted to be zero? (*Structure 1.1*)

Gibbs energy (*Reactivity 1.4.2* and *Reactivity 1.4.3*)

An increase in heat energy (enthalpy) within a reaction system will result in increased movement of particles, leading to greater disorder and an increase in the entropy of the system. Therefore, you need to consider the effects that both changes in enthalpy and entropy have on the spontaneity of a chemical reaction:

- Exothermic reactions are more likely to be spontaneous, as this leads to a decrease in enthalpy, and therefore greater stability of the reaction products.

- An increase in entropy makes reactions more likely to be spontaneous, as greater disorder leads to a more random distribution of energy within the system.

However, reactions that are spontaneous, and therefore **thermodynamically favourable**, can sometimes be **kinetically unfavourable** due to their high activation energies.

The impact that the enthalpy change of a reaction has on the entropy of the surroundings depends on the conditions existing in the system. Imagine if heat energy is transferred into two separate systems, one at low temperature and one at high temperature, such as a block of ice at 0 °C and a bowl of water at 60 °C. Will the transfer of the same amount of energy into each system have the same effect? The ice will begin to melt as the kinetic energy of the water molecules increases, resulting in a significant change in entropy. However, the hot water already has significant disorder compared to the ice, so the additional energy will have a much less marked effect on the system entropy.

The combination of enthalpy, entropy and temperature of the system can be used to define a new state function called **Gibbs energy**, G. The change in Gibbs energy, ΔG, is the maximum amount of energy that can be obtained from a system. The **change in Gibbs energy, ΔG**, is related to the change in enthalpy, ΔH, the change in entropy, ΔS, and the absolute temperature of the system, T, according to the following equation:

$$\Delta G^{\ominus} = \Delta H^{\ominus} - T\Delta S^{\ominus}$$

The units of change in Gibbs energy are $kJ\,mol^{-1}$. ΔG takes into account the direct entropy change of the system resulting from the transformation of the chemicals and the indirect entropy change of the surroundings resulting from the transfer of heat energy.

At constant pressure, a reaction is spontaneous if the change in Gibbs energy has a negative value ($\Delta G^{\ominus} < 0$). Therefore, if we know the enthalpy change, entropy change and temperature, we can determine whether a reaction is spontaneous under standard conditions. This is summarized in table 2.

$\Delta H^{\ominus}$	$\Delta S^{\ominus}$	$\Delta G^{\ominus}$	Spontaneity
positive (> 0): *endothermic*	positive (> 0): *more disorder*	negative (< 0) at high T positive (> 0) at low T	spontaneous only at high temperatures when $T\Delta S^{\ominus} > \Delta H^{\ominus}$
positive (> 0): *endothermic*	negative (< 0): *more order*	always positive (> 0)	non-spontaneous at any temperature
negative (< 0): *exothermic*	positive (> 0): *more disorder*	always negative (< 0)	spontaneous at any temperature
negative (< 0): *exothermic*	negative (< 0): *more order*	negative (< 0) at low T positive (> 0) at high T	spontaneous only at low temperatures when $T\Delta S^{\ominus} > \Delta H^{\ominus}$

▲ **Table 2** Factors affecting ΔG and the spontaneity of a reaction

It is not possible to predict whether every chemical reaction is spontaneous. Exothermic reactions with an increase in disorder will always be spontaneous, with $\Delta G^{\ominus} < 0$. Similarly, endothermic reactions with a decrease in disorder will always be non-spontaneous, with $\Delta G^{\ominus} > 0$. The spontaneity of other reactions depends on the temperature of the system: if $T\Delta S^{\ominus} > \Delta H^{\ominus}$, the value of $\Delta G^{\ominus}$ will be negative, while at $T\Delta S^{\ominus} < \Delta H^{\ominus}$ the value of $\Delta G^{\ominus}$ will be positive.

Calculating the change in Gibbs energy for a reaction

To determine the spontaneity of a reaction under standard conditions, we need to calculate the change in Gibbs energy for the reaction under the same conditions using $\Delta G^{\ominus} = \Delta H^{\ominus} - T\Delta S^{\ominus}$. We can calculate $\Delta H^{\ominus}$ and $\Delta S^{\ominus}$ for the reaction using the thermodynamic data given in the data booklet.

Worked example 3

An equation for the combustion of propane is given below.

$$C_3H_8(g) + 5O_2(g) \rightarrow 3CO_2(g) + 4H_2O(g)$$

a. Determine the standard enthalpy change, $\Delta H^{\ominus}$, in kJ mol^{-1}, for this reaction, using the bond enthalpy data in section 11 of the data booklet.

b   Calculate the standard enthalpy change, $\Delta H^{\ominus}$, in kJ mol^{-1}, for this reaction using the enthalpy of formation data in section 12 of the data booklet.

c. Predict, giving a reason, whether the entropy change, $\Delta S^{\ominus}$, for this reaction is negative or positive.

d. Calculate $\Delta S^{\ominus}$ for the reaction in J K^{-1} mol^{-1}, using the standard entropy values in section 12 of the data booklet. The standard entropy for oxygen gas is 205 J K^{-1} mol^{-1}.

e. Calculate the standard Gibbs energy change, $\Delta G^{\ominus}$, in kJ mol^{-1}, for the reaction at 5 °C, using your answers to (b) and (d).

Solution

a. In the reactants, there are eight C–H bonds, two C–C bonds and five O=O bonds. In the products, there are six C=O bonds and eight O–H bonds. Therefore, the enthalpy change for the reaction is as follows:

$$\Delta H^{\ominus} = \sum(BE \text{ of bonds broken}) - \sum(BE \text{ of bonds formed})$$

$$= [8BE(\text{C–H}) + 2BE(\text{C–C}) + 5BE(\text{O=O})]$$
$$- [6BE(\text{C=O}) + 8BE(\text{O–H})]$$

$$= [(8 \times 414) + (2 \times 346) + (5 \times 498)]$$
$$- [(6 \times 804) + (8 \times 463)]$$

$$= 6494 - 8528$$

$$= -2034 \text{ kJ mol}^{-1}$$

This reaction is exothermic.

b. Use the equation:

$$\Delta H^{\ominus} = \sum(\Delta H_f^{\ominus} \text{ products}) - \sum(\Delta H_f^{\ominus} \text{ reactants})$$

$$= [3\Delta H_f^{\ominus}(\text{CO}_2) + 4\Delta H_f^{\ominus}(\text{H}_2\text{O}(g))]$$
$$- [\Delta H_f^{\ominus}(\text{C}_3\text{H}_8)]$$

$$= [(3 \times (-394)) + (4 \times (-242))] - [-105]$$

$$= -2150 + 105$$

$$= -2045 \text{ kJ mol}^{-1}$$

Alternatively, use the summation of equations method or the enthalpy cycle diagrams method.

c. The change in entropy value for this reaction should be positive ($\Delta S^{\ominus} > 0$) because there is an increase in the number of moles of gaseous species from 6 mol to 7 mol.

d. $\Delta S^{\ominus} = \sum S^{\ominus}(\text{products}) - \sum S^{\ominus}(\text{reactants})$

$$= [3S^{\ominus}(\text{CO}_2(g)) + 4S^{\ominus}(\text{H}_2\text{O}(g))]$$
$$- [S^{\ominus}(\text{C}_3\text{H}_8(g)) + 5S^{\ominus}(\text{O}_2(g))]$$

$$= [(3 \times 214) + (4 \times 189)]$$
$$- [(270) + (5 \times 205)]$$

$$= 1398 - 1295$$

$$= +103 \text{ J K}^{-1} \text{mol}^{-1}$$

e. First, convert temperature from °C into K:

$$T = 5 + 273.15 = 278.15 \text{ K}$$

Then, convert the entropy value into kJ K^{-1} mol^{-1} from J K^{-1} mol^{-1} by dividing by 1000:

$$\Delta S^{\ominus} = \frac{+103 \text{ J K}^{-1}\text{mol}^{-1}}{1000} = +0.103 \text{ kJ K}^{-1}\text{mol}^{-1}$$

Finally, substitute the values obtained into the equation for change in Gibbs energy:

$$\Delta G^{\ominus} = \Delta H^{\ominus} - T\Delta S^{\ominus}$$

$$= -2045 \text{ kJ mol}^{-1} - (278.15 \text{ K} \times 0.103 \text{ kJ K}^{-1}\text{mol}^{-1})$$

$$= -2074 \text{ kJ mol}^{-1}$$

The negative value for the Gibbs energy change indicates that the reaction is spontaneous. You can predict this by looking at table 2 earlier in the topic: the negative value for enthalpy change and the positive value for entropy change suggest that the reaction will be spontaneous at any temperature.

You can practise calculating $\Delta H^{\ominus}$ from bond enthalpy data and enthalpy of formation data using the material in *Reactivity 1.2*.

Practice questions

5. Ethane-1,2-diol, $HOCH_2CH_2OH$, reacts with thionyl chloride, $SOCl_2$, according to the equation below.

$$HOCH_2CH_2OH(l) + 2SOCl_2(l) \rightarrow ClCH_2CH_2Cl(l) + 2SO_2(g) + 2HCl(g)$$

a. Calculate the standard enthalpy change for this reaction using the following enthalpy of formation data.

Substance	$HOCH_2CH_2OH(l)$	$SOCl_2(l)$	$ClCH_2CH_2Cl(l)$	$SO_2(g)$	$HCl(g)$
$\Delta H_f^{\ominus}$ / $kJ\,mol^{-1}$	−453	−246	−165	−297	−92

b. Calculate the standard entropy change for this reaction using the following data.

Substance	$HOCH_2CH_2OH(l)$	$SOCl_2(l)$	$ClCH_2CH_2Cl(l)$	$SO_2(g)$	$HCl(g)$
$S^{\ominus}$ / $J\,K^{-1}\,mol^{-1}$	167	279	209	248	187

c. The standard Gibbs energy change, $\Delta G^{\ominus}$, for the above reaction is −103 kJ mol⁻¹ at 298 K.

Suggest why $\Delta G^{\ominus}$ has a large negative value considering the sign of $\Delta H^{\ominus}$ in part (a).

In a reaction where the change in enthalpy $\Delta H^{\ominus}$ is positive (endothermic) and the change in entropy $\Delta S^{\ominus}$ is also positive (more disorder), the value of Gibbs energy change can be either positive or negative. The spontaneity of this type of reaction is therefore dependent on the temperature of the system:

• At high temperature, $T\Delta S^{\ominus} > \Delta H^{\ominus}$ and $\Delta G^{\ominus}$ is negative (spontaneous reaction)

• At low temperature, $T\Delta S^{\ominus} < \Delta H^{\ominus}$ and $\Delta G^{\ominus}$ is positive (non-spontaneous reaction)

This means that the temperature at which $\Delta G^{\ominus} = 0$ is the temperature above which the reaction becomes spontaneous.

$$0 = \Delta H^{\ominus} - T\Delta S^{\ominus}$$

Rearranging the equation in terms of T gives:

$$T = \frac{\Delta H^{\ominus}}{\Delta S^{\ominus}}$$

This expression can be used to determine the temperature above which an endothermic reaction with positive entropy change becomes spontaneous.

Worked example 4

Hydrogen gas can be produced industrially by the reaction of natural gas with steam. The standard enthalpy change and entropy change of the reaction are given.

$$CH_4(g) + H_2O(g) \rightarrow 3H_2(g) + CO(g) \qquad \Delta H^\ominus = 205\,kJ\,mol^{-1}$$
$$\Delta S^\ominus = 216\,J\,K^{-1}\,mol^{-1}$$

a. Calculate the standard Gibbs energy change, $\Delta G^\ominus$, in $kJ\,mol^{-1}$, for the reaction at 298 K.

b. Determine the temperature, in K, above which the reaction becomes spontaneous.

Solution

a. $\Delta G^\ominus = \Delta H^\ominus - T\Delta S^\ominus$

$\quad = 205\,kJ\,mol^{-1} - (298\,K \times 0.216\,kJ\,K^{-1}\,mol^{-1})$

$\quad = +141\,kJ\,mol^{-1}$

The positive value for the change in Gibbs energy indicates that the reaction is non-spontaneous at 298 K. This can be predicted given the positive value of enthalpy change (endothermic) and the low temperature.

b. Make the assumption that the value for Gibbs energy is 0, and solve for T.

$0 = \Delta H^\ominus - T\Delta S^\ominus$

$0 = 205 - (T \times 0.216)$

$T = \dfrac{205}{0.216}$

$\quad = 949\,K$

The reaction becomes spontaneous at a temperature greater than 949 K.

Practice questions

6. An equation for the hydrogenation of ethene is given below.

$$C_2H_4(g) + H_2(g) \rightarrow C_2H_6(g)$$

a. Determine the standard enthalpy change, $\Delta H^\ominus$, in $kJ\,mol^{-1}$, for this reaction, using section 12 of the data booklet.

b. Calculate the standard enthalpy change, $\Delta H^\ominus$ value, in $kJ\,mol^{-1}$, for this reaction using section 13 of the data booklet.

c. Predict, giving a reason, whether the standard entropy change, $\Delta S^\ominus$, for this reaction is negative or positive.

d. Calculate the $\Delta S^\ominus$ value, in $J\,K^{-1}\,mol^{-1}$, for the reaction, using section 13 of the data booklet. The standard entropy for hydrogen gas is $131\,J\,K^{-1}\,mol^{-1}$.

e. Calculate the standard Gibbs energy change, $\Delta G^\ominus$, in $kJ\,mol^{-1}$, for the reaction at 150 °C, using your answers to (b) and (d). Use section 1 of the data booklet.

(ATL) Thinking skills

The chemistry guide states: "ΔG takes into account the direct entropy change resulting from the transformation of the chemicals and the indirect entropy change of the surroundings resulting from the transfer of heat energy." Explain, as fully as you can, how the reaction described below illustrates this statement.

The reaction between solid ammonium chloride and hydrated barium hydroxide is endothermic:

$$2NH_4Cl(s) + Ba(OH)_2 \cdot 8H_2O(s) \rightarrow 2NH_3(g) + BaCl_2(aq) + 10H_2O(l) \quad \Delta H^\ominus > 0$$

When the two white solid reactants are mixed together, the mixture quickly turns into a slush and becomes so cold that it can easily freeze a few drops of water placed between the reaction flask and the block of wood underneath (figure 4).

▲ Figure 4 The reaction between barium hydroxide and ammonium chloride is endothermic

 Linking question

How can electrochemical data also be used to predict the spontaneity of a reaction? (*Reactivity 3.2*)

The effect of temperature on spontaneity

In this skills task, you will investigate the effect of temperature on the spontaneity of a process where a polymer dissolves in water.

Relevant skills

- Inquiry 1: Formulate hypotheses
- Inquiry 2: Interpret qualitative and quantitative data
- Inquiry 3: Relate the outcomes of an investigation to the stated research question or hypothesis

Poly(N-isopropylacrylamide) (PNIPAM) is a polymer that dissolves in water, producing a clear solution. On heating, PNIPAM precipitates out, causing the solution to become cloudy (figure 5).

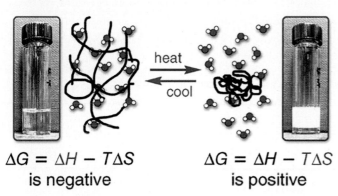

$$\Delta G = \Delta H - T\Delta S$$
is negative

$$\Delta G = \Delta H - T\Delta S$$
is positive

▲ Figure 5 PNIPAM (represented by the black lines) forms hydrogen bonds with water and dissolves. This is shown on the left. At a certain temperature, PNIPAM no longer forms hydrogen bonds and makes an insoluble precipitate. This is shown on the right

The reason for this temperature-dependent change is explained using $\Delta G = \Delta H - T\Delta S$. The polymer dissolves due to the formation of hydrogen bonds between water and the amido groups located along the polymer chain. This dissolution process is exothermic and it has a negative entropy change because of the regular arrangement of water molecules it requires. The process is therefore favourable in terms of enthalpy and unfavourable in terms of entropy. At high temperatures, the entropy term, $T\Delta S$, exceeds the enthalpy term, ΔH, making ΔG positive, so the dissolution becomes non-spontaneous.

Bergbreiter, Mijalis and Fi (2012) explored the effect of varying concentrations of LiCl and LiBr on the temperature at which dissolution of PNIPAM becomes non-spontaneous. They slowly heated a PNIPAM solution and monitored the

temperature, T, at which it went cloudy. The results are shown in table 3.

Solution composition	T / °C
PNIPAM in 0.3 mol dm⁻³ LiCl	29.5
PNIPAM in 0.5 mol dm⁻³ LiCl	27.1
PNIPAM in 0.3 mol dm⁻³ LiBr	31.5
PNIPAM in 0.5 mol dm⁻³ LiBr	30.7
PNIPAM in water	32.5

▲ Table 3 The effect of varying LiCl and LiBr concentration on the temperature at which dissolution becomes non-spontaneous for PNIPAM solutions (Source of data: D. E. Bergbreiter, A. J. Mijalis, and H. Fu, *J. Chem. Educ.*, 89 (5), pp. 675–677, 2012)

Questions

1. State the signs of ΔH and ΔS for the dissolution of PNIPAM.

2. Explain in your own words why the dissolution of PNIPAM is spontaneous at low temperatures, but non-spontaneous at high temperatures.

3. Identify some of the control variables needed for the experiment used to gather the data in table 3.

4. The results in table 3 can be used to derive conclusions about how two different factors affect the temperature at which dissolution becomes non-spontaneous, T:
   - the presence or absence of an aqueous ionic compound
   - the concentration of an aqueous ionic compound.

   Choose one of the two factors to focus on.

   a. Identify which of the results in table 3 allow you to explore this relationship.

   b. With reference to the results you have identified, describe how the factor you chose affects T.

   c. Using your knowledge of ion–dipole interactions, entropy and Gibbs energy, construct a hypothesis of the effect of your chosen factor on T.

   d. Discuss the extent to which the data in table 3 is sufficient to construct your hypothesis.

   e. Give examples of further experiments that could be carried out to explore your hypothesis.

ΔG and equilibrium (*Reactivity 1.4.4*)

The ratio of concentration of products to reactants is called the **reaction quotient**, Q.

$$Q = \frac{\text{concentration of products}}{\text{concentration of reactants}}$$

Chemical equilibrium is achieved when the forward and backward reactions in a reversible reaction occur at the same rate. The equilibrium arrow, $\rightleftharpoons$, is used instead of the normal reaction arrow to show reversible reactions. The Haber process used to make ammonia is an example of a reversible reaction.

$$N_2(g) + 3H_2(g) \rightleftharpoons 2NH_3(g)$$

When a chemical system has reached equilibrium, the ratio of concentrations of products to concentrations of reactants is called the **equilibrium constant**, K.

You can compare the reaction quotient, Q, to the equilibrium constant, K, to determine the progress of the reaction as it moves towards equilibrium and the direction (either forward or backwards) of the reaction that is favoured to establish equilibrium. This is described in table 4.

We have already established in this topic how the change in Gibbs energy, ΔG, can describe the spontaneity and temperature dependence of a reaction. When $\Delta G < 0$, reactions taking place at constant pressure are spontaneous. From the time a reaction begins, the Gibbs energy changes as the ratio of reactants to products changes. For reversible reactions, the minimum Gibbs energy is reached at the point of equilibrium. At this minimum, $\Delta G = 0$. After this point, the Gibbs energy increases (figure 6).

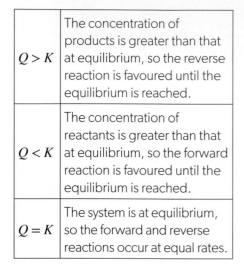

$Q > K$	The concentration of products is greater than that at equilibrium, so the reverse reaction is favoured until the equilibrium is reached.
$Q < K$	The concentration of reactants is greater than that at equilibrium, so the forward reaction is favoured until the equilibrium is reached.
$Q = K$	The system is at equilibrium, so the forward and reverse reactions occur at equal rates.

▲ **Table 4** The relationship between the reaction quotient, Q, and the equilibrium constant, K

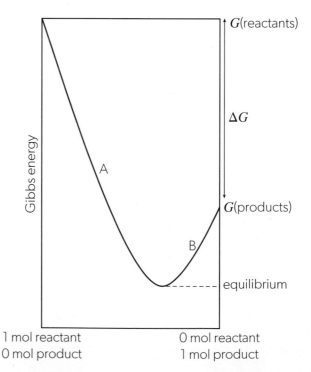

▲ **Figure 6** How the Gibbs energy changes as a reversible reaction proceeds

Examine the shape of the curve and the ratio of reactants to products as the Gibbs energy changes. You will notice that in region A, the forward reaction is favoured, as the amounts of reactants decrease and the amounts of products increase. At the point of equilibrium ($Q = K$), the Gibbs energy reaches a minimum and then increases again. From this point onwards, the forward reaction becomes non-spontaneous (region B), as it is now associated with a positive Gibbs energy change. The reverse reaction in region B is therefore spontaneous until the Gibbs energy reaches a minimum in the same way as during the forward reaction.

Relationship between ΔG, Q, K and T

At any point during a reaction, the ratio of reactants to products is different. The reaction quotient, Q, gives you a snapshot of the ratio of reactants to products at a given point in the reaction pathway. If you know the standard Gibbs energy change for a reaction, $\Delta G^\ominus$, and the temperature, T, you can determine the change in Gibbs energy for the position indicated by the reaction quotient, Q, using the following equation:

$$\Delta G = \Delta G^\ominus + RT \ln Q$$

where R, the gas constant $= 8.31\,\mathrm{J\,K^{-1}\,mol^{-1}}$.

In figure 6, you saw that in a reversible reaction at the point of equilibrium, $\Delta G = 0$. At the point of equilibrium, $Q = K$. We can substitute this information into the equation above to find the relationship between the standard Gibbs energy change, $\Delta G^\ominus$, temperature, T, and the equilibrium constant, K:

$$\Delta G^\ominus = -RT \ln K$$

You will find these expressions in section 1 of the data booklet.

If the equilibrium constant is the unknown value, you can rearrange the last expression in terms of K.

$$K = e^{\frac{-\Delta G^\ominus}{RT}}$$

When the equilibrium constant K is determined for a reversible reaction, its value indicates whether the products or reactants are favoured at equilibrium. The Gibbs energy change for a given reaction is an indication of whether the forward or reverse reaction is favoured. The relationship between the standard Gibbs energy change, $\Delta G^\ominus$, and the equilibrium constant, K, is summarized in table 5.

Equilibrium constant	Description	Standard Gibbs energy change
$K = 1$	neither reactants nor products favoured	$\Delta G^\ominus = 0$
$K > 1$	products favoured	$\Delta G^\ominus < 0$
$K < 1$	reactants favoured	$\Delta G^\ominus > 0$

▲ Table 5 The values of K and $\Delta G^\ominus$ indicate whether products or reactants are favoured at equilibrium

Worked example 5

The synthesis of ammonia is an important industrial process that provides feedstock for the production of fertilizers, plastics, explosives and pharmaceuticals.

$$N_2(g) + 3H_2(g) \rightleftharpoons 2NH_3(g) \quad \Delta H^\ominus = -92 \, kJ \, mol^{-1}$$
$$\Delta S^\ominus = -202 \, J \, K^{-1} \, mol^{-1}$$
$$T = 298 \, K$$

a. Find the standard Gibbs energy change, $\Delta G^\ominus$, for the forward reaction and determine whether the reaction is spontaneous.

b. Determine the equilibrium constant, K, for the reaction and state whether you expect the forward reaction to be favourable.

c. At a given point during the reaction, the reaction quotient, Q, was 1×10^6. Determine the change in Gibbs energy, ΔG, of the forward reaction at this point, and comment on the spontaneity of the reaction.

Solution

a. $\Delta G^\ominus = \Delta H^\ominus - T\Delta S^\ominus$

$= -92 \, kJ \, mol^{-1} - 298 \, K \times (-0.202 \, kJ \, K^{-1} \, mol^{-1})$

$= -31.8 \, kJ \, mol^{-1}$

b. Substitute the $\Delta G^\ominus$ value from (a), and the values of R and T, into $\Delta G^\ominus = -RT \ln K$. Remember that the units of gas constant, R, include J, not kJ, so the $\Delta G^\ominus$ value must be converted from kJ to J.

$$\Delta G^\ominus = -RT \ln K$$

$$-31800 \, J \, mol^{-1} = -(8.31 \, J \, K^{-1} \, mol^{-1} \times 298 \, K) \ln K$$

Rearrange in terms of K:

$$\ln K = \frac{-31800}{-(8.31 \times 298)}$$

$$K = e^{\left[\frac{-31800}{-(8.31 \times 298)}\right]}$$

$$= 3.77 \times 10^5$$

This large positive value of K tells us that the formation of ammonia at room temperature is very favourable.

c. $\Delta G = \Delta G^\ominus + RT \ln Q$

$= -31800 \, J \, mol^{-1} + (8.31 \, J \, K^{-1} \, mol^{-1}) \times 298 \, K$
$\times \ln(1 \times 10^6)$

$= 2410 \, J \, mol^{-1} = 2.41 \, kJ \, mol^{-1}$

ΔG is positive, so the reaction is not spontaneous at this point in the reaction pathway.

Practice questions

7. Consider the reaction between nitrous oxide (N_2O) and oxygen in the formation of nitrogen dioxide (NO_2).

$$2N_2O(g) + 3O_2(g) \rightarrow 4NO_2(g)$$

a. Determine the value of $\Delta H^\ominus$, in $kJ \, mol^{-1}$, for the reaction using the values in the table.

Substance	$N_2O(g)$	$NO_2(g)$
$\Delta H_f^\ominus$ / $kJ \, mol^{-1}$	82	33.2

b. Calculate the standard entropy change, $\Delta S^\ominus$, in $J \, K^{-1} \, mol^{-1}$, for the reaction.

Substance	$N_2O(g)$	$NO_2(g)$	$O_2(g)$
$S^\ominus$ / $J \, K^{-1} \, mol^{-1}$	220	240	205

c. Calculate the standard Gibbs energy change, $\Delta G^\ominus$, in $kJ \, mol^{-1}$, for the reaction at 298 K using your answer to (a) and (b).

d. Calculate the equilibrium constant at 298 K for the reaction.

Graphing the relationship between $\Delta G^\ominus$ and T

In this task, you will explore the relationship between $\Delta G^\ominus$ and T for the following process:

$$H_2O(l) \rightarrow H_2O(g)$$

Relevant skills

- Tool 2: Use spreadsheets to manipulate data
- Tool 3: Use basic arithmetic and algebraic calculations to solve problems
- Tool 3: Sketch graphs with labelled but unscaled axes, to qualitatively describe trends
- Tool 3: Interpret features of graphs including gradient and intercepts
- Inquiry 2: Interpret diagrams, graphs and charts

Method

Part 1: Calculation

1. Predict and explain, with reference to the equation above, the sign of $\Delta S^\ominus$.

2. Predict the sign of $\Delta H^\ominus$. Explain your reasoning.

3. Deduce how the spontaneity of the reaction changes with temperature.

4. With reference to section 13 in the data booklet, calculate $\Delta H^\ominus$, $\Delta S^\ominus$ and $\Delta G^\ominus$ for the reaction at 298 K. Show your working.

5. Calculate the temperature at which the spontaneity of the reaction changes.

6. The temperature value you have just calculated corresponds to the boiling point of water. Calculate the percentage error of your value.

Part 2: Graphing the relationship between $\Delta G^\ominus$ and T

7. Using a spreadsheet, compute the value of $\Delta G^\ominus$ at different temperatures. A suggested layout and formulas are shown in figure 7.

Values:

E5	fx =$B$4−(D5*$B$5/1000)				
	A	B	C	D	E
1	Reaction:	H₂O(l) -> H₂O(g)			
2					
3	Thermodynamic parameters:			Data for graphical analysis:	
4	ΔH	44		T/K	ΔG/kJ mol⁻¹
5	ΔS	119		298	8.538
6	ΔG	8		0	44
7				100	32.1
8				200	20.2
9				300	8.3
10				400	−3.6
11				500	−15.5

Formulas:

E5	fx =$B$4−(D5*$B$5/1000)				
	A	B	C	D	E
1	Reaction:	H₂O(l) -> H₂O(g)			
2					
3	Thermodynamic parameters:			Data for graphical analysis:	
4	ΔH	44		T/K	ΔG/kJ mol⁻¹
5	ΔS	119		298	=$B$4−(D5*$B$5/1000)
6	ΔG	8		0	=$B$4−(D6*$B$5/1000)
7				=D6+100	=$B$4−(D7*$B$5/1000)
8				=D7+100	=$B$4−(D8*$B$5/1000)
9				=D8+100	=$B$4−(D9*$B$5/1000)
10				=D9+100	=$B$4−(D10*$B$5/1000)
11				=D10+100	=$B$4−(D11*$B$5/1000)

▲ Figure 7 A possible spreadsheet arrangement for this task. The values (top) were calculated using the formulas shown at the bottom. The sign $ in the formulas is an absolute cell reference. It tells the spreadsheet to retrieve the same value every time. Note that the input parameters in the yellow boxes are where you would enter the reaction being studied and the values calculated in step 4 above

8. Plot a graph of $\Delta G^{\ominus}$ vs. T. Include a line or curve of best fit, axis labels, units, minor gridlines and a title.

9. With reference to the equation $\Delta G^{\ominus} = \Delta H^{\ominus} - T\Delta S^{\ominus}$, identify the significance of the:
    - gradient
    - y-intercept
    - x-intercept

10. Use your graph to approximate the temperature at which the process becomes spontaneous. How does it compare to the value you calculated in step 5?

11. Explain why the graph:
    - is linear
    - shows a negative relationship (has a negative gradient).

Part 3: Further exploration

12. Below are examples of reactions for each of the three other possible combinations of $\Delta H^{\ominus}$ and $\Delta S^{\ominus}$:
    - positive $\Delta H^{\ominus}$ and negative $\Delta S^{\ominus}$ (for example, formation of ethene, $C_2H_4(g)$, from graphite and diatomic hydrogen)
    - negative $\Delta H^{\ominus}$ and positive $\Delta S^{\ominus}$ (for example, transformation of diamond into graphite)
    - negative $\Delta H^{\ominus}$ and negative $\Delta S^{\ominus}$ (for example, formation of ammonia from diatomic nitrogen and diatomic hydrogen).

Using these examples or different reactions of your choice, search online databases for thermodynamic data that will allow you further explore the relationship between $\Delta G^{\ominus}$ and T for the three other possible combinations of $\Delta H^{\ominus}$ and $\Delta S^{\ominus}$. You will need to repeat the steps in Parts 1 and 2 for each reaction.

13. Sketch the four graphs on one pair of axes.
14. Explain the differences in gradient and y-intercept between the four graphs you sketched above.

(ATL) Self-management skills

This chapter contains concepts that involve different types of calculations. Make a list of the common mistakes, such as forgetting to convert J into kJ, or °C into K and not including negative signs. For each, consider:
- Why might someone make this mistake?
- How could they avoid it?

Write three multiple-choice questions that specifically check for some of the common mistakes you have identified. Then, share your questions with a partner and quiz each other.

 Linking question

What is the likely composition of an equilibrium mixture when $\Delta G^{\ominus}$ is positive? (*Reactivity 2.3*)

End of topic questions

Topic review

1. Using your knowledge from the *Reactivity 1.4* topic, answer the guiding question as fully as possible:

   What determines the direction of chemical change?

Exam-style questions

Multiple-choice questions

2. Which system has the most negative entropy change, ΔS, for the forward reaction?

   A. $N_2(g) + 3H_2(g) \rightleftharpoons 2NH_3(g)$

   B. $CaCO_3(s) \rightarrow CaO(s) + CO_2(g)$

   C. $2S_2O_3^{2-}(aq) + I_2(aq) \rightarrow S_4O_6^{2-}(aq) + 2I^-(aq)$

   D. $H_2O(l) \rightarrow H_2O(g)$

3. Which combinations of values will result in a spontaneous reaction?

	$\Delta H^\ominus$ / kJ mol^{-1}	$\Delta S^\ominus$ / J K^{-1} mol^{-1}	T / K
I	−100	−100	300
II	+100	−100	300
III	+100	+100	1500

   A. I and II only

   B. I and III only

   C. II and III only

   D. I, II and III

4. Which statement is correct?

   A. If $\Delta H < 0$, the reaction is always spontaneous.

   B. If $\Delta H > 0$, the reaction is never spontaneous.

   C. If $\Delta S < 0$, the reaction can be spontaneous if the temperature is low enough.

   D. If $\Delta S < 0$, the reaction can be spontaneous if the temperature is high enough.

Extended-response questions

5. The hydrogenation of propene produces propane.

   $$C_3H_6(g) + H_2(g) \rightarrow C_3H_8(g)$$

   a. Calculate the standard entropy change, $\Delta S^\ominus$, for the hydrogenation of propene. [2]

Substance	$\Delta S^\ominus$ / J K^{-1} mol^{-1}
$H_2(g)$	+131
$C_3H_6(g)$	+267
$C_3H_8(g)$	+270

   b. The standard enthalpy change, $\Delta H^\ominus$, for the hydrogenation of propene is -124.4 kJ mol^{-1}. Predict the temperature above which the hydrogenation reaction is non-spontaneous. [2]

6. A molecule of citric acid, $C_6H_8O_7$, is shown.

   The equation for the first dissociation of citric acid in water is

   $$C_6H_8O_7(aq) + H_2O(l) \rightleftharpoons C_6H_7O_7^-(aq) + H_3O^+(aq)$$

   The value of K for this reversible reaction at 298 K is 5.01×10^{-4}.

   a. Calculate the standard Gibbs energy change, $\Delta G^\ominus$, in kJ mol^{-1}, for this reaction at 298 K, using section 1 of the data booklet. [1]

   b. Comment on the spontaneity of the reaction at 298 K. [1]

7. The reaction for the formation of liquid tetracarbonylnickel is shown below:

   $$Ni(s) + 4CO(g) \rightleftharpoons Ni(CO)_4(l)$$

   a. Calculate the standard entropy change, $\Delta S^\ominus$, of the forward reaction, in J K^{-1} mol^{-1}, using the values given. [2]

Substance	$S^\ominus$ / J K^{-1} mol^{-1}
$Ni(s)$	3
$CO(g)$	198
$Ni(CO)_4(l)$	313

   b. Calculate the standard enthalpy change, $\Delta H^\ominus$, of the forward reaction in kJ mol^{-1}. [1]

Substance	$\Delta H_f^\ominus$ / kJ mol^{-1}
$CO(g)$	−110.5
$Ni(CO)_4(l)$	−633.0

   c. Use your answers to (a) and (b) to determine the temperature, in °C, below which the reaction becomes favourable. [3]

   d. At a given point in the forward reaction pathway, the reaction quotient, Q, is equal to 0.5. Calculate the Gibbs energy change, ΔG, at this point and determine whether the reaction is spontaneous at this point. [2]

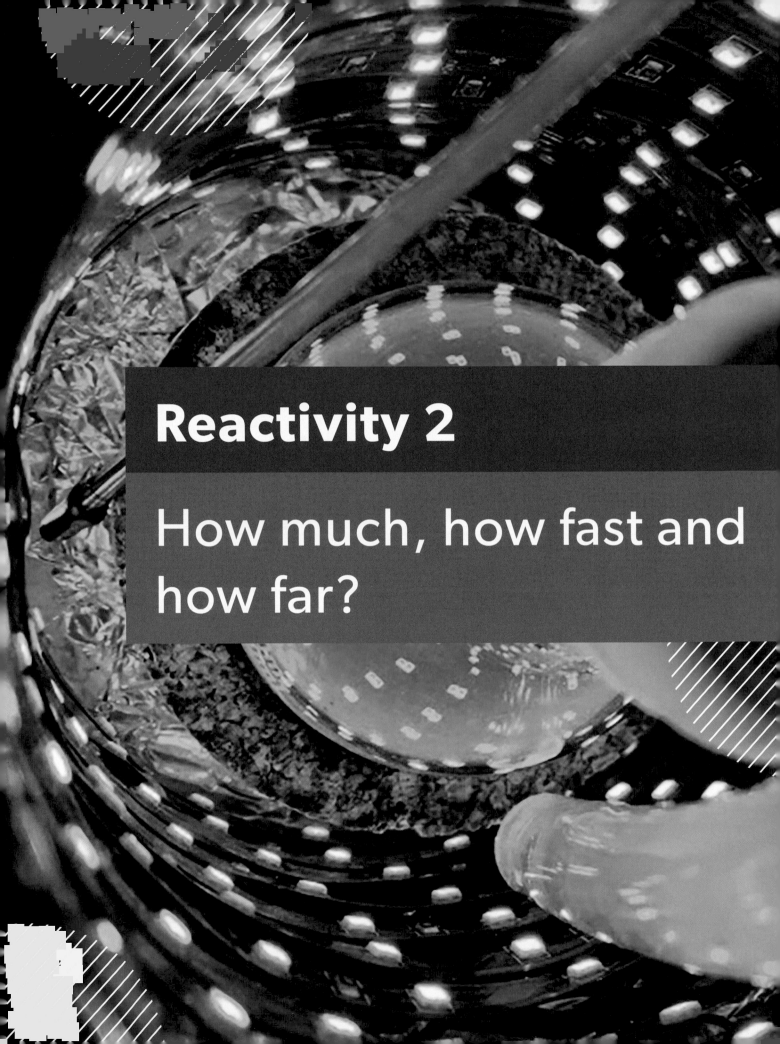

Reactivity 2

How much, how fast and how far?

Reactivity 2.1 — How much? The amount of chemical change

How are chemical equations used to calculate reacting ratios?

Chemical equations contain essential information about the nature of chemical changes, the participating species, the states of these species and the proportions in which they are consumed or produced. In chemical changes, atoms can be combined or rearranged, but never created or destroyed. So, the total number of atoms of each type remains constant when reactants are transformed into products. As a result, all species participating in a chemical reaction are used up or formed in fixed mole ratios, which are described by the stoichiometric coefficients before these species. For gases, the volumes of reacting species are also proportional to their stoichiometric coefficients.

It is important to understand that a chemical equation reflects only the changes in amounts or volumes of reacting species, but not the initial or final quantities of these species. In many cases, one of the reactants may be present in excess, or the reaction may not come to completion. Even in these cases, the mole ratios can be used to calculate any unknown quantity (amount, volume, mass and so on) of a reactant or product from given quantities of other species participating in the reaction.

Understandings

Reactivity 2.1.1—Chemical equations show the ratio of reactants and products in a reaction.

Reactivity 2.1.2—The mole ratio of an equation can be used to determine:

- the masses and/or volumes of reactants and products

- the concentrations of reactants and products for reactions occurring in solution.

Reactivity 2.1.3—The limiting reactant determines the theoretical yield.

Reactivity 2.1.4—The percentage yield is calculated from the ratio of experimental yield to theoretical yield.

Reactivity 2.1.5—The atom economy is a measure of efficiency in green chemistry.

Stoichiometry and the mole ratio (*Reactivity 2.1.1*)

Stoichiometry is the **quantitative** study of the reactants and products involved in a chemical reaction. If you know the reactants and products, you can write a balanced chemical equation by adding **stoichiometric coefficients** before each species to ensure that the number of each type of atom in the reactants and products is identical.

TOK

On the internet, find the Wikipedia article on *Stoichiometry*. Then, select a language that you are not familiar with from the menu on the left of the page. Even though you will not be able to read the text, you will probably recognize many of the diagrams, chemical equations and related mathematical calculations.

Chemical equations describe processes using a universal and internationally adopted set of symbols. To what extent does the use of universal "languages" help knowledge development?

Worked example 1

Ethane gas, C_2H_6, is completely combusted in oxygen, O_2, to form carbon dioxide, CO_2, and water, H_2O, under standard conditions. Write a balanced equation for this reaction.

Solution

First, write the unbalanced equation for the reaction, with the reactants on one side of the reaction arrow and the products on the other:

$$C_2H_6 + O_2 \rightarrow CO_2 + H_2O$$

Then, count the number of atoms of each type on both sides of the equation, adding stoichiometric coefficients where required:

- There are two carbon atoms in the reactants, and one in the products. Write the coefficient "2" before CO_2 to balance the carbon atoms.

$$C_2H_6 + O_2 \rightarrow 2CO_2 + H_2O$$

- There are six hydrogen atoms in the reactants and two in the products. Write the coefficient "3" before H_2O to balance the hydrogen atoms.

$$C_2H_6 + O_2 \rightarrow 2CO_2 + 3H_2O$$

- There are two oxygen atoms in the reactants and now seven in the products. Write the coefficient "3.5" before O_2 to balance the oxygen atoms.

$$C_2H_6 + 3.5O_2 \rightarrow 2CO_2 + 3H_2O$$

- To get a whole-number ratio, multiply each coefficient by 2.

$$2C_2H_6 + 7O_2 \rightarrow 4CO_2 + 6H_2O$$

- If you know the physical states of the species involved in the reaction, you should write the state symbols after each species:

$$2C_2H_6(g) + 7O_2(g) \rightarrow 4CO_2(g) + 6H_2O(l)$$

Practice questions

1. The anaerobic fermentation of glucose, $C_6H_{12}O_6$, forms ethanol, C_2H_5OH, and carbon dioxide, CO_2. Write the balanced equation for this reaction.

State	Symbol
solid	s
liquid	l
gas	g
aqueous solution	aq

▲ Table 1 State symbols commonly used in chemical equations

The stoichiometric coefficients in a balanced chemical equation are proportional to the amounts of reacting species. The ratio of the stoichiometric coefficients is called the **mole ratio**. Reactants will always react in relative amounts equal to the mole ratio of the reactants, to produce relative amounts of the products equal to the mole ratio of the products.

For example, the equation for the photosynthesis of glucose, $C_6H_{12}O_6$, is given below:

$$6CO_2(g) + 6H_2O(l) \rightarrow C_6H_{12}O_6(aq) + 6O_2(g)$$

The mole ratio of the reactants is $1:1$, and the mole ratio of the products is $1:6$. This means that equal amounts of carbon dioxide and water will be consumed in the reaction to produce six times less glucose than oxygen.

 Linking question

When is it useful to use half-equations? (Reactivity 3.2)

Limiting reactants (*Reactivity 2.1.2* and *Reactivity 2.1.3*)

The relative amounts of reactants in a reaction mixture might not be equal to the mole ratio, so some reactants could be left over even if all other reactants are consumed completely. The **limiting reactant** is the reactant used up completely, while other reactants are present in excess. It controls the amount of product formed in the reaction and therefore the **theoretical yield** of the reaction. The **excess reactant** is the reactant that is added to the reaction mixture in a greater proportion than that required by the mole ratio. Some of the excess reactant will be left over when the reaction reaches completion.

The reaction yield will be discussed in *Reactivity 2.1.4*.

For example, imagine that 1.0 mol of hydrogen is mixed with 3.0 mol of oxygen and the mixture is ignited. An explosion occurs, and hydrogen and oxygen react in a 2:1 mole ratio, producing water:

$$2H_2(g) + O_2(g) \rightarrow 2H_2O(l)$$

The reaction stops when all the hydrogen is consumed: 1.0 mol of hydrogen will react with 0.5 mol of oxygen and 2.5 mol of oxygen remains unreacted. Therefore, the final mixture will contain 1.0 mol of water and 2.5 mol of oxygen.

It is helpful to record the amounts of all substances as follows:

$$2H_2(g) + O_2(g) \rightarrow 2H_2O(l)$$

$n_{initial}$ / mol	1.0	3.0	0
Δn / mol	−1.0	−0.5	+1.0
n_{final} / mol	0	2.5	1.0

- The first row under the equation represents the initial mixture, in which the substances can be present in any proportions. We can take any amounts of reactants and mix them together, regardless of the stoichiometric coefficients.

- The second row shows how the amount of each substance changes in the course of the reaction. The sign before each amount shows whether it decreases (for reactants) or increases (for products). These changes must be proportional to the mole ratio shown by the stoichiometric coefficients.

- The last row represents the final mixture, where the amounts of all substances are calculated as $n_{final} = n_{initial} + \Delta n$. As with the initial mixture, the composition of the final mixture is not related to the stoichiometric coefficients.

In this example, the amount of the reaction product (water) is limited by the amount of hydrogen, which is consumed completely in the reaction. Therefore, hydrogen is the limiting reactant and oxygen is in excess.

Practice questions

2. Household lighters often contain pressurized butane, C_4H_{10}, which is combusted upon release:

$$2C_4H_{10}(g) + 13O_2(g) \rightarrow 8CO_2(g) + 10H_2O(l)$$

Determine the limiting reactant in the following mixtures:

a. 20 molecules of C_4H_{10} and 100 molecules of O_2

b. 0.20 mol of C_4H_{10} and 2.6 mol of O_2

c. 8.72 g of C_4H_{10} and 28.8 g of O_2

You can find the amount of a substance, n, from its mass, m, using $n = \dfrac{m}{M}$, where M is the **molar mass** in g mol^{-1}. Molar mass can be calculated using the A_r values given in the data booklet. This was covered in *Structure 1.4*.

The concept of the limiting reactant allows us to determine the **extent of reaction**. Consider the mixture of 1.0 mol of hydrogen and 3.0 mol of oxygen again. If we double the amount of hydrogen, the amount of oxygen that reacts doubles, and the amount of water formed will also double:

$2H_2(g) + O_2(g) \rightarrow 2H_2O(l)$			
$n_{initial}$ / mol	1.0	3.0	0
Δn / mol	−1.0	−0.5	+1.0
n_{final} / mol	0	2.5	1.0

amount of hydrogen doubled ⟹

$2H_2(g) + O_2(g) \rightarrow 2H_2O(l)$			
$n_{initial}$ / mol	2.0	3.0	0
Δn / mol	−2.0	−1.0	+2.0
n_{final} / mol	0	2.0	2.0

However, if we double the amount of oxygen (which was already in excess), the amount of hydrogen that reacts and the amount of water produced in the reaction will not change:

$2H_2(g) + O_2(g) \rightarrow 2H_2O(l)$			
$n_{initial}$ / mol	1.0	3.0	0
Δn / mol	−1.0	−0.5	+1.0
n_{final} / mol	0	2.5	1.0

amount of oxygen doubled ⟹

$2H_2(g) + O_2(g) \rightarrow 2H_2O(l)$			
$n_{initial}$ / mol	1.0	6.0	0
Δn / mol	−1.0	−0.5	+1.0
n_{final} / mol	0	5.5	1.0

Worked example 2

A 16.0 g sample of calcium carbide, $CaC_2(s)$, was placed into a sealed vessel containing 19.3 dm³ of oxygen. The vessel was heated at 1000 °C until all carbide was converted into calcium oxide, $CaO(s)$, and carbon dioxide, $CO_2(g)$. When the vessel was cooled down, the oxides reacted with each other to produce calcium carbonate, $CaCO_3(s)$. Calculate the mass of calcium carbonate and the volumes of individual gases in the remaining mixture. Assume that all reactions have come to completion. All volumes are measured at **standard temperature and pressure** (STP).

Solution

First, write the balanced equation of the combustion reaction:

$$CaC_2(s) + 2.5O_2(g) \rightarrow CaO(s) + 2CO_2(g)$$

Then, find the amounts of the reactants:

$$n(CaC_2) = \frac{m(CaC_2)}{M(CaC_2)}$$
$$= \frac{16.0\,g}{64.10\,g\,mol^{-1}}$$
$$= 0.250\,mol$$

$$n(O_2) = \frac{V(O_2)}{V_m}$$
$$= \frac{19.3\,dm^3}{22.7\,dm^3\,mol^{-1}}$$
$$= 0.850\,mol$$

Avogadro's law states that equal volumes of all gases, at the same temperature and pressure, have the same number of molecules (*Structure 1.4*). At standard temperature and pressure (STP), the **molar volume**, $V_m = 22.7\,dm^3\,mol^{-1}$. This can be used to work out the amount of gas when its volume is known, using $n = \dfrac{V}{V_m}$ (*Structure 1.5*).

Reactivity 2.1 How much? The amount of chemical change

According to the equation, calcium carbide and oxygen react in 1 : 2.5 mole ratio. In our case, the ratio of the reactants in the reaction mixture is 0.250 : 0.850 = 1 : 3.4. Therefore, oxygen is in excess, so calcium carbide is the limiting reactant. All calcium carbide will be consumed, while some oxygen will remain unreacted:

$$CaC_2(s) + 2.5O_2(g) \rightarrow CaO(s) + 2CO_2(g)$$

	CaC_2	O_2	CaO	CO_2
$n_{initial}$ / mol	0.250	0.850	0	0
Δn / mol	−0.250	−0.625	+0.250	+0.500
n_{final} / mol	0	0.225	0.250	0.500

Next, write the balanced equation for the second reaction, in which the oxides react with each other:

$$CaO(s) + CO_2(g) \rightarrow CaCO_3(s)$$

Calcium oxide and carbon dioxide react in a 1 : 1 mole ratio. However, the ratio of CaO to CO_2 in the reaction mixture is 0.250 : 0.500 = 1 : 2. Therefore, carbon dioxide is in excess, so calcium oxide is the limiting reactant. All calcium oxide will be consumed, while some carbon dioxide will remain unreacted:

$$CaO(s) + CO_2(g) \rightarrow CaCO_3(s)$$

	CaO	CO_2	$CaCO_3$
$n_{initial}$ / mol	0.250	0.500	0
Δn / mol	−0.250	−0.250	+0.250
n_{final} / mol	0	0.250	0.250

The final mixture will contain 0.250 mol of $CaCO_3(s)$, 0.250 mol of $CO_2(g)$ and 0.225 mol of $O_2(g)$. Therefore:

$$m(CaCO_3) = 0.250 \text{ mol} \times 100.09 \text{ g mol}^{-1} \approx 25.0 \text{ g}$$

$$V(CO_2) = 0.250 \text{ mol} \times 22.7 \text{ dm}^3 \text{ mol}^{-1} \approx 5.68 \text{ dm}^3$$

$$V(O_2) = 0.225 \text{ mol} \times 22.7 \text{ dm}^3 \text{ mol}^{-1} \approx 5.11 \text{ dm}^3$$

Chemical reactions are often carried out in **aqueous solutions**, which are solutions where water is the **solvent**. Aqueous solutions are easier to handle and mix than solids and gases. When solving quantitative problems involving **concentrations** and **volumes** of solutions, you should focus on the amounts, in mol, of the reacting substances and their mole ratios in the balanced chemical equation. Typical problems involving reactions in aqueous solutions are discussed in the following worked examples.

Practice questions

3. Hydrogen and chlorine react with each other to produce hydrogen chloride, HCl(g). A mixture of 4.54 dm³ of hydrogen, $H_2(g)$, and 2.27 dm³ of chlorine, $Cl_2(g)$, was heated until the reaction was complete. All volumes are measured at STP.

   a. Deduce the balanced equation for this reaction.

   b. Determine the limiting reactant.

   c. Calculate the volumes, in dm³ at STP, of each substance in the final mixture.

Definitions for solutions, solutes, solvents and concentration are given in *Structure 1.4*.

Worked example 3

Hydrochloric acid is an aqueous solution of hydrogen chloride, HCl(aq). The reaction of hydrochloric acid with a sodium hydroxide solution, NaOH(aq), produces sodium chloride, NaCl(aq). A 1.00 dm³ sample of 0.500 mol dm⁻³ hydrochloric acid was mixed with an equal volume of a 0.200 mol dm⁻³ sodium hydroxide solution.

a. Formulate the equation for the reaction of hydrochloric acid with sodium hydroxide.

b. Determine the final concentrations, in mol dm⁻³, of all solutes in the solution once the reaction is complete. Assume that the volumes of solutions are additive.

Solution

a. $HCl(aq) + NaOH(aq) \rightarrow NaCl(aq) + H_2O(l)$

b. First, work out the amounts of each solute using $n = V \times c$:

$$n(HCl) = V(HCl) \times c(HCl)$$

$$= 1.00 \, dm^3 \times 0.500 \, mol \, dm^{-3}$$

$$= 0.500 \, mol$$

$$n(NaOH) = V(NaOH) \times c(NaOH)$$

$$= 1.00 \, dm^3 \times 0.200 \, mol \, dm^{-3}$$

$$= 0.200 \, mol$$

According to the equation, HCl and NaOH react in 1:1 ratio. In the reaction mixture, the amount of HCl (0.500 mol) is greater than that of NaOH (0.200 mol). Therefore, NaOH is the limiting reactant, so it will be consumed completely while some HCl will be left over:

	$HCl(aq)$	$+ NaOH(aq)$	$\rightarrow NaCl(aq)$	$+ H_2O(l)$
$n_{initial}$ / mol	0.500	0.200	0	excess
Δn / mol	−0.200	−0.200	+0.200	+0.200
n_{final} / mol	0.300	0	0.200	excess

The final solution contains 0.300 mol of unreacted HCl(aq) and 0.200 mol of NaCl(aq). The amount of the solvent (water) is not relevant, as it is present in large excess in both the initial and final solutions.

Assuming that the volumes of solutions are additive, $V_{final} = V(HCl) + V(NaOH) = 1.00 \, dm^3 + 1.00 \, dm^3 = 2.00 \, dm^3$. You can use $\frac{n}{V_{final}}$ to calculate the final concentrations:

$$c_{final}(HCl) = \frac{0.300 \, mol}{2.00 \, dm^3} = 0.150 \, mol \, dm^{-3}$$

$$c_{final}(NaCl) = \frac{0.200 \, mol}{2.00 \, dm^3} = 0.100 \, mol \, dm^{-3}$$

You can find the amount of a substance in a solution, n, from the volume of the solution, V, using $n = V \times c$, where is c is the **molar concentration** of that substance in mol dm⁻³. This was covered in *Structure 1.4*.

Worked example 4

Sulfur dioxide, $SO_2(g)$, is a toxic gas with a strong smell of burnt matches. When released into the atmosphere in large quantities, sulfur dioxide contributes to acid rain by reacting with water vapour to produce unstable sulfurous acid, $H_2SO_3(aq)$.

Sulfur dioxide can be absorbed by alkaline solutions, such as a solution of sodium hydroxide, $NaOH(aq)$, to produce salts of sulfurous acid, such as $Na_2SO_3(aq)$.

a. Formulate equations for the reaction of sulfur dioxide with water vapour and for the reaction of sulfur dioxide with excess aqueous sodium hydroxide.

b. Determine the concentrations, in $mol\,dm^{-3}$, of all solutes in a solution produced when $2.84\,dm^3$ of sulfur dioxide is absorbed by $2.00\,dm^3$ of a $0.400\,mol\,dm^{-3}$ solution of sodium hydroxide at STP. Assume that the absorption does not affect the volume of the solution.

Solution

a. $SO_2(g) + H_2O(g) \rightarrow H_2SO_3(aq)$

$SO_2(g) + 2NaOH(aq) \rightarrow Na_2SO_3(aq) + H_2O(l)$

b. Sulfur dioxide is a gas at STP, so use $n = \dfrac{V}{V_m}$ to determine the amount present:

$n(SO_2) = \dfrac{2.84\,dm^3}{22.7\,dm^3\,mol^{-1}} = 0.125\,mol$

Work out the amount of the sodium hydroxide solute using $n = V \times c$:

$n(NaOH) = 2.00\,dm^3 \times 0.400\,mol\,dm^{-3} = 0.800\,mol$

According to the equation, SO_2 and $NaOH$ react in a $1:2$ ratio. The amount of NaOH required to absorb $0.125\,mol$ of SO_2 is therefore $0.250\,mol$. The actual amount of NaOH in the reaction mixture $(0.800\,mol)$ is greater than $0.250\,mol$. Therefore, SO_2 is the limiting reactant, so it will be consumed completely, while some NaOH will be left over:

$$SO_2(g) + 2NaOH(aq) \rightarrow Na_2SO_3(aq) + H_2O(l)$$

	SO_2	$2NaOH$	Na_2SO_3	H_2O
$n_{initial}$ / mol	0.125	0.800	0	excess
Δn / mol	−0.125	−0.250	+0.125	+0.125
n_{final} / mol	0	0.550	0.125	excess

The final solution contains $0.550\,mol$ of unreacted $NaOH(aq)$ and $0.125\,mol$ of $Na_2SO_3(aq)$. The volume of the solution at the end of the reaction will be approximately equal to the volume of the initial solution, $2.00\,dm^3$. Therefore:

$c_{final}(NaOH) = \dfrac{0.550\,mol}{2.00\,dm^3} = 0.275\,mol\,dm^{-3}$

$c_{final}(Na_2SO_3) = \dfrac{0.125\,mol}{2.00\,dm^3} = 0.0625\,mol\,dm^{-3}$

Measurement

All measurements are, to an extent, limited in precision and accuracy. However, sometimes these limitations are negligible. For example, strictly speaking, the volumes of solutions are not additive. If we mix the solutions from worked example 3, the volume of the final solution will be $2.01\,dm^3$, or approximately 0.5% greater than expected $(2.00\,dm^3)$. However, such differences are usually small and comparable with the accuracy of standard volumetric glassware (0.2–0.5%). Therefore, in most practical situations, we can assume that the volume of the solution at the end of the reaction is approximately equal to the sum of volumes of the initial solutions.

Have you come across other negligible limitations to accuracy or precision in chemistry? When do such effects become significant?

Practice questions

4. Equal volumes of $0.100\,mol\,dm^{-3}$ solutions of potassium hydroxide, $KOH(aq)$, and sulfuric acid, $H_2SO_4(aq)$, were mixed to produce potassium sulfate, $K_2SO_4(aq)$, and water.

   a. Formulate the balanced equation for the neutralization.

   b. Determine the concentrations, in $mol\,dm^{-3}$, of all solutes in the final solution.

The volumes of solutions and gases must *never* be added together. The densities of most gases are approximately 1,000 times lower than those of aqueous solutions, so when a gas is absorbed by a solution, the increase in the solution volume is negligible for most practical purposes.

The mole ratio techniques discussed in this topic are commonly used in **titration** calculations. In an acid–base titration, an acid and a base react with each other to form water and a salt. An acid–base indicator is added to the reaction mixture to monitor the reaction progress. The indicator changes colour when the neutralization is complete. The mole ratio is shown by the balanced equation for this neutralization reaction. If the volumes of both acid and base solutions are known, and the concentration of one of the reactants is also known, the unknown concentration of the second reactant can be determined.

The method for performing a titration is discussed in the *Skills* chapter.

▲ **Figure 1** An acid–base titration with a coloured indicator

Worked example 5

20.0 cm³ of sulfuric acid solution, H_2SO_4(aq), of unknown concentration was titrated with a 0.200 mol dm⁻³ solution of sodium hydroxide, NaOH(aq), in the presence of an acid–base indicator. The colour change was observed when 14.3 cm³ of the sodium hydroxide solution was added.

a. Formulate the balanced equation for the neutralization reaction.

b. Determine the concentration, in mol dm⁻³, of sulfuric acid in the analysed solution.

Solution

a. H_2SO_4(aq) + 2NaOH(aq) → Na_2SO_4(aq) + 2H_2O(l)

b. Start by converting the volume of sodium hydroxide solution to dm³:

V(NaOH) = 14.3 cm³ = 0.0143 dm³

Then, find the amount of sodium hydroxide solute using $n = V \times c$:

n(NaOH) = 0.0143 dm³ × 0.200 mol dm⁻³ = 0.00286 mol

According to the equation, H_2SO_4 and NaOH react in 1 : 2 ratio, so $n(H_2SO_4)$ = 0.00143 mol

Convert the volume of sulfuric acid solution to dm³:

$V(H_2SO_4)$ = 20.0 cm³ = 0.0200 dm³

Then, calculate the concentration of sulfuric acid using $c = \dfrac{n}{V}$:

$c(H_2SO_4) = \dfrac{0.00143 \text{ mol}}{0.0200 \text{ dm}^3} = 0.0715$ mol dm⁻³

You can also use the formula $\dfrac{c_1 V_1}{x_1} = \dfrac{c_2 V_2}{x_2}$, where x_1 and x_2 are the stoichiometric coefficients before the reactants in the balanced equation.

This way, you can skip the step where you find the amounts of the reactants used in titration.

Practice question

5. Complete neutralization of 10.0 cm³ of a sodium hydroxide solution required 9.10 cm³ of a 0.500 mol dm⁻³ hydrochloric acid. Determine the concentration, in mol dm⁻³, of sodium hydroxide in the initial solution.

Back titration

When an analyte cannot be easily quantified with a titration, a back titration is employed. In a back titration, the substance being analysed is first neutralized by a reagent of known volume and concentration. The excess of this reagent is then determined by titration.

Eggshells contain calcium carbonate. Calcium carbonate is insoluble in water, so its quantity can be conveniently determined using a back titration method. The eggshells are first crushed and reacted with a known amount of hydrochloric acid taken in excess. The leftover acid is then titrated against a standard sodium hydroxide solution to find out how much of the acid was consumed in the reaction with the eggshells. The amount of calcium carbonate in the eggshells can then be found using stoichiometric calculations.

Relevant skills

- Tool 1: Applying techniques: titration
- Tool 3: Propagate uncertainties in processed data
- Inquiry 2: Collect and record sufficient relevant data
- Inquiry 2: Carry out correct and accurate data processing

Instructions

1. Using the list of equipment shown below, design a method to analyse the mass percentage of calcium carbonate in eggshells. The equipment list is not definitive: you do not have to use all the items in the list, and you can add further items or substances if you wish.

    - Calcium carbonate chips, $CaCO_3(s)$
    - A variety of different types of eggshells
    - 1.00 mol dm^{-3} hydrochloric acid, HCl(aq)
    - 1.00 mol dm^{-3} sodium hydroxide, NaOH(aq)

- Phenolphthalein indicator solution
- Burette
- Burette clamp and stand
- Conical flasks
- Drying oven
- Pestle and mortar
- Top-pan balance
- Volumetric pipette and filler
- Wash bottle containing distilled water
- Waste container(s)
- Standard lab safety equipment

2. Conduct a risk assessment of your method and show it to your teacher for approval.

3. Collect sufficient relevant data, including qualitative observations. You will have to decide how much data will be sufficient.

4. Process the data to determine the mass percentage of calcium carbonate in each of the eggshell samples.

5. Propagate measurement uncertainties to obtain the uncertainties of the calculated mass percentages found above.

6. Graph your data using an appropriate software, including error bars.

7. If possible, compare your data to literature values. Comment on the accuracy and precision of your results.

8. Discuss at least two sources of error in your investigation and propose improvements that could minimize these errors.

 ## Linking questions

How does the molar volume of a gas vary with changes in temperature and pressure? (Structure 1.5)

 In what ways does Avogadro's law help us to describe, but not explain, the behaviour of gases? (Structure 1.4)

The reaction yield (*Reactivity 2.1.4*)

In any chemical reaction, the total mass of the reaction products is equal to the total mass of the consumed reactants. This principle, known as the **law of conservation of mass**, follows from atomic theory. Since atoms cannot be created or destroyed, their total number and therefore mass cannot be affected by chemical changes.

Science as a shared endeavour

"Nothing comes from nothing" was one of the basic principles in ancient Greek philosophy. The first recorded experimental evidence for the conservation of mass was obtained in 1748 by the Russian scientist Mikhail Lomonosov, who carried out chemical reactions in sealed vessels and found that chemical changes did not affect the total mass of the mixtures. However, the works of Lomonosov were almost unknown in western European countries, so it was over 25 years before the French chemist Antoine Lavoisier formulated this principle in 1774 and was commonly credited for its discovery.

▲ **Figure 2** Left: Mikhail Lomonosov; right: Antoine Lavoisier

This example shows the importance of sharing knowledge and ideas: a valuable discovery that remains hidden from the global scientific community may go unnoticed for some time and have little impact on the advancement of science. What conditions facilitate an individual's access to the international scientific community? How can the number of valuable contributions of scientists from around the world be maximized?

Activity

Calculate the masses, in g, of the initial and final mixtures from worked example 2 and verify the validity of the law of conservation of mass.

When chemical reactions are carried out experimentally, the amounts of the reaction products are usually lower than the theoretical amounts predicted by the equation. This can happen for many reasons, including incomplete conversion of the reactants or simply because some product was lost during its isolation and purification. In such cases, we can calculate the **percentage yield**, which is the ratio of the experimental and theoretical amounts of the product.

Since the amount of an individual substance is proportional to its mass ($n = \frac{m}{M}$), the yield can also be found as the ratio of the experimental and theoretical masses of the product:

$$\text{percentage yield} = \frac{n_{\text{experimental}}}{n_{\text{theoretical}}} \times 100\% = \frac{m_{\text{experimental}}}{m_{\text{theoretical}}} \times 100\%$$

The term "yield" can also refer to the amount or mass of the reaction product. In this sense, the **theoretical yield** is the amount or mass of the product that could be obtained from a particular reaction if the reaction went to completion and no product was lost. In contrast, the **experimental yield** is the actual amount or mass of the product isolated from a particular laboratory experiment. The experimental yield can be lower than or equal to the theoretical yield, but it can never exceed it.

The percentage yield, and theoretical and experimental yields of a reaction, are related as follows:

$$\text{percentage yield} = \frac{\text{experimental yield}}{\text{theoretical yield}} \times 100\%$$

Worked example 6

A mixture of 10.0 g of calcium metal and 9.62 g of elemental sulfur was heated for some time to produce 17.0 g of calcium sulfide, CaS(s).

a. Calculate the theoretical yield, in g, of calcium sulfide.

b. Using your answer to part **a**, calculate the percentage yield of calcium sulfide.

Solution

a. First, write the balanced equation for the reaction:

$$Ca(s) + S(s) \rightarrow CaS(s)$$

Then calculate the amounts of the reactants and products using $n = \frac{m}{M}$:

$$n(Ca) = \frac{10.0 \text{ g}}{40.08 \text{ g mol}^{-1}} \approx 0.250 \text{ mol}$$

$$n(S) = \frac{9.62 \text{ g}}{32.07 \text{ g mol}^{-1}} \approx 0.300 \text{ mol}$$

$$n(CaS) = \frac{17.0 \text{ g}}{72.15 \text{ g mol}^{-1}} \approx 0.236 \text{ mol}$$

The limiting reactant is calcium, as calcium and sulfur are consumed in a 1:1 ratio. The theoretical amount of calcium sulfide is 0.250 mol (the same as the initial amount of the limiting reactant, calcium). Therefore, the theoretical yield of calcium sulfide is:

$$m_{\text{theoretical}}(CaS) = 0.250 \text{ mol} \times 72.15 \text{ g mol}^{-1} \approx 18.0 \text{ g}$$

b. The reaction produced 0.236 mol of calcium sulfide, while the theoretical yield of this compound was 0.250 mol. Therefore:

$$\text{percentage yield} = \frac{n_{\text{experimental}}}{n_{\text{theoretical}}} \times 100\%$$

$$= \frac{0.236 \text{ mol}}{0.250 \text{ mol}} \times 100\% = 94.4\%$$

The same result could be obtained through the experimental and theoretical masses of calcium sulfide:

$$\text{percentage yield} = \frac{m_{\text{experimental}}}{m_{\text{theoretical}}} \times 100\%$$

$$= \frac{17.0 \text{ g}}{18.0 \text{ g}} \times 100\% = 94.4\%$$

Practice question

6. Calculate the theoretical and percentage yields of the reaction between 9.443 g of aluminium and 7.945 dm³ (STP) of oxygen that produced 17.13 g of aluminium oxide, Al_2O_3(s).

Data-based questions

Salicylic (2-hydroxybenzoic) acid was isolated from the bark of willow trees in the first half of the 19th century and used as a pharmaceutical drug for pain and fever relief. However, pure salicylic acid caused severe stomach irritation and bleeding. These side effects could be significantly reduced by the use of a chemically modified salicylic acid, which is now known as acetylsalicylic acid or aspirin.

salicylic acid
(2-hydroxybenzoic acid)
M_r 138.13

ethanoic
anhydride
M_r 102.10

H_3PO_4

aspirin
(acetylsalicylic acid)
M_r 180.17

ethanoic
acid
M_r 60.06

The reaction of salicylic acid with ethanoic anhydride can be used in a school laboratory for the preparation of aspirin. In a typical experiment, salicylic acid is mixed with excess ethanoic anhydride and several drops of the catalyst (concentrated phosphoric acid, H_3PO_4). The mixture is heated for a short time, then diluted with water and allowed to cool down slowly, producing crystals of aspirin. The crystals are isolated by filtration and dried in the open air.

The aspirin obtained in this way is usually impure, so it needs to be recrystallized. To do so, impure aspirin is dissolved in hot ethanol, and the solution is allowed to cool down. Pure aspirin precipitates while most of the impurities remain in the solution. The crystals of pure aspirin are filtered and dried until all the ethanol evaporates.

Each step of the procedure outlined causes some loss of the product and therefore reduces the yield of aspirin. In particular, the reaction of salicylic acid with ethanoic anhydride is never 100% complete, some aspirin stays in the solution, small crystals of aspirin can be trapped in the paper filter, and so on. The results of three experiments performed by students are shown in table 2.

Experimental data	Experiment 1	Experiment 2	Experiment 3
mass of salicylic acid / g	2.06	2.57	2.35
mass of ethanoic anhydride / g	2.49	2.85	3.11
mass of aspirin / g	3.42	2.31	1.87
recrystallization from ethanol	none	once	twice
melting point of product / °C	122–128	133–135	135–136

▲ Table 2 Synthesis of aspirin data. The literature value for the melting point of pure aspirin is 136 °C

Impurities usually lower the melting point and widen the melting range of substances. The most common example is sea water, which freezes between -2.5 and $-1.8\,°C$ instead of at 0 °C like pure water.

Questions

1. Calculate the amounts of the reactants for each experiment and determine the limiting reactant in each case.

2. Determine the percentage yield of aspirin in each experiment.

3. Identify which experiment produced an impossible percentage yield of aspirin. Using the data in table 2, suggest a reason for this and provide supporting data for your answer.

4. Suggest which experiment produced the purest sample of aspirin. Explain your answer.

The identity of aspirin can be confirmed by IR spectroscopy, NMR spectroscopy, mass spectrometry (Structure 3.2, AHL) or by determining its melting point (Structure 1.1) and comparing the result with the literature data.

Gravimetric analysis of a precipitation reaction

The reaction between calcium chloride and sodium carbonate solutions produces a precipitate of calcium carbonate, which can be removed by filtration:

$$CaCl_2(aq) + Na_2CO_3(aq) \rightarrow CaCO_3(s) + 2NaCl(aq)$$

In this practical, you will carry out this reaction, isolate the resulting calcium carbonate and calculate the yield.

Relevant skills

- Tool 1: Measuring mass and volume
- Tool 1: Drying to a constant mass
- Tool 3: Applying general mathematics
- Inquiry 2: Identify and record relevant qualitative observations
- Inquiry 2: Collect and record sufficient relevant quantitative data
- Inquiry 2: Carry out relevant and accurate data processing

Safety

- Wear eye protection
- Sodium carbonate solution is an irritant, wear gloves when handling it

Materials

- two $25\,cm^3$ graduated pipettes or burettes
- $100\,cm^3$ beaker
- stirring rod
- vacuum filtration apparatus
- top-pan balance
- filter paper
- wash bottle containing distilled water
- drying oven (if available)
- $0.50\,mol\,dm^{-3}$ calcium chloride solution, $CaCl_2(aq)$
- $1.0\,mol\,dm^{-3}$ sodium carbonate solution, $Na_2CO_3(aq)$

Instructions

1. Write your initials on the filter paper in pencil and pre-weigh it.

2. In a $100\,cm^3$ beaker, add $25\,cm^3$ of $0.50\,mol\,dm^{-3}$ calcium chloride solution to $15\,cm^3$ of $1.0\,mol\,dm^{-3}$ sodium carbonate solution.

3. Stir the reaction mixture. Remember to record relevant qualitative observations throughout.

4. Filter the mixture (figure 3). Rinse the beaker and stirring rod with distilled water to completely transfer all traces of the precipitate onto the filter paper.

5. Allow the filter paper to dry undisturbed, inside a drying oven if possible. Once it looks dry, weigh it. Repeat until the mass remains constant.

6. Clean up as instructed by your teacher.

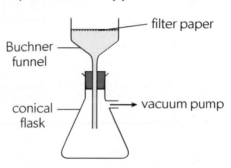

▲ Figure 3 Vacuum filtration

Questions

1. Determine the limiting and excess reactants.

2. Calculate the theoretical yield of calcium carbonate.

3. Calculate the experimental yield of calcium carbonate and hence, the percentage yield.

4. Compare and contrast your results to your classmates' results.

5. Outline two ways to improve the efficiency of this synthesis.

ATL **Self-management skills**

The following steps show a general approach to solving problems involving stoichiometric calculations. To help you remember all the steps, describe each step in three words, and then again in one word.

1. Formulate and balance the chemical equation(s) for any chemical changes mentioned in the problem.

2. Calculate the amounts of as many substances as you can using the formulas $n = \dfrac{m}{M}$ and $n = \dfrac{V}{V_m}$. If you know a mass percentage, convert it first to mass (if possible) and then to amount.

3. Write all known amounts of substances below their formulas in chemical equations. Any missing values may suggest the next step of the solution.

4. Determine the limiting reactant and use the mole ratio shown by stoichiometric coefficients in the balanced equation to calculate the changes in amount for all substances. Remember that the mole ratio does not reflect the initial or final amounts of reactants and products.

5. Check the mass balance. The total mass of products must be equal to the total mass of reactants. If it is not the case, revise your solution.

6. Check that the answer makes sense. All percentages should add up to 100%, and no individual percentage can be greater than 100%. The yield of the final product must not exceed 100%. Finally, treat very low or very high percentages with caution.

In practice, chemists often have to find a balance between yield and purity. For example, most medical drugs must have a purity of 99% or even higher, as any impurities could have unpredicted effects on the patient's health. As a result, the strict safety requirements reduce the yields and increase the prices of medical-grade products. At the same time, the acceptable purity for many industrial and agricultural materials, such as detergents and fertilizers, could be well below 50%, as long as the impurities do not compromise the safety and intended use of the product.

The yields of chemical reactions are particularly important in the chemical industry, where the loss of even a small percentage of the final product could mean a significant drop in profit. Low reaction yields increase the amount of waste, which needs to be disposed of safely or reused. The development of highly efficient synthetic procedures with low environmental impact is reflected in the concept of **green chemistry**, which is now adopted by the majority of commercial and research organizations around the world.

 Linking question

What errors may cause the experimental yield to be i) higher and ii) lower than the theoretical yield? (Tool 1, Inquiry 1, 2, 3)

Green chemistry and atom economy (*Reactivity 2.1.5*)

In traditional chemistry, the efficiency of a synthetic procedure is measured in terms of the product yield and the cost of raw materials. Many other factors, such as the toxicity of reagents and solvents, energy consumption and the amount of waste produced are often ignored. A completely different approach, known as green chemistry, takes into account the environmental impact of the entire technological process. Green chemistry encourages scientists to minimize the use and generation of hazardous chemicals when designing reactions that are used in industrial processes. Common practices of green chemistry include aqueous or solvent-free reactions, renewable starting materials, mild reaction conditions, efficient catalysis and the utilization of any by-products formed during the synthesis.

Global impact of science

The term "green chemistry" was coined in 1991 by American chemists Paul Anastas and John Warner, who formulated 12 principles for their approach to chemical technology. These principles include the use of non-hazardous chemicals and solvents, reduction of energy consumption and waste production: "the best form of waste disposal is not to create it in the first place". Other principles include the use of renewable materials and prevention of accidents. The philosophy of green chemistry has passed into national and international laws, which restrict the use of certain chemical substances and promote environmentally friendly technologies. To what extent can you incorporate the 12 green chemistry principles into your practical work in DP chemistry?

Another key concept of green chemistry is the **atom economy**, which is a measure of the efficiency of a reaction. The atom economy is the ratio of the molar mass of the isolated target product to the combined molar masses of all starting materials, catalysts and solvents used in the reaction:

$$\text{atom economy} = \frac{\text{molar mass of desired product}}{\text{total molar mass of reaction mixture}} \times 100\%$$

For example, the atom economy of a solvent-free reaction $A + B \rightarrow C$ is equal to the experimental yield and can potentially reach almost 100%. However, in the reaction $A + B \rightarrow C + D$, where C is the target product, the atom economy will always be significantly lower than 100% because the unwanted by-product D is formed. The use of solvents and catalysts will further reduce the atom economy because their constituent atoms do not form the target product and must be disposed of or recycled. Therefore, the atom economy is inversely proportional to the percentage of waste generated in an industrial process or laboratory experiment.

> The equation for the calculation of atom economy is given in the data booklet.

Examples of atom-efficient reactions are addition polymerization (*Structure 2.4*) and the hydrogenation of alkenes (*Reactivity 3.2*), which proceed with high yields, require no solvents and form almost no by-products under appropriate conditions. At the same time, many traditional organic reactions, such as the oxidation of alcohols (*Reactivity 3.2*) or electrophilic substitution in aromatic compounds (*Reactivity 3.4, AHL*), are very inefficient. This is because they often require large volumes of solvents, have low yields and in some cases produce mixtures of isomers (*Structure 3.2*) instead of individual target products.

Worked example 7

Dimethyl carbonate, $(CH_3O)_2CO$, is a non-toxic and highly efficient "green" solvent. Calculate the percentage atom economy for the synthesis of dimethyl carbonate by the following reaction:

$$4CH_3OH(l) + 2CO(g) + O_2(g) \rightarrow 2(CH_3O)_2CO(l) + 2H_2O(l)$$

Assume that the percentage yield of the reaction is 100%.

Solution

The total molar mass of the products is equal to the total molar mass of the reactants, so either can be used as the total molar mass of the reaction mixture.

Using the molar mass of reactants in the denominator:

$$\text{atom economy} = \frac{2 \times M((CH_3O)_2CO)}{4 \times M(CH_3OH) + 2 \times M(CO) + M(O_2)} \times 100\%$$

$$= \frac{2 \times 90.09}{4 \times 32.05 + 2 \times 28.01 + 32.00} \times 100\% = 83.33\%$$

Using the molar mass of products in the denominator:

$$\text{atom economy} = \frac{2 \times M((CH_3O)_2CO)}{2 \times M(CH_3O)_2CO) + 2 \times M(H_2O)} \times 100\%$$

$$= \frac{2 \times 90.09}{2 \times 90.09 + 2 \times 18.02} \times 100\% = 83.33\%$$

Practice questions

7. Dimethyl carbonate can be used to convert phenylamine, $C_6H_5NH_2(l)$, into *N*-methylphenylamine, $C_6H_5NHCH_3(l)$. The by-products of this reaction are methanol, $CH_3OH(l)$, and carbon dioxide, $CO_2(g)$.

   a. Formulate the balanced equation for this process and calculate its atom economy.

   b. Suggest, by referring to the chemical equation in worked example 7, how the atom economy of the synthesis of *N*-methylphenylamine can be improved.

The cost of green chemistry

Green technologies vary in efficiency and in many cases involve expensive equipment, raw materials and recycling facilities. However, these initial investments reduce the costs associated with environmental remediation, waste management and energy consumption. Therefore, in the long term, green chemistry is a commercially attractive and sustainable alternative to traditional organic chemistry.

Increasing adoption of green industrial processes in developed countries has significantly reduced the emissions of many hazardous chemicals, such as chlorinated solvents or greenhouse gases. At the same time, some non-hazardous substances branded as "green" or "environmentally friendly" still require toxic chemicals or large amounts of energy for their production. The industrial use of natural products, such as plant oils and starch, takes up agricultural resources, reduces biodiversity and increases the cost of food. This can lead to various ecological and social issues. Therefore, the "greenness" of a product must be assessed using all direct and indirect environmental implications of its entire life cycle. This is one of the most controversial problems in green chemistry.

▲ Figure 4 Left: supercritical carbon dioxide is widely used in food industry as a non-toxic, green alternative to chlorinated solvents; right: a biofuel power plant

(ATL) **Self-management skills**

Create your own practice question involving mole ratios. Depending on how confident you feel about this topic, you could develop a relatively straightforward question based on reacting masses, or a more complex one that brings in your knowledge of concentrations and/or gas volumes. Try to address some of the common mistakes and misconceptions related to mole ratios. Create a booklet or record a video showing the step-by-step solution to the problem and share it with a partner. Give each other feedback in the form of "two stars and a wish": two things they have done well and one constructive suggestion for improvement. Improve your work, then share it with the rest of your class.

Linking question

The atom economy and the percentage yield both give important information about the "efficiency" of a chemical process. What other factors should be considered in this assessment? (Structure 2.4, Reactivity 2.2)

End of topic questions

Topic review

1. Using your knowledge from the *Reactivity 2.1* topic, answer the guiding question as fully as possible:

   How are chemical equations used to calculate reacting ratios?

Exam-style questions

Multiple-choice questions

2. Pyrite, FeS_2, reacts with oxygen according to the following equation:

   $4FeS_2(s) + 11O_2(g) \rightarrow 2Fe_2O_3(s) + 8SO_2(g)$

   How many grams of $SO_2(g)$ will be produced from 0.25 mol of FeS_2?

   A. 16

   B. 32

   C. 64

   D. 128

3. In which mixture is sulfuric acid the limiting reactant?

   A. 0.10 mol H_2SO_4 + 0.20 mol KOH

   B. 0.20 mol H_2SO_4 + 0.10 mol KOH

   C. 0.10 mol H_2SO_4 + 0.20 mol $Ca(OH)_2$

   D. 0.20 mol H_2SO_4 + 0.10 mol $Ca(OH)_2$

4. What is the percentage yield when 18 dm^3 of ethane, $C_2H_6(g)$, is produced from 20 dm^3 of ethene, $C_2H_4(g)$? All volumes are measured at STP.

   $C_2H_4(g) + H_2(g) \rightarrow C_2H_6(g)$

   A. 10%

   B. 20%

   C. 80%

   D. 90%

5. Which of the following expressions gives the atom economy of the synthesis of methanal, HCHO(g), from methanol, $CH_3OH(l)$, if the experimental yield is 100%?

   $CH_3OH(l) + 0.5O_2(g) \rightarrow HCHO(g) + H_2O(l)$

   I $\dfrac{M(HCHO)}{M(CH_3OH) + M(HCHO)}$

   II $\dfrac{M(HCHO)}{M(CH_3OH) + 0.5 \times M(O_2)}$

   III $\dfrac{M(HCHO)}{M(HCHO) + M(H_2O)}$

   A. I and II only

   B. I and III only

   C. II and III only

   D. I, II and III

Extended-response questions

6. Potassium superoxide, $KO_2(s)$, is used to regenerate oxygen from the carbon dioxide exhaled by the crew of a submarine or spacecraft.

   a. Formulate the balanced equation for the reaction between potassium superoxide and carbon dioxide that produces oxygen and potassium carbonate, $K_2CO_3(s)$.

   b. Deduce the limiting reactant in the reaction of 28.44 g of potassium superoxide and 6.81 dm^3 (STP) of carbon dioxide.

   c. Calculate the mass, in g, of potassium carbonate and the volume, in dm^3 at STP, of oxygen produced in this reaction.

7. Chlorine gas is produced in the laboratory by the reaction of hydrochloric acid, HCl(aq), with manganese(IV) oxide, $MnO_2(s)$:

   $MnO_2(s) + 4HCl(aq) \rightarrow MnCl_2(aq) + Cl_2(g) + 2H_2O(l)$

   A 17.4 g sample of manganese(IV) oxide was added to 0.100 dm^3 of 10.0 mol dm^{-3} hydrochloric acid, and the mixture was heated until the reaction was complete.

   a. Deduce the limiting reactant in this experiment.

   b. Calculate the volume, in dm^3 at STP, of chlorine gas produced.

   c. Calculate the molar concentrations, in mol dm^{-3}, of all solutes in the final solution. Assume that the volume of the solution did not change during the reaction.

8. A $10.0\,cm^3$ sample of a solution with unknown concentration of barium hydroxide, $Ba(OH)_2(aq)$, was titrated with $0.100\,mol\,dm^{-3}$ hydrochloric acid, $HCl(aq)$, in the presence of an acid–base indicator. The colour change was observed when $11.7\,cm^3$ of the standard solution was added.

    a. Formulate the balanced equation for the neutralization if the reaction products are barium chloride, $BaCl_2(aq)$, and water.

    b. Determine the concentration, in $mol\,dm^{-3}$, of barium hydroxide in the analysed solution.

9. The thermal decomposition of sodium hydrogencarbonate, $NaHCO_3(s)$, produces sodium carbonate, $Na_2CO_3(s)$, carbon dioxide and water.

    a. Formulate the balanced equation for this reaction.

    b. Calculate the theoretical yield, in g, of carbon dioxide produced by the decomposition of $10.0\,g$ of sodium hydrogencarbonate.

    c. The decomposition of $10.0\,g$ of sodium hydrogencarbonate produced $1.20\,dm^3$ (STP) of carbon dioxide. Calculate the theoretical yield, in dm^3, of carbon dioxide and its percentage yield in this reaction.

    d. Calculate the masses of solid substances in the final mixture obtained in part **c**.

10. Copper(I) sulfide, Cu_2S, is found in nature as the mineral *chalcocite*, which is a major component of copper ores. At high temperatures, chalcocite reacts with oxygen to produce copper metal and sulfur dioxide, $SO_2(g)$.

    a. Formulate the balanced equation for the reaction of copper(I) sulfide with oxygen.

    b. Determine the theoretical yield, in kg, of copper metal that could be obtained from $1.00\,ton$ $(1000\,kg)$ of an ore that contains 75.3% of chalcocite. Assume that other components of the ore are unreactive.

    c. Using your answer to part **b**, determine the percentage yield of copper metal if its practical yield is $506\,kg$.

11. Calculate the atom economy for the following reactions, assuming that the organic compound is the target product in each case.

    a. $CH_3CH_2OH(l) \rightarrow H_2C{=}CH_2(g) + H_2O(l)$

    b. $6CO_2(g) + 6H_2O(l) \rightarrow C_6H_{12}O_6(aq) + 6O_2(g)$

    c. $nH_2C{=}CH_2(g) \rightarrow {+}CH_2{-}CH_2{+}_n(s)$

12. Nitrogen dioxide, $NO_2(g)$, can be synthesized from nitrogen monoxide, $NO(g)$, and oxygen.

    a. Formulate the balanced equation for this reaction.

    b. Calculate the theoretical atom economy for the equation from part **a**.

    c. In a particular experiment, $1.00\,dm^3$ of nitrogen monoxide was mixed with $1.00\,dm^3$ of oxygen to produce $0.700\,dm^3$ of nitrogen dioxide. Calculate the theoretical yield, in dm^3, and the percentage yield of nitrogen dioxide.

    d. Calculate the actual atom economy for the synthesis in part **c**.

How can the rate of a reaction be controlled?

One of the main considerations in designing a synthetic procedure is whether it can be carried out fast enough to be useful. There is no point in starting a reaction if it takes a hundred years before any product could be isolated! Chemists often have to find a balance between the rate and yield of a reaction to make it feasible from both the practical and commercial points of view.

Chemical kinetics is the branch of chemistry that studies the rates of chemical reactions. Kinetics studies provide important information about reaction mechanisms and the ways the reaction rates can be controlled by altering the experimental conditions, such as pressure, temperature, concentration, surface area and the presence of catalysts.

Understandings

Reactivity 2.2.1—The rate of reaction is expressed as the change in concentration of a particular reactant/product per unit time.

Reactivity 2.2.2—Species react as a result of collisions of sufficient energy and proper orientation.

Reactivity 2.2.3—Factors that influence the rate of a reaction include pressure, concentration, surface area, temperature and the presence of a catalyst.

Reactivity 2.2.4—Activation energy, E_a, is the minimum energy that colliding particles need for a successful collision leading to a reaction.

Reactivity 2.2.5—Catalysts increase the rate of reaction by providing an alternative reaction pathway with lower E_a.

AHL

Reactivity 2.2.6—Many reactions occur in a series of elementary steps. The slowest step determines the rate of the reaction.

Reactivity 2.2.7—Energy profiles can be used to show the activation energy and transition state of the rate-determining step in a multistep reaction.

Reactivity 2.2.8—The molecularity of an elementary step is the number of reacting particles taking part in that step.

AHL

Reactivity 2.2.9—Rate equations depend on the mechanism of the reaction and can only be determined experimentally.

Reactivity 2.2.10—The order of a reaction with respect to a reactant is the exponent to which the concentration of the reactant is raised in the rate equation. The order with respect to a reactant can describe the number of particles taking part in the rate-determining step. The overall reaction order is the sum of the orders with respect to each reactant.

Reactivity 2.2.11—The rate constant, k, is temperature dependent and its units are determined from the overall order of the reaction.

Reactivity 2.2.12—The Arrhenius equation uses the temperature dependence of the rate constant to determine the activation energy.

Reactivity 2.2.13—The Arrhenius factor, A, takes into account the frequency of collisions with proper orientations.

The rate of reaction (*Reactivity 2.2.1*)

Chemical reactions proceed at different speeds. Fast reactions, such as explosions, occur within milliseconds. Slow reactions, such as the formation of fossil fuels, take millions of years. However, the terms "fast" and "slow" are very vague, so chemists need to describe the speed of chemical changes more precisely.

Therefore, it is convenient to define the **rate of reaction**, v, as the change in concentration, Δc, of a reactant or product per unit time, Δt:

$$v = \frac{|\Delta c|}{\Delta t}$$

This is the average rate of reaction, that is, the mean reaction rate over a period of time, Δt, rather than the reaction rate at a given moment in time. Typically, reaction rates are measured in $mol\,dm^{-3}\,s^{-1}$, although other units, such as $mmol\,dm^{-3}\,min^{-1}$ or $mol\,m^{-3}\,h^{-1}$, can also be used.

A reaction rate cannot be negative, so it must be calculated using the absolute value (modulus) of the concentration change. For example, if the concentration of species X in a reaction mixture decreases from 0.50 to $0.20\,mol\,dm^{-3}$ over 25 seconds, the rate of reaction with respect to that species is calculated as follows:

$$v(X) = \frac{|0.20 - 0.50|\,mol\,dm^{-3}}{25\,s} = \frac{0.30\,mol\,dm^{-3}}{25\,s} = 0.012\,mol\,dm^{-3}\,s^{-1}$$

When the mole ratio of two or more species participating in a chemical reaction differs from 1 : 1, the reaction rates with respect to these species will also be different. To avoid ambiguity, the rates for individual species can be divided by their stoichiometric coefficients. The resulting **overall rate** of reaction is independent of the species used in calculations.

▲ Figure 1 Fireworks use fast chemical reactions, whereas corrosion of metal constructions takes centuries

Worked example 1

Consider the following reaction:

$$2N_2O(g) \rightarrow 2N_2(g) + O_2(g)$$

Under certain conditions, the concentration of nitrogen(I) oxide in the reaction mixture decreases by $0.20\,mol\,dm^{-3}$ over 10 seconds, producing $0.20\,mol\,dm^{-3}$ of nitrogen and $0.10\,mol\,dm^{-3}$ of oxygen over the same period of time. Calculate:

a. the rate of reaction, in $mol\,dm^{-3}\,s^{-1}$, with respect to each of the three species involved in the reaction

b. the overall rate of reaction, in the same units as above.

Solution

a. Use $v = \dfrac{|\Delta c|}{\Delta t}$ to find the rate of reaction for each species:

$$v(N_2O) = \frac{0.20\,mol\,dm^{-3}}{10\,s} = 0.020\,mol\,dm^{-3}\,s^{-1}$$

$$v(N_2) = \frac{0.20\,mol\,dm^{-3}}{10\,s} = 0.020\,mol\,dm^{-3}\,s^{-1}$$

$$v(O_2) = \frac{0.10\,mol\,dm^{-3}}{10\,s} = 0.010\,mol\,dm^{-3}\,s^{-1}$$

b. To find the overall rate of reaction, you can divide any of the rates for individual species by their stoichiometric coefficients:

$$v_{overall} = \frac{v(N_2O)}{2} = \frac{v(N_2)}{2} = \frac{v(O_2)}{1} = 0.010\,mol\,dm^{-3}\,s^{-1}$$

Practice questions

1. Under certain conditions, ammonia can be oxidized to nitrogen(II) oxide:

   $$4NH_3(g) + 5O_2(g) \rightarrow 4NO(g) + 6H_2O(g)$$

   Over a period of 5 s, the concentration of nitrogen(II) oxide in the reaction mixture increased from 0 to 6.0×10^{-3} mol dm^{-3}. Calculate:

   a. the rate of reaction, in mol dm^{-3} s^{-1}, with respect to NO(g) and O_2(g)

   b. the overall rate of reaction, in the same units as above.

In addition to mass and volume, the following quantitative characteristics are often used to calculate the concentration changes:

- pressure (for reactions involving gases, *Structure 1.5*)

- pH (for acid–base reactions, *Reactivity 3.1*)

- electrical conductivity (for reactions involving electrolytes, *Structure 2.1*)

- colour (for reactions involving transition metals or other coloured compounds, *Structure 3.1, AHL*).

Any of these characteristics can be recorded automatically using a data logger and then related to concentration using an appropriate mathematical expression or a calibration curve (*Structure 1.4*). Alternatively, small samples of the reaction mixture could be taken at certain time intervals and analysed by spectroscopic techniques (*Structure 3.2, AHL*) or titration (*Reactivity 2.1*).

Direct measurement of concentrations can be problematic, so the change in concentration is often calculated from other experimental data, such as changes in mass or volume of the reaction mixture or a particular substance.

For instance, the reaction of magnesium metal with hydrochloric acid can be investigated by measuring the volume of hydrogen gas released from the solution:

$$Mg(s) + 2HCl(aq) \rightarrow MgCl_2(aq) + H_2(g)$$

In a typical experiment, a carefully measured volume of hydrochloric acid with known concentration is placed into a conical flask, and a sample of magnesium is added. The flask is immediately closed with a rubber bung and connected to a gas syringe (figure 2). The gas volume is recorded at regular time intervals until the reaction is complete.

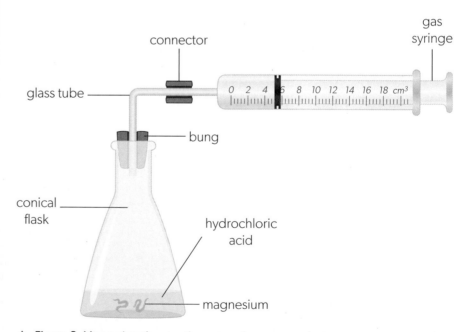

▲ **Figure 2** Measuring the reaction rate using a gas syringe

The change in concentration of hydrochloric acid and the reaction rate can be calculated using gas laws and reaction stoichiometry.

Worked example 2

The experiment shown in figure 2 was carried out using a piece of magnesium ribbon and $0.100\,dm^3$ of $0.250\,mol\,dm^{-3}$ hydrochloric acid. The volume of gas released by the reaction was recorded every 10 seconds until a constant reading was obtained (table 1).

$t\,/\,s$	0	10	20	30	40	50	60	70	80	90	100
$V(H_2)\,/\,cm^3$	0	111	164	192	210	223	234	241	246	248	248

▲ Table 1 Volume of hydrogen gas released by the reaction of hydrochloric acid with magnesium metal at SATP

a. Calculate the rate of reaction, in $mol\,dm^{-3}\,s^{-1}$, with respect to $H_2(g)$, for the following time intervals:

  i. between 0 and 10 s

  ii. between 0 s and the moment when the reaction was complete.

b. Calculate the overall rate of reaction for the time interval in (a)(ii).

Assume that the reaction was carried out at SATP (25 °C and 100 kPa), and the solution volume did not change during the reaction.

Solution

a. i. Use $\Delta n = \dfrac{\Delta V}{V_m}$ to calculate the change in the amount of hydrogen. Note that V_m depends on temperature.

The value of $22.7\,dm^3\,mol^{-1}$ can be used only at $T = 0\,°C\,(273.15\,K)$ and $p = 100\,kPa$ (STP). The V_m at SATP or any other conditions can be calculated using the ideal gas equation (Structure 1.5):

$$\frac{p_1V_1}{T_1} = \frac{p_2V_2}{T_2}$$

$$\frac{100\,kPa \times 22.7\,dm^3\,mol^{-1}}{273.15\,K} = \frac{100\,kPa \times V_m(SATP)}{298.15\,K}$$

$$V_m(SATP) = 24.8\,dm^3\,mol^{-1}$$

Over the first 10 s of the reaction, $111\,cm^3$ $(0.111\,dm^3)$ of hydrogen was released at SATP, so

$$\Delta n(H_2) = \frac{0.111\,dm^3}{24.8\,dm^3\,mol^{-1}} = 0.00448\,mol.$$

To determine the change in concentration, use $\Delta c = \dfrac{\Delta n}{V}$, where V is the volume of the reaction mixture:

$$\Delta c(H_2) = \frac{0.00448\,mol}{0.100\,dm^3} = 0.0448\,mol\,dm^{-3}$$

Note that molecular hydrogen is a gas while the reaction mixture is an aqueous solution. Therefore, the value of $\Delta c(H_2)$ is not the actual concentration of $H_2(g)$ in the solution but only the ratio that shows the amount of the product

released from a unit volume of the solution. An alternative approach could involve the concentration change of another species, such as $HCl(aq)$ or $MgCl_2(aq)$, which remains in the aqueous solution. That approach will be demonstrated in worked example 3.

Finally, use $v = \dfrac{|\Delta c|}{\Delta t}$ to calculate the rate of reaction:

$$v = \frac{0.0448\,mol\,dm^{-3}}{10\,s} = 0.00448\,mol\,dm^{-3}\,s^{-1}$$

ii. According to table 1, the gas volume between 90 and 100 s remained constant, so the reaction was complete at 90 s. Over the same period, $248\,cm^3$ $(0.248\,dm^3)$ of hydrogen was released, so:

$$\Delta n(H_2) = \frac{0.248\,dm^3}{24.8\,dm^3\,mol^{-1}} = 0.0100\,mol$$

$$\Delta c(H_2) = \frac{0.0100\,mol}{0.100\,dm^3} = 0.100\,mol\,dm^{-3}$$

$$v(H_2) = \frac{0.100\,mol\,dm^{-3}}{90\,s} = 0.00111\,mol\,dm^{-3}\,s^{-1}$$

b. The stoichiometric coefficient of $H_2(g)$ is 1, so the overall reaction rate is also $0.00111\,mol\,dm^{-3}\,s^{-1}$.

Data-based questions

Chalk (calcium carbonate) reacts with hydrochloric acid to release carbon dioxide:

$$CaCO_3(s) + 2HCl(aq) \rightarrow CaCl_2(aq) + CO_2(g) + H_2O(l)$$

A conical flask was charged with 50.0 cm³ of 0.500 mol dm⁻³ hydrochloric acid and powdered chalk, plugged with cotton wool to prevent droplets of water from escaping but let the gas through (figure 3), placed on a digital balance and tared. The balance readings were recorded every 30 s using a data logger (table 2).

t/s	$\Delta m/g$
0	0
30	−0.161
60	−0.262
90	−0.326
120	−0.368
150	−0.396
180	−0.415
210	−0.428
240	−0.437
270	−0.440
300	−0.440

◀ Table 2 Balance readings for the flask with the reaction mixture

Calculate the rate of reaction, in mol dm⁻³ s⁻¹, for the following time intervals:

a. between 0 and 30 s

b. between 0 s and the moment when the reaction was complete.

Assume that the solution volume did not change during the reaction.

◀ Figure 3 Monitoring the loss of mass of the reaction mixture

Instantaneous reaction rate

If you examine table 1, you will notice that the rate of the reaction between magnesium metal and hydrochloric acid changes with time. Indeed, the volume of hydrogen released during the first 10 s of the reaction (111 cm³) was more than 50 times greater than that produced during the last 10 s (2 cm³). Therefore, the average reaction rate gives us only a general idea of how fast (or slow) the reaction proceeds over a period of time but tells us nothing about the reaction rate at any given moment.

To obtain this information, we need to introduce the concept of **instantaneous reaction rate**, v_{inst}, which is defined as the concentration change (dc) over an infinitesimally small period of time (dt):

$$v_{inst} = \frac{|dc|}{dt}$$

Like the average rate of reaction, instantaneous reaction rate cannot be negative, so it must be calculated using absolute values of concentration changes. If an instantaneous rate is measured with respect to a particular substance, it must be divided by the stoichiometric coefficient of that substance to give the overall reaction rate. The **initial reaction rate**, v_{init}, is the instantaneous rate measured at $t = 0$. Both initial and instantaneous rates have the same units (typically mol dm⁻³ s⁻¹) as average reaction rate.

The instantaneous reaction rate, v_{inst}, at a given time, t, can be determined by plotting the concentration of a reactant or product against time. If you draw a **tangent line** to the curve at t, the **gradient** of the tangent line will be numerically equal to v_{inst} at t. Similarly, the initial reaction rate, v_{init}, can be determined by measuring the slope of the tangent line at $t = 0$ s.

Worked example 3

Determine the initial rate, in $mol\,dm^{-3}\,s^{-1}$, and the instantaneous rate, in the same units, at $t = 55\,s$ for the reaction of hydrochloric acid with magnesium metal in worked example 2.

Solution

First, we need to find concentrations of hydrochloric acid every 10 seconds:

- At $t = 0\,s$, the initial concentration of the acid ($0.250\,mol\,dm^{-3}$) is given in the problem.

- At $t = 10\,s$, $\Delta n(H_2)$ was calculated to be $0.00448\,mol$ in worked example 2. The mole ratio of $H_2(g)$ to $HCl(aq)$ is $1:2$, so $\Delta n(HCl)$ at $t = 10\,s$ is $0.00448\,mol \times 2 = 0.00896\,mol$. Therefore:

$$\Delta c(HCl) = \frac{0.00896\,mol}{0.100\,dm^3}$$
$$= 0.0896\,mol\,dm^{-3}$$

To find the concentration of HCl at $t = 10\,s$, subtract the change in concentration of HCl from the initial concentration:

$$c(HCl) = 0.250 - 0.0896 \approx 0.160\,mol\,dm^{-3}$$

All other concentrations up to $t = 90\,s$ can be calculated in the same way (table 3).

Now we can plot these concentrations against time and draw tangent lines to the curve at $t = 0\,s$ and $t = 55\,s$ (figure 4).

To determine the gradient, we can select any two points on each tangent line and divide the difference in their y-coordinates (Δc) by the difference in their x-coordinates (Δt).

$$gradient = \frac{y_2 - y_1}{x_2 - x_1}$$

For the tangent at $t = 0\,s$, the most obvious point is $(0, 0.250)$ while the second point could be, for example, $(14, 0.100)$. In this case:

$$gradient(0) = \frac{0.100 - 0.250}{14 - 0} = -0.0107\,mol\,dm^{-3}\,s^{-1}$$

Since the stoichiometric coefficient before HCl(aq) is 2, the overall initial rate of the reaction will be half the absolute slope value:

$$v_{init} = \frac{|gradient(0)|}{2} = \frac{|-0.0107|}{2} = 0.0054\,mol\,dm^{-3}\,s^{-1}$$

For the tangent at $t = 55\,s$, two possible points are $(32, 0.085)$ and $(78, 0.045)$, so:

$$gradient(55) = \frac{0.045 - 0.085}{78 - 32} \approx -8.7 \times 10^{-4}\,mol\,dm^{-3}\,s^{-1}$$

$$v_{inst}(55) = \frac{|gradient(55)|}{2} = \frac{|-8.7 \times 10^{-4}|}{2}$$
$$\approx 4.4 \times 10^{-4}\,mol\,dm^{-3}\,s^{-1}$$

$t\,/\,s$	0	10	20	30	40	50	60	70	80	90
$c(HCl)$ $/\,mol\,dm^{-3}$	0.250	0.160	0.118	0.095	0.081	0.070	0.061	0.056	0.052	0.050

▲ Table 3 Concentration of hydrochloric acid in the reaction with magnesium metal

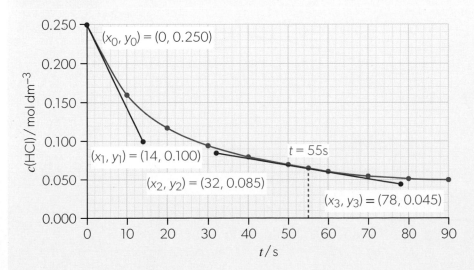

◀ Figure 4 Determining initial and instantaneous reaction rates from concentration changes

Tips for plotting experimental data as a graph, drawing tangent lines and determining the gradient can be found in the *Skills* chapter.

 Activity

For all the values of t listed in table 2, calculate the concentrations of calcium chloride in the reaction mixture and plot these concentrations against time on a graph paper. Using the tangent line method, determine the initial reaction rate, in $mol\,dm^{-3}\,s^{-1}$, and the instantaneous reaction rate, in the same units, at $t = 150\,s$.

By definition, reaction rates should always be expressed through the changes in concentration of reactants and/or products. However, reaction rates are often expressed through changes in volume, mass or other parameters that can be measured directly in the course of the experiment. In such cases, the units of the calculated reaction rate are defined by the chosen parameter.

Worked example 4

a. Determine the initial rate, in $cm^3\,s^{-1}$, for the reaction of hydrochloric acid with magnesium metal in worked example 2.

b. Use your answer to part **a** to calculate the overall initial rate in $mol\,dm^{-3}\,s^{-1}$.

Solution

a. The units $cm^3\,s^{-1}$ suggest that the rate should be expressed through the volume of hydrogen gas from table 1. The plot of $V(H_2)$ against time is shown in figure 5.

To find the instantaneous rate, we need to draw a tangent line to the curve at $t = 0\,s$ and select any two points on that line, for example, (0, 0) and (12, 160). In this case:

$$gradient(0) = \frac{160 - 0}{12 - 0} \approx 13.3\ cm^3\,s^{-1}$$

The stoichiometric coefficient before $H_2(g)$ is 1, so the initial reaction rate is numerically equal to the absolute slope value:

$$v_{init} = |gradient(0)| = 13.3\,cm^3\,s^{-1}$$

b. Initially, the volume of hydrogen gas released per second from $0.100\,dm^3$ of the reaction mixture was $13.3\,cm^3$, or $0.0133\,dm^3$. Therefore:

$$\Delta n(H_2) = \frac{V(H_2)}{V_m} = \frac{0.0133\,dm^3}{24.8\,dm^3\,mol^{-1}} \approx 0.00054\,mol$$

$$\Delta c = \frac{\Delta n}{V(solution)} = \frac{0.00054\,mol}{0.100\,dm^3} = 0.0054\,mol\,dm^{-3}$$

$$v_{init} = \frac{|\Delta c|}{\Delta t} = \frac{0.0054\,mol\,dm^{-3}}{1\,s} = 0.0054\,mol\,dm^{-3}\,s^{-1}$$

Note that the v_{init} value for the overall reaction is the same as that obtained in worked example 3.

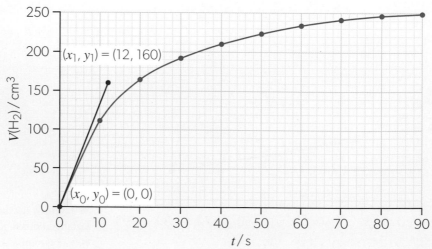

▲ **Figure 5** Determining initial and instantaneous reaction rates from volume changes

Linking questions

Concentration changes in reactions are not usually measured directly. What methods are used to provide data to determine the rate of reactions? (Tool 1, 3, Inquiry 2)

What experiments measuring reaction rates might use time as i) a dependent variable ii) an independent variable? (Tool 1)

Practice questions

2. a. Using the data from table 2, plot on a graph paper the change in mass of the reaction mixture as a function of time.

    b. Using the tangent line method, determine the initial reaction rate in $g\,s^{-1}$.

    c. Use your answer to part (b) to calculate the overall initial rate in $mol\,dm^{-3}\,s^{-1}$.

The collision theory (*Reactivity 2.2.2*)

The **kinetic molecular theory** (*Structure 1.1*) describes an ideal gas as a collection of randomly moving particles that collide elastically. Elastic collisions do not affect the total energy of gas particles, which means that an ideal gas is chemically inert and therefore has a constant molecular composition.

The kinetic molecular theory also states that the average **kinetic energy**, $\bar{E}_k$, of gas particles is proportional to the absolute temperature of the gas. At absolute zero (0 K), the movement of all particles slows down to a point at which they cannot transfer energy to the surroundings and so they cannot get any colder. When the temperature increases, the particles move faster and exchange energy with one another through collisions. If we know the temperature of an ideal gas in kelvin, we can always calculate the average kinetic energy of the gas particles. However, such calculations are not required in DP chemistry.

The **collision theory** expands the kinetic molecular theory by allowing the particles to collide non-elastically and undergo chemical changes. In contrast to the kinetic molecular theory, this theory is not limited to gases and can be applied to reacting species in any state of matter.

According to the collision theory, the **kinetic energy**, E_k, of most particles is insufficient for breaking chemical bonds. Such particles bounce off each other like billiard balls and fly in opposite directions. These are called **unsuccessful collisions**, as they do not affect the chemical nature of the colliding particles.

Collisions redistribute kinetic energy between particles unequally, so some particles accelerate while others slow down. A collision between two fast-moving particles might be violent enough to break or rearrange chemical bonds and transform the reactants into products. Such **successful collisions** lead to chemical changes.

In addition to kinetic energy, the orientation of colliding particles is also important. As an example, consider the following exchange reaction:

$$AB + CD \rightarrow AD + BC$$

For a collision to be successful, existing covalent bonds A–B and C–D must be broken, and new covalent bonds A–D and B–C must be formed. This is only possible if atom A comes into close proximity with atom D, and atom B comes into close proximity with atom C (figure 6, top). Collisions in other orientations will be unsuccessful, even if the kinetic energies of molecules AB and CD are high enough for the reaction to take place (figure 6, bottom).

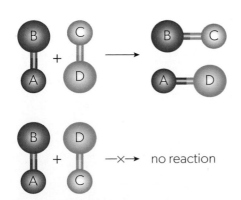

▲ Figure 6 Favourable (top) and unfavourable (bottom) orientations of colliding molecules

Linking question

What is the relationship between the kinetic molecular theory and collision theory? (Structure 1.1)

Some reacting species, such as individual atoms and monatomic ions, are symmetrical, so the results of their collisions do not depend on orientation. Chemical reactions between these species tend to proceed faster than reactions involving large and complex molecules, where the orientation of reactants is particularly important.

In summary, for a chemical reaction to occur, the following conditions must be satisfied:

1.   Two (or more) particles must collide with each other.

2.   The colliding particles must have the correct mutual orientation.

3.   The sum of kinetic energies of the particles must be sufficient to initiate the reaction.

Factors affecting reaction rates (*Reactivity 2.2.3*)

It follows from collision theory that the rate of a chemical reaction is proportional to the frequency of successful collisions in a given volume of the reaction mixture. Any change in reaction conditions that affects the number of collisions per second, or the average kinetic energy of colliding particles, will also affect the reaction rate.

An increase in **concentration** of a reactant leads to more frequent collisions between the particles of that reactant and other species, so the reaction rate increases (figure 7). Conversely, a decrease in concentration of any reactant slows down the reaction.

▶ **Figure 7** Effect of concentration on the frequency of collisions

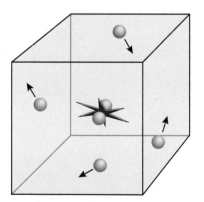 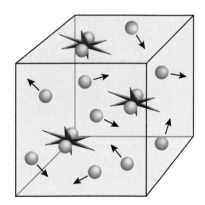

low concentration   high concentration

Pressure affects the rates of reactions with gaseous reactants in the same way as concentration: an increase in pressure increases the reaction rate, while a decrease in pressure has the opposite effect. Unlike gases, liquids and solids are almost incompressible, so pressure has no effect on reactions that do not involve gaseous reactants.

In **heterogeneous mixtures**, the collisions between reactant molecules are possible only at the surface where the different **phases** meet. A phase is an individual substance or mixture that has uniform chemical and physical properties. Although the term "phase" is often used as a synonym for "state of matter", immiscible liquids can form two or more separate phases with different chemical compositions.

Similarly, each solid substance in a heterogeneous mixture is a separate phase. Therefore, the rates of heterogeneous reactions depend on the **surface area** of reacting species. When a solid reactant is broken down into smaller pieces, its surface area increases, and so does the reaction rate (figure 8).

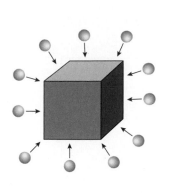 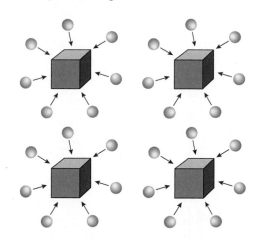

◄ **Figure 8** Effect of surface area on the frequency of collisions (left: one large piece of a solid reactant, smaller surface area, less frequent collisions, lower reaction rate; right: several smaller pieces of the solid reactant, larger surface area, more frequent collisions, higher reaction rate)

Heterogeneous reactions often proceed more slowly than **homogeneous reactions**, in which the reaction mixture is a single phase. However, when the surface area of reactants is extremely large, heterogeneous reactions can also be very fast. Combustion of powdered substances, such as flour or coal dust, can be very violent, which in the past led to many major explosions in grain mills and coal mines. Similarly, fine powders of active metals ignite spontaneously in air and react explosively with water and acids. For example, the reaction between hydrochloric acid and magnesium from worked example 2 would be complete within a few seconds if the magnesium ribbon were replaced with magnesium powder. When a reaction involves a liquid and a gas, the reaction rate increases when the gas is bubbled through the liquid rather than passed over the liquid surface.

Concentration, pressure and surface area affect reaction rates by increasing or decreasing the frequency of collisions between reacting species. However, none of these factors affects the kinetic energy of colliding particles, so the proportion of successful collisions remains the same. This proportion depends on the **temperature** of the reaction mixture.

When temperature increases, the average speed and kinetic energy of particles also increase. As particles move faster, they collide with one another more often. At the same time, a higher percentage of these collisions become successful, as a greater number of particles have sufficient kinetic energy for the reaction to occur (figure 9). Therefore, the rate of almost any reaction increases with temperature. Conversely, low temperature slows down almost all chemical and biochemical reactions.

(ATL) Thinking skills

Bacterial activity in food is slowed down at low temperatures, so refrigerated food remains fresh for much longer than that stored at room temperature. What other examples of chemical activity decreasing at low temperatures can you think of in the world around you?

◄ **Figure 9** Effect of temperature on the frequency of collisions (left: low temperature; right: high temperature)

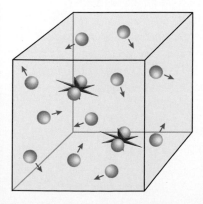 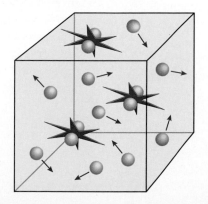

Activity

Hydrogen peroxide in aqueous solutions decomposes as follows:

$$2H_2O_2(aq) \rightarrow 2H_2O(l) + O_2(g)$$

The graph in figure 10 shows the volume of oxygen gas released in this reaction as a function of time.

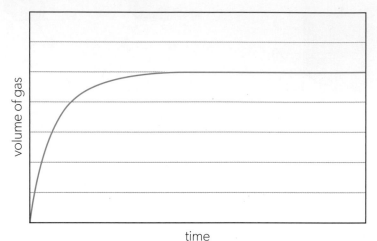

▲ **Figure 10** Volume of oxygen gas released by decomposition of aqueous hydrogen peroxide over time

Copy figure 10 and, on the same graph, sketch and label three curves for the following changes in experimental conditions:

- increased initial concentration of hydrogen peroxide

- lowered temperature of the reaction mixture

- increased atmospheric pressure.

(ATL) Self-management skills

Effective note-taking is a valuable skill. If done well, can help you follow and engage with a particular set of ideas, while also developing a useful set of notes to revise from. You may take notes during class discussion, while reading a textbook, or doing research, for example. One note-taking method, known as the Cornell method, is useful because it reminds you to write in your own words, as well as prompting you to come up with "cues" that identify key ideas and the connections with other areas of the subject. Typically, the Cornell method also includes summarizing the set of notes a short time after the notes were written. Effective note-taking is particularly important when learning concepts such as the ones covered in this chapter, which contain a large amount of interlinked ideas and key vocabulary.

On the internet, search for a reliable explanation of Cornell note-taking, then try it out. You might want to compare how effective this method is for taking notes in class compared to note-taking while you read a certain section in this book. Later, reflect on your experience and adjust your note-taking strategies as needed.

Investigating the effect of concentration on rate of reaction

Dilute hydrochloric acid reacts with calcium carbonate to produce carbon dioxide, water and calcium chloride:

$$2HCl(aq) + CaCO_3(s) \rightarrow CO_2(g) + H_2O(l) + CaCl_2(aq)$$

The rate of the reaction can be monitored by measuring the volume of gas produced over time.

Relevant skills

- Tool 3: Calculate mean
- Tool 3: Plot linear and non-linear graphs
- Tool 3: Draw lines or curves of best fit
- Inquiry 1: Identify independent, dependent and control variables
- Inquiry 2: Identify and justify the removal or inclusion of outliers in data
- Inquiry 2: Assess accuracy and precision
- Inquiry 3: Evaluate the implications of methodological weaknesses, limitations and assumptions on conclusions.
- Inquiry 3: Explain realistic and relevant improvements to an investigation.

A student investigating the effect of the concentration of acid on the rate of production of carbon dioxide, carried out this reaction at five different concentrations of acid using the apparatus shown below. The student recorded the time taken to collect 50 cm³ of gas in the measuring cylinder.

gas collecting in
measuring cylinder

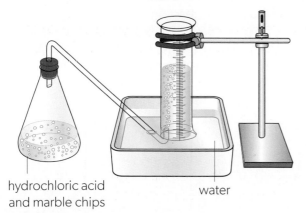

hydrochloric acid
and marble chips

water

▲ **Figure 11** Apparatus for investigating the effect of concentration on the rate of reaction

Calcium carbonate in the form of marble chips was added in large excess. The results are shown in table 4.

Concentration of HCl / mol dm⁻³	Time taken, t, to collect 50 cm³ of CO₂ / s		
	1	2	3
0.2	202	498	215
0.4	118	104	98
0.6	62	58	65
0.8	54	48	52
1.0	30	35	33

▲ Table 4 The student's results

Questions

1. Identify the independent and dependent variables in the investigation.

2. Describe two control variables that were, or should have been, kept constant.

3. Identify the outlier.

4. The rate of reaction is proportional to $\frac{1}{t}$. Calculate $\frac{1}{t}$ for each of the acid concentrations.

5. Using graph-plotting software, plot graphs to show:
   a. time taken, t, to collect 50 cm³ of CO₂ vs concentration of HCl
   b. $\frac{1}{t}$ vs concentration of HCl

6. Draw lines or curves of best fit on your graphs.

7. Explore the effect of including and excluding the outlier on the results. Decide whether it should be discarded from the data set. Justify your choice.

8. By referring to the graphs, comment on the relative impact of systematic and random errors on the outcome of the investigation.

9. Describe and explain three major sources of error in this investigation.

10. Suggest changes to the methodology that would help to eliminate or minimize the sources of error you have identified.

11. Describe and explain one extension which would build on the outcome of this investigation.

12. Outline how you could investigate the rate of the following reactions:
   a. magnesium and hydrochloric acid,
      $$Mg(s) + 2HCl(aq) \rightarrow MgCl_2(aq) + H_2(g)$$
   b. hydrochloric acid and sodium thiosulfate,
      $$2HCl(aq) + Na_2S_2O_3(aq) \rightarrow 2NaCl(aq) + H_2O(l) + SO_2(g) + S(s)$$
   c. bromine and methanoic acid,
      $$Br_2(aq) + HCOOH(aq) \rightarrow 2Br^-(aq) + 2H^+(aq) + CO_2(g)$$

13. Formulate a research question that investigates an aspect of the kinetics of a reaction of your choice. Outline your plan, referring to the Inquiry 1 skills.

Linking questions

What variables must be controlled in studying the effect of a factor on the rate of a reaction? (Tool 1)

How can graphs provide evidence of systematic and random error? (Tool 3, Inquiry 3)

Activation energy (*Reactivity 2.2.4*)

The minimum kinetic energy, E_k, of colliding particles required for the reaction to occur is known as the **activation energy**, E_a.

- When the sum of kinetic energies of colliding particles is less than the activation energy ($E_k < E_a$), the collision between these particles will always be unsuccessful, regardless of their mutual orientation.

- When $E_k \geq E_a$, the collision can be successful if the mutual orientation of the particles is correct.

The concept of activation energy can be illustrated by a simple analogy. Imagine a ball at the bottom of a hill (figure 12). On its own, the ball cannot reach the lower ground at the opposite side of the hill. However, if we spend some energy and push the ball over the top, it will roll down and eventually reach the lowest point. In this analogy, the hill height represents the activation energy, E_a, while the difference in elevation between the initial and final levels of the ball represents the reaction enthalpy, ΔH_r.

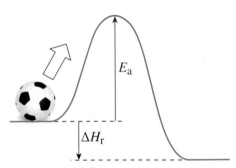

▲ Figure 12 An analogy of activation energy, E_a, and the reaction enthalpy, ΔH_r

The progress of a chemical reaction can be represented by its **energy profile** (figure 13). The **reaction coordinate** on the horizontal axis shows the general direction of the chemical change from reactants to products, while the vertical axis shows the energy at each stage of the reaction. For an exothermic process ($\Delta H_r < 0$), the reactants have a higher energy than the products (figure 13, left), so excess energy is released in the form of heat. In contrast, an endothermic process ($\Delta H_r > 0$) requires energy because the energy of the products is higher than that of the reactants (figure 13, right).

Enthalpy changes and thermal effects of chemical reactions are discussed in *Reactivity 1.1*.

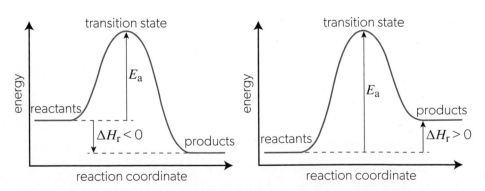

▲ Figure 13 Energy profiles for exothermic (left) and endothermic (right) reactions

The activation energy, E_a, is equal to the difference in energy between the reactants and the **transition state**. The transition state is the least stable arrangement of atoms in the reacting species. In this state, some of the chemical bonds in the reactants are broken or weakened while the new bonds have not been fully formed. This situation is very unfavourable, as the energy used for bond breaking is not compensated by the energy released by bond forming. Therefore, the transition state has the highest energy along the reaction coordinate.

> The enthalpy changes associated with the breaking and formation of chemical bonds are discussed in *Reactivity 1.2*. The structures of transition states for certain reaction types will be discussed in *Reactivity 3.4*.

Practice questions

3. Consider the following reaction that proceeds in a single step:

$$A + B \rightarrow C + D \qquad \Delta H_r = -10\,kJ\,mol^{-1}$$

The activation energy, E_a, for this reaction is $20\,kJ\,mol^{-1}$.

a. Sketch the energy profile for this reaction.

b. Deduce the activation energy, in $kJ\,mol^{-1}$, for the reaction $C + D \rightarrow A + B$ that has the same transition state as the forward reaction.

The effects of temperature on reaction rates can be explained by analysing the distribution of kinetic energies (E_k) of reacting species. According to the kinetic molecular theory, the average kinetic energy, $\bar{E}_k$, of gas particles is proportional to the absolute temperature of the gas. The particles move randomly and collide with one another, so their kinetic energies change constantly. Although it is not possible to predict the E_k value for each individual particle, we can plot the statistical distributions of these energies at various temperatures (figure 14). These plots, known as **Maxwell–Boltzmann energy distribution curves**, show the probability of finding a particle with a certain kinetic energy against that energy. The maximum of the curve shows the kinetic energy of the highest proportion of particles, and the average kinetic energy of all the particles, $\bar{E}_k$, is slightly to the right of this maximum. The area under the curve represents the total number of particles in the sample.

At moderate temperatures (T_1), the majority of particles are unreactive, as their kinetic energies are lower than the activation energy. As the temperature increases to T_2, the Maxwell–Boltzmann curve flattens and broadens, so its maximum shifts to the right. As a result, the proportion of reactive particles with $E_k > E_a$ increases, and so does the reaction rate.

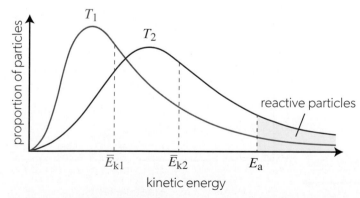

reactive particles

▲ Figure 14 Maxwell–Boltzmann energy distribution curves for gas particles at two temperatures, where $T_2 > T_1$

Activity

Copy the plot shown in figure 14 and sketch a third Maxwell–Boltzmann curve at a temperature T_3, where $T_3 < T_1$. Outline the main features of the new curve (the position and height of its maximum). Predict, with a reference to E_k and E_a, whether the proportion of reactive particles at T_3 will be lower, higher or the same as that at T_1.

Although the Maxwell–Boltzmann distribution of kinetic energies applies to gaseous particles only, the general relationship between reaction rate and temperature remains valid for substances in any state of matter. For most reactions at moderate temperatures (0–100 °C), the rate increases two to four times when the temperature rises by 10 °C.

Catalysts (*Reactivity 2.2.5*)

In addition to raising the temperature, the proportion of successful collisions in a reaction mixture can be increased using a **catalyst**. A catalyst is a substance that increases the rate of a chemical reaction but is not consumed in that reaction. Typically, a catalyst reacts with a reactant to form an unstable compound known as an **intermediate**. An intermediate can exist in a reaction mixture for a certain period of time. In contrast, a transition state is a particular configuration of atoms along the reaction pathway that cannot exist for any prolonged period of time.

A useful analogy of a reaction intermediate is a pencil standing upright on its flat end. Although its balance is precarious, the pencil will not fall until someone taps on the table or disturbs the air in the room. In contrast, a transition state is like a pencil standing upright on its sharpened tip—the pencil will fall at once, even without a disturbance.

The intermediate undergoes further chemical changes and eventually forms the reaction product, releasing the catalyst in an unchanged form. Therefore, the catalyst itself is both a reactant and product of the same reaction.

By providing an alternative pathway for a reaction, a catalyst reduces the activation energy of that reaction (figure 15). A catalysed reaction usually proceeds in several steps and involves at least one intermediate and two or more transition states. Each step requires less energy than the uncatalysed reaction ($E_a' < E_a$), so a greater proportion of collisions between reacting particles are successful.

The overall enthalpy change for the reaction, ΔH_r, is the same with and without a catalyst.

The effect of a catalyst on the reaction rate can also be explained using a Maxwell–Boltzmann distribution curve (figure 16). A lower activation energy means that more particles have enough energy to react with one another, so the reaction rate increases.

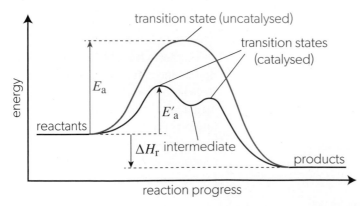

▲ Figure 15 Energy profiles for uncatalysed and catalysed reactions

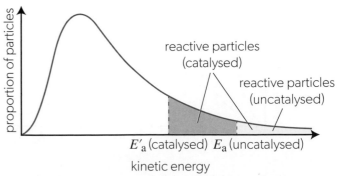

▲ Figure 16 Maxwell–Boltzmann energy distribution curves for uncatalysed and catalysed reactions

Both temperature and catalysts affect reaction rates by altering the proportion of particles with $E_k > E_a$. A rise in temperature does not affect E_a but increases the average E_k. In contrast, a catalyst does not affect the average E_k but reduces E_a. In both cases, the frequency of successful collisions increases, and so does the reaction rate.

Practice questions

4. Sulfur(IV) oxide reacts with molecular oxygen as follows:

$$2SO_2(g) + O_2(g) \rightarrow 2SO_3(g) \qquad \Delta H_r < 0$$

In the presence of nitrogen(IV) oxide, the rate of this reaction increases up to 10,000 times owing to the following two-step process:

$$SO_2(g) + NO_2(g) \rightarrow SO_3(g) + NO(g)$$

$$NO(g) + 0.5O_2(g) \rightarrow NO_2(g)$$

a. State the roles of the nitrogen oxides in this process and explain their effects on the reaction rate.

b. Sketch an energy profile showing the uncatalysed and catalysed reactions, labelling the reactants, products and any intermediates.

Biochemical reactions are catalysed by proteins called **enzymes**. Without enzyme catalysis, many processes in living organisms would be impossible, and life itself would not exist in its present form. The efficiency of enzymes greatly exceeds the catalytic power of synthetic catalysts. Some enzymes can accelerate reactions as much as 10^{16} times, so chemical transformations that would normally take millions of years proceed within milliseconds. At the same time, every enzyme is very specific and will usually catalyse only one or very few chemical reactions.

(ATL) Research skills

Catalysis is one of the green chemistry principles, and a focus of much chemical research. Catalysts can help to lower the environmental impact of a process by reducing its energetic demands, and in some cases, decreasing the amount of waste. For example, the now shorter and more efficient synthesis of the anti-inflammatory drug ibuprofen is due to the use of catalysts. Several chemistry Nobel Prizes have been awarded for work relating to catalysis.

Choose a catalyst to research, for example:

- iron in the production of ammonia
- transition elements in catalytic converters
- zeolites in catalytic cracking
- lactase in the digestion of lactose
- organocatalysts, such as proline, in asymmetric aldol reactions.

Prepare a five-minute explanation for a non-scientist, outlining:

1. The specific problem or issue the catalyst helps to address.
2. How the catalyst helps to address the problem.
3. Social, economic and/or environmental impacts related to the use of the catalyst.

Activity

State the effects of various factors on the reaction rate by copying and completing the table below. In each case, assume that all other reaction conditions remain unchanged. Some cells are already filled as examples.

Factor	Frequency of collisions	Average E_k of reacting species	Activation energy	Rate of forward reaction
decrease in a reactant concentration				
increase in pressure of a gaseous reactant	increases			
decrease in pressure of a gaseous product		no change		
increase in volume of a gaseous reaction mixture			no change	
decrease in temperature				decreases
increase in surface area of a solid reactant				
addition of a catalyst				

 Linking questions

What is the relative effect of a catalyst on the rate of the forward and backward reactions? (Reactivity 2.3)

 AHL What are the features of transition elements that make them useful as catalysts? (Structure 3.1)

Multistep reactions (*Reactivity 2.2.6, 2.2.7 and 2.2.8*)

As mentioned earlier, an intermediate is an unstable compound or other molecular species that can exist in a reaction mixture for a certain period of time. In contrast, a transition state is a particular configuration of atoms along the reaction pathway that cannot exist for any prolonged period of time.

Many reactions involve a series of chemical changes across multiple intermediates and transition states. The sequence of these changes along the reaction pathway is known as the **reaction mechanism**. A reaction mechanism involves one or more **elementary steps**. A reaction mechanism with more than one elementary step is called a **multistep reaction**.

In each elementary step, one, two or (very rarely) three molecular species undergo a chemical change through a single transition state and no intermediates. The number of species (atoms, molecules or ions) involved in an elementary step defines its **molecularity**: a **unimolecular** step involves a single species, a **bimolecular** step involves two colliding species, and a **termolecular** step involves three species colliding at the same time.

AHL

Reactivity 2.2 How fast? The rate of chemical change

AHL

Termolecular reactions are very rare because three particles are very unlikely to collide with each other at exactly the same time and in the correct orientation. This situation is similar to snooker (figure 17), in which two-ball collisions occur all the time, while three-ball collisions are uncommon.

Each elementary step is characterized by its own activation energy, E_a. The slowest step, known as the **rate-determining step**, limits the overall rate of reaction for any given concentrations of the reactants. Typically, the rate-determining step has the highest E_a among all other elementary steps of the reaction.

The concept of the rate-determining step can be illustrated by a simple analogy. Before leaving an international airport, passengers must go through passport control and collect their luggage. A long queue at the passport control desks might hold the passengers at the airport for a long time, even if the luggage has already been delivered. In contrast, if there is no queue at passport control but the luggage has not arrived yet, the rate at which the passengers leave the airport will depend on the rate at which the luggage will appear on the carousel. In the same way, the rate of the slowest elementary step determines the rate of a chemical reaction.

▲ Figure 17 You can think of balls on a snooker table as a model of colliding particles in elementary reactions

Reaction mechanisms, including intermediates, transition states and activation energy, can be represented by energy profiles. For a single-step reaction, the energy profile includes a single transition state and no intermediates. A two-step reaction typically involves one intermediate and two transition states. Use the following rules when sketching an energy profile diagram for multistep reactions:

1. The less stable the species, the higher it appears in the profile.

2. Reactants are less stable than products if the reaction is exothermic ($\Delta H_r < 0$) and more stable than products if the reaction is endothermic ($\Delta H_r > 0$).

3. An intermediate is less stable than both the reactants and products.

4. The rate-determining step usually has the highest E_a value.

5. The number of transition states is equal to the number of elementary steps.

6. The number of intermediates is equal to the number of elementary steps minus one.

7. An annotated diagram must have the labels for reactants, products, intermediate(s), transition states and activation energies.

8. The y-axis should be labelled "energy" and the x-axis should be labelled "reaction coordinate" or "progress of reaction".

TOK

The reaction mechanism is a hypothesis about the sequence of events that transform the reactants into products. A correct hypothesis must not contradict the experimental data. If two or more alternative mechanisms are compatible with the data, further studies of the reaction kinetics are required. Hypotheses are falsifiable by any evidence that contradicts them. However, no amount of evidence can ever prove a hypothesis to be true with complete certitude. To what extent are some things unknowable?

Worked example 5

Consider the following reaction:

$$NO_2(g) + CO(g) \rightarrow NO(g) + CO_2(g)$$

At low temperatures, the reaction proceeds in two steps.

| step 1 | $NO_2(g) + NO_2(g) \rightarrow NO_3(g) + NO(g)$ |
| step 2 | $NO_3(g) + CO(g) \rightarrow NO_2(g) + CO_2(g)$ |

The concentration of CO(g) does not affect the rate of reaction.

    a.   Deduce the rate-determining step for this reaction at low temperatures.

    b.   Identify the reaction intermediate at low temperatures.

The overall reaction has a ΔH_r value of $-126\,kJ\,mol^{-1}$. At high temperatures, the reaction proceeds in a single step.

    c.   Using the kinetics data, sketch the energy profiles for this reaction at low and high temperatures.

Solution

a.   Carbon monoxide, CO(g), participates only in the second step of the reaction. If this step were the rate-determining step (RDS), any change in the concentration of CO(g) would affect the frequency of successful collisions of the reactants and therefore the overall rate of reaction. However, the experimental data show that the concentration of CO(g) has no effect on the reaction rate. Therefore, the second step is fast while the first step is slow (RDS).

b.   The reaction intermediate is $NO_3(g)$, which is formed on the first step and consumed on the second step. As any intermediate, it does not appear in the overall equation for the reaction.

c.   At low temperatures, the reaction proceeds in two steps, so it has a single intermediate, $NO_3(g)$, and two transition states. The reaction is exothermic ($\Delta H_r < 0$), so the products will appear lower in the energy profile. The intermediate is likely to be less stable than both the reactants and products (otherwise, the reaction would probably stop after the first step). Finally, the activation energy of the rate-determining first step is likely to be higher than that of the second (fast) step. This is summarized in figure 18.

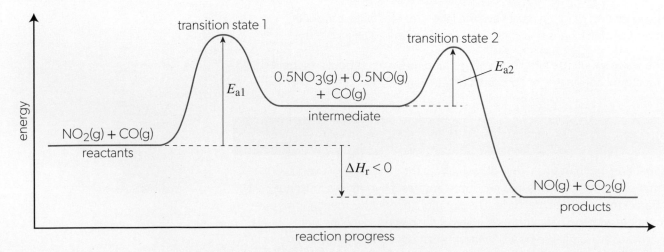

▲ **Figure 18** Energy profile for the reaction at low temperatures

Reactivity 2.2 How fast? The rate of chemical change

AHL

At high temperatures, the reaction has a single transition state and no intermediates. The relative energies of the reactants and products remain unchanged (figure 19).

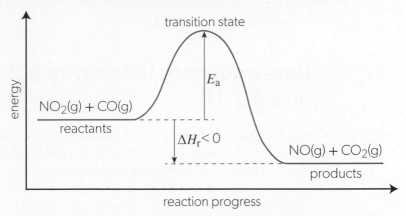

▲ Figure 19 Energy profile for the reaction at high temperatures

Evidence

Scientists make provisional explanations, known as hypotheses, using the patterns that emerge from their observations. Results from further experimentation can then refute or support these hypotheses.

Until recently, many reaction intermediates and all transition states were hypothetical species that could not observed directly. The development of femtosecond lasers (figure 20) allowed scientists to detect these species and analyse their structures, effectively observing chemical reactions at the timescales on which they occur.

▲ Figure 20 A femtosecond (1 fs = 10^{-15} s) laser is used for studying reaction intermediates and transition states

These studies were pioneered by the Egyptian chemist Ahmed Zewail, the "father of femtochemistry". He was awarded the 1999 Nobel Prize in Chemistry for his works in the field of chemical kinetics.

Linking question

Which mechanism in the hydrolysis of halogenoalkanes involves an intermediate? (Reactivity 3.4)

Rate equations (*Reactivity 2.2.9, 2.2.10 and 2.2.11*)

You already know that reaction rates depend on the reactant concentrations. Generally, an increase in concentration of a particular reactant by a certain factor increases the reaction rate by a power of that factor. For most reactions, this power has a small integer value ranging from 0 to 2. In other words, for the reaction $A + B \rightarrow C$, the rate ν depends on the concentrations of reactants A and B as follows:

$$\nu \propto [A]^n$$

$$\nu \propto [B]^m$$

where $n, m = 0, 1$ or 2. Notice that the reaction rate does *not* depend on the concentrations of products of that reaction.

The expressions describing the proportionality of the rate to the reactant concentrations can be combined into a single equation, known as the **rate equation**:

$$\nu = k[A]^n[B]^m$$

where $n, m = 0, 1$ or 2 and k is the **rate constant**. The value of the rate constant is specific for each reaction, and its units depend on the values of n and m. The integer exponents n and m are known as the **reaction orders** with respect to the individual reactants A and B. The sum of these exponents, $n + m$, is the **overall reaction order**.

Molar concentrations are often denoted by square brackets around the substance formulas or names. For example, the molar concentration of reactant A can be denoted as either $c(A)$ or $[A]$. Similarly, a concentration raised to the nth power can be written as $(c(A))^n$ or $[A]^n$. The second expression takes less space, so it is used more commonly.

In many cases, the rate equation for a multistep reaction can be determined from the knowledge of the reaction mechanism and the rate-determining step. Conversely, the reaction mechanism can be evaluated using the experimentally determined rate equation for that reaction.

Reactivity 2.2 How fast? The rate of chemical change

AHL

Worked example 6

Consider the following reaction from worked example 5:

$$NO_2(g) + CO(g) \rightarrow NO(g) + CO_2(g)$$

You saw that at high temperatures, the reaction proceeds in a single step, and at low temperatures, the reaction proceeds in two steps:

| step 1 | $NO_2(g) + NO_2(g) \rightarrow NO_3(g) + NO(g)$ |
| step 2 | $NO_3(g) + CO(g) \rightarrow NO_2(g) + CO_2(g)$ |

a. Deduce the rate equation for this reaction at low temperatures.

b. Determine the order of the reaction with respect to $NO_2(g)$ and $CO(g)$, and the overall order of the reaction, at low temperatures.

c. Deduce the rate equation for this reaction at high temperatures.

d. Determine the order of the reaction with respect to $NO_2(g)$ and $CO(g)$, and the overall order of the reaction, at high temperatures.

Solution

a. In worked example 5, we determined that step 1 was the rate-determining step at low temperatures. Therefore, the reaction rate depends on the frequency of collisions between two $NO_2(g)$ molecules, which is proportional to the square of $[NO_2]$:

$$v = k[NO_2]^2$$

b. The reaction is second order ($n = 2$) with respect to nitrogen dioxide, $NO_2(g)$, and zero order ($m = 0$) with respect to carbon monoxide, $CO(g)$. Remember that any number raised to the zero power equals one, so $[CO]^0 = 1$, and therefore the concentration of carbon monoxide does not appear in the rate equation. Overall, this is a second order reaction ($n + m = 2$).

c. If the reaction proceeds in a single step, then the rate of this bimolecular reaction depends on the frequency of collisions between $NO_2(g)$ and $CO(g)$ molecules, which is proportional to both $[NO_2]$ and $[CO]$:

$$v = k[NO_2][CO]$$

d. The reaction is first order ($n = 1$) with respect to nitrogen dioxide, $NO_2(g)$, and first order ($m = 1$) with respect to carbon monoxide, $CO(g)$. Overall, this is a second order reaction ($n + m = 2$).

It is important to understand that reaction orders depend on the mechanism of the reaction and **cannot** be deduced from the mole ratio of the reactants. In the general case, the reaction $aA + bB \rightarrow cC + dD$ has the rate equation $v = [A]^n[B]^m$, in which n and m might or might not be equal to a and b, respectively.

The orders of reactants, and therefore the exponents in the rate equation can be determined experimentally by measuring the effect that changing the concentrations of the reactants will have on the initial rate of reaction.

For example, consider the hypothetical reaction:

$$A + B + C \rightarrow D$$

Imagine that data showed that doubling the concentration of A doubles the initial rate of reaction. This means that the rate is proportional to the concentration of A, or $v \propto [A]$. Therefore, the reaction is first order with respect to A.

Doubling the concentration of B quadruples the initial rate. This means that the rate is proportional to the square of the concentration of B, or $v \propto [B]^2$. Therefore, the reaction is second order with respect to B.

Doubling the concentration of C has no effect on the initial rate, so the reaction is zero order with respect to C. Therefore, the overall rate equation would be as follows:

$$v = k[A][B]^2$$

The overall reaction order is the sum of the exponents, $1 + 2 = 3$, so the reaction is third order overall. Notice that the exponent of [A] is 1, so it is not shown in the rate equation, but it contributes to the overall reaction order.

Once the exponents have been determined, the value of k can be calculated by substituting known values of concentrations and initial rate into the rate equation. The temperature in all experiments must be kept constant because the rate constant, k, is temperature-dependent.

Worked example 7

Iron(III) cations are reduced by iodide anions as follows:

$$2Fe^{3+}(aq) + 2I^-(aq) \rightarrow 2Fe^{2+}(aq) + I_2(aq)$$

The experimental data obtained for this reaction at 298 K are shown in table 5.

a. Deduce the reaction orders with respect to $Fe^{3+}(aq)$ and $I^-(aq)$.

b. State the rate equation and the overall reaction order.

c. Calculate the value of the rate constant at 298 K and state its units.

d. Determine the initial reaction rate at 298 K and $[Fe^{3+}] = [I^-] = 3.00 \times 10^{-2}$ mol dm^{-3}.

Experiment	$[Fe^{3+}]$ / mol dm^{-3}	$[I^-]$ / mol dm^{-3}	v_{init} / mol dm^{-3} s^{-1}
1	1.00×10^{-2}	1.00×10^{-2}	1.62×10^{-5}
2	2.00×10^{-2}	1.00×10^{-2}	3.24×10^{-5}
3	2.00×10^{-2}	2.00×10^{-2}	1.30×10^{-4}

▲ Table 5 Kinetics data for the reaction between iron(III) and iodide ions at 298 K. Source of data: G. S. Laurence and K. J. Ellis, J. Chem. Soc., Dalton Trans., 1972, pp. 2229–2233

Solution

a. The concentration of iodide ions in experiments 1 and 2 is the same (1.00×10^{-2} mol dm^{-3}), so any change in the reaction rate is caused by the change in the concentration of iron(III) ions. When $[Fe^{3+}]$ doubles (increases from 1.00×10^{-2} to 2.00×10^{-2} mol dm^{-3}), the rate also doubles (increases from 1.62×10^{-5} to 3.24×10^{-5} mol dm^{-3} s^{-1}). Therefore, the reaction is first order with respect to $Fe^{3+}(aq)$.

In experiments 2 and 3, the concentration of iron(III) is constant (2.00×10^{-2} mol dm^{-3}), so the rate is affected by the concentration of iodide ions only. When $[I^-]$ doubles (increases from 1.00×10^{-2} to 2.00×10^{-2} mol dm^{-3}), the rate increases four times (from 3.24×10^{-5} to 1.30×10^{-4} mol dm^{-3} s^{-1}). Therefore, the reaction is second order with respect to $I^-(aq)$.

b. Rate equation: $v = k[Fe^{3+}][I^-]^2$

Overall reaction order: $1 + 2 = 3$

c. The value of k does not depend on concentrations, so we can use data from any row in table 5. The first row gives the following:

$$k = \frac{v_{init}}{[Fe^{3+}][I^-]^2}$$

$$= \frac{1.62 \times 10^{-5} \text{ mol dm}^{-3} \text{ s}^{-1}}{(1.00 \times 10^{-2} \text{ mol dm}^{-3}) \times (1.00 \times 10^{-2} \text{ mol dm}^{-3})^2}$$

$$= 16.2 \text{ dm}^6 \text{ mol}^{-2} \text{ s}^{-1}$$

d. Substitute the values of k, $[Fe^{3+}]$, and $[I^-]$ into the rate equation:

$$v_{init} = k[Fe^{3+}][I^-]^2$$

$$= 16.2 \text{ dm}^6 \text{ mol}^{-2} \text{ s}^{-1} \times 3.00 \times 10^{-2} \text{ mol dm}^{-3} \times (3.00 \times 10^{-2} \text{ mol dm}^{-3})^2$$

$$= 4.37 \times 10^{-4} \text{ mol dm}^{-3} \text{ s}^{-1}$$

Reactivity 2.2 How fast? The rate of chemical change

AHL

Practice questions

5. Consider the following reaction and experimental data from table 6:

$$A(g) + 2B(g) \rightarrow C(g)$$

Experiment	[A]/mol dm^{-3}	[B]/mol dm^{-3}	v_{init}/mol dm^{-3} s^{-1}
1	0.020	0.015	2.0×10^{-4}
2	0.020	0.030	4.0×10^{-4}
3	0.030	0.015	3.0×10^{-4}

▲ Table 6 Kinetics data for the reaction between A(g) and B(g) at 298 K

a. Deduce the reaction orders with respect to A(g) and B(g).

b. State the rate equation and the overall reaction order.

c. Calculate the value of the rate constant at 298 K and state its units.

Graphical representations of reaction order

Orders of chemical reactions can be deduced from their kinetic curves. One type of kinetic curve, a **rate–concentration curve**, represents the relationship between rate and concentration in graphical form (figure 21). Each of these curves can be constructed from experimental data with five or more data points.

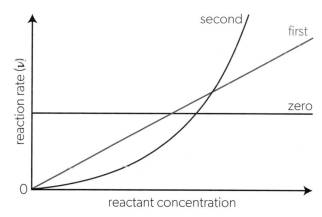

▲ Figure 21 The rate–concentration curves for zero, first and second order reactions

For a zero order reaction, the rate is independent of the reactant concentration, so the rate–concentration curve is a horizontal line that intersects the y-axis at $v = k$. For a first order reaction, the rate is proportional to the reactant concentration, so the curve is a straight line that begins at the origin and has a slope of k. For a second order reaction, the rate is proportional to $[A]^2$, so the curve is parabolic, and the k is numerically equal to the rate at $[A] = 1$ mol dm^{-3}. Interpreting rate–concentration curves is summarized in table 7.

Reaction order	Rate equation	Curve	k
0	$v = k$	horizontal line	y-axis intercept
1	$v = k[A]$	sloped line	slope
2	$v = k[A]^2$	parabola	v_{init} at $[A] = 1$ mol dm^{-3}

▲ Table 7 Interpretation of rate–concentration curves for reactant A

The construction of rate–concentration curves is a time-consuming process, as each curve requires a series of kinetic experiments using different concentrations of the reactant(s). In contrast, a **concentration–time curve** can be obtained from a single experiment, in which the concentration of a particular reactant is recorded at certain time intervals using a data logger (figure 22).

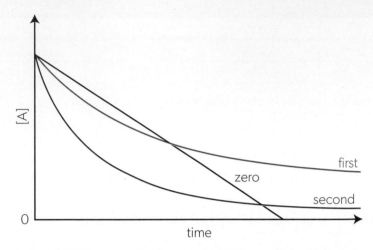

▲ **Figure 22** The concentration–time graph for zero, first and second order reactions

However, the interpretation of these curves is straightforward only in the case of zero order reactions, where the concentration–time plot is a straight line. For a zero order reaction, the y-axis intercept shows the initial concentration, $[A]_0$, and the line slope gives $-k$. The concentration–time curves for the first and second order reactions look similar, so they cannot be distinguished from sketches. These reactions can instead be represented by linear plots (figure 23), which allow you to determine the rate constant from the line slope.

Interpreting concentration–time curves is summarized in table 8.

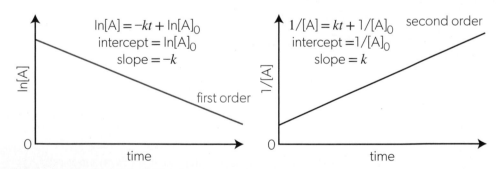

▲ **Figure 23** Linear plots for first order (left) and second order (right) reactions

Reaction order	Curve	Linear function		
		Equation	Slope	y-axis intercept
0	negatively sloped line	$[A] = -kt + [A]_0$	$-k$	$[A]_0$
1	exponential decay	$\ln[A] = -kt + \ln[A]_0$	$-k$	$\ln[A]_0$
2	hyperbolic decay (faster than exponential)	$\dfrac{1}{[A]} = kt + \dfrac{1}{[A]_0}$	k	$\dfrac{1}{[A]_0}$

▲ **Table 8** Interpretation of concentration–time curves for reactant A

Reactivity 2.2 How fast? The rate of chemical change

AHL

Worked example 8

Recall the reaction of magnesium metal with hydrochloric acid:

$$Mg(s) + 2HCl(aq) \rightarrow MgCl_2(aq) + H_2(g)$$

In worked example 3, the following concentrations for HCl(aq) were determined (table 9).

Using the first six data points, determine the order of this reaction with respect to hydrochloric acid.

t/s	0	10	20	30	40	50	60	70	80	90
$c(HCl)/$ $mol\,dm^{-3}$	0.250	0.160	0.118	0.095	0.081	0.070	0.061	0.056	0.052	0.050

▲ Table 9 Concentration of hydrochloric acid in the reaction with magnesium metal

Solution

The kinetic curve shown in figure 5 (worked example 4) may represent either a first or second order reaction. Therefore, we need to process the experimental data and plot the linear functions from table 8 for these two reaction types. The function that produces a linear plot will give us the reaction order.

The initial concentration of HCl(aq) is $0.250\,mol\,dm^{-3}$, so $\ln[HCl]_0 = -1.39$ and $\dfrac{1}{[HCl]_0} = 4.00\,dm^3\,mol^{-1}$. The logarithms and reciprocal values for other data points can be calculated in the same way (table 10).

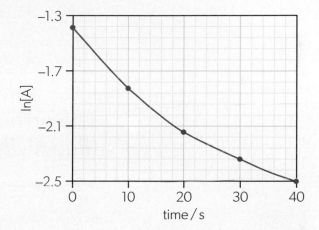

t/s	0	10	20	30	40	50
[HCl] $/mol\,dm^{-3}$	0.250	0.160	0.118	0.095	0.081	0.070
$\ln[HCl]$	−1.39	−1.83	−2.14	−2.35	−2.51	−2.66
$\dfrac{1}{[HCl]_0}$ $/dm^3\,mol^{-1}$	4.00	6.25	8.47	10.5	12.3	14.3

▲ Table 10 The first six data points for the concentration of hydrochloric acid in the reaction with magnesium metal

Now we can use the data from table 10 to plot $\ln[HCl]$ and $\dfrac{1}{[HCl]}$ against time (figure 24).

The second plot is linear, so the reaction is second order with respect to hydrochloric acid. Notice that the points in this plot deviate slightly from the straight line because of random errors in the experimental data.

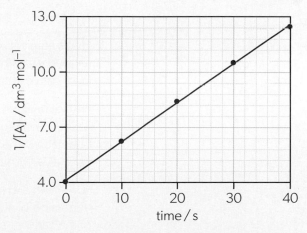

▲ Figure 24 Logarithmic (top) and reciprocal (bottom) concentration–time curves for the reaction of magnesium metal with hydrochloric acid

Practice questions

6. Consider the reaction of magnesium metal with hydrochloric acid, which is discussed in worked examples 3 and 8.

    a. Using the instantaneous reaction rates at $t = 0$ and 55 s from worked example 3, determine the reaction order with respect to hydrochloric acid. Round your answer to the nearest integer number.

    b. Suggest **one** advantage and **one** disadvantage of this method compared to the graphical technique used in worked example 8.

 Linking questions

What measurements are needed to deduce the order of reaction for a specific reactant? (Tool 1, 3, Inquiry 2)

 Why are reaction mechanisms only considered as "possible mechanisms"?

What are the rate equations and units of k for the reactions of primary and tertiary halogenoalkanes with aqueous alkali? (Reactivity 3.4)

The Arrhenius equation (*Reactivity 2.2.12* and *2.2.13*)

You already know that the rate constants are temperature-dependent. This dependence is quantitatively described by the **Arrhenius equation**, which was first derived by the Swedish scientist Svante Arrhenius:

$$k = Ae^{-\frac{E_a}{RT}}$$

In this equation k is the rate constant, A is the **Arrhenius factor**, e is the base of natural logarithms (2.718...), E_a is the activation energy of the reaction, R is the ideal gas constant and T is the absolute temperature.

The value of the Arrhenius factor is specific to each reaction, as it takes into account the frequency of collisions of the reacting particles with proper mutual orientation. The units of A are identical to those of the rate constant, k, so they can be derived from the rate equation.

For simple and highly symmetrical reactant particles, such as individual atoms or diatomic molecules, A values are high, because collisions can lead to reaction of the particles at any mutual orientation. In contrast, reactions between large molecules occur only at specific orientations, so their A values are low.

For a first order reaction, the units of both A and k are s^{-1}, so the Arrhenius factor is sometimes called the **frequency factor** (as frequency is often expressed in the same units, s^{-1}). Both A and E_a in the Arrhenius equation are almost independent of temperature. In contrast, the value of k increases exponentially with temperature, as $k \propto e^{-1/T} = e^T$.

Reactivity 2.2 How fast? The rate of chemical change

AHL

The exact rate of this increase depends on the E_a value.

This means that by analysing the temperature dependence of the rate constant for a particular reaction, you can determine the activation energy. To do so, the Arrhenius equation can be rearranged using the formula $\ln(e^x) = x$:

$$\ln k = -\frac{E_a}{RT} + \ln A$$

If you know two k values for the same reaction at different temperatures, or the ratio of these two k values, you can determine E_a without knowing the value of A.

To do this, the Arrhenius equations for k_1 at T_1 and k_2 at T_2 can be subtracted as follows:

$$\ln k_1 - \ln k_2 = -\frac{E_a}{RT_1} + \ln A - \left(-\frac{E_a}{RT_2} + \ln A\right)$$

By using the formula $\ln x - \ln y = \ln\frac{x}{y}$ and cancelling out the terms $\ln A$ and $-\ln A$, the equation can be rearranged as follows:

$$\ln\frac{k_1}{k_2} = \frac{E_a}{R}\left(\frac{1}{T_2} - \frac{1}{T_1}\right)$$

Now the activation energy of the reaction can be found by substituting k_1, k_2, T_1 and T_2 with experimental data and solving the equation. If the concentrations of all reactants are kept constant, the reaction rate shows the same temperature dependence as the rate constant: $\frac{v_1}{v_2} = \frac{k_1}{k_2}$. Therefore, the E_a value can also be determined from the reaction rates (instead of rate constants) at two different temperatures.

The logarithmic form of the Arrhenius equation is also useful for the analysis of reaction kinetics in graphical form, as the plot of $\ln k$ against $\frac{1}{T}$ is a straight line. The slope of this line gives $-\frac{E_a}{R}$, and the intercept on the y-axis gives $\ln A$.

> The Arrhenius equation in both its linear and logarithmic forms is given in the data booklet.

Practice question

7. The rate of a certain reaction at constant concentrations of all reactants increases 8.2 times when the temperature rises from 25 to 45 °C. Determine the activation energy, in kJ mol⁻¹, of that reaction.

Worked example 9

The decomposition of nitrogen(V) oxide in the gas phase is a first order reaction:

$$2N_2O_5(g) \rightarrow 2N_2O_4(g) + O_2(g)$$

The plot of $\ln k$ against $\frac{1}{T}$ for this reaction is shown in figure 25.

a. Determine the activation energy, E_a, in kJ mol⁻¹, for this reaction.

b. Determine the Arrhenius factor, A, and state its units.

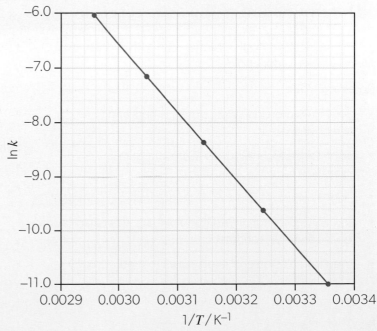

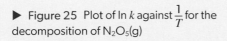
▶ Figure 25 Plot of $\ln k$ against $\frac{1}{T}$ for the decomposition of $N_2O_5(g)$

Solution

a. To determine the activation energy, we need to choose two points on the line as far apart as possible, for example, (0.00302, −6.8) and (0.00334, −10.8). For the linear form of the Arrhenius equation, the line slope gives $-\dfrac{E_a}{R}$, in K. Therefore:

$$-\frac{E_a}{R} = \frac{-6.8 - (-10.8)}{0.00302 - 0.00334}$$

$$= -12\,500\,K$$

You can find E_a by substituting the value of R into the expression for the slope:

$$-\frac{E_a}{8.31\,J\,K^{-1}\,mol^{-1}} = -12\,500\,K$$

$$E_a = -(-12\,500\,K \times 8.31\,J\,K^{-1}\,mol^{-1})$$

$$= 104\,000\,J\,mol^{-1}$$

$$= 104\,kJ\,mol^{-1}$$

The E_a values for many chemical reactions fall into a narrow range (50–100 kJ mol⁻¹), so the rates of different reactions show similar temperature dependence. Remember that for most reactions at moderate temperatures (0–100 °C), the rate increases two to four times when the temperature rises by 10 °C.

b. First, choose any point on the line, for example, the point (0.00296, −6.02). This gives the following values:

$$\ln k = -6.02$$

$$\frac{1}{T} = 0.00296\,K^{-1}$$

From part **a**, we know that the gradient is as follows:

$$-\frac{E_a}{R} = -12\,500\,K$$

Then, substitute these values into the logarithmic form of the Arrhenius equation ($\ln k = -\dfrac{E_a}{RT} + \ln A$) and solve for A:

$$-6.02 = -12\,500\,K \times 0.00296\,K^{-1} + \ln A$$

$$\ln A = 30.98$$

$$A = e^{30.98} = 2.85 \times 10^{13}$$

The Arrhenius factor, A, has the same units as the rate constant, k. Therefore, you should construct the rate equation and find the units of k. The decomposition of $N_2O_5(g)$ is a first order reaction, so it has the following rate equation:

$$v = k[N_2O_5]$$

If the rate is measured in mol dm⁻³ s⁻¹ and $[N_2O_5]$ in mol dm⁻³, then the units for k are s⁻¹, so $A = 2.85 \times 10^{13}\,s^{-1}$.

🧪 Determination of the activation energy of a reaction

In this practical, you will determine the activation energy of a reaction of your choice.

Relevant skills

- Tool 1: Addressing safety of self, others and the environment
- Tool 2: Use technology to represent data in graphical form.
- Tool 3: Interpret features of graphs including gradient
- Inquiry 1: Develop investigations that involve hands-on laboratory experiments
- Inquiry 1: Design and explain a valid methodology

- Inquiry 1: Pilot methodologies
- Inquiry 1: Maintain constant environmental conditions of systems
- Inquiry 2: Assess validity and reliability

Safety

- Wear eye protection
- Conduct a full risk assessment. Consider all the substances and methods used, as well as the behaviour of all substances when heated
- Minimize the amounts of substances used
- Dispose of all substances appropriately

Reactivity 2.2 How fast? The rate of chemical change

AHL

Choose one of the following experimental procedures to carry out:

Reaction	Method for monitoring rate
"Sulfur clock": sodium thiosulfate and hydrochloric acid	Due to the gradual formation of a sulfur precipitate, you can measure the time taken for the reaction mixture to obscure a mark below the flask.
Decomposition of hydrogen peroxide, catalysed by catalase, copper(II) ions, or iodide ions	Due to the formation of oxygen gas, you can measure the rate of oxygen production with a gas syringe, or by collecting the gas over water, in an inverted measuring cylinder.
"Iodine clock": hydrogen peroxide and iodide ions, in acid solution and in the presence of thiosulfate ions and starch	The reaction involves multiple steps. In the last step, the reaction mixture undergoes a sudden change from colourless to blue-black when there is no more thiosulfate left to react with the free iodine produced in an earlier step. You can therefore measure the time taken for the colour change to occur.
Reduction of permanganate ions by ethanedioate ions in acid solution	The solution is decolorized as the purple manganate(VII) ions, MnO_4^-(aq), are consumed, so you can measure the time taken for this colour change to happen.
A chemiluminescent reaction, such as that in a light stick (figure 26)	In the absence of other light sources, the light intensity of the reaction can be measured by a digital sensor (or suitable smartphone app).

Method

1. Conduct background research to determine a basic procedure that you will follow. Do not forget to cite all sources appropriately.
2. Have your risk assessment checked by your teacher, and act on any feedback given.
3. Liaise with your teacher to obtain and prepare the reagents required for your preliminary trials.
4. Conduct the reaction at room temperature to get a sense of how to carry it out. Make note of any adjustments you need to make to the concentrations of solutions, volumes, equipment sizes, etc.
5. Repeat the reaction, this time at a higher temperature (for example, 50 °C, checking whether it is safe to do so first). You will need to devise a way to maintain the reaction mixture at a constant temperature and monitor the temperature to see if it varies during the reaction.
6. Once you have made the necessary changes to your materials and method, collect reaction rate data for at least five different temperatures. Repeat the experiment three times at each temperature.
7. Process your data to compute the rate, or an approximation of the rate of reaction. Depending on your chosen reaction and method, this could involve:
   - Determining the initial rate of reaction by drawing a tangent line at (0,0) on graphs of volume, concentration or absorbance against time

   or
   - Calculating $\frac{1}{t}$ for each of the times measured.

8. Using appropriate software, construct an Arrhenius plot of the data.
9. Draw a line of best fit and determine the gradient.
10. Calculate the experimental value for the activation energy of the reaction you chose.
11. If possible, compare your experimental value to the literature value by calculating the percentage error.
12. Comment on the validity and reliability of the value you have obtained.
13. Outline some of the possible sources of error.

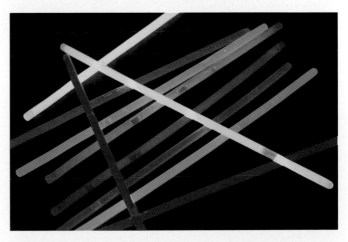

▲ Figure 26 Chemiluminescent reactions produce light. When the thin glass vial inside a light stick is snapped open, the reactants are mixed, and a spontaneous reaction occurs. Some energy of this reaction is released in the form of visible light

End of topic questions

Topic review

1. Using your knowledge from the *Reactivity 2.2* topic, answer the guiding question as fully as possible:
   How can the rate of a reaction be controlled?

Exam-style questions

Multiple-choice questions

2. Excess magnesium powder was added to a beaker containing dilute hydrochloric acid. The mass of the beaker and its contents was recorded and plotted against time (figure 27, curve 1).

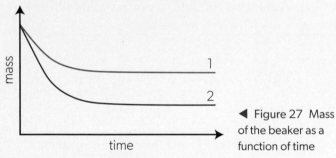

◀ **Figure 27** Mass of the beaker as a function of time

Which change could produce curve 2 in figure 27?

   A. Increasing the temperature
   B. Using the same mass of Mg ribbon
   C. Doubling the mass of powdered Mg
   D. Using the same volume of concentrated HCl(aq)

3. Thiosulfate ions, $S_2O_3^{2-}$(aq), are oxidized by hydrogen peroxide, H_2O_2(aq), as follows:

   $H_2O_2(aq) + 2S_2O_3^{2-}(aq) + 2H^+(aq) \rightarrow S_4O_6^{2-}(aq) + 2H_2O(l)$

   Which graph is consistent with the experimentally determined rate equation, $v = k[H_2O_2][S_2O_3^{2-}]$?

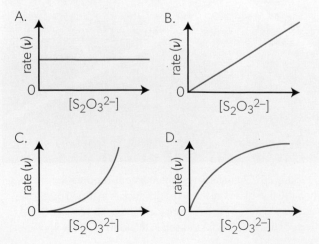

Extended-response questions

4. The questions below refer to the experiments described in question 2.
   a. Explain the shape of curve 1 from figure 27 in terms of the collision theory.
   b. Using the same graph, sketch the third curve that would be produced if the original experiment were run at a lower temperature.
   c. Using a different graph, sketch the concentration of Mg^{2+}(aq) ions against time for experiments 1 and 2.

5. Calcium oxide reacts with carbon dioxide as follows:
   $CaO(s) + CO_2(g) \rightarrow CaCO_3(s)$
   Excess calcium oxide was placed into a pressurized vessel with a volume of $1.00\,dm^3$. The vessel was filled with gaseous carbon dioxide and sealed off. A constant temperature of $25.0\,°C$ was maintained during the whole experiment. The plot of the pressure inside the vessel against time is shown below.

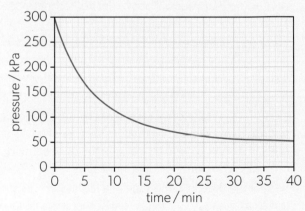

   a. Using the ideal gas law (*Structure 1.5*), calculate the concentrations of CO_2(g) in the vessel at $t = 0$, 5, 10, …, 40 min. Assume that the volume occupied by solid substances is negligible.
   b. Calculate the average reaction rate, in $mol\,dm^{-3}\,min^{-1}$, for the period between 0 and 20 min.
   c. Plot the concentration of carbon dioxide against time on graph paper.
   d. Using the tangent line method, determine the initial reaction rate and the instantaneous reaction rate at $t = 10$ min.

6. Consider the following process:
   $3I^-(aq) + H_2O_2(aq) + 2H^+(aq) \rightarrow I_3^-(aq) + 2H_2O(l)$
   Over a period of 2.0 min, the concentration of iodide ions in the solution decreased from 0.050 to $0.020\,mol\,dm^{-3}$.

AHL

Reactivity 2.2 How fast? The rate of chemical change

AHL

a. Calculate:

   i. average reaction rates, in $mol\,dm^{-3}\,s^{-1}$, with respect to $I^-(aq)$ and $H_2O_2(aq)$

   ii. overall average reaction rate expressed in the same units.

b. Sketch a graph to show the changes in concentrations of $I^-(aq)$, $H_2O_2(aq)$ and $I_3^-(aq)$ over time. Assume that the initial concentrations of $I^-(aq)$ and $H_2O_2(aq)$ were equal, and the initial concentration of $I_3^-(aq)$ was zero.

c. Suggest how the change in concentration of iodine, $I_2(aq)$, could be monitored.

d. The experiment was repeated with the following changes:

   i. the concentration of $H_2O_2(aq)$ was doubled at constant temperature

   ii. the reaction was carried out in a pressurized vessel at 200 kPa instead of 100 kPa

   iii. the temperature of the reaction mixture was lowered by 5 °C

   iv. the solution of iodide ions, $I^-(aq)$, was prepared from a fine powder of sodium iodide, $NaI(s)$, instead of large crystals of this compound.

   For each of the changes, predict, stating a reason, its effect on the rate of reaction.

7. Phosgene, $COCl_2$, is usually produced by the reaction between carbon monoxide, CO, and chlorine, Cl_2, according to the equation:

   $$CO(g) + Cl_2(g) \rightarrow COCl_2(g) \quad \Delta H_r = -108\,kJ\,mol^{-1}$$

   a. Sketch the potential energy profile for the synthesis of phosgene, indicating both the enthalpy of reaction and the activation energy.

   b. This reaction is normally carried out using a catalyst. Sketch a second curve labelled "catalysed" on the diagram to indicate the effect of the catalyst.

   c. Sketch a Maxwell–Boltzmann energy distribution curve for gaseous reactants at a certain temperature.

   d. Using your sketch from part c, explain the effect of catalyst on the reaction rate.

8. Nitrogen(II) oxide reacts with chlorine as follows:

   $$2NO(g) + Cl_2(g) \rightarrow 2NOCl(g)$$

   The following experimental data were obtained for this reaction at 260 K:

[NO] / $mol\,dm^{-3}$	[Cl_2] / $mol\,dm^{-3}$	v_{init} / $mol\,dm^{-3}\,min^{-1}$
1.50×10^{-2}	1.50×10^{-2}	5.30×10^{-3}
1.50×10^{-2}	3.00×10^{-2}	1.06×10^{-2}
3.00×10^{-2}	3.00×10^{-2}	4.24×10^{-2}

   a. Deduce the reaction orders with respect to $Cl_2(g)$ and NO(g).

b. State the rate equation and the overall reaction order.

c. Calculate the value of the rate constant at 260 K.

   The reaction proceeds in two steps:

   step 1 $2NO(g) \rightarrow N_2O_2(g)$

   step 2 $N_2O_2(g) + Cl_2(g) \rightarrow 2NOCl(g)$

d. Using the rate equation deduced in part (b), identify the rate-determining step (RDS) in this reaction.

9. Consider the following two-step reaction mechanism:

   step 1 (slow) $N_2O(g) \rightarrow N_2(g) + O(g)$

   step 2 (fast) $N_2O(g) + O(g) \rightarrow N_2(g) + O_2(g)$

   a. Formulate the equation for the overall reaction.

   b. Identify the role (reactant, product or intermediate) of each species involved in the reaction mechanism.

   c. Identify the molecularity of the rate-determining step.

   d. Deduce the rate equation of the overall reaction and state the overall reaction order.

   e. The overall reaction has a ΔH_r value of $-164.1\,kJ\,mol^{-1}$. Sketch the energy profile for this reaction.

10. The rate constants of a certain reaction at 30 and 60 °C are 0.183 and $5.45\,dm^3\,mol^{-1}\,s^{-1}$, respectively.

    a. Identify the overall order of this reaction.

    b. Determine the activation energy, in $kJ\,mol^{-1}$, of the reaction.

    c. Calculate the rate constant, in $dm^3\,mol^{-1}\,s^{-1}$, of this reaction at 45 °C.

11. Chlorine monoxide, $ClO\bullet$, is an unstable radical that plays an important role in the process of ozone depletion. In the absence of ozone, chlorine monoxide decomposes as follows:

    $$2ClO\bullet(g) \rightarrow Cl_2(g) + O_2(g)$$

    Kinetics studies show that this reaction has the second order with respect to $ClO\bullet(g)$. The plot of $\ln k$ against $1/T$ for this reaction is shown below.

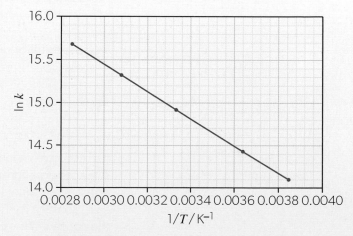

    a. Determine the activation energy, E_a, in $kJ\,mol^{-1}$, for this reaction.

    b. Determine the Arrhenius factor, A, for this reaction and state its units.

How far? The extent of chemical change

Chemical reactions are often reversible, which means that they can proceed both forwards and backwards at the same time. In most practical situations, chemists want to maximize the yield of the target product by facilitating the forward reaction and suppressing the reverse reaction. This goal can be achieved by controlling the temperature, pressure and concentrations of reacting species. The optimum conditions for a chemical process can be determined using Le Châtelier's principle, which enables chemists to predict the effects of specific changes in these conditions on the extent of a reversible reaction.

Understandings

Reactivity 2.3.1—A state of dynamic equilibrium is reached in a closed system when the rates of forward and backward reactions are equal.

Reactivity 2.3.2—The equilibrium law describes how the equilibrium constant, K, can be determined from the stoichiometry of a reaction.

Reactivity 2.3.3—The magnitude of the equilibrium constant indicates the extent of a reaction at equilibrium and is temperature dependent.

Reactivity 2.3.4—Le Châtelier's principle enables the prediction of the qualitative effects of changes in concentration, temperature and pressure to a system at equilibrium.

Reactivity 2.3.5—The reaction quotient, Q, is calculated using the equilibrium expression with non-equilibrium concentrations of reactants and products.

Reactivity 2.3.6—The equilibrium law is the basis for quantifying the composition of an equilibrium mixture.

Reactivity 2.3.7—The equilibrium constant and Gibbs energy change, ΔG, can both be used to measure the position of an equilibrium reaction.

AHL

Dynamic equilibrium (*Reactivity 2.3.1*)

Many physical and chemical changes are reversible, so the interconversion of reactants and products can proceed simultaneously in both directions. In equations that represent reversible changes, the reaction arrow ($\rightarrow$) is replaced with the **equilibrium sign** ($\rightleftharpoons$), which symbolizes the bidirectional nature of the process. For example, the evaporation of liquid bromine, $Br_2(l) \rightarrow Br_2(g)$, and the condensation of gaseous bromine, $Br_2(g) \rightarrow Br_2(l)$, can be represented by a single equation:

$$Br_2(l) \rightleftharpoons Br_2(g)$$

If a reversible change occurs in a closed system, it eventually reaches the state of **dynamic equilibrium**, in which the rates of the forward and reverse processes are equal. For example, if we place a small quantity of liquid bromine into a closed flask, some of the bromine will evaporate (figure 1). The evaporation will increase the concentration of $Br_2(g)$ in the flask and the frequency of collisions between gaseous molecules, some of which will lose energy and return to the condensed phase, $Br_2(l)$.

▲ **Figure 1** The colour of gaseous bromine in a closed flask first darkens and then remains unchanged when the state of dynamic equilibrium between $Br_2(g)$ and $Br_2(l)$ is reached

At some point, the rate of evaporation will equal the rate of condensation, so there will be no further changes in the amounts of $Br_2(g)$ and $Br_2(l)$ present in the flask (figure 2). Both phases will remain in dynamic equilibrium indefinitely unless there is a change in external conditions, such as temperature or pressure.

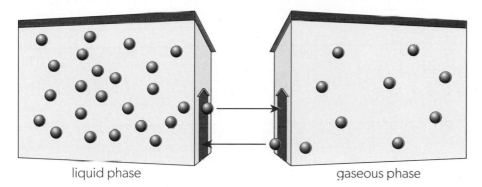

liquid phase   gaseous phase

▲ **Figure 2** In a dynamic equilibrium, the forward and reverse reactions occur at equal rates

Equilibrium is a dynamic state, in which two or more ongoing processes perfectly balance one another. This means that each participating species is consumed and produced at the same rate. Although this might give the illusion that no change is occurring during dynamic equilibrium, all components of the mixture are constantly being transformed into one another. In other words, the changes at equilibrium stop only at the macroscopic level but not at the microscopic level.

The dynamic nature of equilibrium can be observed experimentally. Naturally occurring bromine is a mixture of two stable isotopes, ^{79}Br (51%) and ^{81}Br (49%). Imagine that we have allowed the system shown in figure 1 to reach equilibrium and then replaced some of the liquid bromine with the same quantity of radioactive bromine, $^{80}Br_2(l)$. The concentrations of both liquid and gaseous bromine will not change, but radioactive $^{80}Br_2$ molecules will appear in the gas phase almost immediately. This is only possible if the evaporation and condensation of bromine continues even after the dynamic equilibrium has been established.

Figures 1 and 2 illustrate a **heterogeneous equilibrium**, in which the two participating species, $Br_2(l)$ and $Br_2(g)$, are present in different phases. Another type of heterogeneous equilibrium is established between a solid substance and its solution if the solid substance is present in excess. If we mix excess sodium chloride, $NaCl(s)$, with water, the salt will begin to dissolve and form aqueous ions, $Na^+(aq)$ and $Cl^-(aq)$. Some of these ions will recombine and precipitate out of the solution as solid sodium chloride:

$$NaCl(s) \underset{\text{precipitation}}{\overset{\text{dissolution}}{\rightleftharpoons}} Na^+(aq) + Cl^-(aq)$$

Initially, the rate of dissolution greatly exceeds the rate of precipitation (figure 3, left). However, as the concentration of aqueous ions increases, the rate of precipitation also increases and eventually becomes equal to the rate of dissolution (figure 3, right). At this point, the solution becomes **saturated**, and a dynamic equilibrium is established.

Evidence

Some salts are described as insoluble in water. In practice, minuscule amounts of these salts can still dissolve in water, so traces of respective ions can be detected when analysing these salts in aqueous media. The resolution of detection methods has increased over time, allowing us to detect ever-smaller trace amounts. Is the term "insoluble" incorrect? What type of evidence could cause a revision of scientific terminology? You might want to look up definitions that have evolved over time such as transition element or hydrogen bond.

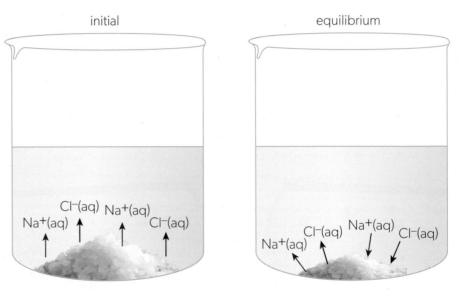

▲ **Figure 3** Dynamic equilibrium in a saturated solution of sodium chloride

A saturated solution has the highest possible concentration of the solute under given conditions. An **unsaturated** solution contains less solute than possible, so it can dissolve an additional quantity of that solute under given conditions. The highest possible concentration of the solute in a solution under given conditions is known as the **solubility** of that solute.

Chemical equilibrium

Reversible chemical reactions can proceed in both directions at the same time. A typical example of a reversible chemical change is the reaction involving gaseous nitrogen, hydrogen and ammonia:

$$N_2(g) + 3H_2(g) \rightleftharpoons 2NH_3(g)$$

In this reversible reaction, two processes take place: the synthesis of ammonia from nitrogen and hydrogen (**forward reaction**) and the decomposition of ammonia into nitrogen and hydrogen (**backward reaction**). The equilibrium sign in this equation tells us that both processes take place simultaneously, so reactant and product molecules are constantly interconverted.

The concept of reactants and products in a reversible process is somewhat ambiguous, as the same substance can be a reactant in the forward reaction but a product in the reverse reaction. However, in DP chemistry, species on the left of the equilibrium sign are referred to as reactants, and species on the right of the equilibrium sign are referred to as products.

The main direction in which a reversible reaction will proceed depends on the initial concentrations of all participating species. For example, if the initial mixture contains only nitrogen and hydrogen, only the forward reaction takes place: some molecules of nitrogen and hydrogen combine to form ammonia. Over time, the concentrations of nitrogen and hydrogen in the reaction mixture decrease while the concentration of ammonia increases (figure 4a).

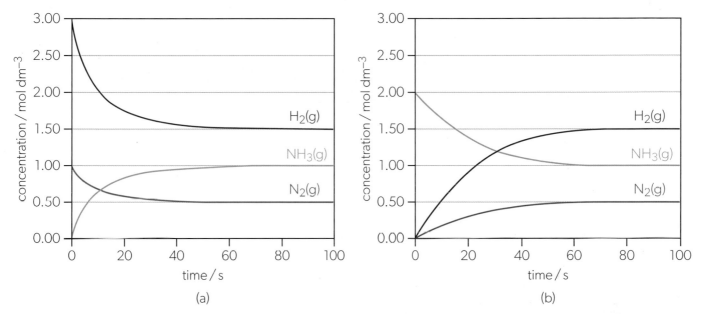

▲ **Figure 4** Concentrations of $N_2(g)$, $H_2(g)$ and $NH_3(g)$ in reaction mixtures approaching equilibrium at 475 K

Once the molecules of ammonia appear in the reaction mixture, the reverse reaction becomes possible, so these ammonia molecules begin to decompose into nitrogen and hydrogen. This process accelerates as the concentration of ammonia increases. At the same time, the forward reaction slows down, as the mixture contains fewer and fewer molecules of nitrogen and hydrogen. Eventually, the rates of the forward and reverse reactions become equal, and the concentrations of all three substances (nitrogen, hydrogen and ammonia) remain constant even though the forward and reverse reactions continue. This state is known as **chemical equilibrium**.

Chemical equilibrium can be achieved from any initial state of the system. In figure 4a, the initial mixture contained $1.00\,\mathrm{mol\,dm^{-3}}$ of nitrogen, $3.00\,\mathrm{mol\,dm^{-3}}$ of hydrogen and no ammonia, while the equilibrium concentrations of these species were 0.50, 1.50 and $1.00\,\mathrm{mol\,dm^{-3}}$, respectively. The same equilibrium concentrations of all three species will be produced if we start the reaction with $2.00\,\mathrm{mol\,dm^{-3}}$ of ammonia only, with no nitrogen or hydrogen in the initial mixture (figure 4b). In this case, some ammonia will decompose into nitrogen and hydrogen:

$$2NH_3(g) \rightleftharpoons N_2(g) + 3H_2(g)$$

As the process continues, the concentration of ammonia will decrease, while the concentrations of nitrogen and hydrogen will increase. Eventually, the rates of forward and reverse reactions will become equal, and the state of equilibrium will be established.

In summary, when a system is at equilibrium:

- chemical and/or physical changes continue at the microscopic level

- forward and reverse reactions proceed at equal rates

- concentrations of reactants and products remain constant

- macroscopic properties of the system, such as colour or density, remain unchanged

- the same equilibrium state can be achieved from either direction.

The equilibrium law (*Reactivity 2.3.2* and *Reactivity 2.3.3*)

When a chemical system is at equilibrium, the ratio of the equilibrium concentrations of reactants and products raised to the power of their stoichiometric coefficients is a constant, if the temperature is kept constant. This observation is known as the **equilibrium law**, and the constant is called the **equilibrium constant**, K.

Consider the general case of a chemical equilibrium involving several species with various stoichiometric coefficients:

$$a\text{A} + b\text{B} + \ldots \rightleftharpoons x\text{X} + y\text{Y} + \ldots$$

In the K expression, the products of the forward process (X, Y, ...) appear in the numerator while the reactants of the forward process (A, B, ...) appear in the denominator:

$$K = \frac{[\text{X}]^x [\text{Y}]^y \cdots}{[\text{A}]^a [\text{B}]^b \cdots}$$

The use of square brackets and the omission of state symbols saves space and focuses our attention on the most important information about the system at equilibrium. For the same reason, K values are treated as unitless quantities, although all equilibrium concentrations must still be expressed using appropriate units (typically $mol\,dm^{-3}$).

For example, here is the K expression for the synthesis of ammonia from nitrogen and hydrogen:

$$N_2(g) + 3H_2(g) \rightleftharpoons 2NH_3(g) \qquad K = \frac{[NH_3]^2}{[N_2][H_2]^3}$$

If a reaction takes place in a solution, the concentration of solvent does not change significantly during the reaction, so it is not included in the K expression. For example, the K expression for the hydrolysis of an ester in an aqueous solution does not include the concentration of water:

$$HCOOCH_3(aq) + H_2O(l) \rightleftharpoons HCOOH(aq) + CH_3OH(aq) \qquad K = \frac{[HCOOH][CH_3OH]}{[HCOOCH_3]}$$

However, if the same reaction takes place in a non-aqueous solution, water is treated as any other reactant, so its concentration must be included in the K expression.

Worked example 1

Calculate the K value for the synthesis of ammonia at 475 K, using the data from figure 4.

Solution

First, write the K expression for the reaction:

$$K = \frac{[NH_3]^2}{[N_2][H_2]^3}$$

Then input the values of concentration at equilibrium into the expression:

$$= \frac{1.00^2}{0.50 \times 1.50^3}$$

$$= 0.59$$

Practice questions

1 a. State the K expression for the following equation:

$$2NH_3(g) \rightleftharpoons N_2(g) + 3H_2(g)$$

   b. Using the data from figure 4, calculate the K value for this process.

For any given process, the value of the equilibrium constant depends on temperature, but not on other reaction conditions, such as pressure or concentration. For example, if we increase the initial concentration of $N_2(g)$ from 1.0 to 2.5 mol dm^{-3}, the state of equilibrium in the reaction $N_2(g) + 3H_2(g) \rightleftharpoons 2NH_3(g)$ will be achieved at different concentrations of nitrogen, hydrogen and ammonia (figure 5). However, these new concentrations (1.85, 1.13 and 1.25 mol dm^{-3}, respectively) will still give the same value of K:

$$K = \frac{1.25^2}{1.85 \times 1.13^3} = 0.59$$

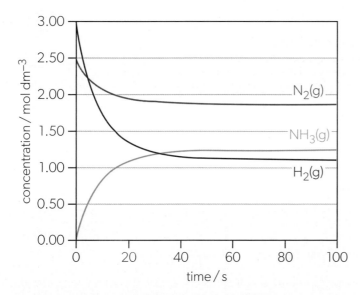

▲ **Figure 5** Concentrations of $N_2(g)$, $H_2(g)$ and $NH_3(g)$ in a reaction mixture approaching equilibrium at 475 K

The value of K provides important information about the direction and extent of the spontaneous process under standard conditions. If $K > 1$, the products are favoured over the reactants, so the forward reaction will proceed spontaneously until most reactants are converted into products.

Conversely, if $K < 1$, the reactants are favoured, so the reverse reaction will be spontaneous, and the reaction mixture at equilibrium will consist predominantly of reactants. Finally, if $K = 1$, the equilibrium will be established at approximately equal concentrations of reactants and products. However, the exact ratio between these concentrations will depend on the stoichiometric coefficients in the equation.

Chemical reactions with very large and very small K values are practically irreversible. If $K >> 1$, the forward reaction is favoured so strongly that the extent of the reverse reaction becomes negligible. In other words, the reaction with $K >> 1$ will proceed from left to right until the reactants are used up completely. In contrast, when $K << 1$, only the reverse reaction will be observed while the forward reaction will not proceed to any noticeable extent.

The relationship between K and $\Delta G^{\ominus}$ will be discussed in the AHL section of this topic.

The value of K for a given reaction depends on the stoichiometric ratio of reactants and products. If all stoichiometric coefficients are halved, the new K will be the square root of the original constant (table 1). Conversely, if the coefficients are doubled, the K will be squared. Finally, if the equation is reversed, the new K will be reciprocal to the original constant.

Change in the equation	Effect on the K value
coefficients halved	$K' = \sqrt{K}$
coefficients doubled	$K' = K^2$
equation reversed	$K' = \dfrac{1}{K}$
two equations added together	$K' = K_1 \times K_2$

▲ Table 1 The equilibrium constant and reaction stoichiometry

Worked example 2

At 475 K, the equilibrium constant K for the reaction
$N_2(g) + 3H_2(g) \rightleftharpoons 2NH_3(g)$ has a value of 0.59.
Calculate the K' values for the following reactions at the same temperature:

a. $2NH_3(g) \rightleftharpoons N_2(g) + 3H_2(g)$

b. $NH_3(g) \rightleftharpoons 0.5N_2(g) + 1.5H_2(g)$

Solution

a. According to table 1, for a reversed reaction $K' = \dfrac{1}{K}$ so:

$$K' = \frac{1}{0.59} \approx 1.7$$

b. The stoichiometric coefficients in reaction (b) are halved, as compared to those in reaction (a). According to table 1, $K' = \sqrt{1.7} \approx 1.3$.

 Linking question

How does the value of K for the dissociation of an acid convey information about its strength? (Reactivity 3.1)

Le Châtelier's principle (*Reactivity 2.3.4*)

Chemical equilibrium is a dynamic process, so it can be easily disturbed by any change in the reaction conditions, such as temperature, pressure or the concentrations of reacting species. The effects of such changes can be predicted by using **Le Châtelier's principle**, which was formulated at the end of the 19th century by the French chemist Henry Le Châtelier:

If a dynamic equilibrium is disturbed by a change in the reaction conditions, the balance between the forward and reverse processes will shift to counteract the change and return the system to equilibrium.

In other words, a system at equilibrium will counteract any changes in concentration, pressure or temperature by altering the rates and spontaneity of the forward and backward reactions until a new equilibrium is established. If the forward reaction becomes more favourable than the reverse reaction, we say that the **position of the equilibrium** shifts to the right. Conversely, if the backward reaction becomes more favourable than the forward reaction, the equilibrium position shifts to the left. Shifting the equilibrium position to the right increases the concentrations of products, and shifting the equilibrium to the left increases the concentrations of reactants.

Effect of concentration on equilibrium

Imagine an equilibrium mixture of nitrogen, hydrogen and ammonia. The temperature is 475 K and the concentrations are as follows: $[N_2] = 0.50\,\text{mol dm}^{-3}$, $[H_2] = 1.50\,\text{mol dm}^{-3}$, $[NH_3] = 1.00\,\text{mol dm}^{-3}$.

$$N_2(g) + 3H_2(g) \rightleftharpoons 2NH_3(g) \qquad K = \frac{1.00^2}{0.50 \times 1.50^3} = 0.59$$

The changes in concentrations of the reacting species over time are shown in figure 6.

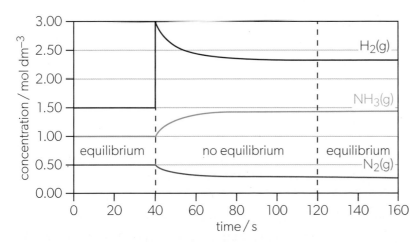

▲ Figure 6 The effect of a concentration change on the equilibrium position

Initially, the concentrations of all three gases remain constant, as the mixture is already at equilibrium, and the rates of the forward and reverse reactions are equal. At $t = 40\,\text{s}$, we disturb the equilibrium by injecting more hydrogen into the reaction mixture, so the concentration of $H_2(g)$ increases from 1.5 to $3.0\,\text{mol dm}^{-3}$.

This change in concentration leads to more frequent collisions between hydrogen and nitrogen molecules and so increases the rate of the forward reaction. As a result, the concentrations of hydrogen and nitrogen in the reaction mixture decrease while the concentration of ammonia increases. In other words, the system counteracts the presence of excess reactant (hydrogen) by converting it into product (ammonia) and shifting the equilibrium position to the right.

The new equilibrium concentrations of nitrogen, hydrogen and ammonia (0.28, 2.33 and 1.45 mol dm^{-3}, respectively) still satisfy the K expression:

$$K = \frac{1.45^2}{0.28 \times 2.33^3} = 0.59$$

The opposite effect would be observed if we increased the concentration of ammonia instead of hydrogen. Since ammonia is the product, its higher concentration would accelerate the reverse reaction and shift the equilibrium position to the left. In this case, the system would counteract the presence of excess product (ammonia) by converting it into reactants (nitrogen and hydrogen).

Practice questions

2. Outline how the following changes will affect the equilibrium position and the K value of the reaction $N_2(g) + 3H_2(g) \rightleftharpoons 2NH_3(g)$:

   a. increase in concentration of nitrogen

   b. decrease in concentration of ammonia.

Case study: the chromate–dichromate equilibrium

Le Châtelier's principle can be studied in a school laboratory using the following reversible reaction:

$$2CrO_4{}^{2-}(aq) + 2H^+(aq) \rightleftharpoons Cr_2O_7{}^{2-}(aq) + H_2O(l)$$

chromate   dichromate
(yellow)   (orange)

The colours of transition metal ions are discussed in *Structure 3.1 (AHL)*. The concentrations of chromate and dichromate ions in the solution can be measured by spectrophotometry (*Structure 1.4*).

Both chromate and dichromate ions contain a transition metal, chromium(VI), so they are brightly coloured (figure 7). Therefore, any change in the equilibrium position can be detected by observing the colour of the solution.

If we add a small amount of acid to a solution of $CrO_4{}^{2-}(aq)$, the concentration of $H^+(aq)$ ions in the solution increases. To counteract this change, some hydrogen ions will react with chromate ions to produce dichromate ions. The equilibrium position will shift to the right, so the solution colour will change from yellow to orange.

The opposite effect can be achieved by the addition of an alkali. The alkali will neutralize the acid, decreasing the concentration of $H^+(aq)$ ions in the solution:

$$H^+(aq) + OH^-(aq) \rightarrow H_2O(l)$$

This change will favour the reverse reaction that will increase the concentration of $H^+(aq)$ ions and therefore counteract the change: the position of equilibrium will shift to the left. Therefore, dichromate ions will turn into chromate ions, and the solution colour will change from orange to yellow.

The chromate–dichromate equilibrium involves water, which acts as both the solvent and the product of the forward reaction. However, water is present in the solution in large excess, so the addition or evaporation of water has very little effect on the equilibrium position. If an orange solution of dichromate is diluted with water, its colour will become paler, but it will still be orange, not yellow. For this reason, the concentration of water is never included in the K expressions for equilibria in aqueous solutions:

▲ **Figure 7** Aqueous chromate, $CrO_4{}^{2-}(aq)$, is yellow, while aqueous dichromate, $Cr_2O_7{}^{2-}(aq)$, is orange

$$K = \frac{[Cr_2O_7{}^{2-}]}{[CrO_4{}^{2-}]^2 [H^+]^2}$$

Effect of pressure on equilibrium

The effect of pressure on the equilibrium position depends on the stoichiometric ratio of gaseous reactants and products. For example, the reaction $N_2(g) + 3H_2(g) \rightleftharpoons 2NH_3(g)$ involves four gaseous molecules (one N_2 and three H_2) on the left-hand side of the equation but only two gaseous molecules of NH_3 on the right-hand side. If the pressure increases, the system will counteract this change by converting the reactants into the products, as this will reduce the number of gaseous molecules in the reaction mixture. Therefore, the equilibrium position will shift to the right. Conversely, a decrease in pressure will shift the equilibrium position to the left, towards the greater number of gaseous molecules. In both cases, the K value of the reaction will not change.

If an equation contains more gaseous molecules on the right-hand side than on the left-hand side, an increase in pressure will shift the equilibrium position to the left, while a decrease in pressure will shift the equilibrium position to the right. If the numbers of gaseous molecules on each side of the equation are equal, a change in pressure will have no effect on the equilibrium position. Once again, the K value will remain unchanged in all cases.

Changes in pressure have no effect on equilibria in condensed phases because solids and liquids are almost incompressible. However, heterogeneous equilibria that involve gaseous species are affected by pressure changes in the same way as homogeneous equilibria in the gas phase. For example, consider the heterogeneous equilibrium between liquid and gaseous bromine that was discussed at the beginning of this topic:

$$Br_2(l) \rightleftharpoons Br_2(g)$$

If the pressure increases, the concentration of $Br_2(l)$ will remain unchanged (because liquids are incompressible) while the concentration of $Br_2(g)$ will increase. According to Le Châtelier's principle, the system will counteract this change by increasing the rate of the reverse reaction (condensation of bromine), so the equilibrium position will shift to the left. Conversely, a decrease in pressure will shift the equilibrium position to the right, making the forward reaction (evaporation of bromine) more favourable than the reverse reaction.

In the general case, when predicting the effects of pressure on heterogeneous equilibria, you can consider only gases and ignore solids, liquids and aqueous species.

According to the ideal gas law (*Structure 1.5*), the pressure of a gas mixture is inversely proportional to its volume. When the volume of a system decreases, the pressure increases, and when the volume increases, the pressure decreases. Therefore, the effects of a change in volume on the equilibrium position are opposite to the effects of pressure. Like pressure, a change in the volume of system has no effect on the value of K.

The chemistry of acidic and alkaline species is detailed in *Reactivity 3.1*.

Practice questions

3. State the effects of increasing pressure on the equilibrium positions and K values for the following reactions:
   a. $N_2O_4(g) \rightleftharpoons 2NO_2(g)$
   b. $SO_2(g) + NO_2(g) \rightleftharpoons SO_3(g) + NO(g)$
   c. $4HCl(g) + O_2(g) \rightleftharpoons 2H_2O(g) + 2Cl_2(g)$
   d. $SO_2(g) + H_2O(l) \rightleftharpoons H_2SO_3(aq)$

Effect of temperature on equilibrium

Temperature is the only factor that affects both the position of equilibrium and the equilibrium constant. The synthesis of ammonia from nitrogen and hydrogen is an exothermic process ($\Delta H_r^\ominus < 0$), so heat is released by the forward reaction and consumed by the reverse reaction:

$$N_2(g) + 3H_2(g) \rightleftharpoons 2NH_3(g) \quad \Delta H_r^\ominus = -91.8\,kJ\,mol^{-1}$$

At equilibrium, the forward and reverse reactions proceed at the same rates, so the total amount of heat in the system remains constant. If we increase the temperature, more heat will be introduced into the system. In accordance with Le Châtelier's principle, the system will counteract this change by consuming excess heat, so the reverse reaction will be favoured, and the equilibrium position will shift to the left. As a result, the concentration of ammonia in the reaction mixture will decrease while the concentrations of both nitrogen and hydrogen will increase.

Now consider the K expression for this reaction:

$$K = \frac{[NH_3]^2}{[N_2][H_2]^3}$$

The numerator contains $[NH_3]$, which decreases with temperature. The denominator contains $[N_2]$ and $[H_2]$, both of which increase with temperature. Therefore, an increase in temperature lowers the K value of this reaction. The same result would be observed for any other exothermic process. Conversely, a decrease in temperature shifts the equilibrium position of an exothermic process to the right and thus increases its K value.

For an endothermic process, the reverse is true: an increase in temperature will shift the equilibrium position to the right and increase the K value. A decrease in temperature will have opposite effects, shifting the equilibrium position to the left and decreasing the K value.

Chemical or physical changes with $\Delta H_r^\ominus = 0$ are very rare but still possible. For example, the isotopic exchange in a mixture of normal water (H_2O, hydrogen oxide) and heavy water (D_2O, deuterium oxide) proceeds as follows:

$$H_2O(l) + D_2O(l) \rightleftharpoons 2DOH(l)$$

The chemical properties of isotopes are nearly identical (*Structure 1.2*), so the $\Delta H_r^\ominus$ value for this reaction is very close to zero. As a result, temperature has no effect on the equilibrium position or the K value for this, or any other reaction with $\Delta H_r^\ominus = 0$.

To predict the effect of a temperature change on the equilibrium position, it is convenient to treat heat as an imaginary substance (Q) participating in the reaction. For an exothermic process, this "substance" is a reaction product, and for an endothermic process, it is a reactant:

$$reactants \rightleftharpoons products + Q \quad \Delta H_r^\ominus < 0\ (exothermic)$$

$$reactants + Q \rightleftharpoons products \quad \Delta H_r^\ominus > 0\ (endothermic)$$

An increase in temperature will increase the "concentration" of heat in the system, so the equilibrium position of the exothermic reaction will shift to the left, and the equilibrium position of the endothermic reaction will shift to the right. Conversely, a decrease in temperature decreases the "concentration" of heat and has the opposite effects on the above equilibria.

The effect of catalysts on equilibrium

A catalyst provides an alternative pathway for the reaction and therefore lower its activation energy. This increases the rate of a chemical reaction (*Reactivity 2.2*). In a reversible process, the forward and reverse reactions follow the same pathway in opposite directions, so the rates of both reactions increase to the same extent. Therefore, in the presence of a catalyst, the equilibrium state of the system is achieved faster, but the position of this equilibrium and the *K* value remain unchanged.

The effect of a catalyst on the equilibrium position can be illustrated by a simple analogy. Imagine two communicating vessels that represent reactants and products (figure 8). The levels of liquid in the vessels represent the relative rates of the forward and reverse reactions.

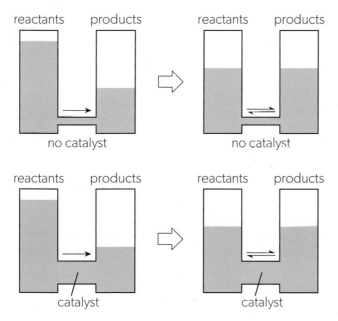

▲ **Figure 8** Communicating vessels as a model of chemical equilibrium

The flow of liquid between the vessels is limited by the diameter of the connecting tube. If we increase this diameter, the levels of liquid in the vessels will become equal faster, but the final state of the system will not be affected. Similarly, a catalyst allows a chemical system to reach the state of equilibrium faster but does not affect the position of that equilibrium.

The same analogy can be used for illustrating Le Châtelier's principle with respect to concentrations. If we add liquid to the left vessel ("increase a reactant concentration"), some of the liquid will flow to the right vessel ("the forward reaction will be favoured") until the levels become equal again ("a new chemical equilibrium will be established").

The effects of reaction conditions on the equilibrium position and K value are summarized in table 2.

Change in condition	Shift of equilibrium	K
concentration of reactant increases	to the right (towards products)	no change
concentration of product decreases		
concentration of reactant decreases	to the left (towards reactants)	
concentration of product increases		
pressure increases	to the side with a smaller number of gas molecules	
volume decreases		
pressure decreases	to the side with a greater number of gas molecules	
volume increases		
temperature increases	$\Delta H_r^\ominus < 0$: to the left $\Delta H_r^\ominus > 0$: to the right $\Delta H_r^\ominus = 0$: no change	$\Delta H_r^\ominus < 0$: decreases $\Delta H_r^\ominus > 0$: increases $\Delta H_r^\ominus = 0$: no change
temperature decreases	$\Delta H_r^\ominus < 0$: to the right $\Delta H_r^\ominus > 0$: to the left $\Delta H_r^\ominus = 0$: no change	$\Delta H_r^\ominus < 0$: increases $\Delta H_r^\ominus > 0$: decreases $\Delta H_r^\ominus = 0$: no change
catalyst is added	no change	no change

▲ Table 2 The effects of reaction conditions on the equilibrium position and K value

Case study: the Haber process

Le Châtelier's principle allows chemists to maximize the yield of the desired product by altering the reaction conditions. In particular, the synthesis of ammonia on an industrial scale becomes profitable only when a fine balance between pressure, temperature and concentrations of reacting species is achieved. This synthesis, developed in early 20th century by the German chemist Fritz Haber, utilizes the reaction that has already been discussed in this topic:

$$N_2(g) + 3H_2(g) \rightleftharpoons 2NH_3(g) + Q$$

Since the number of gas molecules in the forward reaction decreases from four to two, the Haber process is carried out at high pressure (200 atm, or 20 MPa), which pushes the equilibrium position to the right and at the same time increases the reaction rate. Nitrogen and hydrogen are constantly injected into the reaction mixture while ammonia is condensed and removed after each cycle of the process (figure 9). These measures maximize the concentrations of reactants and minimize the concentration of the product, shifting the equilibrium even further to the right.

The forward reaction is exothermic ($\Delta H_r^\ominus < 0$), so it is favoured by a low temperature. However, this reduces the kinetic energy and frequency of collisions between reactant particles. Therefore, a lower temperature will decrease the reaction rate. As a compromise, the reaction is carried out at a moderate temperature (400–450 °C) and in the presence of a catalyst (iron powder with various additives). While the catalyst itself does not affect the equilibrium position, it greatly increases the reaction rate.

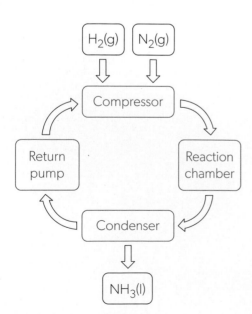

▲ Figure 9 Flow chart of the Haber process

In modern industrial plants, the efficiency of the Haber process is improved by recycling unreacted gases and utilizing the heat released by the forward reaction. Such improvements increase the overall reaction yield to 98%, while reducing the cost and environmental impact of ammonia production.

TOK

The Haber process has had significant impacts – positive and negative – on society. On the one hand, ammonia is widely used for making urea, ammonium nitrate and other fertilizers. Without these, current farming practices could not feed the growing world population. According to some estimates, a half of the nitrogen in our bodies comes from the Haber process, making it arguably the most important invention of the 20th century.

However, the Haber process was also vital to produce explosives and ammunition in both world wars. Haber himself also worked on and promoted the use of poison gases during the First World War, earning him the name "the father of chemical warfare". His story illustrates how advances in scientific knowledge can benefit society or be used for creating weapons of mass destruction.

Is science, or should it be, value-free?

Linking question

Why do catalysts have no effect on the value of K or on the equilibrium composition? (Reactivity 2.2)

Reaction quotient (*Reactivity 2.3.5*)

If a system has not reached equilibrium, the ratio of actual concentrations of products and reactants differs from the K value. This ratio is known as the **reaction quotient**, Q. It can be used for determining the direction of ongoing chemical changes within the system.

The reaction quotient, Q, is calculated just like the K expression, but with non-equilibrium concentrations of all reacting species. So, for the reaction $a\text{A} + b\text{B} + \ldots \rightleftharpoons x\text{X} + y\text{Y} + \ldots$ not at equilibrium, the Q expression is as follows:

$$Q = \frac{[\text{X}]^x\,[\text{Y}]^y \cdots}{[\text{A}]^a\,[\text{B}]^b \cdots}$$

You can make the following conclusions from the relative values of Q and K:

- When $Q < K$, the reaction mixture contains more reactants and less products than needed for equilibrium, so the forward reaction will be favoured.

- When $Q > K$, there are less reactants and more products than needed for equilibrium, so the reverse reaction will be the dominant process.

- When $Q = K$, the system is already at equilibrium, so the forward and reverse reactions will proceed at the same rate.

Practice question

4. Consider the following reaction at 283 K:

   $$2NO_2(g) \rightleftharpoons N_2O_4(g) \quad K = 11.5$$

   The reaction mixture contains 0.025 mol dm^{-3} of nitrogen dioxide and 0.10 mol dm^{-3} of dinitrogen tetroxide. Determine the direction of the spontaneous reaction in this mixture at 283 K.

▲ **Figure 10** Japanese geochemist Katsuko Saruhashi

Worked example 3

Calculate the reaction quotient for a mixture of nitrogen, hydrogen and ammonia at 475 K, where $K = 0.59$ and the concentration of each species is 0.50 mol dm^{-3}. Predict the direction of the spontaneous reaction in this mixture.

$$N_2(g) + 3H_2(g) \rightleftharpoons 2NH_3(g) \quad K = \frac{[NH_3]^2}{[N_2][H_2]^3} = 0.59$$

Solution

To calculate the reaction quotient, we need to use the K expression with the actual concentrations instead of the equilibrium concentrations:

$$Q = \frac{0.5^2}{0.5 \times 0.5^3} = 4$$

Since $Q > K$, the system contains too much ammonia and too little nitrogen and hydrogen, so the reverse reaction is favoured.

 ## Global impact of science

The oceans are an important store of carbon dioxide in the carbon cycle. The determination of $[CO_2]$ in seawater is a complex process because carbon dioxide participates in various equilibria. Japanese geochemist Katsuko Saruhashi explored and modelled the relationship between the total concentration of carbonic acid, pH and temperature. Based on equilibrium constants, she devised a methodology to determine the CO_2 levels in seawater. This information was compiled for easy access in what became known as "Saruhashi's Table". Saruhashi also investigated the effects of nuclear testing on seawater, co-founded the Society of Japanese Women Scientists in 1958 and established the annual Saruhashi Prize in 1981, awarded to female scientists for their contributions to scientific research.

What makes an idea worthy of a scientific prize? What is the purpose of scientific prizes? What is the role of scientific societies?

Calculations involving K (*Reactivity 2.3.6*)

The K value can be used for determining the composition of a reaction mixture at equilibrium if the initial or equilibrium concentrations of some participating species are known. Alternatively, the initial concentrations can be determined from the K value and equilibrium concentrations.

 ## Data-based questions

In esterification reactions, a carboxylic acid reacts with an alcohol to form an ester. The equilibrium constant, K, for esterification reactions can be determined by analysis of the concentrations of the reacting species at equilibrium. Values for K were determined in this way for the esterification of several different alcohols and carboxylic acids, shown in table 3.

Reactivity 2.3 How far? The extent of chemical change

AHL

	ethanoic acid	propanoic acid	butanoic acid
methanol	5.03	4.72	4.08
ethanol	4.86	2.13	2.28
propan-1-ol	4.21	4.10	2.33
butan-1-ol	2.85	1.10	1.34
2-methylpropan-1-ol	2.02	1.23	0.92

▲ Table 3 Equilibrium constants, in $mol\,dm^{-3}$, for esterification reactions at reflux temperature (Source of data: E. Sarlo, P. Svoronos, and P. Kulas, *J. Chem. Ed.*, 1990, **67 (9)**, 796)

1. Prepare a graph of the results.

2. Describe the trends, patterns and relationships you observe in the data.

3. Research ways to determine the equilibrium constant, K, for an esterification reaction. Use this information and your answers to this task to brainstorm possible investigations exploring a factor connected to esterification.

4. Choose one of your ideas and plan an investigation. Remember to:

   • outline the independent and dependent variables
   • state the major control variables
   • state the research question
   • outline the context of the research question
   • outline the methodology

Worked example 4

A mixture of sulfur(IV) oxide and oxygen was heated in a sealed vessel at 1000 K until the following equilibrium was reached:

$$2SO_2(g) + O_2(g) \rightleftharpoons 2SO_3(g) \qquad K = 3.0$$

The equilibrium concentrations of sulfur(IV) oxide and sulfur(VI) oxide in the final mixture were 0.12 and 0.18 $mol\,dm^{-3}$, respectively.

a. Calculate the concentration of oxygen at equilibrium.

b. Determine the initial concentrations of both reactants.

Solution

a. First, we need to write the K expression:

$$K = \frac{[SO_3]^2}{[O_2][SO_2]^2}$$

Since we know three of the four values in this expression, we can find the equilibrium concentration of oxygen:

$$3.0 = \frac{0.18^2}{[O_2] \times 0.12^2}$$

$$[O_2] = 0.75\,mol\,dm^{-3}$$

2. To find the initial concentrations of $SO_2(g)$ and $O_2(g)$, we need to look at their stoichiometric ratio. The formation of each mole of $SO_3(g)$ requires the same quantity of $SO_2(g)$ and twice as little $O_2(g)$.

If 0.18 $mol\,dm^{-3}$ of $SO_3(g)$ was produced, then

0.18 $mol\,dm^{-3}$ of $SO_2(g)$ and $\dfrac{0.18}{2} = 0.090\,mol\,dm^{-3}$ of $O_2(g)$ were consumed:

$2SO_2(g)$	$+ O_2(g)$	$\rightleftharpoons 2SO_3(g)$	
$c_{initial}$ / $mol\,dm^{-3}$	x	y	0
Δc / $mol\,dm^{-3}$	-0.18	-0.09	$+0.18$
c_{eq} / $mol\,dm^{-3}$	0.12	0.75	0.18

Therefore:

$x - 0.18 = 0.12$, so $x = 0.12 + 0.18 = 0.30$

$y - 0.09 = 0.75$, so $y = 0.75 + 0.09 = 0.84$

$[SO_2]_{initial} = 0.30\,mol\,dm^{-3}$

$[O_2]_{initial} = 0.84\,mol\,dm^{-3}$

Practice question

5. Nitrogen dioxide was cooled down to 283 K in a sealed vessel until it reached equilibrium with dinitrogen tetroxide:

$$2NO_2(g) \rightleftharpoons N_2O_4(g) \quad K = 11.5$$

Calculate the equilibrium and initial concentrations of $NO_2(g)$ if the equilibrium concentration of $N_2O_4(g)$ was 0.041 $mol\,dm^{-3}$.

Using a spreadsheet scrollbar to determine equilibrium concentrations

Equilibrium concentrations of reactants and products depend on factors such as the value of the equilibrium constant and the concentrations of the other species. In this task, you will use a method developed by Andrés Raviolo and published in the *Journal of Chemical Education* for analysing an equilibrium mixture by creating a spreadsheet model containing a scrollbar (2012).

Relevant skills

- Tool 2: Generate data from models and simulations
- Inquiry 3: Evaluate methodological limitations
- Tool 2: Use spreadsheets to manipulate data

Nitrogen and hydrogen react reversibly to form ammonia: $N_2(g) + 3H_2(g) \rightleftharpoons 2NH_3(g)$

The equilibrium constant expression for this equation is $K = \dfrac{[NH_3]^2}{[N_2][H_2]^3}$

Instructions

Part 1: Create the spreadsheet model

a. Nitrogen and hydrogen are placed in a 4.00 dm³ container and allowed to reach equilibrium. The initial amounts, in mol, of nitrogen and hydrogen are 0.200 and 0.500, respectively. The equilibrium constant, K, at a certain temperature is 269. Enter this information into a spreadsheet:

	A	B	C	D	E	F
1		V (dm³)	K	x	Q	
2		4.00	269			
3						
4				n(N_2)	n(H_2)	n(NH_3)
5			I (mol)	0.200	0.500	0.000
6			C (mol)			
7			E (mol)			
8						
9						

b. Next, enter the formulas needed to calculate:

- reaction quotient, Q (cell E2),
- changes in concentration (cells D6, E6 and F6)
- equilibrium concentrations (cells D7, E7 and F7).
- parameter labelled x (cell D2) is related to the concentration change and is computed based on a value in cell B10, which is currently empty.

	A	B	C	D	E	F
1		V (dm³)	K	x	Q	
2		4	269	=B10/100000	=F7^2/(D7*E7^3)*B2^2	
3						
4				n(N_2)	n(H_2)	n(NH_3)
5			I (mol)	0.2	0.5	0
6			C (mol)	=−1*D2	=−3*D2	=2*D2
7			E (mol)	=D5+D6	=E5+E6	=F5+F6
8						
9						

Reactivity 2.3 How far? The extent of chemical change

AHL

→

3. Now insert a scroll bar into the space between cells B9 and F9. If you are using Excel, you will need to show the "Developer tab", and select "Scroll bar" from the "Form control" section of this tab. In the scroll bar settings, set the cell link to B10, the minimum value to 1 and maximum value to 10000. For clarity, cells B10 to F10 have been merged and centred in the image below:

	A	B	C	D	E	F
1		$V\,(dm^3)$	K	x	Q	
2		4.00	269	0.08474	269	
3						
4				$n(N_2)$	$n(H_2)$	$n(NH_3)$
5			I (mol)	0.200	0.500	0.000
6			C (mol)	−0.085	−0.254	0.169
7			E (mol)	0.115	0.246	0.169
8						
9		◀				▶
10				8474		

4. The scroll bar allows you to change the value of x, and the spreadsheet computes the corresponding value of the reaction quotient, Q. Use the scroll bar to change x until $Q = K$, which indicates that the reaction mixture is at equilibrium.

Part 2: Use the spreadsheet model

5. Using your knowledge of equilibrium, predict what will happen to the equilibrium concentrations of N_2, H_2 and NH_3 if there are changes to the following:

- volume of the container

- value of K

- initial amounts of reactants

6. Using your spreadsheet model, investigate your ideas above.

7. Explain the formula used to compute Q, in cell E2: "=F7^2/(D7*E7^3)*B2^2".

Part 3: Evaluate the model

8. Identify and explain at least one advantage of this spreadsheet model.

9. Identify and explain at least two limitations of this spreadsheet model.

Part 4: Create a new model

Sulfur dioxide, $SO_2(g)$, reacts with oxygen, $O_2(g)$, to form sulfur trioxide, $SO_3(g)$, according to the following reversible reaction:

$$2SO_2(g) + O_2(g) \rightleftharpoons 2SO_3(g)$$

The value of K at a certain temperature is 7.5×10^{-2}.

10. Create a new spreadsheet model, like the one above, that can be used to determine the equilibrium concentrations of $SO_2(g)$, $O_2(g)$, and $SO_3(g)$, when 1.0 mol of $SO_2(g)$ and 1.4 mol of $O_2(g)$ are placed in a 3.0 dm³ container and allowed to reach equilibrium.

Weak acids and bases in aqueous solutions undergo reversible ionization, producing $H^+(aq)$ and $OH^-(aq)$ ions, respectively. Here are the general equations for these ionizations:

weak acid: $HA(aq) \rightleftharpoons H^+(aq) + A^-(aq)$

weak base: $B(aq) + H_2O(l) \rightleftharpoons BH^+(aq) + OH^-(aq)$

Acid–base equilibria will be discussed in greater detail in *Reactivity 3.1*.

The equilibrium constants for these processes are usually very low, so only a small proportion of the weak acid or base exists as ions, while most of these species remain in molecular form. Therefore, we can assume that the equilibrium concentration of a weak acid or base is approximately equal to its initial concentration:

$$[HA]_{eq} \approx [HA]_{initial}$$

$$[B]_{eq} \approx [B]_{initial}$$

These assumptions simplify the calculations involving K of acid–base equilibria, as shown in the worked example below.

Worked example 5

The weak ethanoic acid, CH_3COOH, dissociates in aqueous solutions as follows:

$$CH_3COOH(aq) \rightleftharpoons CH_3COO^-(aq) + H^+(aq)$$

At 298 K, the K value for this process is 1.74×10^{-5}.

Determine the equilibrium concentrations, in $mol\,dm^{-3}$, of all species in a $0.100\,mol\,dm^{-3}$ solution of ethanoic acid at 298 K.

Solution

We can assume that the initial concentration of ethanoic acid is equal to its equilibrium concentration, as the value of K is very low:

$$[CH_3COOH]_{eq} \approx [CH_3COOH]_{initial}$$

$$= 0.100\,mol\,dm^{-3}$$

Write the expression for K:

$$K = \frac{[CH_3COO^-][H^+]}{[CH_3COOH]}$$

Then substitute in the values for K and $[CH_3COOH]_{eq}$.

$$1.74 \times 10^{-5} = \frac{[CH_3COO^-][H^+]}{0.100}$$

The concentrations of $CH_3COO^-(aq)$ and $H^+(aq)$ are approximately the same, so:

$$1.74 \times 10^{-5} \approx \frac{[CH_3COO^-]^2}{0.100}$$

$$[CH_3COO^-] \approx \sqrt{1.74 \times 10^{-6}}$$

$$[CH_3COO^-]_{eq} \approx [H^+]_{eq} \approx 0.00132\,mol\,dm^{-3}$$

Without assuming that $[CH_3COOH]_{eq} = [CH_3COOH]_{initial}$, we would have to solve a quadratic equation, which is not assessed in DP chemistry examination papers. The exact solution of the quadratic equation would give $[CH_3COO^-]_{eq} \approx 0.0131$, which is very close to the result of our approximate calculations.

Practice question

6. Phenylamine, $C_6H_5NH_2$, also known as aniline, is a weak base. In aqueous solutions, it undergoes a reversible ionization:

$$C_6H_5NH_2(aq) + H_2O(l) \rightleftharpoons C_6H_5NH_3^+(aq) + OH^-(aq)$$

At 298 K, the K value for this process is 7.41×10^{-10}.

Determine the concentration of $OH^-(aq)$ in a $0.100\,mol\,dm^{-3}$ aqueous solution of phenylamine.

Reactivity 2.3 How far? The extent of chemical change

AHL

 Linking question

How does the equilibrium law help us to determine the pH of a weak acid, weak base or a buffer solution? (Reactivity 3.1)

Gibbs energy and equilibrium (*Reactivity 2.3.7*)

In the SL section of this topic, we discussed the relationship between the K value and spontaneity of a reaction under standard conditions. If $K > 1$, the forward reaction proceeds spontaneously while a $K < 1$ favours the reverse reaction. If $K = 1$, neither reaction is favoured under standard conditions, as the equilibrium is already established.

Another measure of the reaction spontaneity is its **Gibbs energy change**, ΔG (*Reactivity 1.3*). For a reversible chemical reaction, a negative ΔG value favours the forward reaction, while a positive ΔG value favours the reverse reaction.

The **standard Gibbs energy change for a reaction**, $\Delta G^\ominus$, refers to the chemical change from reactants to products in their standard states ($p = 100$ kPa for gases or $c = 1$ mol dm^{-3} for aqueous species) at a given temperature. If the temperature is not specified, it is assumed to be 298 K. If the spontaneity is determined under standard conditions, the value of $\Delta G^\ominus$ must be used instead of ΔG.

Equilibrium constant	Standard Gibbs energy change	Spontaneous process under standard conditions
$K > 1$	$\Delta G^\ominus < 0$	forward reaction (products are favoured)
$K < 1$	$\Delta G^\ominus > 0$	reverse reaction (reactants are favoured)
$K = 1$	$\Delta G^\ominus = 0$	neither forward nor reverse (equilibrium)

▲ **Table 3** Equilibrium position and the values of K and $\Delta G^\ominus$

The relationship between K and $\Delta G^\ominus$ is described by the following equation:

$$\Delta G^\ominus = -RT \ln K$$

where $R = 8.31$ J K^{-1} mol^{-1} is the gas constant and T is the absolute temperature in kelvin. Notice that $\Delta G^\ominus$ is usually expressed in kJ mol^{-1}, so it must be converted to J mol^{-1} to match the units of R.

This equation is very useful because standard Gibbs energies of formation, $\Delta G_f^\ominus$, are known for many compounds. These data allow us to calculate first the $\Delta G^\ominus$ value and then the K value for almost any chemical or physical change. Conversely, an unknown $\Delta G^\ominus$ value for a particular compound can be determined from the K value of a reversible reaction involving this compound and the thermodynamic data for other species participating in that reaction.

The expression $\Delta G^\ominus = -RT \ln K$ and the value of R are given in the data booklet.

Calculations involving standard Gibbs energy changes and thermodynamic data for reactants and products are discussed in *Reactivity 1.4*. Thermodynamic data for selected compounds are given in the data booklet.

 Activity

Calculate the equilibrium constant, K, for a reversible reaction with $\Delta G^\ominus = -3.9$ kJ mol^{-1} at 298 K.

Worked example 6

Dinitrogen dioxide, $N_2O_2(g)$, is highly unstable, so its Gibbs energy change of formation, $\Delta G_f^{\ominus}$, cannot be determined directly. However, the equilibrium constant for the following process at 298 K can be calculated from spectroscopic data:

$$2NO(g) \rightleftharpoons N_2O_2(g) \qquad K = 1.39 \times 10^{-5}$$

a. Determine $\Delta G^{\ominus}$, in kJ mol^{-1}, for the forward reaction at 298 K.

b. Using your answer to part a, determine $\Delta G_f^{\ominus}$ for $N_2O_2(g)$, in kJ mol^{-1}, at 298 K if $\Delta G_f^{\ominus}$ for $NO(g)$ at the same temperature is 87.6 kJ mol^{-1}.

Solution

a. $\Delta G^{\ominus} = -RT \ln K$

$\quad = -8.31 \, J \, K^{-1} \, mol^{-1} \times 298 \, K \times \ln(1.39 \times 10^{-5})$

$\quad = 27\,700 \, J \, mol^{-1}$

$\quad = 27.7 \, kJ \, mol^{-1}$

b. Using the enthalpy cycle diagram method from *Reactivity 1.2*:

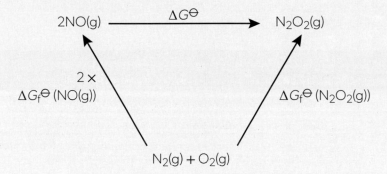

$\Delta G^{\ominus} = -[2 \times \Delta G_f^{\ominus}(NO(g))] + \Delta G_f^{\ominus}(N_2O_2(g))$

Substitute in the values of $\Delta G^{\ominus}$ and $\Delta G_f^{\ominus}(NO(g))$:

$27.7 \, kJ \, mol^{-1} = -[2 \times 87.6 \, kJ \, mol^{-1}] + \Delta G_f^{\ominus}(N_2O_2(g))$

$\Delta G_f^{\ominus}(NO(g)) = 202.9 \, kJ \, mol^{-1}$

🔗 Linking question

How can Gibbs energy be used to explain which of the forward or backward reaction is favoured before reaching equilibrium? (Reactivity 1.4)

Reactivity 2.3 How far? The extent of chemical change

AHL

(ATL) Self-management skills

Graphic-display calculator (GDC) skills that you have learnt in maths, such as using numerical solvers and finding the intersection of two graphs, can also be applied to chemistry.

Consider the following practice question: Calculate the equilibrium constant, K, for a reaction at 298 K. The value of $\Delta G^\ominus$ for the reaction at this temperature is −0.82 kJ mol⁻¹.

This can be done by solving $\Delta G^\ominus = -RT \ln K$ for K, as shown in the text. Two further ways of calculating K are possible, using your GDC:

1. Numerical solver

Enter the equation (remembering to express $\Delta G^\ominus$ in J mol⁻¹, not kJ mol⁻¹) into the numerical solver (usually called nSolve, Nsolve or solveN) and compute K:

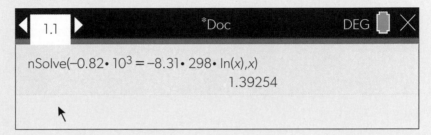

2. Graphing both sides of the equation and finding the intersection

Graph both sides of the equation as two separate functions. In the example above, the two functions are $f_1(x) = -820$ and $f_2(x) = -8.31 \times 298 \times \ln(x)$. Fit the graphs to the screen if necessary, then find the intersection. The x-coordinate of the intersection is the value you are looking for:

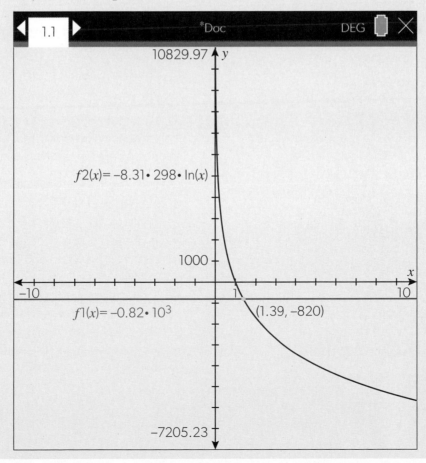

End of topic questions

Topic review

1. Using your knowledge from the *Reactivity 2.3* topic, answer the guiding question as fully as possible:

   How can the extent of a reversible reaction be influenced?

Exam-style questions

Multiple-choice questions

2. Which of the following statements is correct?

   A   The changes at equilibrium stop only at the microscopic level but not at the macroscopic level.

   B   The changes at equilibrium stop at both the macroscopic level and the microscopic level.

   C   The rates of both forward and reverse reactions at equilibrium are equal to zero.

   D   The equilibrium constant expression can be deduced from the reaction stoichiometry.

3. Which factors increase the rates of both forward and reverse reactions in any system at equilibrium?

   I.   Increase in temperature

   II. Increase in pressure

   III. Addition of a catalyst

   A   I and II only

   B   I and III only

   C   II and III only

   D   I, II and III

4. What will happen in the following reaction mixture at equilibrium if the pressure decreases?

   $SO_2(g) + H_2O(l) \rightleftharpoons H^+(aq) + HSO_3^-(aq)$

   A   The equilibrium position will shift to the left and $[H^+]$ will increase

   B   The equilibrium position will shift to the right and $[H^+]$ will increase

   C   The equilibrium position will shift to the left and $[H^+]$ will decrease

   D   The equilibrium position will shift to the right and $[H^+]$ will decrease

5. What is true for any system at equilibrium?

   A $K = 1$

   B $Q = 1$

   C $Q = K$

   D $\Delta G^\ominus = 0$

6. For a certain reversible reaction in a closed system, $Q > K$. Which changes will occur when the system will be approaching equilibrium at constant temperature?

   A Q will decrease

   B K will increase

   C   Concentrations of reactants will decrease

   D   Concentrations of products will increase

7. At 1 000 K, the forward reaction has a negative $\Delta G^\ominus$ value:

   $2NO_2(g) \rightleftharpoons 2NO(g) + O_2(g)$

   Which expression is correct for the equilibrium at this temperature?

   A $[NO_2]^2 > [NO]^2[O_2]$

   B $[NO_2] < [NO]$

   C $[NO] = 2[O_2]$

   D $K > 1$

Extended-response questions

8. Deduce the equilibrium constant expressions for the following reactions.

   a. $3O_2(g) \rightleftharpoons 2O_3(g)$

   b. $2NO_2(g) \rightleftharpoons 2NO(g) + O_2(g)$

   c. $NO_2(g) \rightleftharpoons NO(g) + 0.5O_2(g)$

   d. $CH_3COOH(aq) \rightleftharpoons CH_3COO^-(aq) + H^+(aq)$

   e. $NH_3(aq) + H_2O(l) \rightleftharpoons NH_4^+(aq) + OH^-(aq)$

   f. $CaCO_3(s) \rightleftharpoons CaO(s) + CO_2(g)$

9. At a certain temperature, the K value for reaction **b** from the previous problem is 0.81. Deduce the K value for reaction **c** from the same problem.

10. Consider the following reaction:

    $H_2(g) + I_2(g) \rightleftharpoons 2HI(g)$

    The equilibrium concentrations of hydrogen, iodine and hydrogen iodide in the reaction mixture at 760 K were found to be 0.012, 0.015 and 0.091 mol dm^{-3}, respectively.

    a.   Deduce the equilibrium constant expression and determine the K value for this reaction at 760 K.

    b.   The K value for the same reaction at 715 K is 48.0. Deduce the sign of the standard enthalpy change for the forward reaction.

Reactivity 2.3 How far? The extent of chemical change

AHL

11. Consider the following reaction:

    $2CO(g) + O_2(g) \rightleftharpoons 2CO_2(g)$ $\Delta H^\ominus = -566\,kJ\,mol^{-1}$

    Outline how the following changes will affect the equilibrium position and the K value of this reaction:

    a. decrease in pressure
    b. increase in temperature
    c. increase in concentration of oxygen gas
    d. decrease in concentration of carbon monoxide
    e. addition of a catalyst

12. At high temperature, ammonia decomposes into nitrogen and hydrogen:

    $2NH_3(g) \rightleftharpoons N_2(g) + 3H_2(g)$

    The progress of this reaction under certain conditions without a catalyst is shown in the graph below.

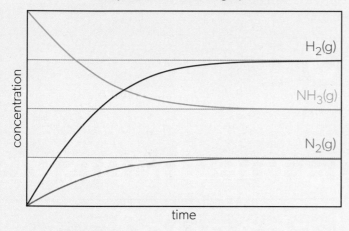

time

    On a copy of the same graph, sketch the changes in concentrations of $NH_3(g)$, $N_2(g)$ and $H_2(g)$ over time for the same reaction under the same conditions in the presence of a catalyst.

13. The esterification reaction between ethanoic acid and ethanol proceeds as follows:

    $CH_3COOH(org) + CH_3CH_2OH(org) \rightleftharpoons$
    $CH_3COOCH_2CH_3(org) + H_2O(org)$

    A mixture of 1.00 mol of ethanoic acid and 2.00 mol of ethanol was dissolved in an inert organic solvent (designated as "org") to produce 1.00 dm³ of the solution and heated at 50 °C in the presence of an acid catalyst until equilibrium was reached. The equilibrium mixture contained 1.40 mol of ethanol.

    a. Calculate the K value for this equilibrium at 50 °C.
    b. Determine the direction of the spontaneous reaction at 50 °C in the mixture containing equal concentrations of all four species participating in this equilibrium.

14. Benzoic acid, C_6H_5COOH, dissociates in aqueous solutions as follows:

    $C_6H_5COOH(aq) \rightleftharpoons C_6H_5COO^-(aq) + H^+(aq)$

    At 298 K, the K value for this equilibrium is 6.31×10^{-5}.

    a. Determine the equilibrium concentration, in mol dm⁻³, of $H^+(aq)$ ions in a 0.0200 mol dm⁻³ solution of benzoic acid at 298 K.
    b. Calculate the $\Delta G^\ominus$ value, in kJ mol⁻¹, for the forward reaction at 298 K.

15. Consider the equilibrium between two oxides of nitrogen:

    $N_2O_4(g) \rightleftharpoons 2NO_2(g)$ $\Delta H^\ominus = +58\,kJ\,mol^{-1}$

    The graph below shows the concentrations of both oxides at 320 K as a function of time.

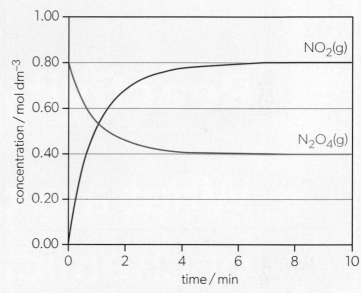

time / min

    a. State the K expression for this equilibrium and determine the K value at 320 K.
    b. Calculate the $\Delta G^\ominus$ value, in kJ mol⁻¹, for the forward reaction at 320 K.
    c. Determine the direction of the spontaneous reaction at 320 K in the mixture containing 0.50 mol dm⁻³ of $N_2O_4(g)$ and 0.10 mol dm⁻³ of $NO_2(g)$.

Reactivity 3

What are the mechanisms of chemical change?

Reactivity 3.1 | Proton transfer reactions

What happens when protons are transferred?

Acids and bases have been known for thousands of years. The term "acid" is derived from the Latin word *acere*, which means "sour" and refers to the characteristic taste of vinegar, lemon juice and other acidic solutions. Basic substances, such as potash (potassium carbonate) and lime water (a solution of calcium hydroxide), were used by ancient Egyptians for making soap and parchment. People who worked with basic solutions noted their slippery, soap-like feeling to the touch, bitter taste and ability to react with acids.

In previous chapters, you learned that **acids** and **bases** are two classes of chemical compounds with opposing properties. One of the most common reaction types, **neutralization**, usually involves an acid and a base as reactants and a **salt** and water as products. At the same time, some salts and even water itself demonstrate acidic or basic properties by reacting with bases or acids, respectively. During acid–base reactions, protons are transferred from one chemical species (acid) to another (base). In this chapter, we will discuss the nature of acids and bases, proton transfer reactions and acid–base equilibria in aqueous solutions.

Understandings

Reactivity 3.1.1—A Brønsted–Lowry acid is a proton donor and a Brønsted–Lowry base is a proton acceptor.

Reactivity 3.1.2—A pair of species differing by a single proton is called a conjugate acid–base pair.

Reactivity 3.1.3—Some species can act as both Brønsted–Lowry acids and bases.

Reactivity 3.1.4—The pH scale can be used to describe the $[H^+]$ of a solution: $pH = -\log_{10}[H^+]$; $[H^+] = 10^{-pH}$

Reactivity 3.1.5—The ion product constant of water, K_w, shows an inverse relationship between $[H^+]$ and $[OH^-]$. $K_w = [H^+][OH^-]$

Reactivity 3.1.6—Strong and weak acids and bases differ in the extent of ionization.

Reactivity 3.1.7—Acids react with bases in neutralization reactions.

Reactivity 3.1.8—pH curves for neutralization reactions involving strong acids and bases have characteristic shapes and features.

Reactivity 3.1.9—The pOH scale describes the $[OH^-]$ of a solution. $pOH = -\log_{10}[OH^-]$; $[OH^-] = 10^{-pOH}$

Reactivity 3.1.10—The strengths of weak acids and bases are described by their K_a, K_b, pK_a or pK_b values.

Reactivity 3.1.11—For a conjugate acid–base pair, the relationship $K_a \times K_b = K_w$ can be derived from the expressions for K_a and K_b.

Reactivity 3.1.12—The pH of a salt solution depends on the relative strengths of the parent acid and base.

Reactivity 3.1.13—pH curves of different combinations of strong and weak monoprotic acids and bases have characteristic shapes and features.

Reactivity 3.1.14—Acid–base indicators are weak acids, where the components of the conjugate acid–base pair have different colours. The pH of the end point of an indicator, where it changes colour, approximately corresponds to its pK_a value.

Reactivity 3.1.15—An appropriate indicator for a titration has an end point range that coincides with the pH at the equivalence point.

Reactivity 3.1.16—A buffer solution is one that resists change in pH on the addition of small amounts of acid or alkali.

Reactivity 3.1.17—The pH of a buffer solution depends on the pK_a or pK_b of its acid or base, and the ratio of the concentration of acid or base to the concentration of the conjugate base or acid.

AHL

Theories of acids and bases (*Reactivity 3.1.1*)

The first rational approach to acids and bases was proposed by the Irish scientist Robert Boyle in the 17th century. According to Boyle, acids and bases can be defined as follows:

- acids taste sour, react with metals, turn litmus red and can be neutralized by bases

- bases feel slippery, turn litmus blue and can be neutralized by acids.

In 1884, the Swedish scientist Svante Arrhenius defined acids and bases in terms of the ions they produced in aqueous solutions:

- an **Arrhenius acid** is a substance that dissociates in water to form hydrogen ions (H^+)

- an **Arrhenius base** is a substance that dissociates in water to form hydroxide ions (OH^-).

For example, hydrogen chloride is an acid, as it produces H^+ ions when dissolved in water:

$$HCl(aq) \rightarrow H^+(aq) + Cl^-(aq)$$

Similarly, sodium hydroxide is a base, as it produces OH^- ions in aqueous solutions:

$$NaOH(aq) \rightarrow Na^+(aq) + OH^-(aq)$$

According to Arrhenius, neutralization is a combination reaction between H^+ and OH^- ions:

$$H^+(aq) + OH^-(aq) \rightarrow H_2O(l)$$

The Arrhenius theory has limitations. First, all Arrhenius acids and bases must be soluble in water, or they will not be able to produce any ions in aqueous solutions. Second, some bases, such as ammonia, NH_3, contain no oxygen and thus cannot produce hydroxide ions by dissociation. When dissolved in water, ammonia does produce some OH^- ions by reacting reversibly with water:

$$NH_3(aq) + H_2O(l) \rightleftharpoons NH_4^+(aq) + OH^-(aq)$$

However, the source of hydroxide ions in this reaction is water, not ammonia itself, so in the absence of water, no OH^- ions would be formed. Nevertheless, gaseous ammonia behaves as a typical base. For example, it readily reacts with gaseous hydrogen chloride, producing a salt, ammonium chloride, in the form of a white smoke (figure 1). This is an example of a neutralization reaction.

$$NH_3(g) + HCl(g) \rightarrow NH_4Cl(s)$$

To overcome the limitations of the Arrhenius approach, a new acid–base theory was proposed in 1923 by two physical chemists, Johannes Brønsted from Denmark and Martin Lowry from England. Working independently of each other, Brønsted and Lowry concluded that both acids and bases could be defined by their roles in the transfer of protons (H^+):

▲ **Figure 1** Reaction of hydrogen chloride with ammonia. "Ammonium hydroxide" is a common name for aqueous ammonia

- a **Brønsted–Lowry acid** is a **proton donor**: a species that can lose an H^+ ion

- a **Brønsted–Lowry base** is a **proton acceptor**: a species that can gain an H^+ ion.

Consider the reaction of hydrogen chloride with gaseous ammonia:

$$NH_3(g) + HCl(g) \rightarrow NH_4Cl(s)$$

Hydrogen chloride formally loses a proton and therefore acts as a Brønsted–Lowry acid:

$$HCl(g) \rightarrow H^+ + Cl^-$$

In turn, ammonia accepts a proton and therefore acts as a Brønsted–Lowry base:

$$NH_3(g) + H^+ \rightarrow NH_4^+$$

The ammonium cation and chloride anion form ammonium chloride:

$$NH_4^+ + Cl^- \rightarrow NH_4Cl(s)$$

If we add the last three equations together, the H^+, NH_4^+ and Cl^- ions will cancel one another, and the resulting equation will represent the overall neutralization reaction:

$$NH_3(g) + HCl(g) \rightarrow NH_4Cl(s)$$

It is important to note that the Brønsted–Lowry approach to acids and bases does not replace the Arrhenius theory but rather expands it by removing any references to the solvent (water) and recognizing a wider range of species as acids and bases. For example, any Arrhenius acid, such as HCl(aq), will be a Brønsted–Lowry acid too, as it acts as a proton donor. At the same time, many Arrhenius bases, such as sodium hydroxide (NaOH), are treated by the Brønsted–Lowry theory as complexes of a base (OH^-) with a metal cation (Na^+). Sodium hydroxide cannot act as a proton acceptor without losing another ion (Na^+), while the hydroxide ion can:

$$OH^-(aq) + H^+(aq) \rightarrow H_2O(l)$$

Theories

Boyle's definitions of acids and bases emphasize the most characteristic properties of these compounds (reactivity towards metals and each other). Boyle's theory also suggested a simple experimental procedure (the colour change of **litmus**, a natural acid–base indicator) for distinguishing between acidic and basic solutions. This approach is a good illustration of the scientific method, which is based on systematic observations and experimental evidence.

At the same time, Boyle was unable to explain why some compounds behaved as acids and others as bases. In the 17th century, the chemical composition of most substances was still unknown, and even the existence of chemical elements was not universally accepted. As a result, Boyle's classification had no theoretical background or predictive power, so it was not a scientific theory but rather a set of empirical rules.

How does the meaning of the word "theory" in science compare to its meaning in other contexts?

Worked example 1

The equation for the neutralization of sulfuric acid, $H_2SO_4(aq)$, with potassium hydroxide, $KOH(aq)$, is shown below:

$$H_2SO_4(aq) + 2KOH(aq) \rightarrow K_2SO_4(aq) + 2H_2O(l)$$

Deduce the Brønsted–Lowry acid and base in this reaction.

Solution

Sulfuric acid loses two protons and therefore acts as a Brønsted–Lowry acid:

$$H_2SO_4(aq) \rightarrow 2H^+(aq) + SO_4^{2-}(aq)$$

In Brønsted–Lowry theory, potassium hydroxide is considered a complex of a base (OH^-) with a metal cation (K^+). In aqueous solution, potassium hydroxide dissociates into potassium and hydroxide ions:

$$KOH(aq) \rightarrow K^+(aq) + OH^-(aq)$$

The hydroxide ion accepts a proton and therefore acts as a Brønsted–Lowry base:

$$OH^-(aq) + H^+(aq) \rightarrow H_2O(l)$$

The potassium cation and sulfate anion form a salt, potassium sulfate:

$$2K^+(aq) + SO_4^{2-}(aq) \rightarrow K_2SO_4(aq)$$

To double check that these assignments match the overall equation, we can add these four equations together. The $H^+(aq)$, $K^+(aq)$, $SO_4^{2-}(aq)$ and $OH^-(aq)$ ions will cancel one another, giving the original neutralization equation:

$$H_2SO_4(aq) + 2KOH(aq) \rightarrow K_2SO_4(aq) + 2H_2O(l)$$

Practice questions

1. Deduce the molecular and net ionic equations for the neutralization of

   a. nitric acid with sodium hydroxide

   b. hydrogen bromide with barium hydroxide.

2. Deduce the equation for the reaction between hydrogen chloride and water.

According to the Brønsted–Lowry theory, the reaction of ammonia with water is also a neutralization process. Water acts as an acid by losing a proton while ammonia acts as a base by accepting a proton:

$$H_2O(l) \rightleftharpoons H^+(aq) + OH^-(aq)$$

$$NH_3(aq) + H^+(aq) \rightleftharpoons NH_4^+(aq)$$

When these two equations are combined, the resulting equation represents the **ionization** of ammonia in water:

$$NH_3(aq) + H_2O(l) \rightleftharpoons NH_4^+(aq) + OH^-(aq)$$

The reaction of ammonia with water is usually referred to as ionization rather than **dissociation**. While these two terms are often used interchangeably, there is a subtle difference between them. Dissociation means that a single molecule (or other species) breaks into two or more species, while ionization refers to any process that produces ions. Since the reaction of ammonia with water involves more than one reactant, it should not be called dissociation.

Hydronium ion

Modern studies show that a free proton, H^+, cannot exist in aqueous solutions because it immediately reacts with water and produces a **hydronium** ion (also known as **hydroxonium**), H_3O^+:

In the above scheme, the oxygen atom donates one of its lone electron pairs to the empty orbital of the hydrogen cation. All three O–H bonds in the hydronium ion have identical lengths, and the overall shape of the ion is a trigonal pyramid (*Structure 2.2*).

In this book, the H^+ symbol will be used as a shorthand equivalent of H_3O^+. However, you should always remember that all acid–base processes in aqueous solutions involve hydronium ions rather than isolated protons.

Alkaline species

"**Alkali**" is a collective name for water-soluble bases, such as hydroxides of alkali metals (group 1 elements) and alkaline earth metals (group 2 elements). The broader term "base" includes all substances that demonstrate basic properties, regardless of their strength and solubility in water. For example, sodium hydroxide, NaOH, is both an alkali and a base while iron(II) oxide, FeO, is a base but not an alkali.

Conjugate acid–base pairs (*Reactivity 3.1.2*)

Another consequence of the Brønsted–Lowry theory is that any Brønsted–Lowry acid that loses a proton produces a Brønsted–Lowry base. In turn, any Brønsted–Lowry base that gains a proton produces a Brønsted–Lowry acid. The acid–base pairs in which the species differ by exactly one proton are called **conjugate acid–base pairs**.

For example, hydrogen cyanide (HCN) acts as a Brønsted–Lowry acid by losing a proton and producing a conjugate base, the cyanide anion (CN^-):

$$HCN(aq) \rightleftharpoons H^+(aq) + CN^-(aq)$$
$$\text{conjugate acid} \qquad\qquad \text{conjugate base}$$

If we expand this equation to include a molecule of water, two different conjugate acid–base pairs will be formed:

$$\begin{array}{cc} \text{conjugate} & \text{conjugate} \\ \text{base 2} & \text{acid 2} \end{array}$$

$$HCN(aq) + H_2O(l) \rightleftharpoons H_3O^+(aq) + CN^-(aq)$$

$$\begin{array}{cc} \text{conjugate} & \text{conjugate} \\ \text{acid 1} & \text{base 1} \end{array}$$

In a conjugate acid–base pair, the acid and the base differ by exactly one proton (the acid has one more proton than the base). For example, in the following equation, sulfuric acid (H_2SO_4) and the sulfate ion (SO_4^{2-}) do not form a conjugate acid–base pair, as they differ by two protons instead of one:

$$H_2SO_4(aq) \rightarrow 2H^+(aq) + SO_4^{2-}(aq)$$

However, if we formulate the equations for stepwise dissociation of sulfuric acid, each of these equations will contain a pair of conjugates:

$$H_2SO_4(aq) \rightarrow H^+(aq) + HSO_4^-(aq)$$
$$\begin{array}{cc} \text{conjugate} & \text{conjugate} \\ \text{acid} & \text{base} \end{array}$$

$$HSO_4^-(aq) \rightarrow H^+(aq) + SO_4^{2-}(aq)$$
$$\begin{array}{cc} \text{conjugate} & \text{conjugate} \\ \text{acid} & \text{base} \end{array}$$

Notice how the same species (HSO_4^-) acts as a Brønsted–Lowry base in one process and as a Brønsted–Lowry acid in another.

Practice questions

3. Hydrogencarbonate ions, $HCO_3^-(aq)$, can participate in the following equilibria:

   $HCO_3^-(aq) + H_2O(l) \rightleftharpoons$
   $CO_3^{2-}(aq) + H_3O^+(aq)$

   $HCO_3^-(aq) + H_2O(l) \rightleftharpoons$
   $H_2CO_3(aq) + OH^-(aq)$

   Identify two conjugate acid–base pairs in each equation and state which species act as Brønsted–Lowry acids and which as Brønsted–Lowry bases.

4. Deduce the formula of the conjugate base of phosphoric acid, H_3PO_4.

5. Deduce the formula of the conjugate acid of the ethanoate ion, CH_3COO^-.

Amphiprotic and amphoteric species (*Reactivity 3.1.3*)

An interesting consequence of the Brønsted–Lowry theory is that the same species can behave as both an acid and a base. For example, water can lose a proton and produce a hydroxide anion, or accept a proton and produce a hydronium cation:

$$H_2O(l) \rightleftharpoons H^+(aq) + OH^-(aq) \qquad \text{(water acts as an acid)}$$

$$H_2O(l) + H^+(aq) \rightleftharpoons H_3O^+(aq) \qquad \text{(water acts as a base)}$$

Moreover, one molecule of water can pass a proton to another molecule of water, in which case the first molecule will act as an acid and the second molecule as a base:

$$H_2O(l) + H_2O(l) \rightleftharpoons H_3O^+(aq) + OH^-(aq)$$

Water and other species that can be both Brønsted–Lowry acids and Brønsted–Lowry bases are often called **amphiprotic** (as they can accept or donate a proton). A broader term, **amphoteric**, refers to species that can react with both acids and bases (and therefore have both acidic and basic properties).

Any amphiprotic species is amphoteric: if it can donate a proton, it is an acid, and if it can accept a proton, it is a base. However, not all amphoteric species are amphiprotic. For example, zinc oxide can react with both acids and bases, so it is amphoteric:

$$ZnO(s) + 2HCl(aq) \rightarrow ZnCl_2(aq) + H_2O(l)$$

$$ZnO(s) + 2NaOH(aq) + H_2O(l) \rightarrow Na_2[Zn(OH)_4](aq)$$

At the same time, ZnO cannot donate a proton (as it has none), so it is not amphiprotic.

2-Amino acids are amphiprotic organic compounds with the general formula $H_2N–CH(R)–COOH$, where R is a hydrocarbon or other organic substituent. In neutral aqueous solutions, 2-amino acids exist as *zwitterions* (from the German *zwitter*, "hybrid"), which have both a positive and a negative charge within the same species:

cation
(acidic solution)

zwitterion
(neutral solution)

anion
(basic solution)

In acidic solutions, zwitterions act as Brønsted–Lowry bases by accepting a proton and producing cations. In basic solutions, zwitterions act as Brønsted–Lowry acids by losing a proton and producing anions.

2-Amino acids are structural units of peptides, proteins and other biologically important macromolecules (*Structure 2.4*). Another type of amphiprotic organic species, anthocyanins, will be discussed later in this topic.

Activity

Consider the following species: HF, F^-, NH_3, NH_4^+, H_3PO_4, $H_2PO_4^-$, HPO_4^{2-}, PO_4^{3-}, Al_2O_3, Zn^{2+}. State, with a reason, which of these species can act as Brønsted–Lowry acids, Brønsted–Lowry bases, or both. Identify amphiprotic and amphoteric species.

The pH scale (*Reactivity 3.1.4*)

The value of [H⁺] in an aqueous solution can be very small, so [H⁺] is often represented by its negative decimal logarithm, known as the **potential of hydrogen**, pH:

$$pH = -\log[H^+]$$

pH is used to characterize the acidity or basicity of aqueous solutions over a broad range of H⁺ concentrations (table 1).

Acidic solutions with high concentration of H⁺ ions have low pH, while basic solutions with low [H⁺] have high pH. This follows from the definition $pH = -\log[H^+]$: the higher the value, the lower its negative logarithm. At 25°C (298 K), pure water and solutions with pH = 7 are called neutral, as they have equal concentrations of H⁺ and OH⁻ ions. The pH of an acidic solution is less than 7, while the pH of a basic solution is greater than 7. These definitions are summarized in table 2.

Solution	pH	[H⁺] and [OH⁻]
acidic	< 7	$[H^+] > [OH^-]$
neutral	7	$[H^+] = [OH^-]$
basic	> 7	$[H^+] < [OH^-]$

▲ Table 2 Acidic, neutral and basic aqueous solutions at 298 K

[H⁺]	pH	Example
1×10^{-14}	14	liquid drain cleaner
1×10^{-13}	13	bleach
1×10^{-12}	12	ammonia solution
1×10^{-11}	11	mild detergent
1×10^{-10}	10	toothpaste
1×10^{-9}	9	baking soda
1×10^{-8}	8	seawater
1×10^{-7}	7	pure water
1×10^{-6}	6	urine
1×10^{-5}	5	black coffee
1×10^{-4}	4	tomato juice
0.001	3	orange juice
0.01	2	vinegar
0.1	1	gastric juice
1	0	battery acid

▲ Table 1 The pH scale

If you know the concentration of H⁺(aq) in solution, you can work out the pH. A 0.100 mol dm⁻³ solution of hydrogen chloride contains 0.100 mol dm⁻³ H⁺(aq) ions. The pH value of this solution will be −log 0.100 = 1.00. As expected, 1.00 < 7, so the solution is acidic.

If the pH of a solution is known, the concentration of H⁺ ions in that solutions can be calculated using the formula $[H^+] = 10^{-pH}$. Hydrochloric acid with pH = 3 will have a hydrogen chloride concentration of $10^{-3} = 0.001$ mol dm⁻³.

The formulas $pH = -\log[H^+]$ and $[H^+] = 10^{-pH}$ are given in the data booklet.

Practice questions

6. Calculate the pH value for 0.0100 mol dm⁻³ aqueous solution of sulfuric acid.

7. Calculate the concentration of nitric acid in its aqueous solution with pH = 4.2.

In a modern chemical laboratory, the pH of a solution can be measured with high accuracy and precision by a digital pH meter (figure 2). Alternatively, the pH can be estimated using **universal indicator**, which gradually changes colour across the whole pH range, as shown in table 1.

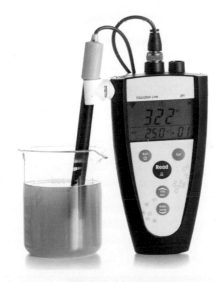

▲ Figure 2 Digital pH meter

The pH scale

In this task, you will explore the logarithmic nature of the pH scale by performing serial dilutions and measuring the pH of the resulting solutions. You will also consider the advantages and disadvantages of measuring pH with a probe compared to pH paper.

Relevant skills

- Tool 1: Measuring pH of a solution
- Tool 1: Carrying out dilutions
- Tool 2: Use sensors
- Tool 2: Represent data in graphical form
- Tool 3: Calculations involving logarithmic functions
- Inquiry 1: Calibrate measuring apparatus, including sensors

Safety

- Wear eye protection
- Wear gloves when handling chemicals (hydrochloric acid and sodium hydroxide at these concentrations are irritants)
- Dispose of all substances appropriately

Materials

- $1.0 \, mol \, dm^{-3}$ hydrochloric acid, HCl(aq)
- $1.0 \, mol \, dm^{-3}$ sodium hydroxide, NaOH(aq)
- graduated pipettes
- distilled water
- test tubes and rack
- pH probe, including calibration solutions
- pH (universal indicator) paper

Instructions

1. Review the section in the *Tools for chemistry* chapter that discusses serial dilutions.

2. Prepare and label five consecutive solutions of HCl(aq), by performing serial dilutions. Start with $1.0 \, mol \, dm^{-3}$ HCl(aq), and carry out a ten-fold dilution, resulting in a $0.10 \, mol \, dm^{-3}$ solution.

3. Create another solution that is ten times more dilute again and so on until you end up with five solutions, each with a concentration different from the next by a factor of 10.

4. Repeat step 2 for sodium hydroxide.

5. Arrange your five HCl(aq) solutions and five NaOH(aq) solutions in a rack. Calculate the concentration of hydrogen ions in each. Calculate the pH.

6. Calibrate the pH probe by following the instructions in the accompanying manual.

7. Using the pH probe, measure the pH of each solution.

8. Measure the pH of each solution again, but this time using pH paper.

Questions

1. Comment on the difference between the pH values measured with the probe and those measured with pH paper.

2. What are the advantages and disadvantages of pH probes and pH paper?

3. Comment on any differences you observe between your calculated and measured pH values.

4. Which pH values have greater reliability: those you measured with a probe or the ones you obtained with pH paper?

5. Table 3 shows the change in pH of a strong acid during the addition of a strong base. Enter the data into a spreadsheet.

6. Plot a graph showing the relationship between pH and volume of NaOH(aq) added.

7. Using the spreadsheet, convert the pH values into concentrations of hydrogen ions.

8. Plot a graph showing the relationship between hydrogen ion concentration and volume of NaOH(aq) added.

9. Based on your graphs, discuss the advantages of using a logarithmic scale such as the pH scale for measuring acidity.

Volume of NaOH(aq) added / cm³	0	1.0	2.0	3.0	4.0	4.5	5.0	5.5	6.0	7.0	8.0
pH	1.30	1.35	1.60	2.15	3.6	10.00	12.80	13.30	13.40	13.50	13.51

▲ Table 3 pH measured on the addition of a strong base to a strong acid

Ionic product of water (*Reactivity 3.1.5*)

Water dissociates into H^+ and OH^- ions:

$$H_2O(l) \rightleftharpoons H^+(aq) + OH^-(aq)$$

Like any other equilibrium, the dissociation of water can be characterized by its equilibrium constant, K:

$$K = \frac{[H^+][OH^-]}{[H_2O]}$$

where $[H^+]$, $[OH^-]$ and $[H_2O]$ are the equilibrium concentrations of participating species. As the density of pure water at room temperature is approximately $1{,}000\,g\,dm^{-3}$, the mass of each dm^3 of water is about $1{,}000\,g$, and the amount of water in $1\,dm^3$ is:

$$\frac{m(H_2O)}{M(H_2O)} = \frac{1000\,g}{18.02\,g\,mol^{-1}} = 55.5\,mol$$

Nearly all water molecules exist in undissociated form. Therefore, in any dilute solution, the equilibrium concentration of water will have approximately the same value, $55.5\,mol\,dm^{-3}$.

The equation for K can be rearranged as follows:

$$K \times [H_2O] = [H^+][OH^-]$$

Both factors on the left, K and $[H_2O]$, are constants, so their product is also a constant. This new constant, $K_w = K \times [H_2O]$, is called the **ionic product of water**:

$$K_w = [H^+][OH^-]$$

At room temperature (25 °C, or 298 K), $K_w = 1.00 \times 10^{-14}$.

The equation for K_w and its value at 298 K are given in the data booklet.

The value of $K_w = 1.00 \times 10^{-14}$ can be used only for dilute aqueous solutions. In a concentrated solution (for example, in battery acid containing 30% H_2SO_4 by mass), the concentration of water becomes significantly less than $55.5\,mol\,dm^{-3}$, which affects the value of K_w. In addition, the K_w value increases with temperature. In this book, we will assume that all solutions are dilute and have a temperature of 25 °C (298 K), so the K_w value remains constant.

If we remove a proton from a water molecule, a hydroxide ion will be left, so the amounts and concentrations of these ions in pure water will always be equal. Therefore, $[H^+] = [OH^-]$ and so each concentration can be found as a square root of K_w:

$$[H^+] = [OH^-] = \sqrt{1.00 \times 10^{-14}} = 1.00 \times 10^{-7}\,mol\,dm^{-3}$$

This means that for each $55.5\,mol$ of water, only $1.00 \times 10^{-7}\,mol$ exists as ions. In other words, only one in approximately 1,800,000,000 water molecules dissociates into $H^+(aq)$ and $OH^-(aq)$ ions, while the rest of the molecules stay in the form of $H_2O(l)$ species. This is the reason why water is represented as $H_2O(l)$ molecules rather than $H^+(aq)$ and $OH^-(aq)$ ions in ionic equations.

Acids and bases can affect the concentrations of $H^+(aq)$ and $OH^-(aq)$ ions in aqueous solutions. For example, if we dissolve hydrogen chloride in water, more $H^+(aq)$ ions will be produced:

$$HCl(aq) \rightarrow H^+(aq) + Cl^-(aq)$$

According to Le Châtelier's principle (*Reactivity 2.3*), the increased concentration of $H^+(aq)$ ions will shift the position of the following equilibrium to the left:

$$H_2O(aq) \rightleftharpoons H^+(aq) + OH^-(aq)$$

As a result, the concentration of $OH^-(aq)$ ions will decrease. Therefore, in an acidic solution, $[H^+]$ will always be greater than $[OH^-]$.

Worked example 2

Calculate the concentration of $OH^-(aq)$ ions in a $0.100 \, mol \, dm^{-3}$ solution of hydrogen chloride at 25 °C.

Solution

A $0.100 \, mol \, dm^{-3}$ HCl solution will contain $0.100 \, mol \, dm^{-3}$ $H^+(aq)$ ions:

$$HCl(aq) \rightarrow H^+(aq) + Cl^-(aq)$$

$$c, mol \, dm^{-3} \qquad 0.100 \qquad 0.100 \qquad 0.100$$

Since $K_w = [H^+][OH^-]$, we can find the concentration of $OH^-(aq)$ ions as follows:

$$1.00 \times 10^{-14} = 0.100 \times [OH^-]$$

$$[OH^-] = \frac{1.00 \times 10^{-14}}{0.100}$$

$$= 1.00 \times 10^{-13} \, mol \, dm^{-3}$$

As expected, $0.100 > 1.00 \times 10^{-13}$, so $[H^+] > [OH^-]$.

Activity

Calculate the concentrations of $H^+(aq)$ and $OH^-(aq)$ ions in a $0.0500 \, mol \, dm^{-3}$ aqueous solution of potassium hydroxide.

An addition of a base will increase the concentration of $OH^-(aq)$ ions in the solution and decrease the concentration of $H^+(aq)$ ions. Therefore, in a basic solution, $[H^+]$ will be lower than $[OH^-]$.

Properties of acids and bases (*Reactivity 3.1.6*)

Properties of acids

All Brønsted–Lowry acids must contain at least one **exchangeable** (weakly bound) hydrogen atom that can detach from the rest of the acid molecule. Exchangeable hydrogen atoms usually form bonds with highly electronegative atoms, such as oxygen, halogens or sulfur. In almost all organic acids, exchangeable hydrogen atoms are bonded to oxygen.

For example, hydrogen chloride, sulfuric acid and ethanoic (acetic) acid have the following structural formulas:

| hydrogen chloride | sulfuric acid | ethanoic acid |

The exchangeable hydrogen atoms are shown in red. In aqueous solutions, these hydrogen atoms dissociate and form $H^+(aq)$ (or $H_3O^+(aq)$) cations, while the remaining part of the acid produces an anion, also known as the **acid residue**. Here are three examples of acid dissociation:

$$HCl(aq) \rightarrow H^+(aq) + Cl^-(aq)$$

$$H_2SO_4(aq) \rightarrow 2H^+(aq) + SO_4^{2-}(aq)$$

$$CH_3COOH(aq) \rightleftharpoons H^+(aq) + CH_3COO^-(aq)$$

Notice that ethanoic acid contains four hydrogen atoms, but only one of them is exchangeable. To explain this, you need to consider the electronegativities of hydrogen ($\chi = 2.2$), carbon ($\chi = 2.6$) and oxygen ($\chi = 3.4$). Hydrogen and carbon have similar electronegativities ($\Delta\chi = 2.6 - 2.2 = 0.4$), so the C–H bond has low polarity and does not break easily. In contrast, the difference in electronegativity between hydrogen and oxygen is significant ($3.4 - 2.2 = 1.2$), so the O–H bond is highly polar. Since the bonding electron pair is shifted towards the more electronegative O atom, the less electronegative H atom develops a partial positive charge. As a result, it dissociates readily to form an $H^+(aq)$ ion.

In inorganic acids containing oxygen (**oxoacids**), all hydrogen atoms are usually bonded to oxygen, and so are exchangeable.

| hypochlorous acid | chlorous acid |

Depending on the number of exchangeable hydrogen atoms, acids are classified as:

- **monoprotic** (one exchangeable hydrogen atom), for example, HCl

- **diprotic** (two exchangeable hydrogen atoms), for example, H_2SO_4

- **triprotic** (three exchangeable hydrogen atoms), for example, H_3PO_4.

In contrast to inorganic acids, organic acids often contain both exchangeable and nonexchangeable hydrogen atoms. For example, both methanoic and ethanoic acids are monoprotic, even though their molecules contain two and four hydrogen atoms, respectively.

The nomenclature, structure and properties of organic acids are discussed in *Structure 3.2*.

Electronegativity and bond polarity are discussed in *Structure 2.2*. The electronegativity values for all elements are given in the data booklet.

Activity

Draw the structural formulas for the following oxoacids: chloric ($HClO_3$), perchloric ($HClO_4$), carbonic (H_2CO_3) and phosphoric (H_3PO_4).

The formulas and names of common acids and their anions are given in table 4. Along with the systematic names, many organic acids have trivial names, which are shown in brackets.

Acid			Anion	
Formula	Name	Strength	Formula	Name
HF	hydrogen fluoride	weak	F^-	fluoride
HCl	hydrogen chloride	strong	Cl^-	chloride
HBr	hydrogen bromide	strong	Br^-	bromide
HI	hydrogen iodide	strong	I^-	iodide
H_2S	hydrogen sulfide	weak	S^{2-}	sulfide
HCN	hydrogen cyanide	weak	CN^-	cyanide
HNO_3	nitric	strong	NO_3^-	nitrate
HNO_2	nitrous	weak	NO_2^-	nitrite
H_2SO_4	sulfuric	strong	SO_4^{2-}	sulfate
H_2SO_3	sulfurous	weak	SO_3^{2-}	sulfite
H_3PO_4	phosphoric	weak	PO_4^{3-}	phosphate
H_3PO_3	phosphorous	weak	PO_3^{3-}	phosphite
$HClO_4$	perchloric	strong	ClO_4^-	perchlorate
$HClO_3$	chloric	strong	ClO_3^-	chlorate
$HClO_2$	chlorous	weak	ClO_2^-	chlorite
$HClO$	hypochlorous	weak	ClO^-	hypochlorite
H_2CO_3	carbonic	weak	CO_3^{2-}	carbonate
HCOOH	methanoic (formic)	weak	$HCOO^-$	methanoate (formate)
CH_3COOH	ethanoic (acetic)	weak	CH_3COO^-	ethanoate (acetate)
$H_2C_2O_4$	ethanedioic (oxalic)	weak	$C_2O_4^{2-}$	ethanedioate (oxalate)

▲ Table 4 Common acids and their anions. Exchangeable protons are shown in red

Although the names *hydrogen chloride* and *hydrochloric acid* refer to the same substance, HCl, they have slightly different meanings in chemistry. When we say, "hydrogen chloride", we mean an individual compound, HCl, which is a gas under normal conditions, while "hydrochloric acid" is a solution of HCl in water. Therefore, it is incorrect to say, "a solution of hydrochloric acid", as "hydrochloric acid" already refers to a solution.

Similar problems may arise when we talk about sulfuric acid, which is often used as an aqueous solution but can also exist in pure form (so-called "100% sulfuric acid"). When this difference is important, we should always say, "aqueous sulfuric acid" when we refer to a solution, or "anhydrous sulfuric acid" when we refer to pure H_2SO_4.

An important characteristic of any acid is its strength. **Strong acids**, such as hydrogen chloride, dissociate completely in aqueous solutions. If we dissolve one mole of HCl in water, the resulting solution will contain one mole of hydrogen cations and one mole of chloride anions but no HCl molecules. In other words, the dissociation of HCl is irreversible, which is represented by the single arrow:

$$HCl(aq) \rightarrow H^+(aq) + Cl^-(aq)$$

In addition to hydrogen chloride, six other strong acids are listed in table 4. You are advised to memorize their formulas and names.

Weak acids, such as ethanoic acid, dissociate only to a small extent when dissolved in water. For example, table vinegar (an aqueous solution of ethanoic acid) contains both CH_3COOH molecules and the products of their dissociation, $H^+(aq)$ and $CH_3COO^-(aq)$ ions. The reversible nature of this process is represented by the equilibrium sign:

$$CH_3COOH(aq) \rightleftharpoons H^+(aq) + CH_3COO^-(aq)$$

Almost all organic and many inorganic acids are weak, so if an acid is not listed in table 4, it is safe to assume that it is weak. There are a few exceptions, but they are not discussed in DP chemistry.

The strengths of acids and bases have no direct relationship to the concentrations of their solutions, so the terms "strong" and "weak" should not be confused with "concentrated" and "dilute" (table 5). The colloquial phrases "strong solution" and "weak solution" will not be accepted in the IB assessments.

	Concentrated	Dilute
Strong	$10\ mol\ dm^{-3}$ HCl	$0.1\ mol\ dm^{-3}$ HCl
Weak	$10\ mol\ dm^{-3}$ CH_3COOH	$0.1\ mol\ dm^{-3}$ CH_3COOH

▲ **Table 5** Examples of solutions of acids with different strength and concentration

The strength of oxoacids generally increases with the oxidation state of the central atom. In turn, a higher oxidation state usually means that the acid molecule contains more oxygen atoms. For example, the nitrogen atom in weak nitrous acid, HNO_2, is bound to two oxygen atoms and has an oxidation state of +3. An addition of another oxygen increases the oxidation state of nitrogen from +3 to +5 and produces strong nitric acid, HNO_3. Similarly, strong sulfuric acid, H_2SO_4, has a higher oxidation state of sulfur and more oxygen atoms than weak sulfurous acid, H_2SO_3.

Binary acids (acids that consist of only two elements) demonstrate clear periodic trends: their strength increases along the period and down the group (figure 3). For example, in the third period, phosphine (PH_3) does not show any acidic properties in aqueous solutions, hydrogen sulfide (H_2S) is a weak acid, and hydrogen chloride (HCl) is a strong acid. Similarly, down group 17, hydrogen fluoride (HF) is a weak acid while the other three hydrogen halides (HCl, HBr and HI) are strong acids.

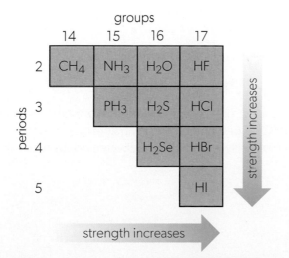

▲ **Figure 3** Periodic trends in the strength of binary acids

 Activity

Formulate the equations for the dissociation of hydrogen bromide, hydrogen cyanide and methanoic acid. Do not forget to use a single arrow for strong acids and an equilibrium sign for weak acids.

Oxidation states of elements were introduced in *Structure 3.1*.

 Activity

Write down the formulas of all oxoacids of chlorine and phosphorus from table 3. Deduce the oxidation states of these elements in each acid and outline how they affect the acid strength.

Diprotic or triprotic weak acids dissociate stepwise, for example:

$$H_2SO_3(aq) \rightleftharpoons H^+(aq) + HSO_3^-(aq)$$

$$HSO_3^-(aq) \rightleftharpoons H^+(aq) + SO_3^{2-}(aq)$$

The second proton dissociates to a much smaller extent than the first, so nearly all H^+ ions produced by a polyprotic acid are formed on its first dissociation step. The reason for that becomes clear if we consider the charges on the ions involved in the above equations. The first step produces $H^+(aq)$ and $HSO_3^-(aq)$ ions. These ions exert electrostatic attraction on each other, so pulling them apart requires some energy. On the second step, the electrostatic attraction between $H^+(aq)$ and $SO_3^{2-}(aq)$ ions is much greater, as the anion $SO_3^{2-}(aq)$ is doubly charged. As a result, the second step requires more energy, which makes this process less likely to occur.

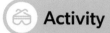

Activity

Formulate the equations for the stepwise dissociation of phosphoric acid.

Properties of bases

Despite the difference in chemical properties, bases show many similarities to acids in terms of their behaviour in aqueous solutions. Where an acid produces an $H^+(aq)$ ion, a base either produces an $OH^-(aq)$ ion (Arrhenius base) or accepts an $H^+(aq)$ ion (Brønsted–Lowry base). However, the general principles behind these processes remain the same and thus can be explained by similar concepts and equations.

Most inorganic bases are metal hydroxides, which contain a metal atom and one or more OH groups. The nature of the chemical bond between the metal and the OH group depends on the metal electronegativity. Alkali metals (Li, Na, K, Rb and Cs) and most group 2 metals (Mg, Ca, Sr and Ba) have very low electronegativities, so they form ionic hydroxides. Such hydroxides consist of a metal cation (M^{n+}) and one or more hydroxide anions (OH^-).

All ionic hydroxides are strong bases. Except for $Mg(OH)_2$ and $Ca(OH)_2$, they are readily soluble in water and fully dissociate into ions, for example:

$$NaOH(aq) \rightarrow Na^+(aq) + OH^-(aq)$$

$$Ba(OH)_2(aq) \rightarrow Ba^{2+}(aq) + 2OH^-(aq)$$

Calcium hydroxide is only slightly soluble in water, and magnesium hydroxide is almost insoluble. If an excess of such hydroxide is added to water, a heterogeneous equilibrium between the solid base and aqueous ions is established:

$$Mg(OH)_2(s) \rightleftharpoons Mg^{2+}(aq) + 2OH^-(aq)$$

$$Ca(OH)_2(s) \rightleftharpoons Ca^{2+}(aq) + 2OH^-(aq)$$

These hydroxides are strong bases, so their solutions contain no undissociated molecules of $Mg(OH)_2$ or $Ca(OH)_2$. The reversible nature of the above processes is caused by low solubility of these bases in water, not by their low strength.

Less active metals, such as beryllium, aluminium and all transition elements, form covalent hydroxides, in which the metal atom and the oxygen of the OH group are linked together by a polar covalent bond. For example, both $Fe(OH)_2$ and $Fe(OH)_3$ are covalent hydroxides. Almost all covalent hydroxides are weak bases. In addition, these hydroxides are virtually insoluble in water, so they only show their basic nature in reactions with acids.

Ammonia (NH_3) is one of the few inorganic bases that does not contain a metal. As discussed earlier, aqueous ammonia acts as a weak Brønsted–Lowry base by accepting a proton from an acid or water:

$$NH_3(aq) + H^+(aq) \rightleftharpoons NH_4^+(aq)$$

$$NH_3(aq) + H_2O(l) \rightleftharpoons NH_4^+(aq) + OH^-(aq)$$

In chemical equations, aqueous ammonia is often represented as ammonium hydroxide, $NH_4OH(aq)$, which is unstable and exists only in solutions:

$$NH_3(aq) + H_2O(l) \rightleftharpoons NH_4OH(aq)$$

The organic derivatives of ammonia, amines (*Structure 3.2*), contain one, two or three hydrocarbon substituents at the nitrogen atom:

| ammonia | methylamine | dimethylamine | trimethylamine |

Instead of methyl groups ($-CH_3$), amines may contain any other hydrocarbon substituents, such as ethyl ($-CH_2CH_3$), phenyl ($-C_6H_5$) and so on. Acid–base properties of amines are similar to those of ammonia, for example:

$$CH_3NH_2(aq) + H^+(aq) \rightleftharpoons CH_3NH_3^+(aq)$$

$$CH_3NH_2(aq) + H_2O(l) \rightleftharpoons CH_3NH_3^+(aq) + OH^-(aq)$$

Practice questions

8. Amines are organic derivatives of ammonia.

   a. Formulate molecular and ionic equations for the reaction of hydrochloric acid with the following amines:

      i. dimethylamine

      ii. trimethylamine.

   b. Identify conjugate acid–base pairs in each ionic equation and state the role (Brønsted–Lowry acid or Brønsted–Lowry base) of each species.

The anions of weak acids can also act as Brønsted–Lowry bases. For example, the ethanoate ion is produced by the dissociation of ethanoic acid:

$$CH_3COOH(aq) \rightleftharpoons H^+(aq) + CH_3COO^-(aq)$$

If we reverse this equation, the basic nature of the ethanoate ion will become obvious:

$$CH_3COO^-(aq) + H^+(aq) \rightleftharpoons CH_3COOH(aq)$$

The driving force of acid–base reactions is the formation of weak conjugates. For example, ethanoic acid, CH_3COOH, is a weak acid. Therefore, the dissociation of ethanoic acid is an unfavourable process, so only a small proportion of the acid exists as ions while most CH_3COOH molecules remain undissociated. In contrast, the reaction of a $CH_3COO^-(aq)$ ion (base) with a proton produces CH_3COOH (weak conjugate acid), so the equilibrium of this process is shifted almost completely to the right.

In aqueous solutions, ethanoate ions react with water, producing $OH^-(aq)$ ions:

$$CH_3COO^-(aq) + H_2O(l) \rightleftharpoons CH_3COOH(aq) + OH^-(aq)$$

This reaction involves two conjugate acid–base pairs: CH_3COO^- (base)/CH_3COOH (acid) and H_2O (acid)/OH^- (base). Water is a weaker acid than ethanoic acid, so the equilibrium of this reaction is shifted to the left.

Anions of polyprotic acids behave as polyprotic bases, for example:

$$CO_3{}^{2-}(aq) + H^+(aq) \rightleftharpoons HCO_3{}^-(aq)$$

$$HCO_3{}^-(aq) + H^+(aq) \rightleftharpoons H_2CO_3(aq)$$

These processes are similar to the stepwise dissociation of weak polyprotic acids, except that all reactions are now reversed.

Practice questions

9. Formulate the equations, showing the states of all species, in which the following ions act as Brønsted–Lowry bases:

    a. cyanide ion, CN^-

    b. phosphate ion, PO_4^{3-}

    c. hydrogenphosphate ion, HPO_4^{2-}.

Patterns and trends

Chemists classify substances based on patterns they observe. Three main acid–base classification systems have evolved over time: Arrhenius, Brønsted-Lowry and **Lewis**. According to Arrhenius theory, acids release H^+ ions in aqueous solution, whereas bases release OH^- ions. Brønsted–Lowry theory is based on the ability of species to donate or accept protons (H^+ ions). Both these theories are relevant in the study of acid–base systems in aqueous media. Lewis theory (which you will learn about in *Reactivity 3.4*) defines acids and bases in terms of their ability to accept or donate electron pairs and is therefore independent of the solvent. These theories represent related, but different, classification systems, each with its own advantages and disadvantages. For instance, Lewis theory covers a broad range of reactions, whereas Brønsted–Lowry theory underpins many of the pH calculations and equilibria you are familiar with.

What other classification systems have you encountered in chemistry? Why do scientists often classify their objects of study?

Reactions of acids and bases (*Reactivity 3.1.7*)

You have already seen how acids react with bases in neutralization reactions. In addition to these reactions, most acids react with metals, metal oxides and salts of weak acids, such as carbonates and hydrogencarbonates. The reactions of acids with **active metals** produce salts and hydrogen gas (figure 4), for example:

$$Mg(s) + 2HCl(aq) \rightarrow MgCl_2(aq) + H_2(g)$$

For strong acids, the actual reacting species are hydrogen ions, $H^+(aq)$, which can be shown by **ionic equations**. In the **total ionic equation**, all ions present in the solution are shown:

$$Mg(s) + 2H^+(aq) + 2Cl^-(aq) \rightarrow Mg^{2+}(aq) + 2Cl^-(aq) + H_2(g) \quad \text{total ionic equation}$$

Ions that do not participate in the reaction are called **spectator ions**. The anions of a strong acid are spectator ions, and hence the chloride anions can be cancelled out to give the **net ionic equation**:

$$Mg(s) + 2H^+(aq) \rightarrow Mg^{2+}(aq) + H_2(g) \quad \text{net ionic equation}$$

In contrast, ionic equations involving weak acids must show the acids in the molecular form, as they are less likely to dissociate:

$$Mg(s) + 2CH_3COOH(aq) \rightarrow Mg(CH_3COO)_2(aq) + H_2(g) \quad \text{molecular equation}$$

$$Mg(s) + 2CH_3COOH(aq) \rightarrow Mg^{2+}(aq) + 2CH_3COO^-(aq) + H_2(g) \quad \text{ionic equation}$$

In the last example, there are no spectator ions, so the total and net ionic equations are identical.

Strong and weak acids can be distinguished by comparing the rates of their reactions with an active metal (figure 4). However, such comparison will be valid only if the concentrations of both acids are equal, as the reaction rate depends on the reactant concentrations (*Reactivity 2.2*).

Reactions of acids with metal oxides produce salts and water. For example:

$$MgO(s) + 2HCl(aq) \rightarrow MgCl_2(aq) + H_2O(l) \quad \text{molecular equation}$$

$$MgO(s) + 2H^+(aq) + 2Cl^-(aq) \rightarrow Mg^{2+}(aq) + 2Cl^-(aq) + H_2O(l) \quad \text{total ionic equation}$$

$$MgO(s) + 2H^+(aq) \rightarrow Mg^{2+}(aq) + H_2O(l) \quad \text{net ionic equation}$$

The last equation shows that magnesium oxide accepts two H^+ ions and thus acts as a Brønsted–Lowry base. Therefore, reactions of acids with metal oxides can be classified as neutralization reactions.

▲ **Figure 4** Reaction of magnesium metal with hydrochloric acid (left) and ethanoic acid (right) of equal concentrations

The term "active metals" refers to elements above hydrogen in the activity series, which is given in section 19 of the data booklet. Copper, silver and other metals after hydrogen in the activity series do not react with most acids and never produce hydrogen gas in such reactions.

Practice questions

10. Formulate the molecular and ionic equations for the reactions of

    a. lithium metal with ethanoic acid

    b. aluminium metal with dilute sulfuric acid.

11. Formulate the molecular and ionic equations for the reaction of iron(III) oxide, $Fe_2O_3(s)$, with:

    a. hydrochloric acid

    b. dilute sulfuric acid.

▲ Figure 5 Reaction of baking soda with an acid

Metal carbonates also react with acids, producing unstable carbonic acid, H_2CO_3:

$$Na_2CO_3(aq) + 2HCl(aq) \rightarrow 2NaCl(aq) + H_2CO_3(aq)$$

Carbonic acid quickly decomposes into water and carbon dioxide, which bubbles out of the solution (figure 5):

$$H_2CO_3(aq) \rightarrow CO_2(g) + H_2O(l)$$

These two reactions are often combined together:

$$Na_2CO_3(aq) + 2HCl(aq) \rightarrow 2NaCl(aq) + CO_2(g) + H_2O(l)$$

Again, the net ionic equation reveals the nature of this process:

$2Na^+(aq)$ $+ CO_3^{2-}(aq) + 2H^+(aq) +$ **$2Cl^-(aq) \rightarrow 2Na^+(aq) + 2Cl^-(aq)$** $+ CO_2(g) + H_2O(l)$ total ionic equation

$$CO_3^{2-}(aq) + 2H^+(aq) \rightarrow CO_2(g) + H_2O(l)$$ net ionic equation

As in the previous example, the anion of weak carbonic acid, CO_3^{2-}, acts as a Brønsted–Lowry base by accepting two protons before decomposing to carbon dioxide and water.

Metal hydrogencarbonates, such as baking soda, $NaHCO_3$, react with acids in the same way as carbonates:

$$NaHCO_3(aq) + HCl(aq) \rightarrow NaCl(aq) + CO_2(g) + H_2O(l)$$ molecular equation

$Na^+(aq)$ $+ HCO_3^-(aq) + H^+(aq) +$ **$Cl^-(aq) \rightarrow Na^+(aq) + Cl^-(aq)$** $+ CO_2(g) + H_2O(l)$ total ionic equation

$$HCO_3^-(aq) + H^+(aq) \rightarrow CO_2(g) + H_2O(l)$$ net ionic equation

In this case, the hydrogencarbonate ion, HCO_3^-, acts as a Brønsted–Lowry base by accepting a proton before decomposing to carbon dioxide and water.

 Global impact of science

Acid deposition, a secondary pollutant, can take many different forms including rain, snow, fog and dry dust. The components of acid deposition (the primary pollutants) may be generated in one country and, depending on climate patterns, may be deposited in neighbouring countries or even different continents. There are no boundaries for acid deposition. The effects of acid rain may occur away from the actual source leading to widespread deforestation and pollution of lakes and river systems. National and regional environmental protection agencies throughout the world collaborate in an effort to better understand and control acid deposition. The US Environmental Protection Agency and the Acid Deposition Monitoring Network in East Asia (EANET) websites provide data that can be used in the discussion of secondary pollutants and their political implications.

How can our understanding of chemistry help to address environmental problems such as acid deposition?

Practice questions

12. Formulate the molecular and ionic equations for the reaction of dilute sulfuric acid with

    a. potassium hydrogencarbonate

    b. calcium carbonate.

Antacids

Heartburn and other symptoms of indigestion are caused by excess hydrochloric acid in the stomach. These symptoms can be alleviated by medicines known as **antacids** (figure 6). The active ingredients in antacids are weak bases, such as metal oxides, hydroxides, carbonates and hydrogencarbonates. All these compounds neutralize the excess acid.

Like any pharmaceutical drugs, antacids have various side effects. In particular, carbon dioxide produced in the body from the reaction of stomach acid with carbonates and hydrogencarbonates causes bloating and belching, while the intake of metal ions disturbs the balance of electrolytes in the body.

▲ Figure 6 "Milk of magnesia", a suspension of magnesium hydroxide in water, is a common antacid

TOK

Pharmaceutical drugs (for example, antacids) are subject to rigorous efficiency tests prior to approval by the relevant health authorities. This process often includes **placebo**-controlled clinical trials, where participants are placed into two groups. Half of the study participants are given the drug and the other half are administered an inactive placebo. Participants do not know which group they are in and therefore do not know whether they have received the active drug or the placebo. The results from the two groups are compared to ascertain that any observed physiological reactions are due to the treatment. If the drug works, it should have the desired therapeutic effect on the participants who received the active drug, but not on the members of the placebo group.

Sometimes, a therapeutic effect is observed in people who are given the placebo. This is known as the **placebo effect**. Statistical and methodological approaches can be used to assess and control the impact of the placebo effect on the results.

How could a participant's (or a doctor's) awareness of the existence and administration of placebos affect the results of the trial? To what extent is bias inevitable in the production of knowledge? The mechanism of the placebo effect is not fully understood. Are some things unknowable?

When balancing acid–base equations, you should use the following steps:

1. Balance all nonmetals except hydrogen and oxygen.

2. Balance all metals. If you need to change any stoichiometric coefficients deduced earlier, return to step 1.

3. Balance hydrogen. Again, if you need to change any coefficients, return to step 1.

4. At this point, the equation should be balanced already. To verify it, count the oxygen atoms in the reactants and products. If their numbers do not match, return to step 1.

In most cases, this strategy produces a balanced equation with the fewest trials. This strategy works well for acid–base reactions, in which none of the elements change their oxidation state. It can also be used for most reactions of acids with metals.

Worked example 3

Deduce the balanced equation for the following chemical reaction:

$$H_3PO_4 + CaCO_3 \rightarrow Ca_3(PO_4)_2 + CO_2 + H_2O$$

Solution

Besides H and O, the equation involves two nonmetals, P and C, which should be balanced first. There are two P atoms on the right but only one on the left, so we need to write "2" in front of H_3PO_4:

$$2H_3PO_4 + CaCO_3 \rightarrow Ca_3(PO_4)_2 + CO_2 + H_2O$$

Carbon seems to be balanced already (one atom on each side), so we can now look at metals. The only metal in the equation is Ca (one atom on the left, three atoms on the right), so we write "3" in front of $CaCO_3$:

$$2H_3PO_4 + 3CaCO_3 \rightarrow Ca_3(PO_4)_2 + CO_2 + H_2O$$

However, this changes the balance of carbon (three atoms on the left, one on the right), so we need to write "3" before CO_2:

$$2H_3PO_4 + 3CaCO_3 \rightarrow Ca_3(PO_4)_2 + 3CO_2 + H_2O$$

Steps 1 and 2 are complete, so the next element is hydrogen. There are six H atoms on the left but only two on the right, so we need to write "3" before H_2O:

$$2H_3PO_4 + 3CaCO_3 \rightarrow Ca_3(PO_4)_2 + 3CO_2 + 3H_2O$$

Step 3 is complete, so we need to check the last remaining element, oxygen. There are $2 \times 4 = 8$ O atoms in $2H_3PO_4$ and $3 \times 3 = 9$ O atoms in $3CaCO_3$, so we have a total of $8 + 9 = 17$ O atoms on the left. On the right, there are $4 \times 2 = 8$ O atoms in $Ca_3(PO_4)_2$, $3 \times 2 = 6$ O atoms in $3CO_2$ and 3 O atoms in $3H_2O$, so the total number of O atoms on the right is also $8 + 6 + 3 = 17$. Therefore, oxygen is balanced and so is the equation.

The balancing of redox equations (those involving changes in the oxidation state of participating elements) will be discussed in *Reactivity 3.2*.

Identification of parent acids and bases

Salts are often produced by neutralization reactions. One way to identify the parent acid and base for a particular salt is to formally split the salt into cation(s) and anion(s). For example, sodium sulfate, Na_2SO_4, consists of two Na^+ cations and one SO_4^{2-} anion:

$$Na_2SO_4 \rightarrow 2Na^+ + SO_4^{2-}$$

Now we can add hydroxide ions to cations and protons to anions according to their charges:

$$Na^+ + OH^- \rightarrow NaOH$$

$$2H^+ + SO_4^{2-} \rightarrow H_2SO_4$$

Therefore, the parent base and acid for Na_2SO_4 are NaOH and H_2SO_4, respectively.

The same result could be obtained by analysing the systematic name of the salt. The word "sodium" in "sodium sulfate" refers to sodium hydroxide, NaOH, while the word "sulfate" refers to sulfuric acid, H_2SO_4 (table 4).

For ammonium salts, such as NH_4Cl (ammonium chloride), the parent base could be either NH_3 (ammonia) or NH_4OH (ammonium hydroxide). Both answers will be accepted in IB assessments.

pH curves (*Reactivity 3.1.8*)

The unknown concentration of an acid or base in a solution can be determined by **titration** (*Reactivity 2.1*) using a standard solution of a base or acid, respectively. The reaction progress can be monitored using a digital pH meter (figure 7) and a data logger, which automatically records the pH of the reaction mixture as the standard solution is added to the analysed solution.

The pH data collected during a titration experiment can be plotted against the added volume of the standard solution, producing a **pH curve** (figures 8 and 9). The overall shape of the pH curve depends on the strengths and concentrations of the reactants and on the addition order.

When a strong acid, such as HCl, is titrated with a strong base, such as NaOH, the curve intercepts the y-axis at a low pH value (figure 8), as the analysed solution is strongly acidic. Typical concentrations of acids and bases used in titration experiments are from 0.01 to 1 mol dm^{-3}, so the initial pH may vary from 2 to 0.

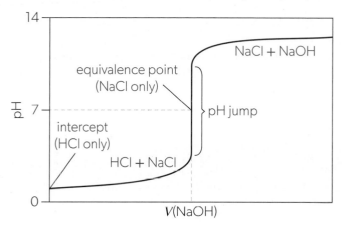

▲ **Figure 8** Typical pH curve for the titration of a strong acid (HCl) with a strong base (NaOH)

At the beginning of the titration, the pH of the mixture increases very slowly, as the solution still contains large excess of acid. For example, if the initial concentration of the acid were 0.1 mol dm^{-3} and half of the acid were neutralized, the pH of the mixture would increase from 1.0 to approximately 1.5.

As the acid concentration decreases, the pH curve becomes progressively steeper. At the equivalence point, the pH raises sharply to 7.0, as the acid is neutralized completely, and the reaction mixture contains only NaCl(aq):

$$HCl(aq) + NaOH(aq) \rightarrow NaCl(aq) + H_2O(l)$$

The pH continues to rise sharply immediately after the equivalence point, as an excess of NaOH(aq) makes the solution basic. When the excess of the titrant becomes very large, the curve flattens out and gradually approaches the pH of the standard NaOH(aq) solution, typically between 12 and 14.

To construct the complete pH curve, we need to continue the titration beyond the equivalence by adding excess titrant. In contrast, a titration experiment with an acid–base indicator must be stopped at or near the equivalence point, when the indicator changes colour. In both cases, the concentration of the analysed acid or base is calculated using the volume of the standard solution at the equivalence point, as explained in *Reactivity 2.1*. The use of acid–base indicators in titrations is discussed in the AHL section of this topic.

▲ **Figure 7** Acid–base titration with a pH meter

Activity

The pH curve shown in figure 8 represents the titration of 0.1 mol dm^{-3} HCl(aq) with 0.1 mol dm^{-3} NaOH(aq). Copy the axes and curve from figure 8 and sketch the second pH curve for the titration of 0.01 mol dm^{-3} HNO$_3$(aq) with 0.01 mol dm^{-3} KOH(aq).

Explain whether the changes in the nature and concentrations of the reactants will affect the following features of the curve:

*   y-axis intercept
*   pH at equivalence
*   pH at which the curve flattens out.

The construction and interpretation of pH curves involving weak acids and bases will not be assessed at SL. The shapes and features of these curves are discussed in the AHL section of this topic.

When a strong base is titrated with a strong acid, the pH curve is inverted (figure 9). Since the initial solution is basic, the curve intercepts the y-axis at a high pH value and declines gradually as the titrant is added. The equivalence is achieved at the same pH value (7.0), as the solution at that point contains only a neutral salt (in our example, NaCl). At the end of the titration, the curve flattens at a low pH, as the solution contains excess strong acid.

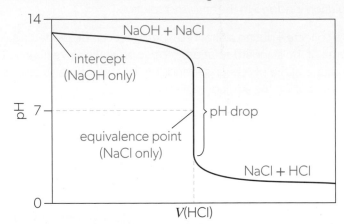

▲ Figure 9 Typical pH curve for the titration of a strong base (NaOH) with a strong acid (HCl)

The main features of pH curves for titration experiments involving strong acids and bases are summarized in table 6.

Analyte	Titrant	pH		
		y-axis intercept	equivalence	flattening out
strong acid	strong base	low	7.0	high
strong base	strong acid	high	7.0	low

▲ Table 6 Summary of pH curves for titrations involving strong acids and bases

Conductometric acid–base titration

The progress of an acid–base reaction can be monitored by measuring the conductivity of the reaction mixture. This is possible because the concentration of aqueous ionic species changes during the reaction. For example, in the reaction between hydrochloric acid and sodium hydroxide, the conductivity of the solution decreases because water is molecular, and all the other species present are fully dissociated into ions:

$$H^+(aq) + Cl^-(aq) + Na^+(aq) + OH^-(aq) \rightarrow Na^+(aq) + Cl^-(aq) + H_2O(l)$$

In this practical, you will perform a conductometric titration. You will analyse and interpret the resulting data in graphical form. Then, you will consider how the strength of the acid affects the shape of the titration curve.

Relevant skills

- Tool 1: Titration
- Tool 2: Use sensors
- Tool 3: Sketch graphs with labelled but unscaled axes
- Tool 3: Extrapolate graphs
- Tool 3: Interpret features of graphs

Safety

- Wear eye protection.
- Dilute hydrochloric acid and sodium hydroxide are irritants.
- Dispose of all substances appropriately.

Materials

- 0.01 mol dm⁻³ hydrochloric acid, HCl(aq)
- 0.1 mol dm⁻³ sodium hydroxide, NaOH(aq)
- 250 cm³ beaker
- burette, burette clamp and stand
- magnetic stir bar and stir plate
- conductivity probe

Instructions

1. You will titrate a known volume of 0.01 mol dm^{-3} HCl(aq) with 0.1 mol dm^{-3} NaOH(aq) and measure how the volume of NaOH(aq) added affects the conductivity. With reference to the ionic equation above, explain why the conductivity should decrease as the reaction approaches the equivalence point.

2. Predict and explain the shape of the graph of conductivity vs volume of NaOH(aq) added that you expect to obtain.

3. Use the conductivity probe, and any other necessary equipment, to prepare and measure the conductivity of:
   a. 0.1 mol dm^{-3} HCl(aq)
   b. 0.01 mol dm^{-3} NaOH(aq)
   c. distilled water

4. Using the measurements you obtained in step 3, make any necessary changes and refinements to the graph you sketched in step 2.

5. Set up the equipment as shown in figure 10:

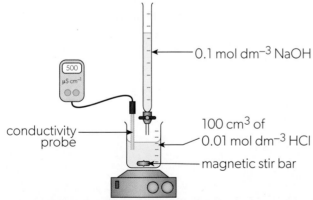

▲ Figure 10 Experimental apparatus for conductometric acid–base titration

6. Record the initial conductivity. Add the NaOH(aq) in small (~1 cm³) increments. Record the exact volume of NaOH(aq) added and the conductivity. Continue adding NaOH(aq) well past the equivalence point, so you can later compare the conductivity changes before and after the equivalence point.

Questions

1. Plot a graph showing conductivity vs. volume of NaOH(aq) added.

2. Draw two lines of best fit (one for the points before the equivalence point and another for the points after). Identify the equivalence point by extrapolating the two lines.

3. Compare and contrast this graph to your initial sketch. Explain any differences between the two plots.

4. Interpret and explain the shape of the graph, noting and explaining the following:
   a. change in conductivity before the equivalence point
   b. conductivity at the equivalence point
   c. change in conductivity after the equivalence point.

5. Interpret your graph to compare and explain the conductivity of H⁺(aq) ions and Na⁺(aq) ions.

6. The total volume of solution in the beaker was kept roughly constant by ensuring that the sodium hydroxide concentration was ten times that of hydrochloric acid. Suggest why the volume in the beaker should be kept constant in this practical.

A conductometric titration of 0.1 mol dm^{-3} ethanoic acid, $CH_3COOH(aq)$, with 0.1 mol dm^{-3} sodium hydroxide, NaOH(aq) added in 2.0 cm³ increments was carried out. The data recorded are shown in figure 11.

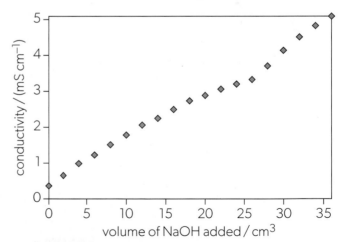

▲ Figure 11 Conductivity against volume of sodium hydroxide for the titration of 0.1 mol dm^{-3} $CH_3COOH(aq)$ with 0.1 mol dm^{-3} NaOH(aq). Source of data: Smith, C. K., Edionwe, E. and Michel, B., *J. Chem. Educ.*, 2010, 87, 11, 1217–1221

7. Compare and contrast the graph in figure 11 with the graph you obtained. Explain any differences you observe.

The pOH scale (*Reactivity 3.1.9*)

The **pOH scale** allows us to represent the basicity of aqueous solutions over a broad range of $OH^-(aq)$ concentrations. The pOH scale is particularly useful for solutions of Arrhenius bases, in which the concentration of hydroxide ions is related to the solution pOH as follows:

$$pOH = -\log[OH^-] \qquad [OH^-] = 10^{-pOH}$$

Basic solutions with high concentrations of $OH^-(aq)$ ions have low pOH values while acidic solutions with low concentrations of $OH^-(aq)$ ions have high pOH values.

The pOH scale is a logarithmic scale, like the pH scale. Working with logarithms requires some practice but greatly simplifies the calculations. The expression "pOH = 14" is more compact than "$[OH^-] = 1.00 \times 10^{-14}\,mol\,dm^{-3}$" and less likely to cause errors when written. Also, logarithms and p-numbers allow us to use addition and subtraction instead of multiplication and division:

$$\log(a \times b) = \log a + \log b$$

$$\log\left(\frac{a}{b}\right) = \log a - \log b$$

As a result, formulas with logarithms or p-numbers are easier to memorize and use in calculations (table 7).

Without p-numbers	With p-numbers
$[H^+][OH^-] = 1.00 \times 10^{-14}$	$pH + pOH = 14$
$[H^+] = \dfrac{1 \times 10^{-14}}{[OH^-]}$	$pH = 14 - pOH$
$[OH^-] = \dfrac{1 \times 10^{-14}}{[H^+]}$	$pOH = 14 - pH$

▲ **Table 7** Useful expressions involving K_w, $[H^+]$, $[OH^-]$, pH and pOH

Worked example 4

Calculate the pOH values for the following solutions:

a. $0.025\,mol\,dm^{-3}$ KOH(aq)

b. $0.025\,mol\,dm^{-3}$ H_2SO_4(aq).

Solution

a. First, write the equation for the dissociation of potassium hydroxide:

$$KOH(aq) \rightarrow K^+(aq) + OH^-(aq)$$

The concentration of KOH is equal to the concentration of hydroxide ions:

$$[OH^-] = [KOH] = 0.025\,mol\,dm^{-3}$$

Then, use the expression $pOH = -\log[OH^-]$ to determine pOH:

$$pOH = -\log 0.025 = 1.60$$

b. Write the equation for the dissociation of sulfuric acid:

$$H_2SO_4(aq) \rightarrow 2H^+(aq) + SO_4^{2-}(aq)$$

When one mole of sulfuric acid dissociates, it produces two moles of H^+ ions. Therefore, the concentration of H^+ ions is double that of sulfuric acid:

$$[H^+] = 2 \times [H_2SO_4] = 2 \times 0.025 \, mol \, dm^{-3} = 0.050 \, mol \, dm^{-3}$$

You can use the expression $K_w = [H^+][OH^-]$ to determine $[OH^-]$:

$$[OH^-] = \frac{K_w}{[H^+]} = \frac{1.00 \times 10^{-14}}{0.050} = 2.00 \times 10^{-13} \, mol \, dm^{-3}$$

Then, use the expression $pOH = -\log[OH^-]$ to determine pOH:

$$pOH = -\log(2.00 \times 10^{-13}) \approx 12.70$$

The same answer could be obtained using the formula $pH + pOH = 14$:

$$pH = -\log 0.050 \approx 1.30$$

$$pOH = 14 - 1.30 = 12.70$$

Practice questions

13. Calculate:

a. the pOH of a $5.0 \times 10^{-3} \, mol \, dm^{-3}$ solution of $Ba(OH)_2(aq)$

b. the concentration of hydroxide ions in a solution with $pOH = 4.70$.

Weak acids and bases (*Reactivity 3.1.10* and *Reactivity 3.1.11*)

As with any other equilibria, the dissociation of a weak acid can be characterized by an equilibrium constant. This equilibrium constant is known as the **acid dissociation constant**, K_a:

$$HA(aq) \rightleftharpoons H^+(aq) + A^-(aq) \qquad K_a = \frac{[H^+][A^-]}{[HA]}$$

Bases can be characterized by the **base dissociation constant**, K_b:

$$B(aq) + H_2O(l) \rightleftharpoons BH^+(aq) + OH^-(aq) \qquad K_b = \frac{[BH^+][OH^-]}{[B]}$$

Notice that the concentration of the solvent (water) is not included in the equilibrium constant expressions.

Stronger acids and bases dissociate to a greater extent and therefore have larger dissociation constants. Therefore, K_a and K_b values can be used to compare the relative strengths of different acids and bases.

Like $[H^+]$, $[OH^-]$ and K_w, the values of K_a and K_b are often expressed as p-numbers (their negative decimal logarithms):

$$pK_a = -\log K_a \qquad K_a = 10^{-pK_a}$$

$$pK_b = -\log K_b \qquad K_b = 10^{-pK_b}$$

In contrast to K_a and K_b, larger values of pK_a and pK_b correspond to weaker acids and bases, respectively (table 8). For example, methanoic acid ($pK_a = 3.75$) is stronger than ethanoic acid ($pK_a = 4.76$), while methylamine ($pK_b = 3.34$) is a stronger base than ammonia ($pK_b = 4.75$).

Acid	K_a	pK_a
HF	6.76×10^{-4}	3.17
HCOOH	1.78×10^{-4}	3.75
CH_3COOH	1.74×10^{-5}	4.76
HCN	6.17×10^{-10}	9.21

strength increases

Base	K_b	pK_b
$(CH_3)_2NH$	5.37×10^{-4}	3.27
CH_3NH_2	4.57×10^{-4}	3.34
NH_3	1.78×10^{-5}	4.75
$C_6H_5NH_2$	7.41×10^{-10}	9.13

▲ Table 8 Dissociation constants of weak acids and bases at 298 K

The dissociation constants for weak acids and bases can be determined experimentally by measuring the pH of their standard solutions. Conversely, if we know the dissociation constant for a weak acid or base and its concentration in the solution, we can calculate the pH of that solution, as shown in the worked examples below.

Worked example 5

A $0.0100 \, mol \, dm^{-3}$ solution of propanoic acid, $CH_3CH_2COOH(aq)$, has a pH of 3.44. Determine the pK_a of propanoic acid.

Solution

First of all, we need to consider all acid–base equilibria in the solution:

$$CH_3CH_2COOH(aq) \rightleftharpoons CH_3CH_2COO^-(aq) + H^+(aq) \qquad K_a = \frac{[CH_3CH_2COO^-][H^+]}{[CH_3CH_2COOH]}$$

$$H_2O(l) \rightleftharpoons H^+(aq) + OH^-(aq) \qquad K_w = [H^+][OH^-] = 1.00 \times 10^{-14}$$

Although $H^+(aq)$ ions are formed by the dissociation of both propanoic acid and water, typical K_a values of organic acids (table 8) are much higher than K_w. Therefore, we can assume that nearly all $H^+(aq)$ ions in the solution are produced by the acid. In such case, $[CH_3CH_2COO^-] \approx [H^+]$, which follows from the first equilibrium.

$$[CH_3CH_2COO^-] \approx [H^+] = 10^{-3.44} \approx 3.63 \times 10^{-4} \, mol \, dm^{-3}$$

Weak acids dissociate to a very small extent, so most of the propanoic acid in the solution exists as undissociated molecules, $CH_3CH_2COOH(aq)$. Therefore, we can assume that the equilibrium concentration of $CH_3CH_2COOH(aq)$ is approximately the same as its initial concentration. In that case:

$$[CH_3CH_2COOH] \approx 0.0100 \, mol \, dm^{-3}$$

When performing equilibrium calculations, always state any approximations you make and explain why these approximations are valid. In many cases, approximations greatly simplify the calculations, which otherwise would involve quadratic equations. The use of quadratic equations will not be required in examination papers, so any reasonable approximations will be accepted.

Substitute the values into the expression for K_a:

$$K_a = \frac{(3.63 \times 10^{-4})(3.63 \times 10^{-4})}{0.0100} \approx 1.32 \times 10^{-5}$$

Then use $pK_a = -\log K_a$ to determine pK_a:

$$pK_a = -\log(1.32 \times 10^{-5}) \approx 4.88.$$

The approximations made in this example could potentially reduce both the accuracy and precision of our calculations. However, the final answer is very close to the actual pK_a value of propanoic acid (4.87), so all the approximations are valid.

Worked example 6

Using the pK_a value from the previous example, calculate the pH of a $0.100\,mol\,dm^{-3}$ solution of propanoic acid.

Solution

We determined K_a to be 1.32×10^{-5} in the previous example. We can assume that $[CH_3CH_2COO^-] \approx [H^+]$, so $[CH_3CH_2COO^-][H^+] \approx [H^+]^2$, giving the following expression for K_a:

$$K_a = \frac{[H^+]^2}{[CH_3CH_2COOH]}$$

The concentration of propanoic acid is given in the question ($0.100\,mol\,dm^{-3}$). Substitute the values of K_a and $[CH_3CH_2COOH]$ into the above expression to determine $[H^+]$:

$$1.32 \times 10^{-5} = \frac{[H^+]^2}{0.100}$$

$$[H^+]^2 = 1.32 \times 10^{-6}$$

$$[H^+] = \sqrt{1.32 \times 10^{-6}} \approx 1.15 \times 10^{-3}\,mol\,dm^{-3}$$

Then use $pH = -\log[H^+]$ to determine pH:

$$pH = -\log(1.15 \times 10^{-3}) = 2.94.$$

Worked example 7

A $0.0100\,mol\,dm^{-3}$ solution of trimethylamine, $(CH_3)_3N(aq)$, has a pH of 10.90. Determine the pK_b of trimethylamine.

Solution

Similar to worked example 5, we need to consider all acid–base equilibria in the solution:

$$(CH_3)_3N(aq) + H_2O(l) \rightleftharpoons (CH_3)_3NH^+(aq) + OH^-(aq) \qquad K_b = \frac{[(CH_3)_3NH^+][OH^-]}{[(CH_3)_3N]}$$

$$H_2O(l) \rightleftharpoons H^+(aq) + OH^-(aq) \qquad K_w = [H^+][OH^-] = 1.00 \times 10^{-14}$$

Amines are stronger bases than water, so we can assume that all $OH^-(aq)$ ions in the solution are formed by the ionization of the amine. In such case, $[(CH_3)_3NH^+] \approx [OH^-]$. Since amines are weak bases, $[(CH_3)_3N] \approx 0.0100\,mol\,dm^{-3}$.

Use $pOH = 14 - pH$ to work out the value of pOH, and then use $[OH^-] = 10^{-pOH}$ to determine the concentration of hydroxide ions:

$$pOH = 14 - pH = 14 - 10.90 = 3.10$$

$$[OH^-] = 10^{-3.10} = 7.94 \times 10^{-4}\,mol\,dm^{-3}$$

Then substitute the values in the expression for K_b:

$$K_b = \frac{(7.94 \times 10^{-4})(7.94 \times 10^{-4})}{0.0100} \approx 6.31 \times 10^{-5}$$

$$pK_b = -\log(6.31 \times 10^{-5}) \approx 4.20$$

Our answer matches the actual pK_b of trimethylamine.

Worked example 8

Using the pK_b value from the previous example, calculate the pH of a $0.100\,mol\,dm^{-3}$ solution of trimethylamine.

Solution

K_b was determined to be 6.31×10^{-5} in the previous example. We can assume that $[(CH_3)_3NH^+]\approx[OH^-]$, so $[(CH_3)_3NH^+][OH^-]\approx[OH^-]^2$, giving the following expression for K_b:

$$K_b = \frac{[OH^-]^2}{[(CH_3)_3N]}$$

Substitute the values into the K_b expression:

$$6.31\times10^{-5} = \frac{[OH^-]^2}{0.100}$$

$$[OH^-]^2 = 6.31\times10^{-6}$$

$$[OH^-] = \sqrt{6.31\times10^{-6}} \approx 2.51\times10^{-3}\,mol\,dm^{-3}$$

Then determine the pOH, and finally the pH:

$$pOH = -\log(2.51\times10^{-3}) \approx 2.60$$

$$pH = 14 - 2.60 = 11.40$$

Practice questions

14. Calculate the pH values for the following solutions:

    a. $0.0200\,mol\,dm^{-3}$ hydrogen cyanide, $HCN(aq)$

    b. $5.00\times10^{-3}\,mol\,dm^{-3}$ phenylamine, $C_6H_5NH_2(aq)$.

    Refer to table 8.

A conjugate acid–base pair can be characterized by a single ionization constant, as the pK_a of the acid and pK_b of the base are related to each other. For example, the acid–base equilibria involving the acid HA and its conjugate base A^- can be represented as follows:

$$HA(aq) \rightleftharpoons H^+(aq) + A^-(aq) \qquad K_a = \frac{[H^+][A^-]}{[HA]}$$

$$A^-(aq) + H_2O(l) \rightleftharpoons HA(aq) + OH^-(aq) \quad K_b = \frac{[HA][OH^-]}{[A^-]}$$

When these two equations are added together, HA(aq) and A^-(aq) cancel out, giving the equation for the ionic product of water:

$$H_2O(l) \rightleftharpoons H^+(aq) + OH^-(aq) \qquad K_w = [H^+][OH^-]$$

According to table 1 from *Reactivity 2.3*, when two chemical equations are added together, the equilibrium constant of the resulting equation is equal to the product of the equilibrium constants of the individual equations. As a result, $K_w = K_a \times K_b$ and $pK_w = pK_a + pK_b$.

At 298 K, $pK_w = 14$, so:

$$pK_a = 14 - pK_b$$

$$pK_b = 14 - pK_a$$

Note that these equations are valid only for conjugate acid–base pairs, as the pK_a and pK_b values of non-conjugated acids and bases are not related to one another in any way.

The equation $K_w = K_a \times K_b$ can be rearranged as follows:

$$K_a = \frac{K_w}{K_b}$$

This equation shows the inverse relationship between the strengths of conjugates: the stronger the acid, the weaker its conjugate base, and the stronger the base, the weaker its conjugate acid.

Acid–base equilibria in solutions of salts (*Reactivity 3.1.12*)

As you already know, any salt can be considered as a product of a neutralization reaction between an acid and a base. The ions produced by the salt in an aqueous solution can react with water and form conjugate acids and/or bases, therefore affecting the solution's pH. The reactions between salts and water are called **hydrolysis** reactions. The direction and extent of hydrolysis reactions depend on the strengths of the acid and base that form the salt. The four possible combinations of parent acids and bases are strong acid–strong base, strong acid–weak base, weak acid–strong base and weak acid–weak base.

Salts of strong acids and strong bases

Both the cation and the anion in salts of this type have strong conjugates (base and acid, respectively), so they do not undergo hydrolysis. For example, sodium chloride dissociates in aqueous solutions as follows:

$$NaCl(aq) \rightarrow Na^+(aq) + Cl^-(aq)$$

The hypothetical reaction of $Na^+(aq)$ with water would produce the strong base $NaOH(aq)$, which cannot exist in aqueous solutions in undissociated form:

$$Na^+(aq) + H_2O(l) \xrightarrow{\quad\times\quad} \text{no reaction}$$

Similarly, the reaction of $Cl^-(aq)$ with water would produce the strong acid $HCl(aq)$, which also does not exist in aqueous solutions in molecular form:

$$Cl^-(aq) + H_2O(l) \xrightarrow{\quad\times\quad} \text{no reaction}$$

Therefore, neither ion is involved in acid–base equilibria.

The only equilibrium in solutions of salts formed by strong acids and strong bases is the dissociation of water:

$$H_2O(l) \rightleftharpoons H^+(aq) + OH^-(aq)$$

The $H^+(aq)$ and $OH^-(aq)$ ions are produced in equal amounts, so $[H^+] = [OH^-]$, and the solution remains neutral (pH = 7). In other words, the presence of a salt formed by a strong acid and a strong base has no effect on the solution's pH.

Salts of strong acids and weak bases

In salts of this type, the hydrolysis involves cations only, as the formation of weak conjugates is a favourable process. For example, ammonium chloride in aqueous solutions produces the following ions:

$$NH_4Cl(aq) \rightarrow NH_4^+(aq) + Cl^-(aq)$$

The conjugate of the $Cl^-(aq)$ ion is a strong acid, $HCl(aq)$, so the chloride anion does not undergo hydrolysis. In contrast, the conjugate of the $NH_4^+(aq)$ ion is a weak base, $NH_3(aq)$, so the ammonium cation itself behaves as a weak acid:

$$NH_4^+(aq) \rightleftharpoons NH_3(aq) + H^+(aq)$$

Practice questions

15. The pK_a and pK_b expressions for the hydrogencarbonate ion, $HCO_3^-(aq)$, are 10.32 and 7.64, respectively.

    a. Formulate the equations that represent the acid–base equilibria characterized by K_a and K_b of the hydrogencarbonate ion.

    b. Calculate the pK_a value for carbonic acid, $H_2CO_3(aq)$, and the pK_b value for the carbonate ion, $CO_3^{2-}(aq)$, at 298 K.

According to table 8, $pK_b(NH_3) = 4.75$, so $pK_a(NH_4^+) = 14 - 4.75 = 9.25$. Therefore, the acidity of $NH_4^+(aq)$ is comparable to that of hydrogen cyanide, $HCN(aq)$, which is a very weak acid with a pK_a of 9.21.

The last equation can be expanded to include a molecule of water and show the true nature of the hydrolysis process:

$$NH_4^+(aq) + H_2O(l) \rightleftharpoons NH_3(aq) + H_3O^+(aq)$$

As mentioned earlier, the H^+ symbol is used as a shorthand equivalent of H_3O^+, so the hydrolysis of the ammonium ion can be represented by either of the two equations. In IB assessments, the use of H^+ or H_3O^+ ions will be equally accepted.

The excess $H^+(aq)$ or $H_3O^+(aq)$ ions means that $[H^+] > [OH^-]$, so the solution becomes acidic (pH < 7).

Salts of weak acids and strong bases

In salts of this type, the hydrolysis involves anions only. For example, sodium ethanoate in aqueous solutions produces the following ions:

$$CH_3COONa(aq) \rightarrow CH_3COO^-(aq) + Na^+(aq)$$

The conjugate of the $CH_3COO^-(aq)$ ion is a weak acid, $CH_3COOH(aq)$, which is formed in the reversible reaction of the ethanoate ion with water:

$$CH_3COO^-(aq) + H_2O(l) \rightleftharpoons CH_3COOH(aq) + OH^-(aq)$$

This reaction produces excess $OH^-(aq)$ ions, so $[H^+] < [OH^-]$, and the solution becomes basic (pH > 7).

Anions of polyprotic weak acids accept a single proton to form their conjugates, for example:

$$CO_3^{2-}(aq) + H_2O(l) \rightleftharpoons HCO_3^-(aq) + OH^-(aq)$$

$$HCO_3^-(aq) + H_2O(l) \rightleftharpoons H_2CO_3(aq) + OH^-(aq)$$

Notice that each equation involves only one molecule of water and produces a single hydroxide anion.

Salts of weak acids and weak bases

In these salts, both the cation and the anion undergo hydrolysis. For example, ammonium ethanoate produces the following ions:

$$CH_3COONH_4(aq) \rightarrow CH_3COO^-(aq) + NH_4^+(aq)$$

Both ions have weak conjugates, so they react reversibly with water:

$$CH_3COO^-(aq) + H_2O(l) \rightleftharpoons CH_3COOH(aq) + OH^-(aq)$$

$$NH_4^+(aq) + H_2O(l) \rightleftharpoons NH_3(aq) + H_3O^+(aq)$$

or

$$NH_4^+(aq) \rightleftharpoons NH_3(aq) + H^+(aq)$$

Practice question

16. Using table 8, calculate the pK_b value for the ethanoate anion, $CH_3COO^-(aq)$, and show that this ion is a weak base. Compare the strength of this base with other basic species listed in table 8.

These reactions produce both $H^+(aq)$ and $OH^-(aq)$ ions, so the acidity (or basicity) of the solution depends on the relative strengths of the parent acid and base. If the conjugate acid for the salt anion is slightly stronger than the conjugate base for the salt cation, the solution will be slightly acidic (pH < 7); otherwise, it will be slightly basic (pH > 7).

In the case of ammonium ethanoate, the pK_a of the parent acid (4.76) and the pK_b of the parent base (4.75) are almost identical, so the solution pH will be 7.0. For many salts, the relative strengths of their parent acid and base are similar to each other, so $[H^+] \approx [OH^-]$ and pH $\approx$ 7.

Hydrolysis of salts: summary

The pH of a salt solution depends on the relative strengths of the parent acid and base, as shown in table 9.

Parent acid	Parent base	Hydrolysis	Ions produced	Solution pH
strong	strong	none	$[H^+] = [OH^-]$	7
strong	weak	cation only	$[H^+] > [OH^-]$	< 7
weak	strong	anion only	$[H^+] < [OH^-]$	> 7
weak	weak	cation and anion	$[H^+] \approx [OH^-]$	$\approx$ 7

▲ **Table 9** Hydrolysis of salts and solution pH

This table can be summarized by the following informal rules:

*   "hydrolysis is for the weak"

*   "the stronger wins".

These rules emphasize the fact that the pH of a salt solution depends on the relative strengths of the parent acid and base: if the acid is stronger, the solution will be acidic, and if the base is stronger, the solution will be basic.

Practice questions

17. Formulate the equations for acid–base equilibria in aqueous solutions of the following salts:

    a. potassium cyanide, KCN

    b. potassium sulfate, K_2SO_4

    c. methylammonium methanoate, $HCOONH_3CH_3$

    d. methylammonium bromide, CH_3NH_3Br.

    For each solution, predict whether it will be neutral, acidic or basic.

pH curves of strong and weak acids and bases (*Reactivity 3.1.13*)

The shape of the pH curve in an acid–base titration depends on the strengths of the acid and the base. There are four distinct shapes that correspond to the following types of acid–base pair: strong acid–strong base, strong acid–weak base, weak acid–strong base and weak acid–weak base.

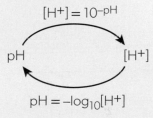
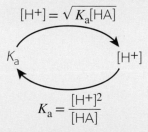

pH curves involving strong acids and strong bases

The pH curves for titration experiments involving strong acids and bases are shown in figures 8 and 9 in the SL section of this topic. At typical concentrations of the analyte and titrant (approximately $0.1 \, mol \, dm^{-3}$ each), these pH curves have the following features:

- y-axis intercept at pH $\approx$ 1 (when the analyte is an acid) or pH $\approx$ 13 (when the analyte is a base)

- gradual rise or fall in pH at the beginning of the titration

- sharp rise or drop in pH near the equivalence point

- equivalence at pH = 7 (no hydrolysis of the salt)

- flattening out at the end of the titration to pH $\approx$ 13 (when the titrant is a base) or pH $\approx$ 1 (when the titrant is an acid).

pH curves involving weak acids and strong bases

A typical pH curve for the titration of a weak acid with a strong base is shown in figure 12. Although the overall shape of this curve is somewhat similar to that in figure 8, there are several important differences:

- The curve intercepts the y-axis at a higher pH. This is because the weak acid dissociates only partially and therefore produces a lower concentration of $H^+(aq)$ ions.

- There is a **buffer region** before the equivalence point. This is when the solution contains both components of a weak conjugate acid–base pair.

- The jump in pH near the equivalence point is smaller than that in figure 8.

- The equivalence is achieved at a pH greater than 7, as the salt anion undergoes hydrolysis and produces $OH^-(aq)$ ions.

Buffer solutions will be discussed later in this topic.

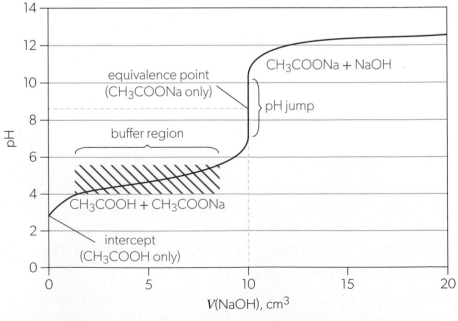

▲ Figure 12 pH curve for the titration of $0.1 \, mol \, dm^{-3}$ $CH_3COOH(aq)$ (weak acid) with $0.1 \, mol \, dm^{-3}$ $NaOH(aq)$ (strong base)

The final part of the curve in figure 12 is very similar to that in figure 8, as in both cases the solution contains excess strong base, NaOH(aq). Therefore, both curves flatten out at pH ≈ 13.

The stage of the titration at which exactly one-half of the acid has been neutralized is known as the **half-equivalence point**. The pH at half-equivalence is equal to the pK_a of the weak acid. In our example, the weak ethanoic acid dissociates as follows:

$$CH_3COOH(aq) \rightleftharpoons CH_3COO^-(aq) + H^+(aq)$$

$$K_a = \frac{[CH_3COO^-][H^+]}{[CH_3COOH]}$$

At half-equivalence, $[CH_3COO^-] = [CH_3COOH]$, so $K_a = [H^+]$ and $pK_a = pH$.

Figure 12 shows that the equivalence is achieved at $V(NaOH) = 10\,cm^3$, so the half-equivalence occurs at $V(NaOH) = 5\,cm^3$. At this point, the solution pH is approximately 4.8, which is very close to the pK_a of ethanoic acid (4.76, table 8).

The pH curve for the titration of a strong base with a weak acid would be almost a mirror image of figure 12, except that the buffer region would be much longer and occur in the second half of the curve. Titrations of this type are uncommon, as the use of a weak acid as a titrant has no practical value.

pH curves involving strong acids and weak bases

A typical pH curve for the titration of a weak base with a strong acid is shown in figure 13. The overall shape of this curve is similar to that in figure 9, with the following differences:

- The curve intercepts the y-axis at a lower pH. This is because the weak base dissociates only partially and thus produces a lower concentration of $OH^-(aq)$ ions.

- There is a buffer region before the equivalence point.

- The drop in pH near the equivalence point is smaller than that in figure 9.

- The equivalence is achieved at pH < 7, as the salt cation undergoes hydrolysis and produces $H^+(aq)$ ions.

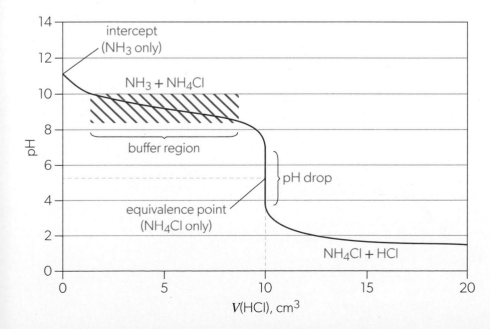

◀ **Figure 13** pH curve for the titration of $0.1\,mol\,dm^{-3}$ $NH_3(aq)$ (weak base) with $0.1\,mol\,dm^{-3}$ HCl(aq) (strong acid)

The final part of the curve in figure 13 is very similar to that in figure 9, as in both cases the solution contains excess strong acid, HCl(aq). Therefore, both curves flatten out at pH ≈ 1.

The half-equivalence point of this titration experiment can be used to determine the pK_b of the weak base. The acid–base equilibrium in the solution of ammonia can be represented as follows:

$$NH_3(aq) + H_2O(l) \rightleftharpoons NH_4^+(aq) + OH^-(aq)$$

$$K_b = \frac{[NH_4^+][OH^-]}{[NH_3]}$$

At half-equivalence, $[NH_4^+] = [NH_3]$, so $K_b = [OH^-]$ and $pK_b = pOH$.

Figure 13 shows that the half-equivalence is achieved at $V(HCl) = 5\ cm^3$. At this point, the solution pH is approximately 9.2, so $pOH = 14 - 9.2 = 4.8$. The last value is very close to the pK_b of ammonia (4.75, table 8).

The pH curve for the titration of a strong acid with a weak base would be almost a mirror image of figure 13, except that the buffer region would be much longer and occur in the second half of the curve. Like the titration of a strong base with a weak acid, titrations of this type are uncommon.

pH curves involving weak acids and weak bases

A typical pH curve for the titration of a weak acid with a weak base is shown in figure 14. This curve has the following features:

- y-axis intercept at pH ≈ 3, as the weak acid dissociates only partially and therefore produces a relatively low concentration of $H^+(aq)$ ions

- a buffer region before the equivalence point

- no sharp change in pH near the equivalence point

- the equivalence is achieved at pH ≈ 7, as both the cation and anion of the salt undergo hydrolysis

- very gradual flattening out above pH ≈ 9 because of the second buffer region after the equivalence point.

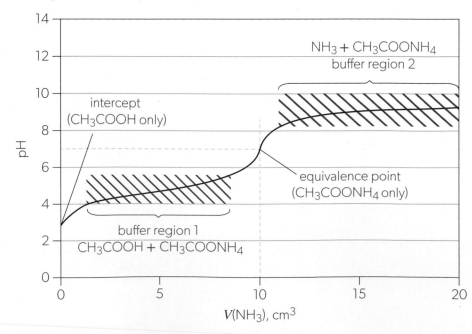

▶ Figure 14 pH curve for the titration of 0.1 mol dm^{-3} CH$_3$COOH(aq) (weak acid) with 0.1 mol dm^{-3} NH$_3$(aq) (weak base)

The two buffer regions of the pH curve in figure 14 are formed by different conjugates. Before the equivalence point, the weak acid and its anion form a buffer. After the equivalence point, the weak base and its cation form another buffer.

The half-equivalence point is achieved at $V(NH_3) = 5\,cm^3$. At this point, the solution pH is equal to the pK_a of the weak ethanoic acid (4.76, table 8). At $V(NH_3) = 20\,cm^3$, the solution contains equal concentrations of $NH_3(aq)$ and $NH_4^+(aq)$, so $pOH = pK_b(NH_3) = 4.75$ and $pH = 14 - 4.75 = 9.25$.

The pH curve for the titration of a weak base with a weak acid would be almost a mirror image of figure 14. Titrations involving both a weak acid and a weak base have little practical value, as the equivalence point is difficult to determine.

Acid–base indicators (*Reactivity 3.1.14*)

Acid–base indicators are weak acids or bases that differ in colour from their conjugates. If the indicator (HInd) is a weak acid, its dissociation scheme can be represented as follows:

$$HInd(aq) \rightleftharpoons H^+(aq) + Ind^-(aq)$$

<table>
<tr><td>conjugate</td><td>conjugate</td></tr>
<tr><td>acid</td><td>base</td></tr>
</table>

When the pH is much lower than pK_a, the indicator exists predominantly in its protonated form, HInd(aq). At pH $\gg pK_a$, the indicator loses a proton and forms the conjugate base $Ind^-(aq)$. If HInd(aq) and $Ind^-(aq)$ absorb visible light at different wavelengths, their solutions have different colours (figure 15).

For each indicator, the colour change occurs within a certain pH range, typically from $pK_a - 1$ to $pK_a + 1$, which is known as the **transition range** of the indicator. However, some indicators have slightly narrower or broader transition ranges (table 10).

Indicator	pK_a	pH transition range	Colour	
			Acid	**Base**
methyl orange	3.7	3.1–4.4	red	yellow
bromothymol blue	7.0	6.0–7.6	yellow	blue
phenolphthalein	9.6	8.3–10.0	colourless	pink

▲ **Table 10** Common acid–base indicators

The universal indicator is a mixture of several acid–base indicators with overlapping transition ranges. As a result, the universal indicator has no specific transition range, and its colour changes gradually over the whole pH range from 0 to 14, as shown in table 1.

Acid–base indicators are often used in titration experiments, as they allow chemists to monitor the reaction progress by observing the solution colour and stop the titration when the colour changes. This moment, known as the **end point** of the titration, depends on the pK_a of the indicator. The pK_a of the indicator approximately corresponds to the pH at the end point.

> ## Activity
>
> - Sketch the pH curve for the titration of $0.1\,mol\,dm^{-3}$ $NH_3(aq)$ with $0.1\,mol\,dm^{-3}$ $CH_3COOH(aq)$.
> - Identify and label the following features of the curve: intercept with the pH axis, equivalence point, buffer regions and the points where $pH = pK_a(CH_3COOH)$ and $pOH = pK_b(NH_3)$.

The absorption spectra of chemical compounds are discussed in *Structure 1.3*.

▲ **Figure 15** Methyl orange indicator is red in acidic solutions and yellow in neutral and alkaline solutions

The colours and transition ranges of common acid–base indicators are given in section 18 of the data booklet.

Identifying appropriate indicators (*Reactivity 3.1.15*)

A suitable indicator for a titration should have a transition range that includes the pH at the equivalence point in a pH curve. When a titration involves a strong acid and a strong base, the equivalence is achieved at pH = 7, so the best indicator for this type of titration is bromothymol blue, which changes colour within the pH range 6.0–7.6 (table 10). However, all three common indicators would produce satisfactory results due to the large pH jump at the equivalence point in strong acid–strong base titrations (figure 16).

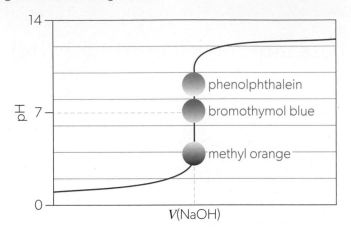

▲ **Figure 16** Titration of a strong acid with a strong base using various indicators

The amount of indicator added to the analysed solution can affect the titration result. Each indicator is a weak Brønsted–Lowry acid or base, so when the indicator changes colour, it reacts with the titrant. The more indicator you add, the greater systematic error you introduce. On the other hand, adding too little indicator to the solution makes it difficult to see the solution colour. Therefore, you should always try to use just enough indicator to make the colour change clearly visible.

In titrations involving a weak acid and a strong base, the equivalence is achieved at pH > 7, so the best indicator for this type of titration is phenolphthalein, which also changes colour at pH > 7 (figure 17). Bromothymol blue could also be used. Methyl orange cannot be used, as it would produce a very large systematic error.

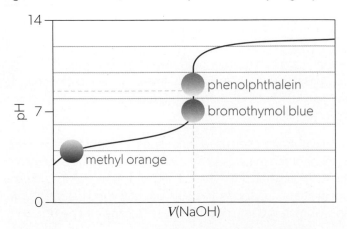

▲ **Figure 17** Titration of a weak acid with a strong base using various indicators

In titrations involving a strong acid and a weak base, the equivalence is achieved at pH < 7, so the best indicator for this type of titration is methyl orange with a transition range closest to the equivalence point (figure 18). Bromothymol blue could also be used, whereas phenolphthalein is not suitable for this type of titration.

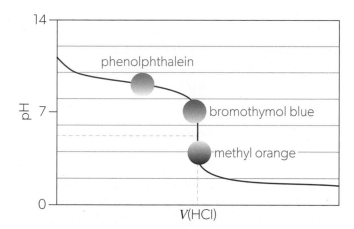

▲ **Figure 18** Titration of a weak base with a strong acid using various indicators

Section 18 of the data booklet lists many other acid–base indicators, including methyl red with the transition range 4.4–6.2. This range includes the equivalence point (pH ≈ 5.5), so methyl red is a better choice than methyl orange.

Worked example 9

An acid–base titration produces sodium methanoate, HCOONa, at the equivalence point. Identify an appropriate indicator for this titration.

Solution

First, you need to identify the parent acid and base for sodium methanoate. You can do this by splitting the salt into the cation and anion:

$$HCOONa \rightarrow HCOO^- + Na^+$$

Then, you formally add OH^- to the cation and H^+ to the anion:

$$Na^+ + OH^- \rightarrow NaOH$$

$$HCOO^- + H^+ \rightarrow HCOOH$$

Therefore, the titration involves a strong base, NaOH, and a weak acid, HCOOH, so the pH at the equivalence point will be greater than 7. The most suitable indicator for this type of titration is phenolphthalein (figure 18). You can also use any other indicator with $pK_a > 7$, such as phenol red (see section 18 of the data booklet).

Titrations involving a weak acid and a weak base cannot be performed using acid–base indicators. In such titrations, the change in pH is very gradual during the whole experiment (figure 14), so the colour of the solution will also change gradually, and the end point will be impossible to determine. The equivalence point in titrations of this type can only be determined using a pH meter.

Practice questions

18. Using table 10, identify the most suitable acid–base indicator for the titration of:

    a. dimethylamine, $(CH_3)_2NH(aq)$, with sulfuric acid, $H_2SO_4(aq)$

    b. barium hydroxide, $Ba(OH)_2(aq)$, with hydrochloric acid, HCl(aq)

    c. methanoic acid, HCOOH(aq), with sodium hydroxide, NaOH(aq).

19. Identify an appropriate indicator for an acid–base titration that produces ammonium bromide, $NH_4Br(aq)$, at the equivalence point. Refer to table 10 and section 18 of the data booklet.

Buffer solutions (*Reactivity 3.1.16*)

In many chemical and biological experiments, it is important to maintain a constant pH of the reaction mixture. This can be achieved using **buffer solutions**. These solutions resist changes in pH when small amounts of acids or bases are added to the solution.

Each buffer solution contains both components of a conjugate acid–base pair, in which both the acid and the base are weak. When a strong acid is added to a buffer, it is neutralized by the weak base. Similarly, when a strong base is added, it is neutralized by the weak acid. As a result, the solution pH remains almost unchanged.

Acid–base buffers are extremely efficient in maintaining their pH. For example, if we add a single drop (about $0.05\,cm^3$) of $0.1\,mol\,dm^{-3}$ HCl(aq) to $100\,cm^3$ of pure water, the pH of the resulting solution will drop by 2.7, from 7.0 to approximately 4.3. However, if we add the same quantity of HCl(aq) to $100\,cm^3$ of a typical phosphate buffer (pH $\approx$ 7.0) used in biological experiments, the pH of the solution will decrease by less than 0.001. This is below the detection limit of most laboratory pH meters.

Buffer solutions are very important components of all living organisms. For example, the pH of human blood is kept within a very narrow range of 7.35–7.45 units by several buffer systems that involve hydrogencarbonate and hydrogenphosphate ions, carbon dioxide and proteins.

A simple buffer solution can be prepared from the weak ethanoic acid and its salt, sodium ethanoate. When these two compounds are dissolved in water, the following processes take place:

$$CH_3COOH(aq) \rightleftharpoons CH_3COO^-(aq) + H^+(aq)$$

$$CH_3COONa(aq) \rightarrow CH_3COO^-(aq) + Na^+(aq)$$

When a strong acid, such as HCl(aq), is added to this solution, it is neutralized by the weak conjugate base of the buffer system:

$$H^+(aq) + CH_3COO^-(aq) \rightarrow CH_3COOH(aq)$$

strong	weak	weak
acid	conjugate base	conjugate acid

Similarly, when a strong base, such as NaOH(aq), is added to this solution, it is neutralized by the weak conjugate acid of the buffer system:

$$OH^-(aq) + CH_3COOH(aq) \rightarrow CH_3COO^-(aq) + H_2O(l)$$

strong	weak	weak
base	conjugate acid	conjugate base

The neutralization reactions of a strong acid and a strong base by different components of the buffer system are known as the **buffer action**. The nature of the buffer action is always the same, regardless of the buffer components. In all cases, a strong acid reacts with the buffer's conjugate base and releases the buffer's conjugate acid, while a strong base reacts with the buffer's conjugate acid and releases the buffer's conjugate base.

In other words, a strong acid is replaced with a weak acid, and a strong base is replaced with a weak base. Since weak acids and bases dissociate only to a small extent, they have very little effect on the solution pH.

An acid–base buffer can work efficiently only if both components of its conjugate pair are weak and present in the solution in sufficient concentrations. This can usually be achieved when the conjugate acid and the conjugate base originate from different compounds. A solution of a single compound, such as a weak acid, a weak base or a salt, may contain a high concentration of only one conjugate, but not both at the same time.

Buffer solutions are often classified according to the acid (or base) and the salt used for their preparation. For example, the ethanoate buffer contains the weak acid $CH_3COOH(aq)$ and its anion $CH_3COO^-(aq)$, and the ammonia buffer contains the weak base $NH_3(aq)$ and its cation $NH_4^+(aq)$. If specific salts, such as $CH_3COONa(s)$ or $NH_4Cl(s)$, were used for preparing the solutions, the same buffers are sometimes called "a weak acid and its salt" and "a weak base and its salt", respectively. Common types of acid–base buffers are listed in table 11.

Type	Example	Conjugate acid	Conjugate base	pK_a
weak acid and its anion	ethanoate buffer	$CH_3COOH(aq)$	$CH_3COO^-(aq)$	4.76
weak base and its cation	ammonia buffer	$NH_4^+(aq)$	$NH_3(aq)$	9.25
anions of two acid salts	phosphate buffer	$H_2PO_4^-(aq)$	$HPO_4^{2-}(aq)$	7.20
anions of an acid salt and a normal salt	carbonate buffer	$HCO_3^-(aq)$	$CO_3^{2-}(aq)$	10.32

▲ Table 11 Common acid–base buffers

In a laboratory, buffer solutions can be prepared by various methods. For example, the ethanoate buffer can be made as follows:

1. solid sodium ethanoate and liquid ethanoic acid are dissolved in water

2. solutions of sodium ethanoate and ethanoic acid are mixed together

3. excess ethanoic acid is mixed with sodium hydroxide

4. excess sodium ethanoate solution is mixed with hydrochloric acid.

Regardless of the method, the resulting buffer will contain the same two components, the weak acid $CH_3COOH(aq)$ and its conjugate base $CH_3COO^-(aq)$. In method 3, the conjugate base will form as follows:

$CH_3COOH(aq) + NaOH(aq) \rightarrow CH_3COONa(aq) + H_2O(l)$ molecular equation

$CH_3COOH(aq) + OH^-(aq) \rightarrow CH_3COO^-(aq) + H_2O(l)$ net ionic equation

In method 4, the conjugate acid will form as follows:

$CH_3COONa(aq) + HCl(aq) \rightarrow CH_3COOH(aq) + NaCl(aq)$ molecular equation

$CH_3COO^-(aq) + H^+(aq) \rightarrow CH_3COOH(aq)$ net ionic equation

Practice question

20. Explain, using ionic equations, the buffer action of a solution containing ammonium cations, $NH_4^+(aq)$, and ammonia, $NH_3(aq)$.

Activity

Suggest **three** methods for the preparation of the ammonia buffer.

If the buffer solutions prepared by different methods contain the same concentrations of both the conjugate acid and the conjugate base, the properties of these buffers will also be the same.

pH of buffer solutions (*Reactivity 3.1.17*)

The pH of a buffer solution depends on the ratio of the components in the conjugate acid–base pair and the dissociation constant of the weak conjugate. The acid–base equilibrium in a buffer solution containing a weak acid HA(aq) and its conjugate base A⁻(aq) is characterized by the K_a of the weak acid:

$$HA(aq) \rightleftharpoons H^+(aq) + A^-(aq) \qquad K_a = \frac{[H^+][A^-]}{[HA]}$$

To find the concentration of H⁺(aq) ions, we need to rearrange the K_a expression as follows:

$$[H^+] = K_a \times \frac{[HA]}{[A^-]}$$

We can then take the logarithm of every factor to determine the pH:

$$\log[H^+] = \log K_a + \log\frac{[HA]}{[A^-]}$$

Add a minus sign to every term in the equation:

$$-\log[H^+] = -\log K_a - \log\frac{[HA]}{[A^-]}$$

Finally, subsitute in pH = $-\log[H^+]$ and $pK_a = -\log K_a$, and use the rule that $-\log \frac{x}{y} = \log\frac{y}{x}$:

$$pH = pK_a + \log\frac{[A^-]}{[HA]}$$

The last equation, known as the **Henderson–Hasselbalch equation**, can be used to calculate the pH of any buffer solution of known composition. Alternatively, it can be used to find the ratio of the concentrations of the conjugate acid and base in the buffer solution at a known pH.

An important consequence from the Henderson–Hasselbalch equation is that the pH of a buffer solution is not affected by dilution. If the solution is diluted by a certain factor, both [A⁻] and [HA] decrease by the same factor. Therefore, the value of $\log\frac{[A^-]}{[HA]}$ remains unchanged, and so does the solution pH. However, a buffer solution cannot be diluted infinitely without changing its pH. An infinitely diluted aqueous solution of any substance at 298 K will have a pH of 7.00.

The ability of a buffer to resist changes in pH on addition of acids and bases is also limited because the amounts of the weak conjugate acid and base in the solution are finite. At the point when either of the weak conjugates is used up, the buffer ceases to function, and the solution pH changes significantly on further addition of an acid or base.

Practice questions

21. Calculate the pH for the buffer solutions containing the following compounds:

    a. 0.50 mol dm⁻³ NaH₂PO₄(aq) and 0.20 mol dm⁻³ Na₂HPO₄(aq)

    b. 0.25 mol dm⁻³ NH₃(aq) and 0.50 mol dm⁻³ NH₄Cl(aq).

The pK_a values for conjugate acids are given in table 10.

Worked example 10

Calculate the pH of an ethanoate buffer containing $0.100\,mol\,dm^{-3}$ $CH_3COOH(aq)$ and $0.200\,mol\,dm^{-3}$ $CH_3COONa(aq)$.

Solution

The conjugate acid is $CH_3COOH(aq)$ and the conjugate base is $CH_3COO^-(aq)$, which is produced by the dissociation of sodium ethanoate:

$$CH_3COONa(aq) \rightarrow CH_3COO^-(aq) + Na^+(aq)$$

All salts are strong electrolytes, so sodium ethanoate dissociates completely, and $[CH_3COO^-] = 0.200\,mol\,dm^{-3}$.

According to table 8, $pK_a(CH_3COOH) = 4.76$, so:

$$pH = 4.76 + \log\frac{0.200}{0.100} = 5.06$$

Worked example 11

A buffer solution with a pH of 11.00 was prepared by the reaction of methylamine, $CH_3NH_2(aq)$, with hydrochloric acid, $HCl(aq)$.

a. Identify the conjugate acid–base pair in this buffer and state the role of each species.

b. State, with a reason, which of the two reactants was in excess.

c. Calculate the mole ratio of the conjugate acid to the conjugate base in the solution.

Solution

a. The reaction between hydrochloric acid and methylamine proceeds as follows:

$$CH_3NH_2(aq) + HCl(aq) \rightarrow CH_3NH_3Cl(aq) \qquad \text{molecular equation}$$
$$CH_3NH_2(aq) + H^+(aq) \rightarrow CH_3NH_3^+(aq) \qquad \text{net ionic equation}$$

The net ionic equation involves two species that differ by a single proton. Therefore, the conjugate acid is methylammonium, $CH_3NH_3^+(aq)$, and the conjugate base is methylamine, $CH_3NH_2(aq)$.

Note that the solution contains two more conjugate acid–base pairs, $H_3O^+(aq)/H_2O(l)$ and $H_2O(l)/OH^-(aq)$. However, none of these pairs can form a buffer, as $H_3O^+(aq)$ is a strong acid, while $OH^-(aq)$ is a strong base.

b. A buffer must contain *both* components of the conjugate acid–base pair. If hydrochloric acid were in excess, all methylamine (weak conjugate base) would be consumed, and the solution could not act as an acid–base buffer. Therefore, methylamine was in excess.

c. According to table 8, $pK_b(CH_3NH_2) = 3.34$, so $pK_a(CH_3NH_3^+) = 14 - 3.34 = 10.66$. Substituting pK_a and pH into $pH = pK_a + \log\frac{[A^-]}{[HA]}$ gives:

$$11.00 = 10.66 + \log\frac{[CH_3NH_2]}{[CH_3NH_3^+]}$$

Rearrange in terms of $\log\frac{[CH_3NH_2]}{[CH_3NH_3^+]}$:

$$\log\frac{[CH_3NH_2]}{[CH_3NH_3^+]} = 0.34$$

Simplify the expression by making each term the exponential of 10:

$$\frac{[CH_3NH_2]}{[CH_3NH_3^+]} = 10^{0.34} = 2.19$$

Therefore, $[CH_3NH_3^+] : [CH_3NH_2] = 1 : 2.19$. Note that $n = c \times V$, so the mole ratio is equal to the ratio of concentrations.

End of topic questions

Topic review

1. Using your knowledge from the *Reactivity 3.1* topic, answer the guiding question as fully as possible:
   What happens when protons are transferred?

Exam-style questions
Multiple-choice questions

2. A solution of acid **X** has a pH of 1 while a solution of acid **Y** has a pH of 3. Which statement is correct?

   A $[X] > [Y]$

   B Acid **X** is stronger than acid **Y**

   C $[H^+]$ in the solution of **X** is three times higher than $[H^+]$ in the solution of **Y**

   D $[OH^-]$ in the solution of **X** is lower than $[OH^-]$ in the solution of **Y**

3. Which pair of species is a conjugate acid–base pair?

   A H^+ and OH^-

   B H_3O^+ and H_2O

   C HCl and NaOH

   B H_2CO_3 and CO_3^{2-}

AHL

4. Which indicators can be used for the titration of ammonia, $NH_3(aq)$, with sulfuric acid, $H_2SO_4(aq)$?

   I   Bromocresol green ($pK_a = 4.7$)

   II Methyl red ($pK_a = 5.1$)

   III Phenol red ($pK_a = 7.9$)

   A I and II only

   B I and III only

   C II and III only

   D I, II and III

Extended-response questions

5. Identify the chemical formulas of parent acids and bases for the following salts:

   a. $Ca(NO_3)_2$

   b. $Fe_2(SO_4)_3$

   c. NH_4HCO_3

6. For each salt from question 5, formulate and balance **one** molecular equation that produces that salt from the parent acid and base.

7. Two unlabelled bottles contain solutions of a strong monoprotic acid, HX(aq), and a weak monoprotic acid, HY(aq), of the same concentration.

   a. Suggest how these acids can be distinguished using solid calcium carbonate, $CaCO_3(s)$.

   b. Formulate molecular and net ionic equations for the reaction of each acid with calcium carbonate.

8. Deduce the formulas of conjugate acids and bases for each species listed in table 12. The first two rows are filled for you as examples.

Species	Conjugate acid	Conjugate base
H_2O	H_3O^+	OH^-
Cl^-	HCl	does not exist
HF		
NH_3		
$(CH_3)_3N$		
HCO_3^-		
CO_3^{2-}		

▲ Table 12 Conjugate acids and bases

9. 2-Amino acids exist as *zwitterions*, which have both a positive and a negative charge within the same species:

$$\underset{\substack{| \\ \text{R} \\ \text{cation}}}{\overset{+}{H_3N}-CH-COOH} \underset{+ H^+}{\overset{- H^+}{\rightleftharpoons}} \underset{\substack{| \\ \text{R} \\ \text{zwitterion}}}{\overset{+}{H_3N}-CH-COO^-} \underset{+ H^+}{\overset{- H^+}{\rightleftharpoons}} \underset{\substack{| \\ \text{R} \\ \text{anion}}}{H_2N-CH-COO^-}$$

Identify **two** conjugate acid–base pairs involved in these equilibria and state the role of each species.

10. The K_w value at 10 °C is 3.47 times lower than that at 25 °C.

   a. Calculate the pH of pure water at 10 °C.

   b. Discuss whether pure water at 10 °C is acidic, basic or neutral.

11. Calculate the pH for the following solutions:

   a. 0.015 mol dm⁻³ HNO_3(aq)

   b. 0.010 mol dm⁻³ H_2SO_4(aq)

   c. 0.020 mol dm⁻³ KOH(aq)

12. A 0.100 dm³ sample of 0.020 mol dm⁻³ KOH(aq) was mixed with 0.900 dm³ of water. Calculate the pH of the final solution. Assume that solution volumes are additive.

13. Calculate:

   a. the pOH of a 0.015 mol dm⁻³ solution of sodium hydroxide, NaOH(aq)

   b. the concentration of hydroxide ions in a solution with pOH = 9.50.

14. A 0.020 mol dm⁻³ solution of benzoic acid, C_6H_5COOH(aq), has a pH of 2.95.

   a. Determine the pK_a of benzoic acid.

   b. Using the pK_a value from part **a**, calculate the pH of a 0.10 mol dm⁻³ solution of benzoic acid.

15. Calculate the pH values for the following solutions:

   a. 0.010 mol dm⁻³ methylamine, CH_3NH_2(aq)

   b. 2.0×10⁻³ mol dm⁻³ ethanoic acid, CH_3COOH(aq). Refer to table 8.

16. The pK_a and pK_b values for the dihydrogenphosphate ion, $H_2PO_4^-$(aq), are 7.20 and 11.88, respectively.

   a. Formulate the equations that represent the acid–base equilibria characterized by K_a and K_b of the dihydrogenphosphate ion.

   b. Calculate the pK_a value for phosphoric acid, H_3PO_4(aq), and the pK_b value for the hydrogenphosphate ion, HPO_4^{2-}(aq).

17. Formulate the equations for acid–base equilibria in aqueous solutions of the following salts:

   a. sodium methanoate, HCOONa

   b. potassium iodide, KI

   c. ammonium cyanide, NH_4CN

   d. trimethylammonium chloride, $(CH_3)_3NHCl$.

   For each solution, predict whether it will be neutral, acidic or basic.

The pK_a and pK_b values for weak acids and bases are given in table 8.

18. Calculate the pH values for buffer solutions containing the following compounds:

   a. 0.25 mol dm⁻³ HCOOH(aq) and 0.50 mol dm⁻³ HCOONa(aq);

   b. 0.50 mol dm⁻³ CH_3NH_2(aq) and 0.20 mol dm⁻³ CH_3NH_3Cl(aq).

19. Explain, using ionic equations, the buffer action of the solutions from the previous question.

20. Using table 10, identify the most suitable acid–base indicator for the titration of:

   a. methylamine, CH_3NH_2(aq), with hydrochloric acid, HCl(aq);

   b. hydrogen cyanide, HCN(aq), with potassium hydroxide, KOH(aq).

What happens when electrons are transferred?

In a reaction where electrons are transferred, one species will lose electrons—the species is **oxidized**—and a different species will gain electrons—the species is **reduced**. It is impossible to have **oxidation** of one species without **reduction** of another. These reactions are referred to as **redox** reactions.

Some redox reactions are spontaneous and exothermic. If the electron transfer from one species to another is made to occur through a wire, such as in electrochemical cells or batteries, the energy released in the process can be used to power appliances. Other redox reactions are not spontaneous and require energy to occur, such as in electroplating, or the reduction of aluminium oxide, Al_2O_3, to produce aluminium metal.

Understandings

Reactivity 3.2.1—Oxidation and reduction can be described in terms of electron transfer, change in oxidation state, oxygen gain/loss or hydrogen loss/gain.

Reactivity 3.2.2—Half-equations separate the processes of oxidation and reduction, showing the loss or gain of electrons.

Reactivity 3.2.3—The relative ease of oxidation and reduction of an element in a group can be predicted from its position in the periodic table. The reactions between metals and aqueous metal ions demonstrate the relative ease of oxidation of different metals.

Reactivity 3.2.4—Acids react with reactive metals to release hydrogen.

Reactivity 3.2.5—Oxidation occurs at the anode and reduction occurs at the cathode in electrochemical cells.

Reactivity 3.2.6—A primary (voltaic) cell is an electrochemical cell that converts energy from spontaneous redox reactions to electrical energy.

Reactivity 3.2.7—Secondary (rechargeable) cells involve redox reactions that can be reversed using electrical energy.

Reactivity 3.2.8—An electrolytic cell is an electrochemical cell that converts electrical energy to chemical energy by bringing about non-spontaneous reactions.

Reactivity 3.2.9—Functional groups in organic compounds may undergo oxidation.

Reactivity 3.2.10—Functional groups in organic compounds may undergo reduction.

Reactivity 3.2.11—Reduction of unsaturated compounds by the addition of hydrogen lowers the degree of unsaturation.

Reactivity 3.2.12—The hydrogen half-cell $H^+(aq) + e^- \rightleftharpoons \frac{1}{2}H_2(g)$ is assigned a standard electrode potential of zero by convention. It is used in the measurement of standard electrode potential, $E^\ominus$.

Reactivity 3.2.13—Standard cell potential, $E^\ominus_{cell}$, can be calculated from standard electrode potentials. $E^\ominus_{cell}$ has a positive value for a spontaneous reaction.

Reactivity 3.2.14—The equation $\Delta G^\ominus = -nFE^\ominus_{cell}$ shows the relationship between standard change in Gibbs energy and standard cell potential for a reaction.

Reactivity 3.2.15—During electrolysis of aqueous solutions, competing reactions can occur at the anode and cathode, including the oxidation and reduction of water.

Reactivity 3.2.16—Electroplating involves the electrolytic coating of an object with a metallic thin layer.

AHL

Oxidation and reduction (*Reactivity 3.2.1*)

Reduction and oxidation can be defined in several ways:

1. in terms of the loss and gain of oxygen
2. in terms of the gain and loss of hydrogen
3. in terms of electron transfer
4. in terms of oxidation state.

1. Oxidation and reduction in terms of oxygen gain/loss

According to the first definition, oxidation is a reaction where a substance combines with oxygen. Examples of this type of oxidation reaction include the combustion of metals to form metal oxides:

$$2Mg(s) + O_2(g) \rightarrow 2MgO(s)$$

$$4Fe(s) + 3O_2(g) \rightarrow 2Fe_2O_3(s)$$

During aerobic respiration, oxygen reacts with glucose to form carbon dioxide and water. This can also be described as an oxidation reaction:

$$C_6H_{12}O_6(aq) + 6O_2(g) \rightarrow 6CO_2(g) + 6H_2O(l)$$

According to the first definition, reduction is a reaction where oxygen is removed from a substance. Examples of this type of reduction reaction include the reduction of nickel(II) oxide by carbon to produce pure nickel and carbon monoxide:

$$NiO(s) + C(s) \rightarrow Ni(s) + CO(g)$$

In all redox reactions, one species is oxidized, and another is reduced. If the substance being oxidized gains oxygen, then the substance being reduced loses oxygen. In the experiment in figure 2, copper(II) oxide is being reduced, losing oxygen, and hydrogen gas is being oxidized, gaining oxygen:

$$CuO(s) + H_2(g) \rightarrow Cu(s) + H_2O(g)$$

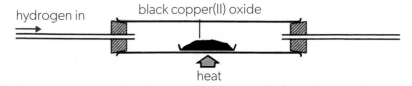

▲ Figure 2 Experimental set-up for the reduction of copper(II) oxide by hydrogen

In some cases, one substance can be simultaneously reduced and oxidized in the same reaction. This is known as **disproportionation**. For example, potassium chlorate, $KClO_3$, decomposes on heating as follows:

$$4KClO_3(s) \rightarrow 3KClO_4(s) + KCl(s)$$

In this reaction, some formula units of $KClO_3$ are reduced to KCl by losing oxygen while other formula units are oxidized to $KClO_4$ by gaining oxygen.

▲ Figure 1 In the oxidation of iron, iron(III) oxide, or rust, is produced

2. Oxidation and reduction in terms of hydrogen loss/gain

Oxidation can also be considered as a loss of hydrogen. For example, in the reaction between hydrogen chloride and oxygen, hydrogen chloride loses hydrogen to form chlorine gas:

$$4HCl(g) + O_2(g) \rightarrow 2Cl_2(g) + 2H_2O(g)$$

Reduction can be considered as the addition of hydrogen to a species. An example of such a reaction is the hydrogenation of ethene:

$$C_2H_4(g) + H_2(g) \rightarrow C_2H_6(g)$$

This reaction typically requires Ni(s) as a heterogeneous catalyst.

The process of heterogeneous catalysis is detailed in *Structure 3.1*.

3. Oxidation and reduction in terms of electron transfer

During a reaction, if a species loses electrons, it is oxidized, and if a species gains electrons, it is reduced. A useful mnemonic for remembering this is **OIL RIG**:

OIL: **O**xidation **I**s **L**oss of electrons

RIG: **R**eduction **I**s **G**ain of electrons

Consider the reaction between sodium metal and chlorine gas:

$$2Na(s) + Cl_2(g) \rightarrow 2NaCl(s)$$

This reaction cannot be described in terms of the gain or loss of hydrogen and oxygen, as these elements are not involved in the reaction. However, we can instead describe the oxidation and reduction occurring in terms of the transfer of electrons between sodium and chlorine.

Sodium metal loses electrons to form sodium cations, so it is oxidized.

$$2Na(s) \rightarrow 2Na^+ + 2e^-$$

These electrons are transferred to chlorine gas, reducing it to chloride anions:

$$Cl_2(g) + 2e^- \rightarrow 2Cl^-$$

The sodium cations and chloride anions are held together by ionic bonds in a three-dimensional lattice structure, NaCl(s) (figure 3).

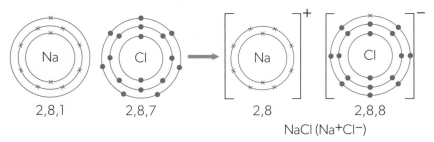

2,8,1   2,8,7   2,8   2,8,8

NaCl (Na^+Cl^-)

The formation of ionic lattice structures is detailed in *Structure 2.1*.

▲ **Figure 3** Sodium atoms are oxidized (lose electrons) and chlorine atoms are reduced (gain electrons) in the formation of sodium chloride

Case study: Redox reactions in optometry

Optometrists often prescribe glasses with photochromic lenses. These lenses darken in the presence of ultraviolet light (from sunlight); this change is based on a redox reaction.

Ordinary glass is composed of silicates while photochromic lenses contain copper(I) chloride, CuCl, and silver chloride, AgCl. The chloride ions are oxidized to chlorine atoms on exposure to ultraviolet light (hf).

$$\overset{hf}{Cl^- \rightarrow Cl + e^-}$$

Electron transfer then takes place, reducing the silver cations to silver atoms.

$$Ag^+ + e^- \rightarrow Ag$$

The silver atoms turn the lenses dark, inhibiting the transmission of light. The darkening process is reversed by copper(I) chloride allowing the lenses to become transparent again. When the lenses are no longer exposed to ultraviolet light, the following reaction takes place:

$$Cu^+ + Cl \rightarrow Cu^{2+} + Cl^-$$

The chlorine atoms formed by the exposure to light are reduced by the Cu^+ ions. In turn, Cu^+ ions are oxidized to Cu^{2+} ions. These Cu^{2+} ions then oxidize silver atoms to Ag^+ ions:

$$Cu^{2+} + Ag \rightarrow Cu^+ + Ag^+$$

As a result, the lenses become transparent again and the silver and chlorine atoms return to the initial species, Ag^+ and Cl^-.

4. Oxidation and reduction in terms of oxidation state change

Some redox reactions do not involve a transfer of electrons between species. For example, in the formation of carbon disulfide, electrons are shared between the carbon and sulfur atoms, forming a covalently bonded molecule:

$$C(s) + 2S(s) \rightarrow CS_2(l)$$

Carbon disulfide has the following Lewis structure:

$$:\!\overset{\bullet\bullet}{\underset{\bullet\bullet}{S}}\!=\!C\!=\!\overset{\bullet\bullet}{\underset{\bullet\bullet}{S}}\!:$$

We can describe oxidation and reduction in terms of the change in oxidation state of the atoms in the reacting species. The **oxidation state** represents the charge that an atom would have in a compound if the compound were composed of ions. In other words, all polar covalent bonds are treated as ionic, so all shared electrons in each bond are formally transferred to the more electronegative atom.

A compound is oxidized if the oxidation state of an atom in that compound increases, and reduced if the oxidation state of an atom in that compound decreases.

▲ **Figure 4** Photochromic lenses

You learned the rules for assigning oxidation states to atoms in covalent compounds in *Structure 3.1*.

In the carbon disulfide example, the oxidation state of carbon increases from 0 in the reactant, C(s), to +4 in the product, $CS_2(l)$, where carbon shares four electrons with the more electronegative sulfur. Carbon is therefore oxidized.

Elemental sulfur has an oxidation state of 0. In carbon disulfide, the oxidation state of each sulfur atom decreases to −2. Sulfur is therefore reduced.

TOK

The underlying assumption when assigning oxidation states is that electrons are transferred between atoms, not shared. However, oxidation states are not real because they do not represent the actual charges on the atoms in a molecule.

If you study the HL course, you learned about formal charges in *Structure 2.2*. Formal charges are also artificial and do not represent actual charges. When assigning formal charges, we assume that the electrons in covalent bonds are shared equally between the atoms involved.

In spite of these simplifying assumptions, oxidation states and formal charges are both useful tools that help us to explain redox reactions and Lewis formulas, respectively.

Worked example 1

Sodium chloride, sulfuric acid, and manganese(IV) oxide react according to the following chemical equation.

$$4NaCl(s) + 2H_2SO_4(aq) + MnO_2(s) \rightarrow 2Na_2SO_4(aq) + MnCl_2(aq) + 2H_2O(l) + Cl_2(g)$$

a. Deduce the change in oxidation states for each atom.

b. State which atom is oxidized, and which atom is reduced.

c. Identify the oxidizing agent and the reducing agent.

Solution

a. First, review the species on each side of the equation and assign oxidation states to each atom:

On the left-hand side	On the right-hand side
In NaCl, Na: +1, Cl: −1 In H_2SO_4, H: +1, S: +6, O: −2 In MnO_2, Mn: +4, O: −2	In Na_2SO_4, Na: +1, S: +6, O: −2 In $MnCl_2$, Mn: +2, Cl: −1 In H_2O, H: +1, O: −2 In Cl_2, Cl: 0

All the oxidation states stay the same, except for manganese, where the oxidation state changes from +4 to +2, and chlorine, where the oxidation state changes from −1 to 0.

b. The oxidation state of manganese decreases, so manganese is reduced. The oxidation state of chlorine increases, so chlorine is oxidized.

c. As manganese has been reduced and caused chlorine to be oxidized, manganese(IV) oxide, $MnO_2(s)$, is the oxidizing agent. Chlorine has been oxidized and caused manganese to be reduced. That makes sodium chloride, NaCl(s), the reducing agent.

Worked example 2

Consider the following balanced equation:

$$Fe(s) + 2HBr(aq) \rightarrow FeBr_2(aq) + H_2(g)$$

a. Deduce the oxidation states of iron and hydrogen in the reactants and products.

b. State which species is oxidized, and which species is reduced.

c. Identify the oxidizing agent and the reducing agent.

Solution

a.

On the left-hand side	On the right-hand side
Fe: 0 H in HBr: +1 Br in HBr: −1	Fe in $FeBr_2$: +2 H in H_2: 0 Br in $FeBr_2$: −1

b. The oxidation state of Fe increases, so Fe(s) is oxidized. The oxidation state of H decreases, so HBr(aq) is reduced.

c. The oxidizing agent is HBr(aq) and the reducing agent is Fe(s)

An **oxidizing agent** causes another species to be oxidized, with the oxidizing agent itself being reduced in the process. A **reducing agent** causes another species to be reduced, with the reducing agent itself being oxidized in the process.

The use of Roman numerals to represent oxidation states of oxyacids and transition element compounds was covered in *Structure 3.1*.

Practice questions

1. Consider the following balanced equation:

   $$Cl_2(aq) + 2KI(aq) \rightarrow 2KCl(aq) + I_2(aq)$$

   a. Deduce the oxidation states of chlorine and iodine in the reactants and products.

   b. State which element is oxidized, and which element is reduced.

   c. Identify the oxidizing agent and the reducing agent.

2. Identify the oxidizing agents and the reducing agents in the following reactions:

   a. $PbO(s) + H_2(g) \rightarrow Pb(s) + H_2O(l)$

   b. $2CuO(s) + C(s) \rightarrow CO_2(g) + 2Cu(s)$

   c. $Mg(s) + Cl_2(g) \rightarrow MgCl_2(s)$

   d. $2KI(aq) + Br_2(aq) \rightarrow I_2(aq) + 2KBr(aq)$

When writing oxidation states, the sign is always placed before the number and not after it. In worked example 2, the oxidation state of hydrogen in HBr is +1 and not 1+.

Remember to describe the *compounds* as being oxidized and reduced, not the individual atoms in the compounds. This also applies to describing the oxidizing and reducing agents. In worked example 2, HBr(aq) is reduced and is the oxidizing agent, not H.

Half-equations (*Reactivity 3.2.2*)

Consider the reaction between sodium metal and chlorine gas, discussed in the previous section:

$$2Na(s) + Cl_2(g) \rightarrow 2NaCl(s)$$

You saw that the processes of oxidation and reduction could be separated into two equations:

$$Na(s) \rightarrow Na^+ + e^-$$

$$Cl_2(g) + 2e^- \rightarrow 2Cl^-$$

These equations are called **half-equations**. Separating a redox process into two half-equations helps to show the transfer of electrons. Half-equations can also make it easier to balance the full equation for the redox reaction. The steps for writing redox reactions in aqueous solutions are:

1. Identify the species being oxidized and reduced.

2. Separate the equation into an oxidation half-equation and reduction half-equation.

3. Balance any atoms being oxidized or reduced.

4. For the oxidation half-equation, write the electrons lost on the right-hand side of the equation. The number of electrons should be equal to the magnitude of the change in oxidation state of the oxidized species.

5. For the reduction half-equation, write the electrons gained on the left-hand side of the equation. The number of electrons should be equal to the magnitude of the change in oxidation state of the reduced species.

6. Balance these half-equations so that the number of electrons lost in oxidation equals the number of electrons gained in reduction.

7. Add the two half-equations together and cancel the electrons.

8. If the reaction is occurring in acidic solution, add $H_2O(l)$ to balance any oxygen atoms and $H^+(aq)$ to balance any hydrogen atoms.

9. For neutral or basic solutions, add $OH^-(aq)$ to balance oxygen atoms and $H_2O(l)$ to balance hydrogen atoms.

10. Finally, add up the charges and check that the sum is equal to zero.

Worked example 3

Iron metal, Fe(s), will react with a solution of silver(I) ions, $Ag^+(aq)$, to form aqueous iron(III) ions, $Fe^{3+}(aq)$, and silver metal, Ag(s).

Write the balanced equation for this redox reaction.

Solution

First, write the unbalanced equation for the reaction:

$$Fe(s) + Ag^+(aq) \rightarrow Fe^{3+}(aq) + Ag(s)$$

The oxidation state of iron increases from 0 to +3, so Fe(s) loses electrons and is therefore oxidized.

The oxidation state of silver decreases from +1 to 0, so $Ag^+(aq)$ gains electrons and is therefore reduced.

Write the oxidation and reduction half-equations, ensuring that the atoms are balanced:

$$\text{oxidation: } Fe \rightarrow Fe^{3+}$$

$$\text{reduction: } Ag^+ \rightarrow Ag$$

To save time, you can omit states of reacting species in all steps except the final answer.

Add the electrons to make sure that the number of electrons in each half-equation is equal to the magnitude of the change in oxidation state:

$$Fe \rightarrow Fe^{3+} + 3e^-$$

$$Ag^+ + e^- \rightarrow Ag$$

Then, multiply the reduction half-equation by three, so that there are three electrons in each half-equation:

$$Fe \rightarrow Fe^{3+} + 3e^-$$

$$3Ag^+ + 3e^- \rightarrow 3Ag$$

Add the two half-equations together and cancel the electrons:

$$Fe + 3Ag^+ + \cancel{3e^-} \rightarrow Fe^{3+} + 3Ag + \cancel{3e^-}$$

$$Fe(s) + 3Ag^+(aq) \rightarrow Fe^{3+}(aq) + 3Ag(s)$$

Finally, check that the charges are balanced: on the left-hand side, there are three silver(I) ions with a 1+ charge each, and on the right, there is one iron ion with a 3+ charge. So, the total charge on each side of the equation is 3+ and therefore the equation is balanced. Do not forget to add states to all species in the final equation.

Worked example 4

Iron(II) ions, $Fe^{2+}(aq)$, and dichromate(VI) ions, $Cr_2O_7^{2-}(aq)$, react in acidic solution to form iron(III) ions, $Fe^{3+}(aq)$, and chromium(III) ions, $Cr^{3+}(aq)$.

Deduce the balanced redox equation for this reaction.

Solution

First, write the unbalanced equation:

$$Fe^{2+}(aq) + Cr_2O_7^{2-}(aq) \rightarrow Fe^{3+}(aq) + Cr^{3+}(aq)$$

The oxidation state of iron increases from +2 to +3, therefore $Fe^{2+}(aq)$ loses electrons and is oxidized.

In $Cr_2O_7^{2-}(aq)$, chromium has an oxidation state of +6. The oxidation state of chromium decreases from +6 to +3, therefore $Cr_2O_7^{2-}(aq)$ gains electrons and is reduced.

Write the oxidation and reduction half-equations, ensuring that all the atoms that change their oxidation states are balanced:

oxidation: $Fe^{2+} \rightarrow Fe^{3+}$

reduction: $Cr_2O_7^{2-} \rightarrow 2Cr^{3+}$

Add the electrons such that number of electrons in each half-equation is equal to the magnitude of the change in oxidation state (remembering that there are two Cr^{3+} ions in the second half-equation):

$$Fe^{2+} \rightarrow Fe^{3+} + e^-$$

$$Cr_2O_7^{2-} + 6e^- \rightarrow 2Cr^{3+}$$

Then, multiply the oxidation half-equation by six, so that there are six electrons in each half-equation:

$$6Fe^{2+} \rightarrow 6Fe^{3+} + 6e^-$$

$$Cr_2O_7^{2-} + 6e^- \rightarrow 2Cr^{3+}$$

Add the two half-equations together and cancel the electrons:

$$6Fe^{2+} + Cr_2O_7^{2-} + \cancel{6e^-} \rightarrow 6Fe^{3+} + 2Cr^{3+} + \cancel{6e^-}$$

$$6Fe^{2+} + Cr_2O_7^{2-} \rightarrow 6Fe^{3+} + 2Cr^{3+}$$

As this reaction is taking place in acidic solution, we need to balance the oxygen and hydrogen atoms. There are seven oxygen atoms on the left-hand side of the equation, so add seven moles of water to the right-hand side of the equation to balance the oxygen atoms:

$$6Fe^{2+} + Cr_2O_7^{2-} \rightarrow 6Fe^{3+} + 2Cr^{3+} + 7H_2O$$

There are now 14 hydrogen atoms on the right-hand side of the equation, so add 14 moles of H^+ to the left-hand side of the equation to balance the hydrogen atoms:

$$6Fe^{2+} + Cr_2O_7^{2-} + 14H^+ \rightarrow 6Fe^{3+} + 2Cr^{3+} + 7H_2O$$

Finally, check that the charges are balanced. On the left-hand side, there are six Fe^{2+}, $Cr_2O_7^{2-}$ and 14 H^+ ions:

$$(6 \times 2) + (-2) + (14 \times 1) = 24$$

On the right-hand side, there are six Fe^{3+} and two Cr^{3+} ions:

$$(6 \times 3) + (2 \times 3) = 24$$

So, the total charge on each side of the equation is 24+ and therefore the equation is balanced. As usual, do not forget to add states to all reacting species in the final equation:

$$6Fe^{2+}(aq) + Cr_2O_7^{2-}(aq) + 14H^+(aq) \rightarrow$$
$$6Fe^{3+}(aq) + 2Cr^{3+}(aq) + 7H_2O(l)$$

Practice questions

3. Write balanced equations for the following reactions that occur in acidic solutions:

    a. $Zn(s) + SO_4^{2-}(aq) \rightarrow Zn^{2+}(aq) + SO_2(g)$

    b. $MnO_4^-(aq) + Br^-(aq) \rightarrow Mn^{2+}(aq) + BrO_3^-(aq)$

    c. $I_2(s) + OCl^-(aq) \rightarrow IO_3^-(aq) + Cl^-(aq)$

    d. $Cr_2O_7^{2-}(aq) + C_2O_4^{2-}(aq) \rightarrow Cr^{3+}(aq) + CO_2(g)$

Oxidation and reduction of metals and halogens (*Reactivity 3.2.3*)

Relative ease of reduction of halogens

Halogens can act as oxidizing agents in redox reactions. In these reactions, halogens in their elemental state gain electrons, being reduced to singly charged halogen anions. In *Structure 3.1*, you learned that the reactivity of halogens increases going up the group.

increasing reactivity

$\longleftarrow$

F_2 Cl_2 Br_2 I_2

This means that fluorine is the strongest oxidizing agent among the halogens, and the most easily reduced, followed by chlorine, and then bromine. For example, chlorine, Cl_2, can oxidize bromide ions, Br^-:

$$Cl_2 + 2Br^- \rightarrow 2Cl^- + Br_2$$

However, chlorine cannot oxidize fluoride ions because fluorine is a stronger oxidizing agent:

$$Cl_2 + 2F^- \longrightarrow\!\!\!\!\times\!\!\!\!\longrightarrow \text{no reaction}$$

Instead, the reverse reaction between fluorine and chloride ions is possible:

$$F_2 + 2Cl^- \rightarrow 2F^- + Cl_2$$

Iodine is the weakest oxidizing agent among the halogens, so it cannot be reduced by the other halogens. However, iodine will oxidize many metals and other strong reducing agents.

Relative ease of oxidation of metals

Group 1 metals can act as reducing agents because they lose their valence electron easily. The reactivity of group 1 metals increases going down the group, and therefore the relative ease of oxidation increases going down the group.

increasing reactivity

$\longrightarrow$

Li   Na   K   Rb   Cs

For other metals, you can deduce their relative ease of oxidation by placing a pure metal into a solution of ions of a different metal. If a reaction occurs, then the pure metal is more easily oxidized and it is a stronger reducing agent. If no reaction occurs, then the metal comprising the ionic solution is more easily oxidized.

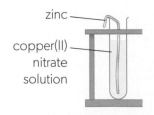

zinc

copper(II)
nitrate
solution

▲ Figure 5 Zinc will displace copper in solution, changing the colour of the solution from blue to colourless and forming a red copper precipitate

Consider the reaction between zinc metal and copper(II) nitrate solution. In the copper nitrate solution, there are copper(II) ions, $Cu^{2+}(aq)$. In this reaction, zinc metal will dissolve to form a zinc nitrate solution, and copper(II) ions will be reduced to copper metal, which will precipitate out as a solid:

$$Zn(s) + Cu(NO_3)_2(aq) \rightarrow Zn(NO_3)_2(aq) + Cu(s)$$

Therefore, zinc is a stronger reducing agent than copper, and is more easily oxidized. Conversely, this means that copper(II) ions are a stronger oxidizing agent and are more easily reduced than zinc ions. You can track the transfer of electrons in this reaction using half-equations:

$$\text{oxidation: } Zn(s) \rightarrow Zn^{2+}(aq) + 2e^-$$

$$\text{reduction: } Cu^{2+}(aq) + 2e^- \rightarrow Cu(s)$$

You can repeat this experiment with several different metals to obtain a **reactivity series**. In a reactivity series, the most easily oxidized metal is listed first and the least easily oxidized metal is listed last.

Worked example 5

Strips of five different metals, zinc, iron, magnesium, copper and silver, were each added to solutions of their metal nitrate counterparts. The mixtures were observed for a period of time to see whether a reaction has occurred or not. These observations were recorded in table 1.

	$Zn(NO_3)_2(aq)$	$Fe(NO_3)_2(aq)$	$Mg(NO_3)_2(aq)$	$Cu(NO_3)_2(aq)$	$AgNO_3(aq)$
Zn(s)	–	Yes	No	Yes	Yes
Fe(s)	No	–	No	Yes	Yes
Mg(s)	Yes	Yes	–	Yes	Yes
Cu(s)	No	No	No	–	Yes
Ag(s)	No	No	No	No	–

▲ Table 1 Summary of reactions between metals and metal ion solutions

Use table 1 to deduce the reactivity series of the five metals.

Solution

Magnesium reacts with all four solutions and is therefore the most easily oxidized and the most reactive. Silver metal does not react with any solution and is the least easily oxidized. However, $Ag^+(aq)$ ions are the most easily reduced, making $AgNO_3(aq)$ the best oxidizing agent on the list.

Completing the list by inspection gives the following activity series, from the most easily oxidized to the least easily oxidized: Mg(s), Zn(s), Fe(s), Cu(s), Ag(s).

Redox reactions of acids and metals (*Reactivity 3.2.4*)

Reactive metals, such as magnesium, zinc, and iron, are readily oxidized by strong acids, such as hydrochloric acid, HCl(aq), and sulfuric acid, H_2SO_4(aq). In dilute solutions, these reactions produce hydrogen gas and a metal salt:

$$Zn(s) + H_2SO_4(aq) \rightarrow ZnSO_4(aq) + H_2(g)$$

$$Zn(s) + 2HCl(aq) \rightarrow ZnCl_2(aq) + H_2(g)$$

In these reactions, the oxidation state of zinc changes from 0 to +2, and the oxidation state of hydrogen changes from +1 to 0. The electron transfer is shown by the following half-equations:

oxidation: $Zn(s) \rightarrow Zn^{2+}(aq) + 2e^-$

reduction: $2H^+(aq) + 2e^- \rightarrow H_2(g)$

Therefore, in the reaction between a metal and an acid, the metal is the reducing agent, and the acid is the oxidizing agent.

Copper and silver are less easily oxidized than magnesium, zinc and iron, so they do not react with dilute solutions of common acids (figure 7).

▲ **Figure 6** The reaction of metals with acids can be detected by the "pop" test: the gas released from the reaction mixture is collected in an inverted test tube, and a lit splint is held close to the test tube opening. A small explosion ("pop") suggests the presence of hydrogen gas that reacts with oxygen in the air

potassium — most reactive
sodium
calcium
magnesium
aluminium
zinc
iron
tin
lead
(hydrogen)
copper
silver
gold
platinum — least reactive

▲ **Figure 7** A reactivity series showing the most easily oxidized metals to the least easily oxidized. Metals above hydrogen on the list can react with common acids, those below can not

▲ Figure 8 Gold is at the bottom of the reactivity series of metals, so it is not oxidized easily. Therefore, it is the most likely to be found in its reduced form, with zero oxidation state: elemental gold. It is impossible to "pan for lithium" because it is at the top of the activity series

(ATL) Thinking skills

A student investigating the reactivity of zinc and copper noted the following qualitative observations:

1. Copper wire was placed in dilute sulfuric acid. No change was observed. A zinc strip was placed in dilute sulfuric acid. Bubbles appeared.

2. Some copper wire was wrapped around one end of the zinc strip. This end of the strip and the surrounding copper were placed in dilute sulfuric acid. Bubbles evolved quickly on the surface of the copper.

3. Strips of copper and zinc were placed in dilute sulfuric acid and connected to each other. Bubbles evolved on the surface of the copper.

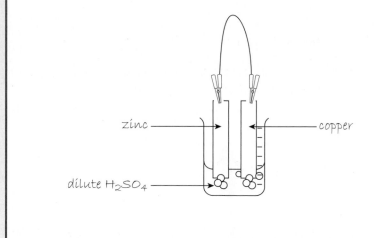

Explain the student's observations, using your knowledge of metal reactivity, electron transfer reactions, reactions of acids and metallic bonding.

Write two or three linking questions relating the concepts illustrated by this experiment.

Redox titration

Iron supplements are used to treat iron deficiency. You will determine the iron content in iron tablets by titrating them with a solution of potassium manganate(VII) of known concentration. Manganate(VII) ions, $MnO_4^-(aq)$, are powerful oxidizing agents. In their reaction with iron(II) in acidic solution, purple manganate(VII) ions are reduced to pale pink manganese(II) ions, $Mn^{2+}(aq)$. In this process, iron(II) ions are oxidized to iron(III) ions.

→

Relevant skills

- Tool 1: Titration
- Tool 3: General mathematics
- Tool 3: Record and propagate uncertainties
- Inquiry 2: Assess reliability and validity of results

Safety

- Wear eye protection
- Sulfuric acid, $H_2SO_4(aq)$, is an irritant
- Dilute potassium manganate(VII), $KMnO_4(aq)$, is an irritant and will stain skin and fabrics

Materials

- iron tablets (or other source of iron(II), $Fe^{2+}(aq)$, ions)
- $1.0\,mol\,dm^{-3}$ sulfuric acid, $H_2SO_4(aq)$
- distilled water
- $0.020\,mol\,dm^{-3}$ potassium manganate(VII) solution, $KMnO_4(aq)$
- pestle and mortar
- top pan balance
- two $250\,cm^3$ conical flasks
- $100\,cm^3$ measuring cylinder
- $250\,cm^3$ volumetric flask
- funnel
- filter paper
- reagent jar
- burette
- $25\,cm^3$ volumetric pipette
- white tile

Instructions

Part A Preparation of acidified iron tablet extract

1. Grind four iron tablets into a fine powder using a pestle and mortar.

2. Weigh the iron tablet powder and transfer it into a conical flask.

3. Add $100\,cm^3$ of $1.0\,mol\,dm^{-3}$ sulfuric acid, $H_2SO_4(aq)$, to the tablet powder and leave for 24 to 48 hours.

4. In the meantime, review the titration and uncertainty propagation sections in the *Tools for chemistry* chapter before starting part B.

Part B Titration against potassium manganate(VII)

5. Filter the iron tablet extract. Transfer the filtrate into a volumetric flask and make up to the $250\,cm^3$ mark with distilled water. Store in a labelled reagent jar.

6. Fill the burette with $0.020\,mol\,dm^{-3}$ potassium manganate(VII), $KMnO_4(aq)$.

7. Transfer $25.0\,cm^3$ of iron tablet solution to a clean conical flask. Place this flask on a white tile under the burette.

8. Perform a rough titration of the iron tablet solution, stopping when the solution in the flask permanently turns pale pink.

9. Repeat the titration several times until you obtain two concordant values.

10. Clear up according to the directions given by your teacher.

Questions

1. Write the oxidation and reduction half-equations for this reaction.

2. Deduce the redox equation for this reaction.

3. Using your results, determine the mass of iron in one tablet.

4. Propagate the measurement uncertainties to obtain the uncertainty of the mass of iron per tablet.

5. Compare your result to the iron content reported on the tablet packaging. Calculate the percentage error.

6. Comment on the reliability and validity of your result.

7. Describe at least two sources of systematic error and suggest improvements that would minimize these sources of error.

8. Explain why you used four iron tablets in this analysis, not just one.

9. Explain why you left the iron tablet powder in dilute acid solution for 24 to 48 hours.

10. Suggest why redox titrations such as this are said to be "self-indicating".

Electrochemical cells (*Reactivity 3.2.5* and *Reactivity 3.2.6*)

An **electrochemical cell** interconverts electrical and chemical energy. There are two types of electrochemical cell:

1. In **primary (voltaic) cells**, **secondary (rechargeable) cells**, and **fuel cells** the energy produced by spontaneous chemical reactions is used to generate electrical energy.

2. In **electrolytic cells**, electrical energy is used to drive forward non-spontaneous chemical reactions.

In redox reactions, electrons move from the substance being oxidized to the substance being reduced. Nearly all spontaneous redox reactions are exothermic, and the energy released in these chemical changes can be used to produce electrical energy. Redox reactions used in primary (voltaic) cells are irreversible while secondary (rechargeable) cells utilize reversible redox reactions.

In the 18th century, the Italian scientist Luigi Galvani discovered accidentally that an electric current could be produced by two dissimilar metals connected by a moist substance. He noticed that he could cause an amputated frog leg to twitch by touching it with two dissimilar metals. Alessandro Volta, another Italian scientist, doubted that animal legs were integral to produce electricity. He showed that chemical reactions can produce electricity and made the first **battery**.

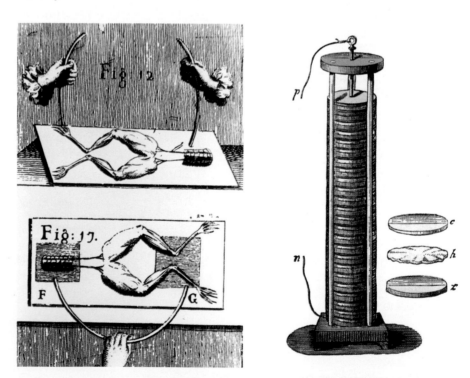

▲ **Figure 9** (a) Galvani's frog legs experiment (b) Volta's voltaic pile, the first modern type of electric battery

Any two dissimilar metals and their ions can participate in redox reactions. The one higher in the activity series will oxidize to ions and the one lower in the series will have its ions reduced to the pure metal. For example, zinc can be oxidized by copper(II) ions, to form zinc ions. The Cu^{2+} ions get reduced, and act as an oxidizing agent.

$$Zn(s) + Cu^{2+}(aq) \rightarrow Zn^{2+}(aq) + Cu(s)$$

Separating this into two half-equations gives:

$$Zn(s) \rightarrow Zn^{2+}(aq) + 2e^-$$

$$Cu^{2+}(aq) + 2e^- \rightarrow Cu(s)$$

These two processes can occur in separate beakers, called **half-cells**.

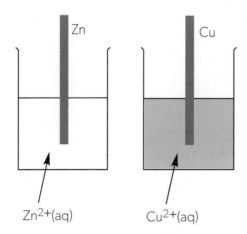

▲ **Figure 10** The zinc and copper half-cells

The two half-cells can be connected with a wire to form an electrochemical cell. In electrochemical cells, oxidation always occurs at the **anode**, and reduction at the **cathode**. **RED CAT** is a useful mnemonic: **RED**uction at **CAT**hode. Therefore, in this case, Zn(s) is the anode and Cu(s) the cathode. This electrochemical cell can be represented by the following equations:

$$Zn(s) \rightarrow Zn^{2+}(aq) + 2e^- \| Cu^{2+}(aq) + 2e^- \rightarrow Cu(s)$$

When the half-cells are connected, the electrons produced by the oxidation of zinc at the anode flow through the wire to the copper half-cell to reduce copper(II) ions. As a result, the zinc half-cell loses two electrons and now takes on a slight positive charge while the copper half-cell gains two electrons, taking on a slightly negative charge. Any further oxidation of zinc is prevented by the slight negative charge on the copper half-cell, which repels the electrons. The electrons therefore remain in the slightly positively charged zinc half-cell. The cell becomes **polarized**, and the redox reaction stops.

To get around this, a salt bridge is used to connect the solutions in the two half-cells and complete the electrical circuit. A salt bridge consists of an ionic salt solution, such as sodium sulfate, Na_2SO_4, or potassium nitrate, KNO_3. It allows the positive ions (cations) to flow toward the slightly negatively charged half-cell with the cathode and negative ions (anions) to flow toward the slightly positively charged half-cell with the anode.

Consider the addition of a sodium sulfate salt bridge to our zinc–copper cell. The slightly negatively charged copper half-cell attracts $Na^+(aq)$ and $Zn^{2+}(aq)$ cations through the salt bridge and solution. The slightly positively charged zinc half-cell attracts $SO_4^{2-}(aq)$ anions through the salt bridge and solution. This neutralizes the charge in each half-cell, so the redox reaction can continue. This is now a complete primary cell (figure 11). This kind of primary cell is known as the Daniell cell, named after its inventor, the British chemist John Frederic Daniell.

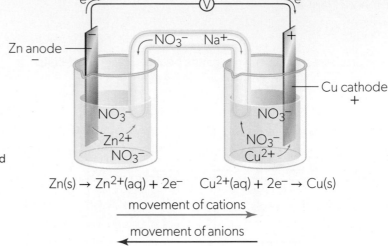

$$Zn(s) \rightarrow Zn^{2+}(aq) + 2e^- \quad Cu^{2+}(aq) + 2e^- \rightarrow Cu(s)$$

movement of cations →

movement of anions ←

▶ Figure 11 A primary cell consisting of a zinc metal anode (labelled as negative because it is the source of electrons) dipped in $ZnSO_4(aq)$, a copper metal cathode (labelled positive as it attracts electrons) dipped in $CuSO_4(aq)$, an electrical connecting wire, a voltmeter and a salt bridge

As the reaction continues, the blue colour of the copper(II) sulfate solution fades, the copper bar increases in size as it becomes coated in more copper, and the zinc bar gets thinner. Once there is a significant build-up of $Zn^{2+}(aq)$ ions on the cathode side, the cell ceases to function.

Cell diagrams are used as a short-hand way to represent primary cells. In this convention, the cathode is always written on the right-hand side and the anode is always written on the left-hand side. The salt bridge is represented by two parallel vertical lines. You can use the following general template to write cell diagrams for metal–ion primary cells:

anode being oxidized | product of oxidation || species being reduced | product of reduction/cathode

Therefore, the cell diagram for the Daniell cell would be written as follows:

$$Zn(s) \mid Zn^{2+}(aq) \parallel Cu^{2+}(aq) \mid Cu(s)$$

The species on the left-hand side, $Zn(s) \mid Zn^{2+}(aq)$, represent the anode, with zinc metal being oxidized to $Zn^{2+}(aq)$. The species on the right-hand side, $Cu^{2+}(aq) \mid Cu(s)$, represent the cathode, with $Cu^{2+}(aq)$ being reduced to $Cu(s)$.

Worked example 6

Manganese metal reacts with nickel(II) ions to form manganese(II) ions and nickel metal.

a. Write the redox reaction that occurs between nickel(II) ions and manganese metal.

b. Assuming that this redox reaction occurs in a manganese–nickel primary cell, write the half-equations that occur in each half-cell.

c. Write the cell diagram to represent the primary cell for this redox reaction.

d. Sketch a primary cell for this reaction and identify the anode, cathode, direction of electron flow and direction of ion flow.

Solution

a. $Mn(s) + Ni^{2+}(aq) \rightarrow Mn^{2+}(aq) + Ni(s)$

b. At the cathode: $Ni^{2+}(aq) + 2e^- \rightarrow Ni(s)$
    At the anode: $Mn(s) \rightarrow Mn^{2+}(aq) + 2e^-$

c. You can use the general template for cell diagrams:

    anode being oxidized | product of oxidation || species being reduced | product of reduction/cathode

    This gives:

    $Mn(s) \,|\, Mn^{2+}(aq) \,||\, Ni^{2+}(aq) \,|\, Ni(s)$

d.

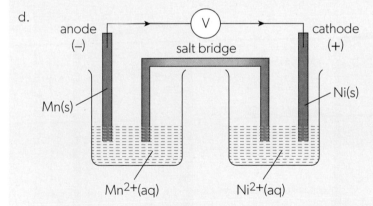

Electrons flow from left to right (from the anode to the cathode) through the electrical wire. $Mn^{2+}(aq)$ and the cations of the salt in the salt bridge also flow from left to right through the salt bridge toward the cathode. The anions of the salt flow from right to left through the salt bridge toward the anode.

Secondary cells (*Reactivity 3.2.7*)

A **battery** is a series of two or more electrochemical cells, typically enclosed in a single container. In a primary electrochemical cell, ultimately the reaction materials will be consumed, and the reaction is not reversible. Either the anode, the electrolyte, or both, need to be replaced, or the battery will be thrown away. Typically, the anode (negative electrode) is oxidized and can no longer be used.

As previously discussed, the ions travelling through the solution and salt bridge can polarize the cell, which causes the chemical reaction to stop. Polarization can also cause a build-up of hydrogen bubbles on the surface of the anode. These can increase the internal resistance of the cell and reduce its electrical output. Primary cells do not operate well under high-current demands, such as in flash photography or electric cars, but are suitable for low-current household devices.

In a **secondary cell**, or a **rechargeable cell**, the chemical reactions that generate electricity can be reversed by applying an electric current to the cell. Secondary cells can satisfy higher current demands than primary cells but have a higher rate of self-discharge. For example, batteries in cell phones are made of secondary cells that can be recharged using electrical energy. When you purchase a replacement battery for a phone, you need to charge it before use because it will have self-discharged during the storage and transportation.

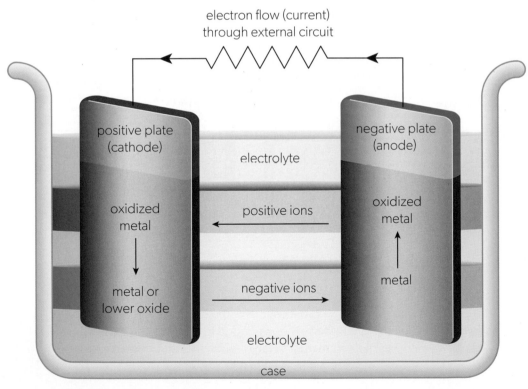

▲ Figure 12 Structure of an electrochemical cell. The negative anode is oxidized and the flow of ions causes polarization. This process cannot be reversed in a primary cell, but it can be reversed in a secondary (rechargeable) battery

Case study: Lead–acid battery

Car batteries are made of secondary cells. The electrical energy from the battery is used to power the motor that starts the engine, and to power any electrical systems in the car. This is known as **discharge**. Some of the chemical energy from the combustion in the engine is used to reverse the chemical reactions that occur during discharge, therefore recharging the battery. An idling engine runs slowly and does not provide enough energy to recharge the battery, and so the battery will gradually discharge.

A typical car uses a lead–acid battery. This battery is composed of a lead anode, $Pb(s)$, a lead(IV) oxide cathode, $PbO_2(s)$, and sulfuric acid, $H_2SO_4(aq)$. Sulfuric acid is a strong acid, so it will ionize into $H^+(aq)$ and $HSO_4^-(aq)$. When the battery is powering the motor and the car's electrical systems, $HSO_4^-(aq)$ will oxidize $Pb(s)$ at the anode, and $H^+(aq)$ will reduce $PbO_2(s)$ at the cathode. This gives rise to the following discharge reactions:

anode (oxidation): $Pb(s) + HSO_4^-(aq) \rightarrow PbSO_4(s) + H^+(aq) + 2e^-$

cathode (reduction): $PbO_2(s) + 3H^+(aq) + HSO_4^-(aq) + 2e^- \rightarrow PbSO_4(s) + 2H_2O(l)$

overall cell reaction: $Pb(s) + PbO_2(s) + 2H_2SO_4(aq) \rightarrow 2PbSO_4(s) + 2H_2O(l)$

Note that the $H^+(aq)$ and $HSO_4^-(aq)$ ions are ultimately produced by the dissociation of sulfuric acid, so in the overall cell reaction they are combined together and shown as $H_2SO_4(aq)$.

Worked example 7

Determine the reactions that occur at the anode and cathode in a car battery cell, and the overall cell reaction, when the engine is charging the battery.

Solution

The charging reactions are the reverse of the discharge reactions:

anode: $PbSO_4(s) + H^+(aq) + 2e^- \rightarrow Pb(s) + HSO_4^-(aq)$

cathode: $PbSO_4(s) + 2H_2O(l) \rightarrow PbO_2(s) + 3H^+(aq) + HSO_4^-(aq) + 2e^-$

overall cell reaction: $2PbSO_4(s) + 2H_2O(l) \rightarrow Pb(s) + PbO_2(s) + 2H_2SO_4(aq)$

The continual charging of a lead–acid battery tends to produce hydrogen and oxygen from water. Therefore, non-sealed car batteries occasionally need to be topped up with distilled water.

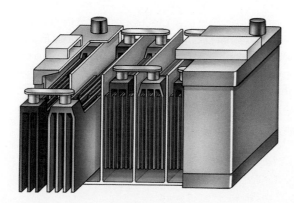

◀ **Figure 13** A lead–acid battery consists of a series of secondary cells with lead(IV) oxide plates, lead plates, and sulfuric acid electrolyte

Case study: Lithium-ion batteries

Lithium-ion rechargeable batteries use lithium atoms embedded in a lattice of graphite electrodes, rather than pure lithium metal for the anode. The cathode is a lithium–cobalt oxide complex, $LiCoO_2$. The lithium atoms are oxidized to lithium ions during discharge.

During charging, the opposite process occurs: the lithium ions in the complex migrate through the battery medium to the anode, where they accept electrons and are reduced to lithium atoms. These atoms become embedded in the graphite lattice. The battery medium must be completely non-aqueous, usually a polymer gel, as lithium reacts vigorously with water. Table 2 summarizes these reactions:

Electrode	Discharging reaction	Charging reaction
anode	$Li(s) \rightarrow Li^+ + e^-$	$Li^+ + e^- \rightarrow Li(s)$
cathode	$Li^+ + e^- + CoO_2(s) \rightarrow LiCoO_2(s)$	$LiCoO_2(s) \rightarrow Li^+ + e^- + CoO_2(s)$

▲ Table 2 Anode and cathode reactions in the lithium-ion battery

The overall cell reaction during discharge is $Li(s) + CoO_2(s) \rightarrow LiCoO_2(s)$.

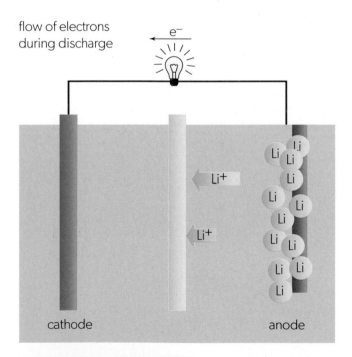

flow of electrons during discharge

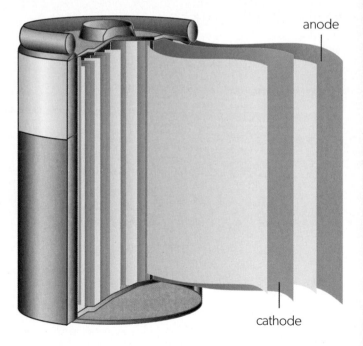

anode

cathode

▲ Figure 14 Structure of a typical lithium-ion rechargeable battery. The battery consists of a series of secondary cells composed of cathodes and anodes with a layer of polymer (yellow) separating them

When the lithium-ion battery is in use, electrons flow from the anode to the cathode through the external circuit while lithium ions flow from the anode to the cathode through the polymer gel inside the cell. When no more lithium ions are left on the anode, the battery is flat. To recharge it, the process is reversed, transferring lithium ions back to the anode.

Practice questions

4. During the discharge of a nickel–cadmium battery, the following reactions occur in the cells:

   anode: $Cd(s) + 2OH^-(aq) \rightarrow Cd(OH)_2(s) + 2e^-$
   cathode: $NiO(OH)(s) + H_2O(l) + e^- \rightarrow Ni(OH)_2(s) + OH^-(aq)$

   a. Write the overall equation for the cell.

   b. Write the cell diagram.

   c. Determine the charging reactions that occur in the cell.

Fuel cells

Fuel cells are a type of electrochemical cell that convert hydrogen, methanol, or ethanol and oxygen into water, carbon dioxide and heat. They cause little pollution and are very efficient. Like primary cells, they are not rechargeable, but unlike primary cells, they require a steady supply of fuel and oxygen, so the reactions in the cell can continue indefinitely.

For example, in a **hydrogen fuel cell**, hydrogen gas is supplied to the anode while oxygen gas is supplied to the cathode. The following reactions occur in a hydrogen fuel cell:

anode: $H_2(g) \rightarrow 2H^+(aq) + 2e^-$

cathode: $O_2(g) + 4H^+(aq) + 4e^- \rightarrow 2H_2O(l)$

overall cell equation: $2H_2(g) + O_2(g) \rightarrow 2H_2O(l)$

In a **direct methanol fuel cell (DMFC)**, methanol is supplied to the anode. The following reactions occur in a DMFC:

anode: $CH_3OH(l) + H_2O(l) \rightarrow CO_2(g) + 6H^+(aq) + 6e^-$

cathode: $\frac{3}{2}O_2(g) + 6H^+(aq) + 6e^- \rightarrow 3H_2O(l)$

overall cell equation: $CH_3OH(l) + \frac{3}{2}O_2(g) \rightarrow CO_2(g) + 2H_2O(l)$

A typical fuel cell has the following key components:

- **Electrolyte** or **separator**: this keeps components from mixing. For example, a **proton exchange membrane (PEM)** is a polymer that allows H^+ ions to diffuse through but prevents the diffusion of other ions, electrons or molecules.

- **Electrodes**: The electrodes are made of a catalyst that allows for the chemical reactions to occur. There is an oxidizing electrode (anode) and reducing electrode (cathode).

- **Bipolar plate**: This conducts the electrical current from cell to cell and ensures uniform distribution of the fuel gas.

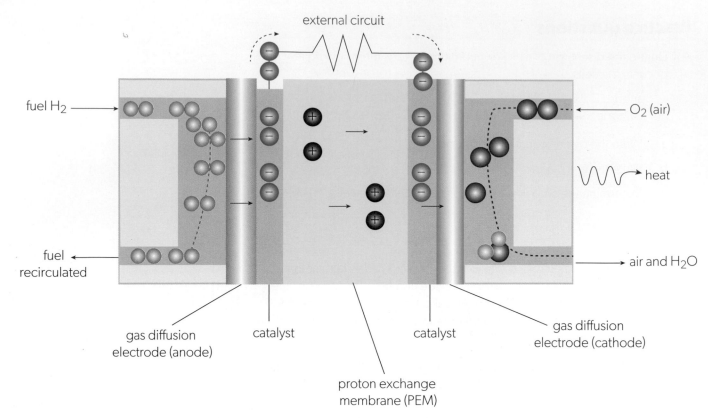

▲ **Figure 15** In a hydrogen fuel cell, $H_2(g)$ is oxidized at the anode and $O_2(g)$ is reduced at the cathode. The proton exchange membrane (PEM) allows H^+ ions to diffuse but prevents the diffusion of other ions, molecules and electrons

Hydrogen fuel cells use hydrogen gas as the fuel. These cells do not produce greenhouse gases. The heat formed in the exothermic reaction $2H_2(g) + O_2(g) \rightarrow 2H_2O(l)$ can be utilized, increasing the efficiency of the cell. The oxygen gas can be obtained from the air.

The catalysts used for the electrodes are often made of platinum or other expensive metals, which makes fuel cells expensive to run on a commercial scale. The hydrogen gas used must be very pure to prevent the poisoning of the catalyst and therefore the reduction of the cell electrical output. There are two main sources of hydrogen gas:

Practice question

5. Outline the function of the proton exchange membrane (PEM) in fuel cells.

The effect of greenhouse gases on atmospheric temperatures is discussed in *Structure 3.2*.

1. Clean hydrogen can be produced by the electrolysis of water. Solar cells or wind generators provide the cleanest form of energy for powering the electrolysis.

2. Hydrogen gas can also be obtained by the reaction of hydrocarbons, especially methane, with steam: $CH_4(g) + H_2O(g) \longrightarrow 3H_2(g) + CO(g)$

DMFCs have the advantage of not needing to extract hydrogen gas because they use methanol as the fuel. However, they produce carbon dioxide, $CO_2(g)$, which is a greenhouse gas.

Electrolytic cells (*Reactivity 3.2.8*)

An electrolytic cell is an electrochemical cell that converts electrical energy to chemical energy. The oxidation and reduction reactions in an electrolytic cell are non-spontaneous, so they require an external source of electricity to bring about the chemical changes. This process is known as **electrolysis**.

An electrolytic cell consists of a single container filled with an **electrolyte**. The electrolyte can be a solution of an ionic salt, or a molten ionic salt, composed of free-moving cations and anions. Two electrodes (the cathode and the anode) are dipped in the electrolyte, and a direct current (DC) power source is connected to the electrodes.

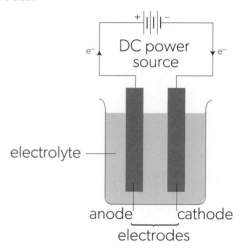

▲ Figure 16 The structure of an electrolytic cell

In a closed circuit, electrons flow from the negative terminal to the positive terminal of the DC power source. The negative terminal of the DC power source is connected to the cathode. Therefore, electrons flow to the cathode and reduce the cations in the electrolyte. The anions in the electrolyte flow to the anode and undergo oxidation, releasing electrons. The electrons flow to the positive terminal of the DC power source to complete the circuit. The flow of electrons and ions in an electrolytic cell comprises electric **current**.

Consider the electrolysis of molten sodium chloride, NaCl(l), shown in figure 17.

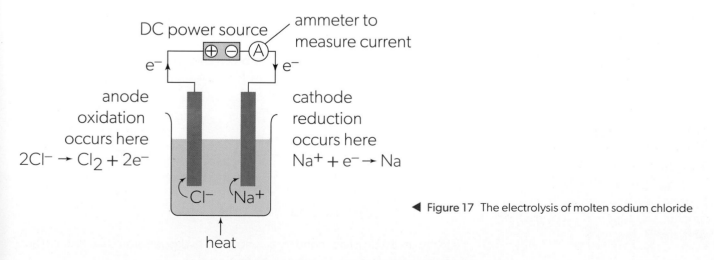

◀ Figure 17 The electrolysis of molten sodium chloride

In the electrolytic cell, molten sodium chloride is the electrolyte, which contains sodium cations, Na^+, and chloride anions, Cl^-. Electrons flow to the cathode, reducing the sodium cations in the electrolyte to form molten sodium metal. The reduction half-equation is as follows:

$$Na^+ + e^- \rightarrow Na(l)$$

At the anode, chloride anions are oxidized, producing chlorine gas and electrons to complete the circuit. The oxidation half-equation is as follows:

$$2Cl^- \rightarrow Cl_2(g) + 2e^-$$

The overall equation for reactions in the electrolytic cell is:

$$2NaCl(l) \rightarrow 2Na(l) + Cl_2(g)$$

This electrolytic cell is therefore useful for the production of sodium metal and chlorine gas.

Reactive metals, such as lithium, magnesium, aluminium, and sodium are all obtained by electrolysis of their molten salts. These metals react with oxygen, so the electrolysis must take place in an inert atmosphere.

Practice question

6. Write the half-equations, and full equation, for the electrolysis of molten lead bromide, $PbBr_2$.

Oxidation of organic compounds (*Reactivity 3.2.9*)

Some **functional groups** in organic compounds can undergo oxidation under certain conditions. For example, in the presence of an oxidizing agent, the **hydroxyl group** in **secondary alcohols** can be oxidized to a **carbonyl group**, forming a **ketone**:

secondary alcohol   ketone

The symbol [O] is used to indicate the oxidizing agent, which provides a source of oxygen atoms. In this oxidation reaction, two hydrogen atoms are lost from the compound, forming water with an oxygen atom from the oxidizing agent. An additional bond forms between carbon and oxygen.

Primary alcohols can be oxidized to **carboxylic acids** in a two-step reaction. In the first step, the hydroxyl group is oxidized to a carbonyl group, forming an **aldehyde**:

primary alcohol   aldehyde

The definitions of primary, secondary and tertiary alcohols are given in *Structure 3.2*. Naming compounds with hydroxyl, carboxyl and carbonyl functional groups is also covered in *Structure 3.2*.

Like in the oxidation of a secondary alcohol, two hydrogen atoms are lost and an additional bond forms between carbon and oxygen.

In the second step, the carbonyl group is oxidized to a **carboxyl group**, forming a carboxylic acid:

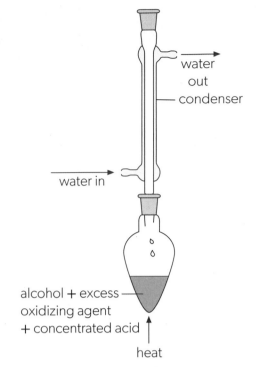

▲ **Figure 18** The experimental set-up for a reflux reaction. Reflux allows vapours to condense back to the boiling reaction mixture for further oxidation

In this oxidation reaction, the aldehyde gains an oxygen atom.

The oxidation of a primary alcohol to a carboxylic acid, or a secondary alcohol to a ketone, can be accomplished by **reflux**. Reflux involves heating the reaction mixture with a reflux condenser, which cools any vapours produced and returns them back to the reaction mixture (figure 18). An oxidizing agent, such as potassium dichromate(VI), $K_2Cr_2O_7$, and a concentrated acid are also added to the reaction mixture.

For the oxidation of a primary alcohol to a carboxylic acid, the oxidizing agent is present in excess to ensure complete two-step oxidation, rather than partial oxidation to an aldehyde after the first oxidation step.

The oxidation of a primary alcohol to an aldehyde can be accomplished by **distillation**. Distillation allows the aldehyde to be isolated before it undergoes further oxidation to carboxylic acid (figure 19). In this case, the alcohol, not the oxidizing agent, is used in excess.

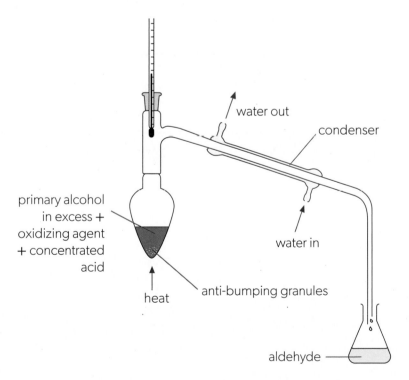

◀ **Figure 19** The experimental set-up for the distillation of an aldehyde obtained by the oxidation of a primary alcohol

Worked example 8

Write the equations for the following oxidation reactions, showing displayed formulas:

a. oxidation of ethanol to an aldehyde

b. oxidation of propan-1-ol to a carboxylic acid

c. oxidation of propan-2-ol.

Use [O] to symbolize the oxidizing agent. In each equation, state which reacting species is in excess and suggest which experimental procedure, reflux or distillation, must be used.

Solution

a.

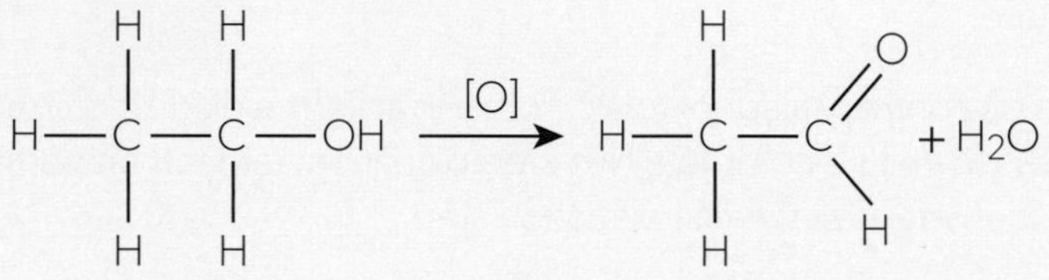

ethanol   ethanal

ethanol in excess

distillation

b.

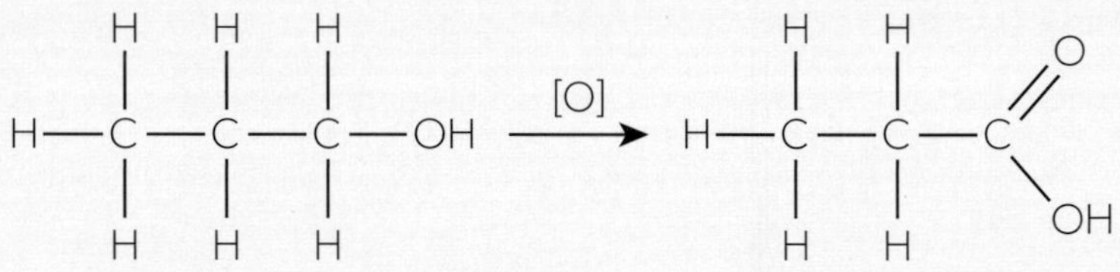

propan-1-ol   propanoic acid

oxidizing agent in excess

reflux

c.

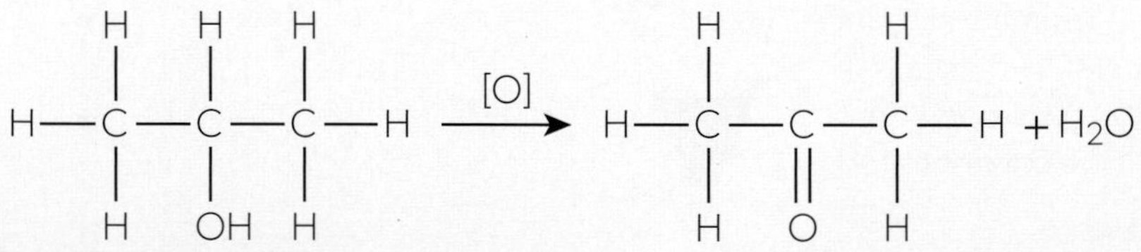

propan-2-ol   propanone

oxidizing agent in excess

reflux

In both primary and secondary alcohols, oxidation of the hydroxyl group to a carbonyl group involves the removal of the hydrogen atom connected to the carbon with the hydroxyl group. Tertiary alcohols do not have this hydrogen atom, so they cannot be oxidized in the same way as primary and secondary alcohols (figure 20).

Reduction of organic compounds (*Reactivity 3.2.10*)

Carboxylic acids can be reduced to primary alcohols via a two-step reaction involving an aldehyde intermediate, and ketones can be reduced to secondary alcohols. The reactions are the opposite of the corresponding oxidation reactions.

In the presence of a reducing agent, the carbonyl group in a ketone can be reduced to a hydroxyl group, forming a secondary alcohol:

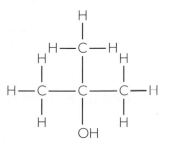

▲ Figure 20 2-methylpropan-2-ol is a tertiary alcohol, so it cannot be oxidized by reflux or distillation in the presence of an oxidizing agent

$$ \text{ketone} \xrightarrow{[H]} \text{secondary alcohol} $$

The symbol [H] is used to indicate the reducing agent, which provides hydrogen atoms. In this reduction reaction, the ketone gains two hydrogen atoms and one of the bonds between carbon and oxygen is broken.

Carboxylic acids can be reduced to primary alcohols in a two-step reaction. In the first step, the carboxyl group is reduced to a carbonyl group, forming an aldehyde:

$$ \text{carboxylic acid} \xrightarrow{[H]} \text{aldehyde} $$

In this reduction reaction, the carboxylic acid loses an oxygen atom.

In the second step, the carbonyl group is reduced to a hydroxyl group, forming a primary alcohol:

$$ \text{aldehyde} \xrightarrow{[H]} \text{primary alcohol} $$

Like in the reduction of a ketone, two hydrogen atoms are gained and one of the carbon–oxygen bonds is broken.

In the reduction of carboxylic acids, the aldehyde is produced as an intermediate product, but in most cases it cannot be isolated.

A common reducing agent used in all the reactions above is lithium aluminium hydride, $LiAlH_4$. Aldehydes and ketones (but not carboxylic acids) can also be reduced by sodium borohydride, $NaBH_4$.

Science as a shared endeavour

Vincent van Gogh used lead and chromium paints to create many of his colourful and well-known paintings. Redox reactions have caused some of the colours to fade over time.

Chemists, art historians and conservators work together to explore and understand these changes, as well as to devise possible preservation methods. Where else does chemistry intersect with the world of art?

▲ **Figure 21** The yellow paint used in Van Gogh's *Bedroom at Arles* was probably made from lead chromate, $PbCrO_4$ (right)

Reduction of alkenes and alkynes (*Reactivity 3.2.11*)

Alkenes and alkynes are unsaturated compounds. Alkenes have a carbon–carbon double bond, and alkynes have a carbon–carbon triple bond. Unsaturated compounds can be reduced by the addition of hydrogen to the multiple bond.

In the presence of a suitable catalyst, such as deactivated palladium, Pd(s), alkynes can be reduced by hydrogen gas to alkenes:

$$R-C\equiv C-R' \quad + \quad H-H \quad \longrightarrow \quad \begin{matrix} H & & H \\ \backslash & & / \\ C & = & C \\ / & & \backslash \\ R & & R' \end{matrix}$$

$$\text{alkyne} \qquad\qquad \text{hydrogen gas} \qquad\qquad \text{alkene}$$

In this equation, the symbol R represents either an alkyl group or a hydrogen atom.

Alkenes are reduced by hydrogen gas to alkanes:

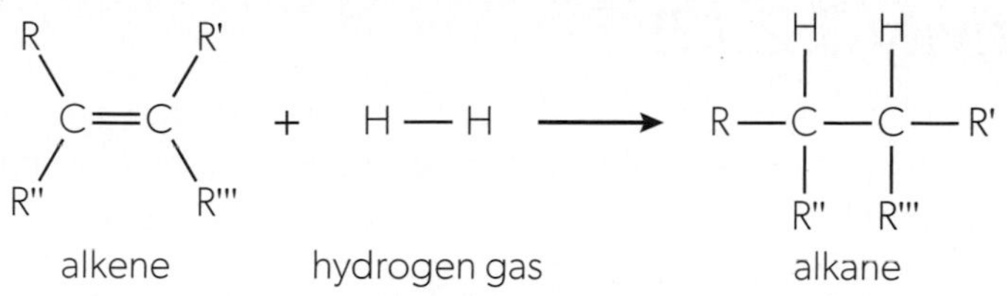

This reaction also requires a transition metal catalyst, such as Ni(s) or Pt(s). The same catalysts can be used to reduce alkynes directly to alkanes using excess hydrogen gas:

$R–C≡C–R' + 2H–H → R–CH_2–CH_2–R'$

alkyne   excess   alkane
          hydrogen

Reduction of alkynes and alkenes decreases the degree of unsaturation of these compounds.

ATL Thinking skills

In this task, you will create a concept map to summarize the material covered in this unit and identify the connections between various aspects of electron transfer reactions. You will need a large sheet of paper and several sticky notes. This task is based on a *Harvard Project Zero Visible Thinking Routine* known as Generate-Sort-Connect-Elaborate.

- Make a list of the key words in this unit and write them on sticky notes. These are the nodes of your concept map.

- Write the title "Electron transfer reactions" at the centre of a large sheet of paper.

- Arrange the nodes (sticky notes) on the paper around the title, from general to more specific.

- Draw lines between pairs of nodes to represent the connections between them.

- Write a brief statement along each connecting line to describe how the key words are linked.

- Share your concept map with your class and expand it once you have received feedback.

Practice questions

7. Predict the products of the reaction of excess hydrogen gas with:

    a. propene

    b. pent-1-yne.

    Write the full equations, including displayed formulas, for these reactions.

Standard electrode potentials (*Reactivity 3.2.12*)

The ease of oxidation and reduction of a species in an electrochemical cell can be described numerically using a **standard electrode potential**, or **standard reduction potential**, $E^{\ominus}$. Standard electrode potentials are defined relative to a hydrogen-based half-cell where the following reaction occurs:

$$H^+(aq) + e^- \rightleftharpoons \frac{1}{2}H_2(g) \qquad E^{\ominus} = 0\,V$$

This half-cell is known as the **standard hydrogen electrode** (**SHE**) and is assigned a standard electrode potential of zero. An inert platinum electrode is used in the SHE because the reduced species is hydrogen gas and not a metal.

Species with a more negative standard electrode potential will have greater ease of oxidation, and greater tendency to reduce other species. They will be higher in the reactivity series. Species with a more positive standard electrode potential will have greater ease of reduction and greater tendency to oxidize other species. They will be lower in the reactivity series. Of all the elements, lithium has the greatest ease of oxidation and fluorine has the greatest ease of reduction.

> The reactivity series and the corresponding standard electrode potentials are given in section 19 of the data booklet. SATP conditions are provided in section 4.

Standard electrode potentials are measured in volts (V). They are correct for standard temperature and pressure (SATP) conditions (temperature = 298 K, pressure = 100 kPa). All aqueous species present in the half-cell equation must also have a concentration of $1.0\,mol\,dm^{-3}$.

Practice question

8. The standard electrode potentials of four metals are given in table 3.

Metal	$E^{\ominus}$ / V
tin, Sn(s)	−0.14
calcium, Ca(s)	−2.87
lithium, Li(s)	−3.04
aluminium, Al(s)	−1.66

▲ Table 3 $E^{\ominus}$ values for selected metals

Order these metals according to their ease of oxidation, with the most readily oxidized first.

Standard cell potentials (*Reactivity 3.2.12*)

The voltage of an electrochemical cell depends on the identity of the electrodes in each half-cell. The further apart the species are on the reactivity series, the greater the voltage. This voltage, known as the **standard cell potential**, $E^{\ominus}_{cell}$, can be calculated by finding the difference between the standard electrode potentials of the half-cells:

$E^{\ominus}_{cell} = E^{\ominus}(\text{reduced species}) - E^{\ominus}(\text{oxidized species})$

For a reaction in an electrochemical cell to be spontaneous, $E^{\ominus}_{cell}$ must have a positive value. In that case, the reduction will occur at the electrode with the more positive value of $E^{\ominus}$ (the cathode) and the oxidation will occur at the electrode with the more negative value of $E^{\ominus}$ (the anode). In other words:

$E^{\ominus}_{cell} = E^{\ominus}(\text{cathode}) - E^{\ominus}(\text{anode})$

Consider an electrochemical cell composed of $Fe^{2+} | Fe$ and $Cu^{2+} | Cu$ half-cells. The data booklet states the half-cell reduction reactions and their standard electrode potentials as follows:

$Fe^{2+}(aq) + 2e^- \rightleftharpoons Fe(s)$ $\qquad E^{\ominus} = -0.45\,V$

$Cu^{2+}(aq) + 2e^- \rightleftharpoons Cu(s)$ $\qquad E^{\ominus} = +0.34\,V$

The copper electrode has the more positive value of $E^{\ominus}$, so $Cu^{2+}(aq)$ will be reduced to $Cu(s)$, and the half-equation at that electrode will proceed in the forward direction.

The iron electrode has the more negative value of $E^{\ominus}$, so $Fe(s)$ will be oxidized to $Fe^{2+}(aq)$, and the half-equation at that electrode will proceed in the reverse direction to that stated above:

$Fe(s) \rightleftharpoons Fe^{2+}(aq) + 2e^-$

Now you can determine the overall standard potential of the cell:

$E^{\ominus}_{cell} = E^{\ominus}(\text{reduced species}) - E^{\ominus}(\text{oxidized species})$

$= (+0.34\,V) - (-0.45\,V)$

$= 0.79\,V$

To find the overall equation for the electrochemical cell, add the two half-equations together, cancelling the electrons and ensuring that the equation is balanced:

$Fe(s) + Cu^{2+}(aq) \rightleftharpoons Fe^{2+}(aq) + Cu(s)$

This reaction is described as being spontaneous in the forward direction. You can therefore use $E^{\ominus}$ data to predict which direction will be spontaneous for a reversible redox reaction in an electrochemical cell.

Worked example 9

An electrochemical cell composed of a $Zn^{2+} \mid Zn$ half-cell and a standard hydrogen electrode (SHE) half-cell is shown in figure 22. The redox reaction in the cell can be written as follows:

$$2H^+(aq) + Zn(s) \rightleftharpoons Zn^{2+}(aq) + H_2(g)$$

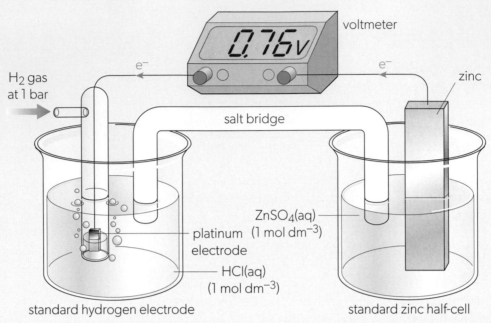

▲ Figure 22 The electrochemical cell with a SHE half-cell and a $Zn \mid Zn^{2+}$ half-cell. The voltmeter is set up to measure the overall cell potential

a. State which half-cell contains the anode and which contains the cathode.

b. Predict whether the forward or backwards reaction is spontaneous.

c. Show that the standard cell potential, $E^{\ominus}_{cell}$, is +0.76 V using section 19 of the data booklet.

Solution

a. First, write the half-cell reduction reactions and standard electrode potentials from the data booklet:

$$Zn^{2+}(aq) + 2e^- \rightleftharpoons Zn(s) \quad E^{\ominus} = -0.76\,V$$
$$H^+(aq) + e^- \rightleftharpoons \tfrac{1}{2}H_2(g) \quad E^{\ominus} = 0.00\,V$$

The SHE half-cell has the more positive value of $E^{\ominus}$, so it will involve a reduction reaction and therefore contains the cathode.

The zinc half-cell has the more negative value of $E^{\ominus}$, so it will involve an oxidation reaction and therefore contains the anode.

b. Reduction occurs at the SHE half-cell, so its half-equation will proceed in the forward direction as given in the data booklet.

Oxidation occurs at the zinc half-cell, so its half-equation will proceed in the reverse direction:

$$Zn(s) \rightleftharpoons Zn^{2+}(aq) + 2e^-$$

Before you add the half-equations together, you first need to double the stoichiometric coefficients in the hydrogen half-equation to ensure that there is an equal number of electrons in each half-equation:

$$2H^+(aq) + 2e^- \rightleftharpoons H_2(g)$$

Now add the half-equations together, cancelling the electrons:

$$2H^+(aq) + Zn(s) \rightarrow Zn^{2+}(aq) + H_2(g).$$

Therefore, the reaction will be spontaneous in the forward direction for the equation given in the question.

c. $E^{\ominus}_{cell} = E^{\ominus}(\text{reduced species}) - E^{\ominus}(\text{oxidized species})$:

$$(0.00\,V) - (-0.76\,V)$$

$$= 0.76\,V$$

Worked example 10

An electrochemical cell comprises a $Cu^{2+}\,|\,Cu$ half-cell and an $Ag^+\,|\,Ag$ half-cell.

a. State which half-cell contains the anode, and which contains the cathode.

b. Write the spontaneous reaction that occurs in the electrochemical cell.

c. Calculate the standard cell potential, $E^{\ominus}{}_{cell}$.

Solution

a. Copy the half-equations and standard reduction potentials for each half-cell from the data booklet:

$$Ag^+(aq) + e^- \rightleftharpoons Ag(s) \qquad E^{\ominus} = +0.80\,V$$

$$Cu^{2+}(aq) + 2e^- \rightleftharpoons Cu(s) \qquad E^{\ominus} = +0.34\,V$$

The silver half-cell has the more positive value, so it contains the cathode, and the copper half-cell contains the anode.

b. The silver half-cell has the more positive $E^{\ominus}$ value, so reduction occurs in this half-cell and the half-equation is already shown in the correct direction. Oxidation occurs in the copper half-cell, so the half-equation needs to be reversed:

$$Cu(s) \rightleftharpoons Cu^{2+}(aq) + 2e^-$$

Before you add the half-equations together, you first need to double the stoichiometric coefficients in the silver half-equation to ensure that there is an equal number of electrons in each half-equation:

$$2Ag^+(aq) + 2e^- \rightleftharpoons 2Ag(s)$$

Now add the half-equations together, cancelling the electrons:

$$2Ag^+(aq) + Cu(s) \rightarrow 2Ag(s) + Cu^{2+}(aq)$$

c. $E^{\ominus}{}_{cell} = E^{\ominus}(\text{reduced species}) - E^{\ominus}(\text{oxidized species})$:

$$= (+0.80\,V) - (+0.34\,V)$$

$$= 0.46\,V$$

It is important to note that the half-cell voltage for silver electrode is not multiplied by two. The electrons lost and gained need to be balanced, but the potential of the cell only depends on the chemical nature of the species involved in the half-equation, and not on the way the half-equation is balanced.

Gibbs energy and standard cell potentials (*Reactivity 3.2.14*)

The **standard change in Gibbs energy**, $\Delta G^\ominus$, over the course of a chemical reaction occurring in an electrochemical cell, can be determined from the standard cell potential, $E^\ominus_{cell}$:

$$\Delta G^\ominus = -nFE^\ominus_{cell}$$

where n is the number of electrons transferred in the balanced redox equation, and F is the Faraday constant, $9.65 \times 10^4\,C\,mol^{-1}$.

You know that an electrochemical reaction will be spontaneous if $E^\ominus_{cell}$ is positive. As the right-hand term in the equation above has a negative sign, a reaction will be spontaneous if ΔG is negative.

F has units of coulombs per mole, $C\,mol^{-1}$, where coulomb is the SI unit of charge. $E^\ominus_{cell}$ has units of volts, V. When the n, F and $E^\ominus_{cell}$ terms are multiplied together, the resulting value has units of $C\,V\,mol^{-1}$. One volt is equivalent to one unit of energy per unit charge; in other words, one joule per coulomb: $1\,V = 1\,J\,C^{-1}$. Substituting this into $C\,V\,mol^{-1}$ gives:

$$C\,J\,C^{-1}\,mol^{-1} = J\,mol^{-1}.$$

Therefore, the units of $\Delta G^\ominus$ are $J\,mol^{-1}$, or more often $\Delta G^\ominus$ is converted to $kJ\,mol^{-1}$ by dividing by 1,000.

> This equation and the value of the Faraday constant are given in sections 1 and 2 of the data booklet. Change in Gibbs energy, ΔG, can also be defined in terms of the enthalpy change, entropy change and temperature, as described in *Reactivity 1.4* (AHL).

Worked example 11

In worked example 9, $E^\ominus_{cell}$ for the reaction $2H^+(aq) + Zn(s) \rightarrow Zn^{2+}(aq) + H_2(g)$ was calculated to be $+0.76\,V$. Calculate $\Delta G^\ominus$ for this reaction.

Solution

Two electrons are transferred in this redox reaction, so $n = 2$.

$$\Delta G^\ominus = -nFE^\ominus_{cell}$$

$$= -2 \times (9.65 \times 10^4\,C\,mol^{-1}) \times 0.76\,J\,C^{-1}$$

$$= -1.47 \times 10^5\,J\,mol^{-1}\,\text{or}\,-147\,kJ\,mol^{-1}$$

Practice questions

9. The standard change in Gibbs energy, $\Delta G^\ominus$, is $-152\,kJ\,mol^{-1}$ for the following electrochemical reaction:

   $$Fe(s) + CuSO_4(aq) \rightarrow FeSO_4(aq) + Cu(s)$$

   a. State whether the reaction will occur spontaneously.

   b. Calculate the value of $E^\ominus_{cell}$. Compare the value obtained with the difference in standard reduction potentials of Fe^{2+}/Fe and Cu^{2+}/Cu electrodes, which are given in section 19 of the data booklet.

Measuring standard cell potentials

The displacement reaction between zinc and copper(II) ions is as follows:

$$Zn(s) + Cu^{2+}(aq) \rightarrow Cu(s) + Zn^{2+}(aq)$$

If the oxidation and reduction processes are separated into two half-cells connected by an external wire and salt bridge, chemical energy from the spontaneous redox reaction is converted into electrical energy. In this practical, you will measure the cell potential of a zinc–copper cell and use this to determine ΔG for the reaction.

You will be using $1.0 \, mol \, dm^{-3}$ solutions, which require large quantities of the corresponding hydrated salts. To minimize waste, very small volumes of the electrolytes will be used. Once finished, you are encouraged to recover the electrolyte solutions instead of discarding them, as they can be used for other experiments.

Relevant skills

- Tool 1: Measuring potential difference
- Tool 1: Constructing electrochemical cells
- Tool 3: Calculate percentage error
- Tool 3: General mathematics

Safety

- Wear eye protection
- Copper(II) sulfate and zinc sulfate are harmful and irritant. Avoid contact with eyes and skin
- Copper(II) sulfate and zinc sulfate are toxic to the environment and must be disposed of safely

Materials

- copper(II) sulfate pentahydrate, $CuSO_4 \bullet 5H_2O$
- zinc sulfate heptahydrate, $ZnSO_4 \bullet 7H_2O$
- distilled water
- zinc electrode
- copper electrode
- potassium chloride
- two weighing bottles (or other small wide-mouth containers)
- sandpaper

- $50 \, cm^3$ beakers
- filter paper
- tweezers
- high-resistance voltmeter
- crocodile clips
- connecting wires

Instructions

1. Prepare a $1.0 \, mol \, dm^{-3}$ copper(II) sulfate solution and transfer $5.0 \, cm^3$ of it into a weighing bottle.

2. Prepare a $1.0 \, mol \, dm^{-3}$ zinc sulfate solution and transfer $5.0 \, cm^3$ of it to a second weighing bottle.

3. Sand the two electrodes to remove any surface contaminants.

4. Prepare a small volume of saturated potassium chloride solution. Cut a strip of filter paper to use as a salt bridge between the two half-cells. Dip the strip into the potassium chloride solution.

5. Construct the voltaic cell:
   - connect the two solutions in weighing bottles with the salt bridge
   - connect the electrodes to the voltmeter using the crocodile clips and connecting wires
   - dip the electrodes into their corresponding solutions in weighing bottles.

6. Measure the potential difference.

7. Clear up according to the directions given by your teacher.

Questions

1. Calculate the theoretical standard cell potential for your zinc–copper cell, using the standard electrode potentials in the data booklet.

2. Compare your experimental result to the theoretical value and calculate the percentage error.

3. Suggest at least two reasons why your measured value differs from the theoretical value.

4. Calculate ΔG from your measured cell potential.

5. Research the relationship between E_{cell} and ΔG under non-standard conditions. Use this to briefly outline two research questions related to voltaic cells, E_{cell} and ΔG.

Electrolysis of aqueous solutions (*Reactivity 3.2.15*)

We have discussed the electrolysis of molten ionic salts where cations are reduced and anions are oxidized. The electrolysis of aqueous solutions of ionic salts introduces oxidation and reduction reactions involving water, which compete with the redox reactions of the anions and cations of the salt at the anode and cathode.

The reduction of water can proceed as follows:

$$H_2O(l) + e^- \rightarrow \frac{1}{2}H_2(g) + OH^-(aq) \qquad\qquad E^{\ominus} = -0.83\,V$$

If the standard electrode potential of the salt cation is more negative than $-0.83\,V$, then water will be preferentially reduced over the salt cation and hydrogen gas will be formed at the cathode.

The other possible competing reaction is the oxidation of water. In the data booklet, the reduction of oxygen is given as follows:

$$\frac{1}{2}O_2(g) + 2H^+(aq) + 2e^- \rightarrow H_2O(l) \qquad\qquad E^{\ominus} = +1.23\,V$$

Reversing the equation gives the half-equation for the oxidation of water. The reduction potential has to be reversed as well, giving an oxidation potential:

$$H_2O(l) \rightarrow \frac{1}{2}O_2(g) + 2H^+(aq) + 2e^- \qquad\qquad E^{\ominus} = -1.23\,V$$

If the oxidation potential of the salt anion is more negative than $-1.23\,V$, then water will be preferentially oxidized over the salt anion and oxygen gas will be formed at the anode.

For example, consider two electrolytic cells: one composed of molten sodium chloride, NaCl(l), and one composed of an aqueous sodium chloride, NaCl(aq). In the electrolysis of NaCl(l), electrical energy is provided to the cell, resulting in the reduction of sodium ions to form sodium metal at the cathode and the oxidation of chloride ions to form chlorine gas at the anode:

cathode: $Na^+ + e^- \rightarrow Na(l)$

anode: $Cl^- \rightarrow \frac{1}{2}Cl_2(g) + e^-$

overall cell equation: $2NaCl(l) \rightarrow 2Na(l) + Cl_2(g)$

In the electrolysis of aqueous sodium hydroxide, NaCl(aq), the two competing oxidation and reduction reactions involving water are introduced. There are now two species that can potentially be reduced at the cathode: sodium ions from the salt, and water. The two competing reduction reactions are as follows:

$$Na^+(aq) + e^- \rightarrow Na(s) \qquad\qquad E^{\ominus} = -2.71\,V$$

$$H_2O(l) + e^- \rightarrow \frac{1}{2}H_2(g) + OH^-(aq) \qquad\qquad E^{\ominus} = -0.83\,V$$

The reduction potential of water is less negative than that of sodium, so water is more easily reduced than the sodium ions. Therefore, the only product formed at the cathode will be hydrogen gas, $H_2(g)$. The sodium ions, $Na^+(aq)$, and hydroxide ions, $OH^-(aq)$, will stay in the solution.

The chloride ions from the salt and water also compete to be oxidized at the anode. The half-equations and standard electrode potentials involving the oxidation of chloride ions and water are shown below.

$$\frac{1}{2}Cl_2(g) + e^- \rightarrow Cl^-(aq) \qquad\qquad E^\ominus = +1.36\,V$$

$$\frac{1}{2}O_2(g) + 2H^+(aq) + 2e^- \rightarrow H_2O(l) \qquad\qquad E^\ominus = +1.23\,V$$

The equations need to be reversed to reflect the oxidation of these species:

$$Cl^-(aq) \rightarrow \frac{1}{2}Cl_2(g) + e^- \qquad\qquad E^\ominus = -1.36\,V$$

$$H_2O(l) \rightarrow \frac{1}{2}O_2(g) + 2H^+(aq) + 2e^- \qquad\qquad E^\ominus = -1.23\,V$$

The chloride ions have a more negative oxidation potential than water, so we can expect water to be oxidized preferentially compared to chloride ions. However, the difference between the two oxidation potentials is small, so in solutions with high concentration of $Cl^-(aq)$ ions the main product formed at the anode will be chlorine gas, $Cl_2(g)$ (figure 23).

The overall cell equation is as follows:

$$Na^+(aq) + Cl^-(aq) + H_2O(l) \rightarrow \frac{1}{2}H_2(g) + \frac{1}{2}Cl_2(g) + Na^+(aq) + OH^-(aq)$$

or

$$NaCl(aq) + H_2O(l) \rightarrow \frac{1}{2}H_2(g) + \frac{1}{2}Cl_2(g) + NaOH(aq)$$

Doubling the stoichiometric coefficients gives the final equation:

$$2NaCl(aq) + 2H_2O(l) \rightarrow H_2(g) + Cl_2(g) + 2NaOH(aq)$$

In dilute solutions of NaCl(aq), the concentration of $Cl^-(aq)$ ions is low, so water is oxidized along with chloride ions. In that case, a mixture of oxygen gas and chlorine gas is produced at the anode. In very dilute solutions, oxygen gas will be the only product at the anode.

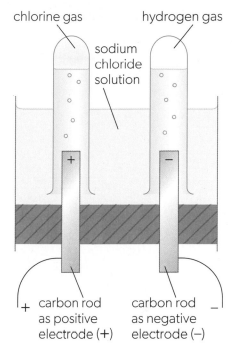

▲ Figure 23 The electrolysis of aqueous sodium chloride

Worked example 12

Deduce the products of the electrolysis of aqueous copper(II) sulfate, $CuSO_4(aq)$, with inert electrodes. Write the overall cell equation.

Solution

First, write the half-equations and standard electrode potentials involving the reduction of copper(II) ions and water.

$$Cu^{2+}(aq) + 2e^- \rightarrow Cu(s) \qquad\qquad E^\ominus = +0.34\,V$$

$$H_2O(l) + e^- \rightarrow \frac{1}{2}H_2(g) + OH^-(aq) \qquad\qquad E^\ominus = -0.83\,V$$

The reduction potential of copper(II) ions is more positive than that of water, so copper(II) ions will be reduced to copper metal at the cathode.

The two competing species for oxidation at the anode are sulfate ions, $SO_4{}^{2-}$ (aq), and water. In this case, the sulfate ions cannot be oxidized because sulfur is in its highest oxidation state of +6. Therefore, removing any more electrons from sulfate would result in an impossible electron configuration for sulfur.

Practice question

10. Deduce the products of the electrolysis of aqueous potassium dichromate, $K_2Cr_2O_7(aq)$, with inert electrodes.

This means that water will be oxidized at the anode and oxygen gas will be released, with hydrogen ions staying in solution:

$$H_2O(l) \rightarrow \frac{1}{2}O_2(g) + 2H^+(aq) + 2e^-$$

The overall cell equation is therefore:

$$Cu^{2+}(aq) + SO_4^{2-}(aq) + H_2O(l) \rightarrow Cu(s) + \frac{1}{2}O_2(g) + 2H^+(aq) + SO_4^{2-}(aq)$$

or

$$CuSO_4(aq) + H_2O(l) \rightarrow Cu(s) + \frac{1}{2}O_2(g) + H_2SO_4(aq)$$

Doubling the stoichiometric coefficients gives the final equation:

$$2CuSO_4(aq) + 2H_2O(l) \rightarrow 2Cu(s) + O_2(g) + 2H_2SO_4(aq)$$

Electroplating (*Reactivity 3.2.16*)

The electrolysis of aqueous copper(II) sulfate described in worked example 12 uses inert carbon electrodes (figure 24, left). However, if the electrodes are made from a reactive metal, the electrolysis of an aqueous solution of an ionic salt will add material to the cathode and take material away from the anode. These processes are known as **plating** and **eroding**, respectively.

For example, consider an electrolytic cell containing a solution of copper sulfate, $CuSO_4(aq)$, with the anode and the cathode each made of copper metal (figure 24, right). When the electric current is applied, the copper anode will erode to form $Cu^{2+}(aq)$ ions while the same ions will be reduced at the cathode to form $Cu(s)$. This process can be used to purify a sample of impure copper metal. The impure copper is used as the anode, which will be eroded to produce $Cu^{2+}(aq)$ ions. These ions will then be reduced at the cathode and plated there as pure copper metal. The impurities will either stay on the anode (if they are less readily oxidized than copper metal) or remain in the solution (if their ions are less readily reduced than copper(II) ions).

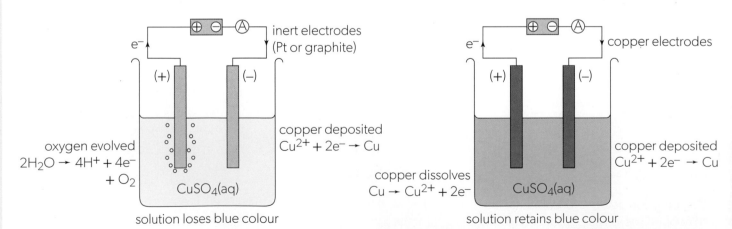

▲ **Figure 24** Electrolysis of copper(II) sulfate with inert electrodes reduces copper(II) ions at the cathode and oxidizes water, producing oxygen gas at the anode. The blue colour of the solution fades as copper(II) ions are replaced with hydrogen ions.

Electrolysis using copper electrodes causes the copper anode to dissolve and copper metal to deposit on the cathode. The amount of copper(II) ions in the solution remains constant, so the blue colour of the solution does not fade

Electroplating involves coating an object with a thin layer of pure metal by electrolysis. A metal anode is oxidized to form cations in the solution. The cations travel through the solution to the cathode, where they are reduced to form a thin layer of metal on the cathode.

For example, in an electrochemical cell comprising a copper anode, a steel cathode, and copper(II) sulfate solution, the steel cathode will be plated with a thin layer of copper (figure 25).

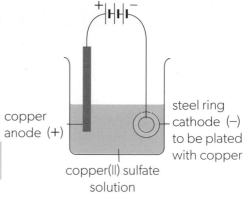

▲ Figure 25 In copper electroplating, copper(II) ions are reduced at the cathode to form a thin layer of $Cu(s)$ on the surface of the cathode (in this case, a steel ring)

Observations

Scientists make inferences from their observations. Observation involves use of the senses. Our knowledge of the behaviour of matter then allows us to infer conclusions from observed data. For example, you can observe gas bubbles being generated at an electrode during electrolysis, and you can infer the identity of the gas from your knowledge of the composition of the electrolyte. Similarly, you may observe a brownish-red solid being deposited at the cathode during the electrolysis of aqueous copper(II) sulfate and, from that, infer that copper is reduced at that electrode. What "counts" as an observation in science?

Worked example 13

Deduce the half-equations at each electrode in the electroplating of a steel electrode with copper in a copper(II) sulfate solution, $CuSO_4(aq)$.

Solution

At the anode, copper metal is oxidized to copper(II) ions:

$$Cu(s) \rightarrow Cu^{2+}(aq) + 2e^-$$

At the cathode, the reverse reaction occurs:

$$Cu^{2+}(aq) + 2e^- \rightarrow Cu(s)$$

Practice question

11. A nickel spoon is used in a silver electroplating experiment as shown below.

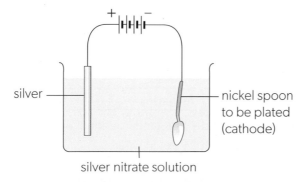

a. Write the half-equation for the reaction occurring at the:

   i. cathode   ii. anode.

b. Describe and explain what would happen to the concentration of the silver(I) nitrate electrolyte.

c. Describe and explain the mass changes at each of the two electrodes.

▲ Figure 26 The trophy for the top prize at the Berlin International Film Festival, the Golden Bear, is made of bronze coated with a thin layer of gold by electroplating

End of topic questions

Topic review

1. Using your knowledge from the *Reactivity 3.2* topic, answer the guiding question as fully as possible:
   What happens when electrons are transferred?

Exam-style questions
Multiple-choice questions

2. Which species contains nitrogen with an oxidation state of +5?
   A. N_2
   B. N_2O
   C. NO_2^-
   D. HNO_3

3. Which element is reduced in the following reaction?
   $2MnO_4^-(aq) + Br^-(aq) + H_2O(l) \rightarrow$
   $$2MnO_2(s) + BrO_3^-(aq) + 2OH^-(aq)$$
   A. Mn
   B. O
   C. Br
   D. H

4. Which of the following is **not** a redox reaction?
   A. $2H_2O_2 \rightarrow 2H_2O + O_2$
   B. $CH_3COOH + LiOH \rightarrow CH_3COOLi + H_2O$
   C. $C_3H_6 + Br_2 \rightarrow C_3H_6Br_2$
   D. $Na + Cl_2 \rightarrow 2NaCl$

5. Which pair reacts most vigorously?
   A. Li and Br_2
   B. Li and F_2
   C. K and Br_2
   D. K and F_2

6. What are the products of the electrolysis of molten lead(II) bromide, $PbBr_2$?

	Cathode product	Anode product
A.	lead	bromine
B.	lead	bromide
C.	lead(II)	bromine
D.	lead(II)	bromide

7. The following reaction occurs in a voltaic cell.
   $$2Ag^+(aq) + Cu(s) \rightarrow 2Ag(s) + Cu^{2+}(aq)$$
   What reaction occurs at each electrode?

	Cathode	Anode
A.	$Ag^+(aq) \rightarrow Ag(s) + e^-$	$Cu(s) \rightarrow Cu^{2+}(aq) + 2e^-$
B.	$Ag^+(aq) + e^- \rightarrow Ag(s)$	$Cu(s) + 2e^- \rightarrow Cu^{2+}(aq)$
C.	$Ag^+(aq) + e^- \rightarrow Ag(s)$	$Cu(s) \rightarrow Cu^{2+}(aq) + 2e^-$
D.	$Cu(s) \rightarrow Cu^{2+}(aq) + 2e^-$	$Ag^+(aq) + e^- \rightarrow Ag(s)$

8. Which of the following classes of compound can be reduced?
   I   alkene
   II   carboxylic acid
   III aldehyde
   A. I and II only
   B. I and III only
   C. II and III only
   D. I, II and III

9. Which class of compound is formed when a secondary alcohol is oxidized?
   A. aldehyde
   B. ether
   C. ketone
   D. carboxylic acid

10. Hydrogen can be added to ethene to produce ethane.
    $$C_2H_4(g) + H_2(g) \rightarrow C_2H_6(g)$$
    Which of the following is correct for this reaction?

	Degree of unsaturation	Species that undergoes reduction
A.	increases	ethene
B.	decreases	ethene
C.	increases	hydrogen
D.	decreases	hydrogen

11. Consider the standard reduction potentials:
    $Cd^{2+}(aq) + 2e^- \rightleftharpoons Cd(s)$ $E^\ominus = -0.40$ V
    $Cr^{3+}(aq) + 3e^- \rightleftharpoons Cr(s)$ $E^\ominus = -0.74$ V
    Which is the strongest oxidizing agent?
    A. Cd^{2+}
    B. Cr^{3+}
    C. Cd
    D. Cr

AHL

AHL

12. What are the major products of the electrolysis of concentrated sodium chloride solution, $NaCl(aq)$?

	Negative electrode	Positive electrode
A.	Na	O_2
B.	H_2	O_2
C.	Na	Cl_2
D.	H_2	Cl_2

Extended-response questions

13. The reaction between ethanedioate ions, $C_2O_4^{2-}$ and manganate(VII) ions, MnO_4^-, in acidic solution is a redox reaction. The incomplete equation for this reaction is shown below:

    $$C_2O_4^{2-} + MnO_4^- \rightarrow CO_2 + Mn^{2+}$$

    a. Deduce the oxidation state of carbon in the ethanedioate ion, $C_2O_4^{2-}$. [1]
    b. The oxidation half-equation is:
    $$C_2O_4^{2-}(aq) \rightarrow 2CO_2(g) + 2e^-$$
    Deduce the reduction half-equation, including state symbols. [2]
    c. Deduce the full redox reaction. [1]

14. Sodium chloride is found in table salt.
    a. Describe how bonding occurs in sodium chloride. [2]
    b. Explain why sodium chloride is not an electrical conductor when solid but can conduct electricity when molten. [1]
    c. Molten sodium chloride can be electrolysed. Identify the half-equation for the reaction that takes place at the
       i. cathode [1]
       ii. anode. [1]
    d. Deduce the half-equation for the oxidation reaction that takes place when dilute aqueous sodium chloride is electrolysed. [1]
    e. A student prepares a concentrated sodium chloride solution and adds a few drops of phenolphthalein. The student then electrolyses the solution. Describe two observations made by the student during the electrolysis. [2]

15. a. $Li^+ + e^- \rightarrow Li(s)$ is a reaction occurring at one of the electrodes in a lithium-ion battery. State at which electrode (anode/cathode) this reaction occurs and whether this is the charging or discharging reaction. [2]
    b. Identify the reaction at the opposite electrode. [1]
    c. Explain why lithium-ion batteries must be sealed. [1]

16. a. Write an equation for the complete combustion of ethanol. [1]
    b. i. Deduce the oxidation state of carbon in carbon dioxide. [1]
       ii. Deduce the mean oxidation state of carbon in ethanol. [1]
       iii. State and explain, with reference to oxidation states, whether carbon is oxidized or reduced when ethanol undergoes complete combustion. [1]
    c. Ethanol can also be oxidized when reacted with acidified potassium dichromate(VI).
       i. Draw the structural formulas and state the names of the two possible organic products of this reaction. [2]
       ii. State the methods used to isolate each of the substances in your answer above. [1]

17. A student carries out the following reactions between three unknown metals (X, Y and Z) and several dilute solutions. The following results were obtained:

Reactants	Observations
$Y + Z(NO_3)_2(aq)$	Y is a shiny grey metal. $Z(NO_3)_2(aq)$ is a clear blue solution. When reacted together, a red-brown solid appeared on the surface of the metal. Over time, the solution's blue colour faded.
$X + Y(NO_3)_2(aq)$	No reaction
$Z + X(NO_3)_2(aq)$	No reaction
$Z + HCl(aq)$	No reaction

   a. List metals X, Y and Z, in order of increasing reactivity. [1]
   b. Suggest, with reference to **two** observations, which metal, X, Y or Z, could be copper. [2]
   c. A standard nickel–copper voltaic cell is set up.
      i. Draw the cell diagram for this voltaic cell and label the following:
         - ions in each solution
         - cathode
         - anode
         - direction of travel of electrons in the external circuit
         - direction of travel of cations in the salt bridge. [2]
      ii. Deduce the equation, including state symbols, for the spontaneous redox reaction that occurs in this cell. [2]
   d. Calculate the standard cell potential, in V, for the redox reaction. Refer to section 19 of the data booklet. [1]

AHL

What happens when a species possesses an unpaired electron?

When an atom or polyatomic species has an unpaired electron, it is called a radical. When a covalent bond between two atoms breaks through the process of homolytic fission, the two electrons involved in the bond move onto the separate atoms. The radicals are highly reactive and can combine with other radicals to form more stable covalent molecules.

Understandings

Reactivity 3.3.1—A radical is a molecular entity that has an unpaired electron. Radicals are highly reactive.

Reactivity 3.3.2—Radicals are produced by homolytic fission, e.g. of halogens, in the presence of ultraviolet (UV) light or heat.

Reactivity 3.2.3—Radicals take part in substitution reactions with alkanes, producing a mixture of products.

Introduction to radicals (*Reactivity 3.3.1*)

Organic reaction **mechanisms** are detailed descriptions of the conversion of reactants to products. These mechanisms are a means of explaining which bonds are broken and formed, and how electrons move over the course of the reaction. One type of chemical species involved in such mechanisms are radicals.

A radical is a chemical entity that has an unpaired electron. Radicals are different from charged species (such as ions) in that a radical can exist independently of any other species. In contrast, cations and anions will always have a corresponding counter-ion. A radical species is indicated by a dot (•). There are two common types of radical:

- a single atom, such as a halogen radical, for example the chlorine radical, Cl•

- a polyatomic species, for example the methyl radical, •CH_3, and the hydroxyl radical, •OH.

When the radical consists of several atoms, the dot in the chemical formula is placed next to the atom with the unpaired electron. For example, in the methyl radical, the dot is placed next to carbon, as carbon has the unpaired electron:

$$
\begin{array}{c}
\text{H} \\
| \\
\cdot\text{C}-\text{H} \\
| \\
\text{H}
\end{array}
$$

Due to their high reactivity, radicals are usually formed as intermediates in a reaction, that is, they are not the ultimate reaction products. Two radicals can react with each other to form a new covalently bonded compound with no unpaired electrons. This process is known as **termination**. However, radical species can also react with non-radical species to create further radical species. This process is known as **propagation**.

Hypotheses

Scientists make provisional explanations for the patterns they observe. A universal observation that has puzzled humans for a long time is ageing. All living things deteriorate physically over time. The free radical hypothesis of ageing (also known as the free radical theory of ageing) was proposed in the mid-20th century and has been widely accepted. Radicals are produced in mitochondria (figure 1) as a by-product of metabolic processes.

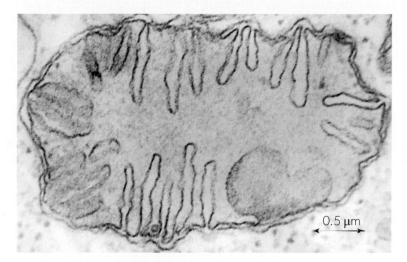

$0.5 \mu m$

▲ **Figure 1** The structure of a mitochondrion

The gradual accumulation of these radical species over time is associated with the oxidation of DNA, proteins and other biomolecules. The free radical hypothesis suggests that ageing is caused by the "oxidative stress" that arises from the build-up of radicals. Correlations have been observed between age, mitochondrial radical production and oxidative damage.

However, critics of this hypothesis challenge the causality of the relationship. The free radical hypothesis of ageing is therefore an area of ongoing research.

What other aspects of nature of science are hypotheses connected to?

Formation of radicals (*Reactivity 3.3.2*)

Curly arrows are used to illustrate the movement of electrons in reaction mechanisms, as bonds are broken and made. A double-barbed arrow shows the movement of a pair of electrons (figure 2).

▲ **Figure 2** The double-barbed arrow represents the movement of an electron pair

A single-barbed, or fish hook, arrow is used to show the movement of a single electron (figure 3).

▲ **Figure 3** The fish hook arrow represents the movement of a single electron

When you draw mechanisms using curly arrows, pay attention to the following:

- The base of the arrow must start at the origin of the moving electrons.

- The arrowhead must finish at the exact destination of the electrons.

- The arrow starts at an electron-rich region and ends at an electron-poor region of the molecule or other species.

Fish hook arrows are commonly used in the reactions involving radicals. A radical is formed when a molecule undergoes **homolytic fission**. The two electrons of a covalent bond are split evenly between two atoms resulting in two radicals that each have a single electron. Halogen radicals are formed this way, where the halogen bond in a diatomic halogen molecule is broken homolytically.

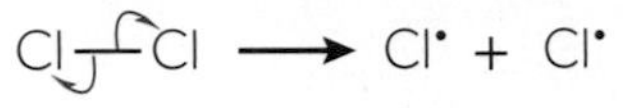

$$Cl \overset{\frown}{\underset{\smile}{}} Cl \longrightarrow Cl^{\bullet} + Cl^{\bullet}$$

▲ Figure 4 Homolytic fission of chlorine, Cl_2, to form chlorine radicals, Cl•

Notice how fish hook curly arrows are used to show the path of each electron that make up the chlorine–chlorine bond.

For homolytic fission of halogens to occur, the reaction mixture must be exposed to ultraviolet (UV) light, or heated. The homolytic fission of halogens to form halogen radicals is the first step in a series of chain reactions initiated by radicals. For this reason, it is known as the **initiation** step.

Radical substitution reactions (*Reactivity 3.3.3*)

One common type of reaction in organic chemistry is substitution reactions. Substitution is the replacement of an atom or a group of atoms in an organic molecule with another atom or group of atoms.

Alkanes are relatively inert due to the strength of the carbon–carbon ($346 \, kJ \, mol^{-1}$) and carbon–hydrogen ($414 \, kJ \, mol^{-1}$) bonds. In addition, the bonds in alkanes are non-polar, which makes them unreactive towards polar reagents. To transform alkanes into more reactive species, some of their non-polar bonds must be replaced with polar bonds. One way of achieving this is to **halogenate** the alkane using radical substitution reactions.

For example, when methane reacts with chlorine in the presence of UV light or heat, chloromethane and hydrogen chloride are formed (figure 5). This is an example of a radical substitution reaction.

$$H-\underset{\underset{H}{|}}{\overset{\overset{H}{|}}{C}}-H \; + \; Cl-Cl \; \xrightarrow[\text{or heat}]{\text{UV light}} \; H-\underset{\underset{H}{|}}{\overset{\overset{H}{|}}{C}}-Cl \; + \; H-Cl$$

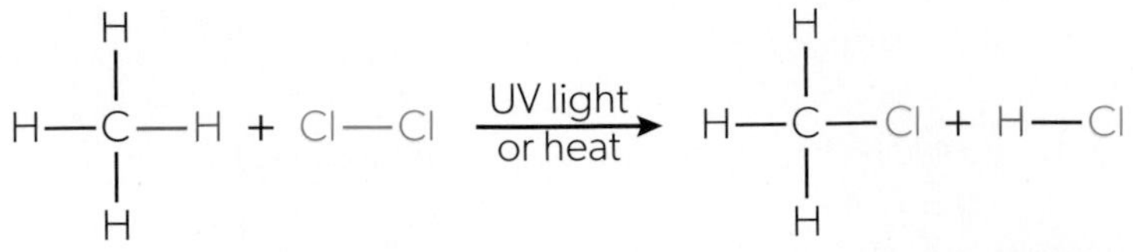

▲ Figure 5 Radical substitution in methane to form chloromethane

There are three stages involved in radical substitution reactions: initiation, propagation and termination.

Initiation

The initiation stage involves the homolytic fission of a molecule to produce radical species. In the presence of UV light, the chlorine molecule splits homolytically, producing two identical chlorine radicals, Cl• (figure 6). The Lewis structures below show the movement of electrons.

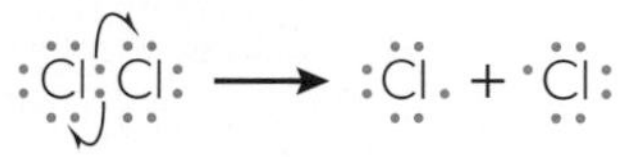

▲ **Figure 6** The homolytic fission of chlorine is the initiation step. This is the same as the reaction in figure 4

Propagation

The propagation stage includes reactions of a radical species with a non-radical species to form a different pair of a non-radical species and a radical species. In this case, the first propagation step occurs between a methane molecule and a chlorine radical, Cl• (figure 7).

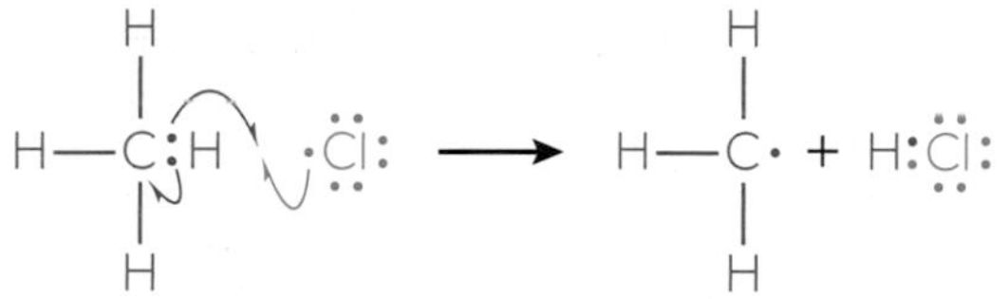

▲ **Figure 7** The first propagation step in the radical substitution of methane

In this step, a methyl radical is formed, which allows the reaction to continue, or **propagate**. Radical substitution is therefore a chain reaction, as reactions in the propagation step produce further reactive radicals.

The methyl radical, •CH_3, reacts with a chlorine molecule, producing the desired halogenoalkane, chloromethane, CH_3Cl, and another chlorine radical, Cl• (figure 8).

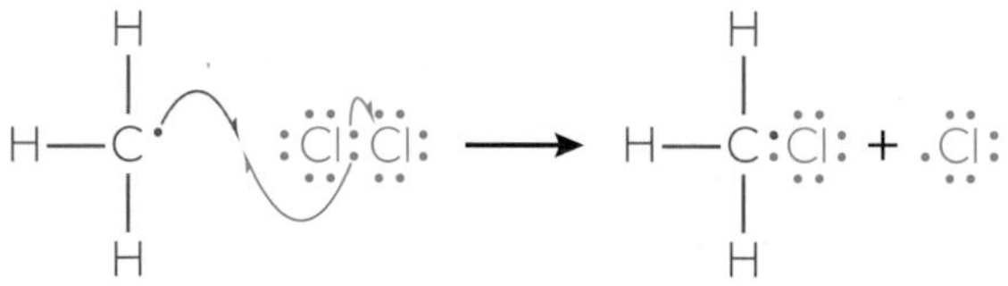

▲ **Figure 8** The second propagation step in the radical substitution of methane

The regenerated chlorine radical can take part again in the first propagation step. This cycle of the two propagation steps will continue until a termination step occurs.

Termination

The termination step includes reactions between two radical species to form a non-radical species. The termination step therefore reduces the concentration of radicals in the reaction mixture. You can see from the propagation steps that these reactions need a constant supply of radicals to continue. Therefore, termination reactions slow down the reaction, eventually stopping it completely.

In the radical substitution of methane, three termination reactions are possible:

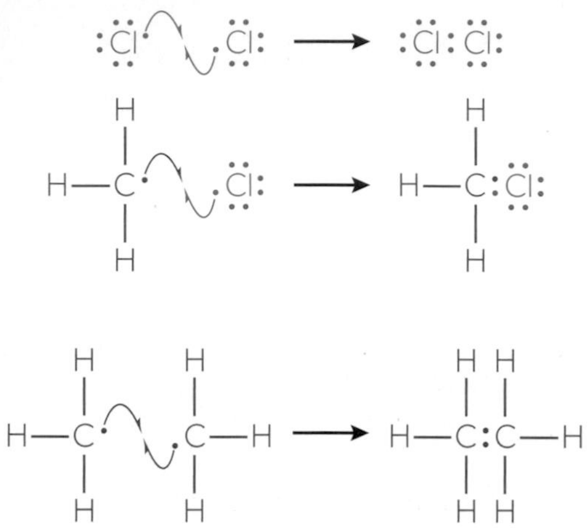

As a result, a mixture of products is formed, including the desired product chloromethane, but also chlorine, which can be recycled for the initiation step. Ethane is also produced, which is a by-product.

Chloromethane has a carbon–chlorine bond, which is polar. Therefore, an organic molecule with greater reactivity has been generated, which can be used in other organic chemistry reactions. In the same way, any other alkane can be halogenated by chlorine or bromine in the presence of UV light or heat. Fluorine is too reactive, so direct fluorination of alkanes often leads to the breaking of carbon–carbon bonds and the formation of a complex mixture of products. In contrast, the reactivity of iodine is too low, so radical iodination of alkanes does not occur.

Molecular modelling

You can build molecular models using software, specialized model kits, or simple materials such as plasticine and toothpicks. Using a medium of your choice, model the free radical mechanism of the reaction between ethane and diatomic bromine, Br_2, under UV light.

Relevant skills

- Tool 2: Physical and digital molecular modelling

Instructions

1. Start by modelling the initiation step. Then, model the propagation steps. Finally, model the termination steps.

2. Share your model with your class. Choose a suitable way to do so. You may decide to create a stop-motion video recording, live explanation, or something else (for example, a flick book).

End of topic questions

Topic review

1. Using your knowledge from the *Reactivity 3.3* topic, answer the guiding question as fully as possible:

   What happens when a species possesses an unpaired electron?

Exam-style questions

Multiple-choice questions

2. What is a propagation step in the radical substitution mechanism of ethane with chlorine?

   A $Cl_2 \rightarrow 2Cl\bullet$

   B $\bullet C_2H_5 + Cl_2 \rightarrow C_2H_5Cl + Cl\bullet$

   C $\bullet C_2H_5 + Cl\bullet \rightarrow C_2H_5Cl$

   D $C_2H_6 + Cl\bullet \rightarrow C_2H_5Cl + \bullet H$

3. Methane reacts with chlorine in sunlight.

   $CH_4(g) + Cl_2(g) \rightarrow CH_3Cl(g) + HCl(g)$

   Which type of reaction occurs?

   A radical substitution

   B electrophilic substitution

   C nucleophilic substitution

   D electrophilic addition

4. Which of these reactions proceeds by a radical mechanism in the presence of UV light?

   A $C_6H_6 + Cl_2 \rightarrow C_6H_5Cl + HCl$

   B $C_6H_6 + 3H_2 \rightarrow C_6H_{12}$

   C $CH_2CH_2 + HBr \rightarrow CH_3CH_2Br$

   D $CH_3CH_3 + Cl_2 \rightarrow CH_3CH_2Cl + HCl$

Extended-response questions

5. a. Define homolytic fission, including the required reaction conditions. [2]

   b. Write an equation to show the movement of electrons during homolytic fission of iodine. [1]

6. This question is about carbon and chlorine compounds.

   a. Ethane, C_2H_6, reacts with chlorine in sunlight. State the type of this reaction and the name of the mechanism by which it occurs. [1]

   b. Formulate equations for the two propagation steps and one termination step in the formation of chloroethane from ethane. [3]

7. Chloromethylbenzene, $C_6H_5CH_2Cl$, is a useful reagent in synthetic reactions in the manufacture of pesticides, medicines and fragrances.

   a. Draw the structural formula of methylbenzene, also known as toluene. [1]

   b. Methylbenzene can undergo chlorination in the presence of UV light to produce chloromethylbenzene. Explain the role of UV light in the initiation step. [1]

   c. Formulate the equation to describe the initiation step. [1]

   d. Write equations for the two propagation steps and one termination step. [3]

   e. Write an equation for the overall reaction, using structural formulas. [3]

Reactivity 3.4 Electron-pair sharing reactions

What happens when reactants share their electron pairs with others?

When heterolytic fission of a molecule occurs, one of the two fragments receives the bonding electron pair, forming a nucleophile. This nucleophile is electron-rich and can share an electron pair with an electron-deficient species (an electrophile), forming a new covalent bond.

Understandings

Reactivity 3.4.1—A nucleophile is a reactant that forms a bond to its reaction partner (the electrophile) by donating both bonding electrons.

Reactivity 3.4.2—In a nucleophilic substitution reaction, a nucleophile donates an electron pair to form a new bond, as another bond breaks producing a leaving group.

Reactivity 3.4.3—Heterolytic fission is the breakage of a covalent bond when both bonding electrons remain with one of the two fragments formed.

Reactivity 3.4.4—An electrophile is a reactant that forms a bond to its reaction partner (the nucleophile) by accepting both bonding electrons from that reaction partner.

Reactivity 3.4.5—Alkenes are susceptible to electrophilic attack because of the high electron density of the carbon–carbon double bond. These reactions lead to electrophilic addition.

Reactivity 3.4.6—A Lewis acid is an electron-pair acceptor and a Lewis base is an electron-pair donor.

Reactivity 3.4.7—When a Lewis base reacts with a Lewis acid, a coordination bond is formed. Nucleophiles are Lewis bases and electrophiles are Lewis acids.

Reactivity 3.4.8—Coordination bonds are formed when ligands donate an electron pair to transition element cations, forming complex ions.

Reactivity 3.4.9—Nucleophilic substitution reactions include the reactions between halogenoalkanes and nucleophiles.

Reactivity 3.4.10—The rate of the substitution reactions is influenced by the identity of the leaving group.

Reactivity 3.4.11—Alkenes readily undergo electrophilic addition reactions.

Reactivity 3.4.12—The relative stability of carbocations in the addition reactions between hydrogen halides and unsymmetrical alkenes can be used to explain the reaction mechanism.

Reactivity 3.4.13—Electrophilic substitution reactions include the reactions of benzene with electrophiles.

Nucleophiles (*Reactivity 3.4.1*)

In *Reactivity 3.3*, you saw the process of activating an alkane by substituting a hydrogen atom with a halogen atom. The resulting halogenoalkane contains a polar carbon–halogen bond, C–X, so the carbon atom is electron-deficient. This means that it is open to attack by a species known as a **nucleophile**.

A nucleophile is an electron-rich species that contains a lone pair of electrons. It can be neutral or carry a full negative charge. A nucleophile (reactant) can donate a pair of electrons to an electron-deficient species called an **electrophile**. This forms a covalent coordination bond between the nucleophile and the electrophile (figure 1).

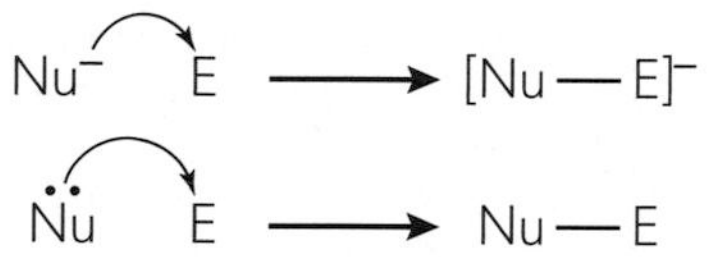

$$Nu^- \quad E \longrightarrow [Nu-E]^-$$

$$\ddot{N}u \quad E \longrightarrow Nu-E$$

▲ Figure 1 A negatively charged or neutral nucleophile (Nu) can attack an electrophile (E) forming a coordination bond

Water, H_2O, is an example of a neutral nucleophile, as it has two lone pairs of electrons on the oxygen atom and no charge. The hydroxide ion, ^-OH, is a negatively charged nucleophile, with three lone pairs of electrons.

$$H \diagdown \overset{\ddots}{\underset{}{O}} \diagup H \qquad \left[:\overset{\ddots}{\underset{\ddots}{O}} - H \right]^-$$

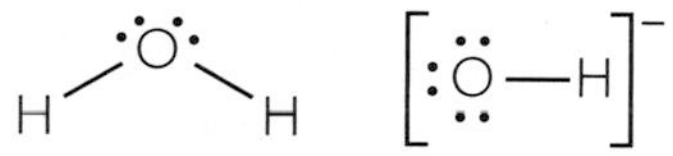

▲ Figure 2 Water and the hydroxide ion are both nucleophiles because they contain at least one lone pair of electrons

The nature and mechanism of the formation of coordination bonds are discussed in *Structure 2.2*.

Other examples of nucleophiles include charged atoms and ions, such as the halogen ions (Cl^-, Br^-, I^-), cyanide ion (CN^-) and hydrogensulfide ion (HS^-), and neutral molecules, such as ammonia (NH_3) and methylamine (CH_3NH_2).

The strength of a nucleophile depends on its ability to donate its electron pair to an electrophile.

Practice questions

1. What must be present in a nucleophile?

   A. Negative charge

   B. Lone pair of electrons

   C. Positive charge

   D. Symmetrical distribution of electrons

2. Which of the following is **not** an example of a nucleophile?

   A. CH_3NH_2

   B. $(CH_3)_2NH$

   C. $(CH_3)_3N$

   D. $(CH_3)_4N^+$

Nucleophilic substitution reactions (*Reactivity 3.4.2*)

In most reactions involving nucleophiles, the nucleophile donates an electron pair to the electrophile, forming a bond. However, in nucleophilic substitution reactions, this also results in the breaking of one of the bonds in the electrophile, producing a small molecule or **leaving group**.

$$Nu^{-} + R—X \longrightarrow R—Nu + X^{-}$$

▲ Figure 3 Mechanism for a nucleophilic substitution reaction, where Nu = nucleophile, R = electron-deficient atom in the electrophile and X = leaving group

An example of nucleophilic substitution is the reaction of chloroethane, a halogenoalkane, with aqueous sodium hydroxide.

$$CH_3CH_2Cl(g) + OH^-(aq) \rightarrow CH_3CH_2OH(aq) + Cl^-(aq)$$

The presence of a highly electronegative chlorine atom polarizes the carbon–chlorine bond in chloroethane. The resulting partial positive charge makes the carbon atom electron-deficient and therefore susceptible to attack by nucleophiles.

The hydroxide nucleophile donates an electron pair to form a new covalent bond between the oxygen and carbon atoms. At the same time, the bonding pair of electrons in the carbon–chlorine bond moves onto the chlorine atom, breaking the bond and creating the leaving group, a chloride ion (Cl⁻). This mechanism is detailed in figure 4.

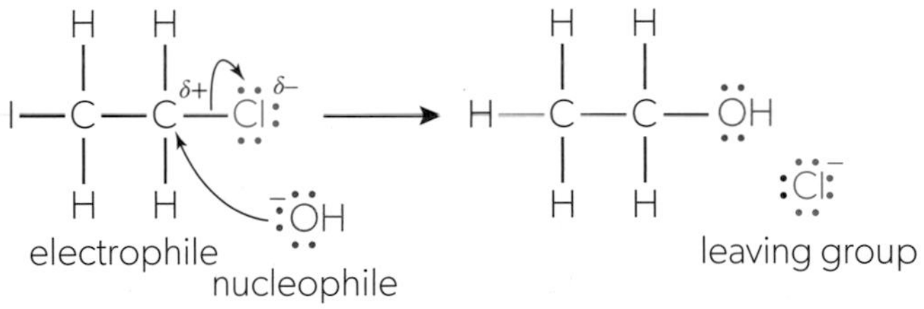

▲ Figure 4 The nucleophilic substitution of chloroethane with a hydroxide ion nucleophile. The electron-deficient nature of the carbon atom is indicated with a partial positive charge ($\delta+$)

Practice questions

3. Deduce the equations for the reactions between the following reactants, identifying the nucleophile and leaving group:

   a. 2-bromo-2-methylpropane and potassium hydroxide

   b. bromopentane and the cyanide ion.

Science as a shared endeavour

For over a century, the journal *Organic Syntheses* has been employing chemists to independently repeat and check every experiment submitted for publication in the journal. There are many peer-reviewed scientific journals in circulation. While not all investigations submitted to journals are verified by replicating the experiment, all articles are thoroughly reviewed. The reviewers are experts who scrutinize the research methods and claims described in the manuscript. The manuscript is then either accepted, or rejected, or sent back to the author for revisions that must be addressed before further review. The peer-review process is widely recognized as a valuable quality control method that ensures that information provided in journal articles is reliable.

Heterolytic fission (*Reactivity 3.4.3*)

When an unsymmetrical cleavage of a covalent bond occurs, the electrons in the bonding pair are distributed unevenly. This is known as **heterolytic fission**. In this process, one of the atoms is left with both bonding electrons while the other atom receives none of the bonding electrons. This results in the formation of a cation that is deficient of an electron and an anion that has an extra electron.

When drawing heterolytic fission, a double-barbed curly arrow is used to show the movement of the electron pair. The arrow starts at the bond and finishes where the electrons are moving to. In the case of figure 5, atom B becomes an anion and atom A becomes a cation.

$$A \! : \! B \longrightarrow A^+ + \; : \! B^-$$
$$\qquad\qquad\quad \text{cation} \quad \text{anion}$$

▲ **Figure 5** Heterolytic fission of a diatomic molecule

Organic compounds, such as halogenoalkanes, can undergo heterolytic fission at the carbon–halogen bond to form a halogen anion and an alkyl cation. Alkyl cations with the positive charge on the carbon atom are called **carbocations**.

Worked example 1

Draw the mechanism for the heterolytic fission of bromomethane and hence deduce the final products.

Solution

First, draw the structure of bromomethane:

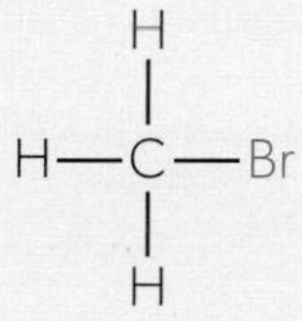

$$H - \overset{\overset{\displaystyle H}{|}}{\underset{\underset{\displaystyle H}{|}}{C}} - Br$$

Identify the partial charges in the molecule:

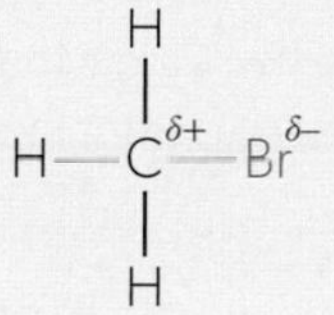

$$H - \overset{\overset{\displaystyle H}{|}}{\underset{\underset{\displaystyle H}{|}}{C^{\delta+}}} - Br^{\delta-}$$

Then draw the double-barbed curly arrow, originating from the carbon–bromine bond and finishing on the bromine atom:

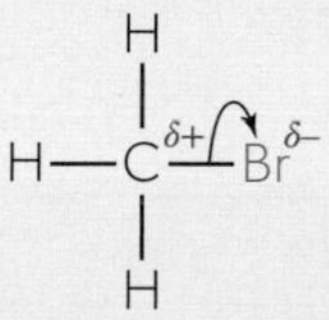

$$H - \overset{\overset{\displaystyle H}{|}}{\underset{\underset{\displaystyle H}{|}}{C^{\delta+}}} \curvearrowright Br^{\delta-}$$

The products are therefore a methyl carbocation and a bromide anion:

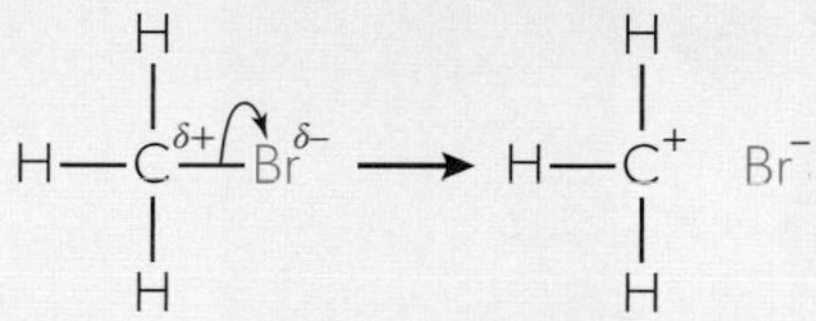

$$H - \overset{\overset{\displaystyle H}{|}}{\underset{\underset{\displaystyle H}{|}}{C^{\delta+}}} \curvearrowright Br^{\delta-} \longrightarrow H - \overset{\overset{\displaystyle H}{|}}{\underset{\underset{\displaystyle H}{|}}{C^{+}}} \quad Br^{-}$$

The two species formed during heterolytic fission are usually unstable, and therefore have a short lifespan. This means that they are usually intermediates in an overall reaction.

Linking question

What is the difference between the bond-breaking that forms a radical and the bond-breaking that occurs in nucleophilic substitution reactions? (Reactivity 3.3)

▲ **Figure 6** Boron trifluoride is an electrophile, with an electron-deficient boron atom

Electrophiles (*Reactivity 3.4.4*)

We defined an electrophile as an electron-deficient species. Electrophiles readily accept a pair of electrons from an electron donor, a nucleophile, to form a covalent bond. Electrophiles are either positively charged ions (cations), or neutral molecules with a partial positive charge ($\delta+$) on one of the atoms. Partial charges are generated by the presence of a highly electronegative species in the molecule resulting in the polarization of a bond.

The methyl cation, $^+CH_3$, is an example of an electrophile with a full positive charge. Boron trifluoride, BF_3, has an electron-deficient boron atom. The boron atom is susceptible to nucleophilic attack. It is an example of an electrophile with a partial positive charge (figure 6).

Compounds with carbonyl or carboxyl groups, such as aldehydes, ketones and carboxylic acids, are also electrophiles. The electron-deficient carbon atom of the carbonyl group is susceptible to nucleophilic attack (figure 7).

Non-polar molecules, such as bromine, Br_2, can also behave as electrophiles. This is covered in detail in the AHL section of this topic.

▲ **Figure 7** Carbon atoms in carbonyl groups or carboxyl groups, such as that in butanoic acid, have a partial positive charge

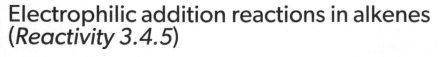

Global impact of science

Alice Ball was a US chemist and researcher working in early 20th century Hawai'i. At the time, it was known that chaulmoogra oil was an effective treatment for leprosy, but it was not suitable for injection. Ball developed a method for converting chaulmoogra oil into ethyl esters that were far easier to administer to patients. This method was used to treat leprosy for decades prior to the introduction of antibiotics.

DP chemistry covers much of the theory underlying Ball's procedure for converting chaulmoogra extract into ethyl ester. Search online for "the chemistry of the Ball Method". You will recognize many of the core chemical concepts involved in this method.

▲ **Figure 8** Alice Ball was the first woman and first African American to earn a master's degree from the University of Hawai'i. She died in 1916 at the age of 24 before her work on the Ball method was published

Electrophilic addition reactions in alkenes (*Reactivity 3.4.5*)

In *Structure 2.3*, we defined alkenes as unsaturated hydrocarbons that contain a carbon–carbon double bond. The presence of the double bond makes alkenes more reactive than the corresponding saturated alkanes. The carbon–carbon double bond is a region of high electron density that is susceptible to electrophilic attack. This reactivity means that alkenes are often used in industrial processes as the **starting molecules** for synthetic reactions. The reactions between alkenes and electrophiles are known as electrophilic addition reactions.

Electrophilic addition of halogens

An example of electrophilic addition is the reaction between an alkene and a diatomic halogen molecule, X_2. In this reaction, the halogen molecule is added across the electron-rich carbon–carbon double bond, resulting in the formation of a disubstituted halogenoalkane with the general formula $C_nH_{2n}X_2$.

For example, the reaction between ethene gas, $C_2H_4(g)$, and bromine water, $Br_2(aq)$, yields a single product, 1,2-dibromoethane, $C_2H_4Br_2$:

| ethene | bromine water (brown) | 1,2-dibromoethane (colourless) |

This reaction can also be used to test for the presence of unsaturated compounds in a mixture of hydrocarbons, as the bromine water will turn colourless in the presence of alkenes or alkynes.

Electrophilic addition of hydrogen halides

Electrophilic addition reactions will also occur between alkenes and hydrogen halides, HX. The process is similar to the addition of halogens: the hydrogen halide molecule is added across the carbon–carbon double bond. This results in the formation of a monosubstituted halogenoalkane with the general formula $C_nH_{2n+1}X$.

For example, consider the electrophilic addition reaction between but-2-ene, $C_4H_8(g)$, and aqueous hydrogen bromide, HBr(aq). But-2-ene is a symmetrical alkene, so the addition of a hydrogen halide molecule will produce only one possible product: 2-bromobutane, C_4H_9Br.

| but-2-ene | hydrogen bromide | 2-bromobutane |

The reaction of an unsymmetrical alkene, such as propene, with a hydrogen halide will yield two possible products:

| propene | hydrogen bromide | 1-bromopropane | 2-bromopropane |

Practice questions

4. a. Determine the product of the reaction of propene gas, $C_3H_6(g)$, with chlorine gas, $Cl_2(g)$. Draw the displayed formulas of all reactants and products.

   b. Iodine dissolves in many solvents to form solutions of various colours, from yellow to brown or purple. Explain how an iodine solution can be used to detect the presence of unsaturated hydrocarbons.

Electrophilic addition of water

Electrophilic addition reactions also occur between water and alkenes. This occurs when an alkene is added to an acidic solution, resulting in the formation of an alcohol. The reaction involves the addition of a water molecule across the carbon–carbon double bond, forming an alcohol with the general formula $C_nH_{2n+1}OH$. This reaction is also known as a hydration reaction.

The selectivity of addition reactions involving unsymmetrical alkenes is discussed in the AHL section of this topic.

TOK

Chemists interested in synthesizing a particular compound use their knowledge of structure and physical properties to determine suitable synthetic pathways to produce that compound. Imagination, intuition and reasoning all play their part in scientific innovation. Imagination transcends the limitations of acquired knowledge and opens up the possibility of new ideas. What are the roles of these ways of thinking when solving chemical problems and applying knowledge in novel situations?

For example, hex-3-ene, C_6H_{12}(l), undergoes electrophilic addition with water, H_2O(l), to form the secondary alcohol hexan-3-ol, $C_6H_{13}OH$(aq). Hex-3-ene is symmetrical, so only one product is formed.

$$H_3CH_2C \diagdown C = C \diagup CH_2CH_3 \quad + \quad H-OH \quad \longrightarrow \quad H_3CH_2C-\overset{H}{\underset{H}{C}}-\overset{H}{\underset{OH}{C}}-CH_2CH_3$$

hex-3-ene   water   hexan-3-ol

As with hydrogen halides, two products will be formed in the electrophilic addition of water to an unsymmetrical alkene.

Practice questions

5. Deduce the equations for the reaction between the following alkenes and electrophiles:

    a. 2-methylbut-2-ene and hydrogen bromide

    b. pent-2-ene and iodine

    c. ethene and water

    d. cyclohexene and hydrogen chloride

    e. methylpropene and hydrogen iodide

 Linking questions

Why is bromine water decolourized in the dark by alkenes but not by alkanes? (*Reactivity 3.3*)

Why are alkenes sometimes known as "starting molecules" in industry? (*Structure 2.4*)

ATL Social skills

Collaboratively develop answers to three of the linking questions in this chapter. Make a list of the key understandings from this chapter, as well as others, that help you address the linking questions. Summarize your responses in a document shared with the rest of your class. Compare and contrast the answers developed by the people in your class. Draw out the common themes in everyone's answers.

Lewis acids and bases (*Reactivity 3.4.6*)

In *Reactivity 3.1*, we defined a **Brønsted–Lowry base** as a substance that can accept a proton (a hydrogen ion, H⁺). The presence of at least one pair of electrons in Brønsted–Lowry bases allows them to form a **coordination bond** with a proton. The hydroxide ion and ammonia are examples of Brønsted–Lowry bases (figure 9).

A **Lewis acid** is defined as an electron-pair acceptor and a **Lewis base** as an electron-pair donor. The Lewis acid–base theory is a more general definition when compared to the Arrhenius and Brønsted–Lowry theories, enabling a wider range of substances to be included.

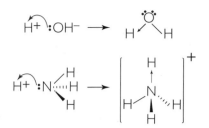

▲ **Figure 9** The lone pair of electrons on Brønsted–Lowry bases forms a coordination bond with a proton

Both ammonia and the hydroxide ion can act as Brønsted–Lowry and Lewis bases, donating a pair of electrons to the hydrogen ion. The hydrogen ion acts as a Lewis acid, as it accepts the electron pair.

We can use Lewis acid–base theory to identify the role of each reacting species in reactions where protons are not involved. For example, consider the reaction between boron trifluoride, BF_3, and ammonia, NH_3:

$$BF_3 + :NH_3 \longrightarrow F_3B \longleftarrow NH_3$$

In this reaction, ammonia donates a lone pair of electrons to boron trifluoride. Therefore, ammonia is a Lewis base and boron trifluoride is a Lewis acid. In this reaction, no proton is involved, so it cannot be described using Brønsted–Lowry acid–base theory.

 Activity

Copy and complete the table for each of the following species: H_2O, Cu^{2+}, CH_3COOH, OH^-, NH_3 and HF. The example of BF_3 has been completed.

Species	BF_3						
Brønsted–Lowry acid	No						
Brønsted–Lowry base	No						
Lewis acid	Yes						
Lewis base	No						

Practice question

6. Which species is a Lewis acid but **not** a Brønsted–Lowry acid?

   A Cu^{2+}

   B NH_4^+

   C   Cu

   D CH_3COOH

Linking question

What is the relationship between Brønsted–Lowry acids and bases and Lewis acids and bases? (*Reactivity 3.1*)

Lewis acid and base reactions (*Reactivity 3.4.7*)

You have learnt that a **nucleophile** is an electron-rich species that possesses at least one lone pair of electrons. It can be either a negatively charged species (anion) or a neutral molecule. A nucleophile can therefore be described as a Lewis base, as it can donate a pair of electrons.

An **electrophile** is an electron-deficient species that will accept a pair of electrons from an electron donor. It can be either a positively charged species (cation) or neutral molecule. An electrophile can therefore be described as a Lewis acid. When a Lewis base reacts with a Lewis acid, a **coordination bond** is formed.

Consider the reaction of boron trifluoride with ammonia. Boron has an electron configuration of $1s^2\,2s^2\,2p^1$ and in boron trifluoride it will form three sp^2 hybrid orbitals, resulting in a trigonal planar geometry and a vacant unhybridized $2p_z$ orbital (figure 10).

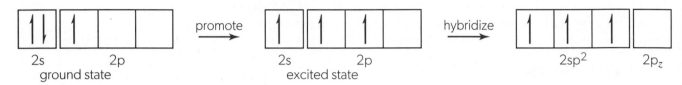

2s   2p
ground state

promote

| 2s   2p |
| excited state |

hybridize

| $2sp^2$ $2p_z$ |

▲ **Figure 10** Hybridization of boron in boron trifluoride

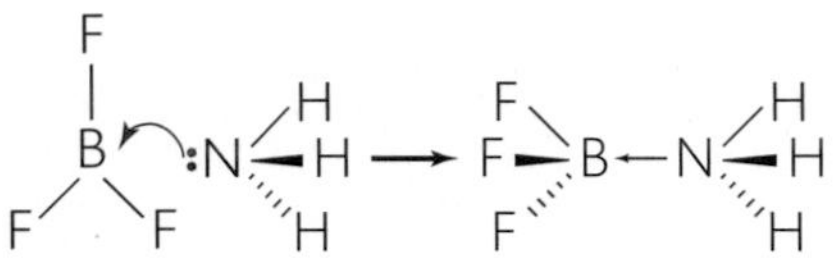

The lone pair of electrons on the nitrogen atom in ammonia is donated to this vacant $2p_z$ orbital, forming a coordination bond with the electron-deficient boron atom and generating a single product (figure 11). Therefore, ammonia acts as a nucleophile and Lewis base, and boron trifluoride acts as an electrophile and Lewis acid.

▲ **Figure 11** Ammonia donates an electron pair to boron trifluoride, forming a coordination bond

Coordination bonds are covalent bonds, so they can be drawn as ordinary bonds or as an arrow from the source of the lone pair to the electron-deficient atom.

Anhydrous aluminium chloride, $AlCl_3$, is another example of a Lewis acid, as the central aluminium atom is electron deficient. Two aluminium chloride molecules can react with each other to form a dimer, where the lone pairs on the chlorine atoms form coordination bonds with adjacent electron-deficient aluminium atoms (figure 12). Therefore, aluminium chloride can act as both a Lewis acid and a Lewis base.

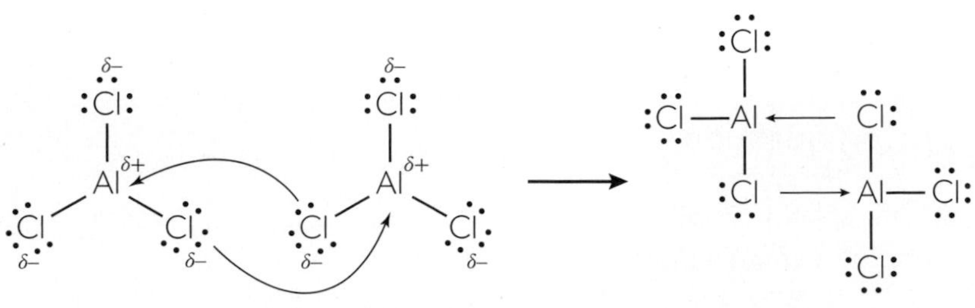

> Coordination bonds and hybrid orbitals were introduced in *Structure 2.2*.

▲ **Figure 12** The formation of the aluminium chloride dimer, Al_2Cl_6

Practice questions

7. Which statements are correct?

    I   Lewis bases can act as nucleophiles.

    II   Electrophiles are Lewis acids.

    III   Lewis acids are electron pair acceptors.

    A   I and II only

    B   I and III only

    C   II and III only

    D   I, II and III

8. Which type of bond is formed when a Lewis acid reacts with a Lewis base?

    A   covalent   C   double

    B   dipole–dipole   D   hydrogen

 Linking question

Do coordination bonds have any different properties from other covalent bonds? (*Structure 2.2*)

Coordination bonds and complex ions (*Reactivity 3.4.8*)

Transition elements are metals that can form ions with a partially filled d-subshell. An example is the chromium(II) ion with an electron configuration of $[Ar]\,3d^4$. Transition element cations are Lewis acids, so they can form several coordination bonds with Lewis bases. A transition element ion bonded to several Lewis bases is called a complex ion.

In the context of complex ions, the surrounding Lewis bases are called **ligands**. Ligand species are normally neutral, such as water, H_2O, and ammonia, NH_3, but can also be anions, such as the cyanide ion, CN^-, chloride ion, Cl^-, and the hydroxide ion, OH^-. They are electron-rich species with at least one lone pair of electrons, and therefore can be considered nucleophiles. The metal cation acts as an electrophile, as it has a positive charge and accepts electron pairs from the ligands. This relationship is summarized in figure 13, where arrows represent coordination bonds.

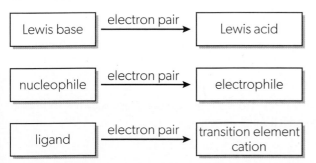

▲ **Figure 13** All Lewis bases are nucleophiles, and all Lewis acids are electrophiles, and vice versa. All ligands are nucleophiles, and all transition element cations are electrophiles, but not vice versa

▲ **Figure 14** The hexaaquacopper(II) complex ion

For example, copper(II) ions in solution, $Cu^{2+}(aq)$, form coordination bonds with water molecules to give the complex ion $[Cu(H_2O)_6]^{2+}$. It has octahedral geometry and is blue in colour (figure 14).

You can deduce the charge on a complex ion if you know the charge on the transition element cation, the charges on the ligands and the number of charged ligands.

Overall charge of complex ion = (charge on transition element cation) + (number of charged ligands × charge on a ligand)

Worked example 2

Consider the equilibrium between two cobalt(II) complex ions, $[Co(H_2O)_6]^n$ and $[CoCl_4]^m$, in a solution containing chloride ions. Their charges are unknown.

$$[Co(H_2O)_6]^n(aq) + 4Cl^-(aq) \rightleftharpoons [CoCl_4]^m(aq) + 6H_2O(l)$$
$$\text{pink} \qquad\qquad\qquad \text{blue}$$

Deduce the charges on each complex ion.

Solution

In each complex ion, the cobalt cation has a charge of 2+. Water is a neutral molecule, therefore the overall charge of the complex ion with water ligands is the same as that of the cobalt cation: $[Co(H_2O)_6]^{2+}$.

Chloride ions have a charge of 1−, and there are four chloride ligands in the second complex ion. The overall charge of the complex ion is equal to the sum of the charges:

Overall charge of complex ion = (charge on transition element cation) + (number of charged ligands × charge on a ligand)

$$= 2 + 4 \times (-1)$$
$$= 2-$$

Therefore, the second complex ion is $[CoCl_4]^{2-}$.

The identity of the ligands also affects the colour of the complex ion. This was discussed in *Structure 3.1*.

You can use the reverse process to deduce the charge on the transition metal cation, that is, you need to know the overall charge of the complex ion, and charge and number of the ligands:

Charge on transition element cation = (overall charge of complex ion) − (number of charged ligands × charge on a ligand)

Worked example 3

Deduce the charge on the transition element cation in the following complexes:

a. $[Fe(OH)(H_2O)_5]^{2+}$

b. $[TiF_6]^{2-}$

Solution

a. The overall charge of the complex ion is 2+. It contains five neutral water ligands, and a hydroxide ligand with a 1− charge:

Charge on transition element cation = (overall charge of complex ion) − (number of charged ligands × charge on a ligand)

$$= 2 - 1 \times (-1)$$
$$= 3+$$

Therefore, the metal cation is Fe^{3+}. You can check your working by doing the reverse process to calculate the overall charge on the complex ion and checking that it equals 2+.

b. The overall charge of the complex ion is 2−. It contains six fluoride ligands, each with a 1− charge:

Charge on transition element cation = (overall charge of complex ion) − (number of charged ligands × charge on a ligand)

$$= (-2) - 6 \times (-1)$$
$$= 4+$$

Therefore, the metal cation is Ti^{4+}.

Practice questions

9. Deduce the charge on the metal ion in the following complexes:

   a. $[Cr(H_2O)_6]^{3+}$

   b. $[NiBr_4]^{2-}$

   c. $[Pt(CN)_6]^{2-}$

   d. $[Fe(H_2O)_2(NH_3)_4]^{3+}$

   e. $[Pd(CN)_4(NH_3)_2]^{2-}$

10. Deduce the total charge (n) on the complex ion in the following complexes:

    a. $[Cr(H_2O)_6]^n$, Cr(III)

    b. $[Ni(OH)_2Br_2]^n$, Ni(II)

    c. $[Pt(CN)_4(H_2O)_2]^n$, Pt(IV)

    d. $[Fe(H_2O)_2(NH_3)_4]^n$, Fe(II)

    e. $[PdCl_6]^n$, Pd(IV)

Element	Electronegativity, χ
carbon	2.6
fluorine	4.0
chlorine	3.2
bromine	3.0
iodine	2.7

▲ Table 1 Halogen atoms have high electronegativity and form polar bonds with carbon

Nucleophilic substitution in halogenoalkanes (*Reactivity 3.4.9*)

Halogenoalkanes contain a carbon–halogen bond, which is polar due to the high electronegativity of the halogen atom compared to that of carbon (table 1).

The electron-deficient carbon atom is susceptible to nucleophilic attack. Halogenoalkanes can therefore undergo nucleophilic substitution reactions, where the halogen atom is displaced by the nucleophile.

There are two types of mechanism that occur in nucleophilic substitution reactions: S_N1 and S_N2. The mechanism that occurs depends on whether the reactant is a **primary**, **secondary** or **tertiary halogenoalkane**.

$$\begin{array}{c} \delta+ \quad \delta- \\ -C - X \end{array}$$

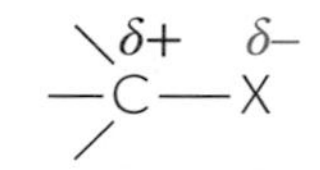

▲ Figure 15 Representation of the partial charges within the polar carbon–halogen bond

Primary, secondary and tertiary halogenoalkanes were defined in *Structure 3.2*.

S_N2 reaction mechanism

Nucleophilic substitution in primary halogenoalkanes follows the S_N2 reaction mechanism. This mechanism is an example of a concerted reaction, which means that reactants are converted directly into products in a single step. Therefore, the S_N2 mechanism does not involve an intermediate.

Reaction order and rate equations are discussed in *Reactivity 2.2*.

The '2' in S_N2 means that there are **two** molecules involved in the rate-determining step (slow step). Therefore, the rate-determining step involves both the halogenoalkane and the nucleophile, so the rate of reaction depends on the concentrations of both reactants. It is described as a second order reaction and has the following rate equation.

$$\text{rate} = k[\text{halogenoalkane}][\text{nucleophile}]$$

Consider the reaction between bromoethane, $C_2H_5Br(l)$, and aqueous hydroxide ions, $^-OH(aq)$, which yields ethanol, $C_2H_5OH(aq)$ and a bromide ion leaving group, $Br^-(aq)$. The mechanism is shown in figure 16.

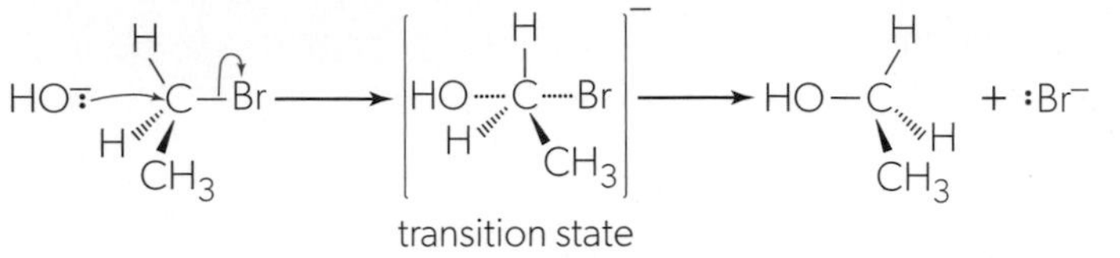

transition state

▲ **Figure 16** The S_N2 mechanism for the reaction between the primary halogenoalkane bromoethane and hydroxide ion

Transition states and intermediates are discussed in greater detail in *Reactivity 2.2*.

The hydroxide nucleophile attacks the electron-deficient carbon atom, forming a **transition state** that includes the halogen and the hydroxyl group. This transition state has a partially formed covalent bond between the nucleophile and the carbon atom, and a weakened carbon–bromine bond that has not completely broken. These partial bonds are represented by dotted lines.

Practice questions

11. Halogenoalkanes can undergo nucleophilic substitution reactions with aqueous potassium hydroxide.

    a. State an equation for the reaction of 1-chlorobutane, $C_4H_9Cl(l)$, with potassium hydroxide, KOH(aq).

    b. Using structural formulas and curly arrows, draw the reaction mechanism for the conversion of 1-chlorobutane to butan-1-ol.

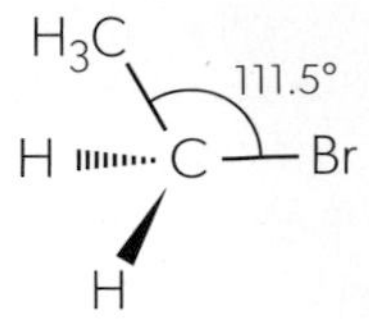

▲ **Figure 17** The tetrahedral arrangement of bromoethane. The bond angle differs slightly from the theoretical value of 109.5° due to the presence of two large substituents, CH_3 and Br

A transition state is not the same as an intermediate. A transition state exists for an infinitesimally small period of time and represents the structure of the reacting species with the highest energy along the reaction pathway. It typically contains bonds that are partially broken and formed, but it does not represent a discrete step of the reaction, as the entire reaction occurs in a single step. In contrast, an intermediate has some degree of stability and does not immediately transform into the ultimate product, so it represents the structure of the reacting species at an intermediate point in a multistep reaction.

When drawing S_N2 reaction mechanisms, pay attention to the following:

1 The curly arrow from the nucleophile originates from its lone pair or negative charge, and terminates at the electron-deficient carbon atom.

2 The curly arrow representing the halogen leaving group originates at the bond between the carbon and halogen atoms. This can be shown either in the reaction substrate or in the transition state.

3 Partial bonds in the transition state are represented by dotted lines, i.e., HO····C····X

4 The transition state is enclosed in square brackets with a single negative charge shown outside the brackets.

5 Both the final product and the leaving group must be shown.

The S_N2 mechanism is **stereospecific**, which means that the product formed will have a specific stereochemistry, rather than be a mixture of isomers. In halogenoalkanes, the electron-deficient carbon atom on the carbon–halogen bond is sp^3 hybridized, and therefore it has a tetrahedral geometry. For example, figure 17 shows the geometry of bromoethane.

In the S_N2 reaction, the nucleophile will attack the carbon atom at 180° to the position of the bromine leaving group. This is because the large halogen atom creates **steric hindrance**, which prevents the nucleophile from attacking the carbon atom from the same side as the halogen atom. Therefore, the nucleophile causes an inversion of the molecule configuration, much like an umbrella turning inside out (figure 18).

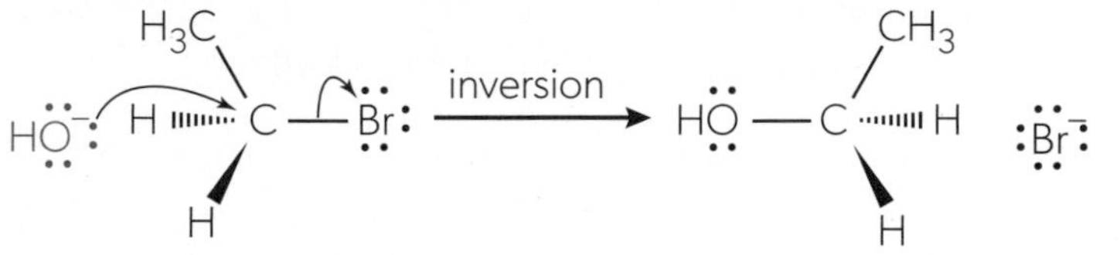

▲ **Figure 18** Inversion of stereochemical configuration in S_N2 reactions

S_N1 reaction mechanism

Tertiary halogenoalkanes undergo nucleophilic substitution in two steps. This is known as the S_N1 mechanism. In S_N1 mechanisms, only **one** molecule is involved in the rate-determining step. In the first step of the S_N1 mechanism, the bond to the leaving group in the halogenoalkane breaks, forming an intermediate carbocation. This is the rate-determining step, and it only involves the halogenoalkane. Therefore, this is a first order reaction, and the rate equation is as follows:

$$\text{rate} = k\,[\text{halogenoalkane}]$$

For example, the reaction between 2-chloro-2-methylpropane, C_4H_9Cl, and aqueous hydroxide ions yields the product 2-methylpropan-2-ol, C_4H_9OH, and the chloride ion leaving group via a carbocation intermediate (figure 19).

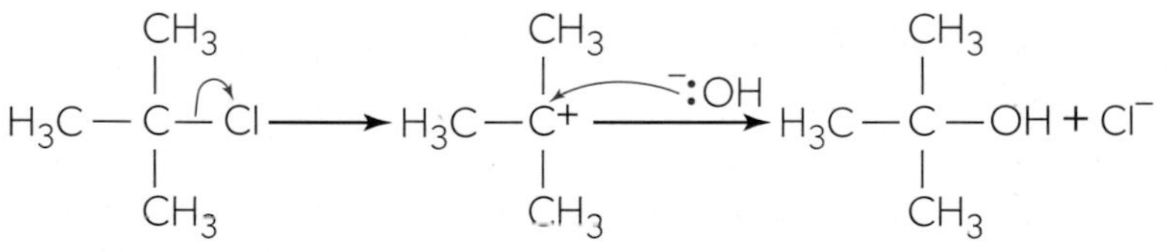

▲ **Figure 19** S_N1 mechanism for the reaction between a tertiary halogenoalkane and aqueous hydroxide ion

When drawing S_N1 reaction mechanisms, pay attention to the following:

1. The curly arrow representing the halogen leaving group originates at the bond between the carbon and the halogen atoms.

2. The carbocation must clearly show a positive charge on the carbon atom.

3. The curly arrow from the nucleophile originates at its lone pair or negative charge and terminates at the positively charged carbon.

4. Both the final product and the leaving group must be shown.

Practice questions

12. Halogenoalkanes can undergo substitution reactions with sodium hydroxide solution.

    a. State an equation for the reaction of 2-iodo-2-methylbutane ($C_5H_{11}I$) and NaOH.

    b. Using structural formulas and curly arrows, draw the reaction mechanism for the conversion of 2-iodo-2-methylbutane to 2-methylbutan-2-ol.

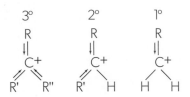

▲ **Figure 20** An arrow is used to represent the movement of electron density caused by the inductive effect

▲ **Figure 21** The inductive effect decreases from tertiary to secondary to primary carbocations

Bond	Bond enthalpy / kJ mol^{-1}
C–F	492
C–Cl	324
C–Br	285
C–I	228

▲ **Table 2** Bond enthalpies of carbon–halogen bonds

Bond enthalpy data are used in calculations in *Reactivity 1.2*.

Practice question

13. In separate reaction vessels, samples of 2-iodopropane, 2-chloropropane and 2-bromopropane are added to an aqueous solution of sodium hydroxide.

    List these halogenoalkanes according to their reaction rates from highest to lowest.

Inductive effects

The different nucleophilic substitution mechanisms of reactions involving halogenoalkanes can be explained by the inductive effects of substituents. In the C–H bond, the carbon atom has a slightly greater electronegativity than hydrogen, creating a weak dipole and a slight shift of bonding electrons towards the carbon atom. As a result, adjacent alkyl groups stabilize the carbocation by donating electron density and reducing the positive charge on the central carbon atom (figure 20). This is known as a **positive inductive effect**.

When several alkyl groups are bonded to the positively charged carbon, their combined positive inductive effect increases. Therefore, a carbocation formed by a tertiary halogenoalkane is more stable than that formed by a primary halogenoalkane. This explains why the nucleophilic substitution of a tertiary halogenoalkane is more likely to proceed according to the S_N1 mechanism (figure 21).

The stability of carbocations formed from secondary halogenoalkanes is intermediate between those formed from primary and tertiary halogenoalkanes, so secondary halogenoalkanes can undergo nucleophilic substitution according to both the S_N1 and S_N2 mechanisms.

 Linking questions

What differences would be expected between the energy profiles for S_N1 and S_N2 reactions? (*Reactivity 2.2*)

What are the rate equations for these S_N1 and S_N2 reactions? (*Reactivity 2.2*)

How useful are mechanistic models such as S_N1 and S_N2? (*Reactivity 2.2*)

Rate of nucleophilic substitution reactions (*Reactivity 3.4.10*)

The rate of a nucleophilic substitution reaction in halogenoalkanes is influenced by the identity of the halogen in the leaving group. In both S_N1 and S_N2 mechanisms, the rate-determining step involves the heterolytic fission of the carbon–halogen bond, in which the two bonding electrons move to the more electronegative atom. The faster this step is completed, the higher the rate of reaction.

The rate of heterolytic fission of the carbon–halogen bond depends on the strength of that bond, which is characterized by the bond enthalpy (table 2). The higher the bond enthalpy, the stronger the bond and therefore the slower the reaction.

Fluoroalkanes are virtually inert due to the high strength of the C–F bond (492 kJ mol^{-1}). As you move down group 17, the strength of the carbon–halogen bond decreases as the size of the halogen atom increases and the electronegativity of the halogen atom decreases. Additionally, the stability of the halide anions also increases down the group.

Data-based questions

Under certain conditions, halide ions can act as nucleophiles and substitute the hydroxyl groups in alcohols. To investigate how the position of the hydroxyl group affects the identity of the reaction product, a series of experiments were carried out. The alcohols used were pentan-1-ol, pentan-2-ol, pentan-3-ol and 2-methylbutan-2-ol. Each of these alcohols was reacted with a mixture of chloride and bromide nucleophiles.

$$\text{R–OH} \xrightarrow[\text{NH}_4\text{Cl, NH}_4\text{Br}]{\text{H}_2\text{O, H}_2\text{SO}_4} \text{R–Cl} + \text{R–Br}$$
1°, 2° or 3°   halogenoalkanes

Aqueous ammonium salts were used to provide chloride and bromide ions for the nucleophilic substitution. Equal amounts of ammonium chloride and ammonium bromide were used as reactants in each experiment. The products, however, did not contain equal amounts of chloroalkanes and bromoalkanes. The relative amounts of the products were determined by ^1H NMR spectroscopy.

Questions

1. For each of the four alcohols used in the experiments, draw the skeletal formula and determine whether it is primary, secondary or tertiary.

2. Identify the independent and dependent variables in the investigation.

3. Answer this part of the question using your knowledge of chemistry and without looking at the results in table 3. Formulate a hypothesis explaining which ion is the stronger nucleophile: the bromide ion or chloride ion. Predict how the ideas behind your hypothesis will be reflected in the amounts of chloroalkane and bromoalkane produced in the reactions with each of the alcohols.

4. Again, answer this part of the question without looking at table 3. Formulate a hypothesis that explains which mechanism, S_N1 and/or S_N2, is favoured for each of the alcohols used. Predict how the ideas behind your hypothesis will be reflected in the amounts of alkyl chloride and alkyl bromide produced in the reactions with each of the alcohols.

5. Select and construct a suitable type of graph or chart to represent the data in table 3.

6. Describe three patterns, trends or relationships you see in the data.

Alcohol reactant	Halogenoalkane products formed	Percentage of products / %
pentan-1-ol	1-bromopentane	87
	1-chloropentane	13
pentan-2-ol	2-bromopentane	50
	2-chloropentane	22
	3-bromopentane	19
	3-chloropentane	9
pentan-3-ol	2-bromopentane	21
	2-chloropentane	9
	3-bromopentane	50
	3-chloropentane	20
2-methylbutan-2-ol	2-bromo-2-methylbutane	53
	2-chloro-2-methylbutane	47

▲ Table 3 Product analysis of nucleophilic substitution reactions. Source of data: K. Herasymchuk, R. Raza, P. Saunders and N. Merbouh, *J. Chem. Ed.*, 2021, **98 (10)**, 3319–3325

7. In certain solvents, such as water, bromide ions are better nucleophiles than chloride ions. Identify and explain the results that support this statement.

8. Using your knowledge of nucleophilic substitution reaction mechanisms, identify and explain which of the alcohols have reacted predominantly via an:

    a. S_N1 mechanism

    b. S_N2 mechanism.

9. Some of the results suggest that a rearrangement has taken place, in which a positive charge of the carbocation initially formed moves to a different carbon atom within the molecule. Identify and explain the results that support this statement.

10. Considering the results, evaluate the hypotheses that you formulated in your answers to questions 3 and 4.

11. Formulate a conclusion to your analysis, which includes:

    • the aim(s) of the investigation

    • a summary of the outcomes of the investigation

    • an appraisal of the hypotheses you proposed

    • any unanswered questions or issues.

Linking question

Why is the iodide ion a better leaving group than the chloride ion? (*Structure 3.1*)

Electrophilic addition mechanisms (*Reactivity 3.4.11*)

Earlier in this topic, you saw that the electron-rich carbon–carbon double bond in alkenes was susceptible to electrophilic attack. This type of attack leads to electrophilic addition reactions. In this section, you will learn about the mechanism of these reactions.

Carbon–carbon double bonds are electron-rich due to the presence of readily accessible pi (π) bonds either side of the bond axis. This is discussed in *Structure 2.2 (AHL)*.

Electrophilic addition of halogens to symmetrical alkenes

Consider the reaction between ethene gas, $C_2H_4(g)$, and bromine water, $Br_2(aq)$, discussed earlier in this chapter. One of the two carbon–carbon bonds in ethene breaks, and two carbon–bromine bonds are formed:

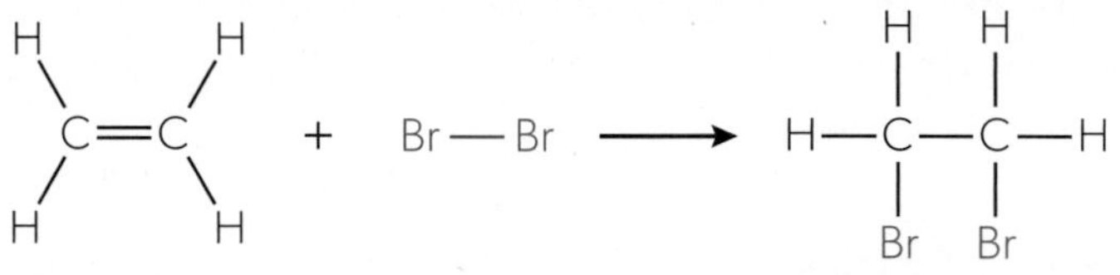

ethene   bromine water   1,2-dibromoethane
                (brown)   (colourless)

The bromine molecule is non-polar, so it must be polarized before it can act as an electrophile. The electrophilic addition proceeds via the following steps:

1. The bromine molecule is polarized as it approaches the electron-rich carbon-carbon double bond of the alkene. The bonding electrons within the bromine molecule are repelled, resulting in an induced, temporary dipole.

2. The electron-rich C=C bond is attacked by the bromine atom with a partial positive charge, and the bromine molecule splits heterolytically to form a bromide anion.

3. When the positively charged bromine atom forms a covalent bond with one of the carbon atoms, another carbon atom becomes positively charged. This produces a carbocation intermediate.

4. Finally, the reaction between the unstable carbocation and the bromide anion results in the formation of the product, 1,2-dibromoethane.

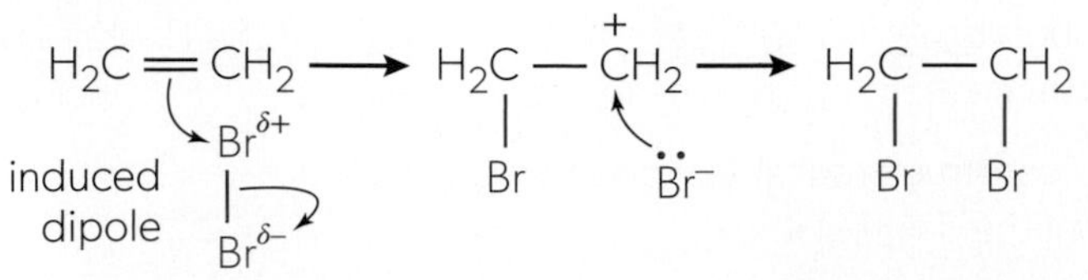

▲ Figure 22 Mechanism for the electrophilic addition of bromine to ethene

When drawing electrophilic addition mechanisms, pay attention to the following:

1. The curly arrow that shows the electrophilic attack originates at the carbon–carbon double bond and finishes at the electron-deficient atom of the electrophile.

2. The curly arrow for heterolytic fission originates at the bond being broken and finishes on the leaving group to give an anion.

3. The last curly arrow originates at the lone pair of electrons or the negative charge on the resulting anion and finishes at the positively charged carbon atom, C^+, in the carbocation.

4. The structural formula of the final product must be shown.

Electrophilic addition of hydrogen halides to symmetrical alkenes

For the electrophilic addition of hydrogen halides to alkenes, the mechanism is similar to that for halogens. The only exception is that the halogen–hydrogen bond is already polar, as the halogen atom is more electronegative than the hydrogen atom.

Consider the electrophilic addition reaction between but-2-ene, $C_4H_8(g)$, and aqueous hydrogen bromide, HBr(aq).

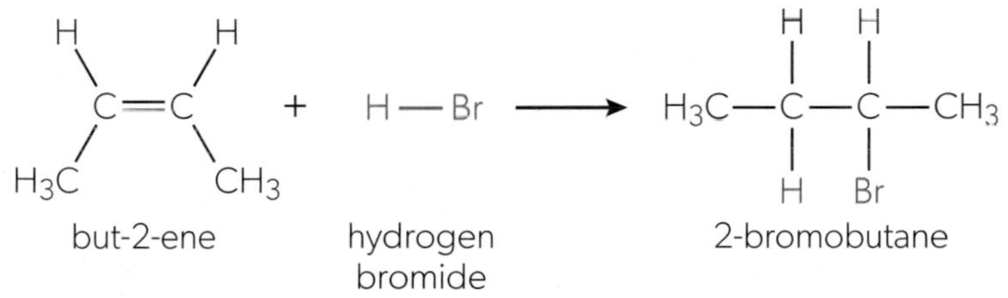

but-2-ene   hydrogen bromide   2-bromobutane

The electrophilic addition proceeds via the following steps:

1. The electron-rich C=C bond is attacked by the partially positive hydrogen atom, and the hydrogen bromide molecule splits heterolytically to form a bromide anion.

2. When the hydrogen atom forms a covalent bond with one of the carbon atoms, another carbon atom becomes positively charged. This produces a carbocation intermediate.

3. The reaction between the unstable carbocation and the bromide anion results in the formation of the product, 2-bromobutane.

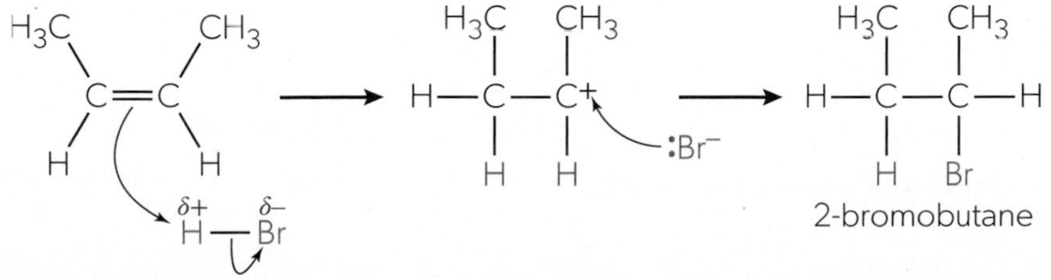

▲ Figure 23 Mechanism for the electrophilic addition of hydrogen bromide to but-2-ene

Electrophilic addition of water to symmetrical alkenes

The third example of electrophilic addition you looked at earlier in the topic was the reaction between alkenes and water, in acidified solution. The mechanism for this reaction involves the protonation of the alkene to form the carbocation, followed by the addition of a water molecule and finally the loss of a proton.

Consider the electrophilic addition reaction between hex-3-ene, $C_6H_{12}(l)$ and water in the presence of an acid.

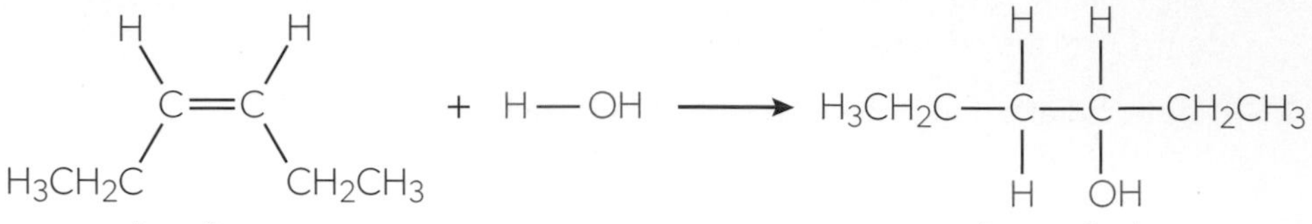

The electrophilic addition proceeds via the following steps:

1. The C=C bond is attacked by a proton present in the acidified reaction mixture, breaking one of the carbon–carbon bonds and producing a carbocation. In this case, the proton is acting as an electrophile.

2. A water molecule, acting as a nucleophile, attacks the positively charged carbon atom in the carbocation.

3. The resulting oxonium ion, a protonated alcohol, is strongly acidic. The oxonium ion deprotonates, forming the alcohol hexan-3-ol and regenerating the proton.

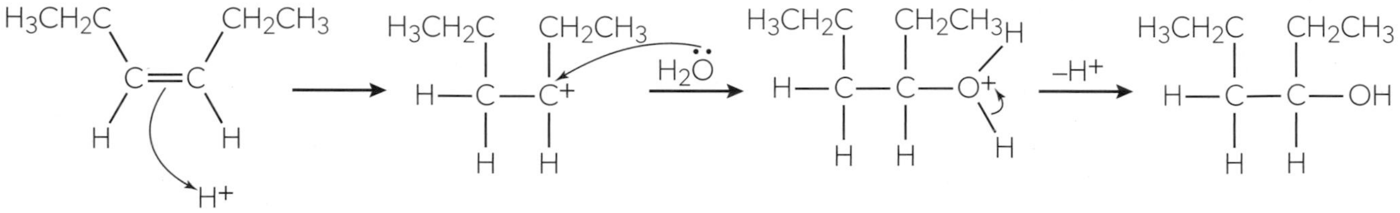

▲ Figure 24 Mechanism for the electrophilic addition of water to hex-3-ene

A proton is consumed at the beginning of the reaction, and a proton is regenerated at the end of the reaction, so it acts as a catalyst.

Practice question

14. Which of the following reactions is an example of electrophilic addition?

    A. $CH_3CH_2CH_2Cl \rightarrow CH_3CHCH_2 + HCl$

    B. $C_3H_7I + KCN \rightarrow C_3H_7CN + KI$

    C. $CH_3CH_2CH_2CHCH_2 + Br_2 \rightarrow CH_3CH_2CH_2CH(Br)CH_2Br$

    D. $CH_4 + Cl_2 \rightarrow CH_3Cl + HCl$

Carbocations in electrophilic addition reactions (*Reactivity 3.4.12*)

To predict the major product of an electrophilic addition reaction involving an unsymmetrical alkene, we need to understand the relative stability of the potential carbocations produced during the reaction. You learned previously that the stability of a carbocation depends on the inductive effects of alkyl groups present in the molecule. A tertiary carbocation has greater stability than a primary carbocation because the positive charge density is offset by the inductive effects of the three alkyl substituents (figure 21).

The major products of the electrophilic addition of hydrogen halides to unsymmetrical alkenes can be predicted using **Markovnikov's rule**.

In an unsymmetrical alkene, there are two possible carbon atoms on the carbon–carbon double bond that are susceptible to electrophilic attack. Markovnikov's rule states that the electropositive part of the polarized electrophile will preferentially bond to the carbon that has the least number of alkyl substituents. This results in a carbocation with the positive charge centred on the most substituted carbon, so the major reaction product will form via the more stable carbocation.

For example, consider the electrophilic addition reaction of hydrogen bromide to the unsymmetrical alkene propene. This reaction has two possible products: 2-bromopropane and 1-bromopropane.

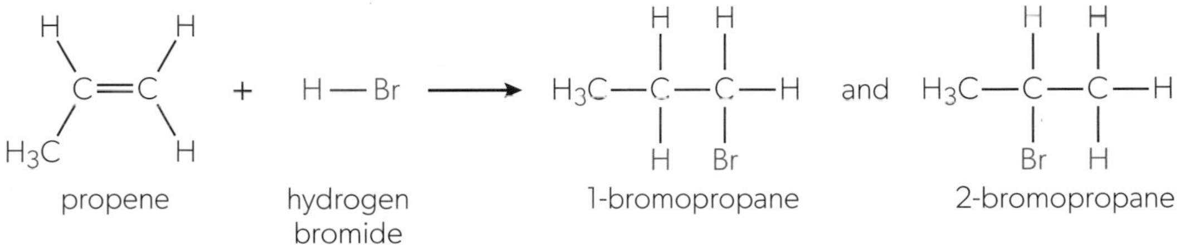

2-bromopropane will form via a secondary carbocation while 1-bromopropane will form via a primary carbocation. The secondary carbocation is more stable, so 2-bromopropane will be the major product (figure 25).

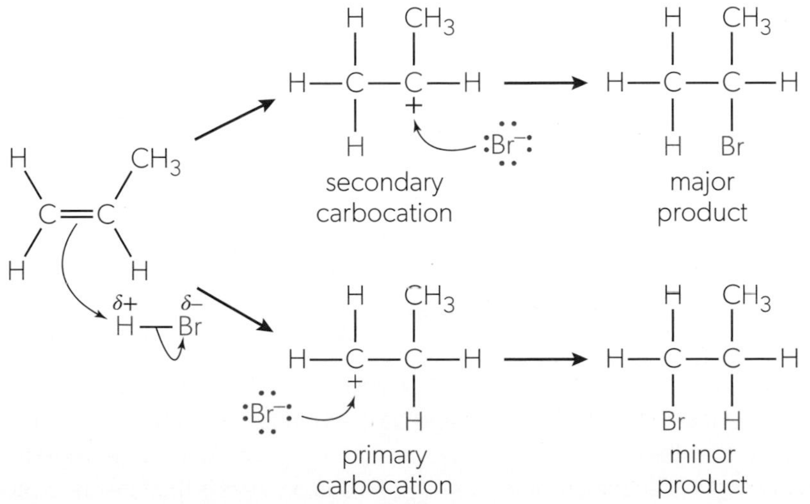

▲ **Figure 25** The major product in this reaction is 2-bromopropane, as the reaction proceeds preferentially via a more stable carbocation

Organic synthesis converts a starting material via a series of reactions into the desired product. Each step produces an intermediate product in quantities less than the theoretical yield, so an efficient synthetic pathway must involve the smallest possible number of steps. Synthetic organic chemists often use a method referred to as retrosynthesis. Starting with knowledge of the structure and properties of the target compound, they think "in reverse" to determine possible synthetic pathways to that compound.

Summarize all the reactions from *Reactivity 3* on one sheet of paper. Use your summary to propose a synthetic route for each of the following:

a.   methanoic acid from bromomethane

b.   propanone from propene

c.   ethyl ethanoate from ethene.

Electrophilic substitution in benzene (*Reactivity 3.4.13*)

The structure of benzene was discussed in *Structure 2.2 (AHL)*.

Benzene does not readily undergo addition reactions because of the stability of its six-electron aromatic ring. Instead, it undergoes substitution reactions. The mechanism of electrophilic substitution in benzene can be illustrated by the nitration reaction.

The first step of benzene nitration is the formation of the nitronium ion, NO_2^+, which acts as the electrophile in this reaction. Pure nitric acid contains only traces of nitronium ions, but in a mixture of sulfuric acid and nitric acid at 50°C the concentration of these ions increases as a result of the following reactions:

$$HNO_3 + H_2SO_4 \rightleftharpoons H_2NO_3^+ + HSO_4^-$$

$$H_2NO_3^+ \rightleftharpoons NO_2^+ + H_2O$$

In turn, the high concentration of nitronium ions increases the rate of the nitration reaction, which proceeds as follows:

1.   The nitronium ion electrophile is attracted to the delocalized pi electrons of the benzene ring.

2.   Two electrons from the benzene ring are donated to the NO_2^+ ion, so a new C–N bond forms while a pi electron from one N–O bond in the nitronium ion moves onto the oxygen atom.

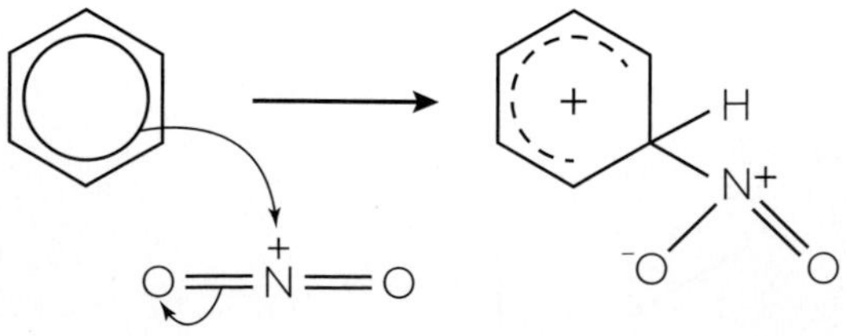

The addition of the nitronium ion to benzene breaks the aromaticity of the ring. This is depicted by the incomplete dashed circle in the ring, which also represents the delocalization of the positive charge. Breaking of the very stable aromatic ring in benzene requires energy, so this process is the rate-determining step of the reaction.

3. Water then acts as a base, deprotonating the carbocation intermediate and restoring the aromaticity of the benzene ring, which gives the final product, nitrobenzene.

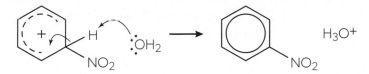

When drawing the mechanism for an electrophilic substitution reaction involving benzene, pay attention to the following:

1. The curly arrow representing the electrophilic attack originates at the ring of delocalized electrons in benzene and terminates at the positive charge on the electrophile.

2. The structure of the carbocation must show an incomplete dashed circle and a positive charge on the ring.

3. The curly arrow representing the hydrogen ion leaving originates at the bond between the carbon and hydrogen atoms and terminates at the benzene ring cation.

4. The last curly arrow originates at a lone electron pair of water and terminates at the hydrogen ion leaving.

5. The structural formula of the substituted benzene must be shown along with the released hydrogen ion, H^+, or hydronium ion, H_3O^+.

TOK

Arrows have many uses in chemistry. Arrows often represent transformations, and chemistry focuses on the transformations of matter. Arrows can also signify movement or imbalance. There are several types of arrows, each with its own specific meaning:

- transformation of reactants into products

- movement of a single electron

- movement of an electron pair

- reversible reaction

- resonance structures

- coordination bond

- bond dipole

How are arrows used as symbols in other areas of knowledge?

Practice questions

15. Benzene is an aromatic hydrocarbon.

    a. State the typical reactions that benzene and cyclohexene will undergo with bromine.

    b. Explain the mechanism for the nitration of benzene, using curly arrows to show the movement of electron pairs.

 Linking questions

What are the features of benzene, C_6H_6, that make it not prone to undergo addition reactions, despite being highly unsaturated? (*Structure 2.2*)

Nitration of benzene uses a mixture of concentrated nitric and sulfuric acids to generate a strong electrophile, NO_2^+. How can the acid/base behaviour of HNO_3 in this mixture be described? (*Reactivity 3.1*)

End of topic questions

Topic review

1. Using your knowledge from the *Reactivity 3.4* topic, answer the guiding question as fully as possible:
   What happens when reactants share their electron pairs with others?

Exam-style questions

Multiple-choice questions

2. Identify and explain why one of the following species cannot act as a nucleophile.

   A.

   B. (structure of $C=N-H$ with H atoms)

   C. (structure of $C=C$ with H atoms)

   D. (structure of $H-C-C-S-H$ with H atoms)

3. Ethene, C_2H_4, reacts with steam in the presence of a strong acid.

   $$C_2H_4 + H_2O \rightarrow C_2H_5OH$$

   What is the name of this type of reaction?

   A. Nucleophilic substitution
   B. Neutralization
   C. Condensation
   D. Electrophilic addition

4. Which statement is correct?

   A. Electrophiles are Brønsted–Lowry acids.
   B. Nucleophiles are Brønsted–Lowry acids.
   C. Electrophiles are Lewis acids.
   D. Nucleophiles are Lewis acids.

5. Which is an example of a Lewis base?

   A. an electrophile
   B. BF_3
   C. CH_4
   D. a nucleophile

6. Which attacking species is matched with its mechanism of reaction?

A.	OH^-	electrophilic substitution
B.	Cl^+	nucleophilic addition
C.	NH_4^+	nucleophilic addition
D.	NO_2^+	electrophilic substitution

7. Which bromoalkane is most likely to hydrolyse via a S_N1 mechanism?

   A. $CH_3CHBrCH_2CH_3$
   B. $(CH_3)_2CHBr$
   C. $(CH_3)_3CBr$
   D. $CH_3CH_2CH_2CH_2Br$

Extended-response questions

8. Organic chemistry can be used to synthesize a variety of products.

   a. Draw the structure of the final product for reaction between but-2-ene and water. [1]
   b. Sketch the mechanism for the reaction of 2-methylbut-2-ene with hydrogen bromide using curly arrows. [3]
   c. Explain why the major organic product of the reaction in part (b) is 2-bromo-2-methylbutane and not 2-bromo-3-methylbutane. [2]

9. Chlorine, Cl_2, undergoes many reactions.

   a. State the type of reaction occurring when ethane reacts with chlorine to produce chloroethane. [1]
   b. Predict, giving a reason, whether ethane or chloroethane is more reactive. [1]
   c. Explain the mechanism of the reaction between chloroethane and aqueous sodium hydroxide, NaOH(aq), using curly arrows to represent the movement of electron pairs. [3]

AHL

AHL

10. Propene, C_3H_6, is an important starting material for many products.

   a. Consider the conversion of propene to a halogenoalkane with the general formula C_3H_7Cl.

      i. State the type of reaction. [1]

      ii. State the IUPAC name of the major product. [1]

      iii. Outline why it is the major product. [1]

      iv. Write an equation for the reaction of the major halogenoalkane product with aqueous sodium hydroxide to produce a compound with the general formula C_3H_8O. [1]

   b. For the reaction between the major halogenoalkane product and aqueous sodium hydroxide, an experiment was carried out to determine the relationship between the rate of reaction and the concentration of the halogenoalkane. The following results were obtained.

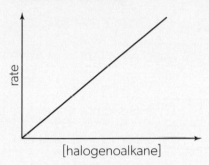

   [halogenoalkane]

      i. State the order of the reaction with respect to the halogenoalkane. [1]

      ii. Under certain conditions, the reaction rate is independent of the concentration of hydroxide ions. Deduce whether the reaction mechanism is S_N1 or S_N2. Explain your answer. [2]

      iii. Sketch the reaction mechanism, using curly arrows to represent the movement of electron pairs. [4]

11. Benzene nitration occurs when benzene reacts with the nitronium ion, NO_2^+.

   a. Write the equation for the production of the nitronium ion from concentrated sulfuric and nitric acids. [1]

   b. Explain the mechanism for the nitration of benzene, using curly arrows to indicate the movement of electron pairs. [4]

12. But-1-ene is an unsymmetrical alkene that can undergo a series of reactions to form an alcohol.

   a. But-1-ene can undergo an electrophilic addition reaction with hydrogen iodide. Deduce the two possible products of this reaction. [2]

   b. Explain which compound is the major product for the reaction. [2]

   c. Draw and explain the mechanism for the reaction using curly arrows. [4]

   d. The major product C_4H_9I can undergo a nucleophilic substitution reaction with aqueous potassium hydroxide. Draw and explain the mechanism for the reaction using curly arrows. [3]

Cross topic exam-style questions

DP exam questions may be topic-specific or refer to content from across different topics. These questions explore the links between various concepts, as well as aspects of NOS and the skills in the study of chemistry.

Below, three exam-style questions have been annotated to show their links to different parts of the course. Next time you do an exam-style question, try to link it to the various course topics, NOS and skills as shown below.

The enhanced greenhouse effect and ozone-layer depletion are two separate atmospheric problems.

Nitrous oxide, N_2O, and carbon dioxide, CO_2, are greenhouse gases. Chlorofluorocarbons (CFCs) are also greenhouse gases, but they are primarily known for their ozone-depleting properties.

Question 1

a. Nitrous oxide, N_2O, is a greenhouse gas.

i. Draw a possible Lewis formula of N_2O, nitrous oxide.	[2]
ii. State and explain the molecular geometry of nitrous oxide.	[2]
iii. Explain why nitrous oxide is IR active.	[2]
iv. Deduce the formal charge of each of the atoms in nitrous oxide.	[1]

← Structure 2.2 Lewis formulas
← Structure 2.2 VSEPR
← Structure 3.2 IR spectroscopy
← Structure 2.2 AHL Formal charge

b. Carbon dioxide is a greenhouse gas. An image of a molecular model of carbon dioxide is shown below:

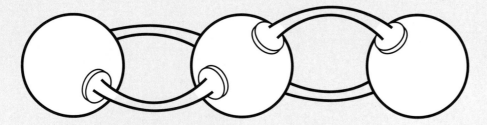

Describe one strength and one limitation of this model's representation of the bonding in carbon dioxide molecules.	[2]

← NOS – models

Question 2

a. Carbon dioxide is produced in the complete combustion of organic compounds such as alcohols.

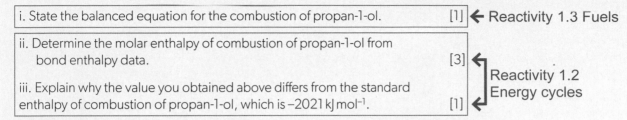

i. State the balanced equation for the combustion of propan-1-ol.	[1] ← Reactivity 1.3 Fuels
ii. Determine the molar enthalpy of combustion of propan-1-ol from bond enthalpy data.	[3]
iii. Explain why the value you obtained above differs from the standard enthalpy of combustion of propan-1-ol, which is $-2021\,kJ\,mol^{-1}$.	[1]

Reactivity 1.2 Energy cycles

b. A student determines the enthalpy of combustion of propan-1-ol by calorimetry using the apparatus shown below:

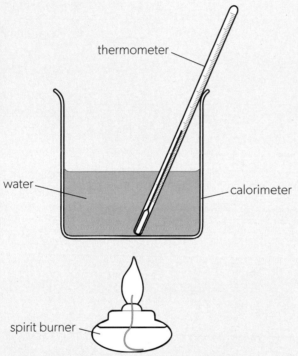

thermometer

water

calorimeter

spirit burner

i. Outline the experimental method employed by the student, identifying the necessary measurements.	[3] ← Tool 1: Experimental techniques – calorimetry
ii. The student's experimental enthalpy of combustion of propan-1-ol was $-894\,kJ\,mol^{-1}$. Calculate the percentage error.	[1] ← Tool 3: Mathematical skills – percentage error
iii. Predict, giving a reason, the sign of the entropy change for the combustion of propan-1-ol.	[1]
iv. The standard entropies of propan-1-ol and oxygen are $193\,J\,K^{-1}\,mol^{-1}$ and $205\,J\,K^{-1}\,mol^{-1}$, respectively. Further standard entropy values are listed in section 13 of the data booklet. Using these data, determine the standard entropy change for the combustion of propan-1-ol, in $J\,K^{-1}\,mol^{-1}$. [1]	
v. Determine the Gibbs energy for the combustion of propan-1-ol at 298 K. Give your answer in $kJ\,mol^{-1}$.	[2]

Reactivity 1.4 AHL Entropy and spontaneity

AHL

653

Question 3

Ozone is a gas found in the upper atmosphere which absorbs harmful UV radiation from the Sun. The following mechanism has been proposed to describe the depletion of ozone:

Step 1 $O_3 + Cl\bullet \rightarrow ClO\bullet + O_2$

Step 2 $ClO\bullet + O_3 \rightarrow Cl\bullet + 2O_2$

<table>
<tr><td>a. A student describes Cl• as a "chloride anion". Outline the student's mistake and suggest the correct term.</td><td>[2]</td><td>← Reactivity 3.3 Radicals</td></tr>
<tr><td>b. Identify, giving a reason, the species in the mechanism that is a catalyst.</td><td>[1]</td><td>← Reactivity 2.2 Catalysts</td></tr>
<tr><td>c. The rate equation for the reaction is found to be rate = k[O₃][Cl•]. Identify, giving a reason, which step in the mechanism is likely to be the rate determining step.</td><td>[1]</td><td>← Reactivity 2.2 AHL Rate equations</td></tr>
<tr><td>d. Chlorofluorocarbons (CFCs) such as CF₂Cl₂ are sources of Cl• when exposed to certain wavelengths of electromagnetic radiation. Complete the diagram below to show the formation of a Cl• species from a CF₂Cl₂ molecule. Include fish-hook arrows and the structural formula of the missing species.</td><td>[2]</td><td>← Reactivity 3.3 Electron sharing reactions</td></tr>
</table>

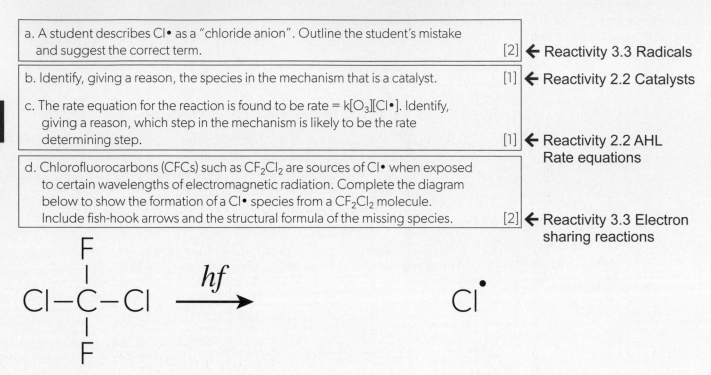

e. CFCs are ozone-depleting substances used as refrigerants. Hydrofluoroolefins, HFOs, are unsaturated organic compounds with the potential to replace CFCs in their application as refrigerants. Since HFOs are more reactive than CFCs, they decompose faster and have shorter lifetimes in the atmosphere. An example of a HFO is 2,3,3,3-tetrafluoropropene.

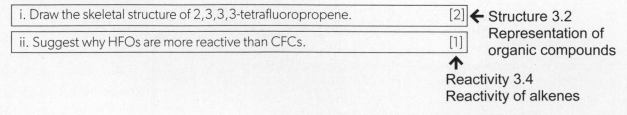

<table>
<tr><td>i. Draw the skeletal structure of 2,3,3,3-tetrafluoropropene.</td><td>[2]</td><td>← Structure 3.2 Representation of organic compounds</td></tr>
<tr><td>ii. Suggest why HFOs are more reactive than CFCs.</td><td>[1]</td><td></td></tr>
</table>

↑
Reactivity 3.4
Reactivity of alkenes

The inquiry process

Introduction

This section clarifies the learning-through-inquiry approach you will use during the DP chemistry course. It will help you to develop the skills in *Tool 1: Experimental techniques*, *Tool 2: Technology* and *Tool 3: Mathematics* during lessons, experiments, the collaborative sciences project, and the internal assessment (IA).

Figure 1 shows how the skills detailed in the *Tools for chemistry* chapter can support the inquiry process.

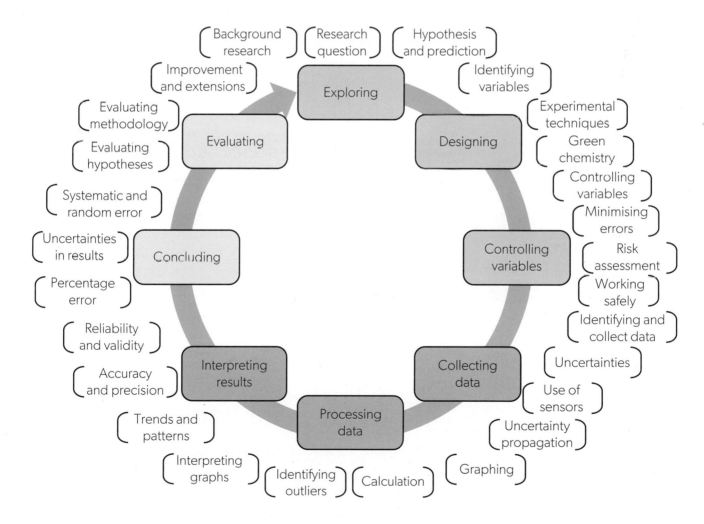

▲ Figure 1 The inquiry cycle with examples of supporting skills

The skills in the approaches to learning (ATL) framework support learning. They are grouped into five categories: thinking skills, communication skills, social skills, research skills and self-management skills. Where do ATL skills fit into the process and tools shown in figure 1?

Theory of knowledge

You will notice that the inquiry process is cyclical: inquiry outcomes feed into further cycles of inquiry. The continuous expansion of science as a body of knowledge relies on this cycle. Scientists use work done previously (by themselves or others) as starting points for their research.

Who owns scientific knowledge? What rights and responsibilities come with owning scientific knowledge?

An inquiry may start in different ways. You may want to go deeper into one or more of the subtopics covered in class. Alternatively, you may have a hobby and want to learn more about its chemistry, such as the materials chemistry used to make equipment for your favourite sport, or the chemical reactions involved in cooking. There also might be a local environmental issue that can be examined using chemistry. In the next section, you will consider one case study that will help you go through the inquiry cycle and identify the skills required to find answers to your questions.

Case study 1: Ocean acidification

The concentration of atmospheric carbon dioxide produced by burning fossil fuels has increased dramatically in recent times. Since the 1980s, it is estimated that oceans have absorbed over 25% of all anthropogenic carbon dioxide. As carbon dioxide dissolves in seawater, it forms carbonic acid, H_2CO_3, a weak acid that dissociates into hydrogencarbonate ions, HCO_3^-, and hydrogen ions, H^+. A higher concentration of H^+ ions increases ocean acidity, resulting in a decrease in pH.

An inquiry might involve testing this hypothesis, but first you need to understand the chemistry behind the theory and find reliable information to support it. You may want to create a list of preliminary questions to form the basis of your inquiry such as:

- What is ocean acidification?

- Is this a local or a global issue?

- Which evidence supports the statement above?

The next step is to do research using reliable sources that provide evidence supporting the claim that increased carbon dioxide emissions has contributed to ocean acidification. For this purpose, you could use journal articles or government data. Figure 2 shows the change in pH of the sea surface from 1700 to 1990, according to the data from the Global Ocean Data Analysis Project (GLODAP).

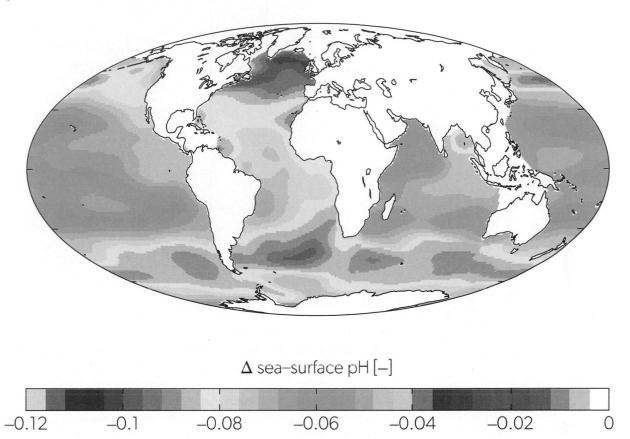

▲ **Figure 2** The change in pH of the sea surface from 1700 to 1990

The map shows that the pH has decreased to a greater extent in regions close to the poles. Can you use any of your learning from the subtopics in DP chemistry to explain this? For example, at lower temperatures, gases will dissolve in water better due to reduced kinetic energy of their molecules. Which subtopic in DP chemistry explains this phenomenon?

In *Structure 2.2*, the polarity of covalent molecules is discussed. From this, you know that carbon dioxide is a non-polar molecule and should not readily dissolve in water. However, you may want to confirm the polarity of carbon dioxide with data. Molecular modelling (*Tool 2: Technology*) is a valuable tool for this purpose. You need to choose molecular modelling software that produces quantitative data, such as WebMO. If you build the carbon dioxide molecule in this software, it will tell you that carbon dioxide has a net dipole moment of 0 D.

You can also check the effect of exposure to carbon dioxide on water pH experimentally. For example, you could set up the experimental apparatus shown in figure 3.

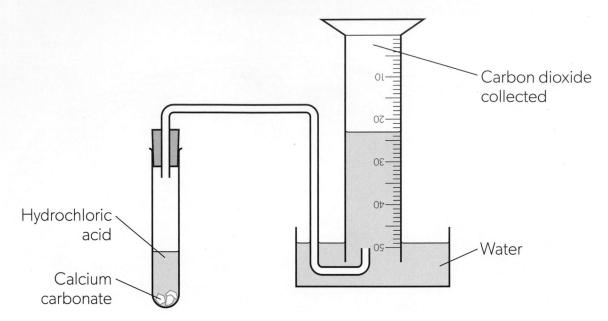

▲ **Figure 3** Experimental apparatus to collect carbon dioxide

A test tube containing calcium carbonate is connected with a tube to a vessel containing a measured volume of deionized water. The initial pH of the water is measured. Hydrochloric acid is added to the test tube containing calcium carbonate, and the carbon dioxide produced in the reaction will be released into the water. You can then record the new pH value. If you change the temperature of the water, you can establish whether there is a correlation between temperature and pH. You can use this data to support your hypothesis that temperature will affect the pH of the water based on the data in figure 2.

You have considered in *Tool 1: Experimental techniques* that not all experiments will yield reliable results. You must critically analyse your experiment: what external factors may affect your results and need to be controlled? Before any experiment, no matter how simple, you should list the variables affecting your results and select instruments with adequate sizes and precisions.

The following are some relevant points to consider:

- How many trials will you do?

- How will you measure the volume of water?

- How will you weigh the sample of calcium carbonate? How much will you use?

- What concentration of hydrochloric acid will you use?

- Will the container of water be open or closed to the surroundings?

- How much time will be required for the reaction to occur? How will you know when the reaction has ended?

- How can you ensure that the carbon dioxide dissolves in the water?

- How will you record the pH of the water?

- What uncertainties are there in your measurements?

You should enter the data you record in a table and produce a graph of your results. Then, interpret the graph (*Tool 3: Mathematics*). Once you have completed the experiment, you should think how you can improve the procedure, or any follow-up experiments you may want to do based on your results.

You may want to consider extensions to your inquiry. While doing your research, you are likely to have found that other gases affect the pH of surface ocean water: sulfur oxides, nitrogen oxides and ammonia. Doing more research, you will find that their impact is significant in coastal areas. Ocean acidification reduces the availability of calcium carbonate for coral reefs and shell-forming marine organisms to build their shells and skeletons. Increased acid concentration also disrupts the equilibrium reactions that occur in the ocean, shifting the positions of equilibria. All these inquiry lines will initiate a new inquiry cycle.

We will now examine the individual steps of the inquiry cycle in figure 1.

Exploring, designing and controlling variables

In the **exploring** step of the inquiry cycle, you should consider the following:

- Find a topic that interests you. Find information that helps you to answer your question and use critical thinking to interpret the data.

- There are many sources for your background reading, and you must carefully select those that are helpful and reliable.

- Produce questions that will help you find the answers you need. Formulating a hypothesis will be helpful and you should make predictions using the knowledge acquired during the course.

In the **designing** step, you should consider the following:

- Carry out a practical experiment, or use databases, simulations or molecular modelling.

- Check that your method will allow you to answer your questions.

- Identify the independent and dependent variables and the variables you should control.

- Do you need to identify the limiting reactant?

- For the independent variables, you must consider the range you investigate and the number of measurements you take.

- Could the storage of the samples affect your results?

- Write the method with enough detail to be able to reproduce it, improve it and be able to justify why it is suitable.

- Your first experiment will always be a pilot that will help you detect areas of improvement.

In the **controlling variables** step, you should consider the following:

- Do you need to calibrate the instrument you are using?

- Which instruments are most suitable in terms of their sizes and precision?

- Keep environmental conditions stable or monitor them if you are unable to control them (such as temperature, pressure and humidity).

- Check if you need to insulate containers to prevent heat loss or gain.

Case study 2: The pH scale

At times your inquiry may be based on the content you are familiar with, but now that you have a better understanding of chemistry, it may look puzzling. For example, the pH scale is defined as values between 0 and 14. You have accepted this range as that was how it was covered in all previous courses. But have you ever considered whether it is correct? In your initial research, you might find out that commercially available concentrated hydrochloric acid has a pH of −1.1, while saturated sodium hydroxide solution has a pH of 15.0.

There are naturally occurring examples as well. Hot springs near Ebeko volcano, Russia, contain naturally occurring hydrochloric and sulfuric acids, and have estimated pH values as low as −1.7. Extremely acidic mine waters with pH values as low as −3.6 have been encountered underground in the Iron Mountain Mine in the US. These are the most acidic waters currently known.

When finding this information, your next step would be to check the conditions under which these results were obtained. At times, you will find it difficult to reproduce them in the school laboratory. For example, it is unusual to find buffers to calibrate your pH probe at the necessary pH values or you would need special probes. You also need to carefully assess the safety risks of an experiment before attempting it, as highly acidic substances are extremely corrosive.

However, some simple experiments may provide initial answers. You could prepare samples with increasing acid concentration and observe the correlation between pH values calculated as $[H^+]$ concentration and the experimentally determined value. This will suggest that DP chemistry is using a simplified approach, as the measured values will deviate from the calculated values. You can then possibly start to consider molecular interactions and suggest an initial hypothesis to explain the deviation. You could also propose a new definition of pH that does not involve $[H^+]$. It also worth noting that pH meters can be unreliable, which may also contribute to the deviation. Therefore, you should consider whether you have enough evidence to support your hypothesis.

Case study 3: The effect of temperature on equilibrium

Equilibrium is open to many investigations. For example, you may be interested in investigating the effect of changes in temperature on the equilibrium constant, K, of the following reaction:

$$Fe^{3+}(aq) + SCN^-(aq) \rightleftharpoons [FeSCN]^{2+}(aq)$$

After some research, you will find that colorimetry is a suitable technique to investigate this. The experiment would involve preparing standard solutions by serial dilution, producing a calibration curve, and then determining the maximum absorbance wavelength (*Tool 1: Experimental techniques*). Due to time constraints, you may need to carry out different parts of the investigation on different days.

Here are some of the challenges you may face in designing the experiment and controlling variables:

- Typical instruments in schools only provide reliable data to absorbances up to 1.

- The optimum wavelength should be determined.

- The complex ion $[FeSCN]^{2+}$ decomposes on standing, so the absorbance values recorded for the solutions prepared more than 1 hour ago will be unreliable.

- The temperature of the solutions will change when they are poured into the cuvette, which will affect the values recorded with the instrument.

You will need to use your mathematics skills to make calculations, and consider and process uncertainties (*Tool 3: Mathematics*).

Before you begin any experiment, you must identify the risks involved and the ways to minimize these risks. You should consider any ethical issues involved and learn how to dispose of any chemicals responsibly. Use the principles of green chemistry whenever possible (*Tool 1: Experimental techniques*).

Collecting data, processing data and interpreting results

In the **collecting data** step of the inquiry cycle, you should consider the following:

- Identify and record relevant qualitative observations. Do not present inferences; these are not raw data.

- Collect and record sufficient relevant quantitative data such that your final results are concordant and you have enough information to support your initial hypothesis.

- Report the uncertainties of the instruments and report all associated data.

- Identify and address issues that arise during data collection.

In the **processing data** step, you should consider the following:

- Can you identify any outliers? If yes, what will you do with those?

- Choose a suitable method for presenting your data.

- Processing that only involves averaging and a bar graph could be an initial step, but you will rarely get useful results with such a simple processing.

In the **interpreting results** step, you should consider the following:

- Interpret both qualitative and quantitative data.

- How can you interpret diagrams, graphs and charts? Qualitative interpretations only provide limited answers.

- If you produced a line or curve of best fit and/or an equation for your graph(s), use it to interpret the trend and correlation.

- Use your knowledge obtained in the DP chemistry course to support your findings. Justify the removal of any outliers. Do not exaggerate your findings.

- Consider whether more experiments are needed to support your findings.

- How have errors affected your results? Are they systematic or random?

- How would you minimize these errors if you were to repeat the experiment?

Case study 4: Caffeine in painkillers

You may be interested in establishing which painkillers include caffeine as a component. You can use this experiment to develop skills in thin-layer chromatography (TLC). You could work with different samples containing aspirin, ibuprofen, paracetamol, and a caffeine standard. In the experiment, the samples are dissolved in a suitable solvent and the TLC chamber is prepared in the fume cupboard. The chamber is covered with a lid and then the plates are loaded (figure 4).

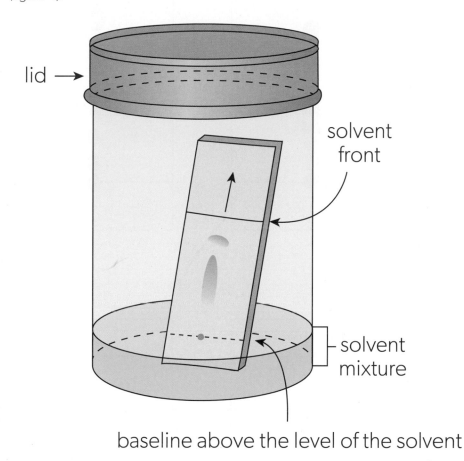

▲ **Figure 4** The experimental set-up for TLC

> Thin-layer chromatography (TLC) is introduced in *Structure 2.2 (AHL)*.

An ultraviolet (UV) lamp is used in this experiment because all the chemicals absorb in the short UV range. In TLC, each dot should be allowed to dry before adding more sample, and you should check that the dots are visible when the lamp is switched on, so that you could detect their positions in the final chromatogram.

Figure 5 shows some possible results for this experiment. You will be unable to reach a valid conclusion if the chromatograms are of poor quality. Identifying the cause of the problems will enable you to fix them in future experiments (table 1).

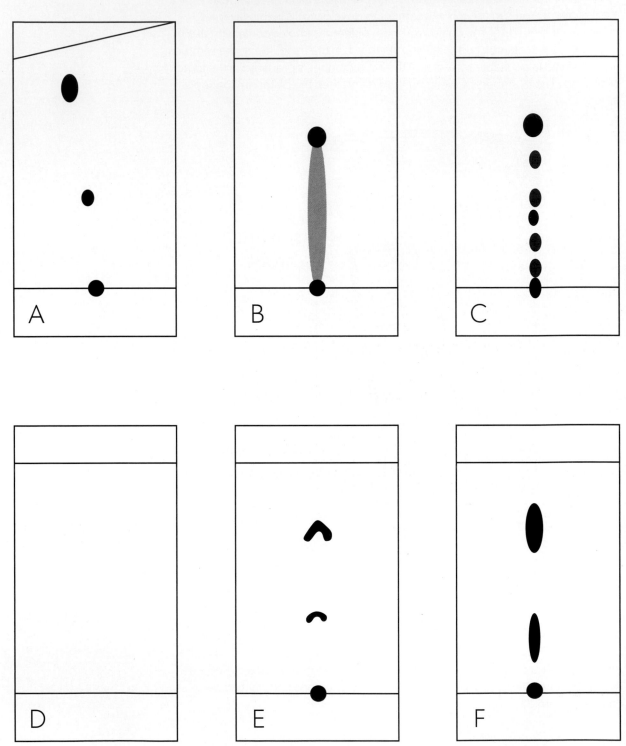

▲ **Figure 5** TLC plates for different painkillers

Appearance	Cause	Fix
A: Curved or skewed solvent front with spots out of lane	Plate touching side of container Plate not lowered level into eluent	Place plate in middle of container Ensure that plate is level when lowering into eluent
B: Streaks, not spots	Too much sample	Dilute sample solution
C: Many blue spots/stripes	Origin marked in pen	Only use pencil on TLC plates
D: No spots on plate	Sample too dilute Wrong visualization used	Remake sample/spot several times Use alternative visualization
E: Crescent-shaped 'spot'	Silica disturbed during spotting	Be gentle when spotting sample
F: Spot(s) smeared out	Acidic/basic groups present in compound	Use an additive in the solvent, such as ammonia or methanoic acid

▲ Table 1 Suggested improvements to the experiment based on the results in figure 5

Case study 5: Antioxidant effects of vitamin C

When considering a topic, try to go for something new and different. For example, there are many studies on the degradation of vitamin C with temperature, light and exposure to air. You might want to investigate its antioxidant effects instead.

For example, you could investigate how temperature affects the nonenzymatic antioxidant activity of vitamin C in oranges. After careful research, you might find that the ferric reducing antioxidant power (FRAP) test is an effective way to find an answer. However, oranges contain phenolic compounds and carotenoids, which are also antioxidants. The FRAP test provides results on all the antioxidants in oranges and not just vitamin C, so this experimental design will not give accurate results for the antioxidant activity of vitamin C. However, you could change the dependent variable to the total antioxidant activity.

Check time constraints when deciding on a method. For example, the reagent in this experiment will need to be freshly prepared: will you have enough time to do this?

Concluding and evaluating

In the **concluding** step of the inquiry cycle, you should consider the following:

- Interpret your processed data and produce an analysis to draw and justify your conclusions. Do they answer your initial question? What about your hypothesis or prediction?

- Compare your results to the accepted scientific context, that is, check whether your results are supported by your theoretical knowledge of science from the DP chemistry topics and your background research, and compare your data to results from existing research.

- How have errors affected your results?

In the **evaluating** step of the inquiry cycle, you should consider the following:

- Did your results support your hypothesis?

- Which errors have affected your results? Have you found R^2? Were the values you obtained aligned with the line of best fit? Do you have any y-intercept that could result from a systematic error? Do you have a reference value that you may use to establish % difference?

- Have you identified any weaknesses in your methodology?

- Did you make any assumptions that were not correct?

- Which improvements can you make to obtain more reliable results if you repeat the experiment?

Case study 6: Enthalpy of solution for metal chlorides

A student wants to investigate how the ionic radii of various metals affect the enthalpy of solution of their chlorides. After three trials of recording the temperature change when different masses of each salt are added to water, the student averages the temperature change for each salt and then calculates the enthalpy of the solution. The student compares their results with enthalpy of hydration data for these metal chlorides from a database, and concludes that the enthalpy of solution results from both the lattice enthalpy and the enthalpy of hydration. However, they do not report the data for lattice enthalpy.

The student has made a common mistake. The calculation of enthalpy should be done for each individual trial rather than the average temperature change. This approach would allow an evaluation of the range of results. The student also needs to include data for the lattice enthalpy of the metal chlorides to support their conclusion.

Case study 7: Solubility of potassium bitartrate

A student wants to determine how the solubility product constant, K_{sp}, of potassium bitartrate is affected by temperature. The experimental method involves using a water bath and adding the potassium bitartrate until it stops dissolving. Then, the solution is filtered and titrated three times with sodium hydroxide, NaOH(aq), using phenolphthalein as an indicator. The overall percentage error in the results is calculated to be 26%, with 20% resulting from the titration stage. When evaluating, the student notes that more trials would improve the reliability of the results and that controlling the temperature of the solution is difficult using the water bath. The student also notes that some potassium bitartrate will precipitate as the temperature decreases during the titration.

Three trials are acceptable in school investigations but having more tests will always reduce random errors. The student fails to note that the temperature would have decreased during filtering as well as the titration. The student has also missed that potassium bitartrate can react with sodium hydroxide even when solid. Some techniques may result in very high error, such as calorimetry, but 20% is too high for titrations.

Summary

Inquiry cycles can be short or involve continuous research over a lifetime. The more skills you develop, the better the quality of your inquiry will be. They will help you to:

* ask relevant questions

* take action to answer them

* collect useful data and find different ways to process them

* reflect on your acquired knowledge and help you move forward in whichever field you plan to work in.

Once you have completed many inquiry cycles, you will be ready to engage meaningfully with the internal assessment (IA).

The internal assessment (IA)

Introduction

The internal assessment (IA) is an opportunity to apply the skills and tools you have learned during the DP chemistry course. In your IA, you are expected to spend ten hours carrying out a scientific investigation to produce a written report. The maximum word count of the report is 3000 words, but this word count does not include charts, diagrams, equations, formulas, calculations, tables, references, bibliography, or headers.

The IA is an inquiry process. Collaboration between up to three students is allowed, and your group must be established before the scientific investigation begins. Even if you collaborate in your IA, you must ensure that you produce an individual and unique research question or title.

The IA is assessed by four criteria, each worth six points and 25% of your final mark (table 1).

Criterion	Number of marks	Weighting
Research design	6	25%
Data analysis	6	25%
Conclusion	6	25%
Evaluation	6	25%

▲ Table 1 Assessment criteria for the IA

Any source used in your scientific investigation should be properly cited as required by the IB's academic integrity policy. However, correct citation is not assessed by the criteria in table 1.

In the research design for your scientific investigation, you are required to

* produce a focused research question or title

* identify the best methodology to answer it

* clearly detail the steps of the chosen methodology

* carry out sufficient research to meet the requirements of each step and be able to justify every decision taken.

Research design

The research question should provide specific and appropriate context for your investigation. In your methodological considerations, you should describe how the chosen data collection methods allow you to answer the research question. The methodology used should be realistic in terms of the time and resources available. It must also be possible to effectively control variables that impact your results. You should present the description of the methodology clearly such that it could be easily reproduced.

You should also make sure that your overall research design allows you to explore the skills detailed in the inquiry process. Two different investigations will rarely follow the same path, but some of the following steps may be involved in developing your research design:

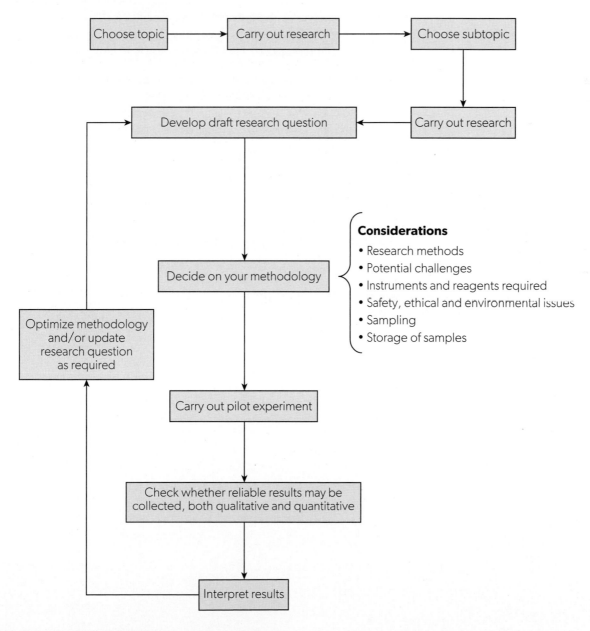

▲ Figure 1 Possible steps in developing your research design

When you are developing your research design, you may find that your chosen methodology does not allow you to collect enough data to answer your research question. In this case, you must ask yourself what makes more sense: changing the methodology or changing the research question. The case studies below give examples of possible research designs and how they link to the inquiry process.

Case study 1 (Inquiry 1: Exploring)

A student wants to devise a research design to investigate how the calcium content of kale changes as it is cooked in water. They suggest the following draft research question: "What is the effect of temperature changes when cooking vegetables?"

This question outlines possible independent and dependent variables for the investigation. However, the question lacks specificity: there are infinite ranges of temperature and many types of vegetables. Readers of the scientific report will not understand the purpose of the investigation with limited information.

The concentration of calcium ions could be measured in the vegetables or the water where they were cooked. Several methods are suitable for a school laboratory. Critical thinking and research skills should be used to determine which is most effective method. Considering a vegetable of local relevance and the range of temperatures typically used in local recipes would also improve the research question. The research question also fails to outline the methodology used, and there are many suitable techniques.

A more specific research question would be: "What is the effect of temperature (35, 45, 55, 65, 75, 85 °C) on the calcium concentration in kale as determined by the titration with EDTA of the water used for cooking it?"

Case study 2 (Inquiry 1: Exploring, Designing, Controlling variables)

A student wants to investigate the effect of citric acid concentration on the antioxidative action of ginger tea. The student decides to measure the change in total phenolic content using an antioxidant assay and colorimetry. The research design requires context, to explain the antioxidant properties of ginger extract and how citric acid enhances these properties.

Natural products are challenging because the results may be affected by the agricultural practices used to farm the ginger, such as soil composition, storage conditions and moisture content, so these factors must be considered. Performing an extraction will also require several decisions regarding the experimental conditions, such as choice of solvent, pH if aqueous, exposure time and temperature.

In developing the research design, the student encounters the following challenges:

- The student decides to use the 2,2-diphenyl-1-picrylhydrazyl (DPPH) assay to test for the phenolic content in the ginger tea, and finds it in the school stock room. However, when doing further research, the student realizes that DPPH is not soluble in water. Therefore, the assay is not suitable for the analysis of tea solutions.

- The student conducts further research and finds out that the Prussian blue assay uses water-soluble reagents. The phenolic compounds in ginger tea will reduce the $Fe(CN)_6^{3-}$ ions in Prussian blue to $Fe(CN)_6^{4-}$ ions, which will react with iron(III) ions to give $Fe_4[(CN)_6]_3$. The colour of $Fe_4[(CN)_6]_3$ is blue, and the intensity of this colour will depend on the antioxidative properties of the ginger tea. The intensity of colour can be measured by colorimetry.

- The student needs to produce a calibration curve, which will require the use of gallic acid as one of the reagents. The student also remembers to take at least three readings and use a blank to make this curve. The student monitors the temperature, as this will affect the absorbance.

When describing the methodology, the student details the Prussian blue test, and provides a brief explanation of Beer's law, both of which add value to the research design. The student also includes the chemical reactions that occur in the test for phenolic content using balanced equations with state symbols.

> The skills required for conducting a titration are discussed in *Tool 1: Experimental techniques*, in the *Tools for chemistry* chapter.

Case study 3 (Inquiry 1: Designing, Controlling variables)

A student wants to investigate how the cooking time affects the concentration of oxalic acid in the spinach. The student outlines their methodology as follows: they will determine the concentration of oxalic acid in the cooking water at different times by titrating it with potassium manganate(VII), $KMnO_4$.

Several control variables need to be considered for this investigation, such as the mass of the spinach, the volume of water, the range of cooking times, and the source and storage of the spinach sample. The following aspects of the methodology also need to be considered:

- Should the leaves, the stem or all the spinach be used?

- Which type of spinach should be used? The concentration of oxalic acid in spinach will vary significantly between different varieties. This information is required to prepare a suitable concentration of the titrant.

- Which range of temperatures is reasonable within the context of the investigation? For example, oxalic acid will decompose if the solution is boiling. Which intervals will provide sufficient and reliable data? For example, figure 2 shows a clear line of best fit with the equation included and a high value of R^2.

- Should the spinach be blended in a food processor first? How long should the sample be blended for? How can the surface area of the spinach sample be kept consistent?

- How will the potassium manganate(VII) solution be prepared? What is a suitable concentration for the solution? The student could run a pilot experiment with a concentration based on their research. The concentration used can be optimized for future experiments based on the results. For instance, the titrant volumes recorded could be very small, which means that the concentration of the titrant needs to be lower.

- Will a hot plate or a water bath be used to reach desired temperatures? A water bath could provide more uniform temperature throughout the sample. Should the temperature be monitored?

- How long should the spinach samples remain at each temperature? Should they be cooled down when removed? If the samples are not immediately cooled down, the cooking process will continue, affecting the results.

- Has appropriate experimental equipment been selected, considering their size and precision? Volumetric flasks and volumetric pipettes should be used for preparing solutions or dilutions. The golden rule is to use the highest precision instruments available.

- Is the titrant solution stable? This must be considered if data are collected in several sessions. Potassium manganate(VII) reacts with water, and this reaction is sped up in the presence of sunlight.

- Are there any specific requirements for the titration? It is a slow reaction, and therefore a gentle heating is advisable. The most suitable temperature can be ascertained in a pilot experiment. If the temperature is too high, the oxalate will decompose.

- Which safety issues must be considered? How can green chemistry come into the picture? How should the leftover spinach samples and reagents be disposed of? Are there any ethical issues? To mention a few: safety gear should always be worn even when the experiment does not involve serious risks. You must consider risks associated with chemical reagents, try to use as little as possible of them and state the environmental hazards involved.

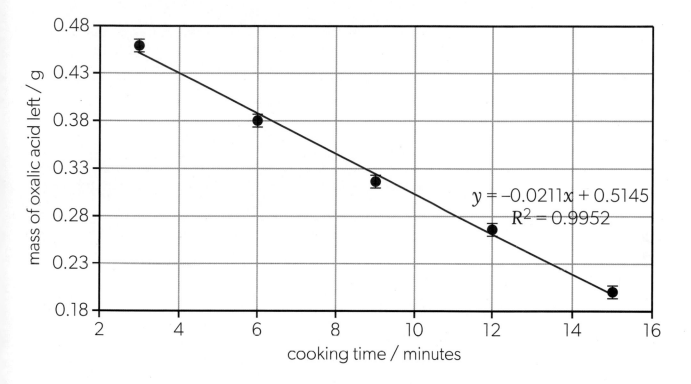

▲ **Figure 2** Graph of measured mass of oxalic acid in spinach vs cooking time

Each decision must be based on a solid rationale and should be justified in the methodological considerations. The student decides to cut the leaf samples with scissors and uses a ruler to measure the pieces and ensure that the surface area of the spinach sample is consistent. The student opts for this method as a pilot experiment shows the sample gets warmer when using a food processor to chop the spinach. The student also decides to monitor the temperature using a thermometer and cool the samples immediately after heating, so the exposure time is the same for each sample. This shows thorough control of variables and adds value to the investigation.

The student also checks that there are high precision instruments available while planning their research design, therefore minimizing of uncertainties affecting the results.

The student conducts the experiment in the fume hood using a lab coat, safety goggles, mask and gloves due to the risks that potassium manganate(VII) presents. These hazards are briefly described in the research design. The student makes every effort to use the minimum quantities and the leftover chemical waste is disposed in a special bin for a company to collect. The report clarifies that the waste should not be disposed through the sink as this compound can cause damage to the environment. The leftovers of the spinach are used for composting. This approach shows that the student adheres to safety, ethical and environmental good practice.

Case study 4 (Inquiry 1: Designing)

A student designs a research method to investigate how the time of chlorine exposure to UV light affects the extent of the photolysis reaction, using a mixture of sodium thiosulfate, $Na_2S_2O_3$, and starch as an indicator.

The following methodological considerations need to be made:

- reagents involved

- concentrations and storage of solutions

- chosen wavelength of UV light

- distance of chlorine sample from lamp

- elimination of other sources of light

- range of exposure times

- the best way to establish the endpoint of the titration and the number of trials.

Each will require several decisions, especially if the pilot shows that extended time periods are needed to obtain reliable results. For example, the temperature needs to be controlled as this will affect the reaction system. This is usually a challenge for longer investigations. If you the temperature cannot be kept constant in the experimental environment, the temperature of the reaction system should be monitored using a temperature probe.

The student's report includes the details of calibrating the pH probe with two buffers and they prepare a blank, showing good experimental technique. The buffers cover the optimum pH needed, to avoid systematic errors. The student covers the container with glass transparent to a wavelength of 365 nm, which gives a high rate of photolysis. This prevents evaporation and changes in concentration that would make the results unreliable.

Sodium thiosulfate is stable for 24 hours, so the student prepares a fresh solution each day that they carry out the experiment. The starch solution is also freshly prepared to avoid degradation.

The processing of errors must be considered before implementing the procedure. If the idea is to use a standard deviation, five trials are the minimum required. However, if any of the trials produce outlier results, then these results need to be excluded and more trials will be necessary. Diagrams or photographs of the experimental set-up could also add value to the description of the methodology.

Data analysis

The data analysis for your scientific investigation requires you to:

- communicate your recording and processing of data in a clear and precise manner

- consider uncertainties and their propagation

- adequately process the relevant data to allow you to answer the research question.

Communication is essential when addressing this criterion. Make sure that you follow these rules:

1. Present clear tables and graphs with adequate titles.

2. Produce well designed tables that allow easy comparison of data.

3. Include the uncertainties of the instruments used.

4. Use correct symbols for physical quantities and their units. Remember to use SI units.

5. Show your processing clearly, but do not add unnecessary steps and descriptions.

6. Report decimal places consistently.

7. Do not use images to replace qualitative data.

8. Make sure that you report raw data, not inferences.

9. Conduct repeats for any trials where the results are inconsistent.

Consider point 8: inferences are the conclusions you make based on your recorded data. For example, if you were testing the rate of a reaction with and without a catalyst, your raw data might be the time taken for the reaction to reach to completion, and the inference would be that the catalyst increases the rate of reaction. The raw data should be in your data analysis, and the inference should be in your conclusion.

The case studies below give examples of data analysis in three scientific investigations and how these link to the inquiry process.

Case study 5 (Inquiry 2: Collecting data, Processing data)

A student investigates the effect of changing the concentration of glucose on the rate of carbon dioxide production during fermentation. Four samples with different concentrations of glucose are prepared by adding different masses of glucose to a fixed volume of water and yeast. The student plans to heat the solutions with a hotplate at a temperature slightly above the expected one to account for the cooling that occurs while data are collected. The concentration of CO_2 will be measured with a probe for 700 seconds.

In this experiment, the temperature must be controlled with a water bath and monitored with a thermometer. To establish a trend, a minimum of five data points is required for the independent variable.

The student's report includes four graphs. Figure 3 shows the results from the pilot experiment.

Figure 3 shows that carbon dioxide concentration changes non-linearly with respect to time, so a line of best fit is not appropriate for this graph. Figure 4 shows two trials that include the other glucose samples, which also do not show a linear trend.

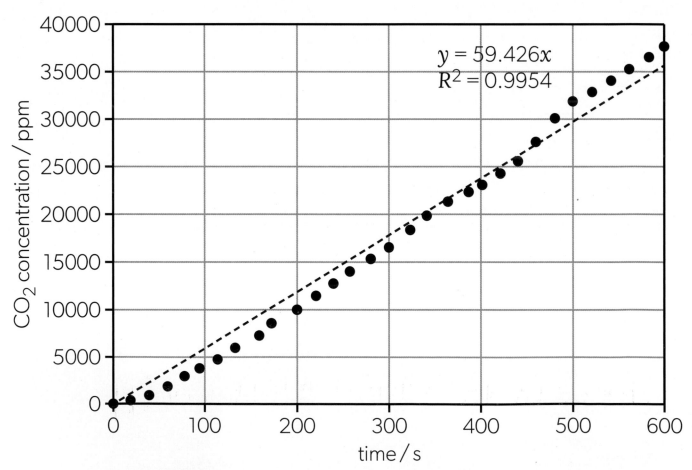

$$y = 59.426x$$
$$R^2 = 0.9954$$

▲ **Figure 3** Graph of carbon dioxide evolved *vs* time during fermentation for one glucose sample

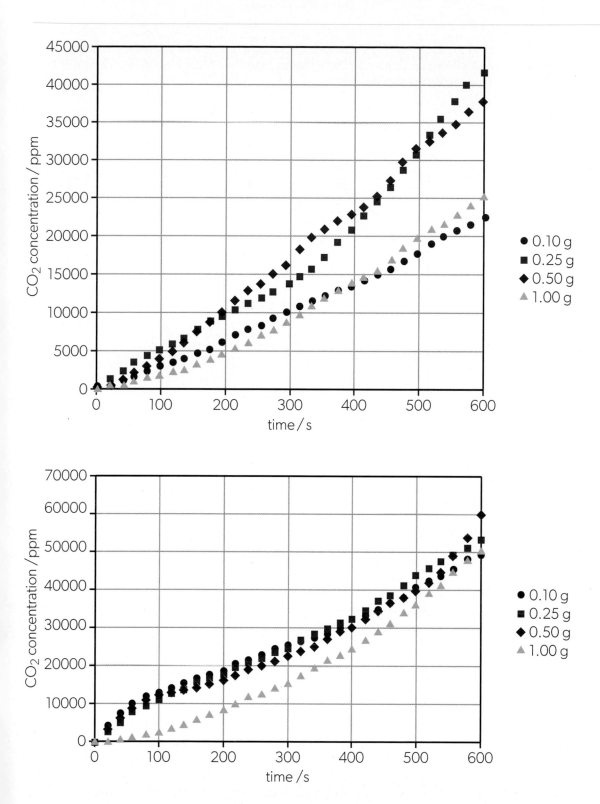

▲ **Figure 4** Graphs showing two trials for the fermentation experiment

The two trials show inconsistent values for all samples, especially 0.10 and 1.00 g. In this case, a repeat is required.

In the final graph in the report, the student has tried to produce a Michaelis–Menten plot (figure 5). The ability to produce a Michaelis–Menten plot is not assessed in DP chemistry.

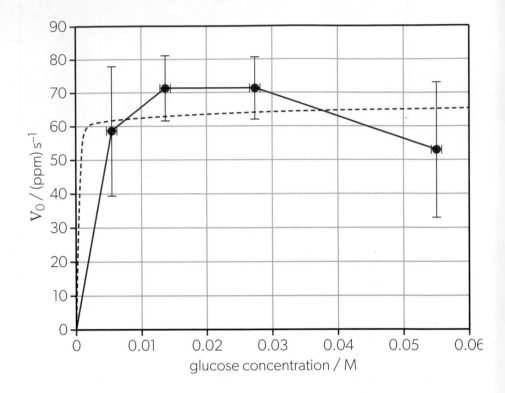

▲ **Figure 5** Michaelis–Menten plot for the fermentation experiment

For each data point, the range of the error bars overlaps with the error bar range for other data points. For example, the first and last points in the graph have error bar ranges that include all of the data points. Therefore, the data are not reliable. More data are required to produce a reliable Michaelis–Menten plot. The student also assumes that the graph starts at the origin but does not have data to support this.

Case study 6 (Inquiry 2: Collecting data, Processing data)

A student conducts an investigation where the independent variable is the length of the carbon chain in primary alcohols, and the dependent variable is the standard enthalpy of combustion. Two databases and the DP chemistry data booklet are used instead of conducting a traditional hands-on experiment. One of the databases includes experimental values, and the other contains predicted values. The data were processed in the form of a scatter graph with the number of carbon atoms as the independent variable.

You should aim to use at least three reliable databases in investigations using secondary data. However, if you have an interesting idea for a database investigation and there are fewer than three databases available, this should not discourage you from pursuing this investigation. You should try to think of other ways to support your hypotheses if limited secondary data is unavailable. The reasons for selecting the chosen databases should be included.

The DP chemistry data booklet is reliable, but it is not a database. Using scientific papers is also not a good idea as the authors have already selected the values from secondary data, and this is a task you must perform.

Many databases do not include uncertainties. In this situation, an estimate through reported precision is too simplistic. An analysis of the differences in reported values in different databases would provide a more realistic uncertainty.

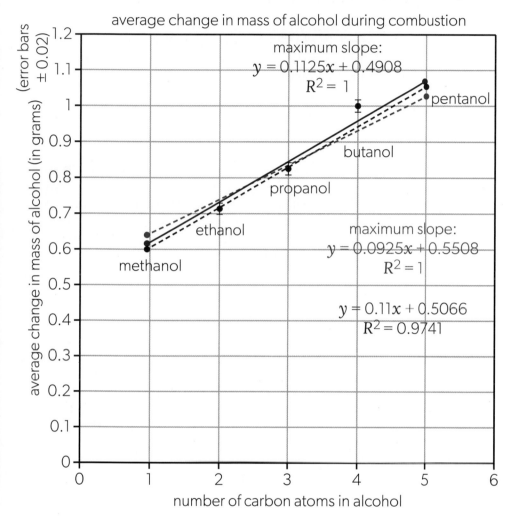

▲ **Figure 6** Graph of change of mass in combustion of primary alcohol *vs* number of carbon atoms in chain

The student processes their data to produce the graph shown in figure 6.

Scatter graphs should not be used to display data where the independent variable is discrete and quantitative with regular intervals (as in figure 7). This is because the line of best fit incorrectly implies that alcohols with a fractional number of carbon atoms exist. Therefore scatter graphs should only be used for continuous data. The student will however receive marks for demonstrating data handling skills such as plotting the graph, identifying the line of best fit, extracting the R^2 value and line of best fit equation. You should use a bar chart when the independent variable is discrete and qualitative.

Case study 7 (Inquiry 2: Collecting data, Processing data)

A student wants to investigate how the temperature of copper(II) sulfate solution in a voltaic cell affects the potential difference of the cell. In their investigation, they use copper and iron electrodes, and sodium nitrate, $NaNO_3$, for the salt bridge. They want to use temperatures of 10, 20, 30, 40, 50 and 60 °C.

In their procedure, the student reports the overall surface area of electrodes in the electrolyte, but not the distance between them. A picture included in the report shows that the electrodes were not parallel and touched the bottom of the beaker. The distance between electrodes needs to be controlled, and the electrolyte flow will be affected if the electrodes touch the container. The thermometers should not be touching it either but be placed in the bulk of the electrolyte. The student has not reported the temperature of the surroundings, nor pressure for the three trials.

The student prepares the electrolyte with $CuSO_4 \bullet 5H_2O$ but they use the molar mass of the anhydrous salt to find the concentration. This is a common mistake that will affect the accuracy of their methodology. Many salts in school stock rooms are hydrates, as they are cheap and easy to store. In this case, the student will need to factor in the molar mass of water in the hydrated salt when making the calculations for the concentrations.

The student prepares a large enough volume of the electrolyte to use in all the trials. The student also decides to change the salt bridge and sand the electrodes for each trial.

In experiments involving electrochemical cells, a water bath should be used to ensure that the temperature does not change while collecting data and the voltmeter reading needs to stabilize before data is recorded.

The student includes qualitative observations on the solutions at the start of the experiment. However, qualitative observations on changes that occur during the process are most useful.

Their results are shown in table 2 below.

◀ Table 2 The student's results

Temperature /K ± 0.1 K	Potential difference / mV ± 1 mV					
	Trial 1	Trial 2	Trial 3	Trial 4	Trial 5	Mean
323	619	614	627	611	627	620
313	631	634	642	638	633	636
303	642	639	639	640	649	642
293	616	622	602	597	619	611
283	579	573	582	589	586	582

Always reflect on collected data and decide whether you need to repeat any trial. Also look for outliers when calculating the mean and use your critical thinking to keep or exclude them. In either case, you should provide a brief rationale for your decision. If you find the standard deviation using your graphical calculator or computer software, you should include a picture of the calculations with the values used.

The student also calculates theoretical data for the voltaic cell using the Nernst equation. The use of the Nernst equation is not assessed in DP chemistry. Both theoretical and experimental data are presented in a graph with a line of best fit and an equation for the line of best fit (figure 7). The R^2 is relatively low for the experimental data, so the student cannot report that there is a strong correlation.

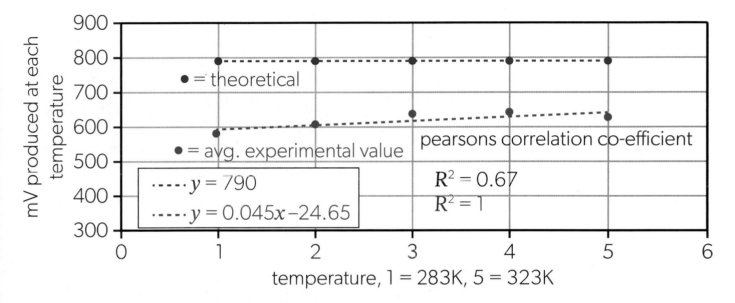

▲ Figure 7 Graph of potential difference *vs* temperature for an iron–copper voltaic cell

You must remember to consider the limitations of the theory you plan to use when designing an experiment. The Nernst equation only applies to solutions of concentrations up to 10^{-3} mol dm^{-3}, and the solutions that the student prepares have a concentration of 0.1 mol dm^{-3}. In their report, the student mentions that the error bars were too small to be shown. The use of error bars is not mandatory in your report. However, if they are used, you should use them correctly. The report should include the error bar calculations and a brief analysis of their significance.

Conclusion

The conclusion for your scientific investigation requires you to:

- present a conclusion that is relevant to the research question and justified by the data

- include scientific context to support your conclusion.

The case study below gives an example of conclusion in a scientific investigation and shows how this links to the inquiry process.

Case study 8 (Inquiry 3: Concluding)

A student wants to investigate how different ratios of active ingredients in antacids impact their neutralizing effect on hydrochloric acid, $HCl(aq)$. Hydrochloric acid is used to simulate stomach acid.

The student chooses to investigate mixtures of five different mass ratios of two antacids, magnesium carbonate, $MgCO_3(s)$, and calcium carbonate, $CaCO_3(s)$. The range of mass ratios used is 1:1 to 1:5, and the overall mass of each mixture is 0.2 g. The method involves adding the antacid mixtures to $0.1 \, mol \, dm^{-3}$ hydrochloric acid and titrating the resulting solution with sodium hydroxide, $NaOH(aq)$.

Their results are shown in table 3 below.

Mass ratio of $MgCO_3$ to $CaCO_3 \pm 0.001$	Volume of NaOH / cm³			
	Trial 1	Trial 2	Trial 3	Mean
1:1	27.6	30.8	24.1	27.5
1:2	28.3	29.2	28.6	28.7
1:3	29.1	30.5	33.5	31.0
1:4	35.2	34.1	32.7	34.0
1:5	37.5	37.2	37.2	37.3

▲ Table 3 Table 3 The student's results

The student includes the two graphs shown in figures 8 and 9 in their report.

The choice of scatter graph is a good one as it allows the student to demonstrate skills in data-handling, such as identifying the line of best fit and extracting information from this. However, the volume on the y-axis is erroneously referred to as "amount".

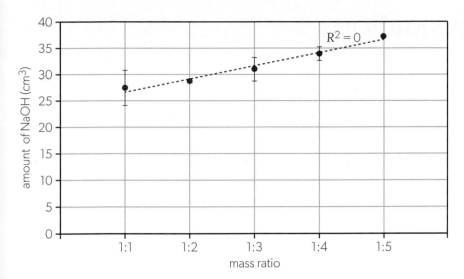

▲ **Figure 8** Graph of volume of sodium hydroxide required to neutralize hydrochloric acid with different mixtures of antacids

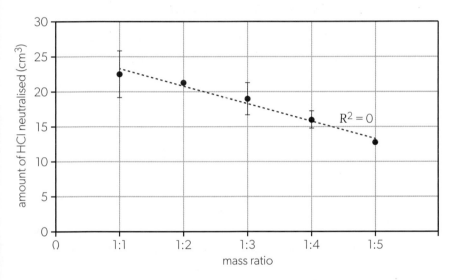

▲ **Figure 9** Graph of volume of hydrochloric acid with different mixtures of antacids neutralized by sodium hydroxide

In their conclusion, the student states that as the relative mass of calcium carbonate in the antacid increases, the neutralizing effect of the mixture decreases. The student also mentions that the 0.97 value of R^2 justifies this trend. However, both graphs report the value as 0 and the origin of the 0.97 value is not reported.

The main drawback of the experimental design is that the concentration for hydrochloric acid used poorly mimics stomach conditions. The student has also not considered the temperature of the stomach, and the time given for the antacid mixture to react with the acid.

As presented, this investigation is more related to stoichiometry calculations. Therefore, mole ratios rather than mass ratios should be used for the antacid mixtures.

Evaluation

The evaluation for your scientific investigation requires you to:

- identify methodological weaknesses and limitations

- suggest realistic and relevant improvements, addressing previously identified weaknesses and limitations to the methodology.

Your methodology is assessed against this criterion, so any significant mistakes made in the experimental design will affect your marks. Identifying issues at this stage that you should have addressed before while collecting data will earn minimal credit. Please remember that suggestions such as 'doing more trials' are considered too simple and will not allow you to reach the second band.

The case study below gives an example of an evaluation in a scientific investigation and shows how this links to the inquiry process.

Case study 9 (Inquiry 3: Evaluating)

A student investigates the effect of changing the mole ratio of a mixture of cyclohexane and cyclohexene on the volatility of the mixture. After several pilots, the student finds that the mole ratios 1:0, 3:1, 1:1, 1:3 and 0:1 produce reasonable data.

The student decides to soak a piece of filter paper in the solvent mixture, and then allows the sample to evaporate in a fume cupboard due to the hazardous nature of the solvents. The temperature and pressure in the fume hood are monitored. The student measures the length and width of the filter papers to ensure that the surface area is the same for each mixture. They decide to control the amount of the solvents by carefully measuring their volumes with a 2.00 cm^3 graduated pipette.

The student attaches the filter paper to a temperature probe using a rubber band, and submerges both through a hole in a test tube filled with the solvent mixture. The student collects data until the temperature remains constant. When the value remains stable for two minutes, they remove the probe with the paper and record the decreasing values until they stabilize for two minutes using a data logger.

The student propagates the uncertainties (*Tool 3: Mathematics*) to find an estimate of the systematic error. The overall percentage error is 8.96%, and random errors range from 3.38 to 5.72%.

Their results are shown in table 4 below:

▶ Table 4 The student's results

Mole fraction of cyclohexane	Average rate of cooling / °C s^{-1}	Total random error / %	Total systematic error / %	Total percentage error / %	Absolute uncertainty / °C s^{-1}
1.00	0.1200	3.63	5.33	8.96	±0.0044
0.75	0.0840	5.72	3.24	8.96	±0.0048
0.50	0.1050	4.89	4.07	8.96	±0.0051
0.25	0.0970	5.60	3.36	8.96	±0.0054
0.00	0.1000	3.38	5.58	8.96	±0.0028

The student uses their results to create the graph shown in figure 10.

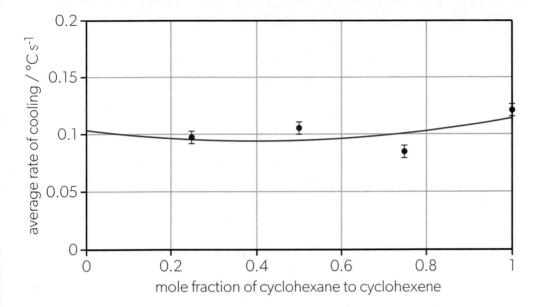

▲ **Figure 10** Graph of rate of cooling of solvent mixture *vs* mole fraction of cyclohexane in solvent mixture

Vapour pressure is a measure of the volatility of a liquid. The student knows that the change in vapour pressure is proportional to the change in temperature, so can use their result to predict the relative volatility of the different mixtures. This is an example of an inference.

In evaluating the results in their report, the student considers that the rubber band introduces a systematic error, as they cannot confirm its position is always the same. They suggest using a thin contact film strip with a well-established surface to minimize this weakness in the design. The student also thinks that the paper needs more contact with the temperature probe, which introduces another systematic error. The student suggests using cylindrical filter paper instead.

The student also notes that the entire filter paper may not always absorb the same volume of the solvent mixture, so they recommend using a more professional device to ensure that the same volume is absorbed every time. The report also mentions that instead of measuring the evaporation rate, a pressure sensor could be used to measure the vapour pressure.

The identified weaknesses are valid and deserve credit as the student has made effort to minimize errors in their methodology earlier in the process, and they identify some useful improvements.

Index

3D representation of molecules 129, 259, 260
absolute uncertainty 355
absolute zero 17, 389
absorbance 76, 339
absorption spectra 34, 35, 39
accident prevention, experiments 309
accuracy of measurements 311–12, 467
acetic acid (ethanoic acid) 526, 549
acetylsalicylic acid (aspirin) 472
achirality 286
acid–base classification systems 552
acid–base equations, balancing of 555–6
acid–base equilibria, in salt solutions 565–7
acid–base indicators 318–19, 571–3
acid–base titration 468, 557–9, 567–73
acid deposition 242, 554
acid dissociation constant 561–5
acid rain 242–3, 554
acid residue 547
acids
    anions 548
    Arrhenius acids 538, 539
    binary acids, periodic trends 549
    Brønsted–Lowry acids 239, 539, 546–7
    buffer region 568
    buffer solutions 574–7
    classification 547
    conjugate acids 541, 571
    dissociation constant 561–5
    half-equivalence point 569, 571
    Lewis acids 241, 635–6
    oxidation states 549
    oxide reactions 241–2
    parent acids 556
    periodic trends 549
    pH curves 557–9, 568, 569–70
    pH scale 543–4
    properties 537, 538, 546–50
    reactions 241–2, 552, 553–6, 591
    redox reactions 552, 591
    strong acids 548
        pH curves 568, 569–70
    theories 538–41
    weak acids 530, 549–50
        anions 552
        dissociation constant 561–5
        reversible ionization 530
        titration 572–3
activation energy 391, 492–4, 507–8
active metals 553
addition polymers 217–19

addition reactions, electrophilic 632–4, 644–7
ageing, free radical theory of 623
air pollution 242, 433
alcohols 261
    aqueous solubility 152
    boiling points 269
    classification 281
    combustion 398, 427
    homologous series 264, 265
    naming 276
    oxidation 604–6
    primary alcohols 604–6, 607
    reduction 607
    secondary alcohols 604, 605, 607
    tertiary alcohols 607
aldehydes 261, 264, 276
    functional group isomers 283
    homologous series 265
    oxidation 604–6
    reduction 607, 608
    terminal position 276
aliphatic 262
alkali metals 550
    metallic character of 238
    periodic table 230
    reactions 239
alkaline species 541
alkanes 262
    boiling points 268, 269
    branched-chain alkanes 271
    combustion 425–6, 428
    homologous series 263
    melting points 268
    naming 272
    radical substitution 624
    root names 272
    straight-chain alkanes 268
    see also halogenoalkanes
alkenes 262
    addition polymerization 217–19
    boiling points 269
    cis–trans isomerism 284–5
    electrophilic addition 632–4, 644–6
    homologous series 264
    naming 274
    reduction 608–9
    straight-chain alkenes 264
alkoxy group 261, 264, 283
alkyl group 260, 272
alkynes 264
    homologous series 264

reduction 608–9
allotropes, carbon 135–6
alloys 205–9
aluminium
boiling and melting points 192
recycling 190
aluminium chloride 204, 636
aluminium nitrate 107
amides 127, 261, 266
amido group 261
amines
classification 281–2
hydrocarbon substituents 551
naming 282
primary amines 261, 266
secondary amines 266
suffix 282
tertiary amines 266
amino acids 222, 542
amino group 261
ammeters 319
ammonia 551
bond angle 129
combustion 413
Haber process 525
ionization 540
reverse reaction 515
van der Waals parameters 84
ammonium ion 104
ammonium nitrate 104
amphiprotic species 542
amphoteric species 241, 542
analyte 335, 469
analytical techniques
colorimetric analysis 253
colorimetry 77, 339–40
combining of 302–3
gas–liquid chromatography 155
infrared spectroscopy 293–6
liquid column chromatography 155
mass spectrometry 29–32, 290–2
paper chromatography 10, 11, 155, 156, 331
proton nuclear magnetic resonance spectroscopy 296–302
spectrophotometry 76, 339–40
thin layer chromatography 157, 158–9, 331
titration 334–6, 400, 468, 469, 557–9, 567–73, 592–3
Anastas, Paul 475
anions 97, 98, 102, 105
acids 548
electrical conductivity 111
oxyanions 245
primary (voltaic) cells 596
radii of 233–4
solubility 112, 113, 114

weak acids 552
anodes 437
cell diagrams 596
electrochemical cells 598
electrolysis 618
electrolytic cells 603
hydrogen fuel cell 602
lithium-ion batteries 600
RED CAT mnemonic 595
antacids 555
anthocyanins 318
antibiotics 127
aqueous solutions 13, 72, 465
electrolysis 616–18
area under a curve, graphs 375–8
aromatic compounds 163, 164, 261
Arrhenius acids 538, 539
Arrhenius bases 538, 539, 560
Arrhenius equation 506–8
Arrhenius factor 506
Arrhenius, Svante 538
Arrhenius theory 552
Āruni, Uddālaka 5
aspirin 472
asymmetric centre 286
atmosphere, Earth's 268, 429
atom economy 475–6
atomic number 24–6
atomic orbitals 44–54, 176–8
Aufbau principle 50, 53–4
d orbitals 47, 251, 252
degenerate orbitals 49, 50
f orbitals 47
Hund's rule 50
orbital diagrams 48–54
p orbitals 45, 47
Pauli exclusion principle 48
s orbitals 44–7
atomic radii, periodicity 232–3
atomic theory 3, 5–6, 46
atomic volume, periodicity 237
atomization, enthalpy of 418
atoms/atomic structure 3, 5–6, 20–6
Bohr model 40–2, 43
diameter of atoms 22
electron configurations 34–62
empirical formula 68–71
excited state 41
ground state 41
ionization energy 54–60
isotopes 26–9
mole unit 63–4, 66, 72
"plum pudding" model 21
quantum mechanical model 43–6
relative atomic charge 23
relative atomic mass 23, 65, 66

Rutherford model 20–1, 22
Aufbau principle, atomic orbital filling 50, 53–4
aurora borealis (Northern Lights) 35
Avogadro's constant 64
Avogadro's law 78
axial bonds 169

back titration 469
backward reactions 514, 522
        see also reversible reactions
Ball, Alice 632
ball-and-stick models 341
balloon models, molecular geometry 169, 170
bar charts 365, 366
barium hydroxide 104
base dissociation constant 561–5
bases
        alkalis 541
        Arrhenius bases 538, 539, 560 ·
        Brønsted–Lowry bases 239, 539, 542, 554, 635
        buffer region 568
        buffer solutions 574–7
        conjugate bases 105, 541, 571
        half-equivalence point 570
        Lewis bases 241, 635–6
        oxide reactions 241–2
        parent bases 556
        pH curves 557–9, 568–71
        pH scale 543–4
        properties 537, 538, 550–2
        reactions 241–2, 552, 553–6
        strong bases 558
                pH curves 568–9
        theories 538–41
        weak bases
                dissociation constants 561–5
                pH curves 569–71
                reversible ionization 530
                titration 573
batteries
        lead–acid batteries 599
        lithium-ion batteries 112, 191, 439, 600
        primary (voltaic) cell 594, 598
        recycling 111
        secondary (rechargeable) cells 598
        voltaic pile 338
bent (V-shaped) geometry, molecules 128, 129
benzene
        1, 2-disubstituted benzene compounds 166
        carbon–carbon bonds in 165
        electron density 165
        electrophilic substitution 648–9
        hydrogenation 166
        isomers 166
        resonance energy 165–6

structure 162–7
beryllium
        electron configuration 56
        ionization energy 246
beta-lactam ring 127
bimolecular reactions 496
Binnig, Gerd 23
biochemical reactions 495
biodegradable materials/products 213, 215, 216
biofuels 427, 433–6, 477
        production by world region 436
        pros and cons of 435
biological carbon fixation 434
bioplastics 216
bipolar plate, fuel cells 601
blocks (periodic table) 229–31
blood, microplastics in 216
Bohr model, atomic structure 40–2, 43
boiling points
        and bonding type 198
        ethanal 150
        ethanol 150
        graphical model of 270
        group 1 metals 191
        halogens 141
        and homologous series 269, 270
        hydrides 145
        methanoic acid 150
        pentane 142
        period 3 metals 192
        propane 150
        straight-chain alkanes 268, 269
Boltzmann, Ludwig 443
bomb calorimeter 401
bond axis (internuclear axis) 175
bond dissociation energy 391
bond enthalpy 125, 293, 405–7
        average bond enthalpy 405–6
        carbon–halogen bonds 642
        definition 405
        positive enthalpy value 405
bonding
        axial bonds 169
        bond angle 128, 129
        bond-breaking and bond-forming 404–7
        bond length 125
        bond order 124–5
        bonding continuum 187, 198–200, 201
        bonding electrons 120
        coordination bonds 126, 251–2, 629, 635, 636, 637–9
        electron domains 127–30
        equatorial bonds 169
        metallic bonds 95, 96, 187, 191–3
        metallic–covalent bonding continuum 187

multiple bonds 130
pi bonds 174, 175–6
polarity 132
sigma bonds 174–5, 176
transition elements 193–4, 251–2
triangular bonding diagrams 198, 201–5
valence shell electron pair repulsion model 127–9, 131
see also covalent bonds; ionic bonds
bonding triangles (diagrams) 198, 201–5
Born–Haber cycles 418–21
boron
electron configuration 56
ionization energy 246
mass spectrum 30
orbital diagram 50
boron trifluoride 632
Boyle, Robert 81, 87, 539
Boyle's law 81, 82, 87
branched-chains alkanes 271
brass 207
brittleness, and bonding type 198, 199
bromine
electrophilic addition 644
reversible reactions 512–14
bromopentane 279
Brønsted, Johannes 538
Brønsted–Lowry acids 239, 539, 546–7
Brønsted–Lowry bases 239, 539, 542, 554, 635
Brønsted–Lowry theory 540, 552
bronze 207
buckminsterfullerenes (buckyballs) 135, 136
buffer action 574
buffer region 568
buffer solutions 574–7
pH of 576–7
burettes 314, 335
measurement uncertainty/error 352, 383
butane 273, 274
boiling and melting points 268
chain isomers 279
empirical and molecular formula 68
enthalpy change of formation 413
molecular formula 263
structural formulas 263
butanoic acid 294, 526

calcium 52, 53
calcium carbonate 104, 469, 473
calcium fluoride 109
calcium hydroxide 550
calcium nitride 106
calcium oxide 105
calculator skills 533
calibration curves 76–8, 339
calorimeters 391, 395, 401

calorimetry 333–4, 399, 401
car batteries 599
carbocations 631, 641, 642, 647
carbon
alkyl groups 260
allotropes of 135–6
biological fixation 434
carbon–carbon bonds 124, 125, 165
catenation 257
chirality 286, 287, 288
hybrid orbitals 177–8
hybridization 176–8
and steel hardness 207
see also hydrocarbons
carbon dioxide 5
atmospheric concentration 429, 430–1
food industry use of 477
Lewis formula 121
molecular polarity 133
seawater concentration 526
carbon–halogen bonds 642
carbon monoxide 5, 428
carbon nanotubes 135, 136
carbonate ion 104, 122–3
carbonates
acid reactions 554
aqueous solubility 114
carbonic acid 242, 526
carbonyl group 261
naming of compounds in 276–7
oxidation 604
reduction 607
carboxyl group 261, 265
functional group isomers 283
oxidation 604
carboxylic acids 261, 265
boiling points 269
naming 277–8
oxidation 604–6
reduction 607
suffix 277
catalysts 247, 494–5
and clean energy 434
equilibrium position 523
Haber process 524
hydrogen fuel cell 602
catalytic converters 194, 247–8
catenation 257
cathodes 437
cell diagrams 596
electrolysis 618
electrolytic cells 603
hydrogen fuel cell 602
lithium-ion batteries 600
RED CAT mnemonic 595

cations 29, 97, 98, 99, 101, 102
    coordination bonds 251
    electrical conductivity 111
    primary (voltaic) cell 596
    radii of 233–4
    solubility 112, 113, 114
    see also carbocations
cell diagrams 596
cell phones 342, 598
cellulose 211, 212
Celsius temperature scale 17
CERN 90
chain isomers 279
changes of state 13–14, 15
charge
    effective nuclear charge 233
    formal charge 172–4
    ionic charge 109
    ions 98–100
    notation of 102
    nuclear charge 234
charts 365–6
    see also graphs; tables
chaulmoogra oil 632
chemical bonds see bonding
chemical compounds 6
    oxidation states, deduction of 243–5
chemical equations see equations
chemical equilibrium see equilibrium
chemical formulas see formulas
chemical industry, reaction yields 474
chemical kinetics see kinetics
chemical reactions see reactions
chemical shift 297, 298, 300
chemical symbols see symbols
chemical weapons 525
chemiluminescent reactions 509
chirality 286, 287, 288
chloride ion 98
chlorides
    aqueous solubility 114
    lattice enthalpy 110
chlorine
    homolytic fission 624, 625
    isotopes 26, 27
    redox reactions 582
chloroethane 297, 630
chloromethane 627
chromate–dichromate equilibrium 520
chromatograms 156, 157, 158
chromatography 155–60
    classification of 156
    experimental technique/apparatus 331–3
    gas–liquid chromatography 155
    liquid column chromatography 155

    locating agents 157
    mobile phase 155
    paper chromatography 10, 11, 155, 156, 331
    retardation factor 157–8
    stationary phase 155
    thin layer chromatography 157, 158–9, 331
chromium
    discovery of 193
    electron configuration 54, 101
    in oil paints 608
    rust prevention 206
    variable oxidation states 249, 250
cis–trans isomers 284–6
citations, academic practice 311
climate change 431, 434
    see also global warming
closed systems 387, 388
cobalt
    electron configuration 101
    mass spectrum 31
coefficient of determination 379
collision theory 487–8
colorimeters 342
colorimetric analysis 253
colorimetry 77, 339–40
colour wheel 251–2, 340
colourless substances, chromatography 157
column chromatography 155
combined gas law 88–9
combustion 424–8
    alcohols 427
    complete combustion 425–7
    enthalpies of 334
    greenhouse gas emissions 429–31
    incomplete combustion 428
    metals 424–5
    non-metals 425
    phlogiston theory 426
    primary alcohols 398
    standard enthalpy change 398, 412–17
complementary colours 340
complex ions 248, 251, 637–9
complexes, transition elements 251–3, 637
compounds 6
    oxidation states, deduction of 243–5
concentrated solutions 73
concentration/s 72–8
    calibration curves 76–8
    colorimetric analysis 253
    colorimetry 77
    equilibrium position 519–20
    mass concentration 74
    molar concentration 73–5
    problem-solving 465
    reaction rate 488, 491

spectrophotometry 76
standard solutions, measurement uncertainty 77–8
titration technique/apparatus 334–6
concentration–time curve 504–5
concordant values, measurement 312
condensation 15, 443
condensation polymers 220–4
condensation between two different monomers 221
condensation of the same monomer 222–3
hydrolysis 224
condensed structural formula 258
condensers 324
configurational isomers 284
conformational isomers 284
conjugate acid–base pairs 541, 564–5
conjugate acids 571
conjugate bases 105, 571
conservation of energy, law of 404
conservation of mass, law of 470
constitutional isomers 279
continuous spectrum 35
coordination bonds 126, 251–2, 629, 635
and complex ions 637–9
Lewis acid and base reactions 636
copper
ancient artefact 186
antimicrobial properties 207
electron configuration 54, 101
electroplating 619
mass spectrum 30
variable oxidation states 249, 250
copper(II) sulfate
crystals 96
electrolysis 618
relative molecular mass 65, 66
Cornell method, note-taking 490
correlation, graphs 371
corrosion 198, 199
cotton 212
covalent bonds 95, 96, 117–34
bond angle 128, 129
bond enthalpy 125
bond length 125
bond order 124–5
coordination bonds 126, 636
electron domains 127–30
electronegativity 119
formation of 118
Lewis formulas 120–3
multiple bonds 130
polarity 132
polarization 632
valence shell electron pair repulsion model 127–9, 131
covalent network structures 117, 135–8
covalent substances 149–54

electrical conductivity 151
intermolecular forces, relative strength of 149–50
properties 151–4, 198
solubility 152
volatility 151
crosslinking 213
crystallization 9, 11, 327
crystals 96, 104
current, electric 603
cycloalkanes, cis–trans isomerism 285

d atomic orbitals 47, 251, 252
d-block elements 230
Dalton, John 5
Daniell cell 596
data collection see measurement
databases 345–6
decimal places, measurement 356–7
deductive reasoning 38
degenerate atomic orbitals 49, 50, 251
delocalization 162, 187, 188, 194
in benzene molecule 164
and hybridization 180
Democritus 5
density, gases 468
deposition 14, 15
detergents 134, 152–3
diamond 135–6
diamorphine (heroin) 153
digital balance 313
digital pH meters 543
digital sensors 342
digitalin 260
dilute solutions 73
dimensional analysis 22
dipole–dipole forces 143, 147, 148, 149
dipole-induced forces 142, 147, 148
dipole moment 132, 133
dipoles, temporary 139, 140
diprotic acids 547, 550
direct-methanol fuel cell (DMFC) 438–9, 601, 602
direct proportionality, graphs 371
discharge 599
disposal of chemicals, safe practice 309
disproportionation 581
dissociation
ionization comparison 540
of water 545
dissociation constant, acid 561–5
dissolution (solvation) 11, 112
distillation 10, 11
experimental technique/apparatus 271, 324–5
fractional distillation 271, 325, 329
oxidation of primary alcohols 605
simple distillation 324

DMFC see direct-methanol fuel cell
DNA (deoxyribonucleic acid), hydrogen bonds 147
double bonds, molecules 124
doublets 300
drugs/medicines
    antacids 555
    aspirin 472
    chaulmoogra oil 632
    clinical trials 555
    diamorphine (heroin) 153
    ibuprofen 278
    impurities 472, 474
    penicillamine 286, 287
dry ice 13, 17
drying samples to a constant mass 326
ductility, metals 189, 190
dynamic equilibrium 512–16, 519–25

Earth
    atmosphere
        carbon dioxide concentration 429, 430–1
        composition of 266
        ozone layer 161
    magnetic field 35
    oceans, carbon dioxide concentration 526
effective nuclear charge 233
Einstein, Albert 43
elasticity 199
electric current 319, 603
electric potential difference 320
electrical conductivity 137
    and bonding type 198
    covalent substances 151
    ionic compounds 111
    measurement of 319
electrical conductors 190
electrical resistance, metals 189
electrochemical cells 336–8, 594–602
electrode potential, standard 610
electrodes 437, 601
    see also anodes; cathodes
electrolysis 603–4
    aqueous solutions 616–18
    copper sulfate 618
    sodium chloride, molten 603–4, 617
    water 438
electrolyte 601, 603
electrolytic cells 336–7, 594, 603–4
electromagnetic (EM) radiation 34, 35, 38, 39
electron affinity 235–6, 418
electron configurations 34–62
    for beryllium atom 56
    for boron atom 56
    for calcium atom 52, 53
    for chromium atom 54

condensed electron configurations 52–3
    for copper atom 54
    emission spectra 34–40
    full electron configurations 52
    isoelectronic 98
    for nitrogen atom 56–7
    noble gas configuration 98, 99
    orbital diagrams 48–54
    transition elements 101, 248
electron deficient molecules 123
electron domain geometry 169–71, 179–80
electron domains, covalent bonds 127–30
electron microscope 23
electron-pair sharing reactions 628–51
    coordination bonds and complex ions 637–9
    electrophiles 632
    electrophilic addition 632–4
    Lewis acids and bases 635–6
    nucleophiles 629
    nucleophilic substitution 630
electron sharing reactions 622–7
electron shielding 233
electron transfer reactions 580–621
    electrochemical cells 594–602
    electrolysis of aqueous solutions 616–18
    electrolytic cells 603–4
    electroplating 618–19
    Gibbs energy and standard cell potential 614
    half-equations 586–8
    oxidation 580–5
        metals 589–90
        organic compounds 604–7
    redox reactions of acids and metals 591
    reduction 580–5
        alkenes and alkynes 608–9
        halogens 589
        organic compounds 607–8
    secondary cells 598–602
    standard cell potential 611
    standard electrode potential 610
electronegativity 187, 201
    covalent bonds 119
    halogens 639
    ionic bonds 102–3, 104
    periodicity 236
electrons 20, 21, 24–5
    atomic orbitals 44–54, 176–8
    bonding electrons 120
    delocalization 164, 180, 187, 188, 194
    electron affinity 235–6, 418
    electron energy 40, 41
    energy levels 37, 40–2, 44, 47, 50–2
    inner core electrons 52
    ionization energy 54–60
    relative mass and charge 23

spin 48–50, 174
valence electrons 52
wave–particle duality 43
see also electron…
electrophiles 629, 632
electrophilic addition 632–4, 644–7
carbocations in 647
halogens 644–5
hydrogen halides 645
water 646
electrophilic substitution, benzene 648–9
electroplating 618–19
electrostatic attraction 102, 187
elementary charge 23
elementary steps, reaction mechanism 496
elements 5, 6, 24
EM see electromagnetic radiation
emission spectra 34–40
flame tests 37
ionization energy calculations 57–8
observation of 36, 37
empirical formula 68–71, 257, 296
enantiomers 286, 287, 289
end point, titration 571
endothermic reactions 14, 15, 109, 391, 392, 404, 442
energy profile 492
equilibrium position 522
energy
of electromagnetic radiation 39
and matter 3–4
energy cycles 404–23
energy density 439, 440
energy distribution curves 493–4
energy levels, electrons 37, 40–2, 44, 48, 50–2
energy profiles 391–3, 492
energy transfers 387–90, 391
enrichment, uranium 28
enthalpy of atomization 418
enthalpy change 387–403
activation energy 391
bond enthalpy 125, 406
Born–Haber cycles 418–21
combustion 334, 398, 412–17
definition 390
and entropy 447
Hess's law 411
hydrogenation 166
standard enthalpies 394–401, 412–17
enthalpy of combustion 334, 398, 412–17
enthalpy of hydrogenation 166
entropy 389, 442–7
and enthalpy change 447
and physical changes 443–4
entropy change 443–6
calculation of 445–6

Gibbs energy change 448, 452
standard entropy change 445, 446
standard entropy values 445, 446
environmental issues
acid rain 242–3, 554
air pollution 242, 433
carbon dioxide concentrations, atmospheric and oceanic 429, 430–1, 526
global warming 427, 430
plastic waste 215–16
see also green chemistry
enzymes 495
equations 26, 461, 462
acid–base equations 555–6
Arrhenius equation 506–8
half-equations 586–8
Henderson–Hasselbalch equation 576
ideal gas equation 90–1
ionic equations 553
net ionic equation 553
rate equations 500–6
Schrödinger equation 43
total ionic equation 553
equatorial bonds 169
equilibrium 442, 514–16
acid–base equilibria, in salt solutions 565–7
chemical equilibrium 514–16, 523
chromate–dichromate equilibrium 520
dynamic equilibrium 512–16, 519–25
effect of catalysts on 523
effect of concentration on 519–20
effect of pressure on 521
effect of temperature on 522–3
Gibbs energy change 454–6, 531–2
Haber process 524–5
heterogeneous equilibrium 514
Le Châtelier's principle 519–25
model of 523
position of the equilibrium 519–24
equilibrium constant 516–18
calculations 526–30
effect of reaction conditions on 524
and Gibbs energy change 454–6
reversible ionization 530
water 545
equilibrium law 516–18
equilibrium sign 512
equivalence point 335
eroding 618
error bars, graphs 369, 370
errors
and graphs 385
and processed results 383–4
random errors 354, 383, 385
systematic errors 383, 385, 395, 399

ester group 261, 283
esterification reaction 220, 526
esters 220–1, 261, 266
ethanal 150
ethane 262
    boiling and melting points 268
    molecular formula 263
    molecular polarity 133
    structural formulas 263
ethanoate anion 180
ethanoic acid (acetic acid) 526, 549
ethanol
    aqueous solubility 152
    biofuels 427, 435
    boiling point 150
    equilibrium constants 526
    intermolecular forces 150
    structural formula 152
ethene 262
    addition polymerization 217–19
ethers 261, 265
ethical issues 28
ethyl ethanoate 220
evaporation 9, 11, 328
Excel spreadsheets 343
excess reactant 395, 463
exchangeable hydrogen atoms 546–7
excited state, atoms 41
exothermic reactions 14, 15, 101, 391, 392, 404, 442
    energy profile 492
    equilibrium position 522
expanded octets 120, 167–71
    Lewis formulas for 167–8
    molecular geometry 169–71
experimental yield 471
experiments 309–41
    academic integrity 311
    accident prevention 309
    empirical formula, determination of 71
    isolation techniques
        drying to a constant mass 326
        recrystallization 329
        separation of mixtures 326–9
    measurement of variables 311–20
    melting point data 8
    molar mass of a gas, determination of 86–7
    non-Newtonian fluids 14
    planning and risk assessments 11
    preparation techniques
        dilution of standard solutions 322–3
        distillation 324–5
        reflux 324
        standard solutions 321–2
    reaction rate 509
    repeatability of results 312
    reproducibility of results 313
    risk assessments 309
    safety 309–11
    variables, control of 89
    see also analytical techniques
extrapolation, graph data 380–1

f atomic orbitals 47
f-block elements 230
falsification 21, 426
femtosecond lasers 499
fermentation, glucose 435
fertilizers 393
filter paper, fluting of 327
filtrate 9, 326
filtration 9, 11, 326–7
first ionization energy, of hydrogen 57–8
flame tests 37
fluorine 589, 627
fluoroalkanes 642
formal charge 172–4
formation, standard enthalpy change of 413–16, 419, 420
formic acid 278
formula units 65
formulas
    condensed structural formula 258
    empirical formula 257, 296
    ionic compounds 104, 105–7, 108
    Lewis formulas 120–3, 167–8
    molecular formula 257
    organic compounds 257–9
    skeletal formula 163, 258
    stereochemical formula 287–8
    structural formula 258, 341
forward reactions 514, 522
    see also reversible reactions
fossil fuels 429–33
fractional distillation 271, 325, 329
fragmentation 290
free radical hypothesis of ageing 623
freezing 15
frequency factor, particle collisions 506
fuel cells 437–9, 594, 601–2
    direct-methanol fuel cell 438–9
    hydrogen fuel cell 437–8
fuels 424–41
    biofuels 427, 433–6, 477
    carbon-neutral fuels 413
    complete combustion 425–7
    energy density 440
    fossil fuels 429–33
    incomplete combustion 428
    specific energy 432, 440
    storage and transportation 440
fullerenes 136

functional group isomers 283
functional groups 260–78
    classes 261
    formulas 261
    homologous series 263–70
    suffixes 261
    see also specific functional groups
funnel, separating 329

gallium 231
Galvani, Luigi 594
galvanizing 206
gas–liquid chromatography 155
gases 13
    Boyle's law 81, 82, 87
    combined gas law 88–9
    density of 468
    ideal gas equation 90–1
    ideal gases 80–93
    molar mass 86–7
    molar volume of an ideal gas 85–7
    pressure–volume relationship 81–2, 83, 85, 87–91
    real gases versus ideal gases 82–5
    Van der Waals parameters 84
GDC see graphic-display calculator
giant covalent structures 117, 135–8
Gibbs energy change 447–58, 531–2
    calculation of 448–51
    entropy change 452
    and equilibrium 454–6, 531–2
    reversible reactions 454–5
    spontaneous reactions 448
    and standard cell potential 614
    temperature 455, 457–8
glassware, volumetric 321
global warming 427, 430
    see also climate change
glucose
    empirical and molecular formula 68
    fermentation 435
    molecular structure 257
gold
    electroplating 619
    nuclear symbol notation 25
    reactivity of 592
gradient
    graphs 373–4
    tangent line 484
graphene 135, 136
graphic-display calculator (GDC) skills 533
graphite 135, 136, 137
graphs
    area under a curve 375–8
    boiling points and homologous series 270
    coefficient of determination 379

concentration–time curve 504–5
correlation 371
direct proportionality 371
error bars 369, 370
and errors 385
extrapolation 380–1
gas laws 88
Gibbs energy change/temperature relationship 457–8
gradient (slope) 373–4
intercepts 374–5
interpolation 380–1
interpretation of 371–2
inverse proportionality 372
line or curve of best fit 369–70
logarithmic scale 367
maximum and minimum values 375–8
non-linear 372
outliers 367
plotting of 367
potential energy/distance between hydrogen atoms 125
proportionality 371–2
rate–concentration curve 503–4
reaction order 503–5
sketching of 363–4
tangent line 374
    see also charts; tables, quantitative data
gravimetric analysis 313, 473
gravity filtration 327
green chemistry 216, 310, 474–7
    atom economy 475–6
    catalysts 495
    cost of 476–7
    principles of 475
greenhouse effect 429, 430
greenhouse gases 429–31
ground state, atoms 41
group 1 metals, boiling and melting points 191
groups (periodic table) 54, 229–31

Haber, Fritz 393
Haber process 524–5
half-cells 595, 596, 612
half-equations 586–8
half-equivalence point 569–71
halide ions
    halogen reactions 239
    nucleophilic substitution 643
halides, melting points 200
halite 113
halogenate 624
halogenoalkanes 261, 263
    classification 281
    heterolytic fission 631
    homologous series 264

naming 275
nucleophilic substitution 630, 639–42
halogens
  boiling points 141
  electronegativity 639
  electrophilic addition 633, 644–5
  halide ion reactions 239
  non-metallic character of 238
  periodic table 230
  reduction 589
  substituents 275
hardness test, Vickers 207
hazard symbols, chemistry labs 309
heat 388, 390, 392, 394
  and temperature 389
  see also enthalpy...
heavy water 28
Heisenberg, Werner 43
helium
  emission spectrum 37
  Van der Waals parameters 84
Henderson–Hasselbalch equation 576
heroin (diamorphine) 153
Hess's law 408–11
  enthalpy change of combustion 414–7
  enthalpy change, determination of 411
  enthalpy change of formation 414–17
  enthalpy cycle diagram method 410, 416
  summation of equations method 409, 416
heterogeneous catalysts 247
heterogeneous composition 6
heterogeneous equilibrium 514
heterogeneous mixtures 488–9
heterolytic fission 631, 642
hexane 274
  boiling and melting points 268
  molecular formula 263
  structural formulas 263
  structural isomers 273
histograms 365, 366
Hofmann voltameter 337
homogeneous composition 6
homogeneous reactions 489
homologous series 141, 263–70
  physical trends in 268–70
homolytic fission 405, 624
Hund's rule, degenerate orbitals 50
hybrid orbitals 177–9
hybridization 176–80
  and delocalization 180
  and molecular geometry 179–80
hydrates 65
hydride anions 99
hydrides, boiling points 145
hydrocarbons
  combustion 412–13
  complete combustion 425–6
  fractional distillation 271
  saturated/unsaturated hydrocarbons 262
  steam reforming of 438
  substituents 551
  see also alkanes; alkenes
hydrochloric acid 548
hydrogels 214–15
hydrogen
  covalent bonding 118
  electron distribution 140
  electron energy 40, 41
  electron transitions 42
  emission spectrum 40–3, 57
  exchangeable hydrogen atoms 546–7
  first ionization energy 57–8
  ion formation 99
  isotopes of 26, 28
  nuclear spin 297
  "pop" test 591
  potential of hydrogen 543
  redox reactions 582
  specific energy 440
hydrogen bonding 143–6, 148, 149
hydrogen chloride 143, 293, 547, 548
hydrogen fluoride 132
hydrogen fuel cell 437–8, 601, 602
hydrogen halides 633, 645
hydrogen iodide 293
hydrogenation, of benzene 166
hydrogencarbonates, acid reactions 554
hydrolysis reactions 224
  amides 127
  direction and extent 565–7
  salt solutions 567
hydronium ion 540–1
  coordination bond 126
  Lewis formula 122–3
hydrophilic molecules 152–3
hydrophobic molecules 152–3
hydroxide ion
  formula 104
  Lewis formula 122–3
hydroxides
  aqueous solubility 114
  properties 550–1
  water reactions 241
hydroxonium 540
hydroxyl group 261
  cellulose 212
  functional group isomers 283
  morphine molecule 153
  nucleophilic substitution 643
  oxidation 604, 607

hypotheses 499

ibuprofen 278
ice
    changes of state 13, 15
    hydrogen bonding 145–6
    molecular structure 341
ideal gases 80–93, 139
    assumptions of ideal gas model 80–2
    combined gas law 88–9
    ideal gas equation 90–1
    molar mass 86–7
    molar volume 85–7
    and real gases 82–5
immiscible liquids 152
impurities, drug production 472, 474
indicators, acid–base 318–19, 571–3
indigenous peoples 431
inductive effect 642
inductive reasoning 38
infrared (IR) spectroscopy 293–6
initial reaction rate 484, 485, 486
initiation 625
inner core electrons 52
inquiry process 665
instantaneous dipoles 139, 140
instantaneous reaction rate 484–5, 486
instruments, measurement uncertainty 351–2
integers 70
integration traces 297, 298
intensive properties, specific heat capacity 394
intercepts, graphs 374–5
intermediate compounds 494
intermolecular forces 138–50
    dipole–dipole forces 143, 147, 148, 149
    dipole-induced forces 142, 147, 148
    hydrogen bonding 143–6, 148, 149
    London (dispersion) forces 136, 139–42, 147, 148, 149, 269, 432
internal assessment 668
internuclear axis (bond axis) 175
interpolation, graph data 380–1
ionic bonds 26, 95, 96, 102–7
    electronegativity 102–3, 104
    lattice enthalpy 109–10
    lattice structure 108
    non-directionality 108
    periodic table position 103–4
    polyatomic ions 104
ionic charge 109
ionic compounds 95, 96
    dissolution of 112
    electrical conductivity 111
    formulas 104, 105–7, 108
    lattice enthalpy 109–10

    lattice structure 108
    naming 104–5
    periodicity 103–4
    properties 110–14, 154, 198
    redox reactions 98
    solubility 112–14
    standard enthalpy of formation 419, 420
    volatility 111
ionic–covalent bonding continuum 201
ionic equations 553
ionic lattices 108
ionic product of water 545–6
ionic radii 109, 233–4
ionic salts, solubility of 114
ionization
    of ammonia 540
    dissociation comparison 540
ionization energy 54–60, 101, 418
    calculation of from spectral data 57–8
    data collection 60
    discontinuities in 246
    first ionization energy 57–8
    periodicity 234–5
    successive ionization energies 58–9, 60
ions 95, 96–102
    alloys 205
    charge 98–100
    complex ions 248, 251, 637–9
    hydrolysis of salt solutions 567
    hydronium ion 540–1
    polyatomic ions 104, 122
    spectator ions 553
    see also anions; cations
IR spectrum 294
    see also infrared spectroscopy
iron 206
    electron configuration 101
    ion formation 100
    ionization energies 101
    oxidation 581
iron disulfide ("fool's gold") 137
isoelectronic 98, 234
isolated systems 387
isomers
    benzene 166
    cis–trans isomers 284–6
    configurational isomers 284
    conformational isomers 284
    functional group isomers 283
    optical isomers 286–90
    stereoisomers 284–90
    structural isomers 273, 279–83
isotope labelling 28
isotopes 26–9
    relative atomic mass 30, 31

IUPAC nomenclature 271

journals, scientific 630

Kekulé, Friedrich August von 164
Kelvin temperature scale 15–17
ketones 261, 264
    functional group isomers 283
    homologous series 265
    oxidation 604
    reduction 607, 608
    suffix 277
Kevlar 211
kilogram, standard measurement 16
kinetic energy 189, 389, 487, 493–4
kinetics 480–511
    activation energy 492–4
    Arrhenius equation 506–8
    catalysts 494–5
    collision theory 487–8
    femtochemistry 499
    Maxwell–Boltzmann energy distribution curves 493–4
    multistep reactions 496–9
    rate equations 500–6
    reaction rate 480–6, 488–91, 509
Kwolek, Stephanie 211

laboratory work see analytical techniques; experiments
lattice enthalpy 109–10, 204, 418, 420
lattice structure 108
Lavoisier, Antoine 470
law of conservation of energy 404
law of conservation of mass 470
LDFs see London (dispersion) forces
Le Châtelier's principle 519–25
lead–acid batteries 599
lead chromate 608
lead(II) chromate 113
leaving group 630
length, measurement of 316
leprosy 632
Lewis acids 241, 635–6
    Lewis base reactions 636–7
Lewis bases 635–6
    Lewis acid reactions 636–7
Lewis formulas 120–3, 167–8
Lewis theory 552
ligands 126, 251–2, 637
limiting reactant 395, 428, 463–8
line or curve of best fit, graphs 369–70
line graphs 365, 366
linear geometry, molecules 128, 130, 169
lipids 153
liquefied petroleum gas (LPG) 426
liquid column chromatography 155

liquids 10, 13, 152
lithium 191
lithium-ion batteries 112, 439, 600
litmus 538, 539
locants 272
logarithmic scale, graphs 367
Lomonosov, Mikhail 470
London (dispersion) forces (LDFs) 136, 139–42, 147, 148, 149, 269, 432
lone pairs 127
Lowry, Martin 538
LPG see liquefied petroleum gas
lysozyme, 3D structure of 223

m/z ratio 29, 30, 31
macromolecules 209
magnesium 6
    boiling and melting points 192
    nuclear symbol notation 25
magnesium hydroxide 555
magnesium iodide 204
magnesium oxide 26
magnesium sulfide 6
magnetic field, Earth's 35
magnetite 69
malleability, metals 189, 190
manganese 101
Markovnikov's rule 647
mass
    conservation of mass 470
    drying samples to a constant mass 326
    measurement of 72, 313
    mole unit 63–4, 66
    molecular mass 141
    subatomic particles 23
mass concentration 74
mass percentage 69
mass spectra 29–32, 290–2
mass spectrometry 29–32, 290–2
materials science 197–227
    alloys 205–9
    aluminium chloride 204
    biodegradable materials 213, 215, 216
    bonding continuum 198–200
    brittleness 199
    corrosion 199
    elasticity 199
    green chemistry 216
    hydrogels 214–15
    magnesium iodide 204
    plasticity 199
    polymers 209–24
    product life cycle 208
    silicon 203
mathematics 350–85

experimental error, sources of 383–5
graphs and tables 363–82
SI units 350
uncertainties 351–62
matter
    changes of state 13–14, 15
    characteristics 4
    composition 3–12
    and energy 3–4
    observations of 37
    pure substances and mixtures 6–12
    states of matter 13–15
Maxwell–Boltzmann energy distribution curves 493–4
mean values, measurement 362
measurement
    accuracy 311–12, 467
    concordant values 312
    decimal places 356–7
    decimal prefixes 64
    electric current 319
    electric potential difference 320
    electrical conductivity 319
    length 316
    mass 72, 313
    mean values 362
    pH of solution 318–19
    precision 311–12, 467
    reaction rate 482
    reactions 395
    reliability 312–13
    SI units 15, 16, 66, 355
    significant figures 356–7
    standard solutions 77–8
    temperature 316–17, 351, 352
    time 315
    uncertainties 351–62
        error bars 369, 370
        expression of 355–6
        human reaction time 353
        instrument uncertainty 351–2
        mean values 362
        propagation of 358–61
        random errors 354
        reaction mixtures 353
        value fluctuation 353
    validity 312–13
    volume 314–15
Meitner, Lise 28
melting 15
melting points
    and bonding type 198
    determination of 8, 330
    group 1 metals 191
    period 3 metals 192
    potassium halides 200

    silver halides 200
    straight-chain alkanes 268
memory metals 206
Mendeleev, Dmitri 231
meniscus, and measurement 314, 315, 383
metal oxides 241–3
metallic bonds 95, 96, 187
    non-directionality 189
    strength of 191–3
    transition elements 193–4
metallic–covalent bonding continuum 187, 201
metallic structures 186–91
    ductility 189, 190
    electrical conductivity 190
    electrical resistance 189
    malleability 189, 190
    properties 188–90, 198
    superconductors 189
    thermal conductivity 189, 190
metalloids, periodic table 230
metals
    boiling and melting points; group 1 metals 191
    combustion 424–5
    electrical resistance 189
    flame tests 37
    oxidation 589–90
    periodicity 229, 230, 238–40
    product life cycle 208
    redox reactions 591
    standard electrode potentials 610
    see also alkali metals; transition elements
methane
    boiling and melting points 268
    bond angle 129
    molecular formula 263
    radical substitution 624, 625, 626
    structural formulas 263
    tetrahedral geometry 177
    Van der Waals parameters 84
methanoic acid 150, 278
methanol 276, 438–9, 526
methylpropane 279, 526
microbeads 216
microplastics 216
Milley–Urey experiment 268
mirrors 188
miscible liquids 10
mitochondria 623
mixtures 6–12
    alloys 205
    heterogeneous mixtures 488–9
    separation techniques 9–12, 326–9
mnemonics
    OIL RIG 582
    RED CAT 595

models/modelling 22, 46, 81, 347–8
    see also molecular models
molar concentration (molarity) 73–5
molar mass 66–7, 86–7, 463
molar volume of an ideal gas 85–7
mole ratio 68–9, 461–2, 468, 477, 481
mole unit 63–4, 66, 72
molecular formula 68, 70, 257
molecular geometry 127–31
    expanded octets 169–71
    and hybridization 179–80
molecular ion peak 291
molecular models 283–4, 303, 341, 348–9, 626
molecules
    3D representation of 129, 259, 260
    electron deficient molecules 123
    intermolecular forces 138–50
    Lewis formulas 120–3
    mass 141
    molar mass 66–7
    mole unit 63–4, 66, 72
    orbital theory 176
    polarity 133–4
    polarizability of 141
    relative molecular mass 65, 66
    resonance structures 160–7
    temporary dipoles 139, 140
Molina, Mario 161
monomers 209
monoprotic acids 547
monosaccharides 224
morphine 153
multiple bonds 130
multiplicity 300
multistep reactions 496–9

naming
    of alcohols 276
    of alkanes 272
    of alkenes 274
    of amines 282
    of carbonyl group compounds 276–7
    of carboxylic acids 277
    of halogenoalkanes 275
    of ionic compounds 104–5
    IUPAC nomenclature 271
    of organic compounds 271–8
    of oxyanions 245
    in periodic table 245
    symbols 245
nanotechnology 136
natural abundance, isotopes 26, 27
natural compounds 260
natural polymers 209, 212
net ionic equation 553

neutralization reactions 241, 468, 537
neutrons 20, 21, 23, 24–5
nickel
    electron configuration 101
    mass spectrum 31
nicotine 287
nitrate ion 104
nitrates, aqueous solubility 114
nitrogen 95
    electron configuration 56–7
    Haber process 393
    ionization energy 246
nitrogen dioxide 162
nitrogen oxides 242
nitrogen trichloride 121, 172
NMR see nuclear magnetic resonance spectroscopy
noble gases
    electron configurations 98, 99
    periodic table 230
non-metal oxides 241–3
non-metals
    combustion 425
    periodic table 229
non-Newtonian fluids 14
non-polar solvents 112
non-spontaneous reactions 442
Northern Lights (aurora borealis) 35
note-taking method 490
nuclear charge 234
nuclear fission 28
nuclear magnetic resonance (NMR) spectroscopy 296–302
nuclear reactors 28
nuclear spin 297
nuclear spin quantum number 296
nuclear symbol notation 24–6
nucleophiles 629, 637
nucleophilic substitution 630
    in halogenoalkanes 639–42
    reaction rate 642
nucleus, atomic 20–6
nylon 220, 221

octahedral geometry, molecules 170, 171
octet rule 99, 120, 123
oil industry 269
oil paints 608
OIL RIG mnemonic 582
open systems 387, 388
opiates 153
optical activity 286
optical isomers 286–90
optometry, redox reactions 583
orbital diagrams 48–54
organic chemistry 257
organic compounds

3D models of 259, 260
classification 260, 281–2
complete combustion 425–7
formulas 257–9
functional groups 260–78
incomplete combustion 428
naming 271–8
oxidation 604–7
reduction 607–8
organic reaction mechanisms 622
organic synthesis 648
outliers, graph data 367
overall reaction order 500
oxidation 98, 580–5
definitions 581–4
electrolytic cells 603–4
electron transfer 582
half-equations 586–8
hydrogen loss/gain 582
metals 424, 589–90
organic compounds 604–7
oxidation state change 583–5
oxygen gain/loss 581
reactivity series 590, 592
oxidation states (oxidation number) 98, 101, 243–5, 583–5
acids 549
transition elements 248–50
variable oxidation states 248–50
oxidizing agents 585, 589, 590
oxoacids 547, 549
oxyanions 245
oxygen
bonding 160
covalent bonding 118
empirical and molecular formula 68
hybridization 179
ionization energy 246
oxygen–oxygen bonds 160, 161
ozone 160, 161
redox reactions 581
ozone (diatomic oxygen) 160, 161

p atomic orbitals 45, 47
p-block elements 230
painkillers 153, 278
paper chromatography 10, 11, 155, 156, 331
parent acids 556
parent bases 556
pascal unit 81
Pascal's triangle 301
Pauli exclusion principle 48
Pauling scale 102
peak ratio 301
peer-review process 630
PEM see proton exchange membrane

penicillamine 286, 287
pentane 273
boiling and melting points 142, 268
molecular formula 263
space-filling molecular model 142
structural formulas 263
peptide bonds 222
percentage composition 69
percentage uncertainty 355
percentage yield 470
period 3 elements
acid–base properties 242
properties 187
period 3 metals, boiling and melting points 192
periodic table/periodicity 229–55, 698
acids 549
alkali metals 230
alternative representations of 240
atomic radii 232–3
atomic volume 237
blocks 229–31
electron affinity 235–6
electronegativity 236
groups 229–31
and ionic charge 99
ionic compounds 103–4
ionic radii 233–4
ionization energy 54–7, 234–5
discontinuities in 246
Mendeleev's work on 231
metal oxides 241–3
metalloids 230
metals 229, 230, 238–40
naming conventions 245
non-metal oxides 241–3
non-metals 229
oxidation states 243–5
periods 229–31
transition elements
complexes 251–3
properties 247–8
variable oxidation states 248–50
see also group 1 metals, boiling and melting points
periods (periodic table) 229–31
personal protective equipment (PPE) 309
PET (polyethene terephthalate) 209
petrochemical industry 269
pH
of buffer solutions 576–7
measurement of 318–19
pH curves 557–9
of strong acids and strong bases 568
of strong acids and weak bases 569–70
of weak acids and strong bases 568–9
of weak acids and weak bases 570–1

pH scale 543–4
pharmaceutical drugs see drugs/medicines
phases, reaction rate 488–9
phenyl group 261
phlogiston theory 426
phosphate ion 104
photochromic lenses 583
photons 37, 40, 43
pi bonds 174, 175–6
pie charts 365, 366
pipettes 314, 321
placebo effect 555
plane-polarized light, rotation of 289–90
planetary model, atomic structure 21
plant pigments, thin layer chromatography 159
plasticity 199
plastics 213
    bioplastics 216
    microplastics 216
    pollution issue 215–16
plating 618
platinum 247, 248
"plum pudding" model, atomic structure 21
pOH scale 560–1
polar covalent bonds 132
polar solvents 112
polarity
    bond polarity 132
    molecular polarity 133–4
polarization 595, 598, 632
polarized light, rotation of 289–90
polyamides 221
polyatomic ions 104, 122
polyatomic molecules 294
poly(chloroethene) 218
polyester 221
polyethene 211
polyethene terephthalate (PET) 209
poly(isoprene) (natural rubber) 211
polymerization 209
polymers 209–24
    addition polymers 217–19
    condensation polymers 220–4
    and the environment 215–16
    examples of 211
    natural polymers 209, 212
    properties 212–13
    repeating units 209–11
    synthetic polymers 209, 212–13
polypeptides 222
polypropene 210, 218
polyprotic acids 552
polysaccharides 224
polystyrene 213
polyvinyl chloride (PVC) 218

"pop" test, hydrogen 591
positional isomers 279
positive inductive effect 642
potassium
    boiling and melting points 191
    electron orbital filling diagram 51
potassium bromide 420
potassium fluoride 104, 109
potassium halides, melting points 200
potassium permanganate 73, 76
potential difference, measurement of 320
potential of hydrogen 543
PPE see personal protective equipment
precipitate 114
precipitation reactions
    aspirin production 472
    gravimetric analysis 473
precision of measurements 311–12, 467
prefixes, decimal 64
pressure
    equilibrium position 521
    pascal unit 81
    reaction rate 488
pressure–volume relationships, gases 81–2, 83, 85, 87–91
primary alcohols
    aqueous solubility 152
    combustion 398
    oxidation 604–6
    reduction 607
primary amines 261, 266
primary compounds 281–2
primary (voltaic) cell 338, 437, 596
    batteries 594, 598
principal quantum number 40, 47, 230
product life cycle 208, 477
propagation 622, 625
propanal 258
propane
    boiling and melting points 150, 268
    intermolecular forces 150
    molecular formula 263
    structural formulas 258, 263
propanoic acid 526
propanone 258
propene 258
proteins 222
proton acceptors 539
proton donors 539
proton exchange membrane (PEM) 437, 601
proton nuclear magnetic resonance (^{1}H NMR) spectroscopy 296–302
    high-resolution ^{1}H NMR 300–1
    low-resolution ^{1}H NMR 297–9
proton transfer reactions 537–79
    acid–base equilibria in salt solutions 565–7

acid–base indicators 571–3
acids and bases
    properties 546–52
    reactions 553–6
    theories 538–41
    weak acids and bases 561–5
amphiprotic and amphoteric species 542
buffer solutions 574–7
conjugate acid–base pairs 541
ionic product of water 545–6
pH of buffer solutions 576–7
pH curves 557–9, 567–71
pOH scale 560–1
protons 20, 21, 23, 24–5
public understanding of science 440
pure covalent bonds 132
pure substances 6, 7
PVC (polyvinyl chloride) 218

quantization 40, 43
quantum mechanical atomic model 43–6
quantum numbers 40
quartz 137

racemic mixtures 289
radical substitution reactions 624–6
radicals 622–7
    formation 623–4
Raman scattering effect 298
random errors 354, 383, 385
rate–concentration curve 503–4
rate constant 500
rate-determining step 497
rate equations 500–6
rate of reaction 480–6
    average rate of reaction 481
    concentration effect 488, 491
    definition 480
    experiments 509
    factors affecting 488–91
    heterogeneous mixtures 488–9
    homogeneous reactions 489
    initial reaction rate 484, 485
    instantaneous reaction rate 484–5, 486
    measurement 482
    overall rate of reaction 481, 483
    phases 488–9
    pressure effect 488
    surface area of reactants 489
    temperature effect 489
    units 486
reaction coordinate 492
reaction kinetics see kinetics
reaction mechanism 496, 497
reaction order 500

graphical representations of 503–5
reaction quotient 525–6
reaction systems 387
reaction yield 470–4
reactions 4
    backward reactions 514, 522
    bimolecular reactions 496
    bond-breaking and bond-forming 404–7
    energy cycles 404–23
    energy profiles 391–3
    energy transfers 387–90
    enthalpy change 387–403
    entropy 389, 442–7
    excess reactant 395, 463
    experimental yield 471
    extent of a reaction 464
    forward reactions 514, 522
    Gibbs energy change 447–58
    Hess's law 408–11
    limiting reactant 395, 463–8
    metal oxides 241–2
    multistep reactions 496–9
    non-metal oxides 241–2
    non-spontaneous reactions 442
    percentage yield 470
    reversible reactions 512–16, 520, 530
    spontaneous reactions 442
        entropy change 443
        Gibbs energy change 448
        temperature effect 453
    standard enthalpy change 394–401
    termolecular reactions 496–7
    theoretical yield 463, 471
    unimolecular reactions 496
    water 239
    see also electron sharing reactions; electron transfer reactions; electron-pair sharing reactions; endothermic reactions; exothermic reactions; kinetics; rate of reaction
reactivity series 590, 591, 592, 610
real gases 82–5, 139
reasoning, types of 38
rechargeable cells 437, 594, 598–602
recrystallization 329
recycling
    batteries 111
    metals 190, 208
    plastics 215
RED CAT mnemonic 595
redox reactions 424, 580–5
    acids 591
    electron transfer 594–7
    half-equations 586–8
    ionic compounds 98
    metals 591
    in optometry 583
    oxidizing agents 585, 589, 590

reducing agents 585, 589, 590, 608
redox titration 335, 592–3
reducing agents 585, 589, 590, 608
reduction 98, 580–5
    alkenes 608–9
    alkynes 608–9
    definitions 581–4
    electrolytic cells 603–4
    electron transfer 582
    half-equations 586–8
    halogens 589
    hydrogen loss/gain 582
    organic compounds 607–8
    oxidation state change 583–5
    oxygen gain/loss 581
reference plane 285
referencing style, academic practice 311
reflux 324, 605
relative abundance of isotopes 29
relative atomic charge 23
relative atomic mass 23, 26, 27, 30, 31, 65, 66
relative molecular mass 65, 66
relative (or fractional) uncertainty 356
reliability of measurements 312–13
renewable energy 433–6
repeatability, experimental results 312
repeating units, polymers 209–11
reproducibility, experimental results 313
residue 326
resonance energy, benzene 165–6
resonance structures 160–7
    benzene 162–7
    delocalization 162
retardation factor, chromatography 157–8
retrosynthesis 648
reverse osmosis, seawater 11
reversible reactions 530
    chromate–dichromate equilibrium 520
    dynamic equilibrium 512–16
    Gibbs energy change 454–5
RGB analyser, smartphones 342
risk assessments, experiments 12, 309
Rohrer, Heinrich 23
root names, alkanes 272
rotary evaporation 328
rubber 199, 211
rulers 316
rusting/rust prevention 199, 206
Rutherford model, atomic structure 20–1, 22
Rydberg constant 40

s atomic orbitals 44–7
s-block elements 230
safety, experiments 309–11
salicylic acid 472

salt bridge 595–6, 612
salts
    acid–base equilibria 565–7
    parent acids and bases 556
    of strong acids and strong bases 565
    of strong acids and weak bases 565–6
    of weak acids and strong bases 566
    of weak acids and weak bases 566–7
Saruhashi, Katsuko 526
saturated hydrocarbons 262
saturated solutions 514
scale resolution, length measurement 316
scandium 101, 250
scanning tunnelling microscope (STM) 23
scatter graphs 365, 366
Schrödinger equation 43
scientific journals 630
scientific knowledge 218
    falsifiability of 21, 426
    public understanding of 440
    sharing of 470
scientific laws 444
scientific models 22, 46, 81
scientific theories 46, 539
seawater
    carbon dioxide concentration 526
    reverse osmosis 11
secondary alcohols, oxidation 604, 605, 607
secondary amines 266
secondary compounds 281–2
secondary (rechargeable) cells 437, 594, 598–602
seesaw geometry, molecules 169
semiconductors 203
sensors 342
separating funnel 329
serial dilution 78, 323
SHE see standard hydrogen electrode
SI system, defining constants 355
SI units 15, 16, 66, 350, 355
sigma bonds 174–5, 176
significant figures, measurement 316, 356–7
silicon 137, 203
silicon dioxide (silica) 137
silver chloride 204
silver halides 200
silver sulfide 104
skeletal formula 163, 258
smartphones 342, 598
S_N1 reaction mechanism 641
S_N2 reaction mechanism 639–41
snowflakes 146
sodium
    boiling and melting points 191
    emission spectrum 36
    reactions 239, 582

successive ionization energies 59
sodium carbonate 106, 473
sodium chloride
    crystals 96
    dynamic equilibrium 514
    electrolysis 603, 617
    lattice enthalpy 204
    lattice structure 108
    mass concentration 74
    molar concentration 74
sodium fluoride 109
sodium hydroxide 616–17
solids 13
solubility 514
    and bonding type 198
    covalent substances 152
    ionic compounds 112–14
solubility rules 114
solute 72, 514
solutions 72–8
    aqueous solutions 13, 72, 465, 616–18
    concentrated solutions 73
    dilute solutions 73
    pH measurement 318–19
    saturated/unsaturated solutions 514
    serial dilution 78, 323
    standard solutions 75, 76, 77–8, 321–3
    stock solutions 75
solvation (dissolution) 11, 112
solvent 72
    chromatography 155, 156, 158
sp hybrid orbitals 178
sp^2 hybrid orbitals 178
sp^3 hybrid orbitals 177
space-filling models 127, 341
specific energy 432, 440
specific heat capacity 394–5
spectator ions 553
spectrophotometers 339
spectrophotometry 76, 339–40
spectroscopes 36, 37
spectroscopy
    infrared spectroscopy 293–6
    proton nuclear magnetic resonance spectroscopy 296–302
    spin resonance spectroscopy 296
    vibrational spectroscopy 293
speed of light 39
spin, electrons 48–50, 174
spin resonance spectroscopy 296
spin–spin coupling 300
spontaneous reactions 442
    entropy change 443
    Gibbs energy change 448–51
    temperature effect 453
spreadsheets 343–5

equilibrium concentrations, determination of 528–9
    functions/operators 344
    modelling 347–8
square planar geometry, molecules 170
square pyramidal geometry, molecules 170
standard cell potential 611, 614, 615
standard change in Gibbs energy see Gibbs energy change
standard electrode potential 610
standard enthalpy change of combustion 398, 412–17
standard enthalpy change of formation 413–16, 419, 420
standard enthalpy change for a reaction 394–401
standard entropy change 445, 446
standard entropy values 445, 446
standard hydrogen electrode (SHE) 610, 612
standard reduction potential 610
standard solutions 75, 76, 77–8
    dilution technique 322–3
    preparation of 321–2
starch 212
starting molecules 632
state function 389
states of matter 13–15
    entropy change 443
    symbols 462
steels 206, 207
stereocentre (asymmetric centre) 286
stereochemical formula 287–8
stereoisomers 284–90
    cis–trans isomers 284–6
stereospecific 640
steric hindrance 641
STM see scanning tunnelling microscope
stock solutions 75
stoichiometric coefficient 65, 461
stoichiometry 461–2
straight-chain alkanes 268, 269
straight-chain alkenes 264
strong acids 548
    pH curves 568, 569–70
strong bases 558
    pH curves 568–9
structural formula 258, 341
structural isomers 273, 279–83
    chain isomers 279
    positional isomers 279
    primary, secondary and tertiary compounds 281–2
sublimation 13, 15
substituents 271, 272, 275, 551
substitution reactions
    electrophilic substitution 648–9
    nucleophilic substitution 639–42
        reaction rate 642
    radical substitution 624–6
sucrose 68
sulfate ion 104, 173

sulfates 114
sulfur 6
sulfur dioxide 127, 242
sulfur hexafluoride 167–8
sulfuric acid 73, 425, 548
Sun, absorption spectrum 40
superconductors 189
surface-active agents (surfactants) 152–3
symbols 6
     hazard symbols 309
     naming conventions 245
     states of matter symbols 462
synthetic polymers 209, 212–13
systematic errors 383, 385, 395, 399

T-shaped geometry, molecules 169
tables (periodic table) 54
tables, quantitative data 364–5
     see also charts; graphs
tangent line 374, 484
technology 342–9
     databases 345–6
     modelling 347–9
     and science 218
     sensors 342
     spreadsheets 343–5
temperature
     equilibrium position 522–3
     gases 83, 85, 87–91, 139
     Gibbs energy change 455, 457–8
     and heat 389
     measurement 316–17, 351, 352
     reaction rate 489
     and spontaneous reactions 453
temperature gradient 389
temperature scales 15–17
terminal position 276
termination 622, 626
termolecular reactions 496–7
tertiary alcohols 607
tertiary amines 266
tertiary compounds 281–2
tests
     flame tests, metals 37
     "pop" test, hydrogen 591
     Vickers hardness test 207
tetrahedral geometry, molecules 129, 130, 177
tetravlent 176
theoretical yield 463, 471
theories, scientific 46, 539
thermal conductivity/conductors 189, 190, 198
thermal energy 389
thermochemistry 390, 399
thermodynamics 404
     second law of 444

thermography 388
thermometric titration 400
Thiele tube 330
thin layer chromatography (TLC) 157, 158–9, 331
three-dimensional representation of molecules 129, 259, 260
time, measurement of 315
titanium 101
titration
     acid–base titration 468, 557–9, 567–73
     back titration 469
     end point 571
     experimental technique/apparatus 334–6
     half-equivalence point 569–71
     pH curves 557–8
     redox titration 335, 592–3
     of strong acid with strong base 572
     thermometric titration 400
     of weak acid with strong base 572–3
     of weak base with strong acid 573
TLC see thin layer chromatography
total ionic equation 553
transition elements 100–1
     atomic orbital filling 53, 55
     bonding 193
     complexes 251–3, 637
     coordination bonds 126
     electron configurations 101, 248
     oxidation states 101, 248–50
     periodic table 230
     properties 247–8
     variable oxidation states 248–50
transition range, acid–base indicators 571
transition state 493, 640
triangular bonding diagrams 198, 201–5
trichloromethane 133
trigonal bypyramidal geometry, molecules 169–70, 171
trigonal planar geometry, molecules 128, 130
trigonal pyramidal geometry, molecules 129
triiodide ion 167–8
triplets 300
triprotic acids 547, 550
tritium 26
tungsten 194

uncertainties, measurement 351–62
     absolute uncertainty 355
     decimal places 356–7
     error bars 369, 370
     expression of 355–6
     human reaction time 353
     instrument uncertainty 351–2
     least count 351
     mean values 362
     percentage uncertainty 355
     propagation of 358–61

reaction mixtures 353
relative (or fractional) uncertainty 356
significant figures 356–7
value fluctuation 353
uncertainty principle, Heisenberg's 43
unimolecular reactions 496
universal indicator 543
unsaturated hydrocarbons 262
unsaturated solutions 514
uranium 28

vacuum filtration 327
valence bond theory 174, 176
valence electrons 52
valence shell electron pair repulsion model (VSEPR) 127–9, 131
validity of measurements 312–13
van Arkel-Ketelaar triangular bonding diagrams 198, 201–5
van der Waals forces 147
van der Waals parameters, gases 84
vanadium 101
vaporization 15
variable oxidation states 248–50
variables, measurement of 311–20
Vauquelin, Nicolas-Louis 193
vernier caliper 317
vibrational spectroscopy 293
Vickers hardness test 207
visible light, wavelengths 34, 35, 38, 39
volatility
    and bonding type 198
    covalent substances 151
    ionic compounds 111
Volta, Alessandro 594
voltaic cell 338, 437, 596
    batteries 594, 598
voltaic pile 338
voltmeters 320
volume
    measurement of 314–15
    problem-solving 465
volumetric analyses 314
volumetric glassware 321
VSEPR see valence shell electron pair repulsion model

Warner, John 475
waste disposal, laboratory chemicals 309
water
    alkali metal reactions 239
    bond angle 129
    dissociation 545
    electrolysis 438
    electrophilic addition 634, 646
    empirical formula 68
    equilibrium constant 545
    heating curve graph 15
    heavy water 28

hydrogen bonding 145–6
intermolecular forces 138
ionic product of water 545–6
Lewis formula 121
metal oxide reactions 241–2
molecular formula 68
molecular models 348, 349
molecular polarity 133
non-metal oxide reactions 241
as nucleophile 629
oxidation state 243
percentage composition 69
as polar solvent 112
reduction 616
reverse osmosis 11
sodium reaction 239
space-filling molecular model 127
states of matter 13
Van der Waals parameters 84
see also aqueous solutions; hydrolysis
wave functions, Schrödinger's 44, 45
wavelengths, electromagnetic radiation 34, 35, 39
wavenumber 293
wave–particle duality 43
weak acids 549–50
    anions 552
    dissociation constant 561–5
    reversible ionization 530
    titration 572–3
weak bases
    dissociation constant 561–5
    pH curves 569–71
    reversible ionization 530
    titration 573
work 389

X-ray diffraction 165
xenon 84
xenon trioxide 167–8

Zewail, Ahmed 499
zinc 247, 590

The periodic table

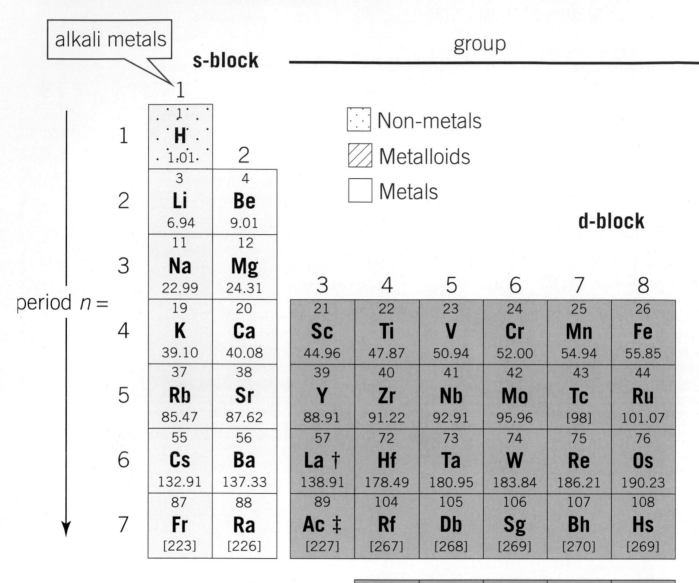

alkali metals

s-block

group

Non-metals

Metalloids

Metals

d-block

period n =

	1	2		3	4	5	6	7	8
1	1 **H** 1.01								
2	3 **Li** 6.94	4 **Be** 9.01							
3	11 **Na** 22.99	12 **Mg** 24.31							
4	19 **K** 39.10	20 **Ca** 40.08		21 **Sc** 44.96	22 **Ti** 47.87	23 **V** 50.94	24 **Cr** 52.00	25 **Mn** 54.94	26 **Fe** 55.85
5	37 **Rb** 85.47	38 **Sr** 87.62		39 **Y** 88.91	40 **Zr** 91.22	41 **Nb** 92.91	42 **Mo** 95.96	43 **Tc** [98]	44 **Ru** 101.07
6	55 **Cs** 132.91	56 **Ba** 137.33		57 **La** † 138.91	72 **Hf** 178.49	73 **Ta** 180.95	74 **W** 183.84	75 **Re** 186.21	76 **Os** 190.23
7	87 **Fr** [223]	88 **Ra** [226]		89 **Ac** ‡ [227]	104 **Rf** [267]	105 **Db** [268]	106 **Sg** [269]	107 **Bh** [270]	108 **Hs** [269]

†lanthanoids

f-block

‡actinoids

58 **Ce** 140.12	59 **Pr** 140.91	60 **Nd** 144.24	61 **Pm** [145]	62 **Sm** 150.36
90 **Th** 232.04	91 **Pa** 231.04	92 **U** 238.03	93 **Np** [237]	94 **Pu** [244]

p-block

noble gases

halogens

				13	14	15	16	17	18
									2 **He** 4.00
				5 **B** 10.81	6 **C** 12.01	7 **N** 14.01	8 **O** 16.00	9 **F** 19.00	10 **Ne** 20.18
9	10	11	12	13 **Al** 26.98	14 **Si** 28.09	15 **P** 30.97	16 **S** 32.06	17 **Cl** 35.45	18 **Ar** 39.95
27 **Co** 58.93	28 **Ni** 58.69	29 **Cu** 63.55	30 **Zn** 65.38	31 **Ga** 69.72	32 **Ge** 72.63	33 **As** 74.92	34 **Se** 78.96	35 **Br** 79.90	36 **Kr** 83.80
45 **Rh** 102.91	46 **Pd** 106.42	47 **Ag** 107.87	48 **Cd** 112.41	49 **In** 114.82	50 **Sn** 118.71	51 **Sb** 121.76	52 **Te** 127.60	53 **I** 126.90	54 **Xe** 131.29
77 **Ir** 192.22	78 **Pt** 195.08	79 **Au** 196.97	80 **Hg** 200.59	81 **Tl** 204.38	82 **Pb** 207.20	83 **Bi** 208.98	84 **Po** [209]	85 **At** [210]	86 **Rn** [222]
109 **Mt** [278]	110 **Ds** [281]	111 **Rg** [281]	112 **Cn** [285]	113 **Nh** [286]	114 **Fl** [289]	115 **Mc** [288]	116 **Lv** [293]	117 **Ts** [294]	118 **Og** [294]

63 **Eu** 151.96	64 **Gd** 157.25	65 **Tb** 158.93	66 **Dy** 162.50	67 **Ho** 164.93	68 **Er** 167.26	69 **Tm** 168.93	70 **Yb** 173.05	71 **Lu** 174.97
95 **Am** [243]	96 **Cm** [247]	97 **Bk** [247]	98 **Cf** [251]	99 **Es** [252]	100 **Fm** [257]	101 **Md** [258]	102 **No** [259]	103 **Lr** [262]